$ 795

D1356948

physical sciences data 25

thermodynamic data for pure compounds

Part A. hydrocarbons and ketones

physical sciences data

Other titles in this series:

physical sciences data 25

thermodynamic data for pure compounds

Part A
hydrocarbons and ketones

buford d. smith, director

and

rakesh srivastava, research associate

Thermodynamics Research Laboratory, Chemical Engineering Department, Washington University, St. Louis, MO 63130, U.S.A.

ELSEVIER
Amsterdam — Oxford — New York — Tokyo 1986

ELSEVIER SCIENCE PUBLISHERS B.V.
Sara Burgerhartstraat 25
P.O. Box 211, 1000 AE Amsterdam, The Netherlands

Distributors for the United States and Canada:

ELSEVIER SCIENCE PUBLISHING COMPANY INC.
52, Vanderbilt Avenue
New York, N.Y. 10017, U.S.A.

ISBN 0-444-42576-4 (Vol. 25A)
ISBN 0-444-41689-7 (Series)
ISBN 0-444-42579-9 (Set)

© Elsevier Science Publishers B.V., 1986

All rights reserved. No part of this publication may be reproduced, stored in a retrieval system or transmitted in any form or by any means, electronic, mechanical, photocopying, recording or otherwise, without the prior written permission of the publisher, Elsevier Science Publishers B.V./Science & Technology Division, P.O. Box 330, 1000 AH Amsterdam, The Netherlands.

Special regulations for readers in the USA — This publication has been registered with the Copyright Clearance Center Inc. (CCC), Salem, Massachusetts. Information can be obtained from the CCC about conditions under which photocopies of parts of this publication may be made in the USA. All other copyright questions, including photocopying outside of the USA, should be referred to the publisher.

Printed in The Netherlands

PREFACE

The project from which this two-volume data compilation was drawn is a good example of the "one thing leads to another" approach to project planning. It began in the early 1970's with the development of a correlation formalism for correlation of liquid mixture properties. The formalism was unique in that it was thermodynamically consistent, i.e., it could simultaneously correlate vapor-liquid equilibrium, heat of mixing and volume change of mixing data.

A thermodynamically consistent formalism will not, of course, fit inconsistent data. Hence, it was necessary to begin a systematic evaluation of VLE, H^E and V^E data from the literature in order to identify mutually consistent sets which could be used as a data base for the correlation. As confidence in the capabilities of the correlation grew, the mixture data base changed from a limited one for testing purposes to a comprehensive data base which would contain all of the useable VLE, H^E and V^E data from the literature.

From the beginning, the mixture data evaluation work was hindered by the lack of readily available, reliable pure compound data. Finally, it became obvious that the accurate evaluation of mixture data on a large scale required the prior development of a comprehensive, evaluated pure compound data file which could be accessed by the computer programs used to process the mixture data. The file had to be comprehensive because it had to furnish data for every compound which had been or would be included in a reported set of VLE, H^E or V^E data. The data in the file had to be evaluated because of the sensitivity of the mixture results to inaccuracies in the pure compound data used.

The decision was made to extend the data retrieval and evaluation effort to pure compounds. By 1975 enough funding became available to permit an energetic attack on both the mixture data and compound data evaluation problems. The objective was to retrieve, process, evaluate, and store in computer data banks all the useful thermodynamic data in the literature for organic compounds and mixtures. That objective was to be reached by 1985.

A massive literature retrieval project retrieved copies of about 28,000 documents. Large, sophisticated computer program libraries were developed to process and list the documents, and to process and store the compound and mixture data. Efficient data processing and evaluation methods were developed and

documented. The capability to combine intensive evaluation with high-volume
data processing was demonstrated in successive large-scale evaluation projects
for both pure compounds and mixtures. All the preliminary objectives were
achieved, and it was just necessary to continue "turning the crank" to produce
the desired pure compound and mixture data banks.

Many things can change over a ten-year period, and two major developments
in the 1975-1985 period eventually made it impossible to complete the two pro-
posed data banks. Inflation increased the cost of doing the work by a factor of
about three but it was not possible to increase the project funding propor-
tionally. Other, less comprehensive, projects in the thermodynamic data area
were organized and their success drained away the increased industrial funding
which was needed. As a result, the gap between the actual budget and that
necessary to complete the banks within the ten-year period widened rapidly
after 1980.

By 1982, it was no longer possible to do both pure compound and mixture
evaluation work simultaneously because it was not possible to maintain experi-
enced evaluators in both areas. Work on one or the other had to be dropped.
Unfortunately, the correlation and evaluation projects were inextricably linked.
The pure compound evaluation work had to be done before the mixture data could
be evaluated. The correlation work needed both the pure compound and mixture
data. Adequate funding was available to do only one part at a time, but doing
the three parts in series would have extended the overall project by at least
another ten years and that was not a practical option for several reasons.
Finally, the goal of creating a comprehensive thermodynamic data system had to
be dropped and the data evaluation projects were shut down.

This two-volume compilation makes available in hardback most of the
information which had been stored in the pure compound data bank. Only four
compound groups are covered – hydrocarbons (C_1-C_9), ketones (all carbon numbers),
alcohols (all carbon numbers) and halogenated hydrocarbons (C_1, C_2 and a few
C_3's). When work stopped, the other halogenated hydrocarbons were being processed and the data for all the nitrogen-containing compounds (amines, amides,
nitriles, nitrated compounds, heterocyclics, etc.) had been transcribed from
the literature onto computer input forms. The final data bank of about 3000
compounds could have been completed in 1985 or 1986.

The user of this compilation must keep in mind the dual-purpose for which
the pure compound data bank was created. The initial and primary objective was
to provide the best possible pure compound information to the mixture evaluation and correlation programs. Consequently, the data for compounds for which
extensive measurements have been made were carefully evaluated and correlated
to provide the most reliable values possible. A secondary objective was to make

the data bank a repository for all the useful data in the literature. i.e., the objective was to make all the useful information immediately available to the user through a computer terminal. The quality of the output from such a system can vary widely from compound depending upon the quality of the information obtained from the literature.

Buford D. Smith

Rakesh Srivastava

CONTENTS

In Part B thermodynamic data of halogenated hydrocarbons and alcohols are given.

Chapter 1

INTRODUCTION

This two-volume data compilation provides values for three experimentally-determined properties (vapor pressure, saturated liquid density and second virial coefficient) and two calculated thermodynamic properties (saturated vapor volume and heat of vaporization). The first volume covers the C_1 through C_9 hydrocarbons and ketones of all carbon numbers. The second volume provides data for halogenated hydrocarbons and alcohols. Alcohols of all carbon numbers are covered, but the halogenated hydrocarbons are restricted to the C_1 and C_2 compounds plus only a few C_3's.

An attempt was made to extract and store all useful experimental data existing in the open literature for the compound classes covered. However, not all of the information stored in the original data bank has been tabulated here. The criteria used to retain or delete compounds were as follows:

1. All deutero and deuteroxy compounds were eliminated.

2. All compounds with a 20 K or greater temperature range for either a vapor pressure or a liquid density correlation were retained, if there were at least four useable experimental data point values for the correlation.

3. In a few cases, compounds were retained with less than a 20 K temperature range because of their similarity to other retained compounds. For example the trans form might be retained with four data points and a 15 K temperature range if the cis form met the standard criteria.

The data values presented for any given compound-property combination are what appeared to be the best available values as determined by a careful, intensive screening process. Unfortunately, thermodynamics does not provide any method for the independent verification of the accuracy of individual sets of vapor pressure, liquid density or second virial coefficient data. Accuracy can be substantiated only by multiple measurements by independent workers. When multiple, high-quality determinations appear in the literature, the selected data for the compound-property combination will be reliable. When they are not available, no guarantee of accuracy can be given to the user. All that can be done is to tabulate the available experimental evidence in the hope that those data will be better than no data. This problem is discussed further in Section 4.3, Assessment of Probable Accuracy.

The experimental property values for each compound are presented in two ways. First, the selected literature data values for the three experimental properties are represented by correlation equations. Second, those correlation equations

1

are used to produce short tabulations of the three experimental property values
as a function of temperature. Those tabulations are extended to include the two
calculated properties by assuming an equation of state.

The presentation of the selected literature data in the form of temperature-
dependent correlation equations permits the easy use of the data in computer
programs. The tabulation of property values shows the magnitude of those values
and makes obvious at a glance the extent of the available data for the given
compound.

The vapor pressure and liquid density data for each compound are represented
by up to four correlation equations, and that redundancy requires explanation.
The pure compound data bank from which this data compilation was produced was
developed to support mixture calculations. The mixture correlation formalism
being used required rationing the configurational Gibbs function of each mixture
component to that of a reference compound in order to calculate the corresponding states parameters as a function of temperature. The vapor pressure
(represented by an equation) is a numerically important term in the configurational Gibbs function, and the ratio of the two Gibbs functions was sensitive
to the behavior of the two correlation polynomials. To include the heat of
mixing in the mixture correlation, it was necessary to differentiate the corresponding states parameters with respect to temperature, and those derivatives
were very sensitive to any misbehavior in the slopes of the calculated vapor
pressure curves.

To obtain well-behaved temperature derivatives of the corresponding states
parameters, it was necessary to use vapor pressure correlation equations with
no more than five constants (and preferably four) and to limit the temperature
range covered by the correlation. Hence, up to four vapor pressure correlations
--one over the maximum temperature range and up to three correlations with
narrower ranges--were stored for each compound in the pure compound data bank.
The narrower range correlations usually fit the selected data points slightly
better than the maximum-range correlation but that difference will not be
important for most users.

The mixture correlation procedure also required the ratioing of two liquid
density correlations. The results were much less sensitive to the temperature
derivatives of the density correlations but, nevertheless, narrower-range
correlations were also stored for the saturated liquid density.

The number of property correlations stored for each compound depends upon the
amount of data available. In any case, those literature documents which provided at least one of the selected data points for the correlation(s) are always
identified for the user. That is done through the Master Reference List (MRL)
numbers. The MRL number is the identification number assigned to a retrieved

2

literature document when its citation was listed on the MRL tapes. The MRL numbers of the related documents are listed with the correlation equations for each compound. The citations for those documents appear in the Bibliography in Chapter 6.

The reader should note that the bibliography for each volume and the two indexes provided for each major compound class appear before--rather than after--the data tables. It should also be noted that the bibliography contains citations only for those documents referenced on the data tables. Those literature documents cited in the pages preceding the data tables are listed separately in Chapter 5, Literature Cited.

Finally, the data tables are sequenced according to the compound sequence numbers which appear on the first line of both the first and second pages for each compound.

Chapter 2

DESCRIPTION OF TABLES

A two-page tabulation is presented for each compound. The first page pre-
sents the compound constants, the constants for the correlations chosen for the
three experimental properties, and the MRL (Master Reference List) numbers for
the relevant literature documents. The second page provides a tabulation of
the properties as a function of temperature.

2.1 FIRST PAGE

2.1.1 Compound identification

The first line of the first page of each compound's tabulations contains the
compound sequence number, the compound name (usually in the IUPAC nomenclature
[1]), the chemical formula and the formula weight. The data tables are ordered
according to the compound sequence number which serves the same basic purpose as
a page number.

2.1.2 Compound constants

The symbols used for the compound constants are as follows:

MP	Melting point, K
NBP	Normal boiling point, K
TC	Critical temperature, K
PC	Critical pressure, MPa
VC	Critical volume, ml/mol
DC	Critical density, g/ml
ZC	Critical compressibility factor
OM	Acentric factor (omega)
MU	Dipole moment, Debye
RG	Radius of gyration, angstrom

Note that the critical pressure is always given in megapascals whereas the vapor
pressure correlations (tabulated below the compound constants) use the kilopascal for all compounds except methane, ethane, ethene and propene; the correlations for those four compounds use the megapascal. Also note that those
four compounds have density correlations expressed in mol/dm^3 instead of the
g/ml used for the critical density or the ml/mol used for the critical volume.
The milliliter used is the new milliliter and is equivalent to the cubic
centimeter.

5

2.1.3 Vapor pressure correlations

Up to four vapor pressure correlations may be tabulated for each compound. Those four fits are for the following general temperature ranges:

Range	Temperature range, K
1	T_m to T_c
2	T_m to $(T_b + 20)$
3	T_m to T_d
4	$(T_b - 20)$ to T_c

The T_m, T_b and T_c are the melting, normal boiling and critical temperatures respectively. The T_d is defined by $T_d = 0.85T_c$ if T_c is available, or by $T_d = 1.45T_b$ if not.

The range 1, 2, 3 and 4 fits are presented in order from left to right on the page. The accuracies with which the four fits represent their respective sets of data points fall in the following order (from highest to lowest): 2, 4, 3, 1. In most cases, the small differences in accuracy will not be important for most users for most compounds.

The first line in each block of vapor pressure correlation constants identifies the correlation equation. The vapor pressure correlations used to fit the data during the evaluation procedures are given in Table 2.1.

The second line of each correlation constant block shows the actual temperature range covered by the data points fitted. Those actual ranges fall within the general temperature ranges defined above.

The third and fourth lines in each block show the number of data points supporting the correlation, and the root-mean-square-deviation of the fit. The RMSD value is in kilopascals.

The correlation constants are presented in order from top to bottom; i.e., the first constant is A, the second B, etc., as defined by the respective equation in Table 2.1. The units are kelvin and kilopascal for all compounds except methane, ethene, ethane and propane; for those four compounds the vapor pressure correlation uses the megapascal.

In order to increase the amount of literature data made available to the user, vapor pressure correlations are sometimes presented when only two or three useable data points were found. In such cases the Riedel-Plank-Miller-2 equation was used with the last two constants set equal to zero; i.e., the first two terms of the RPM2 equation were used to provide a linear fit.

Often, all the selected experimental vapor pressure points for a compound will fall within the range 2 temperature interval. In such cases, the range 2 fit was always stored also in the range 1 and range 3 fit fields in the data bank, and in some cases was also duplicated in the range 4 field. If that was done, the table for the compound will list the same fit three or four times. The reasons for that duplication concern the various uses to which the data bank was put.

In a few cases--generally for high-boiling compounds--there was no range 2 fit because no data points fell within that temperature range. For those compounds the range 3, or range 1, fit was duplicated in the range 2 position.

In many cases, a different range 4 fit will be given even though all of the selected experimental data points fall within the range 2 interval. The two ranges overlap by 40 K and often there were enough points within that 40 K region to justify the separate range 4 fit in order to provide a more accurate representation of the property values around the boiling point. The additional fit was often particularly helpful when the selected data points ran all the way down to the melting point. Vapor pressures at such low temperatures are usually measured with techniques different from those used at high temperatures, and it was sometimes difficult to correlate simultaneously the low temperature data points and those around the normal boiling point. Hence a range 4 fit which did not have to accommodate the lower temperature data often could give a more accurate representation of the data points around the normal boiling point than could the range 2 fit.

All compounds for which vapor pressure data are given in this two-volume compilation are represented by relatively simple four or five constant vapor pressure equations, except for the following hydrocarbons: methane, ethane, ethene and propene. The compound constant values, property correlations and equations of state stored in the original data bank for those four compounds were taken from intensive data evaluation projects summarized in IUPAC and NBS reports [6, 7, 8, 9]. The Wagner-2 and Wagner-3 equations in Table 2-1 are the Wagner equation with three exponents treated as arbitrary constants; rather than list seven constants for the methane and propene fits, the Wagner equation has been repeated in Table 2-1 with the special exponent values used for methane and propene.

7

TABLE 2.1

Vapor pressure equations used.

Equation	$\ln P' =$
Riedel [2]	$A + B\ln T + \dfrac{C}{T} + DT^6$
Frost-Kalkwarf [3]	$A + B\ln T + \dfrac{C}{T} + \dfrac{DP'}{T^2}$
Riedel-Plank-Miller-2 [4]	$A + \dfrac{B}{T} + CT + DT^2$
Vapres-2	$A + \dfrac{B}{T} + CT + DT^2 + E\ln T$
Wagner [5]	$\dfrac{1}{T_r}\left[A(1-T_r) + B(1-T_r)^{1.5} + C(1-T_r)^3 + D(1-T_r)^6 \right]$
Wagner-2 [6]	$\dfrac{1}{T_r}\left[A(1-T_r) + B(1-T_r)^{1.5} + C(1-T_r)^{2.5} + D(1-T_r)^5 \right]$
Wagner-3 [9]	$\dfrac{1}{T_r}\left[A(1-T_r) + B(1-T_r)^{1.5} + C(1-T_r)^{4.0} + D(1-T_r)^{4.5} \right]$
Goodwin [7, 8]	$A + BX + CX^2 + DX^3 + EX^4 + FX(1-X)^G$ where $X = (1-T_t/T)/(1-T_t/T_c)$ and T_t and T_c are the triple point and critical point temperatures.

8

2.1.4 Liquid density correlations

The format for the saturated liquid density correlations is the same as for the vapor pressure.

The general temperature ranges defined for the four density fits vary with the correlation equation used. The Martin equation can be fitted to the same temperature ranges defined above for the vapor pressure. The Francis-1 equation was fitted to the following ranges because it cannot represent data close to T_c.

Range	Temperature Range, K
1	T_m to $(T_c - 10)$
2	T_m to $(T_b + 20)$
3	T_m to $(T_c - 30)$
4	$(T_b - 20)$ to $T_c - 10$

The Francis-2 equation was fitted to the range from $(T_c - 50)$ to T_c. In the early projects before the Martin equation became available, the Rackett equation was used more frequently than in later projects; it was useful in those instances when d_c and T_c were known but only limited amounts of density data were available at lower temperatures. After the Martin program became operative, the Rackett equation was used only for the $(T_c - 50)$ to T_c range where it provided a seldom used alternative to the Francis-2 and Martin fits.

The Francis-1 equation is a very reliable equation when its fit is terminated 10 to 20 K below T_c. However, when T_c and d_c values are available, the Martin equation is generally preferred for ranges 1 and 4 because it can "reach" the critical point. The Martin equation was not in use for some of the earlier compound projects and hence Francis-1 and Francis-2 fits will appear for some range 1 and range 4 fits where the Martin equation would have been selected if it had been in use at the time.

As for the vapor pressure, a range 1 fit was always loaded into the original data bank even when it was a duplicate of the range 2 or range 3 fit. Also, the range 2 fit is sometimes repeated as a range 3 fit, or vice versa. Those repetitions were made necessary by previous uses of the original data file but affect the tabulations in this compilation by sometimes causing the same correlation to be repeated two or three times.

The situation described above for the vapor pressure where a separate range 4 fit is provided in the upper part of the range 2 fit can exist also for the liquid density when the Martin equation is used.

As described above for the vapor pressure, the density correlations for methane, ethane, ethene and propene were taken from data compilation projects by the Union of Pure and Applied Chemistry [6, 9] and by the U. S. Bureau of

9

Standards [7, 8]. Special equations were used for those four compounds, and the density units are mol/dm^3 instead of g/ml.

Finally, the Francis-1 equation is sometimes used with only two constants in order to provide a linear fit of two or three useable density points.

TABLE 2.2

Liquid density equations.

Equation	$d = FW/V^L =$
Francis-1 [10]	$A - BT - \dfrac{C}{E-T}$
Francis-2 [10]	$d_c + [A(T_c-T)]^{1/B}$
Rackett [11, 12]	$\dfrac{FW\ P_c}{R\ T_c}\left[Z_{RA}\right]^{-(1 + (1-T_r)^{2/7})}$
Martin [13]	$d_c\left[1 + A(1-T_r)^{0.35} + B(1-T_r)^{2/3} + C(1-T_r)\right.$
	$\left. + E(1-T_r)^{4/3} + F(1-T_r)^{5/3}\right]$
McCarty [14]	$d_c + (d_{tp}-d_c)\ \exp\left[A_1\ \ln x + \displaystyle\sum_{i=2}^{4} A_i\left(1-x^{(i-5)/3}\right)\right.$
	$\left. + \displaystyle\sum_{i=5}^{7} A_i\left(1-x^{(i-4)/3}\right)\right]$
Goodwin [8]	$d_c + (d_{tp}-d_c)\left[x + (x^E-x)(A + Bx^{1-E} + cx)\right]$
Angus [6]	$d_c\left[\exp\left(\displaystyle\sum_{i=1}^{4} A_i(1-T_r)^{i.A_7} + A_5(1-T_r)^{9.A_7} + \right.\right.$
	$\left.\left. + A_6\ \ln\ (T_r)\right)\right]$
Angus-2 [9]	$d_c\left[\exp\left(\displaystyle\sum_{i=1}^{6} A_i(1-T_r)^{(i+2)/6}\right.\right.$
	$\left.\left. + A_7(1-T_r)^{13/6}\right)\right]$

Nomenclature:

d_{tp} saturated liquid density at the triple point.

FW   formula weight.

V^L liquid molar volume.

T_{tp} triple point temperature.

x $(T-T_c)/(T_{tp}-T_c)$.

Z_{RA} adjustable constant, theoretically equal to T_c.

2.1.5 Second virial coefficient correlations

The second virial coefficient data retrieved from the literature were not fitted in the same manner as the vapor pressure and liquid density data. Instead, the general correlation which best represented the better sets of available data for each compound was selected to represent that property. The chosen correlation and the correlation constants appear below the liquid density correlation constants.

The following five correlation equations were used in the data processing for all the pure compound projects: Pitzer-Curl [15], O'Connell-Prausnitz [16], Nothnagel-Abrams-Prausnitz [17], Kreglewski [18, 19, 20], Tsonopoulos [21, 22]. The Hayden-O'Connell correlation [23] was added when data processing for non-hydrocarbons began.

The Tsonopoulos correlation was chosen most often for the compound classes covered (hydrocarbons, alcohols, ketones, halogenated hydrocarbons). The Pitzer-Curl equation was chosen often for hydrocarbon systems. For polar systems, the Hayden-O'Connell equation was sometimes competitive with the Tsonopoulos correlation. The Kreglewski equation often fits well the data to which it has been fitted [19, 20] but, because of the lack of mixture mixing rules, was usually not chosen if another equation could be used. In those relatively few cases where T_c and T_b values are available but P_c is not known, the Nothnagel et al. equation could be used.

Because of their length and complexity, the second virial coefficient correlations are not reproduced here. If the user needs to program the correlations, it will be necessary in any case to utilize the original literature source documents.

In the tabulation of the correlation constant values for the second virial coefficient, the compound constants (T_b, T_c, P_c, V_c, acentric factor, dipole moment and radius of gyration) are repeated as needed. That is necessary because some of the correlations (e.g., O'Connell-Hayden) use and list values of the compound constants different than the ones selected for storage in our data bank. The values listed with the correlation should be used with that equation. The symbols used for the compound constants have been defined in Section 2.1.2. The symbols used for the other arbitrary constants for each correlation equation correspond closely to those used in the source literature documents to which the user must refer if use is to be made of the correlation equations.

2.1.6 Literature documents

The literature documents related to the properties covered are identified by their Master Reference List (MRL) numbers. The citations for those documents appear in Chapter 6.

The documents listed for the vapor pressure and liquid density correlations are only those which contributed at least one data point to the final correlation. The lists do not include documents whose data were excluded from the final correlation.

The literature documents for the second virial coefficient data were not separated into "used" and "not used" categories in the evaluation and correlation procedures used for that property. Hence, the numbers listed for that property include all the available documents reporting original data.

2.2 SECOND PAGE
2.2.1 Property values

The property correlations tabulated on the first page--plus an appropriate equation of state--permit the calculation of the property values tabulated in the second page of the two-page tabulation for each compound. The saturated vapor volume provided by the chosen equation of state was used with the liquid density and vapor pressure correlations to calculate the heat of vaporization using the Clapeyron equation.

$$\frac{dP'}{dT} = \frac{\Delta H^{vap}}{T(V^G - V^L)}$$

The differentiated forms of the vapor pressure correlation equations were used to provide the dP'/dT values.

The second page tabulations are usually based on the range 1 correlations given on the first pages. The lowest temperature in each second page tabulation coincides with the lower temperature limit of either the vapor pressure or liquid density correlation, whichever is lower. Similarly, the highest temperature coincides with the upper limit of one of those two property correlations, whichever is higher. The interval between those two temperatures has been used to calculate a temperature increment which provides the maximum number of lines which can be tabulated on one page. The resulting tabulation shows at a glance the temperature ranges over which useable experimental vapor pressure and liquid density data are available, and how the magnitudes of the three experimental properties and the related two calculated properties vary with temperature.

The temperature increment calculated as described in the last paragraph was seldom an integer value. Instead of rounding to the next highest integer and then using that constant value throughout the compound's tabulation, the real number increment was used to calculate each succeeding temperature and then that temperature was rounded to the nearest integer value. That procedure

13

provides more lines per page and often facilitates interpolation, but the user must be aware that the temperature increment often varies within a given tabulation.

The range of useable experimental data for the second virial coefficient is not shown by the second page tabulation. If a correlation equation could be established for the compound, that equation was used to calculate second virial coefficient values over the temperature interval determined from the vapor pressure and/or liquid density correlations, regardless of the temperature range actually covered by the useable experimental B values.

Unfortunately, when the second page tabulations cover a wide temperature range, the temperature increment becomes too large to permit accurate linear interpolations and one must resort to the equation constants on the first page.

In those instances where the temperature ranges of the vapor pressure and liquid density correlations do not overlap, there will be a discontinuity in the temperature values at the point where the values for one property ends and the values for the other one starts. Also, it should be noted that the heat of vaporization values must often terminate short of the critical point because the Francis-1 equation was used for the range 1 density fit. However, that is not as great a loss as it first appears. Even when a range 1 density fit all the way to the critical point is available, the heat of vaporization values often become erratic as they decrease rapidly toward zero because of the difficulties in providing accurate--and mutually consistent--correlations for the other three properties in the Clapeyron equation.

As discussed further in the next section, the ideal gas equation of state must be used for compounds when the critical point values required by other equations of state are not available. The calculated saturated vapor volume values are then the ideal gas values. Those ideal gas values, interacting with the dP'/dT values calculated from the vapor pressure values, sometimes cause the heat of vaporization values to terminate prematurely. The heat of vaporization, and the saturated vapor volume, should decrease with temperature. If the calculated value at any given temperature is larger than the value calculated at the preceding temperature, the calculation of the property is terminated to prevent the tabulation of grossly inaccurate values. This safeguard was installed initially to protect against misbehavior of any equation of state around the critical point. Unfortunately, the cutoff is sometimes triggered at lower temperatures when the ideal gas equation of state must be used.

14

2.2.2 Equation of state

The original data bank from which these tables were produced contained one of the following equations of state for each compound:

> Ideal gas
>
> Virial:
> > Through B term
> > Through C term
>
> Redlich-Kwong:
> > Unmodified form [24]
> > Chueh-Prausnitz modification [25]
> > Lu modification [26]
> > Peng-Robinson [27]
>
> Benedict-Webb-Rubin:
> > Unmodified [28]
> > Sood-Haselden modification [29]
> > Starling modification [30, 31]

The Peng-Robinson equation is not actually a modification of the Redlich-Kwong equation but was grouped with the Redlich-Kwong forms because of its similarity to that equation.

The ideal gas equation of state was not stored in the compound record. It was the default when no other equation of state was possible. That happens quite often because all other equations of state require the critical properties (or at least T_c) for the compound in question.

For most compounds, when the use of an equation of state other than the ideal gas equation was possible, the computer program which produced the second page tabulations used the following logic. When $(P/T) < (P_c/3T_c)$, the virial equation of state

$$\frac{PV_i}{RT} = 1 + \frac{B_{ii}}{V_i} + \frac{C_{iii}}{V_i^2} + \cdots$$

truncated to the second (B_{ii}) coefficient was used until the temperature was reached where the inequality sign reversed. For the rest of the tabulation, the program used the Peng-Robinson equation [27]:

$$P = \frac{RT}{V-b} - \frac{a(T)}{V(V+b) + b(V-b)}$$

$$a(T) = a(T_c)\, \alpha(T_r, \omega)$$

$$a(T_c) = 0.45724\, \frac{R^2 T_c^2}{P_c}$$

$$\alpha^{\frac{1}{2}} = 1 + \kappa(1 - T_r^{\frac{1}{2}})$$

$$\kappa = 0.37464 + 1.54226\omega - 0.26992\omega^2$$

$$b = 0.07780 \frac{RT_c}{P_c}$$

ω = acentric factor.

The Peng-Robinson equation of state, along with the Chueh-Prausnitz and the Lu modifications of the Redlich-Kwong equation, requires T_c, P_c and ω values. In those few cases where T_c and P_c values are available but vapor pressure data are not available at T_r = 0.7 for the calculation of the acentric factor, the unmodified Redlich-Kwong [24] had to be used:

$$P = \frac{RT}{V-b} - \frac{a}{T^{\frac{1}{2}} V(V+b)}$$

$$a = 0.42748023 \frac{R^2 T_c^{2.5}}{P_c}$$

$$b = 0.08664035 \frac{RT_c}{P_c}$$

The Chueh-Prausnitz modification [25] of the Redlich-Kwong equation of state uses individual a and b constants instead of those calculated only from the T_c and P_c values with the general equations above. Those constants are available for only a few compounds and, of the compounds covered in these compilations, BWR constants were available for all of those compounds except cyclohexane and benzene. If a BWR form was available in the data bank, it was selected instead of any stored Redlich-Kwong form. The Peng-Robinson equation was used for both benzene and cyclohexane.

The Lu modofication [26] of the Redlich-Kwong equation involves correlating the a and b constants as a function of temperature along the saturated vapor curve. The Lu constants are available only for a few compounds and, for the compounds covered in these compilations, BWR constants were available whenever the Lu constants were. Hence, like the Chueh-Prausnitz modification, the Lu modification was not involved in any of second page tabulations.

The switch from one equation of state to another is not necessary if the third (C_{iii}) virial coefficient is available, or if wide-range constants are available for one of the BWR equations. Good C_{iii} values are available for very few compounds (usually relatively simple compounds or elements) and no attempt was made to use C_{iii} coefficients for any of the compounds covered in these compilations. BWR constants are available for only a few hydrocarbons, halogenated hydrocarbons and inorganic compounds. See Cooper and Goldfrank [32],

Sood and Haselden [29] and Starling et al. [30, 31] for listings of most of the available constants. Those compounds in these compilations which used some form of the BWR equation in the second page tabulations are as follows:

Compound	BWR Form
Propyne	Unmodified
Cyclopropane	Unmodified
Propane	Starling modification
1,3-Butadiene	Unmodified
1-Butene	Unmodified
cis-2-Butene	Unmodified
Isobutene	Unmodified
Butane	Starling modification
Isobutane	Starling modification
Pentane	Starling modification
2-Methylbutane	Starling modification
2,2-Dimethylpropane	Unmodified
Hexane	Starling modification
Heptane	Starling modification
Octane	Starling modification
Trichlorofluoromethane	Starling modification
Dichlorodifluoromethane	Starling modification
Chlorotrifluoromethane	Starling modification
Tetrafluoromethane	Starling modification
Chlorodifluoromethane	Starling modification
Trifluoromethane	Starling modification
1,1,2-Trichloro-1,2,2-trifluoroethane	Starling modification
1,2-Dichloro-1,1,2,2-tetrafluoroethane	Starling modification
1,1-Difluoroethene	Starling modification
1-Chloro-1,1-difluoroethane	Starling modification

More complex equations of state were used for the following four hydrocarbons: methane [6], ethene [7], ethane [8] and propene [9].

Chapter 3

LITERATURE COVERAGE

The objective of the literature retrieval effort was to obtain copies of all literature documents which contained pure compound or mixture thermodynamic data for all compounds containing only the following elements: C, H, O, N, S, F, Cℓ, Br, I and Si. Data were also retrieved for all those elements (except C and Si) plus He, Ne, Ar, Kr and Xe. The relevant documents were identified in the following four ways:

1. Thirty-five journals, which at some period in their history have been rich with thermodynamic data articles, were "clean-swept". A trained searcher leafed through each issue scanning each article for relevant information. The title and abstract were not assumed to be sufficient to establish relevancy or nonrelevancy.

2. Over one hundred compilations, bibliographies, review articles, etc., dealing with the properties and compounds of interest were searched for relevant literature citations.

3. The relevant sections in all the Chemical Abstract volumes from Volume 1 (1907) through Volume 64 (1965) were searched for relevant documents. After 1965, the Bulletin of Chemical Thermodynamics (one of the compilations covered in method 2 above) was used instead of Chemical Abstracts.

4. The literature references in each retrieved document were inspected to identify other relevant documents.

By the end of 1984, about 28,000 documents had been retrieved and their citations listed on the Master Reference List (MRL) tapes.

It was not possible to retrieve 100% of all the published data for every compound-property or mixture-property combination of interest to this project. However, it is believed that over 95% of all the existing relevant documents, and over 98% of all the useable data, were in-hand before the start of any particular compound data project.

Once a compound data project was finished and the selected results stored in the data bank, new data of course continued to appear in the literature. The list below indicates the approximate date to which the literature coverage should be essentially complete for each class of compounds covered in these compilations. After that date, the user may need to search the literature to see if any new important data have been published for compounds of interest.

Compound Class	Literature Covered
C_4 Hydrocarbons	December, 1976
C_3 Hydrocarbons (ex propene)	December, 1977
C_5 Hydrocarbons	March, 1978
C_6 Hydrocarbons	December, 1978
C_7 Hydrocarbons	July, 1979
C_8 Hydrocarbons	December, 1979
C_9 Hydrocarbons	March, 1980
Ketones	December, 1981
C_1-C_6 Alcohols	December, 1981
C_7-C_9 Alcohols	March, 1982
$\geq C_{10}$ Alcohols	December, 1982
Halogenated Hydrocarbons	March, 1984

As discussed previously, the raw literature data for the C_1 and C_2 hydrocarbons were not processed. Instead, the property correlations and equations of state recommended in data compilations by the International Union of Pure and Applied Chemistry [6] for methane, and by the U. S. National Bureau of Standards [7,8] for ethene and ethane were used. Also, the information originally stored in the data bank for propene was later replaced with the equations recommended by the IUPAC compilation for that compound [9].

Chapter 4

DATA EVALUATION

The objectives of the compound data evaluation process were (a) to select
those experimental data points which best defined the true property-temperature
relationship, and (b) to correlate those selected data points with the precision
necessary to support the mixture data evaluation and correlation work which the
pure compound data bank supported. The emphasis was not on the evaluation of
individual data sets per se, but rather the development of property correlation
equations which were the best available definition of the true property-temperature curve.

4.1 DATA SCREENING

All of the retrieved data--regardless of its source or apparent quality--for
the given property for the compound class being processed were transcribed from
the literature documents and keyed into a computer disk file. Up to 200 compounds could be processed in one project. Editing programs were used to identify obvious errors and to make sure the structure of the original data file was
correct. A listing of the file was then proofed against the literature documents to eliminate transcription and keying errors.

The corrected original data file was then used as input to a unit conversion
program which created an output file with the property and temperature values
all in SI units. At this point, the procedure used for the second virial coefficient data deviated from that used for the vapor pressure and liquid density
data. The subsequent processing for the B values is described in Section 4.2.2
while the rest of this section and Section 4.2.1 discusses the procedures for
the vapor pressure and liquid density data.

The unit conversion programs for the vapor pressure and liquid density data
fitted the data points for each compound to a simple equation, and calculated
and stored a residual value (deviation from the correlation value) for each data
point. Those residual values for each compound were plotted versus temperature
by a computer-driven plotter. The compound plots were nine inches high and up
to thirty inches long. Each set of data had its own symbol, and those symbols
were keyed to the MRL number for the literature document from which the data set
came. The plotting program selected the ordinate and abscissa scale factors for
each compound-property plot to give the maximum possible magnification of the
scatter of the data points.

The residual plots were used to make most of the data point deletions. Previously selected values for the compound constants (normal boiling and critical

points for vapor pressure, and the 20°, 25° and critical point values for density) were used to help locate the correct track of the property curves on each compound's residual plots. The literature documents for all the plotted data sets for each compound were studied in order to identify those sets which had been measured most carefully and expertly. (The evaluator must be wary of placing too much reliance on stated accuracies.) Sometimes a data set from a source which had an established record of producing reliable data could be used to help define the correct track. With whatever information could be gathered, the evaluator used his experience and judgment to "carve" away points in order to define the property-temperature relationship.

After making obvious deletions, new residual versus temperature plots were made for all compounds with more than a few data points. Deletion of outlying points permitted expansion of the next plot which in turn helped define the next set of deletions. Compounds with a large amount of data (e.g., benzene, acetone, etc.) usually required three or four "trim and plot" steps before the evaluator was ready to start correlating the selected data points.

When the evaluator felt the track of each property curve had been established as well as possible using the residual plots, the remaining data points in each property's converted file were used as input to a sorting program which merged all of the remaining data sets for each compound into one set of data points ordered with respect to temperature. The source of each data point in the sorted file was still identified by the MRL number of the literature document from which the data point came.

4.2 DATA CORRELATION
4.2.1 Vapor pressure and liquid density

The correlation step had two objectives. First, the equation which worked best for each set of data points being fitted had to be determined. Second, the screening process had to be completed by looking for additional points to be deleted.

The equations used to correlate the vapor pressure and liquid density data are listed in Tables 2.1 and 2.2. As explained previously, the last three equations (Wagner-2, Wagner-3 and Goodwin) in Table 2.1, and the last four equations (McCarty, Goodwin, Angus and Angus-2) in Table 2.2, were taken from the IUPAC and NBS compilations for methane, ethene, ethane and propene and, therefore, were not involved in the iterative correlation procedure used for all the other compounds.

The vapor pressure equations found to be most useful were the Wagner, Riedel-Plank-Miller-2 (RPM2), Vapres-2 and Riedel. The Frost-Kalkwarf equation did not offer any advantages sufficient to balance its inconvenience. The Wagner

equation was preferred whenever T_c and P_c values were available; if critical
point values were available, it was always selected for the range 1 and range 4
fits and almost always for the range 2 and range 3 fits. When T_c and P_c values
were not available, the RPM2 equation was usually preferred over the Riedel and
Vapres-2 equations. The Vapres-2 equation was sometimes useful for the range 2
fit when the experimental data points extended down close to the melting point.
As mentioned earlier, it was often difficult to fit simultaneously the very low
pressure data and the data around the normal boiling point. The extra constant
in the Vapres-2 equation was sometimes helpful in such cases but, otherwise,
that extra constant often made the temperature derivatives misbehave when fitting scattered data points.

The behavior of the liquid density equations has been discussed in Section
2.1.4 in connection with the definitions of the fitting ranges. Whenever critical point values are available, the Martin equation is preferred because it can
reach the critical point. The Francis-1 equation is a very good equation, and
often fits data below the critical region better than the Martin equation but
its fits must be terminated 10 to 20 K below the critical point.

The (T_c-50) to T_c range used for the Francis-2 equation is a difficult region
to fit and the Francis-2 fits for that region--as do some of the Martin fits
close to the critical--often have large RMSD values. In that region, however,
the large RMSD is usually due to deficiencies in the data points rather than
deficiencies in the equation. An equation which is forced to go through the
selected critical point will not correlate well other nearby data points which
are not consistent with that critical point value. It was often found to be
desirable to modify the selected critical point values slightly (within experimental error) to take some of the strain off the correlating equation and reduce
the RMSD somewhat.

Subject to these general constraints, the determination of the best correlation equation for each fitting range for each compound-property combination was
an iterative procedure. The evaluator would specify what equations were to be
applied to each fitting range and then use the information supplied by the
fitting program to select the best equation for each set of data points. That
information included the deviation and percent error for each data point, the
RMSD, the maximum deviation, the deviations from the selected compound constant
values, the number of points with positive deviations and the number of points
with negative deviations, the first and second temperature derivatives calculated from the correlation at spaced temperature values, and a graphical display
which located each data point on a scale from a minus two times the RMSD value
to a plus two times the RMSD value. The latter display magnified the scatter in
the data points and was used to complete the point deletion process begun on the

23

residual plots. It also graphically illustrated how the equation was interacting with the data points; a perfect display shows the data points randomly
scattered around the zero RMSD (center) point of the display with half the
points to the left and half to the right.

Based on the results from each of the equations specified for trial, the
evaluator chose one equation for each fitting range. If needed, additional
points would be deleted, possibly to allow the equation to come closer to firm
compound constant values or to just cause the equation to move closer to the
evaluator's concept of the correct temperature track. Simple reduction of the
RMSD value was usually not a primary goal. Often, conflicting points would be
retained in order to force the temperature track to fall between them, rather
than eliminate one "side of the argument" in order to materially reduce the RMSD
value.

If points were deleted from a particular fitting range, the fit would have to
be repeated--with only one equation if a clearcut choice of equation was possible based on the first round of fits. Deletion of points usually reduced the
RMSD and that magnified the scatter of the points on the graphical display. For
some compounds, it was necessary to go through two or three "trim and fit"
rounds before the evaluator had the property-temperature relationship established as well as possible.

4.2.2 Second virial coefficient

As for the vapor pressure and liquid density data, all the retrieved second
virial coefficient data for all the compounds being processed were keyed into
an original literature data file. That file was edited and, when structurally
correct, the stored values were proofed against the literature documents from
which the data had been transcribed. A unit conversion program was then used to
convert all the units to the SI units, and to convert all the B values to those
defined by the virial equation of state in Section 2.2.2 [33]. However, instead
of using only one correlation equation to calculate data point residuals as done
for the other two experimental properties, the conversion program for the B
values calculated residuals for each of the five correlation equations listed
in Section 2.1.5. It also provided an RMSD value for each literature data set
and for all the combined literature data sets for each compound for each correlation equation.

After the first few data projects, the residuals thus provided were not
plotted versus temperature. The objective for the second-virial coefficient
data was just to determine which correlation equation (with its existing arbitrary constants) was to be chosen--not to define the correct temperature track
by deletion of data points and then fit an equation to the selected points. The

24

evaluator checked the compilation by Dymond and Smith [33] to see if those authors had expressed a preference for individual data sets. The individual literature documents reporting the data sets were studied. The evaluator would thus identify those data sets for each compound which appeared to be best and the performances of the various correlation equations on those data sets were then compared in order to select the best available correlation equation for each compound.

4.3 ASSESSMENT OF PROBABLE ACCURACY

As stated in the Introduction, thermodynamics does not provide any method for the independent verification of the accuracy of individual sets of vapor pressure or liquid density data. Hence, the assessment of accuracy of any particular correlation appearing in these compilations must be based on an evaluation of the literature data in the experimental data base.

Table 4.1 lists the definitions of eleven quality ratings which were applied to correlations previously published for the ketones [34]. It is recommended that users of these compilations use the same approach in their assessments of the probable accuracies of the reported correlations.

In Table 4.1, ratings A through F all involve experimental data from a reliable source. A "reliable source" is one which has reported high-quality data for a large number of compounds and those data consistently agree with other good data. An example of such a source can be drawn from the ketone vapor pressure data--the National Physical Laboratory, Teddington, Middlesex, United Kingdom [35,36,37]. Whenever a data set from that source was available, the compound's vapor pressure correlation was given one of the ratings from A through F. For ratings A through C, the data from the reliable source must cover the entire correlation range while the amount of corroboration from other sources ranges from substantial for A through partial for B to none for C. For ratings D through F, the reliable source data set covers only part of the correlation's temperature range and the three ratings differ only in the amount of corroborating data.

Ratings G through K do not involve a reliable source data set. A G rating means that at least two data sets from different sources showed considerable agreement, whereas an H rating says that all the sources showed substantial disagreement. Both the G and H ratings indicate a better situation than the I and J ratings which indicate that only one source of data was available in any given temperature range.

The K rating indicates the correlation is just a straight-line fit of two or three data points. The only justification for including such correlations is the desire to provide the user with as much information as possible. Two

25

elementary precautions were taken. First, the points fitted were usually required to be from the same source; points from different sources were used only if there were other corroborating measurements which supported the general accuracy of the two or three points fitted to the straight line. Second, the slope of the linear fit was checked to make sure it had the right sign and a reasonable numerical value. The possible pitfalls in the use of such a correlation are obvious and users are warned to be cautious in their use of property values obtained from such correlations.

In general, the quality of the experimental data base decreases as one goes from A to K in Table 4.1, but there can be exceptions to that rule. For example, for the D through F quality ratings, only part of the correlation is supported by a reliable source data set and it is possible for the other part to be less accurate than another correlation given a G rating. Another example concerns the I rating which presumably is superior to the J rating because the I data points are smooth whereas the J data are scattered. However, the I data may be smooth because it was reported only in equation form; it may be that the original raw data which the reporting authors smoothed were more scattered than other data sets which were given a J rating.

It is not possible to characterize precisely all the experimental circumstances which can arise with a limited number of single-character quality rating symbols, and the user may feel that the definitions in Table 4.1 lack precision in some instances. That is inevitable with only eleven ratings, but the addition of more soon becomes self-defeating. Despite any such sortcomings, data used in a critical situation will have to be evaluated by the user in some way, and that must be done for the covered properties by evaluating the quantity and quality of the literature data on which the tabulated correlations are based. The ratings in Table 4.1 define a quality spectrum on which any compound's data can be located.

TABLE 4.1

Definition of quality rating symbols for the individual property correlations.

Rating	Definition
A	Correlation is based on extensive experimental data from a reliable source, and those data are substantiated by measurements from other sources over all or most of the range.
B	Correlation is based on extensive experimental data from a reliable source, but those data are substantiated by measurements from other sources over only part of the range.

TABLE 4.1 (continued)

Rating	Definition
C	Correlation is based on extensive experimental data from a reliable source, but the correlation is not corroborated by other measurements in any major part of the range.
D	Correlation is based on extensive experimental data from a reliable source over a major part of the range and on a less reliable data set in the other part. Data from other sources support the general validity of the correlation in both parts of the range.
E	Correlation is based on extensive experimental data from a reliable source over a major part of the range and on a less reliable data set in the other part. Data from other sources corroborate the data from the reliable source but supporting data are not available for the less reliable data set.
F	Correlation is based on extensive experimental data from a reliable source over a major part of the range and on a less reliable data set in the other part. Neither part of the correlation is corroborated by measurements from other sources.
G	Correlation is based on data from two or more overlapping sources which through their mutual agreement support the validity of the correlation through an important part of its range.
H	Correlation is based on a group of points from a single source or from multiple sources which has been selected from among various disagreeing data sets or scattered data points from several sources.
I	Correlation is based on relatively smooth multiple data points from one source, or from two or more sources which do not overlap much, and there is little or no data available from other sources.
J	Correlation is based on scattered data points from one or two sources.
K	Correlation is just a straight line fit of only two or more data points from one or more sources.

Chapter 5

LITERATURE CITED

   Table 5.1 lists the documents which have been cited in Chapters 1 through 4.
In general, those documents are not part of the bibliographies for the thermodynamic data presented in the compound tables. The literature sources for the
data--as identified by the Master Reference List (MRL) numbers on the first page
tabulation for each compound--are listed in Table 6.1.

TABLE 5.1

Literature cited in Chapters 1 through 4.

1   International Union of Pure and Applied Chemistry, Nomenclature of Organic
    Chemistry, Butterworths, London, 1979.
2   L. Riedel, Chemie Ing. Techn., 26 (1954) 83.
3   A.A. Frost and D.R. Kalkwarf, Chem. Phys., 21 (1953) 264.
4   D.G. Miller, J. Phys. Chem., 68 (1964) 1399.
5   W. Wagner, Cryogenics, 13 (1973) 470.
6   S. Angus, B. Armstrong and K.M. de Reuck, International Thermodynamic Tables
    of the Fluid State--7 Methane, Chemical Data Series No. 17, International
    Union of Pure and Applied Chemistry, Pergamon Press, Oxford, 1978.
7   R.D. McCarty and R.T. Jacobsen, An Equation of State for Fluid Ethylene, NBS
    Technical Note 1045, U. S. Department of Commerce, 1981.
8   R.D. Goodwin, H.M. Roder and G.C. Straty, Thermophysical Properties of
    Ethane, from 90 to 600 K at Pressures to 700 Bar, NBS Technical Note 684,
    U. S. Department of Commerce, 1976.
9   S. Angus, B. Armstrong and K.M. de Reuck, International Thermodynamic Tables
    of the Fluid State--7 Propylene (Propene), Chemical Data Series No. 25,
    International Union of Pure and Applied Chemistry, Pergamon Press, Oxford,
    1980.
10 A.W. Francis, Ind. Eng. Chem., 49 (1957) 1779.
11 H.G. Rackett, J. Chem. Eng. Data, 15 (1970) 514.
12 C.F. Spencer and R.P. Danner, J. Chem. Eng. Data, 17 (1972) 236.
13 R.D. McCarty, J. Chem. Thermodynamics, 14 (1982) 837.
14 B.A. Younglove, J. Phys. Chem. Ref. Data, 11 (1982) Supplement 1.
15 K.S. Pitzer and R.F. Curl, J. Amer. Chem. Soc., 79 (1957) 2369.
16 J.P. O'Connell and J.M. Prausnitz, Ind. Eng. Chem., Process Des. Develop.,
    6 (1967) 245.
17 K.H. Nothnagel, D.S. Abrams and J.M. Prausnitz, Ind. Eng. Chem., Process
    Des. Develop., 12 (1973) 25.
18 A. Kreglewski, J. Phys. Chem., 73 (1969) 608.
19 API 44 Tables, Selected Values of Properties of Hydrocarbons and Related
    Compounds, Thermodynamics Research Center, Texas A&M University, College
    Station, Texas.
20 TRC Tables, Selected Values of Properties of Chemical Compounds,
    Thermodynamics Research Center, Texas A&M University, College Station,
    Texas.
21 C. Tsonopoulos, AIChE J., 20 (1974) 263.
22 C. Tsonopoulos, AIChE J., 21 (1975) 827.
23 J.C. Hayden and J.P. O'Connell, Ind. Eng. Chem., Process Des. Develop.,
    14 (1975) 221.
24 O. Redlich and N.S. Kwong, Chem. Rev., 44 (1949) 233.
25 P.L. Chueh and J.M. Prausnitz, Ind. Eng. Chem., Fundam., 6 (1967) 492.

TABLE 5.1 (continued)

26 S.E.M. Haman, W.K. Chung, I.M. Elshayal and B.C.Y. Lu, Ind. Eng. Chem.,
   Process Des. Develop., 16 (1977) 51.
27 D.Y. Peng and D.B. Robinson, Ind. Eng. Chem., Fundam., 15 (1976) 59.
28 M. Benedict, G.B. Webb and L.C. Rubin, Chem. Eng. Progr., 47 (1951) 419,
   449, 571, 609.
29 S.K. Sood and G.G. Haselden, AIChE J., 16 (1970) 891.
30 K.W. Cox, J.L. Bono, Y.C. Kwok and K.E. Starling, Ind. Eng. Chem., Fundam.,
   10 (1971) 245.
31 K.E. Starling, Fluid Thermodynamic Properties of Light Petroleum Systems,
   Gulf Publishing Company, Houston, Texas, 1973.
32 H.W. Cooper and J.C. Goldfrank, Hydrocarbon Process., 46(12) (1967) 141.
33 J.H. Dymond and E.B. Smith, The Virial Coefficients of Pure Gases and
   Mixtures, Clarendon Press, Oxford, 1980.
34 B.D. Smith and O. Muthu, Definition of Recommended Values of Certain
   Thermodynamic Properties for the Ketones, NBSIR 84-2811, Final Report to
   the Office of Standard Reference Data, National Bureau of Standards, U. S.
   Department of Commerce, Washington, D. C., 1984.
35 R.R. Collerson, J.F. Counsell, R. Handley, J.F. Martin and C.H.S. Sprake,
   J. Chem. Soc., London (1965) 3697.
36 D. Ambrose, C.H.S. Sprake and R. Townsend, J. Chem. Thermodyn.
   6 (1974) 693.
37 D. Ambrose, J.H. Ellender, E.B. Lees, C.H.S. Sprake and R. Townsend, J.
   Chem. Thermodyn., 7 (1975) 453.

Chapter 6

BIBLIOGRAPHY

Table 6.1 lists the documents which contributed data for the compound
tabulations in this volume. The identifying numbers for the citations in Table
6.1 are the Master Reference List (MRL) numbers used to identify the relevant
literature documents for the individual property correlations for each compound.

TABLE 6.1

Bibliography for tabulated data in part A.

3	W.E. Ehrett and J.H. Weber, J. Chem. Eng. Data, 4 (1959) 142.
4	R.L. Nielsen and J.H. Weber, J. Chem. Eng. Data, 4 (1959) 145.
10	S. Weissman and S.E. Wood, J. Chem. Phys., 32 (1960) 1153.
25	S. Wen, Z.-K. Mo, H.-C. Chien, H.-T. Ching and W.-K. Kuang, Hua Kung Hsueh Pao, 2 (1960) 150.
33	B.V. Subbarao and C. Venkatarao, Can. J. Chem. Eng., 42 (1964) 266.
39	H. Kilian and H.-J. Bittrich, Z. Phys. Chem. (Leipzig), 230 (1965) 383.
47	R.P. Rastogi, J. Nath and J. Misra, J. Phys. Chem., 71 (1967) 1277.
70	R.L. Schmidt, J.C. Randall and H.L. Clever, J. Phys. Chem., 70 (1966) 3912.
71	H.W. Prengle, E.G. Felton and M.A. Pike, J. Chem. Eng. Data, 12 (1967) 193.
72	H. Loiseleur, J.-C. Merlin and R.A. Paris, J. Chim. Phys. Physicochim. Biol., 64 (1967) 634.
76	K.C. Reddy, S.V. Subrahmanyam and J. Bhimasenachar, J. Phys. Soc. Jap., 19 (1964) 559.
78	J. Gomez-Ibanez and C.-T. Liu, J. Chem. Phys., 65 (1961) 2148.
90	H.A. Beatty and G. Calingaert, Ind. Eng. Chem., 26 (1934) 504.
91	T. Bell and R. Wright, J. Phys. Chem., 31 (1927) 1884.
102	I. Brown, Aust. J. Sci. Res., 5 (1952) 530.
119	E. Gelus, S. Marple and M.E. Miller, Ind. Eng. Chem., 41 (1949) 1757.
122	S.V. Gorbachev and I.V. Kudryashov, Zh. Fiz. Khim., 28 (1954) 902.
123	S.V. Gorbachev and I.V. Kudryashov, Zh. Fiz. Khim., 29 (1955) 589.
124	H.G. Grimm, Z. Phys. Chem., 140 (1929) 321.
131	W. Herz, Z. Phys. Chem., 87 (1914) 63.
138	J.C. Hubbard, Phys. Rev., 30 (1910) 740.
146	R. Kremann and R. Meingast, Monatsh. Chem., 35 (1914) 1323.
147	R. Kremann, R. Meingast and F. Gugl, Monatsh. Chem., 35 (1914) 1235.
162	C. Mangold, K. Akad. Wiss., 102A (1893) 1071.
175	H.S. Myers, Ind. Eng. Chem., 47 (1955) 2215.
182	W.H. Perkin, J. Chem. Soc., 69 (1896) 1025.
189	M.A. Rosanoff, C.W. Bacon and J.F.W. Schulze, J. Amer. Chem. Soc., 36 (1914) 1993.
193	G. Scatchard, S.E. Wood and J.M. Mochel, J. Phys. Chem., 43 (1939) 119.
194	G. Scatchard, S.E. Wood and J.M. Mochel, J. Amer. Chem. Soc., 61 (1939) 3206.
196	G.C. Schmidt, Z. Phys. Chem., 99 (1921) 71.
197	G.C. Schmidt, Z. Phys. Chem., 121 (1926) 221.
212	P.E. Tahvonen, Commentat. Phys.-Math., 10 (1938) 1.
224	S. Young and E.C. Fortey, J. Chem. Soc., 83 (1903) 45.
228	T. Boublik, Collect. Czech. Chem. Commun., 28 (1963) 1771.
236	J.L. Cruetzen, R. Haase and L. Sieg, Z. Naturforsch. A, 5 (1950) 600.
241	M.B. Donald and K. Ridgway, J. App. Chem., 8 (1958) 403.
243	K. Dvorak and T. Boublik, Collect. Czech. Chem. Commun., 28 (1963) 1249.
254	Fr. Gothard and I. Minea, Rev. Chim. (Bucharest), 14 (1963) 520.
260	H.H. Reamer, B.H. Sage and W.N. Lacey, Ind. Eng. Chem., 45 (1953) 1805.
262	L. Sieg, F.I.A.T., Final Report No. 1095, Office of Military Government for Germany (U.S.), 1947.
269	G. Kortum and H.-J. Freier, Chem. - Ing. - Tech., 26 (1954) 670.
287	H.W. Prengle and G.F. Palm, Ind. Eng. Chem., 49 (1957) 1769.

TABLE 6.1 (continued)

295 J. Sameshima, J. Fac. Sci., 1 (1925) 63.
304 A.V. Storonkin and A.G. Morachevsku, Zh. Fiz. Khim., 30 (1956) 1297.
314 L.M. Watson and B.F. Dodge, Chem. Eng. Progr., 48 (1952) 73.
336 S.E. Wood and O. Sandus, J. Phys. Chem., 60 (1956) 801.
359 O. Maass and C.H. Wright, J. Amer. Chem. Soc., 43 (1921) 1098.
360 S.P. Vohra, T.L. Kang, K.A. Kobe and J.J. Mcketta, J. Chem. Eng. Data,
     7 (1962) 150.
361 J.A. Beattie, W.C. Kay and J. Kaminsky, J. Amer. Chem. Soc.,
     59 (1937) 1589.
362 B.J. Cherney, H. Marchman and R. York, Ind. Eng. Chem., 41 (1949) 2653.
363 W.W. Deschner and G.G. Brown, Ind. Eng. Chem., 32 (1940) 836.
364 P.P. Dawson and J.J. McKetta, Petrol. Refiner, 39 (1960) 151.
366 H.H. Reamer, B.H. Sage and W.N. Lacey, Ind. Eng. Chem., 41 (1949) 482.
371 W.B. Kay, Ind. Eng. Chem., 32 (1940) 353.
372 R.H. Olds, H.H. Reamer, B.H. Sage and W.N. Lacey, Ind. Eng. Chem.,
     36 (1944) 282.
374 B.H. Sage, D.C. Webster and W.N. Lacey, Ind. Eng. Chem., 29 (1937) 1188.
375 R.C. Wackher, C.B. Linn and A.V. Grosse, Ind. Eng. Chem., 37 (1945) 464.
376 J.A. Beattie, D.G. Edwards and S. Marple, J. Chem. Phys., 17 (1949) 576.
379 B.H. Sage and W.N. Lacey, Ind. Eng. Chem., 30 (1938) 673.
391 J.A. Beattie and S. Marple, J. Amer. Chem. Soc., 72 (1950) 1449.
392 C.C. Coffin and O. Maass, J. Amer. Chem. Soc., 50 (1928) 1427.
393 R.H. Olds, B.H. Sage and W.N. Lacey, Ind. Eng. Chem., 38 (1946) 301.
395 J.B. Garner, Petrol. Refiner, 24 (1945) 99.
410 J. Brewer, J. Chem. Eng. Data, 10 (1965) 113.
415 F.-W. Seemann and M. Urban, Erdoel Kohle, 16 (1963) 117.
420 D.W. Morecroft, J. Inst. Petrol., 44 (1958) 433.
424 J.A. Beattie, N. Poffenberger and C. Hadlock, J. Chem. Phys.,3 (1935) 96.
425 F.R. Morehouse and O. Maass, Can. J. Res., 5 (1931) 306.
427 W.M. Morris, B.H. Sage and W.N. Lacey, Trans. Am. Inst. Min.,
     136 (1940) 158.
429 J.A. Beattie, H.G. Ingersoll and W.H. Stockmayer, J. Amer. Chem. Soc.,
     64 (1942) 546.
432 A.J. Davenport, J.S. Rowlinson and G. Saville, Trans. Faraday Soc.,
     62 (1966) 322.
452 G.G. Haselden and P. Snowden, Trans. Faraday Soc., 58 (1962) 1515.
457 W.B. Kay, Ind. Eng. Chem., 32 (1940) 358.
477 NGAA, Ind. Eng. Chem., 34 (1942) 1240.
515 B.H. Sage, H.H. Reamer, R.H. Olds and W.N. Lacey, Ind. Eng. Chem.,
     34 (1942) 1108.
538 J. Shim and J.P. Kohn, J. Chem. Eng. Data, 7 (1962) 3.
543 H.H. Reamer, B.H. Sage and W.N. Lacey, Chem. Eng. Data Ser., 1 (1956) 29.
555 W.B. Kay, Ind. Eng. Chem., 30 (1938) 459.
556 J.H. Weber, J. Chem. Eng. Data, 4 (1959) 301.
558 G.H. Hanson, R.J. Hogan, F.N. Ruehlen and M.R. Cines, Chem. Eng. Progr.,
     49 (1953) 37.
577 G.H. Hanson, R.J. Hogan, W.T. Nelson and M.R. Cines, Ind. Eng. Chem.,
     44 (1952) 604.
578 R.J. Burch and M.W. Leeds, Chem. Eng. Data Ser., 2 (1957) 3.
579 L.C. Kahre, J. Chem. Eng. Data, 18 (1973) 267.
583 D.V.S. Jain and O.P. Yadav, J. Chem. Thermodyn., 5 (1973) 541.
587 H.W. Scheeline and E.R. Gilliland, Ind. Eng. Chem., 31 (1939) 1050.
589 G.H. Goff, P.S. Farrington and B.H. Sage, Ind. Eng. Chem., 42 (1950) 735.

TABLE 6.1 (continued)

590 G. Calingaert and L.B. Hitchcock, J. Amer. Chem. Soc., 49 (1927) 750.
599 W.B. Kay, Ind. Eng. Chem., 33 (1941) 590.
600 H.H. Reamer, B.H. Sage and W.N. Lacey, Ind. Eng. Chem., 38 (1946) 986.
608 W.B. Kay and R.E. Albert, Ind. Eng. Chem., 48 (1956) 422.
612 J.D. Kemp and C.J. Egan, J. Amer. Chem. Soc., 60 (1938) 1521.
614 L.I. Dana, A.C. Jenkins, J.N. Burdick and R.C. Timm, Refrig. Eng.,
     12 (1926) 387.
616 J.G. Aston and G.H. Messerly, J. Amer. Chem. Soc., 62 (1940) 1917.
617 J.G. Aston, R.M. Kennedy and S.C. Schumann, J. Amer. Chem. Soc.,
     62 (1940) 2059.
620 J.A. Beattie and D.G. Edwards, J. Amer. Chem. Soc., 70 (1948) 3382.
624 A.B. Lamb and E.E. Roper, J. Amer. Chem. Soc., 62 (1940) 806.
625 G.B. Kistiakowsky, J.R. Ruhoff, H.A. Smith and W.E. Vaughan, J. Amer.
     Chem. Soc., 57 (1935) 876.
679 H. Hipkin, AIChE J., 12 (1966) 484.
687 G. Scatchard, S.E. Wood and J.M. Mochel, J. Amer. Chem. Soc.,
     68 (1946) 1957.
691 G. Scatchard and F.G. Satkiewicz, J. Amer. Chem. Soc., 86 (1964) 130.
692 C.B. Kretschmer and R. Wiebe, J. Amer. Chem. Soc., 71 (1949) 3176.
694 T.D. Ling and M. Van Winkle, Ind. Eng. Chem., 3 (1958) 88.
695 L.W.T. Cummings, F.W. Stones and M.A. Volante, Ind. Eng. Chem.,
     25 (1933) 728.
697 H. Arm, F. Huegli and R. Signer, Helv. Chim. Acta., 40 (1957) 1200.
705 J.F. Connolly, J. Phys. Chem., 66 (1962) 1082.
707 C.P. Yang and M. Van Winkle, Ind. Eng. Chem., 47 (1955) 293.
708 H. Brandt and H. Roeck, Chem. - Ing. - Tech., 29 (1957) 397.
709 I. Brown and F. Smith, Aust. J. Chem., 10 (1957) 423.
719 A.R. Mathieson, J. Chem. Soc., London, (1958) 4444.
727 G. Werner and H. Schuberth, J. Prakt. Chem., 31 (1966) 225.
742 V. Mathot and A. Desmyter, J. Chem. Phys., 21 (1953) 782.
757 G. Schneider, Z. Phys. Chem. (Frankfurt am Main), 27 (1961) 171.
764 J.H. Weber, Ind. Eng. Chem., 47 (1955) 454.
774 I.A. McLure and F.L. Swinton, Trans. Faraday Soc., 61 (1965) 421.
776 E.L. Washington and R. Battino, J. Phys. Chem., 72 (1968) 4496.
777 I. Brown and F. Smith, Aust. J. Chem., 15 (1962) 9.
779 R.K. Nigam and P.P. Singh, Trans. Faraday Soc., 65 (1969) 950.
791 W.A. Duncan, J.P. Sheridan and F.L. Swinton, Trans. Faraday Soc.,
     66 (1966) 1090.
860 A.E. Jones and W.B. Kay, AIChE J., 13 (1967) 717.
905 E.J. Partington, J.S. Rowlinson and J.F. Weston, Trans. Faraday Soc.,
     56 (1960) 479.
914 W. Beyer, H. Schuberth and E. Leibnitz, J. Prakt. Chem., 27 (1965) 276.
922 M.K.D. Sanghvi and W.B. Kay, Chem. Eng. Sci., 6 (1956) 10.
929 A.A. Mamedov and G.M. Panchenkov, Zh. Fis. Khim., 29 (1955) 1204.
932 F.G. Waelbroeck, J. Chem. Phys., 23 (1955) 749.
939 S.R.M. Ellis and R.M. Contractor, Birmingham Univ. Chem. Eng.,
     15 (1964) 10.
948 A.L. Henne and K.W. Greenlee, J. Amer. Chem. Soc., 67 (1945) 484.
960 R. Rothe, Ph. D. Dissertation, Gottingen, (1958).
965 J. Brewer, AFOSR No. 67-2795, Air Force Office of Scientific Research
     (U.S.), 1967.
974 C.S. Carlson, Ph. D. Dissertation, Pennsylvania State College, University
     Park, Pennsylvania, 1939.

TABLE 6.1 (continued)

985 F. Danusso, Atti Accad. Naz. Lincei, 12-13 (1952) 131.
986 F. Danusso, Atti Accad. Naz. Lincei, 16-17 (1954) 109.
1002 D.H. Knoebel and W.C. Edmister, J. Phys. Chem., 13 (1968) 312.
1005 D.R. Swami, V.N. Kumarkrishna Rao and M. Narasinga Rao, Trans. Indian
      Inst. Chem. Eng., 9 (1958) 47.
1043 G.J. Pierotti, C.H. Deal and E.L. Derr, Ind. Eng. Chem., 51 (1959) 95.
1052 I. Brown, W. Fock and F. Smith, J. Chem. Thermodyn., 1 (1969) 273.
1088 P.G. Francis and M.L. McGlashan, J. Chem. Phys., 20 (1952) 1341.
1089 R.B. Scott and F.G. Brickwedde, J. Res. Nat. Bur. Stand., 35 (1945) 501.
1091 J.C. Rintelen, J.H. Saylor and P.M. Gross, J. Amer. Chem. Soc.,
      59 (1937) 1129.
1095 F. Glaser and H. Rueland, Chem. - Ing. - Tech., 29 (1957) 772.
1098 K.S. Pitzer and D.W. Scott, J. Amer. Chem. Soc., 65 (1943) 803.
1099 R.E. Donaldson and O.R. Quayle, J. Amer. Chem. Soc., 72 (1950) 35.
1146 D.R. Stull, Ind. Eng. Chem., 39 (1947) 517.
1173 M. Dutta, Ph. D. Dissertation, Ohio State University, Columbus,
      Ohio, 1968.
1176 G.W.A. Kahlbaum, Z. Phys. Chem., 26 (1898) 577.
1179 C.F. Muendel, Z. Phys. Chem., 85 (1913) 435.
1211 J.A. Dixon, J. Chem. Eng. Data, 4 (1959) 289.
1257 J.D. Lambert, G.A.H. Roberts, J.S. Rowlinson and V.J. Wilkinson, Proc.
      Royal Soc., 196 (1949) 113.
1266 J. Dojcansky, L. Sakalosova and J. Surovy, Chem. Zvesti, 28 (1974) 160.
1290 C.P. Smyth and E.W. Engel, J. Amer. Chem. Soc., 51 (1929) 2646.
1291 C.P. Smyth and W.N. Stoops, J. Amer. Chem. Soc., 51 (1929) 3312.
1294 C.P. Smyth and R.W. Dornte, J. Amer. Chem. Soc., 53 (1931) 545.
1339 M. Benedict, G.B. Webb and L.C. Rubin, Chem. Eng. Progr., 47 (1951) 419.
1348 R.V. Orye and J.M. Prausnitz, Trans. Faraday Soc., 61 (1965) 1338.
1359 C.B. Kretschmer, J. Nowakowska and R. Wiebe, Ind. Eng. Chem.,
      38 (1946) 506.
1360 R.R. Dreisbach and R.A. Martin, Ind. Eng. Chem., 41 (1949) 2875.
1361 R.R. Dreisbach and S.A. Shrader, Ind. Eng. Chem., 41 (1949) 2879.
1415 J.O. Hirschfelder, F.T. McClure and I.F. Weeks, J. Chem. Phys.,
      10 (1942) 201.
1485 C.B. Willingham, W.J. Taylor, J.M. Pignocco and F.D. Rossini, J. Res.
      Nat. Bur. Stand., 35 (1945) 219.
1486 A.F. Forziati, W.R. Norris and F.D. Rossini, J. Res. Nat. Bur. Stand.,
      43 (1949) 555.
1487 J.M. Geist and M.R. Cannon, Ind. Eng. Chem., 18 (1946) 611.
1488 E.A. Kelso and W.A. Felsing, Ind. Eng. Chem., 34 (1942) 161.
1489 A.F. Forziati, A.R. Glasgow, C.B. Willingham and F.D. Rossini, J. Res.
      Nat. Bur. Stand., 36 (1946) 129.
1490 A.F. Forziati and F.D. Rossini, J. Res. Nat. Bur. Stand., 43 (1949) 473.
1492 J.P. McCullough, R.E. Pennington, J.C. Smith, I.A. Hossenlopp and
      G. Waddington, J. Amer. Chem. Soc., 81 (1959) 5880.
1494 H.M. Huffman, M.E. Gross, D.W. Scott and J.P. McCullough, J. Phys. Chem.,
      65 (1961) 495.
1496 W.B. Kay, J. Amer. Chem. Soc., 69 (1947) 1273.
1497 R.J.L. Andon, J.D. Cox, E.F.G. Herington and J.F. Martin, Trans. Faraday
      Soc., 53 (1957) 1074.
1499 J.M. Stuckey and J.H. Saylor, J. Amer. Chem. Soc., 62 (1940) 2922.
1503 K.N. Marsh, J. Chem. Thermodyn., 2 (1970) 359.
1514 V. Bekarek and E. Hala, Collect. Czech. Chem. Commun., 33 (1968) 2598.

TABLE 6.1 (continued)

1520 C.B. Kretschmer, J. Nowakowska and R. Wiebe, J. Amer. Chem. Soc.,
      70 (1948) 1785.
1521 C.B. Kretschmer and R. Wiebe, J. Amer. Chem. Soc., 71 (1949) 1793.
1524 J.D. Gomez-Ibanez and T.C. Wang, J. Phys. Chem., 70 (1966) 391.
1526 K. Hlousek and E. Hala, Collect. Czech. Chem. Commun., 35 (1970) 1030.
1529 M. Diaz Pena and M. Haya, An. Real Soc. Espan. Fis. Quim., 60 (1964) 423.
1534 M.V. Prabhakara Rao and P.R. Naidu, Can. J. Chem., 52 (1974) 788.
1540 J.F. Connolly and G.A. Kandalic, J. Chem. Eng. Data, 7 (1962) 137.
1541 R.W. Stephenson and M. Van Winkle, J. Chem. Eng. Data, 7 (1962) 510.
1548 R.S. Ramalho and J. Delmas, J. Chem. Eng. Data, 13 (1968) 161.
1549 K.M. Sumer and A.R. Thompson, J. Chem. Eng. Data, 13 (1968) 30.
1559 H. Klapproth, Nova Acta Leopold., 9 (1940) 305.
1602 R.P. Rastogi, J. Nath and J. Misra, J. Phys. Chem., 71 (1967) 2524.
1613 M.B. Ewing, B.J. Levien, K.N. Marsh and R.H. Stokes, J. Chem. Thermodyn.,
      2 (1970) 689.
1631 H. Renon and J.M. Prausnitz, Ind. Eng. Chem., 7 (1968) 220.
1636 F. Schouteden and J. Deveux, Natuurwetensch. Fijdschr. (Ghent),
      18 (1936) 242.
1648 M.W. Cook, Rev. Sci. Instrum., 29 (1958) 399.
1660 M.L. McGlashan and A.G. Williamson, Trans. Faraday Soc., 57 (1961) 588.
1688 H.H. Reamer and B.H. Sage, Chem. Eng. Data Ser., 2 (1957) 9.
1690 K.N. Marsh, Trans. Faraday Soc., 64 (1968) 883.
1698 A.J.B. Cruickshank and A.J.B. Cutler, J. Chem. Eng. Data, 12 (1967) 326.
1711 H.M. Renon, Ph. D. Dissertation, University of California, Berkeley,
      California, 1966.
1716 J.B. Edwards, Ph. D. Dissertation, Georgia Institute of Technology,
      Atlanta, Georgia, 1962.
1732 R.V. Orye, Ph. D. Dissertation, University of California, Berkeley,
      California, 1965.
1740 J.J. Ljunglin, Ph. D. Dissertation, Rensselear Polytechnic Institute,
      Troy, New York, 1961.
1744 K.R. Harris and P.J. Dunlop, J. Chem. Thermodyn., 2 (1970) 813.
1745 K.R. Harris and P.J. Dunlop, J. Chem. Thermodyn., 2 (1970) 805.
1758 O. Vilim, E. Hala, J. Pick and V. Fried, Chem. Listy, 48 (1954) 989.
1769 O.R. Quayle, R.A. Day and G.M. Brown, J. Amer. Chem. Soc., 66 (1944) 938.
1770 R.W. Hermsen, Ph. D. Dissertation, University of California, Berkeley,
      California, 1962.
1771 W.B. Kay, J. Amer. Chem. Soc., 68 (1946) 1336.
1773 D.V.S. Jain, R.K. Dewan and K.K. Tewari, Indian J. Chem., 6 (1968) 511.
1776 R.P. Rastogi and K.T. Rama Varma, J. Chem. Soc., London, (1957) 2257.
1788 J. Surovy and J. Heinrich, Bratislava Slovenska Vysoka Skola Tech. Chem.
      Fakulta, 4 (1966) 201.
1792 M. Diaz Pena and D. Rodriguez Cheda, An. Quim., 66 (1970) 721.
1803 R.F. Hajjar, W.B. Kay and G.F. Leverett, J. Chem. Eng. Data,
      14 (1969) 377.
1809 J.G. Aston, H.L. Fink and S.C. Schumann, J. Amer. Chem. Soc.,
      65 (1943) 341.
1832 K. Katz and M. Newman, Ind. Eng. Chem., 48 (1956) 137.
1837 D.McA. Mason and B.E. Eakin, Chem. Eng. Data Ser., 6 (1961) 499.
1859 E.H. Amick, M.A. Weiss and M.S. Kirshenbaum, Ind. Eng. Chem.,
      43 (1951) 969.
1883 D.V.S. Jain, V.K. Gupta and B.S. Lark, Indian J. Chem., 8 (1970) 815.
1894 A.V. Grosse and C.B. Linn, J. Amer. Chem. Soc., 61 (1939) 751.

TABLE 6.1 (continued)

1898   E.E. Roper, J. Phys. Chem., 44 (1940) 835.
1899   A.W. Tickner and F.P. Lossing, J. Phys. Chem., 55 (1951) 733.
1900   H.W. Pfennig and J.J. McKetta, Petrol. Refiner, 36 (1957) 309.
1901   M.L. McGlashan and C.J. Wormald, Trans. Faraday Soc., 60 (1964) 646.
1906   L.S. Echols and E. Gelus, Anal. Chem., 19 (1947) 668.
1911   F.R. Morehouse and O. Maass, Can. J. Res., 11 (1934) 637.
1912   F.W. Jessen and J.H. Lightfoot, Ind. Eng. Chem., 30 (1938) 312.
1915   G.B. Heisig and C.D. Hurd, J. Amer. Chem. Soc., 55 (1933) 3485.
1917   M.M. Hicks-Bruun and J.H. Bruun, J. Amer. Chem. Soc., 58 (1936) 810.
1922   R.A. Ruehrwein and T.M. Powell, J. Amer. Chem. Soc., 68 (1946) 1063.
1923   C.B. Kretschmer and R. Wiebe, J. Amer. Chem. Soc., 73 (1951) 3778.
1937   G.A. Bottomley, D.S. Massie and R. Whytlaw-Gray, Proc. Royal Soc., 200 (1949) 201.
1938   M.L. McGlashan and D.J.B. Potter, Proc. Royal Soc., 267 (1962) 478.
1939   S.D. Hamann and J.F. Pearse, Trans. Faraday Soc., 48 (1952) 101.
1940   S.D. Hamann, W.J. McManamey and J.F. Pearse, Trans. Faraday Soc., 49 (1953) 351.
1941   M.Y. Shana'a and F.B. Canfield, Trans. Faraday Soc., 64 (1968) 2281.
1948   P. Harteck and R. Edse, Z. Phys. Chem., 182 (1938) 220.
1949   W. Kappallo and K. Schaefer, Z. Elektrochem., 66 (1962) 508.
1950   W. Kappallo, N. Lund and K. Schaefer, Z. Phys. Chem. (Frankfurt am Main), 37 (1963) 196.
1951   P. Sliwinski, Z. Phys. Chem. (Frankfurt am Main), 63 (1969) 263.
1952   R.N. Lichtenthaler and K. Schaefer, Ber. Bunsenges. Phys. Chem., 73 (1969) 42.
1972   B.R. Carney, Petrol. Refiner, 21 (1942) 84.
1978   A. Schulze, Ann. Phys. (Leipzig), 59 (1919) 73.
1981   J.H. Vera and J.M. Prausnitz, J. Chem. Eng. Data, 16 (1971) 149.
1985   D.D. Hanson, Ph. D. Dissertation, The University of Texas, Austin, Texas, 1966.
1991   A.F. Forziati, D.L. Camin and F.D. Rossini, Nat. Bur. Stand. (U.S.), 45 (1950) 406.
2011   J.H. Dymond and E.B. Smith, The Virial Coefficients of Pure Gases and Mixtures, Oxford University Press, New York, 1980.
2026   T. Hammerich and A. Schmitz, Erdoel Kohle, 14 (1961) 1021.
2027   C.C. Coffin and O. Maass, Proc. Trans. Roy. Soc. Can., 21 (1927) 33.
2045   V.V.G. Krishnamurty and C. Venkata Rao, J. Sci. Ind. Res., 14 (1955) 55.
2060   J.B. Conn, G.B. Kistiakowsky and E.A. Smith, J. Amer. Chem. Soc., 61 (1939) 1868.
2065   H.P. Clegg and J.S. Rowlinson, Trans. Faraday Soc., 51 (1955) 1333.
2097   S. Young, Proc. Roy. Irish Acad., 38 (1928) 65.
2098   G.B. Martinenghi, Olli Miner. Grassi Saponi, 23 (1943) 9.
2123   C. McAuliffe, J. Phys. Chem., 70 (1966) 1267.
2187   D.L. Schindler, G.W. Swift and F. Kurata, Hydrocarbon Process., 45 (1966) 205.
2209   H.G. David, S.D. Hamann and R.B. Thomas, Aust. J. Chem., 12 (1959) 309.
2238   G.B. Heisig, J. Amer. Chem. Soc., 55 (1933) 2304.
2244   M. Hirata and T. Hakuta, Mem. Fac. Technol., 18 (1968) 1595.
2281   J.W. Glanville and B.H. Sage, Ind. Eng. Chem., 41 (1949) 1272.
2283   S. Young, J. Chem. Soc., 55 (1889) 486.
2284   E.J. Gornowski, E.H. Amick and A.N. Hixson, Ind. Eng. Chem., 39 (1947) 1348.
2288   W.B. Nichols, H.H. Reamer and B.H. Sage, Ind. Eng. Chem., 47 (1955) 2219.

TABLE 6.1 (continued)

2297 S.F. Shakhova and G.E. Braude, Khim. Prom. (Moscow), 47 (1963) 436.
2299 S.F. Shakhova, G.E. Braude, Khim. Prom. (Moscow), (1964) 906.
2306 M. Hirata, T. Hakuta and T. Onoda., Sekiyu Gakkai Shi, 10 (1967) 440.
2314 J. Timmermans, Bull, 30 (1921) 62.
2320 W. Thomas, M. Zander and F. Peter, Erdoel Kohle, 16 (1963) 764.
2344 D.C.-K. Lin, I.H. Silberberg and J.J. McKetta, J. Chem. Eng. Data,
     15 (1970) 483.
2369 P. Pomerantz, A. Fookson, T.W. Mears, S. Rothberg and F.L. Howard,
     J. Res. Nat. Bur. Stand., 52 (1954) 59.
2402 G.A. Bottomley and T.A. Remmington, J. Chem. Soc., London, (1958) 3800.
2403 A. Perez Masia, M. Diaz Pena and J.A. Burriel Lluna, An. Real Soc. Espan.
     Fis. Quim., 60 (1964) 229.
2404 J.A. Beattie, S.W. Levine and D.R. Douslin, J. Amer. Chem. Soc.,
     74 (1952) 4778.
2405 M.D.G. Garner and J.C. McCoubrey, Trans. Faraday Soc., 55 (1959) 1524.
2407 M. Ratzsch and H.-J. Bittrich, Z. Phys. Chem. (Leipzig), 228 (1965) 81.
2409 J.F. Connolly and G.A. Kandalic, Phys. Fliuds, 3 (1960) 463.
2410 J.D. Cox and R.J.L. Andon, Trans. Faraday Soc., 54 (1958) 1622.
2411 D.W. Scott, G.B. Gutherie, J.F. Messerly, S.S. Todd, W.T. Berg,
     I.A. Hossenlopp and J.P. McCullough, J. Phys. Chem., 66 (1962) 911.
2412 G.A. Bottomley and I.H. Coopes, Nature (London), 193 (1962) 268.
2413 A. Eucken and L. Meyer, Z. Phys. Chem., 5 (1929) 452.
2415 J.H. Baxendale, B.V. Enustun and J. Stern, Phil. Trans. Roy. Soc. London,
     243 (1950 171.
2416 J.H. Baxendale and B.V. Enustun, Phil. Trans. Roy. Soc. London,
     243 (1950) 176.
2418 P.W. Allen, D.H. Everett and M.F. Penney, Proc. Royal Soc.,
     212 (1952) 149.
2420 G.A. Bottomley, C.G. Reeves and R. Whytlaw-Gray, Proc. Royal Soc.,
     246 (1958) 504.
2422 Sh.D. Zaalishvili and Z.S. Belousova, Russ. J. Phys. Chem.,
     38 (1964) 269.
2423 Sh.D. Zaalishvili, Z.S. Belousova and L.E. Kolysko, Russ. J. Phys. Chem.,
     39 (1965) 232.
2426 J. Timmermans and Mm. Hennault-Roland, J. Chim. Phys. Physicochim. Biol.,
     27 (1930) 401.
2475 J. Timmermans and Mm. Hennault-Roland, J. Chim. Phys. Physicochim. Biol.,
     32 (1935) 501.
2478 J. Timmermans and F. Martin, J. Chim. Phys. Physicochim. Biol.,
     25 (1928) 411.
2523 J. Timmermans and Mm. Hennault-Roland, J. Chim. Phys. Physicochim. Biol.,
     52 (1955) 223.
2560 R.F. Brunel, J. Amer. Chem. Soc., 45 (1923) 1334.
2573 J. Timmermans and F. Martin, J. Chim. Phys. Physicochim. Biol.,
     23 (1926) 747.
2575 A.R. Gordon and W.G. Hines, Can. J. Res., 24 (1946) 254.
2645 M.R. Dean and T.W. Legatski, Ind. Eng. Chem., 16 (1944) 7.
2649 S.W. Benson, Ind. Eng. Chem., 14 (1942) 189.
2650 J.G. Aston, H.L. Fink, A.B. Bestul, E.L. Pace and G.J. Szasz, J. Amer.
     Chem. Soc., 68 (1946) 52.
2651 J.A. Beattie and W.H. Stockmayer, J. Chem. Phys., 10 (1942) 473.
2653 L. Guttman and K.S. Pitzer, J. Amer. Chem. Soc., 67 (1945) 324.
2656 J.A. Beattie, G.L. Simard and G.-J. Su, J. Amer. Chem. Soc.,

TABLE 6.1 (continued)

61 (1939) 24.

2657 J.A. Beattie, H.G. Ingersoll and W.H. Stockmayer, J. Amer. Chem. Soc.,
64 (1942) 548.

2678 G.A. Bottomley and T.H. Spurling, Aust. J. Chem., 20 (1967) 1789.

2737 J.G. Aston and G.J. Szasz, J. Amer. Chem. Soc., 69 (1947) 3108.

2751 J.A. Beattie and S. Marple, J. Amer. Chem. Soc., 72 (1950) 4143.

2752 J.G. Aston, S.V.R. Mastrangelo and G.W. Moessen, J. Amer. Chem. Soc.,
72 (1950) 5287.

2768 R.N. MacCallum and J.J. McKetta, Hydrocarbon Process. Petrol. Refiner,
42 (1963) 191.

2777 W. Kozicki, R.F. Cuffel and B.H. Sage, J. Chem. Eng. Data, 7 (1962) 173.

2778 T.B. Tripp and R.D. Dunlap, J. Phys. Chem., 66 (1962) 635.

2790 J.F. Connolly, Ind. Eng. Chem., 48 (1956) 813.

2792 L.B. Petty and J.M. Smith, Ind. Eng. Chem., 47 (1955) 1258.

2814 P.B. Ayscough, K.J. Ivin and J.H. O'Donnell, Trans. Faraday Soc.,
61 (1965) 1601.

2815 J. Huisman and B.H. Sage, J. Chem. Eng. Data, 9 (1964) 536.

2817 G.B. Heisig, J. Amer. Chem. Soc., 63 (1941) 1698.

2824 B. Kalinowska and M. Woycicka, Bull. Acad. Pol. Sci., 21 (1973) 845.

2835 Anon., Oil Gas J., 44 (1945) 115.

2840 E. Kuss, Z. Angew. Phys., 7 (1955) 372.

2850 T.S. Akhundov and F.G. Abdullaev, Izv. Vyssh. Ucheb. Zaved.,
12 (1969) 44.

2851 A.M. Mamedov, T.S. Akhundov and Sh.Yu. Imanov, Russ. J. Phys. Chem.,
44 (1960) 877.

2852 A.M. Mamedov, T.S. Akhundov and N.N. Asadullaeva, Teploenergetika,
14 (1967) 81.

2894 T.W. Richards and J.W. Shipley, J. Amer. Chem. Soc., 38 (1916) 989.

2902 S. Young, Sci. Proc. Roy. Dublin Soc., 12 (1910) 374.

2920 J. Timmermans, Bull, 24 (1910) 244.

2925 D. Ambrose, B.E. Broderick and R. Townsend, J. Chem. Soc., London,
(1967) 633.

2959 H. Landolt and H. Jahn, Z. Phys. Chem., 10 (1892) 289.

2975 L.M. Heil, Phys. Rev., 39 (1932) 666.

2976 E. Cohen and J.S. Buij, Z. Phys. Chem., 35 (1937) 270.

2977 P. Bender, G.T. Furukawa and J.R. Hyndman, Ind. Eng. Chem.,
44 (1952) 387.

2979 N.W. Krase and J.B. Goodman, Ind. Eng. Chem., 22 (1930) 13.

2981 S. Young, J. Chem. Soc., 77 (1900) 1145.

2982 W.B. Kay and F.M. Warzel, Ind. Eng. Chem., 43 (1951) 1150.

2983 E.R. Smith, J. Res. Nat. Bur. Stand., 24 (1940) 229.

2985 S. Young, J. Chem. Soc., 71 (1897) 446.

2986 C.P. Smyth and W.N. Stoops, J. Amer. Chem. Soc., 50 (1928) 1883.

2987 G.H. Messerly and R.M. Kennedy, J. Amer. Chem. Soc., 62 (1940) 2988.

2991 D.E. Stewart, B.H. Sage and W.N. Lacey, Ind. Eng. Chem., 46 (1954) 2529.

2994 B.H. Sage and W.N. Lacey, Ind. Eng. Chem., 34 (1942) 730.

2996 W. Seitz and G. Lechner, Ann. Phys. (Leipzig), 49 (1916) 93.

3016 K.K. McMillin, K.A. Kobe, J.J. McKetta and M. Van Winkle, Chem. Eng. Data
Ser., 3 (1958) 96.

3020 H. Tanneberger, Ber. Deut. Chem. Ges., 66b (1933) 484.

3023 M. Lecat, Rec. Trav. Chim. Pays-Bas, 45 (1926) 620.

3024 G.B. Heisig and H.M. Davis, J. Amer. Chem. Soc., 57 (1935) 339.

3025 R.B. Scott, C.H. Meyers, R.D. Rands, F.G. Brickwedde and N. Bekkedahl,

TABLE 6.1 (continued)

J. Res. Nat. Bur. Stand., 35 (1945) 39.

3026 R.B. Scott, W.J. Ferguson and F.G. Brickwedde, J. Res. Nat. Bur. Stand., 33 (1944) 1.

3036 G.W. Rathjens and W.D. Gwinn, J. Amer. Chem. Soc., 75 (1953) 5629.

3047 J. Dojcansky, J. Heinrich and J. Surovy, Chem. Zvesti, 22 (1968) 514.

3062 P.W. Bridgman, Daedalus (Boston), 66 (1931) 185.

3067 G.A. Bottomley and T.H. Spurling, Aust. J. Chem., 17 (1964) 501.

3087 T. Tonomura, Sci. Rep. Tohoku Univ., 22 (1933) 104.

3101 N. Barrulescu, J. Chim. Phys. Physicochim. Biol., 29 (1932) 418.

3118 H.H. Gilmann and P. Gross, J. Amer. Chem. Soc., 60 (1938) 1525.

3144 L.A. Wood and C.F. Higgins, India Rubber World, 107 (1943) 475.

3249 A. Giacalone, Gazz. Chim. Ital., 72 (1942) 370.

3256 M. Hirata, S. Suda, T. Hakuta and K. Nagahama, Sekiyu Gakkai Shi, 12 (1969) 773.

3262 M.S. Rozhnov, Khim. Prom. (Moscow), 43 (1967) 288.

3263 J. Timmermans, Physico-Chemical Constants of Pure Organic Compounds, Elsevier Publishing Company, Amsterdam, 1965.

3269 B.J. Zwolinski (Ed.), Selected Values of Properties of Hydrocarbons and Related Compounds, American Petroleum Institute Project 44, Thermodynamics Research Center, Texas A & M University, College Station, Texas, 1970.

3317 S. Young, J. Chim. Phys. Physicochim. Biol., 4 (1906) 425.

3318 M. Hirata and S. Suda, Bull. Jap. Petrol. Inst., 10 (1968) 20.

3383 J. Timmermans and Hennault-Roland, J. Chim. Phys. Physicochim. Biol., 56 (1959) 984.

3386 M. Raja Rao and C. Venkata Rao, J. App. Chem., 7 (1957) 659.

3396 M. Lecat, Ann. Soc. Sci. Bruxelles, 45 (1926) 169.

3491 S.W. Benson, Ind. Eng. Chem., 13 (1941) 502.

3569 J.L.H. Wang and B.C.-Y. Lu, J. App. Chem., 21 (1971) 297.

3571 E.W. Funk, F.-C. Chai and J.M. Prausnitz, J. Chem. Eng. Data, 17 (1972) 24.

3610 D. Tyrer, J. Soc. Chem. Ind., 105 (1914) 2534.

3646 W.A. Scheller, A.R. Torres-Soto and K.J. Daphtary, J. Chem. Eng. Data, 14 (1969) 17.

3760 M. Lecat, Ann. Soc. Sci. Bruxelles, 50 (1930) 21.

3773 T. Boublik, V.T. Lam, S. Murakami and G.C. Benson, J. Phys. Chem., 73 (1969) 2356.

3807 T.G. Hartmann, M. S. Thesis, Washington University, St. Louis, Missouri, 1972.

3822 A.N. Gorbunov, M.P. Susarev and I.M. Balashova, Zh. Prikl. Khim. (Leningrad), 41 (1968) 312.

3855 L.J. Hirth, H.G. Harris and J.M. Prausnitz, AIChE J., 14 (1968) 812.

3894 G.F. Meehan, Ph. D. Dissertation, Virginia Polytechnic Institute, Blacksburg, Virginia, 1965.

3907 V.H. Khan and S.V. Subrahmanyam, Trans. Faraday Soc., 67 (1971) 2282.

3910 G.A. Bottomley, and C.G. Reeves, J. Chem. Soc., London, (1958) 3794.

3963 W.B. Kay, J. Chem. Eng. Data, 15 (1970) 46.

3965 J.M. Lenoir, C.J. Rebert and H.G. Hipkin, J. Chem. Eng. Data, 16 (1971) 401.

4015 J.F. Skinner, E.L. Cussler and R.M. Fuoss, J. Phys. Chem., 72 (1968) 1057.

4033 J.L. Hales, E.B. Lees and D.J. Ruxton, Trans. Faraday Soc., 63 (1967) 1876.

TABLE 6.1 (continued)

4060 I.B. Rabinovich, Akad. Nauk SSSR Stroenie Veshchestva I Spectroskopia, (1960) 62.
4076 H. Arm, D. Bankay, K. Strub and M. Walti, Helv. Chim. Acta., 50 (1967) 1013.
4089 K. Strein, R.N. Lichtenthaler, B. Schramm and Kl. Schafer, Ber. Bunsenges. Phys. Chem., 75 (1971) 1308.
4092 C.D. Hurd and R.N. Meinert, J. Amer. Chem. Soc., 53 (1931) 289.
4094 A.W. Francis, Ind. Eng. Chem., 49 (1957) 1779.
4114 D.V.S. Jain, O.P. Yadav and K.N. Kamra, Indian J. Chem., 9 (1971) 1262.
4125 K. Hachmuth, Ind. Eng. Chem., 24 (1932) 82.
4127 C. Holloway and S.H. Thurber, Ind. Eng. Chem., 36 (1944) 980.
4139 A.-L. Vierk, Z. Anorg. Chem., 261 (1950) 283.
4184 K. Quitzsch, J. Prakt. Chem., 35 (1967) 49.
4216 E. Kuss and M. Talsimi, Chem. - Ing. - Tech., 42 (1970) 1073.
4220 V.C. Maripuri and G.A. Ratcliff, J. App. Chem., 22 (1972) 899.
4233 I.P.C. Li, Y-W. Wong, S-D. Chang and B.C.-Y. Lu, J. Chem. Eng. Data., 17 (1972) 492.
4234 V.G. Skripka, I.E. Nikitina, L.A. Zhdanovich, A.G. Sirotin and O.A. Ben'yaminovich, Gazov. Prom., 15 (1970) 35.
4274 A. Carli, S. DiCave and E. Sebastiani, Chem. Eng. Sci., 27 (1972) 993.
4287 M. Hirata, S. Suda, T. Hakuta and K. Nagahama, Mem. Fac. Technol., 19 (1969) 103.
4298 K.S. Pitzer, L. Guttman and E.F. Westrum, J. Amer. Chem. Soc., 68 (1946) 2209.
4328 B. Ringel, Wiss. Z. Univ. Rostock, 18 (1969) 919.
4343 W.B. Kay, J. Chem. Eng. Data, 16 (1971) 137.
4362 S. Bywater, J. Polym. Sci., 9 (1952) 417.
4372 F. Straus and L. Kollek, Ber. Deut. Chem. Ges., 59b (1926) 1664.
4392 E.C. Linder, J. Phys. Chem., 35 (1931) 531.
4418 R.R. Collerson, J.F. Counsell, R. Handley, J.F. Martin and C.H.S. Sprake, J. Chem. Soc., London, (1965) 3697.
4422 T. Hakuta, K. Nagahama and M. Hirata, Bull. Jap. Petrol. Inst., 11 (1969) 10.
4423 C. Drucker, E. Jimeno and W. Kangro, Z. Phys. Chem., 90 (1915) 513.
4450 J. Huisman, Ph. D. Dissertation, California Institute of Technology, Pasadena, California, 1964.
4452 A. Hopfner, U.T. Kreibich and Kl. Schafer, Ber. Bunsenges. Phys. Chem., 74 (1970) 1016.
4455 J. Jose, R. Philippe and P. Clechet, Bull. Soc. Chim. Fr., 74 (1971) 2860.
4457 H.S. Myers, Ind. Eng. Chem., 47 (1955) 1659.
4462 D.C.-K. Lin, Ph. D. Dissertation, University of Texas, Austin, Texas, 1969.
4463 S.W. Chun, Ph. D. Dissertation, Ohio State University, Columbus, Ohio, 1964.
4470 M. Diaz Pena and C.P. Sotomayor, An. Quim., 67 (1971) 233.
4514 D.R. Laurance and G.W. Swift, J. Chem. Eng. Data, 17 (1972) 333.
4515 D.B. Manley and G.W. Swift, J. Chem. Eng. Data, 16 (1971) 301.
4519 V.V. Udovenko and L.G. Fatkulina, Zh. Fiz. Khim., 26 (1952) 719.
4522 J.A. Oxley, M.S. Thesis, Ohio State University, Columbus, Ohio, 1962.
4565 D.D. Deshpande and L.G. Bhatgadde, Aust. J. Chem., 24 (1971) 1817.
4576 E. Nicolini and P. Laffitte, C. R. Acad. Sci., 229 (1949) 757.
4604 T.M. Lesteva and E.I. Khrapkova, Zh. Fis. Khim., 45 (1971) 3113.

TABLE 6.1 (continued)

4606 J. Timmermans, Bull, 26 (1912) 205.
4612 P.S. Puri and K.S.N. Raju, J. Chem. Eng. Data, 15 (1970) 480.
4614 K.S. Howard, L.W. Hammond, R.A. McAllister and F.P. Pike, J. Phys. Chem.,
      62 (1958) 1597.
4651 G.A. Nowak, Seifen, 87 (1961) 405.
4670 M.J. Timmermans and Mme. Hennaut-Roland, J. Chim. Phys. Physicochim.
      Biol., 34 (1937) 693.
4724 A.P. Van der Vet, Density, Compressibility and Expansion of Light
      Hydrocarbons and Hydrocarbon Blends, Presented at the Second World
      Petroleum Congress, Sect. 2, Phys. Chim. Raffinage, 1937, pp. 515-521.
4816 G.F. Carruth and R. Kobayashi, J. Chem. Eng. Data, 18 (1973) 693.
4818 D.V.S. Jain, V.K. Gupta and B.S. Lark, J. Chem. Thermodyn., 5 (1973) 693.
4821 T. Treszczanowicz and H. Kehiaian, Bull. Acad. Pol. Sci., 21 (1973) 693.
4842 J.B. Ferguson, M. Freed and A.C. Morris, J. Phys. Chem., 37 (1933) 87.
4844 L. Boublikova and B.C-Y. Lu, J. App. Chem., 19 (1969) 89.
4856 I. Klesper, Z. Phys. Chem. (Frankfurt am Main), 51 (1966) 1.
4867 H. Wolff and H.-E. Hoeppel, Ber. Bunsenges. Phys. Chem., 72 (1968) 710.
4870 M. Scheller, H. Schuberth and H.G. Koennecke, J. Prakt. Chem.,
      311 (1969) 974.
4886 A.N. Campbell and S.C. Anand, Can. J. Chem., 50 (1972) 1109.
4891 L.L. Lee and W.A. Scheller, J. Chem. Eng. Data, 12 (1967) 497.
4893 S.M.K.A.Gurukul and B.N. Raju, J. Chem. Eng. Data, 11 (1966) 501.
4895 M.S. Reddy and C. Venkata Rao, J. Chem. Eng. Data, 10 (1965) 309.
4896 D.E.G. Jones, I.A. Weeks, S.C. Anand, R.W. Wetmore and G.C. Benson,
      J. Chem. Eng. Data, 17 (1972) 501.
4908 G.O. Oakeson and J.H. Weber, J. Chem. Eng. Data, 5 (1960) 279.
4909 J.M. Skaates and W.B. Kay, Chem. Eng. Sci., 19 (1964) 431.
4938 G.J. Besserer and D.B. Robinson, J. Chem. Eng. Data, 18 (1973) 301.
4939 I.P.-C. Li and B.C.-Y. Lu, J. Chem. Eng. Data, 18 (1973) 305.
4946 N.A. Smirnova and L.M. Kurtynina, Russ. J. Phys. Chem., 43 (1969) 1059.
4949 N.P. Markuzin and V.N. Baidin, Vestn. Leningrad. Univ., 2 (1973) 77.
4972 A.N. Marinichev and M.P. Susarev, Zh. Prikl. Khim. (Leningrad),
      38 (1965) 1619.
4983 K. Sosnkowska-Kehiaian, W. Recko and W. Woycicki, Bull. Acad. Pol. Sci.,
      14 (1966) 475.
5017 M.B. Ewing and K.N. Marsh, J. Chem. Thermodyn., 5 (1973) 651.
5041 E.R. Hopke and G.W. Sears, J. Amer. Chem. Soc., 70 (1948) 3801.
5043 R. Isaac, K. Li and L.N. Canjar, Ind. Eng. Chem., 46 (1954) 199.
5045 F.S. Fawcett, Ind. Eng. Chem., 38 (1946) 338.
5048 H.O. Day, D.E. Nicholson and W.A. Felsing, J. Amer. Chem. Soc.,
      70 (1948) 1784.
5051 J.A. Beattie, S.W. Levine and D.R. Douslin, J. Amer. Chem. Soc.,
      73 (1951) 4431.
5054 J.A. Beattie, D.R. Douslin and S.W. Levine, J. Chem. Phys.,
      19 (1951) 948.
5056 W.A. Wright, J. Phys. Chem., 37 (1933) 233.
5059 J.G. Aston and G.H. Messerly, J. Amer. Chem. Soc., 58 (1936) 2354.
5063 E.R. Nicolini, Ann. Chim. (Paris), 6 (1951) 582.
5065 M.W. Lister, J. Amer. Chem. Soc., 63 (1941) 143.
5067 K.A. Kobe, H.R. Crawford and R.W. Stephenson, Ind. Eng. Chem.,
      47 (1955) 1767.
5068 M. Diaz Pena and D. Rodriguez Cheda, An. Quim., 66 (1970) 737.
5070 M. Diaz Pena and D. Rodriquez Cheda, An. Quim., 66 (1970) 747.

TABLE 6.1 (continued)

5101 M. Schmitt, The Vaporization of Hydrocarbons and Mixtures of Hydrocarbons, Publications Scientifiques et Techniques du Ministere de l'Air, No. 54, Paris, 1934.

5113 L. Boublikova, Collect. Czech. Chem. Commun., 38 (1973) 2033.

5115 J. Linek and I. Wichterle, Collect. Czech. Chem. Commun., 38 (1973) 1846.

5124 P.G. McCracken, T.S. Storvick and J.M. Smith, J. Chem. Eng. Data, 5 (1960) 130.

5173 H.A. Clarke and R.W. Missen, J. Chem. Eng. Data, 19 (1974) 343.

5182 L.S. Kudryavtseva and M.P. Susarev, J. Appl. Chem. USSR, 36 (1963) 1471.

5199 W. Seitz, H. Alterthum and G. Lechner, Ann. Phys. (Leipzig), 49 (1916) 85.

5203 J.A.A. Ketelaar and N. Van Meurs, Rec. Trav. Chim. Pays-Bas, 76 (1957) 437.

5204 J. Timmermans, Sci. Proc. R. Dublin Soc., 13 (1912) 310.

5214 Sh.D. Zaalishvili and L.E. Kolysko, Russ. J. Phys. Chem., 34 (1960) 1223.

5215 W.J. Boyne, Ph. D. Dissertation, Purdue University, Lafayette, Indiana, 1952.

5231 G.J. Quintanilla, Rev. Quim. Ing. Quim., 2 (1956) 23.

5233 C. Koeppel, Gas Wasserfach, 83 (1940) 73.

5235 R.S. Myers and H.L. Clever, J. Chem. Thermodyn., 2 (1970) 53.

5283 M.M. Abbott, Ph. D. Dissertation, Rensselaer Polytechnic Institute, Troy, New York, 1965.

5310 M.Y., Shana'a, Ph. D. Dissertation, University of Oklahoma, Norman, Oklahoma, 1966.

5311 D.P. Tassios, Ph. D. Dissertation, University of Texas, Austin, Texas, 1967.

5321 S.N. Subbanna, Ph. D. Dissertation, University of Florida, Gainesville, Florida, 1969.

5364 L. Massart, Bull. Soc. Chim. Belg., 45 (1936) 76.

5572 E. Kordes, Z. Elektrochem. Angew. Phys. Chem., 58 (1954) 76.

5635 L.M. Besley and G.A. Bottomley, J. Chem. Thermodyn., 6 (1974) 577.

5636 T. Boublik and G.C. Benson, Can. J. Chem., 50 (1972) 1978.

5638 M.B. Ewing and K.N. Marsh, J. Chem. Thermodyn., 6 (1974) 35.

5640 M.B. Ewing and K.N. Marsh, J. Chem. Thermodyn., 6 (1974) 395.

5641 T.M. Letcher and F. Marsicano, J. Chem. Thermodyn., 6 (1974) 509.

5718 A.G. Osborn and D.R. Douslin, J. Chem. Eng. Data, 19 (1974) 114.

5719 D.R. Laurance and G.W. Swift, J. Chem. Eng. Data, 19 (1974) 61.

5736 L. Deffet, Bull, 40 (1931) 385.

5759 W.H. Perkin, J. Chem. Soc., 45 (1884) 421.

5772 F.V. Grimm and W.A. Patrick, J. Amer. Chem. Soc., 45 (1923) 2794.

5775 F.H. Stross, C.M. Gable and G.C. Rounds, J. Amer. Chem. Soc., 69 (1947) 1629.

5839 M.J. Timmermans and Mme. Hennault-Roland, J. Chim. Phys. Physicochim. Biol.,29 (1932) 529.

5842 J.W. Bruhl, Justus Liebigs Ann. Chem., 203 (1880) 1.

5846 K.G. Falk, J. Amer. Chem. Soc., 31 (1909) 806.

5858 K.G. Falk, J. Amer. Chem. Soc., 31 (1909) 86.

5866 R. Schiff, Justus Liebigs Ann. Chem., 220 (1883) 71.

5868 M.A. von Reis, Ann. Phys. (Leipzig), 13 (1881) 447.

5881 Y. Sassa, R. Konishi and T. Katayama, J. Chem. Eng. Data, 19 (1974) 44.

5882 N.N. Nagornov and L.A. Rotinyantz, Izv. Inst. Fiz.-Khim. Anal., 3 (1926) 162.

5936 S. Cabani and N. Ceccanti, J. Chem. Thermodyn., 5 (1973) 9.

TABLE 6.1 (continued)

5948 D.M. Knoebel, Ph. D. Dissertation, Oklahoma State University, Stillwater, Oklahoma, 1967.
5990 P. Walden and R. Swinne, Z. Phys. Chem. (Leipzig), 79 (1912) 700.
5992 S.A. Sanni and P. Hutchison, J. Chem. Eng. Data, 18 (1973) 317.
5994 S.A. Sanni, C.J.D. Fell and H.P. Hutchison, J. Chem. Eng. Data, 16 (1971) 424.
5995 L.C. Kahre, J. Chem. Eng. Data, 18 (1973) 267.
6041 J.B. Rodosevich and R.C. Miller, AIChE J., 19 (1973) 729.
6194 R.M. Gibbons and G.P. Kuebler, Technical Report AFML-TR-68-370, Air Force Materials Laboratory, Wright-Patterson Air Force Base, Ohio, 1968.
6237 G.J. Ostling, J. Chem. Soc., 101 (1912) 457.
6253 W.B. Kay and H.A. Fisch, AIChE J., 4 (1958) 293.
6284 D.W. Scott, G. Waddington, J.C. Smith and H.M. Huffman, J. Amer. Chem. Soc., 71 (1949) 2767.
6298 J.N. Friend and W.D. Hargreaves, Phil. Mag., 35 (1944) 57.
6309 V. Fried, J. Pick, E. Hala and O. Vilim, Collect. Czech. Chem. Commun., 22 (1956) 1535.
6310 R. Willstaetter and J. Bruce, Ber. Deut. Chem. Ges., 46 (1907) 3979.
6351 W. Kozicki, Ph. D. Dissertation, California Institute of Technology, Pasadena, California, 1962.
6359 K. Strein, R.N. Lichtenthaler, B. Schramm and Kl. Schaefer, Ber. Bunsenges. Phys. Chem., 75 (1971) 1308.
6363 G.L. Thomas and S. Young, J. Chem. Soc., 67 (1895) 1071.
6372 J. Rose-Innes and S. Young, Phil. Mag., 47 (1899) 353.
6392 J.A. Beattie, D.R. Douslin and S.W. Levine, J. Chem. Phys., 20 (1952) 1619.
6397 J.E. Kilpatrick and K.S. Pitzer, J. Amer. Chem. Soc., 68 (1946) 1066.
6408 L.W. Hammond, K.S. Howard and R.A. McAllister, J. Phys. Chem., 62 (1958) 637.
6450 D.W. Scott, H.L. Finke, J.P. McCullough, M.E. Gross, J.F. Messerly, R.E. Pennington and G. Waddington, J. Amer. Chem. Soc., 77 (1955) 4993.
6458 G. Waddington, J.C. Smith and H.M. Huffman, J. Amer. Chem. Soc., 71 (1949) 3902.
6465 S.C. Schumann, J.G. Aston and M. Sagenkahn, J. Amer. Chem. Soc., 64 (1942) 1039.
6471 D.W. Scott, M.E. Gross, G.D. Oliver and H.M. Huffman, J. Amer. Chem. Soc., 71 (1949) 1634.
6504 I.H. Silberberg, J.J. McKetta and K.A. Kobe, J. Chem. Eng. Data, 4 (1959) 323.
6516 L.T. Carmichael, B.H. Sage and W.N. Lacey, Ind. Eng. Chem., 45 (1953) 2697.
6518 D.L. Camin and F.D. Rossini, J. Chem. Eng. Data, 5 (1960) 368.
6520 P. Chaiyavech and M. Van Winkle, J. Chem. Eng. Data, 4 (1959) 53.
6525 H.R. Heichelheim and J.J. McKetta, Chem. Eng. Progr., 59 (1963) 23.
6541 J.F. Lemons and W.A. Felsing, J. Amer. Chem. Soc., 65 (1943) 46.
6550 H.L. Finke, D.W. Scott, M.W. Gross, J.F. Messerly and G. Waddington, J. Amer. Chem. Soc., 78 (1956) 5469.
6552 H.O. Day and W.A. Felsing, J. Amer. Chem. Soc., 74 (1952) 1951.
6553 S. Young and E.C. Fortey, J. Chem. Soc., 77 (1900) 1126.
6554 S. Young and E.C. Fortey, J. Chem. Soc., 75 (1899) 873.
6588 D.G. Elliot, R.J.J. Chen, P.S. Chappelear and R. Kobayashi, J. Chem. Eng. Data, 19 (1974) 71.

TABLE 6.1 (continued)

6627 D.W. Scott and G. Waddington, J. Amer. Chem. Soc., 72 (1950) 4310.
6629 L.I. Smith and L.J. Spillane, J. Amer. Chem. Soc., 62 (1940) 2639.
6632 L.S. Kassel, J. Amer. Chem. Soc., 58 (1936) 670.
6633 F.J. Soday and C.E. Boord, J. Amer. Chem. Soc., 55 (1933) 3293.
6634 A. Smith and A.W. Menzies, J. Amer. Chem. Soc., 32 (1910) 1448.
6646 N. Bekkedahl, L.A. Wood and M. Wojciechowski, J. Res. Nat. Bur. Stand., 17 (1936) 883.
6655 K.N. Campbell and L.T. Eby, J. Amer. Chem. Soc., 63 (1941) 2683.
6656 D.B. Brooks, F.L. Howard and H.C. Crafton, J. Res. Nat. Bur. Stand., 24 (1940) 33.
6659 L. Guttman, E.F. Westram and K.S. Pitzer, J. Amer. Chem. Soc., 65 (1943) 1246.
6660 G. Calingaert, H. Soroos, V. Hnizda and H. Shapiro, J. Amer. Chem. Soc., 66 (1944) 1389.
6662 H.E. Clements, K.V. Wise and S.E.J. Johnsen, J. Amer. Chem. Soc., 75 (1953) 1593.
6664 J.L. Finck and R.M. Wilhelm, J. Amer. Chem. Soc., 47 (1925) 1577.
6668 D.L. Camin and F.D. Rossini, J. Phys. Chem., 59 (1955) 1173.
6669 D.L. Camin and F.D. Rossini, J. Phys. Chem., 60 (1956) 1446.
6737 J.P. Wibaut, H. Hoog, S.L. Langedijk, J. Overhoff and J. Smittenberg, Rec. Trav. Chim. Pays-Bas, 58 (1939) 329.
6740 H. Tschamler and F. Kohler, Monatsh. Chem., 81 (1950) 463.
6745 F. Wiegner, Z. Elektrochem., 47 (1941) 163.
6751 K. Li and L.N. Canjar, Chem. Eng. Progr., 49 (1953) 147.
6763 E.P. Sokolova and A.G. Morachevskii, Vestn. Leningrad. Univ., 22 (1967) 98.
6764 A.G. Morachevskii and E.G. Komarova, Vestn. Leningrad. Univ., 12 (1957) 118.
6776 J. Jakubicek, V. Fried and J. Vahala, Chem. Listy, 51 (1957) 1422.
6800 P.K. York and W.A. Felsing, Tex. J. Sci., 4 (1952) 261.
6802 G. Milazzo, Ann. Chim. (Rome), 46 (1956) 1105.
6817 O.A. Nelson and C.E. Senseman, Ind. Eng. Chem., 14 (1922) 58.
6838 K. Auwers, Justus Liebigs Ann. Chem., 419 (1919) 92.
6856 Sh.D. Zaalishvili, Russ. J. Phys. Chem., 34 (1960) 46.
6878 F.J. Krieger and H.H. Wenzke, J. Amer. Chem. Soc., 60 (1938) 2115.
6886 P. Pomerantz, A. Fookson, T.W. Mears, S. Rothberg and F.L. Howard, J. Res. Nat. Bur. Stand., 52 (1954) 51.
6893 E.R. Smith, J. Res. Nat. Bur. Stand., 26 (1941) 129.
6894 E.R. Smith and H. Matheson, J. Res. Nat. Bur. Stand., 20 (1938) 641.
6947 B. Woringer, Z. Phys. Chem., 34 (1900) 257.
6951 J.T. Barker, Z. Phys. Chem., 71 (1910) 235.
6954 A. Neckel and H. Volk, Monatsh. Chem., 88 (1957) 925.
6962 C.W. Hack and M. Van Winkle, Ind. Eng. Chem., 46 (1954) 2392.
6970 E.G. Foehr and M.R. Fenske, Ind. Eng. Chem., 41 (1949) 1956.
6986 I. Wichterle and R. Kobayashi, J. Chem. Eng. Data, 17 (1972) 4.
7040 R.L. Denyer, F.A. Fidler and R.A. Lowry, Ind. Eng. Chem., 41 (1949) 2727.
7091 D.A. Zanolini, M. S. Thesis, Pennsylvania State University, University Park, Pennsylvania, 1964.
7093 R.D. Gunn, M. S. Thesis, University of California, Berkeley, California, 1958.
7148 J.R. Tomlinson, Liquid Densities of Ethane, Propane, and Ethane + Propane Mixtures, Technical Publication I, Natural Gas Processors Association, Tulsa, Oklahoma, 1971.

TABLE 6.1 (continued)

7195 E.A. Bried and G.F. Hennion, J. Amer. Chem. Soc., 59 (1937) 1310.
7234 S.M. Dubrovskii and K.V. Afonina, Ukr. Khim. Zh., 40 (1974) 465.
7291 A.W. Crossley and N. Renouf, J. Chem. Soc., 89 (1906) 26.
7333 D.R. Douslin, Proc. Div. Refining, 50 (1970) 189.
7342 J.-L. Gustin and H. Renon, Bull. Soc. Chim. Fr., 50 (1974) 2719.
7343 B. Gredy, Bull. Soc. Chim. Fr., 2 (1935) 1951.
7362 G. Chiurdoglu, Bull. Soc. Chim. Belg., 60 (1951) 39.
7372 J. Livingston, R. Morgan and E.C. Stone, J. Amer. Chem. Soc.,
      35 (1913) 1505.
7404 A. Schulze, Z. Phys. Chem., 97 (1921) 417.
7406 E.C. Baughan, A.L. Jones and K. Stewart, Proc. Royal Soc.,
      225 (1954) 478.
7439 D.M. Cowan, G.H. Jeffery and A.I. Vogel, J. Chem. Soc., (1940) 171.
7452 A.N. Campbell, E.M. Kartzmark and W.E. Falconer, Can. J. Chem.,
      36 (1958) 1475.
7459 S. Young, J. Chem. Soc., 73 (1898) 675.
7465 A.ElH.N.Mousa, W.B. Kay and A. Kreglewski, J. Chem. Thermodyn.,
      4 (1972) 301.
7470 R.H. Cole, J. Chem. Phys., 9 (1941) 251.
7481 J.-M. Crafts, J. Chim. Phys. Physicochim. Biol., 13 (1915) 105.
7483 G. Chancel, C. R. Acad. Sci., 99 (1884) 1053.
7697 R.M. Varushchenko and L.L. Bulgakova, Russ. J. Phys. Chem.,
      47 (1973) 1491.
7706 L.E. Kolysko, Z.S. Belousova, T.D. Sulimova, L.V. Mozhginskaya and
      V.M. Prokhorov, Russ. J. Phys. Chem., 47 (1973) 1067.
7746 D.D. Kalafati, D.S. Rasskazov and E.K. Petrov, Russ. J. Phys. Chem.,
      41 (1967) 720.
7812 B.H.G. Brady and J.H. O'Donnell, Trans. Faraday Soc., 64 (1968) 23.
7814 S.G. Canagaratna, D. Margerison and J.P. Newport, Trans. Faraday Soc.,
      62 (1966) 3058.
7815 D.H. Everett and F.L. Swinton, Trans. Faraday Soc., 59 (1963) 2476.
7820 W.J. Gaw and F.L. Swinton, Trans. Faraday Soc., 64 (1968) 2023.
7828 A.V. Anantaraman, S.N. Bhattacharyya and S.R. Palit, Trans. Faraday Soc.,
      59 (1963) 1101.
7862 G.F. Carruth, Ph. D. Dissertation, Rice University, Houston, Texas, 1970.
7867 G.H. Findenegg, Monatsh. Chem., 101 (1970) 1081.
7869 G. Miksch, E. Liebermann and F. Kohler, Monatsh. Chem., 100 (1969) 1574.
7870 A. Neckel and H. Volk, Monatsh. Chem., 95 (1964) 822.
7872 L. Ebert, H. Tschamler and F. Kohler, Monatsh. Chem., 82 (1957) 63.
7873 A. Neckel and F. Kohler, Monatsh. Chem., 87 (1956) 176.
7879 G. Geiseler and H. Kohler, Ber. Bunsenges. Phys. Chem., 72 (1968) 697.
7881 H. Wolff and A. Hopfner, Ber. Bunsenges. Phys. Chem., 71 (1967) 461.
7888 U.T. Kreibich, K. Schafer and A. Hopfner, Ber. Bunsenges. Phys. Chem.,
      74 (1970) 1020.
7890 H. Wolff and H.E. Hoppel, Ber. Bunsenges. Phys. Chem., 72 (1968) 722.
7892 H. Wolff and H.E. Hoppel, Ber. Bunsenges. Phys. Chem., 70 (1966) 874.
7904 F. Neubeck, Z. Phys. Chem., 1 (1887) 649.
7924 K. Quitzsch, D. Strittmatter and G. Geiseler, Z. Phys. Chem. (Leipzig),
      240 (1969) 107.
7926 G. Schneider, Z. Phys. Chem. (Frankfurt am Main), 24 (1960) 165.
7927 H. Rock and G. Schneider, Z. Phys. Chem. (Frankfurt am Main),
      8 (1956) 154.
7929 H. Rock and L. Sieg, Z. Phys. Chem. (Frankfurt am Main), 3 (1955) 355.

TABLE 6.1 (continued)

7931 D.F. Saunders and A.J.B. Spaull, Z. Phys. Chem. (Frankfurt am Main), 28 (1961) 332.

7948 H. Loiseleur, J.C. Merlin and R.A. Paris, J. Chim. Phys. Physicochim. Biol., 61 (1964) 1231.

7952 A. Zmaczynski, J. Chim. Phys. Physicochim. Biol., 27 (1930) 503.

8031 V.J. Johnson (Ed.), A Compendium of the Properties of Materials at Low Temperatures (Phase 1): I. Properties of Fluids, WADD Technical Report 60-56. Part I. National Bureau of Standards (U.S.), 1960.

8069 A. Iguchi, Kagaku Sochi, 10 (1968) 74.

8098 S.C. Anand, J.-P.E. Grolier, O. Kiyohara, C.J. Halpin and G. C. Benson, J. Chem. Eng. Data, 20 (1975) 184.

8111 S.M. Khodeeva, E.S. Lebedeva and Z.S. Belousova, Russ. J. Phys. Chem., 40 (1966) 1669.

8125 J.W.M. Boelhouwer, G.W. Nederbragt and G. Verberg, App. Sci. Res., 2 (1950) 249.

8148 L. Clarke and E.R. Riegel, J. Amer. Chem. Soc., 34 (1912) 674.

8149 L. Clarke, J. Amer. Chem. Soc., 33 (1911) 520.

8226 F. Krafft and H. Weilandt, Ber. Deut. Chem. Ges., 29 (1896) 1316.

8310 D.W. Scott, J.P. McCullough, K.D. Williamson and G. Waddington, J. Amer. Chem. Soc., 73 (1951) 1707.

8330 G.A. Bottomley and T.H. Spurling, Nature (London), 195 (1962) 900.

8335 A.R. Glasgow, J. Res. Nat. Bur. Stand., 24 (1940) 509.

8354 K.N. Campbell and M.J. O'Connor, J. Amer. Chem. Soc., 61 (1939) 2897.

8367 F. Cortese, J. Amer. Chem. Soc., 51 (1929) 2266.

8383 J. Miksovsky and I. Wichterle, Collect. Czech. Chem. Commun., 40 (1975) 365.

8397 N.D. Zelinsky and B.A. Kasansky, Ber. Deut. Chem. Ges., 60 (1927) 1101.

8404 B.K. Merezhkovskii, Zh. Russ. Fiz. - Khim. Obshchest, 45 (1913) 1940.

8405 W.J. Michkelson, A.A. Elwelt, L.S. Kudrjawzewa and O.G. Eisen, Monatsh. Chem., 105 (1974) 1379.

8406 A. Hoepfner, N. Parekh, C. Hoerner and A. Abdel-Hamid, Ber. Bunsenges. Phys. Chem., 79 (1975) 216.

8418 G. Ribaud, Thermodynamic Constants of Gases at High Temperatures, No. 266, Publications Scientifiques et Techniques du Ministere de l'Air, Paris, 1952.

8421 L.V. Mozhginskaya and L.E. Kolysko, Russ. J. Phys. Chem., 48 (1974) 1094.

8460 A.C. Bratton, W.A. Felsing and J.R. Bailey, Ind. Eng. Chem., 28 (1936) 424.

8545 D.B. Brooks, J. Res. Nat. Bur. Stand., 21 (1938) 847.

8546 M. Wojciechowski, J. Res. Nat. Bur. Stand., 19 (1937) 347.

8549 J.C. Houck, J. Res. Nat. Bur. Stand., 78 (1974) 617.

8577 R.E. Pennington and K.A. Kobe, J. Amer. Chem. Soc., 79 (1957) 300.

8580 R.W. Dornte and C.P. Smythe, J. Amer. Chem. Soc., 52 (1930) 3546.

8581 W.F. Seyer, J. Amer. Chem. Soc., 53 (1931) 3588.

8628 W.J. Gaw and F.L. Swinton, Trans. Faraday Soc., 64 (1968) 637.

8639 L. Schmerling, B.S. Friedman and V.N. Ipatieff, J. Amer. Chem. Soc., 62 (1940) 2446.

8647 J.P. Kohn and W.F. Bradish, J. Chem. Eng. Data, 9 (1964) 5.

8654 J.E. Troyan, J. Amer. Chem. Soc., 64 (1942) 3056.

8669 W. Wagner, Cryogenics, 13 (1973) 470.

8717 L.S. Mason and E.R. Washburn, J. Phys. Chem., 40 (1956) 481.

8801 F.A.H. Schreinemakers, Z. Phys. Chem., 48 (1904) 257.

8853 R.E. Chaddock, Ph. D. Dissertation, University of Michigan, Ann Arbor,

TABLE 6.1 (continued)

Michigan, 1940.
8869   J.G. Aston and S.C. Schumann, J. Amer. Chem. Soc., 64 (1942) 1034.
8899   J.A. Dixon and R.W. Schiessler, J. Amer. Chem. Soc., 76 (1954) 2197.
8900   A.W. Rytina, R.W. Schiessler and F.C. Whitmore, J. Amer. Chem. Soc.,
       71 (1949) 751.
8902   A.K. Doolittle and R.H. Peterson, J. Amer. Chem. Soc., 73 (1951) 2145.
8912   C. Hoerner, A. Hoepfner and B. Schmeiser, Ber. Bunsenges. Phys. Chem.,
       79 (1975) 222.
8958   S.G. Sayegh and G.A. Ratcliff, J. Chem. Eng. Data, 21 (1976) 71.
8967   J. Hust and R.E. Schramm, J. Chem. Eng. Data, 21 (1976) 7.
8973   A.K. Doolittle, Chem. Eng. Progr., 59 (1963) 1.
8990   N.N. Nagornov and L. Rotinyanz, Izv. Inst. Fiz.-Khim. Anal.,
       2 (1924) 371.
9002   J.-P.E. Grolier, G.C. Benson and P. Picker, J. Chem. Eng. Data,
       20 (1975) 243.
9008   K.-Y. Hsu and H.L. Clever, J. Chem. Eng. Data, 20 (1975) 268.
9027   L.N. Anderson, A.P. Kudchadker and P.T. Eubank, J. Chem. Eng. Data,
       13 (1968) 321.
9039   J.A. Guzman, Ph. D. Dissertation, Ohio State University, Columbus,
       Ohio, 1973.
9050   J.H. McMicking, Ph. D. Dissertation, Ohio State University, Columbus,
       Ohio, 1961.
9053   E.L. Washington, Ph. D. Dissertation, Illinois Institute of Technology,
       Chicago, Illinois, 1966.
9061   F.G. Clark, Ph. D. Dissertation, University of Illinois, Urbana-
       Champaign, Illinois, 1973.
9068   J. Kuhara, Kogyo Kagaku Zasshi, 50 (1947) 164.
9087   G. Chavanne and L.-J. Simon, C. R. Acad. Sci., 168 (1919) 1324.
9132   D.E.L. Dyke, J.S. Rowlinson and R. Thacker, Trans. Faraday Soc.,
       55 (1959) 903.
9138   A.E. Karr, W.M. Bowes and E.G. Scheibel, Anal. Chem., 23 (1951) 459.
9176   S. Peter and K. Reinhartz, Z. Phys. Chem. (Frankfurt am Main),
       24 (1960) 103.
9277   K.T. Thomas and R.A. McAllister, AIChE J., 3 (1957) 161.
9352   C.R. Mueller and J.E. Lewis, J. Chem. Phys., 26 (1957) 286.
9508   H. Wolff, A. Hoepfner and H.-M. Hoepfner, Ber. Bunsenges. Phys. Chem.,
       68 (1964) 410.
9566   R.W. Benoliel, M. S. Thesis, Pennsylvania State University, University
       Park, Pennsylvania, 1941.
9580   P. LeBeau and M. Picon, C. R. Acad. Sci., 156 (1913) 1077.
9617   E.J. Blat, M.J. Gerber and M.B. Newmann, Acta Physicochim. USSR,
       10 (1939) 273.
9623   N.N. Nagornov and L. Rotinjanz, Izv. Inst. Fiz.-Khim. Anal.,
       3 (1925) 562.
9706   G. Chavanne and B. Lejeune, Bull. Soc. Chim. Belg., 31 (1922) 98.
9707   G. Chavanne and P. Becker, Bull. Soc. Chim. Belg., 36 (1927) 591.
9708   G. Chiurdoglu, Bull. Soc. Chim. Belg., 53 (1944) 45.
9709   G. Chavanne and G. Tock, Bull. Soc. Chim. Belg., 41 (1932) 630.
9712   P. Ceuterick, Bull. Soc. Chim. Belg., 45 (1936) 545.
9714   H. de Graef, Bull. Soc. Chim. Belg., 34 (1925) 427.
9717   G. Chiurdoglu, Bull. Soc. Chim. Belg., 43 (1934) 35.
9734   O. Miller, Bull. Soc. Chim. Belg., 44 (1935) 513.
9746   A.M. Mamedov, T.S. Akhundov and N.N. Asadullaeva, Teploenergetika,

TABLE 6.1 (continued)

14 (1967) 83.

9760 T.D. Nevens, Ph. D. Dissertation, Ohio State University, Columbus, Ohio, 1950.

9851 J.F. Norris and R. Reuter, J. Amer. Chem. Soc., 49 (1927) 2624.

9852 M.L. Sherrill and G.F. Walter, J. Amer. Chem. Soc., 58 (1936) 742.

9854 K.N. Campbell and L.T. Eby, J. Amer. Chem. Soc., 63 (1941) 216.

9863 I. Schurman and C.E. Boord, J. Amer. Chem. Soc., 55 (1933) 4930.

9865 F.C. Whitmore and A.H. Homeyer, J. Amer. Chem. Soc., 55 (1933) 4555.

9903 E.W. Funk, Ph. D. Dissertation, University of California, Berkeley, California, 1970.

9918 S.-S. Chen and B.J. Zwolinski, J. Chem. Soc., 70 (1974) 1133.

9920 T.E. Thorpe and L.M. Jones, J. Chem. Soc., 63 (1893) 273.

9922 E.P. Carr and H. Stuecklen, J. Chem. Phys., 6 (1938) 55.

9923 A.I. Vogel, J. Chem. Soc., (1938) 1323.

9943 E. Dickinson, I.A. McLure and B.H. Powell, J. Chem. Soc., 70 (1974) 2321.

9961 M. Hirata and S. Suda, Sekiyu Gakkai Shi, 9 (1966) 885.

9988 S.D. Hamann, J.A. Lambert and R.B. Thomas, Aust. J. Chem., 8 (1955) 149.

10015 K.P. Murphy, J. Chem. Eng. Data, 9 (1964) 259.

10044 H.C. Van Ness and N.K. Kochar, J. Chem. Eng. Data, 12 (1967) 38.

10054 J.P. Kohn and J.H.S. Haggin, J. Chem. Eng. Data, 12 (1967) 313.

10070 R.E. Rondeau and L.A. Harrah, J. Chem. Eng. Data, 10 (1965) 84.

10135 D. Ambrose, C.H.S. Sprake and R. Townsend, J. Chem. Thermodyn., 1 (1969) 499.

10163 L. Boje and A. Hvidt, J. Chem. Thermodyn., 3 (1971) 663.

10165 G.C. Benson, S. Murakami and D.E.G. Jones, J. Chem. Thermodyn., 3 (1971) 719.

10193 J.L. Hales and R. Townsend, J. Chem. Thermodyn., 4 (1972) 763.

10216 P.R. Garrett and J.M. Pollock, J. Chem. Thermodyn., 5 (1973) 569.

10219 B.J. Levien, J. Chem. Thermodyn., 5 (1973) 679.

10241 I.P.-C. Li, J. Polak and B.C.-Y. Lu, J. Chem. Thermodyn., 6 (1974) 417.

10246 T.M. Letcher and F. Marsicano, J. Chem. Thermodyn., 6 (1974) 501.

10253 D. Ambrose, C.H.S. Sprake and R. Townsend, J. Chem. Thermodyn., 6 (1974) 693.

10255 R.P. Tomlins and M. Adamson, J. Chem. Thermodyn., 6 (1974) 757.

10258 E. Rajagopal and S.V. Subrahmanyam, J. Chem. Thermodyn., 6 (1974) 873.

10264 R.S. Myers and H.L. Clever, J. Chem. Thermodyn., 6 (1974) 949.

10271 M.B. Ewing and K.N. Marsh, J. Chem. Thermodyn., 6 (1974) 1087.

10284 W. Woycicki, J. Chem. Thermodyn., 7 (1975) 77.

10288 T.G. Bissell and A.G. Williamson, J. Chem. Thermodyn., 7 (1975) 131.

10290 D.P. Deshpande and S.L. Oswal, J. Chem. Thermodyn., 7 (1975) 155.

10296 T.M. Letcher, J. Chem. Thermodyn., 7 (1975) 205.

10315 A. Baghdoyan and V. Fried, J. Chem. Thermodyn., 7 (1975) 409.

10317 K.-Y. Hsu and H.L. Clever, J. Chem. Thermodyn., 7 (1975) 435.

10318 D. Ambrose, J.H. Ellender, E.B. Lees, C.H.S. Sprake and R. Townsend, J. Chem. Thermodyn., 7 (1975) 453.

10335 D.V.S. Jain and O.P. Yadav, Indian J. Chem., 12 (1974) 721.

10351 A.G. Osborn and D.R. Douslin, J. Chem. Eng. Data, 14 (1969) 208.

10394 J.M. Lenoir, K.E. Hayworth and H.G. Hipkin, J. Chem. Eng. Data, 15 (1970) 474.

10403 A. Baghdoyan, J. Malik and V. Fried, J. Chem. Eng. Data, 16 (1971) 96.

10413 T.M. Letcher and J.M. Bayles, J. Chem. Eng. Data, 16 (1971) 266.

10416 J.M. Lenoir, K.E. Hayworth and H.G. Hipkin, J. Chem. Eng. Data, 16 (1971) 280.

10428 C. Eon, C. Pommier and G. Guiochon, J. Chem. Eng. Data, 16 (1971) 408.

10430 C.C. Chappelow, P.S. Snyder and J. Winnick, J. Chem. Eng. Data,

TABLE 6.1 (continued)

16 (1971) 440.

10432 L.Y. Sadler, D.W. Luff and M.D. McKinley, J. Chem. Eng. Data, 16 (1971) 446.

10483 F.J. Wright, J. Chem. Eng. Data, 6 (1961) 454.

10493 S. Curtice, E.G. Felton and H.W. Prengle, J. Chem. Eng. Data, 17 (1972) 192.

10508 P.P. Dawson, I.H. Silberberg and J.J. McKetta, J. Chem. Eng. Data, 18 (1973) 7.

10552 R.C. Mitra, S.C. Guhaniyogi and S.N. Bhattacharyya, J. Chem. Eng. Data, 18 (1973) 147.

10561 J.A. Ellis and K.-C. Chao, J. Chem. Eng. Data, 18 (1973) 264.

10569 E.F. Meyer and R.D. Hotz, J. Chem. Eng. Data, 18 (1973) 359.

10574 D.J. Subach and C.L. Kong, J. Chem. Eng. Data, 18 (1973) 403.

10592 J.W. Moore and R.M. Wellek, J. Chem. Eng. Data, 19 (1974) 136.

10622 R.S. Murray and M.L. Martin, J. Chem. Thermodyn., 7 (1975) 839.

10665 T.M. Lesteva, S.K. Ogorodnikov and A.I. Morozova, Zh. Prikl. Khim. (Leningrad), 40 (1967) 891.

10697 R.A. Murogova, G.L. Tudorovskaya, I.D. Gridin, A.P. Yurchuk, N.I. Pleskach, V.D. Kozlova, T.V. Bazyleva, N.A. Safonova and L.A. Serafimov, Zh. Prikl. Khim. (Leningrad), 45 (1972) 824.

10719 S. Young, Sci. Proc. R. Dublin Soc., 12 (1909) 374.

10722 W.R. Ham, J.C. Churchill and H.M. Ryder, J. Franklin Inst., 186 (1918) 15.

10749 G.A.McA. Cummings and E. McLaughlin, J. Chem. Soc., (1955) 1391.

10754 E.J. Partington, J.S. Rowlinson and J.F. Weston, Trans. Faraday Soc., 56 (1960) 479.

10757 R. Jacquemain, C. R. Acad. Sci., 215 (1942) 179.

10804 J.A. Neff and J.B. Hickman, J. Phys. Chem., 59 (1955) 42.

10808 E.B. Evans, J. Inst. Petrol. Technol., 24 (1938) 321.

10809 E.B. Evans, J. Inst. Petrol. Technol., 24 (1938) 537.

10834 V.T. Zharov, N.D. Malegina and A.G. Morachevskii, J. Appl. Chem. USSR, 38 (1965) 2089.

10841 A.N. Marinichev and M.P. Susarev, J. Appl. Chem. USSR, 38 (1965) 371.

10894 E.E. Tucker, S.B. Farnham and S.D. Christian, J. Phys. Chem., 73 (1969) 3820.

10901 J.K. Nickerson, K.A. Kobe and J.J. McKetta, J. Phys. Chem., 65 (1961) 1037.

10917 F.E. Francis and S. Young, J. Chem. Soc., 73 (1898) 920.

10919 P. Sabatier and A. Mailhe, C. R. Acad. Sci., 137 (1903) 240.

10940 C.L. Young, Aust. J. Chem., 25 (1972) 1625.

10968 V. Zharov, T. Vitman, Kh. Viit and L. Kudryavtseva, Eesti NSV Tead. Akad. Toim., 20 (1971) 206.

10972 S.F. Di Zio, Ph. D. Dissertation, Rensselaer Polytechnic Institute, Troy, New York, 1964.

11017 T. Tonomura and K. Uehara, Bull. Chem. Soc. Jap., 6 (1931) 255.

11019 A.N. Campbell and R.M. Chatterjee, Can. J. Chem., 46 (1968) 575.

11023 A.N. Campbell and S.C. Anand, Can. J. Chem., 50 (1972) 479.

11040 T.S. Patterson, J. Chem. Soc., 81 (1902) 1097.

11049 V. Fried and J. Pick, Collect. Czech. Chem. Commun., 26 (1961) 954.

11064 R.T. Lagemann, D.R. McMillan and M. Woolsey, J. Chem. Phys., 16 (1948) 247.

11081 J. Kraus and J. Linek, Collect. Czech. Chem. Commun., 36 (1971) 2547.

11100 S.N. Bhattacharyya and A. Mukherjee, J. Phys. Chem., 72 (1968) 56.

11125 G. Opel, B. Zorn and K.-D. Zwerschke, Z. Phys. Chem. (Leipzig),

TABLE 6.1 (continued)

255 (1974) 997.

11128   F. Kuschel, H. Kehlen and H. Sackmann, Z. Phys. Chem. (Leipzig), 255 (1974) 432.

11157   S. Takagi, K. Furukawa and R. Fujishiro, Bull. Chem. Soc. Jap., 41 (1968) 1313.

11160   F. Rivenq, Bull. Soc. Chim. Fr., 41 (1969) 3034.

11161   L. Abello, B. Servais, M. Kern and G. Pannetier, Bull. Soc. Chim. Fr., 41 (1968) 4360.

11166   M. Kern, L. Abello, D. Caceres and G. Pannetier, Bull. Soc. Chim. Fr., 41 (1970) 3849.

11193   J. Quintanilla G., Rev. Quim. Ing. Quim., 2 (1956) 23.

11231   J.L.H. Wang, L. Boublikova and B.C.-Y. Lu, J. Appl. Chem., 20 (1970) 172.

11234   P. Ramachandra Rao, C. Chiranjivi and O.J. Dasarao, J. Appl. Chem., 18 (1968) 166.

11243   K.V. Kurmanadha Rao, V.V.G. Krishnamurty and C. Venkata Rao, J. Appl. Chem., 7 (1957) 535.

11246   R. Satapathy, M. Raja Rao, N.S.R. Anjaneyulu and C. Venkata Rao, J. Appl. Chem., 6 (1956) 261.

11248   J.D. Gomez-Ibanez and F.T. Wang, J. Chem. Thermodyn., 3 (1971) 811.

11250   R.F. Hudson and I. Stelzer, J. Chem. Soc., (1966) 775.

11265   Z. Maczynska, Bull. Acad. Pol. Sci., 11 (1963) 225.

11295   I.B. Rabinovich, Russ. J. Phys. Chem., 34 (1960) 198.

11316   K. Quitzsch, R. Huettig, H.-G. Vogel, H.-J. Gesemann and G. Geiseler, Z. Phys. Chem. (Leipzig), 223 (1963) 225.

11336   S.M. Hosseini and G. Schneider, Z. Phys. Chem. (Frankfurt am Main), 36 (1963) 137.

11475   P.N. Nikolaev, I.B. Rabinovich, V.A. Gal'perin and V.G. Tsvetkov, Russ. J. Phys. Chem., 40 (1966) 586.

11493   Z.S. Belousova and Sh.D. Zaalishvili, Russ. J. Phys. Chem., 41 (1967) 1290.

11556   Sh.D. Zaalishvili, Z.S. Belousova and V.P. Verkhova, Russ. J. Phys. Chem., 45 (1971) 149.

11733   P. Gross, J.C. Rintelen and J.H. Saylor, J. Phys. Chem., 43 (1939) 197.

12038   E.F. Meyer and R.E. Wagner, J. Phys. Chem., 70 (1966) 3162.

12089   D.D. Deshpande and L.G. Bhatgadde, J. Phys. Chem., 72 (1968) 261.

12146   J.C. Shieh and P.A. Lyons, J. Phys. Chem., 73 (1969) 3258.

12169   H. Wolff and R. Wurtz, J. Phys. Chem., 74 (1970) 1600.

12195   M.E. Baur, C.M. Knobler, D.A. Horsma and P. Perez, J. Phys. Chem., 74 (1970) 4594.

12358   E. Bergmann and A. Weizmann, Trans. Faraday Soc., 32 (1936) 1318.

12385   A. Weizmann, Trans. Faraday Soc., 36 (1940) 329.

12387   A. Weizmann, Trans. Faraday Soc., 35 (1940) 978.

12455   R.K. Hind, E. McLaughlin and A.R. Ubbelohde, Trans. Faraday Soc., 55 (1959) 21.

12507   D.J. Coumou and E.L. Mackor, Trans. Faraday Soc., 60 (1964) 1726.

12536   D.D. Deshpande and M.V. Pandya, Trans. Faraday Soc., 63 (1967) 2149.

12551   D. Ambrose and R. Townsend, Trans. Faraday Soc., 64 (1668) 2622.

12559   E. Wilhelm, R. Schano, G. Becker, G.H. Findenegg and F. Kohler, Trans. Faraday Soc., 65 (1969) 1443.

12590   P.J. Flory and H. Hocker, Trans. Faraday Soc., 67 (1971) 2258.

12591   H. Hocker and P.J. Flory, Trans. Faraday Soc., 67 (1971) 2270.

12707   J.L.R. Morgan and O.M. Lammert, J. Amer. Chem. Soc., 46 (1924) 881.

12709   T.W. Richards, C.L. Speyers and E.K. Carver, J. Amer. Chem. Soc., 46 (1924) 1196.

12764   C.P. Smyth and S.O. Morgan, J. Amer. Chem. Soc., 50 (1928) 1547.

TABLE 6.1 (continued)

12849   W.F. Seyer and E.G. King, J. Amer. Chem. Soc., 55 (1933) 3140.
12856   J.M.A. de Bruyne, R.M. Davis and P.M. Gross, J. Amer. Chem. Soc., 55 (1933) 3936.
12993   R.E. Gibson and J.F. Kincaid, J. Amer. Chem. Soc., 60 (1938) 511.
13016   L.S. Mason and H. Paxton, J. Amer. Chem. Soc., 61 (1939) 67.
13053   J.S. Burlew, J. Amer. Chem. Soc., 62 (1940) 690.
13065   P.M. Ginnings, D. Plonk and E. Carter, J. Amer. Chem. Soc., 62 (1940) 1923.
13074   J.M. Stuckey and J.H. Saylor, J. Amer. Chem. Soc., 62 (1940) 2922.
13138   K. Owen, O.R. Quayle and W.J. Clegg, J. Amer. Chem. Soc., 64 (1942) 1294.
13155   J.H. Saylor, V.J. Baxt and P.M. Gross, J. Amer. Chem. Soc., 64 (1942) 2742.
13236   F.H. Stross, J.M. Monger and H. De V. Finch, J. Amer. Chem. Soc., 69 (1947) 1627.
13276   M. Goldsmith and G.W. Wheland, J. Amer. Chem. Soc., 70 (1948) 2632.
13292   R. Van Volkenburgh, K.W. Greenlee, J.M. Derfer and C.E. Boord, J. Amer. Chem. Soc., 71 (1949) 172.
13352   R.G. Charles and H. Freiser, J. Amer. Chem. Soc., 72 (1950) 2233.
13364   P.W. Selwood and R.M. Dobres, J. Amer. Chem. Soc., 72 (1950) 3860.
13365   B.H. Eccleston, H.J. Coleman and N.G. Adams, J. Amer. Chem. Soc., 72 (1950) 3866.
13373   A. Weissler and V.A. Del Grosso, J. Amer. Chem. Soc., 72 (1950) 4209.
13380   E.D. Bergmann, A. Weizmann and E. Fischer, J. Amer. Chem. Soc., 72 (1950) 5009.
13422   H.O. Day and W.A. Felsing, J. Amer. Chem. Soc., 73 (1951) 4839.
13464   R.L. Custer and C.E. Waring, J. Amer. Chem. Soc., 74 (1952) 5726.
13621   A.J. Petro and C.P. Smyth, J. Amer. Chem. Soc., 80 (1958) 73.
13714   R.A. Orwoll and P.J. Flory, J. Amer. Chem. Soc., 89 (1967) 6814.
13729   S. Chang, D. McNally, S. Shary-Tehrany, M.J. Hickey and R.H. Boyd, J. Amer. Chem. Soc., 92 (1970) 3109.
13799   M.G. Mayberry and J.G. Aston, J. Amer. Chem. Soc., 56 (1934) 2682.
13803   T.E. Thorpe, J. Chem. Soc., 37 (1880) 141.
13832   J.H. Gladstone, J. Chem. Soc., 59 (1891) 290.
13838   W.H. Perkin, J. Chem. Soc., 61 (1892) 287.
13839   W.H. Perkin, J. Chem. Soc., 61 (1892) 800.
13877   W.H. Perkin, J. Chem. Soc., 77 (1900) 267.
13929   A.W. Crossley and N. Renouf, J. Chem. Soc., 87 (1905) 1487.
13945   J.S. Lumsden, J. Chem. Soc., 91 (1907) 24.
13997   F. Chick and N.T.M. Wilsmore, J. Chem. Soc., 97 (1910) 1978.
14001   D. Tyrer, J. Chem. Soc., 97 (1910) 2620.
14049   R.P. Worley, J. Chem. Soc., 105 (1914) 273.
14102   T.W. Price, J. Chem. Soc., 115 (1919) 1116.
14116   W.B. Parker and G. Thompson, J. Chem. Soc., 121 (1922) 1341.
14192   F.B. Garner and S. Sugden, J. Chem. Soc., London, (1927) 2877.
14269   E.H. Farmer and F.L. Warren, J. Chem. Soc., London, (1933) 1297.
14292   F. Fairbrother, J. Chem. Soc., London, (1934) 1846.
14299   F.R. Goss, J. Chem. Soc., London, (1935) 727.
14373   J. Kenyon and B.C. Platt, J. Chem. Soc., London, (1939) 633.
14380   D.M. Cowan, G.H. Jeffery and A.I. Vogel, J. Chem. Soc., London, (1939) 1862.
14407   A.I. Vogel, J. Chem. Soc., London, (1946) 133.
14416   A.I. Vogel, J. Chem. Soc., London, (1948) 607.
14417   A.I. Vogel, J. Chem. Soc., London, (1948) 610.
14419   G.H. Jeffery and A.I. Vogel, Chem. Soc., London, (1948) 658.
14422   G.H. Jeffery and A.I. Vogel, J. Chem. Soc., London, (1948) 1804.
14448   F.R. Buck, K.F. Coles, G.T. Kennedy and F. Morton, J. Chem. Soc.,

TABLE 6.1 (continued)

London, (1949) 2377.

14487   H.E. Eduljee, D.M. Newitt and K.E. Weale, J. Chem. Soc., London,
(1951) 3086.

14586   J.Y.H. Chau, R.J.W. Le Fevre and J. Tardif, J. Chem. Soc., London,
(1957) 2293.

14634   R. Grzeskowiak, G.H. Jeffery and A.I. Vogel, J. Chem. Soc., London,
(1960) 4719.

14703   R.J.W. Le Fevre and K.M.S. Sundaram, J. Chem. Soc., London, (1964) 3518.

14905   G.B. Rathmann, A.J. Curtis, P.L. McGeer and C.P. Smyth, J. Chem. Phys.,
25 (1956) 413.

15029   D.L. Hogenboom, W. Webb and J.A. Dixon, J. Chem. Phys., 46 (1967) 2586.

15034   A.C. Plaush and E.L. Pace, J. Chem. Phys., 47 (1967) 44.

15062   F.I. Mopsik, J. Chem. Phys., 50 (1969) 2559.

15090   R. Prydz, G.C. Straty and K.D. Timmerhaus, J. Chem. Phys.,
53 (1970) 2359.

15130   R.B. Heady and J.W. Cahn, J. Chem. Phys., 58 (1973) 896.

15272   A.N. Roy, Proc. Nat. Inst. Sci. India, 12 (1946) 137.

15276   B.N. Raju, R. Ranganathan and M. Narasinga Rao, Indian Chem. Eng.,
5 (1963) 82.

15281   P.D. Murthy and M.V. Raghavacharya, Indian Chem. Eng., 10 (1968) 26.

15290   P. Dakshinamurty, V.V.G. Krishnamurty, C.S. Rao and C. Venkata Rao,
Indian J. Technol., 1 (1963) 196.

15291   H.P. Khadilkar and G. Narsimhan, Indian J. Technol., 1 (1963) 299.

15306   K. Zieborak, Bull. Acad. Pol. Sci., 6 (1958) 443.

15307   K. Zieborak, Bull. Acad. Pol. Sci., 6 (1958) 449.

15380   W.J. Kerns, R.G. Anthony and P.T. Eubank, Chem. Eng. Progr.,
70 (1974) 14.

15411   P.R. Naidu, Aust. J. Chem., 23 (1970) 967.

15485   S.K. Ogorodnikov, V.B. Kogan and M.S. Nemtsov, J. Appl. Chem. USSR,
33 (1960) 1581.

15514   K.P. Mishchenko and G.M. Poltoratskii, J. Appl. Chem. USSR,
35 (1962) 1572.

15521   L.S. Kudryavtseva and M.P. Susarev, J. Appl. Chem. USSR, 36 (1963) 1180.

15532   Y.N. Garber and R.A. Bovkun, J. Appl. Chem. USSR, 37 (1964) 831.

15542   A.N. Gudkov, N.A. Fermor and N.I. Smirnov, J. Appl. Chem. USSR,
37 (1964) 2179.

15650   B.I. Konobeev and V.V. Lyapin, J. Appl. Chem. USSR, 43 (1970) 806.

15688   A.A. Ennan, V.G. Samoilenko, A.N. Chobotarev and V.A. Anikeev, J. Appl.
Chem. USSR, 44 (1971) 1449.

15698   I.L. Krupatkin and M.F. Glagoleva, J. Appl. Chem. USSR, 44 (1971) 2433.

15728   N.N. Slutsman, V.O. Reikhsfel'd and B.V. Mamontov, J. Appl. Chem. USSR,
45 (1972) 2148.

15748   A.N. Marinichev and L.A. Vertsman, J. Appl. Chem. USSR, 46 (1973) 368.

15755   A.A. Terent'eva, B.S. Krumgal'z and Y.I. Gerzhberg, J. Appl. Chem. USSR,
46 (1973) 1213.

15758   V.M. Komarov, A.V. Boldyrev and R.S. Pelevina, J. Appl. Chem. USSR,
46 (1973) 1465.

15791   V.F. Mironenko, Y.N. Garber and B.K. Bormatov, J. Appl. Chem. USSR,
47 (1974) 836.

15843   V.F. Belugin and L.K. Yakushina, J. Appl. Chem. USSR, 48 (1975) 1432.

15872   J.F. Tooke, Ph. D. Dissertation, Louisiana State University, Baton
Rouge, Louisiana, 1959.

15880   F. Gothard, T. Messinger, G. Musca, G. Panaitescu and P. Tomi, Chim.
Ind., 104 (1971) 1454.

15904   B.V. Subba Rao and C. Venkata Rao, Can. J. Chem. Eng., 42 (1964) 266.

16004   K.L. Butcher, K.R. Ramasubramanian and M.S. Medani, J. Appl. Chem.

TABLE 6.1 (continued)

Biotechnol., 22 (1972) 1139.
16006 P. Dakshinamurty, V. Subrahmanyam and M. Narasinga Rao, J. Appl. Chem. Biotechnol., 23 (1973) 323.
16059 J. Chevalley, C. R. Acad. Sci., 250 (1960) 3326.
16160 J.R. Barber, Ph. D. Dissertation, Ohio State University, Columbus, Ohio, 1968
16162 J.V. Fox, Ph. D. Dissertation, University of Houston, Houston, Texas, 1968.
16166 R.A. Orwoll, Ph. D. Dissertation, Stanford University, Palo Alto, California, 1967.
16244 H. Buchowski, B. Janaszewski and J. Teperek, Bull. Acad. Pol. Sci., 14 (1966) 403.
16255 S. Glowka and A.C. Zawisza, Bull. Acad. Pol. Sci., 17 (1969) 365.
16358 H. Enokido, T. Shinoda and Y. Mashiko, Bull. Chem. Soc. Jap., 42 (1969) 84.
16363 O. Kiyohara and K. Arakawa, Bull. Chem. Soc. Jap., 43 (1970) 3037.
16370 K. Nakanishi and O. Toyama, Bull. Chem. Soc. Jap., 45 (1972) 3210.
16414 F.I. Mopsik, J. Res. Nat. Bur. Stand., 71 (1967) 287.
16466 M. Katz, P.W. Lobo, A.S. Minano and H. Solimo, Can. J. Chem., 49 (1971) 2605.
16488 J. Nyvlt and E. Erdos, Collect. Czech. Chem. Commun., 26 (1961) 500.
16497 T. Boublik and K. Aim, Collect. Czech. Chem. Commun., 37 (1972) 3513.
16589 E.F. Viktorova and A.P. Toropov, Izv. Vyssh. Ucheb. Zaved., 10 (1967) 1209.
16662 T.S. Akhundov, Izv. Vyssh. Ucheb. Zaved., 16 (1973) 42.
16664 N.F. Otpushchennikov, B.S. Kir'yakov and P.P. Panin, Izv. Vyssh. Ucheb. Zaved., 17 (1974) 73.
16695 O.V. Skvortsova, A.M. Chashchin and M.S. Perinykh, Gidroliz. Lesokhim. Prom., 25 (1972) 16.
16837 D.S. Tsiklis and L.I. Shenderei, Khim. Prom. (Moscow), 40 (1964) 194.
17031 H.M. Ashton and E.S. Halberstadt, Proc. Royal Soc., 245 (1958) 373.
17056 M.A. Azim, S.S. Bhatnagar and R.N. Mathur, Phil. Mag., 16 (1933) 580.
17061 J.N. Friend and W.D. Hargreaves, Phil. Mag., 35 (1944) 136.
17129 S.V. Subrahamanyam and E. Rajagopal, Z. Phys. Chem. (Frankfurt am Main), 85 (1973) 256.
17144 J. Murto, A. Kivinen and E. Lindell, Suom. Kemistilehti B, 43 (1970) 28.
17145 A. Nissema and L. Koskenniska, Suom. Kemistilehti B, 45 (1972) 203.
17270 I. Matsunaga and T. Katayama, J. Chem. Eng. Jap., 6 (1973) 397.
17342 F. Burriel, An. Real Soc. Espan. Fis. Quim., 29 (1931) 89.
17415 J. Timmermans, H. Van der Horst and H. Kamerlingh-Onnes, Arch. Neer. Sci. Exactes Natur., 6 (1923) 180.
17428 L.S. Kudryavtseva, Kh. Viit and O. Eisen, Eesti NSV Tead. Akad. Toim., 17 (1968) 242.
17432 O. Eisen and A. Orav, Eesti NSV Tead. Akad. Toim., 19 (1970) 202.
17433 O. Eisen, A. Elvelt and L.S. Kudryavtseva, Eesti NSV Tead. Akad. Toim., 20 (1971) 287.
17434 L.S. Kudryavtseva, Kh. Viit and C. Eisen, Eesti NSV Tead. Akad. Toim., 20 (1971) 292.
17437 A. Aarna and T. Kaps, Eesti NSV Tead. Akad. Toim., 23 (1974) 16.
17453 R. Kandiyoti and J.M.L. Penninger, Middle East Tech. Univ. J. Pure Appl. Sci., 5 (1972) 157.
17509 F. Schwers, J. Chim. Phys. Physicochim. Biol., 9 (1911) 15.
17517 G.D. Gal'pern, L.A. Konovalova and M.M. Kusakov, Tr. Inst. Nefti., 1 (1950) 217.
17521 G. Hanumantharao and B.V. Subbarao, Indian J. Technol., 12 (1974) 292.

TABLE 6.1 (continued)

17544   H. Pines and V.N. Ipatieff, J. Amer. Chem. Soc., 61 (1939) 1076.
17545   W.F. Seyer, R.B. Bennett and F.C. Williams, J. Amer. Chem. Soc.,
        71 (1949) 3447.
17569   R.M. Varushchenko, N.A. Belikova, S.M. Skuratov and A.F. Plate, Russ. J.
        Phys. Chem., 44 (1970) 1722.
17612   Sh.D. Zaalishvili, Z.S. Belousova and V.P. Verkhova, Russ. J. Phys.
        Chem., 45 (1971) 902.
17615   N.G. Shlenkina, Russ. J. Phys. Chem., 45 (1971) 954.
17827   M. Grutzmacher, Z. Phys., 28 (1924) 342.
17885   Yu.S. Shoitov, G.M. Pan'kevich and N.F. Otpushchennikov, Teploenergetika,
        15 (1968) 76.
17936   D.V.S. Jain, A.M. North and R.A. Pethrick, J. Chem. Soc., 1 (1974) 1292.
17937   E. Reisler, H. Eisenberg and A.P. Minton, J. Chem. Soc., 2 (1972) 1001.
17943   H.S. Hull, A.F. Reid and A.G. Turnbull, Aust. J. Chem., 18 (1965) 249.
17966   V.R. Deitz, J. Amer. Chem. Soc., 55 (1933) 472.
17974   L. Massart, Bull. Soc. Chim. Belg., 45 (1936) 76.
17986   D.V.S. Jain and O.P. Yadav, Indian J. Chem., 9 (1971) 342.
17992   A. Iguchi, Kagaku Sochi, 16 (1974) 63.
18019   K. Nagahama, H. Konishi, D. Hoshino and M. Hirata, J. Chem. Eng. Jap.,
        7 (1974) 323.
18044   A. Gandek, Ph. D. Dissertation, Columbia University, New York, 1961.
18079   J. Bellm, W. Reineke, K. Schaefer and B. Schramm, Ber. Bunsenges. Phys.
        Chem., 78 (1974) 282.
18080   R. Hahn, K. Schaefer and B. Schramm, Ber. Bunsenges. Phys. Chem.,
        78 (1974) 287.
18101   A. Kirrmann, Bull. Soc. Chim. Fr., 39 (1926) 988.
18172   R.M. Varushchenko, A.I. Druzhinina and G.L. Gal'chenko, Tr. Khim. Khim
        Tekhnol., 39 (1974) 93.
18196   F. Billes and G. Varasnyi, Acta Chim. Acad. Sci. Hung., 35 (1963) 147.
18206   J.N. Bronsted and J. Koefoed, Mat.-Fys. Medd.-K. Dan. Vidensk. Selsk.,
        22 (1946) 1.
18208   H. Kirss, L. Kudryavtseva and O. Eisen, Eesti NSV Tead. Akad. Toim.,
        24 (1975) 15.
18235   J.M. Genco, Ph. D. Dissertation, Ohio State University, Columbus,
        Ohio, 1965.
18243   F. Kohler and E. Rott, Monatsh. Chem., 85 (1954) 703.
18268   K. Schmoll and E. Jenckel, Z. Elektrochem., 60 (1956) 756.
18270   G. Kortum and W. Vogel, Z. Elektrochem., 62 (1958) 40.
18271   Kl. Schafer and W. Rall, Z. Elektrochem., 62 (1958) 1090.
18274   W. Rall and Kl. Schafer, Z. Elektrochem., 63 (1959) 1019.
18314   H. Grunert, Z. Anorg. Allg. Chem., 164 (1927) 256.
18355   A.Z. Golik and I.I. Adamenko, Ukr. Fiz. Zh. (Ukr. Ed.), 10 (1965) 443.
18392   M. Lecat, Bull. Cl. Sci. Acad. Roy. Belg., 29 (1943) 273.
18395   M. Lecat, Ann. Soc. Sci. Bruxelles, 45 (1925) 284.
18421   P. Hennings and U. Von Weber, J. Prat. Chem., 18 (1962) 91.
18439   G. Geiseler and M. Raetzsch, Ber. Bunsenges. Phys. Chem., 69 (1965) 485.
18474   H. Tschamler, F. Wettig and E. Richter, Monatsh. Chem., 80 (1949) 572.
18654   G. Wolf, Helv. Chim. Acta., 55 (1972) 1446.
18668   A.M. Mamedov, T.S. Akhundov and Sh.Yu. Imanov, Teplofiz. Svoistva
        Veshchestv Mater., 55 (1972) 47.
18837   A.P. Toropov, J. Gen. Chem. USSR, 26 (1956) 3635.
18913   D.V.S. Jain, V.K. Gupta and B.S. Lark, Indian J. Chem., 9 (1971) 465.
18956   P. Sellers and S. Sunner, Acta Chem. Scand., 16 (1962) 46.
19014   F. Krafft, Chem. Ber., 15 (1882) 1711.
19056   R.L. Schmidt and H.L. Clever, J. Colloid Interface Sci., 26 (1968) 19.

TABLE 6.1 (continued)

19270   W. Herz, Z. Elektrochem., 25 (1919) 145.
19380   V.A. Atoyan and I.A. Mamedov, Izv. Vyssh. Ucheb. Zaved., 18 (1975) 74.
19427   R. Dreyer, W. Martin and U. von Weber, J. Prakt. Chem., 1 (1954) 324.
19429   K. Engelmann and H.J. Bittrich, J. Prakt. Chem., 19 (1963) 106.
19433   K. Quitzsch, J. Prakt. Chem., 28 (1965) 69.
19435   K. Quitzsch, V. Wunderlich and G. Geiseler, J. Prakt. Chem.,
        30 (1965) 119.
19467   A. Klages and R. Keil, Chem. Ber., 36 (1903) 1632.
19516   D.R. Stull, G.C. Sinke, R.A. McDonald, W.E. Hatton and D.L. Hildenbrand,
        Pure Appl. Chem., 2 (1961) 315.
19531   E.B. Evans, J. Inst. Petrol., 24 (1938) 38.
19532   E.B. Evans, J. Inst. Petrol., 24 (1938) 321.
19607   R.S. Myers, B.A. Berenbach and H.L. Clever, J. Chem. Eng. Data,
        14 (1969) 91.
19662   E.G. Kanakbaeva, Z.Kh. Anisimova and M.I. Shakhparonov, Moscow Univ.
        Chem. Bull., 29 (1974) 20.
19687   M.V. Prabhakara Rao and P.R. Naidu, J. Chem. Thermodyn., 6 (1974) 1195.
19712   I. Wichterle and L. Boublikova, Ind. Eng. Chem., 8 (1969) 585.
19746   G.M. Pan'kevich and Yu.S. Shoitov, Uch. Zap., 54 (1969) 36.
19840   C. Podder, Z. Phys. Chem. (Frankfurt am Main), 39 (1963) 79.
19948   A. Bygden, Z. Phys. Chem. (Leipzig), 90 (1915) 248.
20015   T. Batuecas and F.L. Casado, Z. Phys. Chem., 181 (1937) 197.
20064   E. Bergmann, L. Engel and H.A. Wolff, Z. Phys. Chem., 17 (1931) 81.
20065   E. Bergmann, L. Engel and H. Hoffmann, Z. Phys. Chem., 17 (1931) 92.
20066   E. Bergmann and M. Tschundnowsky, Z. Phys. Chem., 17 (1931) 107.
20074   E. Bergmann and W. Schutz, Z. Phys. Chem., 19 (1932) 389.
20075   E. Bergmann and W. Schutz, Z. Phys. Chem., 19 (1932) 395.
20076   E. Bergmann and W. Schutz, Z. Phys. Chem., 19 (1932) 401.
20090   J. Wellm, Z. Phys. Chem., 28 (1935) 119.
20129   H. Rock and L. Sieg, Z. Phys. Chem. (Frankfurt am Main), 3 (1955) 355.
20206   O. Eisen, L. Kudrjawzewa and A. Elwelt, Z. Phys. Chem. (Frankfurt am
        Main), 86 (1973) 33.
20232   C.R. Koppany, Ph. D. Dissertation, University of Southern California,
        Los Angeles, California, 1972.
20290   M. Diaz Pena and M. Benitez de Soto, An. Real Soc. Espan. Fis. Quim.,
        61 (1965) 1163.
20315   A. Dessart, Bull. Soc. Chim. Belg., 35 (1926) 9.
20318   F.B. Marti, Bull. Soc. Chim. Belg., 39 (1930) 590.
20360   V. Los, Chem. Prum., 24 (1974) 286.
20436   D.V.S. Jain, B.S. Lark, S.S. Chamak and P. Chander, Indian J. Chem.,
        8 (1970) 66.
20442   D.V.S. Jain and O.P. Yadav, Indian J. Chem., 9 (1971) 342.
20447   R.K. Nigam and B.S. Mahl, Indian J. Chem., 9 (1971) 1250.
20452   D.V.S. Jain and O.P. Yadav, Indian J. Chem., 11 (1973) 28.
20469   G.F. Leverett, Ph. D. Dissertation, Ohio State University, Columbus,
        Ohio, 1959.
20472   A.K. Abas-Zade, T.A. Apaev and A.M. Kerimov, Izv. Akad. Nauk Azerb. SSR,
        11 (1970) 70.
20486   A.P. Toropov, J. Gen. Chem. USSR, 26 (1956) 1453.
20518   K. Noda, K. Fukawa, M. Yanagisawa and K. Ishida, Kagaku Kogaku,
        35 (1971) 245.
20540   V.G. Ben'kovskii, T.M. Bogoslovskaya and M.Kh. Hauruzov, Khim. Tekhnol.
        Topl. Masel, 11 (1966) 22.
20636   N. Barbulescu and R. Moldovan-Velniceriu, Rev. Roum. Phys.,

TABLE 6.1 (continued)

13 (1968) 275.

20654 A. Jackowski, Rocz. Chem., 48 (1974) 491.

20670 I. Nitta, S. Seki, H. Chihara and K. Suzuki, Sci. Pap. Osaka Univ., 29 (1951) 1.

20700 S.V. Mainkar and P.S. Mene, Trans. Indian Inst. Chem. Eng., 10 (1968) 169.

20723 L.M. Artyukhovskaya, E.T. Shimanskaya and Yu.I. Shimanskiy, Sov. Phys.-JETP (Engl. Transl.), 59 (1970) 375.

20752 T.D. Sokolova, N.K. Prokof'eva and L.A. Nisel'son, J. Appl. Chem. USSR, 45 (1972) 1685.

20784 W. Markownikoff, J. Prakt. Chem., 45 (1892) 561.

20809 K. Von Auwers and F. Eisenlohr, J. Prakt. Chem., 82 (1910) 65.

20811 K. Von Auwers and F. Eisenlohr, J. Prakt. Chem., Ser. 2, 84 (1911) 37.

20895 R. Siedler and H.-J. Bittrich, J. Prakt. Chem., 311 (1969) 721.

20934 Von Auwers and Muller, Chem. Ber., 44 (1911) 1595.

20949 H. Azimi-Pour, Rev. Inst. Petrole Ann. Combust. Liquides, 15 (1960) 1.

20954 A. Iguchi, Kagaku Sochi, 13 (1971) 76.

20957 P.S. Pavlova, D.M. Popov and V.M. Olevskii, Tr. Gos. Nauchno-Issled. Proektn. Inst. Azotn. Prom-sti, 13 (1972) 89.

21010 V.M. Zakurenov and V.P. Konyakhin, Sint., 5 (1973) 126.

21026 Y. Morino, Kagaku Kenkyusho Hokoku, 23 (1933) 49.

21028 A. Naccari and S. Pagliani, Atti. Accad. Sci. Torino, 16 (1881) 407.

21080 B. Pesce, Gazz. Chim. Ital., 65 (1935) 440.

21109 W. Ramsay and S. Young, Phil. Trans. Roy. Soc. London, 175 (1884) 37.

21121 A. Nissema, Ann. Acad. Sci. Fenn., 153 (1970) 1.

21202 S.D. Burd, Jr., Ph. D. Dissertation, Pennsylvania State University, University Park, Pennsylvania, 1968.

21213 Y.L. Rastorgiev, B.A. Grigorev and R.M. Murdaev, Izv. Vyssh. Ucheb. Zaved., 18 (1975) 66.

21292 D.R. Hafemann and S.L. Miller, J. Phys. Chem., 73 (1969) 1392.

21293 K. Schafer, B. Schramm and J.S. Urieta Navarro, Z. Phys. Chem. (Frankfurt am Main), 93 (1974) 203.

21304 V.G. Benkovskii, M.Kh. Nauruzov, T.M. Bogoslovskaya and Zh. Serikov, Tr. Inst. Khim. Nefti Prir. Solei, 1 (1970) 16.

21306 F. Danusso and E. Fadigati, Atti Accad. Naz. Lincei, 14 (1953) 81.

21313 A.Z. Golik, I.I. Adamenko and V.V. Borovik, Ukr. Fiz. Zh. (Ukr. Ed.), 17 (1972) 2075.

21388 D.W. Brazier and G.R. Freeman, Can. J. Chem., 47 (1969) 893.

21432 M.S. Roshnov, Khim. Prom. (Moscow), 43 (1967) 288.

21441 M. Azizbekova, Izv. Vyssh. Ucheb. Zaved., 18 (1975) 75.

21510 P. Walden and R. Swinne, Z. Phys., 82 (1913) 271.

21533 J. Schmelzer and K. Quitzsch, Z. Phys. Chem. (Leipzig), 252 (1973) 280.

21537 H.-J. Bittrich, D. Klemm and D. Stephan, Z. Phys. Chem. (Leipzig), 256 (1975) 465.

21562 A.M. Kerimov and T.A. Apaev, Teplofiz. Svoistva Veshchestv Mater., 256 (1972) 26.

21583 H.-J. Gumpert, H. Kohler, W. Schille and H.-J. Bittrich, Wiss. Z. Tech. Hochsch. Chem. " Carl Schorlemmer" Leuna-Merseburg, 15 (1973) 179.

21584 Y.C. Chang, H.T. Wang and T. Huang, K'o Hsueh T'ung Pao, 5 (1965) 437.

21601 T.W. Mears, A. Fookson, P. Pomerantz, E.H. Rich, C.S. Dussinger and F.L. Howard, J. Res. Nat. Bur. Stand., 44 (1950) 299.

21602 F.L. Howard, T.W. Mears, A. Fookson, P. Pomerantz and D.B. Brooks, J. Res. Nat. Bur. Stand., 38 (1947) 365.

21607 J.W.M. Boelhouwer, Physica (Utrecht), 26 (1960) 1021.

21718 G. Dejardin, Ann. Phys. (Paris), 11 (1919) 253.

TABLE 6.1 (continued)

21804 A.V. Grosse and R.C. Wackher, Ind. Eng. Chem., 11 (1939) 614.
21805 J. Myers, Chem. Ber., 6 (1873) 439.
21809 F.L. Howard, T.W. Mears, A. Fookson, P. Pomerantz and D.B. Brooks,
       J. Res. Nat. Bur. Stand., 38 (1947) 365.
21810 B.V. Subbarao and C. Venkatarao, Can. J. Chem. Eng., 42 (1964) 266.
21816 L.S. Shraiber and N.G. Pechenyuk, Russ. J. Phys. Chem., 39 (1965) 219.
21835 D.W. Brazier and G.R. Freeman, Can. J. Chem., 47 (1969) 893.
21852 M. Kominek-szczepanik, Rocz. Chem., 33 (1959) 553.
21855 J. Sameshima, J. Amer. Chem. Soc., 40 (1918) 1503.
21870 W.F. Seyer and A.F. Gallaugher, Trans. Roy. Soc. Can., 20 (1926) 343.
21948 J.H. de Wilde, Z. Anorg. Allg. Chem., 233 (1937) 411.
21969 H.W. Deinum, Rec. Trav. Chim. Pays-Bas, 53 (1934) 1061.
22180 T. Tomiyama and C.L.A. Schmidt, J. Gen. Physiol., 19 (1935) 379.
22201 R.D. Goodwin, D.E. Diller, H.M. Roder and L.A. Weber, Cryogenics,
       2 (1961) 81.
22370 A.G. Tokaev and P.P. Pugachevich, Russ. J. Phys. Chem., 47 (1973) 90.
22447 M. Godchot and G. Cauquil, C. R. Acad. Sci., 191 (1930) 1326.
22448 T.E. Thorpe, J. Chem. Soc., 35 (1879) 296.
22457 M. Godchot, Bull. Soc. Chim. Fr., 1 (1934) 1153.
22499 K. Matsuno and K. Han, Bull. Chem. Soc. Jap., 11 (1936) 321.
22600 R. Schiff, Justus Liebigs Ann. Chem., 220 (1884) 278.
22720 J.D. White and F.W. Rose, J. Res. Nat. Bur. Stand., 9 (1932) 711.
22762 S.W. Benson and G.B. Kistiakowsky, J. Amer. Chem. Soc., 64 (1942) 80.
22902 H.C. Eckstrom, J.E. Berger and L.R. Dawson, J. Phys. Chem.,
       64 (1960) 1458.
23080 A.E.H. Mousa, Ph. D. Dissertation, Ohio State University, Columbus,
       Ohio, 1970.
23180 B.S. Chandak, G.D. Nageshwar and P.S. Mene, Can. J. Chem. Eng.,
       54 (1976) 647.
23270 J.-P.E. Grolier, E. Wilhelm and M.H. Hamedi, Ber. Bunsenges. Phys. Chem.,
       82 (1978) 1282.
23278 E. McMahon, J.N. Roper, W.P. Utermohlen, R.H. Hasek, R.C. Harris and
       J.H. Brant, J. Amer. Chem. Soc., 70 (1948) 2971.
23434 A.V. Nikolaev, A.A. Kolesnikov, I.P. Golentovskaya and M.A. Shmeleva,
       Izv. Sib. Otd. Akad. Nauk SSSR, Ser. Khim. Nauk, 2 (1971) 78.
23672 L. Haar and J.S. Gallagher, J. Phys. Chem. Ref. Data, 7 (1978) 635.
23676 S. Angus , B. Armstrong and K.M. De Reuck, International Thermodynamic
       Tables of the Fluid State, Carbon Dioxide, Vol. 3, Pergamon Press,
       Oxford, England, 1976.
23678 S. Angus, B. Armstrong, and K. De Reuck, International Thermodynamic
       Tables of the Fluid State, Methane, Vol. 5, Pergamon Press, Oxford,
       England, 1978.
23679 S. Angus, K.M. De Reuck and B. Armstrong, International Tables of the
       Fluid State, Nitrogen, Vol. 6, Pergamon Press, Oxford, England, 1979.
23680 S. Angus, B. Armstrong and K.M. De Reuck, International Thermodynamic
       Tables of the Fluid State Propylene (Propene), Vol. 7, Pergamon Press,
       Oxford, England, 1980.
23681 B.A. Younglove, J. Phys. Chem. Ref. Data, 11 (1982) 1.
40032 M. Inoue, K. Azumi and N. Suzuki, Ind. Eng. Chem., 14 (1975) 312.
40131 H. Wolff, O. Bauer, R. Goetz, H. Landeck, O. Schiller and L. Schimpf,
       J. Phys. Chem., 80 (1976) 131.
40156 M.V. Prabhakara Rao and P.R. Naidu, J. Chem. Thermodyn., 8 (1976) 73.
40158 M.V. Prabhakara Rao and P.R. Naidu, J. Chem. Thermodyn., 8 (1976) 96.
40166 A.J. Ashworth and D.M. Hooker, J. Chem. Soc., 72 (1976) 2240.
40176 J.-L. Fortier, G.C. Benson and P. Picker, J. Chem. Thermodyn.,

TABLE 6.1 (continued)

8 (1976) 289.

40204 M.L. Lakhanpal, H.G. Mandal and G. Lal, Indian J. Chem., 13 (1975) 1309.

40222 R.S. DePablo, J. Chem. Eng. Data, 21 (1976) 141.

40228 M. Hafez and S. Hartland, J. Chem. Eng. Data, 21 (1976) 179.

40231 T. Katayama and T. Nitta, J. Chem. Eng. Data, 21 (1976) 194.

40253 Z.S. Belousova and T.D. Sulimova, Russ. J. Phys. Chem., 50 (1976) 292.

40270 D.V.S. Jain and R.K. Wadi, J. Chem. Thermodyn., 8 (1976) 493.

40278 D. Ambrose and C.H.S. Sprake, J. Chem. Thermodyn., 8 (1976) 601.

40284 E.F. Meyer and C.A. Hotz, J. Chem. Eng. Data, 21 (1976) 274.

40292 A.H. Absood, M.S. Tutunji, K.-Y. Hsu and H.L. Clever, J. Chem. Eng. Data, 21 (1976) 304.

40329 S. Sundaram and D.S. Viswanath, J. Chem. Eng. Data, 21 (1976) 448.

40337 W. Wagner, J. Ewers and W. Pentermann, J. Chem. Thermodyn., 8 (1976) 1049.

40349 R.P. Tomlins and K.N. Marsh, J. Chem. Thermodyn., 8 (1976) 1185.

40353 K. Subramanyam Reddy and P.R. Naidu, J. Chem. Thermodyn., 8 (1976) 1208.

40360 Y. Jagannadha Rao and D.S. Viswanath, J. Chem. Eng. Data, 22 (1977) 36.

40405 H. Wolff and R. Gotz, Z. Phys. Chem. (Frankfurt am Main), 100 (1976) 25.

40418 G.C. Straty and R. Tsumura, J. Res. Nat. Bur. Stand., 80 (1976) 35.

40425 M.V. Prabhakara Rao and P.R. Naidu, Can. J. Chem., 54 (1976) 2280.

40435 C.R. McClune, Cryogenics, 16 (1976) 289.

40461 J. Aftienjew and A. Zawisza, J. Chem. Thermodyn., 9 (1977) 153.

40463 W.M. Haynes and M.J. Hiza, J. Chem. Thermodyn., 9 (1977) 179.

40489 O. Dusart, C. Piekarski, S. Piekarski and A. Viallard, J. Chim. Phys. Physicochim. Biol., 73 (1976) 837.

40535 Z.S. Belousova and T.D. Sulimova, Russ. J. Phys. Chem., 50 (1976) 585.

40563 J.R. Goates, J.B. Ott and J.F. Moellmer, J. Chem. Thermodyn., 9 (1977) 249.

40566 O. Kiyohara, G.C. Benson and J.-P.E. Grolier, J. Chem. Thermodyn., 9 (1977) 315.

40652 M.S. Medani and M.A. Hasan, J. Appl. Chem. Biotechnol., 27 (1977) 80.

40683 F.V. Kerchove and M.D. Vijider, J. Chem. Eng. Data, 22 (1977) 333.

40685 R.P. Tomlins and K.N. Marsh, J. Chem. Thermodyn., 9 (1977) 651.

40686 T.M. Letcher, J. Chem. Thermodyn., 9 (1977) 661.

40688 J.-P. Grolier, O. Kiyohara and G.C. Benson, J. Chem. Thermodyn., 9 (1977) 697.

40698 R.D. Goodwin, H.M. Roder and G.C. Straty, Thermophysical Properties of Ethane from 90 to 600 K at Pressures to 700 Bar, NBS TN-684, National Bureau of Standards, Cryogenics Division, Boulder, Colorado, 1976.

40724 J. Juza, V. Svoboda, R. Holub and J. Pick, Collect. Czech. Chem. Commun., 42 (1977) 1453.

40733 N.S. Laevskaya, I.V. Bagrov, L.L. Dobroserdov and Yu.L. Bondarenko, J. Appl. Chem. USSR, 49 (1976) 2460.

40772 R.H. Stokes and M. Adamson, J. Chem. Soc., 73 (1977) 1232.

40775 J. Lnenickova and I. Wichterle, Collect. Czech. Chem. Commun., 42 (1977) 1907.

40796 K.L. Young, R.A. Mentzer, R.A. Greenkorn and K.C. Chao, J. Chem. Thermodyn., 9 (1977) 979.

40802 O.P. Bagga, R.C. Katyal and K.S.N. Raju, J. Chem. Eng. Data, 22 (1977) 416.

40813 I.N. V'yunnik, A.M. Zholnovach and A.M. Shkodin, J. Gen. Chem. USSR, 47 (1977) 1541.

40816 A.C. Gupta and R.W. Hanks, Thermochim. Acta, 21 (1977) 143.

40823 A.H.N. Mousa, J. Chem. Thermodyn., 9 (1977) 1063.

40865 V. Mihkelson, H. Kirss, L. Kudrjavzeva and O. Eisen, Fluid Phase

TABLE 6.1 (continued)

Equilibria, 1 (1978) 201.

40895 M. Diaz Pena and G. Tardajos, J. Chem. Thermodyn., 10 (1978) 19.

40900 J.A. Hugill and M.L. McGlashan, J. Chem. Thermodyn., 10 (1978) 95.

40903 S.A. Wieczorek and J. Stecki, J. Chem. Thermodyn., 10 (1978) 177.

40928 H.L. Clever and K.-Y. Hsu, J. Chem. Thermodyn., 10 (1978) 213.

40936 M. Diaz Pena, A. Crespo Colin and A. Compositizo, J. Chem. Thermodyn., 10 (1978) 337.

41156 M. Sreenivasulu and P.R. Naidu, J. Chem. Thermodyn., 10 (1978) 1019.

41167 W. Pentermann and W. Wagner, J. Chem. Thermodyn., 10 (1978) 1161.

41188 M.N. Chandrashekara and D.N. Seshadri, J. Chem. Eng. Data, 24 (1979) 6.

41221 M. Diaz Pena, A. Compositizo and A. Crespo Colin, J. Chem. Thermodyn., 11 (1979) 447.

41226 J.R. Goates, J.B. Ott and R.B. Grigg, J. Chem. Thermodyn., 11 (1979) 497.

41354 P.J. Maher and B.D. Smith, J. Chem. Eng. Data, 24 (1979) 363.

41379 M. Takeo, K. Nishii, T. Nitta and T. Katayama, Fluid Phase Equilibria, 3 (1979) 123.

41407 K.S. Reddy and P.R. Naidu, Aust. J. Chem., 32 (1979) 687.

41430 M. Kato, Ind. Eng. Chem., 19 (1980) 253.

41545 P.J. Maher and B.D. Smith, J. Chem. Eng. Data, 25 (1980) 61.

41622 R. Rigglo, M.H. Ubeda, J.F. Ramos and H.E. Martinez, J. Chem. Eng. Data, 25 (1980) 318.

41656 J.H. Kennan, F.G. Keyes, P.G. Hill and J.G. Moore, Steam Tables - Thermodynamic Properties of Water Including Vapor, Liquid and Solid Phases (SI Units), John Wiley and Sons, Inc., New York, 1978.

41685 J.D. Olson, J. Chem. Eng. Data, 26 (1981) 58.

41757 B. Armstrong, J. Chem. Eng. Data, 26 (1981) 168.

42045 R.D. McCarty and R.T. Jacobsen, An Equation of State for Fluid Ethylene, NBS Technical Note 1045, National Bureau of Standards, Boulder, Colorado, 1981.

42132 H. Poll, H. Huemer and F. Moser, Monatsh. Chem., 111 (1980) 1159.

Chapter 7

INDEXES

Separate indexes are provided for each compound class--hydrocarbons and
ketones in Part A, and halogenated hydrocarbons and alcohols in Part B. In
addition, two kinds of indexes are provided for each compound class--a chemical
formula index and a compound name index.

7.1 CHEMICAL FORMULA INDEXES

The chemical formula indexes all contain four columns in the following order:
the chemical formula, the compound name, the Chemical Abstracts Service
registry number (CAS reg. nr) and the compound sequence number (CSN). The
compound sequence number appears at the left end of the first line on all the
compound tables (both first and second pages) and serves the same basic purpose
as a page number.

All of the chemical formula indexes are divided into carbon number sets. The
order within each carbon number set varies from one compound class to another.

7.1.1 Hydrocarbons

Each carbon number set in the formula index for the hydrocarbons is ordered
with respect to a decreasing hydrogen number. Compounds with the same number of
hydrogen atoms are ordered further by increasing branching, increasing location
numbers and increasing unsaturation.

TABLE 7.1

Formula index for hydrocarbons

(CSN = compound sequence number)		CAS reg. nr	CSN
C(1)H(4)	METHANE	000074-82-8	1
C(2)H(6)	ETHANE	000074-84-0	2
C(2)H(4)	ETHENE	000074-85-1	107
C(3)H(8)	PROPANE	000074-98-6	3
C(3)H(6)	CYCLOPROPANE	000075-19-4	52
C(3)H(6)	PROPENE	000115-07-1	108
C(3)H(4)	PROPADIENE	000463-49-0	194
C(3)H(4)	PROPYNE	000074-99-7	225
C(4)H(10)	BUTANE	000106-97-8	4
C(4)H(10)	ISOBUTANE	000075-28-5	5
C(4)H(8)	CYCLOBUTANE	000287-23-0	53
C(4)H(8)	METHYLCYCLOPROPANE	000594-11-6	54
C(4)H(8)	1-BUTENE	000106-98-9	109
C(4)H(8)	CIS-2-BUTENE	000590-18-1	110
C(4)H(8)	TRANS-2-BUTENE	000624-64-6	111
C(4)H(8)	ISOBUTENE	000115-11-7	112
C(4)H(6)	CYCLOBUTENE	000822-35-5	170
C(4)H(6)	1,2-BUTADIENE	000590-19-2	195
C(4)H(6)	1,3-BUTADIENE	000106-99-0	196
C(4)H(6)	1-BUTYNE	025339-57-5	226
C(4)H(6)	2-BUTYNE	000503-17-3	227
C(4)H(4)	1-BUTEN-3-YNE	000689-97-4	239
C(4)H(2)	1,3-BUTADIYNE	000460-12-8	238
C(5)H(12)	PENTANE	000109-66-0	6
C(5)H(12)	2-METHYLBUTANE	000078-78-4	7
C(5)H(12)	2,2-DIMETHYLPROPANE	000463-82-1	8
C(5)H(10)	CYCLOPENTANE	000287-92-3	55
C(5)H(10)	1-PENTENE	000109-67-1	113
C(5)H(10)	CIS-2-PENTENE	000627-20-3	114
C(5)H(10)	TRANS-2-PENTENE	000646-04-8	115
C(5)H(10)	2-METHYL-1-BUTENE	000563-46-2	116
C(5)H(10)	3-METHYL-1-BUTENE	000563-45-1	117
C(5)H(10)	2-METHYL-2-BUTENE	000513-35-9	118
C(5)H(8)	SPIROPENTANE	000185-94-4	56
C(5)H(8)	CYCLOPENTENE	000142-29-0	171
C(5)H(8)	METHYLENECYCLOBUTANE	001120-56-5	172
C(5)H(8)	1,2-PENTADIENE	000591-95-7	197
C(5)H(8)	1-CIS-3-PENTADIENE	001574-41-0	198
C(5)H(8)	1-TRANS-3-PENTADIENE	002004-70-8	199
C(5)H(8)	1,4-PENTADIENE	000591-93-5	200
C(5)H(8)	2,3-PENTADIENE	000591-96-8	201
C(5)H(8)	3-METHYL-1,2-BUTADIENE	000598-25-4	202
C(5)H(8)	2-METHYL-1,3-BUTADIENE	000078-79-5	203
C(5)H(6)	1,3-CYCLOPENTADIENE	000542-92-7	217

TABLE 7.1 (continued)   CAS reg. nr   CSN

C(6)H(14)	HEXANE	000110-54-3	9
C(6)H(14)	2-METHYLPENTANE	000107-83-5	10
C(6)H(14)	3-METHYLPENTANE	000096-14-0	11
C(6)H(14)	2,2-DIMETHYLBUTANE	000075-83-2	12
C(6)H(14)	2,3-DIMETHYLBUTANE	000079-29-8	13
C(6)H(12)	CYCLOHEXANE	000110-82-7	57
C(6)H(12)	METHYLCYCLOPENTANE	000096-37-7	58
C(6)H(12)	1-HEXENE	000592-41-6	119
C(6)H(12)	CIS-2-HEXENE	007688-21-3	120
C(6)H(12)	TRANS-2-HEXENE	004050-45-7	121
C(6)H(12)	CIS-3-HEXENE	007642-09-3	122
C(6)H(12)	TRANS-3-HEXENE	013269-52-8	123
C(6)H(12)	2-METHYL-1-PENTENE	000763-29-1	124
C(6)H(12)	3-METHYL-1-PENTENE	000760-20-3	125
C(6)H(12)	4-METHYL-1-PENTENE	000691-37-2	126
C(6)H(12)	2-METHYL-2-PENTENE	000625-27-4	127
C(6)H(12)	3-METHYL-CIS-2-PENTENE	000922-62-3	128
C(6)H(12)	3-METHYL-TRANS-2-PENTENE	000616-12-6	129
C(6)H(12)	4-METHYL-CIS-2-PENTENE	000691-38-3	130
C(6)H(12)	4-METHYL-TRANS-2-PENTENE	000674-76-0	131
C(6)H(12)	2-ETHYL-1-BUTENE	000760-21-4	132
C(6)H(12)	2,3-DIMETHYL-1-BUTENE	000563-78-0	133
C(6)H(12)	3,3-DIMETHYL-1-BUTENE	000558-37-2	134
C(6)H(12)	2,3-DIMETHYL-2-BUTENE	000563-79-1	135
C(6)H(10)	CYCLOHEXENE	000110-83-8	173
C(6)H(10)	1-METHYLCYCLOPENTENE	000693-89-0	174
C(6)H(10)	1,2-HEXADIENE	042296-74-2	204
C(6)H(10)	1,TRANS-3-HEXADIENE	020237-34-7	205
C(6)H(10)	1,TRANS-4-HEXADIENE	007319-00-8	206
C(6)H(10)	TRANS-2,TRANS-4-HEXADIENE	005194-51-4	208
C(6)H(10)	1,5-HEXADIENE	000592-42-7	207
C(6)H(10)	2,3-DIMETHYL-1,3-BUTADIENE	000513-81-5	209
C(6)H(10)	1-HEXYNE	000693-02-7	228
C(6)H(10)	3-HEXYNE	000928-49-4	229
C(6)H(8)	1,3-CYCLOHEXADIENE	000592-57-4	218
C(6)H(8)	1,4-CYCLOHEXADIENE	000628-41-1	219
C(6)H(8)	CIS-1,3,5-HEXATRIENE	002612-46-6	210
C(6)H(6)	1,5-HEXADIEN-3-YNE	000821-08-9	240
C(6)H(6)	BENZENE	000071-43-2	241
C(7)H(16)	HEPTANE	000142-82-5	14
C(7)H(16)	2-METHYLHEXANE	000591-76-4	15
C(7)H(16)	3-METHYLHEXANE	000589-34-4	16
C(7)H(16)	3-ETHYLPENTANE	000617-78-7	17
C(7)H(16)	2,2-DIMETHYLPENTANE	000590-35-2	18
C(7)H(16)	2,3-DIMETHYLPENTANE	000565-59-3	19
C(7)H(16)	2,4-DIMETHYLPENTANE	000108-08-7	20
C(7)H(16)	3,3-DIMETHYLPENTANE	000562-49-2	21
C(7)H(16)	2,2,3-TRIMETHYLBUTANE	000464-06-2	22
C(7)H(14)	CYCLOHEPTANE	000291-64-5	59
C(7)H(14)	METHYLCYCLOHEXANE	000108-87-2	60
C(7)H(14)	ETHYLCYCLOPENTANE	001640-40-6	61
C(7)H(14)	1,1-DIMETHYLCYCLOPENTANE	001638-26-2	62
C(7)H(14)	1,CIS-2-DIMETHYLCYCLOPENTANE	001192-18-3	63

TABLE 7.1 (continued)   CAS reg. nr   CSN

C(7)H(14)	1,TRANS-2-DIMETHYLCYCLOPENTANE	000822-50-4	64
C(7)H(14)	1,CIS-3-DIMETHYLCYCLOPENTANE	002532-58-3	65
C(7)H(14)	1,TRANS-3-DIMETHYLCYCLOPENTANE	001759-58-6	66
C(7)H(14)	1-HEPTENE	000592-76-7	136
C(7)H(14)	CIS-2-HEPTENE	006443-92-1	137
C(7)H(14)	TRANS-2-HEPTENE	014686-13-6	138
C(7)H(14)	CIS-3-HEPTENE	007642-10-6	139
C(7)H(14)	TRANS-3-HEPTENE	014686-14-7	140
C(7)H(14)	3-METHYL-CIS-3-HEXENE	004914-89-0	141
C(7)H(14)	3-METHYL-TRANS-3-HEXENE	003899-36-3	142
C(7)H(14)	2,4-DIMETHYL-1-PENTENE	002213-32-3	143
C(7)H(14)	4,4-DIMETHYL-1-PENTENE	000762-62-9	144
C(7)H(14)	3-ETHYL-2-PENTENE	000816-79-5	145
C(7)H(14)	2,4-DIMETHYL-2-PENTENE	000625-65-0	146
C(7)H(14)	4,4-DIMETHYL-CIS-2-PENTENE	000762-63-0	147
C(7)H(14)	4,4-DIMETHYL-TRANS-2-PENTENE	000690-08-4	148
C(7)H(14)	2-ETHYL-3-METHYL-1-BUTENE	007357-93-9	149
C(7)H(14)	2,3,3-TRIMETHYL-1-BUTENE	000594-56-9	150
C(7)H(12)	BICYCLO(2.2.1)HEPTANE	000279-23-2	67
C(7)H(12)	BICYCLO(4.1.0)HEPTANE	000286-08-8	68
C(7)H(12)	1-METHYLBICYCLO(3.1.0)HEXANE	004625-24-5	69
C(7)H(12)	CYCLOHEPTENE	000628-92-2	175
C(7)H(12)	1-METHYL-1-CYCLOHEXENE	000591-49-1	176
C(7)H(12)	3-METHYL-1-CYCLOHEXENE	000591-48-0	177
C(7)H(12)	METHYLENECYCLOHEXANE	001192-37-6	178
C(7)H(12)	2,CIS-4-HEPTADIENE	054354-36-8	211
C(7)H(12)	2,TRANS-4-HEPTADIENE	002384-94-3	212
C(7)H(12)	2,4-DIMETHYL-1,3-PENTADIENE	001000-86-8	213
C(7)H(12)	2,4-DIMETHYL-2,3-PENTADIENE	001000-87-9	214
C(7)H(12)	1-HEPTYNE	000628-71-7	230
C(7)H(12)	2-HEPTYNE	001119-65-9	231
C(7)H(12)	3-HEPTYNE	002586-89-2	232
C(7)H(10)	TRICYCLO(2.2.1.0)HEPTANE	036120-90-8	70
C(7)H(10)	TRICYCLO(4.1.0.0)HEPTANE	000187-26-8	71
C(7)H(10)	BICYCLO(2.2.1)HEPT-2-ENE	000498-66-8	179
C(7)H(10)	BICYCLO(4.1.0)HEPT-3-ENE	016554-83-9	180
C(7)H(8)	TETRACYCLO(2.2.1.0.0)HEPTANE	000278-06-8	72
C(7)H(8)	BICYCLO(2.2.1)HEPTA-2,5-DIENE	000121-46-0	221
C(7)H(8)	1,3,5-CYCLOHEPTATRIENE	002196-23-8	220
C(7)H(8)	TOLUENE	000108-88-3	242
C(8)H(18)	OCTANE	000111-65-9	23
C(8)H(18)	2-METHYLHEPTANE	000592-27-8	24
C(8)H(18)	3-METHYLHEPTANE	000589-81-1	25
C(8)H(18)	4-METHYLHEPTANE	000589-53-7	26
C(8)H(18)	3-ETHYLHEXANE	000619-99-8	27
C(8)H(18)	2,2-DIMETHYLHEXANE	000590-73-8	28
C(8)H(18)	2,3-DIMETHYLHEXANE	000584-94-1	29
C(8)H(18)	2,4-DIMETHYLHEXANE	000589-43-5	30
C(8)H(18)	2,5-DIMETHYLHEXANE	000592-13-2	31
C(8)H(18)	3,3-DIMETHYLHEXANE	000563-16-6	32
C(8)H(18)	3,4-DIMETHYLHEXANE	000583-48-2	33
C(8)H(18)	3-ETHYL-2-METHYLPENTANE	000609-26-7	34
C(8)H(18)	3-ETHYL-3-METHYLPENTANE	001067-08-9	35

TABLE 7.1 (continued)   CAS reg. nr   CSN

C(8)H(18)	2,2,3-TRIMETHYLPENTANE	000564-02-3	36
C(8)H(18)	2,2,4-TRIMETHYLPENTANE	000540-84-1	37
C(8)H(18)	2,3,3-TRIMETHYLPENTANE	000560-21-4	38
C(8)H(18)	2,3,4-TRIMETHYLPENTANE	000565-75-3	39
C(8)H(18)	2,2,3,3-TETRAMETHYLBUTANE	000594-82-1	40
C(8)H(16)	CYCLOOCTANE	000292-64-8	73
C(8)H(16)	ETHYLCYCLOHEXANE	001678-91-7	74
C(8)H(16)	1,1-DIMETHYLCYCLOHEXANE	000590-66-9	75
C(8)H(16)	CIS-1,2-DIMETHYLCYCLOHEXANE	002207-01-4	76
C(8)H(16)	TRANS-1,2-DIMETHYLCYCLOHEXANE	006876-23-9	77
C(8)H(16)	CIS-1,3-DIMETHYLCYCLOHEXANE	000638-04-0	78
C(8)H(16)	TRANS-1,3-DIMETHYLCYCLOHEXANE	002207-03-6	79
C(8)H(16)	CIS-1,4-DIMETHYLCYCLOHEXANE	000624-29-3	80
C(8)H(16)	TRANS-1,4-DIMETHYLCYCLOHEXANE	002207-04-7	81
C(8)H(16)	PROPYLCYCLOPENTANE	002040-96-2	82
C(8)H(16)	ISOPROPYLCYCLOPENTANE	003875-51-2	83
C(8)H(16)	1-ETHYL-1-METHYLCYCLOPENTANE	016747-50-5	84
C(8)H(16)	CIS-1-ETHYL-2-METHYLCYCLOPENTANE	000930-89-2	85
C(8)H(16)	TRANS-1-ETHYL-2-METHYLCYCLOPENTANE	000930-90-5	86
C(8)H(16)	1,1,2-TRIMETHYLCYCLOPENTANE	004259-00-1	87
C(8)H(16)	1,1,3-TRIMETHYLCYCLOPENTANE	004516-69-2	88
C(8)H(16)	1,CIS-2,TRANS-4-TRIMETHYL CYCLOPENTANE	004850-28-6	89
C(8)H(16)	1,TRANS-2,CIS-4-TRIMETHYL CYCLOPENTANE	013398-35-1	90
C(8)H(16)	1-OCTENE	000111-66-0	151
C(8)H(16)	CIS-2-OCTENE	007642-04-8	152
C(8)H(16)	TRANS-2-OCTENE	013389-42-9	153
C(8)H(16)	TRANS-3-OCTENE	014919-01-8	154
C(8)H(16)	CIS-4-OCTENE	007642-15-1	155
C(8)H(16)	TRANS-4-OCTENE	014850-23-8	156
C(8)H(16)	2-METHYL-2-HEPTENE	000627-97-4	157
C(8)H(16)	2,2-DIMETHYL-CIS-3-HEXENE	000690-92-6	158
C(8)H(16)	2,2-DIMETHYL-TRANS-3-HEXENE	000690-93-7	159
C(8)H(16)	3-ETHYL-2-METHYL-1-PENTENE	019780-66-6	160
C(8)H(16)	2,4,4-TRIMETHYL-1-PENTENE	000107-39-1	161
C(8)H(16)	2,4,4-TRIMETHYL-2-PENTENE	000107-40-4	162
C(8)H(14)	CIS-BICYCLO(3.3.0)OCTANE	001755-05-1	91
C(8)H(14)	TRANS-BICYCLO(3.3.0)OCTANE	005597-89-7	92
C(8)H(14)	CIS-BICYCLO(4.2.0)OCTANE	028282-35-1	93
C(8)H(14)	CIS-BICYCLO(5.1.0)OCTANE	016526-90-2	94
C(8)H(14)	CYCLOOCTENE	000931-88-4	181
C(8)H(14)	1-ETHYL-1-CYCLOHEXENE	001453-24-3	182
C(8)H(14)	4,4-DIMETHYL-1-CYCLOHEXENE	014072-86-7	183
C(8)H(14)	1-METHYL-3-METHYLENECYCLOHEXANE	003101-50-6	184
C(8)H(14)	1-METHYL-4-METHYLENECYCLOHEXANE	002808-80-2	185
C(8)H(14)	2,3,3-TRIMETHYL-1-CYCLOPENTENE		186
C(8)H(14)	2-BUTYL-1,3-BUTADIENE	001189-53-3	215
C(8)H(14)	1-OCTYNE	000629-05-0	233
C(8)H(14)	2-OCTYNE	002809-67-8	234
C(8)H(14)	3-OCTYNE	015232-76-5	235
C(8)H(14)	4-OCTYNE	001942-45-6	236
C(8)H(12)	4-VINYLCYCLOHEXENE	000100-40-3	187
C(8)H(12)	CIS-1,CIS-5-CYCLOOCTADIENE	001552-12-1	222
C(8)H(10)	ETHYLBENZENE	000100-41-4	243

TABLE 7.1 (continued)   CAS reg. nr   CSN

C(8)H(10)	O-XYLENE	000095-47-6	244
C(8)H(10)	M-XYLENE	000108-38-3	245
C(8)H(10)	P-XYLENE	000106-42-3	246
C(8)H(8)	1,3,5,7-CYCLOOCTATETRAENE	000629-20-9	223
C(8)H(8)	STYRENE	000100-42-5	247
C(9)H(20)	NONANE	000111-84-2	41
C(9)H(20)	2,6-DIMETHYLHEPTANE	001072-05-5	42
C(9)H(20)	2,2,3-TRIMETHYLHEXANE	016747-25-4	43
C(9)H(20)	2,2,4-TRIMETHYLHEXANE	016747-26-5	44
C(9)H(20)	2,2,5-TRIMETHYLHEXANE	003522-94-9	45
C(9)H(20)	2,4,4-TRIMETHYLHEXANE	016747-30-1	46
C(9)H(20)	3,3-DIETHYLPENTANE	001067-20-5	47
C(9)H(20)	2,2,3,3-TETRAMETHYLPENTANE	007154-79-2	48
C(9)H(20)	2,2,3,4-TETRAMETHYLPENTANE	001186-53-4	49
C(9)H(20)	2,2,4,4-TETRAMETHYLPENTANE	001070-87-7	50
C(9)H(20)	2,3,3,4-TETRAMETHYLPENTANE	016747-38-9	51
C(9)H(18)	PROPYLCYCLOHEXANE	001678-92-8	95
C(9)H(18)	ISOPROPYLCYCLOHEXANE	000696-29-7	96
C(9)H(18)	1-ETHYL-CIS-3-METHYLCYCLOHEXANE	019489-10-2	97
C(9)H(18)	1,1,3-TRIMETHYLCYCLOHEXANE	003073-66-3	98
C(9)H(18)	1,CIS-3,CIS-5-TRIMETHYLCYCLOHEXANE	001795-27-3	99
C(9)H(18)	BUTYLCYCLOPENTANE	002040-95-1	100
C(9)H(18)	1-NONENE	000124-11-8	163
C(9)H(18)	CIS-2-NONENE	006434-77-1	164
C(9)H(18)	TRANS-2-NONENE	006434-78-2	165
C(9)H(18)	CIS-3-NONENE	020237-46-1	166
C(9)H(18)	TRANS-3-NONENE	020063-92-7	167
C(9)H(18)	CIS-4-NONENE	010405-84-2	168
C(9)H(18)	TRANS-4-NONENE	010405-85-3	169
C(9)H(16)	CIS-BICYCLO(4.3.0)NONANE	004551-51-3	101
C(9)H(16)	TRANS-BICYCLO(4.3.0)NONANE	003296-50-2	102
C(9)H(16)	CIS-BICYCLO(6.1.0)NONANE	013757-43-2	103
C(9)H(16)	2-ETHYLBICYCLO(2.2.1)HEPTANE	002146-41-0	104
C(9)H(16)	1,4-DIMETHYLBICYCLO(2.2.1)HEPTANE	020454-81-3	105
C(9)H(16)	TRANS-2,3-DIMETHYLBICYCLO(2.2.1) HEPTANE		106
C(9)H(16)	1-METHYL-2-PROPYL-1-CYCLOPENTENE	053366-60-2	188
C(9)H(16)	1,2-DIETHYL-1-CYCLOPENTENE		189
C(9)H(16)	7-METHYL-2,4-OCTADIENE		216
C(9)H(16)	1-NONYNE	003452-09-3	237
C(9)H(14)	2-METHYLBICYCLO(2.2.2)OCT-2-ENE	004893-13-4	190
C(9)H(14)	2-ETHENYLBICYCLO(2.2.1)HEPTANE	002146-39-6	191
C(9)H(12)	CIS-5-ETHYLIDENEBICYCLO(2.2.1) HEPT-2-ENE	016219-75-3	192
C(9)H(12)	5-ETHENYLBICYCLO(2.2.1)HEPT-2-ENE	003048-64-4	193
C(9)H(12)	CIS-BICYCLO(4.3.0)NONA-3,7-DIENE		224
C(9)H(12)	PROPYLBENZENE	000103-65-1	248
C(9)H(12)	ISOPROPYLBENZENE	000098-82-8	249
C(9)H(12)	1-ETHYL-2-METHYLBENZENE	000611-14-3	250
C(9)H(12)	1-ETHYL-3-METHYLBENZENE	000620-14-4	251
C(9)H(12)	1-ETHYL-4-METHYLBENZENE	000622-96-8	252
C(9)H(12)	1,2,3-TRIMETHYLBENZENE	000526-73-8	253
C(9)H(12)	1,2,4-TRIMETHYLBENZENE	000095-63-6	254

TABLE 7.1 (continued)   CAS reg. nr   CSN

C(9)H(12)	1,3,5-TRIMETHYLBENZENE	000108-67-8	255
C(9)H(10)	CYCLOPROPYLBENZENE	000873-49-4	256
C(9)H(10)	1-PROPENYLBENZENE	000637-50-3	257
C(9)H(10)	ISOPROPENYLBENZENE	000098-83-9	258
C(9)H(10)	1-ETHENYL-2-METHYLBENZENE	000611-12-3	259
C(9)H(10)	1-ETHENYL-3-METHYLBENZENE	000100-80-1	260
C(9)H(10)	1-ETHENYL-4-METHYLBENZENE	000622-97-9	261
C(9)H(10)	INDAN	000496-11-7	262
C(9)H(8)	INDENE	000095-13-6	263

7.1.2 <u>Ketones</u>

Each carbon number set in the formula index for the ketones is further divided into four subsets arranged in the following order: ketones with no other active group, halogenated ketones, hydroxy ketones and ether ketones. All four subsets are in the order of decreasing hydrogen number.

After the hydrogen number, the ketone subset is arranged in the order of increasing oxygen number. For compounds with the same hydrogen and oxygen numbers, further ordering is related to increasing branching, increasing location numbers or increasing unsaturation.

The formulas for the halogenated ketone subset compounds are ordered alphabetically according to the halogens contained.

Whenever more than one hydroxy ketone appears for given carbon and hydrogen numbers, further ordering is based on the location numbers.

Most of the ether ketones contain halogens and have been included in the halogenated ketone subsets. Only two compounds--one C_{11} and one C_{16}--are listed separately as "ether" ketones.

TABLE 7.2

Formula index for ketones

(CSN = compound sequence number)   CAS reg. nr   CSN

C(3)H(6)O(1)	ACETONE	000067-64-1	264
C(3)H(5)CL(1)O(1)	1-CHLORO-2-PROPANONE	000078-95-5	369
C(3)H(4)CL(2)O(1)	1,1-DICHLORO-2-PROPANONE	000513-88-2	370
C(3)H(4)CL(2)O(1)	1,3-DICHLORO-2-PROPANONE	000534-07-6	371
C(3)CL(1)F(5)O(1)	1-CHLORO-1,1,3,3,3-PENTAFLUORO-2-PROPANONE	000079-53-8	372
C(3)F(6)O(1)	1,1,1,3,3,3-HEXAFLUORO-2-PROPANONE	000684-16-2	373
C(4)H(8)O(1)	2-BUTANONE	000078-93-3	265
C(4)H(6)O(1)	CYCLOBUTANONE	001191-95-3	301
C(4)H(6)O(1)	3-BUTEN-2-ONE	000078-94-4	323
C(4)H(6)O(2)	2,3-BUTANEDIONE	000431-03-8	266
C(4)H(4)O(2)	1,3-CYCLOBUTANEDIONE	015506-53-3	302
C(4)H(7)BR(1)O(1)	1-BROMO-2-BUTANONE	000816-40-4	374
C(4)H(7)BR(1)O(1)	3-BROMO-2-BUTANONE	000814-75-5	375
C(4)H(7)CL(1)O(1)	3-CHLORO-2-BUTANONE	004091-39-8	376
C(4)H(8)O(2)	3-HYDROXY-2-BUTANONE	000513-86-0	384
C(5)H(10)O(1)	2-PENTANONE	000107-87-9	267
C(5)H(10)O(1)	3-PENTANONE	000096-22-0	268
C(5)H(10)O(1)	3-METHYL-2-BUTANONE	000563-80-4	269
C(5)H(8)O(1)	CYCLOPENTANONE	000120-92-3	303
C(5)H(8)O(1)	ACETYLCYCLOPROPANE	000765-43-5	304
C(5)H(8)O(1)	3-METHYL-3-BUTEN-2-ONE	000814-78-8	324
C(5)H(8)O(2)	2,4-PENTANEDIONE	000123-54-6	270
C(5)H(9)BR(1)O(1)	4-BROMO-2-PENTANONE		377
C(5)H(10)O(2)	1-HYDROXY-3-METHYL-2-BUTANONE	036960-22-2	385
C(5)H(10)O(2)	3-HYDROXY-3-METHYL-2-BUTANONE	000115-22-0	386
C(5)H(10)O(2)	4-HYDROXY-3-METHYL-2-BUTANONE	003393-64-4	387
C(5)H(6)O(2)	5-HYDROXY-3-PENTYN-2-ONE	015441-65-3	388
C(6)H(12)O(1)	2-HEXANONE	000591-78-6	271
C(6)H(12)O(1)	3-HEXANONE	000589-38-8	272
C(6)H(12)O(1)	4-METHYL-2-PENTANONE	000108-10-1	273
C(6)H(12)O(1)	2-METHYL-3-PENTANONE	000565-69-5	274
C(6)H(12)O(1)	3,3-DIMETHYL-2-BUTANONE	000075-97-8	275
C(6)H(10)O(1)	CYCLOHEXANONE	000108-94-1	305
C(6)H(10)O(1)	5-HEXEN-2-ONE	000109-49-9	325
C(6)H(10)O(1)	4-METHYL-3-PENTEN-2-ONE	000141-79-7	326
C(6)H(10)O(1)	4-METHYL-4-PENTEN-2-ONE	003744-02-3	327

TABLE 7.2 (continued)

		CAS reg. nr	CSN
C(6)H(10)O(2)	3-METHYL-2,4-PENTANEDIONE	000815-57-6	276
C(6)H(4)O(2)	2,5-CYCLOHEXADIENE-1,4-DIONE	000106-51-4	330
C(6)H(11)BR(1)O(1)	6-BROMO-2-HEXANONE	010226-29-6	378
C(6)H(12)O(2)	4-HYDROXY-4-METHYL-2-PENTANONE	000123-42-2	389
C(7)H(14)O(1)	2-HEPTANONE	000110-43-0	277
C(7)H(14)O(1)	3-HEPTANONE	000106-35-4	278
C(7)H(14)O(1)	4-HEPTANONE	000123-19-3	279
C(7)H(14)O(1)	2-METHYL-3-HEXANONE	007379-12-6	280
C(7)H(14)O(1)	2,4-DIMETHYL-3-PENTANONE	000565-80-0	281
C(7)H(12)O(1)	CYCLOHEPTANONE	000502-42-1	306
C(7)H(12)O(1)	2-METHYL-1-CYCLOHEXANONE	000583-60-8	307
C(7)H(12)O(1)	3-METHYL-1-CYCLOHEXANONE	000591-24-2	308
C(7)H(12)O(1)	4-METHYL-1-CYCLOHEXANONE	000589-92-4	309
C(7)H(12)O(2)	3-ETHYL-2,4-PENTANEDIONE	001540-34-7	282
C(7)H(12)O(2)	3,3-DIMETHYL-2,4-PENTANEDIONE	003142-58-3	283
C(7)H(10)O(3)	2,4,6-HEPTANETRIONE	000626-53-9	284
C(7)H(6)O(2)	2-METHYL-2,5-CYCLOHEXADIENE-1,4-DIONE	000553-97-9	331
C(8)H(16)O(1)	2-OCTANONE	000111-13-7	285
C(8)H(16)O(1)	3-OCTANONE	000106-68-3	286
C(8)H(16)O(1)	4-OCTANONE	000589-63-9	287
C(8)H(16)O(1)	2-METHYL-3-HEPTANONE	013019-20-0	288
C(8)H(16)O(1)	2,5-DIMETHYL-3-HEXANONE	001888-57-9	289
C(8)H(16)O(1)	2,2,4-TRIMETHYL-3-PENTANONE	005857-36-3	290
C(8)H(14)O(1)	CYCLOOCTANONE	000502-49-8	310
C(8)H(14)O(1)	2-PROPYL-1-CYCLOPENTANONE	001193-70-0	311
C(8)H(14)O(1)	ACETYLCYCLOHEXANE	000823-76-7	312
C(8)H(12)O(2)	3-ACETYL-5-HEXEN-2-ONE	003508-78-9	328
C(8)H(8)O(1)	ACETOPHENONE	000098-86-2	335
C(8)H(7)CL(1)O(1)	4-CHLOROACETOPHENONE	000099-91-2	379
C(9)H(18)O(1)	2-NONANONE	000821-55-6	291
C(9)H(18)O(1)	3-NONANONE	000925-78-0	292
C(9)H(18)O(1)	5-NONANONE	000502-56-7	293
C(9)H(18)O(1)	2,6-DIMETHYL-4-HEPTANONE	000108-83-8	294
C(9)H(16)O(1)	1-CYCLOHEXYL-1-PROPANONE	001123-86-0	314
C(9)H(16)O(1)	CYCLONONANONE	003350-30-9	313
C(9)H(14)O(1)	2,6-DIMETHYL-2,5-HEPTADIEN-4-ONE	000504-20-1	329
C(9)H(10)O(1)	1-PHENYL-1-PROPANONE	000093-55-0	336
C(9)H(10)O(1)	1-PHENYL-2-PROPANONE	000103-79-7	337
C(9)H(10)O(1)	4-METHYLACETOPHENONE	000122-00-9	338
C(9)H(8)O(1)	1-INDANONE	000083-33-0	339

TABLE 7.2 (continued)   CAS reg. nr   CSN

C(10)H(20)O(1)	3-DECANONE	000928-80-3	295
C(10)H(18)O(1)	CYCLODECANONE	001502-06-3	315
C(10)H(18)O(1)	2-ISOPROPYL-5-METHYL-1-CYCLOHEXANONE	014073-97-3	316
C(10)H(18)O(1)	2,2,5,5-TETRAMETHYL-1-CYCLOHEXANONE	015189-14-7	317
C(10)H(16)O(1)	1,3,3-TRIMETHYLBICYCLO(2.2.1)HEPTAN-2-ONE	000126-21-6	318
C(10)H(16)O(1)	CAMPHOR	000076-22-2	319
C(10)H(16)O(1)	DECAHYDRO-2-NAPHTHALENONE	016021-08-2	320
C(10)H(14)O(1)	5-ISOPROPENYL-2-METHYL-2-CYCLOHEXEN-1-ONE	000099-49-0	332
C(10)H(12)O(1)	1-PHENYL-1-BUTANONE	000495-40-9	340
C(10)H(12)O(1)	1-PHENYL-2-BUTANONE	001007-32-5	341
C(10)H(12)O(1)	4-PHENYL-2-BUTANONE	002550-26-7	342
C(10)H(12)O(1)	3-ETHYLACETOPHENONE	022699-70-3	345
C(10)H(12)O(1)	4-ETHYLACETOPHENONE	000937-30-4	346
C(10)H(10)O(1)	1-OXO-1,2,3,4-TETRAHYDRONAPHTHALENE	000529-34-0	344
C(10)H(10)O(2)	1-PHENYL-1,3-BUTANEDIONE	000093-91-4	343
C(10)H(10)O(2)	1,3-DIACETYLBENZENE	006781-42-6	347
C(10)H(10)O(2)	1,4-DIACETYLBENZENE	001009-61-6	348
C(10)H(15)BR(1)O(2)	5-BROMO-3-METHOXY-2,6-DIMETHYL-2,5-HEPTADIEN-4-ONE		380
C(11)H(22)O(1)	2-UNDECANONE	000112-12-9	296
C(11)H(22)O(1)	6-UNDECANONE	000927-49-1	297
C(11)H(20)O(1)	CYCLOUNDECANONE	000878-13-7	321
C(11)H(14)O(1)	1-PHENYL-1-PENTANONE	001009-14-9	349
C(11)H(14)O(1)	1-PHENYL-2-PENTANONE	006683-92-7	350
C(11)H(16)O(3)	3-ACETOXY-2,6-DIMETHYL-2,5-HEPTADIEN-4-ONE		393
C(12)H(22)O(1)	CYCLODODECANONE	000830-13-7	322
C(12)H(16)O(1)	1-PHENYL-1-HEXANONE	000942-92-7	351
C(12)H(10)O(1)	1-ACETYLNAPHTHALENE	000941-98-0	352
C(13)H(26)O(1)	2-TRIDECANONE	000593-08-8	298
C(13)H(20)O(1)	4-(2,6,6-TRIMETHYL-1-CYCLOHEXEN-1-YL)3-BUTEN-2-ONE	014901-07-6	333
C(13)H(20)O(1)	4-(2,6,6-TRIMETHYL-2-CYCLOHEXEN-1-YL)3-BUTEN-2-ONE	000127-41-3	334
C(13)H(18)O(1)	1-PHENYL-1-HEPTANONE	001671-75-6	353
C(13)H(12)O(1)	1-NAPHTHYL-1-PROPANONE	002876-63-3	354
C(13)H(10)O(1)	BENZOPHENONE	000119-61-9	355

TABLE 7.2 (continued)   CAS reg. nr   CSN

C(14)H(14)O(1)	1-(1-NAPHTHALENYL)-1-BUTANONE	002876-62-2	356
C(14)H(10)O(2)	DIPHENYLETHANEDIONE	000134-81-6	357
C(14)H(8)O(2)	9,10-ANTHRACENEDIONE	000084-65-1	358
C(14)H(8)O(4)	1,4-DIHYDROXY-9,10-ANTHRACENEDIONE	000081-64-1	390
C(14)H(8)O(4)	1,5-DIHYDROXY-9,10-ANTHRACENEDIONE	000117-12-4	391
C(14)H(8)O(4)	1,8-DIHYDROXY-9,10-ANTHRACENEDIONE	000117-10-2	392
C(15)H(30)O(1)	8-PENTADECANONE	000818-23-5	299
C(15)H(22)O(1)	1-PHENYL-1-NONANONE	006008-36-2	359
C(15)H(16)O(1)	1-(1-NAPHTHALENYL)-1-PENTANONE	002876-60-0	360
C(15)H(14)O(1)	1,3-DIPHENYL-2-PROPANONE	000102-04-5	361
C(16)H(18)O(1)	1-(1-NAPHTHALENYL)-1-HEXANONE	002876-61-1	362
C(16)H(19)BR(1)O(2)	3-(4-BROMOBENZYLOXY)-2,6-DIMETHYL-2,5-HEPTADIEN-4-ONE		381
C(16)H(18)BR(2)O(2)	5-BROMO-3(4-BROMOBENZYLOXY)-2,6-DIMETHYL-2,5-HEPTADIEN-4-ONE		382
C(16)H(17)BR(1)O(3)	3-BENZOYLOXY-5-BROMO-2,6-DIMETHYL-2,5-HEPTADIEN-4-ONE		383
C(16)H(18)O(3)	3-BENZOYLOXY-2,6-DIMETHYL-2,5-HEPTADIEN-4-ONE		394
C(17)H(20)O(1)	1-(1-NAPHTHALENYL)-1-HEPTANONE	025897-44-3	363
C(17)H(10)O(1)	7H-BENZ(DE)ANTHRACEN-7-ONE	000082-05-3	364
C(18)H(28)O(1)	1-PHENYL-1-DODECANONE	001674-38-0	365
C(18)H(22)O(1)	1-NAPHTHYL-1-OCTANONE		366
C(20)H(32)O(1)	1-PHENYL-1-TETRADECANONE	004497-05-6	367
C(22)H(36)O(1)	1-PHENYL-1-HEXADECANONE	006697-12-7	368
C(31)H(62)O(1)	16-HENTRIACONTANONE	000502-73-8	300

7.2 COMPOUND NAME INDEXES

The compound name indexes contain the same four columns as the formula
indexes but in the following order: compound name, chemical formula, Chemical
Abstracts Service registry number (CAS reg. nr) and the compound sequence
number (CSN). The compound sequence number is used to sequence the compound
data tables; it appears at the left end of the first line on all the compound
tables.

Because of the large numbers of compounds, and the complexity of many of the
compound names, a straight alphabetical listing of compound names is not very
user friendly in three of the four compound name indexes in these two volumes.
Modified alphabetical ordering is used for the halogenated hydrocarbons but
other ordering procedures were found to provide easier access to the data tables
for the other three compound classes.

7.2.1 Hydrocarbons

The hydrocarbon compound name index is divided into the following compound-
type groups: alkanes, cycloalkanes, alkenes, cycloalkenes, alkadienes, cycloalkadienes, alkynes, alkadiynes, en-yne compounds, and aromatics. Each
compound-type group is further divided into carbon number sets. Order within
each carbon number set depends first upon decreasing hydrogen number, and then
upon increasing branching and increasing location numbers.

TABLE 7.3

Compound name index for hydrocarbons

(CSN = compound sequence number)

		CAS reg. nr	CSN
ALKANES			
METHANE	C(1)H(4)	000074-82-8	1
ETHANE	C(2)H(6)	000074-84-0	2
PROPANE	C(3)H(8)	000074-98-6	3
BUTANE	C(4)H(10)	000106-97-8	4
ISOBUTANE	C(4)H(10)	000075-28-5	5
PENTANE	C(5)H(12)	000109-66-0	6
2-METHYLBUTANE	C(5)H(12)	000078-78-4	7
2,2-DIMETHYLPROPANE	C(5)H(12)	000463-82-1	8
HEXANE	C(6)H(14)	000110-54-3	9
2-METHYLPENTANE	C(6)H(14)	000107-83-5	10
3-METHYLPENTANE	C(6)H(14)	000096-14-0	11
2,2-DIMETHYLBUTANE	C(6)H(14)	000075-83-2	12
2,3-DIMETHYLBUTANE	C(6)H(14)	000079-29-8	13
HEPTANE	C(7)H(16)	000142-82-5	14
2-METHYLHEXANE	C(7)H(16)	000591-76-4	15
3-METHYLHEXANE	C(7)H(16)	000589-34-4	16
3-ETHYLPENTANE	C(7)H(16)	000617-78-7	17
2,2-DIMETHYLPENTANE	C(7)H(16)	000590-35-2	18
2,3-DIMETHYLPENTANE	C(7)H(16)	000565-59-3	19
2,4-DIMETHYLPENTANE	C(7)H(16)	000108-08-7	20
3,3-DIMETHYLPENTANE	C(7)H(16)	000562-49-2	21
2,2,3-TRIMETHYLBUTANE	C(7)H(16)	000464-06-2	22
OCTANE	C(8)H(18)	000111-65-9	23
2-METHYLHEPTANE	C(8)H(18)	000592-27-8	24
3-METHYLHEPTANE	C(8)H(18)	000589-81-1	25
4-METHYLHEPTANE	C(8)H(18)	000589-53-7	26
3-ETHYLHEXANE	C(8)H(18)	000619-99-8	27
2,2-DIMETHYLHEXANE	C(8)H(18)	000590-73-8	28
2,3-DIMETHYLHEXANE	C(8)H(18)	000584-94-1	29
2,4-DIMETHYLHEXANE	C(8)H(18)	000589-43-5	30
2,5-DIMETHYLHEXANE	C(8)H(18)	000592-13-2	31
3,3-DIMETHYLHEXANE	C(8)H(18)	000563-16-6	32
3,4-DIMETHYLHEXANE	C(8)H(18)	000583-48-2	33
3-ETHYL-2-METHYLPENTANE	C(8)H(18)	000609-26-7	34
3-ETHYL-3-METHYLPENTANE	C(8)H(18)	001067-08-9	35
2,2,3-TRIMETHYLPENTANE	C(8)H(18)	000564-02-3	36
2,2,4-TRIMETHYLPENTANE	C(8)H(18)	000540-84-1	37
2,3,3-TRIMETHYLPENTANE	C(8)H(18)	000560-21-4	38
2,3,4-TRIMETHYLPENTANE	C(8)H(18)	000565-75-3	39
2,2,3,3-TETRAMETHYLBUTANE	C(8)H(18)	000594-82-1	40

TABLE 7.3 (continued)   CAS reg. nr   CSN

NONANE	C(9)H(20)	000111-84-2	41
2,6-DIMETHYLHEPTANE	C(9)H(20)	001072-05-5	42
2,2,3-TRIMETHYLHEXANE	C(9)H(20)	016747-25-4	43
2,2,4-TRIMETHYLHEXANE	C(9)H(20)	016747-26-5	44
2,2,5-TRIMETHYLHEXANE	C(9)H(20)	003522-94-9	45
2,4,4-TRIMETHYLHEXANE	C(9)H(20)	016747-30-1	46
3,3-DIETHYLPENTANE	C(9)H(20)	001067-20-5	47
2,2,3,3-TETRAMETHYLPENTANE	C(9)H(20)	007154-79-2	48
2,2,3,4-TETRAMETHYLPENTANE	C(9)H(20)	001186-53-4	49
2,2,4,4-TETRAMETHYLPENTANE	C(9)H(20)	001070-87-7	50
2,3,3,4-TETRAMETHYLPENTANE	C(9)H(20)	016747-38-9	51

CYCLOALKANES

CYCLOPROPANE	C(3)H(6)	000075-19-4	52
CYCLOBUTANE	C(4)H(8)	000287-23-0	53
METHYLCYCLOPROPANE	C(4)H(8)	000594-11-6	54
CYCLOPENTANE	C(5)H(10)	000287-92-3	55
SPIROPENTANE	C(5)H(8)	000185-94-4	56
CYCLOHEXANE	C(6)H(12)	000110-82-7	57
METHYLCYCLOPENTANE	C(6)H(12)	000096-37-7	58
CYCLOHEPTANE	C(7)H(14)	000291-64-5	59
METHYLCYCLOHEXANE	C(7)H(14)	000108-87-2	60
ETHYLCYCLOPENTANE	C(7)H(14)	001640-40-6	61
1,1-DIMETHYLCYCLOPENTANE	C(7)H(14)	001638-26-2	62
1,CIS-2-DIMETHYLCYCLOPENTANE	C(7)H(14)	001192-18-3	63
1,TRANS-2-DIMETHYLCYCLOPENTANE	C(7)H(14)	000822-50-4	64
1,CIS-3-DIMETHYLCYCLOPENTANE	C(7)H(14)	002532-58-3	65
1,TRANS-3-DIMETHYLCYCLOPENTANE	C(7)H(14)	001759-58-6	66
BICYCLO(2.2.1)HEPTANE	C(7)H(12)	000279-23-2	67
BICYCLO(4.1.0)HEPTANE	C(7)H(12)	000286-08-8	68
1-METHYLBICYCLO(3.1.0)HEXANE	C(7)H(12)	004625-24-5	69
TRICYCLO(2.2.1.0)HEPTANE	C(7)H(10)	036120-90-8	70
TRICYCLO(4.1.0.0)HEPTANE	C(7)H(10)	000187-26-8	71
TETRACYCLO(2.2.1.0.0)HEPTANE	C(7)H(8)	000278-06-8	72
CYCLOOCTANE	C(8)H(16)	000292-64-8	73
ETHYLCYCLOHEXANE	C(8)H(16)	001678-91-7	74
1,1-DIMETHYLCYCLOHEXANE	C(8)H(16)	000590-66-9	75
CIS-1,2-DIMETHYLCYCLOHEXANE	C(8)H(16)	002207-01-4	76
TRANS-1,2-DIMETHYLCYCLOHEXANE	C(8)H(16)	006876-23-9	77
CIS-1,3-DIMETHYLCYCLOHEXANE	C(8)H(16)	000638-04-0	78
TRANS-1,3-DIMETHYLCYCLOHEXANE	C(8)H(16)	002207-03-6	79
CIS-1,4-DIMETHYLCYCLOHEXANE	C(8)H(16)	000624-29-3	80
TRANS-1,4-DIMETHYLCYCLOHEXANE	C(8)H(16)	002207-04-7	81
PROPYLCYCLOPENTANE	C(8)H(16)	002040-96-2	82
ISOPROPYLCYCLOPENTANE	C(8)H(16)	003875-51-2	83
1-ETHYL-1-METHYLCYCLOPENTANE	C(8)H(16)	016747-50-5	84
CIS-1-ETHYL-2-METHYLCYCLOPENTANE	C(8)H(16)	000930-89-2	85
TRANS-1-ETHYL-2-METHYLCYCLOPENTANE	C(8)H(16)	000930-90-5	86

TABLE 7.3 (continued)   CAS reg. nr   CSN

1,1,2-TRIMETHYLCYCLOPENTANE	C(8)H(16)	004259-00-1	87
1,1,3-TRIMETHYLCYCLOPENTANE	C(8)H(16)	004516-69-2	88
1,CIS-2,TRANS-4-TRIMETHYL CYCLOPENTANE	C(8)H(16)	004850-28-6	89
1,TRANS-2,CIS-4-TRIMETHYL CYCLOPENTANE	C(8)H(16)	013398-35-1	90
CIS-BICYCLO(3.3.0)OCTANE	C(8)H(14)	001755-05-1	91
TRANS-BICYCLO(3.3.0)OCTANE	C(8)H(14)	005597-89-7	92
CIS-BICYCLO(4.2.0)OCTANE	C(8)H(14)	028282-35-1	93
CIS-BICYCLO(5.1.0)OCTANE	C(8)H(14)	016526-90-2	94
PROPYLCYCLOHEXANE	C(9)H(18)	001678-92-8	95
ISOPROPYLCYCLOHEXANE	C(9)H(18)	000696-29-7	96
1-ETHYL-CIS-3-METHYLCYCLOHEXANE	C(9)H(18)	019489-10-2	97
1,1,3-TRIMETHYLCYCLOHEXANE	C(9)H(18)	003073-66-3	98
1,CIS-3,CIS-5-TRIMETHYLCYCLOHEXANE	C(9)H(18)	001795-27-3	99
BUTYLCYCLOPENTANE	C(9)H(18)	002040-95-1	100
CIS-BICYCLO(4.3.0)NONANE	C(9)H(16)	004551-51-3	101
TRANS-BICYCLO(4.3.0)NONANE	C(9)H(16)	003296-50-2	102
CIS-BICYCLO(6.1.0)NONANE	C(9)H(16)	013757-43-2	103
2-ETHYLBICYCLO(2.2.1)HEPTANE	C(9)H(16)	002146-41-0	104
1,4-DIMETHYLBICYCLO(2.2.1)HEPTANE	C(9)H(16)	020454-81-3	105
TRANS-2,3-DIMETHYLBICYCLO(2.2.1) HEPTANE	C(9)H(16)		106

ALKENES

ETHENE	C(2)H(4)	000074-85-1	107
PROPENE	C(3)H(6)	000115-07-1	108
1-BUTENE	C(4)H(8)	000106-98-9	109
CIS-2-BUTENE	C(4)H(8)	000590-18-1	110
TRANS-2-BUTENE	C(4)H(8)	000624-64-6	111
ISOBUTENE	C(4)H(8)	000115-11-7	112
1-PENTENE	C(5)H(10)	000109-67-1	113
CIS-2-PENTENE	C(5)H(10)	000627-20-3	114
TRANS-2-PENTENE	C(5)H(10)	000646-04-8	115
2-METHYL-1-BUTENE	C(5)H(10)	000563-46-2	116
3-METHYL-1-BUTENE	C(5)H(10)	000563-45-1	117
2-METHYL-2-BUTENE	C(5)H(10)	000513-35-9	118
1-HEXENE	C(6)H(12)	000592-41-6	119
CIS-2-HEXENE	C(6)H(12)	007688-21-3	120
TRANS-2-HEXENE	C(6)H(12)	004050-45-7	121
CIS-3-HEXENE	C(6)H(12)	007642-09-3	122
TRANS-3-HEXENE	C(6)H(12)	013269-52-8	123
2-METHYL-1-PENTENE	C(6)H(12)	000763-29-1	124
3-METHYL-1-PENTENE	C(6)H(12)	000760-20-3	125
4-METHYL-1-PENTENE	C(6)H(12)	000691-37-2	126
2-METHYL-2-PENTENE	C(6)H(12)	000625-27-4	127
3-METHYL-CIS-2-PENTENE	C(6)H(12)	000922-62-3	128
3-METHYL-TRANS-2-PENTENE	C(6)H(12)	000616-12-6	129

TABLE 7.3 (continued)   CAS reg. nr   CSN

4-METHYL-CIS-2-PENTENE	C(6)H(12)	000691-38-3	130
4-METHYL-TRANS-2-PENTENE	C(6)H(12)	000674-76-0	131
2-ETHYL-1-BUTENE	C(6)H(12)	000760-21-4	132
2,3-DIMETHYL-1-BUTENE	C(6)H(12)	000563-78-0	133
3,3-DIMETHYL-1-BUTENE	C(6)H(12)	000558-37-2	134
2,3-DIMETHYL-2-BUTENE	C(6)H(12)	000563-79-1	135
1-HEPTENE	C(7)H(14)	000592-76-7	136
CIS-2-HEPTENE	C(7)H(14)	006443-92-1	137
TRANS-2-HEPTENE	C(7)H(14)	014686-13-6	138
CIS-3-HEPTENE	C(7)H(14)	007642-10-6	139
TRANS-3-HEPTENE	C(7)H(14)	014686-14-7	140
3-METHYL-CIS-3-HEXENE	C(7)H(14)	004914-89-0	141
3-METHYL-TRANS-3-HEXENE	C(7)H(14)	003899-36-3	142
2,4-DIMETHYL-1-PENTENE	C(7)H(14)	002213-32-3	143
4,4-DIMETHYL-1-PENTENE	C(7)H(14)	000762-62-9	144
3-ETHYL-2-PENTENE	C(7)H(14)	000816-79-5	145
2,4-DIMETHYL-2-PENTENE	C(7)H(14)	000625-65-0	146
4,4-DIMETHYL-CIS-2-PENTENE	C(7)H(14)	000762-63-0	147
4,4-DIMETHYL-TRANS-2-PENTENE	C(7)H(14)	000690-08-4	148
2-ETHYL-3-METHYL-1-BUTENE	C(7)H(14)	007357-93-9	149
2,3,3-TRIMETHYL-1-BUTENE	C(7)H(14)	000594-56-9	150
1-OCTENE	C(8)H(16)	000111-66-0	151
CIS-2-OCTENE	C(8)H(16)	007642-04-8	152
TRANS-2-OCTENE	C(8)H(16)	013389-42-9	153
TRANS-3-OCTENE	C(8)H(16)	014919-01-8	154
CIS-4-OCTENE	C(8)H(16)	007642-15-1	155
TRANS-4-OCTENE	C(8)H(16)	014850-23-8	156
2-METHYL-2-HEPTENE	C(8)H(16)	000627-97-4	157
2,2-DIMETHYL-CIS-3-HEXENE	C(8)H(16)	000690-92-6	158
2,2-DIMETHYL-TRANS-3-HEXENE	C(8)H(16)	000690-93-7	159
3-ETHYL-2-METHYL-1-PENTENE	C(8)H(16)	019780-66-6	160
2,4,4-TRIMETHYL-1-PENTENE	C(8)H(16)	000107-39-1	161
2,4,4-TRIMETHYL-2-PENTENE	C(8)H(16)	000107-40-4	162
1-NONENE	C(9)H(18)	000124-11-8	163
CIS-2-NONENE	C(9)H(18)	006434-77-1	164
TRANS-2-NONENE	C(9)H(18)	006434-78-2	165
CIS-3-NONENE	C(9)H(18)	020237-46-1	166
TRANS-3-NONENE	C(9)H(18)	020063-92-7	167
CIS-4-NONENE	C(9)H(18)	010405-84-2	168
TRANS-4-NONENE	C(9)H(18)	010405-85-3	169

CYCLOALKENES

CYCLOBUTENE	C(4)H(6)	000822-35-5	170
CYCLOPENTENE	C(5)H(8)	000142-29-0	171
METHYLENECYCLOBUTANE	C(5)H(8)	001120-56-5	172
CYCLOHEXENE	C(6)H(10)	000110-83-8	173
1-METHYLCYCLOPENTENE	C(6)H(10)	000693-89-0	174

TABLE 7.3 (continued)   CAS reg. nr   CSN

CYCLOHEPTENE	C(7)H(12)	000628-92-2	175
1-METHYL-1-CYCLOHEXENE	C(7)H(12)	000591-49-1	176
3-METHYL-1-CYCLOHEXENE	C(7)H(12)	000591-48-0	177
METHYLENECYCLOHEXANE	C(7)H(12)	001192-37-6	178
BICYCLO(2.2.1)HEPT-2-ENE	C(7)H(10)	000498-66-8	179
BICYCLO(4.1.0)HEPT-3-ENE	C(7)H(10)	016554-83-9	180
CYCLOOCTENE	C(8)H(14)	000931-88-4	181
1-ETHYL-1-CYCLOHEXENE	C(8)H(14)	001453-24-3	182
4,4-DIMETHYL-1-CYCLOHEXENE	C(8)H(14)	014072-86-7	183
1-METHYL-3-METHYLENECYCLOHEXANE	C(8)H(14)	003101-50-6	184
1-METHYL-4-METHYLENECYCLOHEXANE	C(8)H(14)	002808-80-2	185
2,3,3-TRIMETHYL-1-CYCLOPENTENE	C(8)H(14)		186
4-VINYLCYCLOHEXENE	C(8)H(12)	000100-40-3	187
1-METHYL-2-PROPYL-1-CYCLOPENTENE	C(9)H(16)	053366-60-2	188
1,2-DIETHYL-1-CYCLOPENTENE	C(9)H(16)		189
2-METHYLBICYCLO(2.2.2)OCT-2-ENE	C(9)H(14)	004893-13-4	190
2-ETHENYLBICYCLO(2.2.1)HEPTANE	C(9)H(14)	002146-39-6	191
CIS-5-ETHYLIDENEBICYCLO(2.2.1) HEPT-2-ENE	C(9)H(12)	016219-75-3	192
5-ETHENYLBICYCLO(2.2.1)HEPT-2-ENE	C(9)H(12)	003048-64-4	193

ALKADIENES

PROPADIENE	C(3)H(4)	000463-49-0	194
1,2-BUTADIENE	C(4)H(6)	000590-19-2	195
1,3-BUTADIENE	C(4)H(6)	000106-99-0	196
1,2-PENTADIENE	C(5)H(8)	000591-95-7	197
1-CIS-3-PENTADIENE	C(5)H(8)	001574-41-0	198
1-TRANS-3-PENTADIENE	C(5)H(8)	002004-70-8	199
1,4-PENTADIENE	C(5)H(8)	000591-93-5	200
2,3-PENTADIENE	C(5)H(8)	000591-96-8	201
3-METHYL-1,2-BUTADIENE	C(5)H(8)	000598-25-4	202
2-METHYL-1,3-BUTADIENE	C(5)H(8)	000078-79-5	203
1,2-HEXADIENE	C(6)H(10)	042296-74-2	204
1,TRANS-3-HEXADIENE	C(6)H(10)	020237-34-7	205
1,TRANS-4-HEXADIENE	C(6)H(10)	007319-00-8	206
1,5-HEXADIENE	C(6)H(10)	000592-42-7	207
TRANS-2,TRANS-4-HEXADIENE	C(6)H(10)	005194-51-4	208
2,3-DIMETHYL-1,3-BUTADIENE	C(6)H(10)	000513-81-5	209
CIS-1,3,5-HEXATRIENE	C(6)H(8)	002612-46-6	210
2,CIS-4-HEPTADIENE	C(7)H(12)	054354-36-8	211
2,TRANS-4-HEPTADIENE	C(7)H(12)	002384-94-3	212
2,4-DIMETHYL-1,3-PENTADIENE	C(7)H(12)	001000-86-8	213
2,4-DIMETHYL-2,3-PENTADIENE	C(7)H(12)	001000-87-9	214
2-BUTYL-1,3-BUTADIENE	C(8)H(14)	001189-53-3	215

TABLE 7.3 (continued)   CAS reg. nr   CSN

7-METHYL-2,4-OCTADIENE   C(9)H(16)   216

CYCLOALKADIENES

1,3-CYCLOPENTADIENE	C(5)H(6)	000542-92-7	217
1,3-CYCLOHEXADIENE	C(6)H(8)	000592-57-4	218
1,4-CYCLOHEXADIENE	C(6)H(8)	000628-41-1	219
1,3,5-CYCLOHEPTATRIENE	C(7)H(8)	002196-23-8	220
BICYCLO(2.2.1)HEPTA-2,5-DIENE	C(7)H(8)	000121-46-0	221
CIS-1,CIS-5-CYCLOOCTADIENE	C(8)H(12)	001552-12-1	222
1,3,5,7-CYCLOOCTATETRAENE	C(8)H(8)	000629-20-9	223
CIS-BICYCLO(4.3.0)NONA-3,7-DIENE	C(9)H(12)		224

ALKYNES

PROPYNE	C(3)H(4)	000074-99-7	225
1-BUTYNE	C(4)H(6)	025339-57-5	226
2-BUTYNE	C(4)H(6)	000503-17-3	227
1-HEXYNE	C(6)H(10)	000693-02-7	228
3-HEXYNE	C(6)H(10)	000928-49-4	229
1-HEPTYNE	C(7)H(12)	000628-71-7	230
2-HEPTYNE	C(7)H(12)	001119-65-9	231
3-HEPTYNE	C(7)H(12)	002586-89-2	232
1-OCTYNE	C(8)H(14)	000629-05-0	233
2-OCTYNE	C(8)H(14)	002809-67-8	234
3-OCTYNE	C(8)H(14)	015232-76-5	235
4-OCTYNE	C(8)H(14)	001942-45-6	236
1-NONYNE	C(9)H(16)	003452-09-3	237

ALKADIYNES

1,3-BUTADIYNE	C(4)H(2)	000460-12-8	238

ALKEN-YNES

1-BUTEN-3-YNE	C(4)H(4)	000689-97-4	239
1,5-HEXADIEN-3-YNE	C(6)H(6)	000821-08-9	240

TABLE 7.3 (continued)   CAS reg. nr   CSN

AROMATICS

BENZENE	C(6)H(6)	000071-43-2	241
TOLUENE	C(7)H(8)	000108-88-3	242
ETHYLBENZENE	C(8)H(10)	000100-41-4	243
O-XYLENE	C(8)H(10)	000095-47-6	244
M-XYLENE	C(8)H(10)	000108-38-3	245
P-XYLENE	C(8)H(10)	000106-42-3	246
STYRENE	C(8)H(8)	000100-42-5	247
PROPYLBENZENE	C(9)H(12)	000103-65-1	248
ISOPROPYLBENZENE	C(9)H(12)	000098-82-8	249
1-ETHYL-2-METHYLBENZENE	C(9)H(12)	000611-14-3	250
1-ETHYL-3-METHYLBENZENE	C(9)H(12)	000620-14-4	251
1-ETHYL-4-METHYLBENZENE	C(9)H(12)	000622-96-8	252
1,2,3-TRIMETHYLBENZENE	C(9)H(12)	000526-73-8	253
1,2,4-TRIMETHYLBENZENE	C(9)H(12)	000095-63-6	254
1,3,5-TRIMETHYLBENZENE	C(9)H(12)	000108-67-8	255
CYCLOPROPYLBENZENE	C(9)H(10)	000873-49-4	256
1-PROPENYLBENZENE	C(9)H(10)	000637-50-3	257
ISOPROPENYLBENZENE	C(9)H(10)	000098-83-9	258
1-ETHENYL-2-METHYLBENZENE	C(9)H(10)	000611-12-3	259
1-ETHENYL-3-METHYLBENZENE	C(9)H(10)	000100-80-1	260
1-ETHENYL-4-METHYLBENZENE	C(9)H(10)	000622-97-9	261
INDAN	C(9)H(10)	000496-11-7	262
INDENE	C(9)H(8)	000095-13-6	263

7.2.2 Ketones

The ketone compound name index is divided into the following major compound-type groups: ketones with no other active group, halogenated ketones, hydroxy ketones and ether ketones. The first of those four major groups is further divided into the following five groups: alkanones, cycloalkanones, alkenones, cycloalkenones and aromatic ketones. Because of the smallness of the groups, the alkadiene ketones have been included with alkenones, and the cycloalkadiene compounds have been grouped with the cycloalkenones.

Each of the compound-type groups is divided into carbon number sets which are then put in the order of increasing branching, increasing location numbers or increasing unsaturation.

Those few ether ketones which contain one or more halogen atoms are listed as halogenated ketones.

TABLE 7.4

Compound name index for ketones

(CSN = compound sequence number)

		CAS reg. nr	CSN

ALKANONES

ACETONE	C(3)H(6)O(1)	000067-64-1	264
2-BUTANONE	C(4)H(8)O(1)	000078-93-3	265
2,3-BUTANEDIONE	C(4)H(6)O(2)	000431-03-8	266
2-PENTANONE	C(5)H(10)O(1)	000107-87-9	267
3-PENTANONE	C(5)H(10)O(1)	000096-22-0	268
3-METHYL-2-BUTANONE	C(5)H(10)O(1)	000563-80-4	269
2,4-PENTANEDIONE	C(5)H(8)O(2)	000123-54-6	270
2-HEXANONE	C(6)H(12)O(1)	000591-78-6	271
3-HEXANONE	C(6)H(12)O(1)	000589-38-8	272
4-METHYL-2-PENTANONE	C(6)H(12)O(1)	000108-10-1	273
2-METHYL-3-PENTANONE	C(6)H(12)O(1)	000565-69-5	274
3,3-DIMETHYL-2-BUTANONE	C(6)H(12)O(1)	000075-97-8	275
3-METHYL-2,4-PENTANEDIONE	C(6)H(10)O(2)	000815-57-6	276
2-HEPTANONE	C(7)H(14)O(1)	000110-43-0	277
3-HEPTANONE	C(7)H(14)O(1)	000106-35-4	278
4-HEPTANONE	C(7)H(14)O(1)	000123-19-3	279
2-METHYL-3-HEXANONE	C(7)H(14)O(1)	007379-12-6	280
2,4-DIMETHYL-3-PENTANONE	C(7)H(14)O(1)	000565-80-0	281
3-ETHYL-2,4-PENTANEDIONE	C(7)H(12)O(2)	001540-34-7	282
3,3-DIMETHYL-2,4-PENTANEDIONE	C(7)H(12)O(2)	003142-58-3	283
2,4,6-HEPTANETRIONE	C(7)H(10)O(3)	000626-53-9	284
2-OCTANONE	C(8)H(16)O(1)	000111-13-7	285
3-OCTANONE	C(8)H(16)O(1)	000106-68-3	286
4-OCTANONE	C(8)H(16)O(1)	000589-63-9	287
2-METHYL-3-HEPTANONE	C(8)H(16)O(1)	013019-20-0	288
2,5-DIMETHYL-3-HEXANONE	C(8)H(16)O(1)	001888-57-9	289
2,2,4-TRIMETHYL-3-PENTANONE	C(8)H(16)O(1)	005857-36-3	290
2-NONANONE	C(9)H(18)O(1)	000821-55-6	291
3-NONANONE	C(9)H(18)O(1)	000925-78-0	292
5-NONANONE	C(9)H(18)O(1)	000502-56-7	293
2,6-DIMETHYL-4-HEPTANONE	C(9)H(18)O(1)	000108-83-8	294
3-DECANONE	C(10)H(20)O(1)	000928-80-3	295
2-UNDECANONE	C(11)H(22)O(1)	000112-12-9	296
6-UNDECANONE	C(11)H(22)O(1)	000927-49-1	297
2-TRIDECANONE	C(13)H(26)O(1)	000593-08-8	298
8-PENTADECANONE	C(15)H(30)O(1)	000818-23-5	299
16-HENTRIACONTANONE	C(31)H(62)O(1)	000502-73-8	300

TABLE 7.4 (continued)   CAS reg. nr   CSN

CYCLOALKANONES

CYCLOBUTANONE	C(4)H(6)O(1)	001191-95-3	301
1,3-CYCLOBUTANEDIONE	C(4)H(4)O(2)	015506-53-3	302
CYCLOPENTANONE	C(5)H(8)O(1)	000120-92-3	303
ACETYLCYCLOPROPANE	C(5)H(8)O(1)	000765-43-5	304
CYCLOHEXANONE	C(6)H(10)O(1)	000108-94-1	305
CYCLOHEPTANONE	C(7)H(12)O(1)	000502-42-1	306
2-METHYL-1-CYCLOHEXANONE	C(7)H(12)O(1)	000583-60-8	307
3-METHYL-1-CYCLOHEXANONE	C(7)H(12)O(1)	000591-24-2	308
4-METHYL-1-CYCLOHEXANONE	C(7)H(12)O(1)	000589-92-4	309
CYCLOOCTANONE	C(8)H(14)O(1)	000502-49-8	310
2-PROPYL-1-CYCLOPENTANONE	C(8)H(14)O(1)	001193-70-0	311
ACETYLCYCLOHEXANE	C(8)H(14)O(1)	000823-76-7	312
CYCLONONANONE	C(9)H(16)O(1)	003350-30-9	313
1-CYCLOHEXYL-1-PROPANONE	C(9)H(16)O(1)	001123-86-0	314
CYCLODECANONE	C(10)H(18)O(1)	001502-06-3	315
2-ISOPROPYL-5-METHYL-1- CYCLOHEXANONE	C(10)H(18)O(1)	014073-97-3	316
2,2,5,5-TETRAMETHYL-1-CYCLOHEXANONE	C(10)H(18)O(1)	015189-14-7	317
1,3,3-TRIMETHYLBICYCLO(2.2.1) HEPTAN-2-ONE	C(10)H(16)O(1)	000126-21-6	318
CAMPHOR	C(10)H(16)O(1)	000076-22-2	319
DECAHYDRO-2-NAPHTHALENONE	C(10)H(16)O(1)	016021-08-2	320
CYCLOUNDECANONE	C(11)H(20)O(1)	000878-13-7	321
CYCLODODECANONE	C(12)H(22)O(1)	000830-13-7	322

ALKENONES

3-BUTEN-2-ONE	C(4)H(6)O(1)	000078-94-4	323
3-METHYL-3-BUTEN-2-ONE	C(5)H(8)O(1)	000814-78-8	324
5-HEXEN-2-ONE	C(6)H(10)O(1)	000109-49-9	325
4-METHYL-3-PENTEN-2-ONE	C(6)H(10)O(1)	000141-79-7	326
4-METHYL-4-PENTEN-2-ONE	C(6)H(10)O(1)	003744-02-3	327
3-ACETYL-5-HEXEN-2-ONE	C(8)H(12)O(2)	003508-78-9	328
2,6-DIMETHYL-2,5-HEPTADIEN-4-ONE	C(9)H(14)O(1)	000504-20-1	329

TABLE 7.4 (continued)   CAS reg. nr   CSN

CYCLOALKENONES

2,5-CYCLOHEXADIENE-1,4-DIONE	C(6)H(4)O(2)	000106-51-4	330
2-METHYL-2,5-CYCLOHEXADIENE-1,4-DIONE	C(7)H(6)O(2)	000553-97-9	331
5-ISOPROPENYL-2-METHYL-2-CYCLOHEXEN-1-ONE	C(10)H(14)O(1)	000099-49-0	332
4-(2,6,6-TRIMETHYL-1-CYCLOHEXEN-1-YL)3-BUTEN-2-ONE	C(13)H(20)O(1)	014901-07-6	333
4-(2,6,6-TRIMETHYL-2-CYCLOHEXEN-1-YL)3-BUTEN-2-ONE	C(13)H(20)O(1)	000127-41-3	334

AROMATIC KETONES

ACETOPHENONE	C(8)H(8)O(1)	000098-86-2	335
1-PHENYL-1-PROPANONE	C(9)H(10)O(1)	000093-55-0	336
1-PHENYL-2-PROPANONE	C(9)H(10)O(1)	000103-79-7	337
4-METHYLACETOPHENONE	C(9)H(10)O(1)	000122-00-9	338
1-INDANONE	C(9)H(8)O(1)	000083-33-0	339
1-PHENYL-1-BUTANONE	C(10)H(12)O(1)	000495-40-9	340
1-PHENYL-2-BUTANONE	C(10)H(12)O(1)	001007-32-5	341
4-PHENYL-2-BUTANONE	C(10)H(12)O(1)	002550-26-7	342
1-PHENYL-1,3-BUTANEDIONE	C(10)H(10)O(2)	000093-91-4	343
1-OXO-1,2,3,4-TETRAHYDRONAPHTHALENE	C(10)H(10)O(1)	000529-34-0	344
3-ETHYLACETOPHENONE	C(10)H(12)O(1)	022699-70-3	345
4-ETHYLACETOPHENONE	C(10)H(12)O(1)	000937-30-4	346
1,3-DIACETYLBENZENE	C(10)H(10)O(2)	006781-42-6	347
1,4-DIACETYLBENZENE	C(10)H(10)O(2)	001009-61-6	348
1-PHENYL-1-PENTANONE	C(11)H(14)O(1)	001009-14-9	349
1-PHENYL-2-PENTANONE	C(11)H(14)O(1)	006683-92-7	350
1-PHENYL-1-HEXANONE	C(12)H(16)O(1)	000942-92-7	351
1-ACETYLNAPHTHALENE	C(12)H(10)O(1)	000941-98-0	352
1-PHENYL-1-HEPTANONE	C(13)H(18)O(1)	001671-75-6	353
1-NAPHTHYL-1-PROPANONE	C(13)H(12)O(1)	002876-63-3	354
BENZOPHENONE	C(13)H(10)O(1)	000119-61-9	355
1-(1-NAPHTHALENYL)-1-BUTANONE	C(14)H(14)O(1)	002876-62-2	356
DIPHENYLETHANEDIONE	C(14)H(10)O(2)	000134-81-6	357
9,10-ANTHRACENEDIONE	C(14)H(8)O(2)	000084-65-1	358
1-PHENYL-1-NONANONE	C(15)H(22)O(1)	006008-36-2	359
1-(1-NAPHTHALENYL)-1-PENTANONE	C(15)H(16)O(1)	002876-60-0	360
1,3-DIPHENYL-2-PROPANONE	C(15)H(14)O(1)	000102-04-5	361
1-(1-NAPHTHALENYL)-1-HEXANONE	C(16)H(18)O(1)	002876-61-1	362

TABLE 7.4 (continued)

		CAS reg. nr	CSN
1-(1-NAPHTHALENYL)-1-HEPTANONE	C(17)H(20)O(1)	025897-44-3	363
7H-BENZ(DE)ANTHRACEN-7-ONE	C(17)H(10)O(1)	000082-05-3	364
1-PHENYL-1-DODECANONE	C(18)H(28)O(1)	001674-38-0	365
1-NAPHTHYL-1-OCTANONE	C(18)H(22)O(1)		366
1-PHENYL-1-TETRADECANONE	C(20)H(32)O(1)	004497-05-6	367
1-PHENYL-1-HEXADECANONE	C(22)H(36)O(1)	006697-12-7	368

HALOGENATED KETONES

		CAS reg. nr	CSN
1-CHLORO-2-PROPANONE	C(3)H(5)CL(1)O(1)	000078-95-5	369
1,1-DICHLORO-2-PROPANONE	C(3)H(4)CL(2)O(1)	000513-88-2	370
1,3-DICHLORO-2-PROPANONE	C(3)H(4)CL(2)O(1)	000534-07-6	371
1-CHLORO-1,1,3,3,3-PENTAFLUORO-2-PROPANONE	C(3)CL(1)F(5)O(1)	000079-53-8	372
1,1,1,3,3,3-HEXAFLUORO-2-PROPANONE	C(3)F(6)O(1)	000684-16-2	373
1-BROMO-2-BUTANONE	C(4)H(7)BR(1)O(1)	000816-40-4	374
3-BROMO-2-BUTANONE	C(4)H(7)BR(1)O(1)	000814-75-5	375
3-CHLORO-2-BUTANONE	C(4)H(7)CL(1)O(1)	004091-39-8	376
4-BROMO-2-PENTANONE	C(5)H(9)BR(1)O(1)		377
6-BROMO-2-HEXANONE	C(6)H(11)BR(1)O(1)	010226-29-6	378
4-CHLOROACETOPHENONE	C(8)H(7)CL(1)O(1)	000099-91-2	379
5-BROMO-3-METHOXY-2,6-DIMETHYL-2,5-HEPTADIEN-4-ONE	C(10)H(15)BR(1)O(2)		380
3-(4-BROMOBENZYLOXY)-2,6-DIMETHYL-2,5-HEPTADIEN-4-ONE	C(16)H(19)BR(1)O(2)		381
5-BROMO-3(4-BROMOBENZYLOXY)-2,6-DIMETHYL-2,5-HEPTADIEN-4-ONE	C(16)H(18)BR(2)O(2)		382
3-BENZOYLOXY-5-BROMO-2,6-DIMETHYL-2,5-HEPTADIEN-4-ONE	C(16)H(17)BR(1)O(3)		383

HYDROXY KETONES

		CAS reg. nr	CSN
3-HYDROXY-2-BUTANONE	C(4)H(8)O(2)	000513-86-0	384
1-HYDROXY-3-METHYL-2-BUTANONE	C(5)H(10)O(2)	036960-22-2	385
3-HYDROXY-3-METHYL-2-BUTANONE	C(5)H(10)O(2)	000115-22-0	386
4-HYDROXY-3-METHYL-2-BUTANONE	C(5)H(10)O(2)	003393-64-4	387
5-HYDROXY-3-PENTYN-2-ONE	C(5)H(6)O(2)	015441-65-3	388
4-HYDROXY-4-METHYL-2-PENTANONE	C(6)H(12)O(2)	000123-42-2	389

TABLE 7.4 (continued)   CAS reg. nr   CSN

1,4-DIHYDROXY-9,10-ANTHRACENEDIONE	C(14)H(8)O(4)	000081-64-1	390
1,5-DIHYDROXY-9,10-ANTHRACENEDIONE	C(14)H(8)O(4)	000117-12-4	391
1,8-DIHYDROXY-9,10-ANTHRACENEDIONE	C(14)H(8)O(4)	000117-10-2	392

ETHER KETONES

3-ACETOXY-2,6-DIMETHYL-2,5- HEPTADIEN-4-ONE	C(11)H(16)O(3)	393
3-BENZOYLOXY-2,6-DIMETHYL-2,5- HEPTADIEN-4-ONE	C(16)H(18)O(3)	394

Chapter 8

COMPOUND DATA TABLES

This volume contains the tables for the C_1-C_9 hydrocarbons and the ketones.
The hydrocarbon tables precede those for the ketones. A two-page tabulation is
presented for each compound.

The order of the hydrocarbon tables matches the order in which the compounds
are listed in the hydrocarbon compound name index, Table 7.3. The various
compound-type groupings and their compound sequence number ranges are as
follows:

Alkanes	1-51
Cycloalkanes	52-106
Alkenes	107-169
Cycloalkenes	170-193
Alkadienes	194-216
Cycloalkadienes	217-224
Alkynes	225-237
Alkadiynes	238
Alken-ynes	239-240
Aromatics	241-263

The tables for each of the compound-type groups are presented in the order of
increasing carbon number. Further ordering within a carbon number sub-group is
based on increasing branching, increasing location numbers and increasing unsaturation (decreasing hydrogen number).

The order of the ketone tables matches the order in the compound name index
for the ketones, Table 7.4. The ketones are divided into compound-type groups
which partially parallel those for the hydrocarbons:

Alkanones	264-300
Cycloalkanones	301-322
Alkenones	323-329
Cycloalkenones	330-334
Aromatic ketones	335-368
Halogenated ketones	369-383
Hydroxy ketones	384-392
Ether ketones	393-394

The one alkadiene ketone for which tables are presented is listed in the
alkenone group, and the cycloalkadienes are listed with the cycloalkenones.

87

Halogenated compounds containing an ether linkage are listed with the halogenated ketones.

As for the hydrocarbon tables, each ketone compound-type group is ordered by increasing carbon number, and generally the order in each carbon number subgroup is related to increasing branching, increasing location numbers and increasing unsaturation. In general, hydrogen numbers decrease and oxygen numbers increase within a carbon number subgroup, but that order is sometimes broken in order to list similar compounds together. For example, in the aromatic ketone carbon number groups, phenyl alkanones are usually listed before the alkyl phenones.

EXPLANATORY SAMPLE CHART

EXPLANATORY SAMPLE CHART: FIRST PAGE

compound sequence number   molecular weight

(3) PROPANE   C(3)H(8)   (44.096)

(COMPOUND CONSTANTS:)   → for abbreviations see p. 5

MP 85.460 K	TC 370.020 K	VC 197.48 ML/MOL	ZC 0.2735
NBP 231.100 K	PC 4.261 MPA	DC 0.2233 G/ML	OM 0.1514
MU 0.0830 DEBYE	RG 2.4255 ANGSTROM		

(VAPOR PRESSURE CORRELATION CONSTANTS, K AND KPA:) → see pp. 6—8

RIEDEL	VAPRES-2	VAPRES-2	WAGNER
95. - 370. K	95. - 249. K	95. - 304. K	212. - 370. K
155 POINTS	73 POINTS	99 POINTS	125 POINTS
RMSD = 3.8530	RMSD = 0.1780	RMSD = 3.4100	RMSD = 1.8310
0.36917395E+02	-0.16440408E+03	-0.92359737E+02	-0.66876847E+01
-0.34299700E+01	-0.69997894E+03	-0.15264804E+04	0.12488338E+01
-0.31528395E+04	-0.24259399E+00	-0.14721327E+00	-0.19745461E+01
0.94259287E-16	0.21942752E-03	0.12666922E-03	-0.20538574E+01
	0.39758043E+02	0.24039841E+02	

(LIQUID DENSITY CORRELATION CONSTANTS, K AND G/ML:) → see pp. 9—11

RACKETT	FRANCIS1	FRANCIS1	RACKETT
91. - 367. K	91. - 249. K	91. - 314. K	213. - 367. K
150 POINTS	60 POINTS	128 POINTS	109 POINTS
RMSD = 0.00167	RMSD = 0.00040	RMSD = 0.00050	RMSD = 0.00058
0.27681024E+00	0.83095551E+00	0.83616287E+00	0.27702026E+00
	0.93440502E-03	0.87599433E-03	
	0.59999990E+01	0.97111998E+01	
	0.41072729E+03	0.41625684E+03	

(SECOND VIRIAL COEFFICIENT CORRELATION CONSTANTS, ML/MOL:) → see p. 12

PITZER-CURL: TC=370.020   PC=4.261   VC=197.48   OM=0.1514

LITERATURE DOCUMENTS FOR CORRELATED VAPOR PRESSURE VALUES: → see p. 12, 31—56

362	363	366	424	558	577	587	612	614	679	1917	1948	2065
2187	2244	2306	3963	4234	4287	4343	4422	4463	4514	4515	4816	6986
7091	8383	8853	9061	16160	18019							

LITERATURE DOCUMENTS FOR CORRELATED LIQUID DENSITY VALUES:

359	363	366	415	424	477	579	1941	1951	1972	2320	4463	4724
5310	6041	7148	9061	40435	40463							

LITERATURE DOCUMENTS REPORTING VIRIAL COEFFICIENT DATA:

361	363	364	366	965	1912	1923	1937	1938	1949	1950	1952	2011
4089	4463	7093	9061	16160	18080	21293						

90

③ PROPANE compound sequence number see pp. 13—17

T,K	VAPOR PRESSURE, KPA	SATURATED LIQUID VOLUME, ML/MOL	SECOND VIRIAL COEFFICIENT, ML/MOL	HEAT OF VAPORIZATION, J/MOL	SATURATED VAPOR VOLUME, ML/MOL
91		61.11	-64143.		
97	0.0000127	61.56	-39710.	23448.	0.6371E+11
103	0.0000684	62.02	-25593.	23277.	0.1251E+11
110	0.0003831	62.57	-16089.	23078.	0.2387E+10
116	0.001406	63.05	-11243.	22907.	0.6858E+09
122	0.004504	63.55	-8134.	22737.	0.2252E+09
129	0.01512	64.15	-5813.	22538.	0.7094E+08
135	0.03834	64.67	-4505.	22368.	0.2928E+08
141	0.08923	65.21	-3591.	22198.	0.1314E+08
148	0.2176	65.86	-2845.	22000.	5653751.
154	0.4356	66.43	-2387.	21831.	2939135.
160	0.8237	67.02	-2042.	21663.	1615061.
167	1.625	67.73	-1737.	21423.	852644.
173	2.773	68.36	-1536.	21234.	517195.
179	4.546	69.01	-1374.	21038.	325904.
186	7.740	69.79	-1222.	20799.	198474.
192	11.80	70.49	-1115.	20584.	134065.
198	17.48	71.20	-1024.	20358.	93039.
205	26.76	72.08	-934.5	20079.	62644.
211	37.59	72.86	-868.7	19827.	45695.
217	51.68	73.66	-810.9	19560.	33992.
224	73.14	74.65	-751.7	19231.	24611.
230	96.63	75.53	-706.8	18934.	18985.
236	125.6	76.45	-666.3	18622.	14858.
243	167.5	77.57	-623.9	18239.	11347.
249	211.3	78.59	-591.0	17895.	9121.
255	263.2	79.65	-560.9	17537.	7410.
262	335.3	80.97	-528.8	17101.	5886.
268	407.9	82.16	-503.5	16712.	4877.
274	491.6	83.42	-480.1	16310.	4073.
281	604.5	85.00	-454.8	15822.	3329.
287	715.3	86.44	-434.7	15388.	2819.
293	840.2	87.99	-415.9	14938.	2400.
300	1005.	89.95	-395.4	14392.	2001.
306	1164.	91.79	-379.0	13902.	1719.
312	1340.	93.79	-363.6	13389.	1482.
319	1570.	96.38	-346.6	12755.	1249.
325	1788.	98.88	-333.0	12172.	1080.
331	2029.	101.71	-320.1	11542.	933.7
338	2338.	105.55	-305.9	10728.	786.1
344	2631.	109.50	-294.3	9934.	675.4
350	2951.	114.36	-283.4	9010.	575.7
357	3363.	121.99	-271.3	7668.	468.9
363	3752.	132.13	-261.4	6078.	379.5
369	4176.		-252.0		261.7

HYDROCARBONS

COMPOUND CONSTANTS:

MP   90.680 K   TC   190.555 K   VC   98.92 ML/MOL   ZC 0.2869
NBP   111.633 K   PC   4.595 MPA   DC 0.1622 G/ML   OM 0.0110
MU   0.0000 DEBYE   RG   1.1234 ANGSTROM

   VAPOR PRESSURE CORRELATION CONSTANTS, K AND MPA:

   WAGNER-2
    91. - 190. K
  -0.60468521E+01
   0.13456848E+01
  -0.66075063E+00
  -0.13040196E+01

   LIQUID DENSITY CORRELATION CONSTANTS, K AND MOL/DM^3:

   ANGUS   MARTIN
    91. - 190. K   105. - 160. K
   0.93028716E+00   0.18356299E+01
   0.76919405E+00   0.66063290E+00
  -0.66015053E+00   -0.93228500E-01
   0.59845760E-01
  -0.87440672E-01
  -0.10608186E+00
   0.27

   SECOND VIRIAL COEFFICIENT CORRELATION CONSTANTS, ML/MOL:

   HAYDEN-O'CONNELL: TC=191.000 PC=4.641 RG=1.1230 MU=0.000 ETA=0.000

LITERATURE DOCUMENTS FOR CORRELATED VAPOR PRESSURE VALUES:

  23678

LITERATURE DOCUMENTS FOR CORRELATED LIQUID DENSITY VALUES:

  23678 40463

LITERATURE DOCUMENTS REPORTING VIRIAL COEFFICIENT DATA:

1 METHANE

T,K	VAPOR PRESSURE, KPA	SATURATED LIQUID VOLUME, ML/MOL	SECOND VIRIAL COEFFICIENT, ML/MOL	HEAT OF VAPORIZATION, J/MOL	SATURATED VAPOR VOLUME, ML/MOL
91	12.21	35.59	-528.2	8720.	61456.
93	15.67	35.80	-498.7	8673.	48854.
95	19.89	36.01	-472.2	8625.	39252.
97	24.97	36.23	-448.2	8576.	31853.
100	34.49	36.56	-416.4	8502.	23687.
102	42.30	36.79	-397.5	8451.	19649.
104	51.44	37.03	-380.1	8398.	16425.
106	62.06	37.27	-364.2	8345.	13831.
109	81.12	37.64	-342.5	8262.	10822.
111	96.15	37.89	-329.4	8205.	9260.
113	113.2	38.15	-317.2	8146.	7969.
115	132.6	38.42	-305.9	8085.	6895.
118	166.2	38.83	-290.2	7991.	5601.
120	191.9	39.12	-280.5	7926.	4905.
122	220.5	39.41	-271.4	7858.	4313.
124	252.2	39.71	-262.8	7789.	3808.
127	306.1	40.19	-250.9	7680.	3181.
129	346.4	40.51	-243.4	7604.	2834.
131	390.6	40.85	-236.3	7525.	2532.
133	438.7	41.20	-229.6	7444.	2269.
136	518.9	41.75	-220.1	7316.	1935.
138	578.1	42.13	-214.2	7227.	1746.
140	642.0	42.53	-208.5	7134.	1579.
142	710.8	42.95	-203.0	7038.	1431.
145	823.9	43.60	-195.3	6886.	1239.
147	906.2	44.06	-190.4	6779.	1128.
149	994.2	44.54	-185.7	6668.	1028.
151	1088.	45.05	-181.2	6552.	938.8
154	1241.	45.85	-174.8	6368.	821.0
156	1351.	46.43	-170.7	6238.	751.8
158	1468.	47.04	-166.7	6102.	688.9
160	1592.	47.68	-162.9	5959.	631.8
163	1791.	48.73	-157.5	5731.	555.3
165	1934.	49.49	-154.0	5568.	509.7
167	2085.	50.31	-150.6	5397.	467.8
169	2244.	51.20	-147.3	5215.	429.2
172	2499.	52.68	-142.6	4920.	376.7
174	2680.	53.80	-139.6	4706.	344.8
176	2871.	55.05	-136.6	4476.	315.2
178	3071.	56.46	-133.8	4226.	287.3
181	3391.	59.00	-129.7	3802.	248.5
183	3619.	61.11	-127.1	3475.	224.0
185	3858.	63.77	-124.5	3093.	200.0
187	4110.	67.40	-122.0	2618.	175.6
190	4516.	79.40	-118.4	1320.	128.5

2 ETHANE   C(2)H(6)   30.070

COMPOUND CONSTANTS:

MP   90.348 K   TC   305.330 K   VC   147.06 ML/MOL   ZC 0.2822
NBP   184.550 K   PC   4.871 MPA   DC   0.2045 G/ML   OM 0.0990
MU   0.0000 DEBYE   RG   1.8314 ANGSTROM

     VAPOR PRESSURE CORRELATION CONSTANTS, K AND MPA:

     GOODWIN
       91. - 305. K
     -0.67767270E+01
      0.10673240E+02
      0.83378200E+01
     -0.30848900E+01
     -0.65857000E+00
      0.60495500E+01

     LIQUID DENSITY CORRELATION CONSTANTS, K AND MOL/DM^3:

     GOODWIN   MARTIN
       91. - 305. K   100. - 270. K
      0.71968450E+00   0.18297449E+01
      0.28186618E+00   0.19581399E+01
     -0.28993731E+00   -0.27703513E+01
      0.33333333E+00   0.16147464E+01
      0.65192000E+00

     SECOND VIRIAL COEFFICIENT CORRELATION CONSTANTS, ML/MOL:

     HAYDEN-O'CONNELL: TC=305.600   PC=4.894   RG=1.8310   MU=0.000   ETA=0.000

LITERATURE DOCUMENTS FOR CORRELATED VAPOR PRESSURE VALUES:

  40418

LITERATURE DOCUMENTS FOR CORRELATED LIQUID DENSITY VALUES:

  40698

LITERATURE DOCUMENTS REPORTING VIRIAL COEFFICIENT DATA:

T,K	VAPOR PRESSURE, KPA	SATURATED LIQUID VOLUME, ML/MOL	SECOND VIRIAL COEFFICIENT, ML/MOL	HEAT OF VAPORIZATION, J/MOL	SATURATED VAPOR VOLUME, ML/MOL
91	0.001351	46.18	-6322.	17825.	0.5599E+09
95	0.003634	46.49	-4995.	17719.	0.2174E+09
100	0.01111	46.89	-3831.	17574.	0.7487E+08
105	0.03025	47.29	-3025.	17419.	0.2886E+08
110	0.07462	47.70	-2450.	17257.	0.1226E+08
115	0.1689	48.13	-2028.	17092.	5661528.
120	0.3544	48.56	-1711.	16926.	2814242.
125	0.6965	49.00	-1469.	16759.	1491411.
129	1.146	49.36	-1314.	16627.	934763.
134	2.040	49.81	-1156.	16462.	545429.
139	3.465	50.29	-1029.	16299.	332831.
144	5.648	50.77	-924.3	16137.	211313.
149	8.875	51.26	-837.8	15975.	138968.
154	13.49	51.77	-765.1	15812.	94298.
159	19.92	52.30	-703.2	15649.	65800.
163	26.68	52.73	-660.1	15516.	50243.
168	37.62	53.29	-612.6	15348.	36601.
173	51.89	53.86	-571.1	15175.	27216.
178	70.15	54.46	-534.5	14997.	20613.
183	93.12	55.07	-501.9	14813.	15873.
188	121.6	55.71	-472.8	14621.	12406.
193	156.4	56.38	-446.6	14421.	9828.
197	189.3	56.93	-427.5	14255.	8229.
202	237.7	57.65	-405.5	14037.	6656.
207	295.0	58.41	-385.4	13808.	5439.
212	362.0	59.20	-367.0	13567.	4484.
217	439.9	60.04	-350.0	13313.	3728.
222	529.7	60.92	-334.3	13045.	3121.
227	632.3	61.85	-319.7	12761.	2631.
232	748.8	62.85	-306.1	12461.	2230.
236	852.9	63.69	-295.9	12207.	1961.
241	997.3	64.82	-283.9	11872.	1677.
246	1159.	66.03	-272.6	11516.	1439.
251	1339.	67.35	-262.0	11136.	1239.
256	1538.	68.79	-252.0	10729.	1070.
261	1758.	70.38	-242.6	10292.	925.2
266	2000.	72.16	-233.7	9819.	800.7
270	2210.	73.74	-226.9	9410.	713.3
275	2495.	75.97	-218.8	8856.	616.6
280	2806.	78.57	-211.1	8240.	531.4
285	3145.	81.70	-203.8	7543.	455.7
290	3515.	85.61	-196.8	6734.	387.1
295	3917.	90.88	-190.1	5749.	323.2
300	4355.	99.08	-183.8	4423.	260.0
304	4736.	113.48	-178.9	2631.	200.7

COMPOUND CONSTANTS:

MP   85.460 K   TC   370.020 K   VC 197.48 ML/MOL   ZC 0.2735
NBP 231.100 K   PC   4.261 MPA   DC 0.2233 G/ML   OM 0.1514
MU   0.0830 DEBYE   RG   2.4255 ANGSTROM

VAPOR PRESSURE CORRELATION CONSTANTS, K AND KPA:

RIEDEL   VAPRES-2   VAPRES-2   WAGNER
 95. - 370. K   95. - 249. K   95. - 304. K   212. - 370. K
155 POINTS   73 POINTS   99 POINTS   125 POINTS
RMSD = 3.8530   RMSD = 0.1780   RMSD = 3.4100   RMSD = 1.8310
 0.36917395E+02   -0.16440408E+03   -0.92359737E+02   -0.66876847E+01
-0.34299700E+01   -0.69997894E+03   -0.15264804E+04   0.12488338E+01
-0.31528395E+04   -0.24259399E+00   -0.14721327E+00   -0.19745461E+01
 0.94259287E-16   0.21942752E-03   0.12666922E-03   -0.20538574E+01
                      0.39758043E+02   0.24039841E+02

LIQUID DENSITY CORRELATION CONSTANTS, K AND G/ML:

RACKETT   FRANCIS1   FRANCIS1   RACKETT
 91. - 367. K   91. - 249. K   91. - 314. K   213. - 367. K
150 POINTS   60 POINTS   128 POINTS   109 POINTS
RMSD = 0.00167   RMSD = 0.00040   RMSD = 0.00050   RMSD = 0.00058
 0.27681024E+00   0.83095551E+00   0.83616287E+00   0.27702026E+00
                      0.93440502E-03   0.87599433E-03
                      0.59999990E+01   0.97111998E+01
                      0.41072729E+03   0.41625684E+03

SECOND VIRIAL COEFFICIENT CORRELATION CONSTANTS, ML/MOL:

PITZER-CURL: TC=370.020   PC=4.261   VC=197.48   OM=0.1514

LITERATURE DOCUMENTS FOR CORRELATED VAPOR PRESSURE VALUES:

   362   363   366   424   558   577   587   612   614   679   1917   1948   2065
  2187   2244   2306   3963   4234   4287   4343   4422   4463   4514   4515   4816   6986
  7091   8383   8853   9061 16160 18019

LITERATURE DOCUMENTS FOR CORRELATED LIQUID DENSITY VALUES:

   359   363   366   415   424   477   579   1941   1951   1972   2320   4463   4724
  5310   6041   7148   9061 40435 40463

LITERATURE DOCUMENTS REPORTING VIRIAL COEFFICIENT DATA:

   361   363   364   366   965   1912   1923   1937   1938   1949   1950   1952   2011
  4089   4463   7093   9061 16160 18080 21293

T,K	VAPOR PRESSURE, KPA	SATURATED LIQUID VOLUME, ML/MOL	SECOND VIRIAL COEFFICIENT, ML/MOL	HEAT OF VAPORIZATION, J/MOL	SATURATED VAPOR VOLUME, ML/MOL
91		61.11	-64143.		
97	0.0000127	61.56	-39710.	23448.	0.6371E+11
103	0.0000684	62.02	-25593.	23277.	0.1251E+11
110	0.0003831	62.57	-16089.	23078.	0.2387E+10
116	0.001406	63.05	-11243.	22907.	0.6858E+09
122	0.004504	63.55	-8134.	22737.	0.2252E+09
129	0.01512	64.15	-5813.	22538.	0.7094E+08
135	0.03834	64.67	-4505.	22368.	0.2928E+08
141	0.08923	65.21	-3591.	22198.	0.1314E+08
148	0.2176	65.86	-2845.	22000.	5653751.
154	0.4356	66.43	-2387.	21831.	2939135.
160	0.8237	67.02	-2042.	21663.	1615061.
167	1.625	67.73	-1737.	21423.	852644.
173	2.773	68.36	-1536.	21234.	517195.
179	4.546	69.01	-1374.	21038.	325904.
186	7.740	69.79	-1222.	20799.	198474.
192	11.80	70.49	-1115.	20584.	134065.
198	17.48	71.20	-1024.	20358.	93039.
205	26.76	72.08	-934.5	20079.	62644.
211	37.59	72.86	-868.7	19827.	45695.
217	51.68	73.66	-810.9	19560.	33992.
224	73.14	74.65	-751.7	19231.	24611.
230	96.63	75.53	-706.8	18934.	18985.
236	125.6	76.45	-666.3	18622.	14858.
243	167.5	77.57	-623.9	18239.	11347.
249	211.3	78.59	-591.0	17895.	9121.
255	263.2	79.65	-560.9	17537.	7410.
262	335.3	80.97	-528.8	17101.	5886.
268	407.9	82.16	-503.5	16712.	4877.
274	491.6	83.42	-480.1	16310.	4073.
281	604.5	85.00	-454.8	15822.	3329.
287	715.3	86.44	-434.7	15388.	2819.
293	840.2	87.99	-415.9	14938.	2400.
300	1005.	89.95	-395.4	14392.	2001.
306	1164.	91.79	-379.0	13902.	1719.
312	1340.	93.79	-363.6	13389.	1482.
319	1570.	96.38	-346.6	12755.	1249.
325	1788.	98.88	-333.0	12172.	1080.
331	2029.	101.71	-320.1	11542.	933.7
338	2338.	105.55	-305.9	10728.	786.1
344	2631.	109.50	-294.3	9934.	675.4
350	2951.	114.36	-283.4	9010.	575.7
357	3363.	121.99	-271.3	7668.	468.9
363	3752.	132.13	-261.4	6078.	379.5
369	4176.		-252.0		261.7

COMPOUND CONSTANTS:

MP 134.790 K	TC 425.180 K	VC 255.00 ML/MOL	ZC 0.2740
NBP 272.650 K	PC 3.797 MPA	DC 0.2279 G/ML	OM 0.2000
MU 0.0200 DEBYE	RG 2.8885 ANGSTROM		

VAPOR PRESSURE CORRELATION CONSTANTS, K AND KPA:

RPM2	VAPRES-2	RPM2	WAGNER
139. - 425. K	139. - 283. K	139. - 379. K	253. - 423. K
106 POINTS	45 POINTS	88 POINTS	81 POINTS
RMSD = 1.7880	RMSD = 0.0980	RMSD = 1.1010	RMSD = 1.5650
0.23749036E+02	-0.28567404E+03	0.23687475E+02	-0.70268598E+01
-0.37738688E+04	0.68881788E+03	-0.37693058E+04	0.15198848E+01
-0.26160618E-01	-0.32184518E+00	-0.25898512E-01	-0.28095502E+01
0.24825265E-04	0.24895764E-03	0.24470959E-04	-0.19569847E-02
	0.63658675E+02		

LIQUID DENSITY CORRELATION CONSTANTS, K AND G/ML:

FRANCIS1	FRANCIS1	FRANCIS1	RACKETT
213. - 394. K	213. - 292. K	213. - 379. K	253. - 422. K
72 POINTS	35 POINTS	69 POINTS	67 POINTS
RMSD = 0.00039	RMSD = 0.00022	RMSD = 0.00038	RMSD = 0.00106
0.87617052E+00	0.86983806E+00	0.87795979E+00	0.27300918E+00
0.79437206E-03	0.85553573E-03	0.76593761E-03	
0.12206570E+02	0.60392160E+01	0.14978459E+02	
0.48233862E+03	0.44497070E+03	0.49388330E+03	

SECOND VIRIAL COEFFICIENT CORRELATION CONSTANTS, ML/MOL:

TSONOPOULOS: TC=425.160 PC=3.797 VC=255.00 OM=0.2000 A= 0.000000 B=0.0000

LITERATURE DOCUMENTS FOR CORRELATED VAPOR PRESSURE VALUES:

    371   372   374   375   457   590   599   614   616   705 2656 2792 3249
   3256 3318 3807 4287 4452 4816 5719 6253 6588 8418 9566 9617 9961

LITERATURE DOCUMENTS FOR CORRELATED LIQUID DENSITY VALUES:

    392   457   477   579   600 1911 1951 2026 2790 3016 4724 9566

LITERATURE DOCUMENTS REPORTING VIRIAL COEFFICIENT DATA:

    705   860 1415 1938 1950 1952 2651 2768 2778 3067 6359

T,K	VAPOR PRESSURE, KPA	SATURATED LIQUID VOLUME, ML/MOL	SECOND VIRIAL COEFFICIENT, ML/MOL	HEAT OF VAPORIZATION, J/MOL	SATURATED VAPOR VOLUME, ML/MOL
139	0.001419		-19741.		0.8144E+09
145	0.003891		-14828.		0.3098E+09
152	0.01131		-10908.		0.1117E+09
158	0.02600		-8570.		0.5054E+08
165	0.06307		-6625.		0.2175E+08
171	0.1264		-5418.		0.1124E+08
178	0.2665		-4375.		5554006.
184	0.4799		-3703.		3187895.
191	0.9044		-3102.		1755888.
197	1.495		-2702.		1093479.
204	2.575		-2333.		656636.
210	3.972		-2079.		437775.
217	6.360	88.36	-1837.	25198.	282014.
223	9.268	89.15	-1667.	24933.	198507.
230	13.97	90.10	-1500.	24617.	135410.
236	19.43	90.94	-1380.	24337.	99639.
243	27.87	91.95	-1260.	24002.	71234.
249	37.27	92.84	-1171.	23706.	54365.
256	51.27	93.92	-1081.	23350.	40399.
262	66.35	94.87	-1013.	23034.	31774.
269	88.14	96.03	-942.8	22655.	24379.
275	111.0	97.06	-889.2	22320.	19659.
282	143.1	98.31	-833.2	21918.	15488.
288	175.9	99.43	-789.8	21563.	12750.
295	221.3	100.79	-743.9	21136.	10267.
301	266.7	102.01	-708.1	20759.	8598.
308	328.2	103.52	-669.8	20306.	7050.
314	389.0	104.87	-639.7	19905.	5987.
321	470.1	106.55	-607.2	19423.	4982.
327	549.0	108.07	-581.5	18995.	4279.
334	653.1	109.96	-553.6	18477.	3603.
340	753.4	111.70	-531.4	18015.	3122.
347	884.3	113.89	-507.1	17452.	2652.
353	1009.	115.93	-487.7	16944.	2313.
360	1171.	118.52	-466.4	16317.	1977.
366	1324.	120.96	-449.3	15745.	1730.
373	1521.	124.13	-430.4	15026.	1483.
379	1707.	127.19	-415.2	14356.	1299.
386	1945.	131.25	-398.4	13494.	1112.
392	2168.	135.28	-384.8	12667.	969.7
399	2452.		-369.7		821.8
405	2718.		-357.5		706.6
412	3056.		-343.9		580.7
418	3371.		-332.8		472.8
425	3771.		-320.5		292.4

COMPOUND CONSTANTS:

MP 113.550 K	TC 408.140 K	VC 263.00 ML/MOL	ZC 0.2830
NBP 261.430 K	PC 3.648 MPA	DC 0.2210 G/ML	OM 0.1836
MU 0.1320 DEBYE	RG 2.8962 ANGSTROM		

VAPOR PRESSURE CORRELATION CONSTANTS, K AND KPA:

RPM2	VAPRES-2	RIEDEL	WAGNER
122. - 408. K	122. - 278. K	122. - 367. K	246. - 408. K
97 POINTS	31 POINTS	85 POINTS	80 POINTS
RMSD = 4.6050	RMSD = 0.2340	RMSD = 3.0590	RMSD = 2.8130
0.24560233E+02	-0.40094492E+03	0.47902945E+02	-0.64915350E+01
-0.36867359E+04	0.20695749E+04	-0.50565381E+01	0.16195465E+00
-0.30161706E-01	-0.47343790E+00	-0.39655664E+04	0.10003693E+01
0.29930569E-04	0.38781306E-03	0.99718581E-16	-0.19035639E+02
	0.88914485E+02		

LIQUID DENSITY CORRELATION CONSTANTS, K AND G/ML:

FRANCIS1	FRANCIS1	FRANCIS1	RACKETT
213. - 373. K	213. - 278. K	213. - 363. K	244. - 398. K
44 POINTS	17 POINTS	42 POINTS	39 POINTS
RMSD = 0.00035	RMSD = 0.00034	RMSD = 0.00034	RMSD = 0.00066
0.86629736E+00	0.85609049E+00	0.86621267E+00	0.27530753E+00
0.87856059E-03	0.82108285E-03	0.86905202E-03	
0.77421513E+01	0.74298372E+01	0.83093929E+01	
0.44347607E+03	0.41900537E+03	0.44647241E+03	

SECOND VIRIAL COEFFICIENT CORRELATION CONSTANTS, ML/MOL:

TSONOPOULOS: TC=408.130 PC=3.648 VC=263.00 OM=0.1836 A= 0.000000 B=0.0000

LITERATURE DOCUMENTS FOR CORRELATED VAPOR PRESSURE VALUES:

   375   376   379   427   579   614   617   705 1899 1906 3256 3318 3807
   4125 4287 4938 9566 9961

LITERATURE DOCUMENTS FOR CORRELATED LIQUID DENSITY VALUES:

   375   376   477 1951 4274 5995 9566

LITERATURE DOCUMENTS REPORTING VIRIAL COEFFICIENT DATA:

   705 2768 6359

T,K	VAPOR PRESSURE, KPA	SATURATED LIQUID VOLUME, ML/MOL	SECOND VIRIAL COEFFICIENT, ML/MOL	HEAT OF VAPORIZATION, J/MOL	SATURATED VAPOR VOLUME, ML/MOL
122	0.0001373		-35105.		0.7387E+10
128	0.0004941		-24904.		0.2154E+10
135	0.001882		-17224.		0.5963E+09
141	0.005276		-12889.		0.2222E+09
148	0.01563		-9459.		0.7872E+08
154	0.03635		-7429.		0.3523E+08
161	0.08901		-5750.		0.1504E+08
167	0.1794		-4713.		7740992.
174	0.3791		-3819.		3815813.
180	0.6832		-3245.		2190613.
187	1.286		-2731.		1207466.
193	2.120		-2389.		755225.
200	3.638		-2072.		455673.
206	5.586		-1854.		305285.
213	8.891	90.03	-1647.	23924.	197938.
219	12.88	90.90	-1499.	23640.	140168.
226	19.29	91.95	-1355.	23302.	96329.
232	26.64	92.87	-1250.	23004.	71354.
239	37.92	93.98	-1145.	22649.	51418.
245	50.36	94.97	-1067.	22337.	39512.
252	68.74	96.16	-987.6	21963.	29598.
258	88.34	97.21	-927.5	21633.	23438.
265	116.4	98.50	-865.1	21240.	18124.
271	145.5	99.64	-817.3	20894.	14710.
278	186.2	101.04	-767.1	20480.	11675.
284	227.4	102.29	-728.1	20116.	9670.
291	283.9	103.82	-686.7	19681.	7840.
297	340.1	105.20	-654.3	19298.	6602.
304	415.6	106.90	-619.6	18839.	5446.
310	489.7	108.45	-592.1	18433.	4647.
317	587.9	110.38	-562.5	17945.	3886.
323	683.1	112.15	-539.0	17512.	3350.
330	807.9	114.38	-513.5	16985.	2832.
336	927.5	116.46	-493.0	16513.	2460.
343	1083.	119.13	-470.8	15932.	2095.
349	1231.	121.66	-452.8	15403.	1829.
356	1422.	124.99	-433.2	14739.	1563.
362	1603.	128.24	-417.4	14121.	1367.
369	1835.	132.65	-400.0	13326.	1168.
375	2053.		-385.9		1019.
382	2333.		-370.3		864.0
388	2595.		-357.7		744.6
395	2930.		-343.7		616.1
401	3243.		-332.3		509.5
408	3643.		-319.6		353.0

COMPOUND CONSTANTS:

MP   143.420 K   TC   469.810 K   VC   311.00 ML/MOL   ZC   0.2690
NBP   309.220 K   PC   3.375 MPA   DC   0.2320 G/ML   OM   0.2506
MU   0.3700 DEBYE   RG   3.3850 ANGSTROM

VAPOR PRESSURE CORRELATION CONSTANTS, K AND KPA:

WAGNER	RIEDEL	WAGNER	WAGNER
144. - 469. K	144. - 325. K	144. - 398. K	290. - 469. K
177 POINTS	136 POINTS	152 POINTS	105 POINTS
RMSD = 1.9400	RMSD = 0.2360	RMSD = 0.7170	RMSD = 2.1660
-0.72429989E+01	0.59768840E+02	-0.73290032E+01	-0.71766718E+01
0.14401906E+01	-0.67266640E+01	0.16307495E+01	0.12591478E+01
-0.28955580E+01	-0.51580818E+04	-0.31237385E+01	-0.24414220E+01
-0.20862068E+01	0.11257891E-15	-0.18433996E+01	-0.51820521E+01

LIQUID DENSITY CORRELATION CONSTANTS, K AND G/ML:

FRANCIS1	FRANCIS1	FRANCIS1	FRANCIS2
150. - 457. K	150. - 323. K	150. - 439. K	423. - 469. K
164 POINTS	128 POINTS	159 POINTS	15 POINTS
RMSD = 0.00044	RMSD = 0.00038	RMSD = 0.00036	RMSD = 0.00077
0.90024590E+00	0.89427972E+00	0.90014148E+00	0.47694199E-03
0.77327527E-03	0.80343452E-03	0.77310181E-03	0.25762424E+01
0.10687443E+02	0.60030041E+01	0.10655088E+02	
0.51812622E+03	0.47717236E+03	0.51774756E+03	

SECOND VIRIAL COEFFICIENT CORRELATION CONSTANTS, ML/MOL:

TSONOPOULOS: TC=469.810 PC=3.375 VC=311.00 OM=0.2506 A= 0.000000 B=0.0000

LITERATURE DOCUMENTS FOR CORRELATED VAPOR PRESSURE VALUES:

   515 1485 2985 2987 3569 3963 4233 4522 4576 4816 5051 5124 5718
  6751 7091 7888 8912 9918 18274 20895 21583 40461

LITERATURE DOCUMENTS FOR CORRELATED LIQUID DENSITY VALUES:

   260 1972 2475 2985 2994 2996 3101 4452 5204 5321 6372 6737 6751
  8549 8580 8902 9087 10394 10493 13373 14407 16664 19270 20723 40166 40435

LITERATURE DOCUMENTS REPORTING VIRIAL COEFFICIENT DATA:

  1803 1837 1938 1950 2404 2405 2407 4452 6856

T,K	VAPOR PRESSURE, KPA	SATURATED LIQUID VOLUME, ML/MOL	SECOND VIRIAL COEFFICIENT, ML/MOL	HEAT OF VAPORIZATION, J/MOL	SATURATED VAPOR VOLUME, ML/MOL
144	0.0000781		-45296.		0.1534E+11
151	0.0002998	95.64	-32286.	34527.	0.4187E+10
158	0.001006	96.41	-23585.	34086.	0.1306E+10
166	0.003482	97.30	-16952.	33591.	0.3963E+09
173	0.009266	98.10	-13005.	33167.	0.1552E+09
180	0.02259	98.91	-10190.	32752.	0.6625E+08
188	0.05695	99.86	-7901.	32289.	0.2745E+08
195	0.1190	100.71	-6451.	31895.	0.1362E+08
203	0.2571	101.71	-5225.	31458.	6566106.
210	0.4767	102.61	-4418.	31088.	3662561.
217	0.8439	103.52	-3789.	30729.	2138012.
225	1.540	104.60	-3229.	30273.	1212187.
232	2.506	105.56	-2841.	29912.	767656.
240	4.195	106.69	-2485.	29504.	473693.
247	6.376	107.71	-2232.	29148.	320215.
254	9.433	108.76	-2020.	28793.	222143.
262	14.32	110.00	-1817.	28384.	150502.
269	20.15	111.11	-1667.	28024.	109491.
276	27.77	112.26	-1538.	27658.	81184.
284	39.19	113.62	-1410.	27233.	58904.
291	52.02	114.86	-1313.	26853.	45236.
299	70.55	116.32	-1216.	26407.	34033.
306	90.73	117.65	-1140.	26007.	26896.
313	115.2	119.04	-1073.	25595.	21503.
321	149.1	120.69	-1003.	25108.	16866.
328	184.6	122.21	-948.5	24668.	13776.
336	232.9	124.02	-891.8	24147.	11046.
343	282.6	125.70	-846.7	23674.	9182.
350	339.9	127.47	-805.2	23183.	7685.
358	415.6	129.61	-761.7	22598.	6319.
365	491.7	131.61	-726.7	22063.	5356.
372	577.9	133.75	-694.1	21504.	4562.
380	689.6	136.38	-659.6	20833.	3817.
387	799.8	138.89	-631.5	20212.	3279.
395	941.3	142.03	-601.6	19460.	2766.
402	1080.	145.07	-577.2	18756.	2388.
409	1233.	148.44	-554.2	18003.	2065.
417	1427.	152.83	-529.5	17067.	1750.
424	1615.	157.25	-509.2	16165.	1512.
432	1853.	163.20	-487.4	15013.	1275.
439	2082.	169.48	-469.4	13860.	1093.
446	2332.	177.19	-452.3	12516.	928.0
454	2647.	188.62	-433.8	10608.	753.2
461	2950.		-418.5		600.2
468	3283.		-403.9		351.6

COMPOUND CONSTANTS:

MP 113.250 K	TC 460.430 K	VC 305.72 ML/MOL	ZC 0.2700
NBP 300.993 K	PC 3.381 MPA	DC 0.2360 G/ML	OM 0.2278
MU 0.1200 DEBYE	RG 3.3130 ANGSTROM		

VAPOR PRESSURE CORRELATION CONSTANTS, K AND KPA:

WAGNER	VAPRES-2	VAPRES-2	WAGNER
187. - 448. K	187. - 303. K	187. - 373. K	282. - 448. K
34 POINTS	27 POINTS	30 POINTS	18 POINTS
RMSD = 2.6240	RMSD = 0.0520	RMSD = 0.7690	RMSD = 1.7210
-0.69266095E+01	0.10054755E+04	0.41273906E+03	-0.70751992E+01
0.85514811E+00	-0.20774039E+05	-0.11247862E+05	0.12546723E+01
-0.14853190E+01	0.73132239E+00	0.24221346E+00	-0.22168415E+01
-0.49915486E+01	-0.46574364E-03	-0.13355591E-03	-0.37127403E+01
	-0.19445416E+03	-0.75617655E+02	

LIQUID DENSITY CORRELATION CONSTANTS, K AND G/ML:

FRANCIS1	FRANCIS1	FRANCIS1	FRANCIS2
115. - 448. K	115. - 313. K	115. - 423. K	423. - 460. K
82 POINTS	74 POINTS	81 POINTS	2 POINTS
RMSD = 0.00049	RMSD = 0.00041	RMSD = 0.00048	RMSD = 0.00000
0.90053451E+00	0.89918488E+00	0.90222555E+00	0.29075516E-03
0.83946739E-03	0.85423281E-03	0.82881679E-03	0.29549918E+01
0.72778625E+01	0.60025387E+01	0.86762562E+01	
0.50029565E+03	0.49849365E+03	0.51023706E+03	

SECOND VIRIAL COEFFICIENT CORRELATION CONSTANTS, ML/MOL:

PITZER-CURL: TC=460.430   PC=3.381   VC=305.72   OM=0.2278

LITERATURE DOCUMENTS FOR CORRELATED VAPOR PRESSURE VALUES:

1485   2902   5043   6465   6504   8869

LITERATURE DOCUMENTS FOR CORRELATED LIQUID DENSITY VALUES:

432   2573   2902   4606   4700   5043   5199   5759   6737   8549   9087   9920   15062
40435

LITERATURE DOCUMENTS REPORTING VIRIAL COEFFICIENT DATA:

1837   6504   8310

T,K	VAPOR PRESSURE, KPA	SATURATED LIQUID VOLUME, ML/MOL	SECOND VIRIAL COEFFICIENT, ML/MOL	HEAT OF VAPORIZATION, J/MOL	SATURATED VAPOR VOLUME, ML/MOL
115		91.90	-132220.		
122		92.63	-84419.		
130		93.49	-52645.		
137		94.25	-36016.		
145		95.14	-24209.		
152		95.94	-17635.		
160		96.87	-12695.		
167		97.70	-9790.		
175		98.67	-7495.		
183		99.66	-5908.		
190	0.1319	100.55	-4904.	31149.	0.1197E+08
198	0.2904	101.58	-4053.	30548.	5664192.
205	0.5448	102.52	-3490.	30048.	3125093.
213	1.051	103.60	-2993.	29506.	1681925.
220	1.781	104.58	-2651.	29054.	1024510.
228	3.100	105.73	-2337.	28574.	609374.
236	5.158	106.91	-2085.	28105.	378527.
243	7.790	107.97	-1903.	27710.	257595.
251	12.06	109.21	-1727.	27273.	171337.
258	17.23	110.34	-1597.	26901.	122970.
266	25.20	111.66	-1468.	26483.	86348.
273	34.39	112.85	-1369.	26123.	64653.
281	48.00	114.26	-1271.	25714.	47408.
289	65.59	115.72	-1184.	25303.	35447.
296	84.84	117.05	-1116.	24940.	27883.
304	112.0	118.62	-1046.	24518.	21507.
311	140.9	120.05	-990.5	24141.	17337.
319	180.7	121.75	-932.8	23697.	13714.
326	222.2	123.31	-886.6	23295.	11276.
334	278.2	125.18	-838.1	22816.	9104.
342	344.3	127.15	-793.6	22315.	7418.
349	411.4	128.96	-757.5	21853.	6243.
357	499.7	131.17	-719.2	21297.	5161.
364	588.1	133.22	-688.0	20781.	4393.
372	702.9	135.74	-654.7	20154.	3672.
379	816.4	138.12	-627.4	19567.	3152.
387	962.2	141.08	-598.2	18848.	2656.
395	1127.	144.35	-570.8	18068.	2245.
402	1287.	147.54	-548.2	17326.	1941.
410	1490.	151.65	-523.9	16398.	1644.
417	1687.	155.80	-503.8	15500.	1420.
425	1934.	161.37	-482.0	14348.	1198.
432	2173.	167.28	-464.0	13197.	1027.
440	2472.	175.78	-444.4	11648.	851.1
447	2758.	185.60	-428.1	9976.	708.7

COMPOUND CONSTANTS:

MP   256.610 K   TC   433.770 K   VC   311.12 ML/MOL   ZC   0.2758
NBP   282.650 K   PC   3.196 MPA   DC   0.2319 G/ML   OM   0.1961
MU   0.0200 DEBYE   RG   3.1530 ANGSTROM

VAPOR PRESSURE CORRELATION CONSTANTS, K AND KPA:

WAGNER	WAGNER	WAGNER	WAGNER
257. - 433. K	257. - 298. K	257. - 368. K	263. - 433. K
73 POINTS	43 POINTS	55 POINTS	65 POINTS
RMSD = 0.4470	RMSD = 0.0830	RMSD = 0.1040	RMSD = 0.4800
-0.69274470E+01	-0.84402658E+01	-0.69453955E+01	-0.69209280E+01
0.13671699E+01	0.50891973E+01	0.14159933E+01	0.13449179E+01
-0.27240762E+01	-0.90838085E+01	-0.28309755E+01	-0.26442160E+01
-0.10021750E+01	0.16207686E+02	-0.59340447E+00	-0.15681787E+01

LIQUID DENSITY CORRELATION CONSTANTS, K AND G/ML:

RACKETT	FRANCIS1	FRANCIS1	FRANCIS2
258. - 433. K	258. - 293. K	258. - 403. K	388. - 433. K
32 POINTS	13 POINTS	26 POINTS	10 POINTS
RMSD = 0.00119	RMSD = 0.00063	RMSD = 0.00050	RMSD = 0.00094
0.27566716E+00	0.88378024E+00	0.88820922E+00	0.38240741E-03
	0.87586371E-03	0.77005173E-03	0.27025296E+01
	0.62133236E+01	0.14999805E+02	
	0.46501123E+03	0.50157520E+03	

SECOND VIRIAL COEFFICIENT CORRELATION CONSTANTS, ML/MOL:

PITZER-CURL: TC=433.770   PC=3.196   VC=311.12   OM=0.1961

LITERATURE DOCUMENTS FOR CORRELATED VAPOR PRESSURE VALUES:

4452 5054 5059 5718 7888 8406 9566 10508 16358

LITERATURE DOCUMENTS FOR CORRELATED LIQUID DENSITY VALUES:

742   4452 9566 10508

LITERATURE DOCUMENTS REPORTING VIRIAL COEFFICIENT DATA:

2403 4089 4452 6392 6525 9988 17031 18079

T,K	VAPOR PRESSURE, KPA	SATURATED LIQUID VOLUME, ML/MOL	SECOND VIRIAL COEFFICIENT, ML/MOL	HEAT OF VAPORIZATION, J/MOL	SATURATED VAPOR VOLUME, ML/MOL
257	36.10		-1313.		57738.
261	43.06	115.51	-1263.	23779.	49001.
265	51.04	116.28	-1216.	23577.	41819.
269	60.16	117.06	-1172.	23369.	35877.
273	70.52	117.87	-1131.	23159.	30934.
277	82.22	118.69	-1092.	22944.	26798.
281	95.39	119.53	-1056.	22724.	23319.
285	110.1	120.40	-1021.	22501.	20378.
289	126.6	121.29	-988.5	22272.	17878.
293	144.9	122.20	-957.6	22039.	15745.
297	165.2	123.14	-928.2	21802.	13916.
301	187.5	124.11	-900.2	21559.	12340.
305	212.1	125.11	-873.5	21311.	10978.
309	239.1	126.13	-848.1	21059.	9796.
313	268.6	127.19	-823.8	20801.	8766.
317	300.7	128.29	-800.6	20537.	7866.
321	335.6	129.42	-778.4	20268.	7075.
325	373.4	130.59	-757.1	19993.	6380.
329	414.4	131.80	-736.7	19713.	5765.
333	458.6	133.06	-717.1	19425.	5221.
337	506.2	134.36	-698.3	19132.	4737.
341	557.3	135.72	-680.2	18831.	4306.
345	612.1	137.14	-662.8	18522.	3921.
349	670.9	138.62	-646.0	18206.	3576.
353	733.6	140.16	-629.9	17881.	3265.
357	800.6	141.78	-614.3	17547.	2986.
361	871.9	143.48	-599.2	17202.	2733.
365	947.8	145.27	-584.7	16847.	2505.
369	1028.	147.16	-570.6	16479.	2297.
373	1114.	149.16	-557.1	16098.	2109.
377	1205.	151.29	-543.9	15701.	1936.
381	1301.	153.56	-531.2	15288.	1779.
385	1402.	155.99	-518.9	14855.	1634.
389	1510.	158.61	-506.9	14401.	1501.
393	1623.	161.46	-495.4	13921.	1378.
397	1743.	164.57	-484.2	13411.	1265.
401	1869.	167.99	-473.3	12866.	1159.
405	2002.	171.81	-462.7	12279.	1060.
409	2142.	176.14	-452.5	11640.	967.8
413	2290.	181.12	-442.5	10933.	880.2
417	2445.	187.02	-432.8	10139.	796.5
421	2609.	194.29	-423.4	9220.	715.1
425	2782.	203.82	-414.2	8105.	633.7
429	2964.	218.05	-405.3		546.4
433	3157.	251.89	-396.7		237.9

COMPOUND CONSTANTS:

MP   177.830 K   TC   507.680 K   VC   366.53 ML/MOL   ZC 0.2640
NBP   341.886 K   PC   3.040 MPA   DC   0.2351 G/ML   OM 0.3018
MU   0.0500 DEBYE   RG   3.8120 ANGSTROM

VAPOR PRESSURE CORRELATION CONSTANTS, K AND KPA:

RIEDEL	VAPRES-2	VAPRES-2	WAGNER
177. - 508. K	177. - 354. K	177. - 424. K	322. - 508. K
237 POINTS	214 POINTS	226 POINTS	101 POINTS
RMSD = 1.2220	RMSD = 0.1010	RMSD = 0.4670	RMSD = 0.7700
0.53334115E+02	-0.22919783E+03	-0.10761380E+03	-0.75346817E+01
-0.55933264E+01	-0.63811025E+03	-0.26597879E+04	0.14152605E+01
-0.55098512E+04	-0.20871521E+00	-0.11426998E+00	-0.22906092E+01
0.22596241E-16	0.12931528E-03	0.68782069E-04	-0.11694111E+02
	0.50034351E+02	0.25887165E+02	

LIQUID DENSITY CORRELATION CONSTANTS, K AND G/ML:

FRANCIS1	FRANCIS1	FRANCIS1	RACKETT
183. - 494. K	183. - 354. K	183. - 478. K	460. - 508. K
122 POINTS	110 POINTS	120 POINTS	7 POINTS
RMSD = 0.00027	RMSD = 0.00019	RMSD = 0.00024	RMSD = 0.00088
0.91968477E+00	0.91393524E+00	0.92026401E+00	0.26336655E+00
0.75910310E-03	0.77406992E-03	0.74911886E-03	
0.98481293E+01	0.60000496E+01	0.11006895E+02	
0.55327661E+03	0.50989258E+03	0.55941040E+03	

SECOND VIRIAL COEFFICIENT CORRELATION CONSTANTS, ML/MOL:

TSONOPOULOS: TC=507.680 PC=3.040 VC=366.53 OM=0.3018 A= 0.000000 B=0.0000

LITERATURE DOCUMENTS FOR CORRELATED VAPOR PRESSURE VALUES:

```
    3    102    175    254    538    914   1052   1179   1485   1631   1660   1716   1745
 1792   1883   1985   3317   4233   4423   4452   4816   4867   4870   4939   5101   5182
 5311   5641   6363   6947   6954   7870   7872   7881   7890   7892   7931   8958   9508
 9918   9943  10288  10622  10968  12169  15306  15521  15728  15748  18271  19712  20452
21432  40131  40405  40652  40823  40903
```


LITERATURE DOCUMENTS FOR CORRELATED LIQUID DENSITY VALUES:

```
   70     78   1487   1489   1769   2478   2991   4216   5641   6363   6541   6954   7867
12146  12764  13714  14269  14407  14487  15650  16414  19056  20290  21304  21432  40231
40563  40895
```


LITERATURE DOCUMENTS REPORTING VIRIAL COEFFICIENT DATA:

```
 1257   1803   1938   3910   4463   4949   5283  10972  11556  18235
```

T,K	VAPOR PRESSURE, KPA	SATURATED LIQUID VOLUME, ML/MOL	SECOND VIRIAL COEFFICIENT, ML/MOL	HEAT OF VAPORIZATION, J/MOL	SATURATED VAPOR VOLUME, ML/MOL
177	0.001175				
184	0.003093	114.39	-19612.	37262.	0.4947E+09
192	0.008491	115.41	-14971.	36893.	0.1880E+09
199	0.01908	116.33	-12040.	36570.	0.8674E+08
207	0.04463	117.39	-9573.	36203.	0.3856E+08
214	0.08854	118.34	-7961.	35882.	0.2010E+08
222	0.1825	119.45	-6560.	35517.	0.1012E+08
229	0.3277	120.45	-5614.	35198.	5810400.
237	0.6097	121.61	-4768.	34835.	3232120.
244	1.010	122.65	-4180.	34520.	2008390.
252	1.729	123.87	-3638.	34073.	1208723.
259	2.681	124.96	-3253.	33726.	800276.
267	4.284	126.25	-2889.	33321.	515523.
274	6.288	127.40	-2623.	32957.	359773.
282	9.487	128.75	-2366.	32529.	244796.
289	13.30	129.97	-2175.	32142.	178451.
297	19.13	131.41	-1988.	31685.	127065.
304	25.80	132.70	-1845.	31271.	96050.
312	35.61	134.23	-1702.	30781.	71032.
319	46.48	135.61	-1593.	30338.	55345.
327	61.99	137.24	-1481.	29812.	42245.
334	78.67	138.72	-1394.	29338.	33760.
342	101.9	140.48	-1305.	28779.	26461.
350	130.0	142.31	-1225.	28201.	20997.
357	159.3	143.99	-1161.	27681.	17308.
365	198.7	145.99	-1095.	27071.	14013.
372	239.0	147.82	-1042.	26523.	11737.
380	292.1	150.03	-985.8	25882.	9662.
387	345.6	152.07	-940.9	25307.	8200.
395	415.3	154.54	-893.4	24635.	6841.
402	484.6	156.85	-854.8	24031.	5867.
410	573.9	159.67	-813.8	23323.	4947.
417	661.7	162.33	-780.4	22684.	4276.
425	774.0	165.63	-744.7	21927.	3633.
432	883.6	168.79	-715.4	21237.	3157.
440	1023.	172.79	-684.0	20407.	2694.
447	1158.	176.70	-658.1	19896.	2375.
455	1329.	181.78	-630.3	18904.	2024.
462	1494.	186.90	-607.3	17953.	1757.
470	1703.	193.81	-582.5	16737.	1488.
477	1905.	201.12	-561.9	15518.	1279.
485	2160.	211.59	-539.7	13867.	1064.
492	2406.	223.55	-521.2	12072.	890.1
500	2717.		-501.1		712.6
507	3018.		-484.3		476.4

COMPOUND CONSTANTS:

MP 119.479 K	TC 497.502 K	VC 366.71 ML/MOL	ZC 0.2674
NBP 333.388 K	PC 3.017 MPA	DC 0.2350 G/ML	OM 0.2787
MU 0.0300 DEBYE	RG 3.8090 ANGSTROM		

VAPOR PRESSURE CORRELATION CONSTANTS, K AND KPA:

RPM2	RIEDEL	RIEDEL	WAGNER
283. - 498. K	283. - 349. K	283. - 404. K	314. - 498. K
32 POINTS	21 POINTS	25 POINTS	21 POINTS
RMSD = 0.9890	RMSD = 0.0180	RMSD = 0.2340	RMSD = 1.3720
0.25229767E+02	0.71060654E+02	0.47821100E+02	-0.72842451E+01
-0.48380665E+04	-0.83540915E+01	-0.48205424E+01	0.11879084E+01
-0.24865628E-01	-0.60224657E+04	-0.50779957E+04	-0.24439739E+01
0.19705206E-04	0.11144443E-15	0.23826152E-16	-0.67000701E+01

LIQUID DENSITY CORRELATION CONSTANTS, K AND G/ML:

FRANCIS1	FRANCIS1	FRANCIS1	RACKETT
115. - 484. K	115. - 344. K	115. - 464. K	443. - 498. K
38 POINTS	31 POINTS	37 POINTS	4 POINTS
RMSD = 0.00040	RMSD = 0.00030	RMSD = 0.00041	RMSD = 0.00168
0.92119229E+00	0.91508019E+00	0.92130184E+00	0.26592526E+00
0.76870690E-03	0.80373790E-03	0.76810503E-03	
0.10952058E+02	0.60010643E+01	0.11044655E+02	
0.54841138E+03	0.51831372E+03	0.54891113E+03	

SECOND VIRIAL COEFFICIENT CORRELATION CONSTANTS, ML/MOL:

TSONOPOULOS: TC=497.502 PC=3.017 VC=366.71 OM=0.2787 A= 0.000000 B=0.0000

LITERATURE DOCUMENTS FOR CORRELATED VAPOR PRESSURE VALUES:

   1485   1716   3571   4463   8958   9918

LITERATURE DOCUMENTS FOR CORRELATED LIQUID DENSITY VALUES:

   432   1489   1771   4463   6541 10808 21809

LITERATURE DOCUMENTS REPORTING VIRIAL COEFFICIENT DATA:

   4463   6458

T,K	VAPOR PRESSURE, KPA	SATURATED LIQUID VOLUME, ML/MOL	SECOND VIRIAL COEFFICIENT, ML/MOL	HEAT OF VAPORIZATION, J/MOL	SATURATED VAPOR VOLUME, ML/MOL
115		106.72			
123		107.60	-293029.		
132		108.61	-170275.		
141		109.65	-103416.		
149		110.59	-68708.		
158		111.67	-44974.		
167		112.78	-30526.		
175		113.79	-22263.		
184		114.95	-16096.		
193		116.14	-11999.		
202		117.36	-9204.		
210		118.47	-7434.		
219		119.76	-5980.		
228		121.08	-4917.		
236		122.29	-4199.		
245		123.68	-3574.		
254		125.12	-3089.		
262		126.44	-2743.		
271		127.97	-2426.		
280		129.56	-2168.		
289	19.08	131.20	-1953.	30349.	123946.
297	26.93	132.72	-1792.	29920.	89884.
306	38.68	134.49	-1637.	29421.	64096.
315	54.24	136.33	-1503.	28905.	46734.
323	71.91	138.03	-1399.	28431.	35888.
332	96.92	140.04	-1296.	27877.	27120.
341	128.3	142.15	-1206.	27301.	20825.
350	166.9	144.36	-1125.	26701.	16222.
358	208.3	146.44	-1061.	26144.	13132.
367	263.9	148.91	-995.7	25490.	10464.
376	329.9	151.55	-936.6	24801.	8421.
384	398.6	154.05	-888.6	24154.	6992.
393	488.1	157.08	-839.1	23382.	5711.
402	591.8	160.37	-793.8	22550.	4692.
410	697.2	163.56	-756.6	21748.	3953.
419	832.0	167.50	-717.9	20753.	3267.
428	985.4	171.93	-682.0	20170.	2769.
437	1159.	176.97	-648.8	19094.	2308.
445	1332.	182.11	-621.2	18045.	1962.
454	1550.	188.90	-592.1	16732.	1630.
463	1793.	197.18	-564.9	15234.	1348.
471	2034.	206.34	-542.3	13688.	1128.
480	2333.	219.77	-518.2	11587.	907.2
489	2667.		-495.6		700.6
497	2994.		-476.7		483.1

COMPOUND CONSTANTS:

MP	110.252 K	TC	504.654 K	VC	366.71 ML/MOL	ZC	0.2723
NBP	336.398 K	PC	3.116 MPA	DC	0.2350 G/ML	OM	0.2692
MU	0.0300 DEBYE	RG	3.6797 ANGSTROM				

VAPOR PRESSURE CORRELATION CONSTANTS, K AND KPA:

WAGNER	VAPRES-2	VAPRES-2	WAGNER
283. - 505. K	283. - 349. K	283. - 424. K	317. - 505. K
26 POINTS	19 POINTS	23 POINTS	17 POINTS
RMSD = 1.1390	RMSD = 0.0350	RMSD = 0.4400	RMSD = 1.4880
-0.75220392E+01	-0.30092190E+04	0.31403084E+03	-0.75316697E+01
0.19717397E+01	0.54161810E+05	-0.10756933E+05	0.19940911E+01
-0.42592158E+01	-0.19294971E+01	0.14373164E+00	-0.42384813E+01
0.21374453E+01	0.10579897E-02	-0.66329442E-04	0.77457681E+00
	0.58130169E+03	-0.54703379E+02	

LIQUID DENSITY CORRELATION CONSTANTS, K AND G/ML:

FRANCIS1	FRANCIS1	FRANCIS1	FRANCIS2
273. - 494. K	273. - 349. K	273. - 474. K	463. - 505. K
14 POINTS	7 POINTS	13 POINTS	4 POINTS
RMSD = 0.00047	RMSD = 0.00013	RMSD = 0.00033	RMSD = 0.00035
0.92706662E+00	0.91903067E+00	0.92613357E+00	0.28876861E-03
0.75379666E-03	0.73185912E-03	0.70994440E-03	0.29488428E+01
0.11013167E+02	0.93420839E+01	0.14999959E+02	
0.55564844E+03	0.52467065E+03	0.57148608E+03	

SECOND VIRIAL COEFFICIENT CORRELATION CONSTANTS, ML/MOL:

PITZER-CURL: TC=504.654   PC=3.116   VC=366.71   OM=0.2692

LITERATURE DOCUMENTS FOR CORRELATED VAPOR PRESSURE VALUES:

   1485   1771   3571   4463   6552   8958

LITERATURE DOCUMENTS FOR CORRELATED LIQUID DENSITY VALUES:

   1771   4463   6552 10054 10808 21809

LITERATURE DOCUMENTS REPORTING VIRIAL COEFFICIENT DATA:

   4463   6458

T,K	VAPOR PRESSURE, KPA	SATURATED LIQUID VOLUME, ML/MOL	SECOND VIRIAL COEFFICIENT, ML/MOL	HEAT OF VAPORIZATION, J/MOL	SATURATED VAPOR VOLUME, ML/MOL
273		126.30	-2235.		
278		127.13	-2115.		
283	12.94	127.98	-2006.	30927.	179847.
288	16.29	128.85	-1907.	30707.	145074.
294	21.22	129.91	-1799.	30431.	113347.
299	26.22	130.82	-1717.	30190.	93079.
304	32.12	131.74	-1641.	29941.	77003.
309	39.07	132.68	-1571.	29682.	64149.
315	48.94	133.85	-1494.	29358.	51972.
320	58.62	134.84	-1434.	29078.	43906.
325	69.75	135.85	-1379.	28788.	37308.
330	82.50	136.90	-1327.	28488.	31875.
336	100.1	138.18	-1268.	28115.	26568.
341	117.0	139.28	-1223.	27793.	22948.
346	135.9	140.42	-1180.	27462.	19910.
352	161.7	141.82	-1132.	27050.	16885.
357	185.9	143.03	-1094.	26696.	14782.
362	212.9	144.27	-1058.	26331.	12989.
367	242.6	145.56	-1024.	25956.	11452.
373	282.4	147.16	-984.9	25490.	9887.
378	319.2	148.55	-954.2	25089.	8776.
383	359.4	149.98	-925.0	24675.	7811.
388	403.3	151.48	-897.1	24248.	6969.
394	461.2	153.36	-865.2	23717.	6094.
399	514.0	155.00	-839.8	23256.	5462.
404	571.1	156.71	-815.5	22777.	4903.
410	645.7	158.88	-787.7	22175.	4315.
415	713.3	160.80	-765.5	21646.	3884.
420	785.9	162.82	-744.2	21089.	3498.
425	863.9	164.96	-723.7	20498.	3151.
431	965.0	167.72	-700.1	20527.	2882.
436	1056.	170.19	-681.3	19974.	2612.
441	1153.	172.84	-663.1	19397.	2367.
446	1256.	175.72	-645.6	18794.	2146.
452	1390.	179.50	-625.3	18029.	1906.
457	1509.	182.99	-609.1	17351.	1726.
462	1636.	186.85	-593.4	16629.	1561.
468	1799.	192.09	-575.3	15693.	1380.
473	1945.	197.08	-560.7	14840.	1242.
478	2100.	202.81	-546.6	13905.	1113.
483	2264.	209.48	-532.9	12861.	992.7
489	2476.	219.16	-517.1	11412.	855.4
494	2666.	229.17	-504.3	9958.	744.1
499	2868.		-492.0		629.5
504	3086.		-480.0		480.6

COMPOUND CONSTANTS:

MP 173.302 K	TC 488.778 K	VC 359.07 ML/MOL	ZC 0.2708
NBP 322.879 K	PC 3.065 MPA	DC 0.2400 G/ML	OM 0.2305
MU 0.0200 DEBYE	RG 3.4846 ANGSTROM		

VAPOR PRESSURE CORRELATION CONSTANTS, K AND KPA:

WAGNER	VAPRES-2	VAPRES-2	WAGNER
211. - 489. K	211. - 324. K	211. - 414. K	303. - 489. K
46 POINTS	34 POINTS	41 POINTS	25 POINTS
RMSD = 1.0830	RMSD = 0.0320	RMSD = 0.5730	RMSD = 0.9700
-0.69198640E+01	0.75214002E+03	0.31713368E+03	-0.69795717E+01
0.83583142E+00	-0.18074824E+05	-0.10427449E+05	0.98033240E+00
-0.15644990E+01	0.46276213E+00	0.14824183E+00	-0.16626995E+01
-0.57208842E+01	-0.26013358E-03	-0.68496146E-04	-0.78454462E+01
	-0.14086888E+03	-0.55552948E+02	

LIQUID DENSITY CORRELATION CONSTANTS, K AND G/ML:

FRANCIS1	FRANCIS1	FRANCIS1	RACKETT
273. - 464. K	273. - 319. K	273. - 444. K	403. - 489. K
16 POINTS	9 POINTS	15 POINTS	4 POINTS
RMSD = 0.00022	RMSD = 0.00015	RMSD = 0.00022	RMSD = 0.00084
0.91018236E+00	0.89190286E+00	0.90942097E+00	0.27047673E+00
0.69200271E-03	0.69793616E-03	0.68898150E-03	
0.15000000E+02	0.69251318E+01	0.15000000E+02	
0.55313257E+03	0.47689282E+03	0.55249634E+03	

SECOND VIRIAL COEFFICIENT CORRELATION CONSTANTS, ML/MOL:

TSONOPOULOS: TC=488.778 PC=3.065 VC=359.07 OM=0.2305 A= 0.000000 B=0.0000

LITERATURE DOCUMENTS FOR CORRELATED VAPOR PRESSURE VALUES:

   1485 1771 3571 4463 4576 6397 8958 9918

LITERATURE DOCUMENTS FOR CORRELATED LIQUID DENSITY VALUES:

   1771 5063

LITERATURE DOCUMENTS REPORTING VIRIAL COEFFICIENT DATA:

   4463

T,K	VAPOR PRESSURE, KPA	SATURATED LIQUID VOLUME, ML/MOL	SECOND VIRIAL COEFFICIENT, ML/MOL	HEAT OF VAPORIZATION, J/MOL	SATURATED VAPOR VOLUME, ML/MOL
211	0.2556		-5746.		6858804.
217	0.4296		-5007.		4194853.
223	0.6974		-4405.		2654145.
229	1.097		-3911.		1731723.
236	1.795		-3440.		1089796.
242	2.661		-3106.		752902.
248	3.855		-2824.		532092.
255	5.780		-2546.		364257.
261	8.008		-2344.		268620.
267	10.90		-2168.		201491.
274	15.30	129.23	-1990.	28869.	146915.
280	20.12	130.27	-1857.	28526.	113795.
286	26.11	131.34	-1739.	28188.	89287.
293	34.83	132.63	-1618.	27799.	68293.
299	44.01	133.76	-1525.	27468.	54913.
305	55.03	134.93	-1441.	27138.	44593.
312	70.51	136.34	-1353.	26752.	35381.
318	86.34	137.59	-1285.	26418.	29278.
324	104.8	138.88	-1223.	26081.	24413.
331	130.1	140.44	-1156.	25680.	19924.
337	155.3	141.84	-1104.	25328.	16856.
343	184.2	143.28	-1056.	24967.	14344.
349	217.0	144.79	-1011.	24595.	12273.
356	260.6	146.62	-962.3	24144.	10295.
362	303.1	148.27	-923.7	23742.	8898.
368	350.7	150.00	-887.6	23321.	7721.
375	413.1	152.13	-848.2	22805.	6574.
381	472.9	154.06	-816.7	22337.	5746.
387	539.0	156.09	-786.9	21843.	5037.
394	624.7	158.63	-754.3	21225.	4331.
400	705.8	160.95	-728.0	20656.	3812.
406	794.5	163.44	-703.1	20043.	3359.
413	908.2	166.58	-675.7	19660.	2957.
419	1015.	169.51	-653.4	19016.	2619.
425	1131.	172.69	-632.2	18331.	2321.
432	1278.	176.81	-608.8	17474.	2017.
438	1415.	180.74	-589.7	16683.	1787.
444	1563.	185.13	-571.4	15830.	1582.
451	1749.	190.99	-551.2	14743.	1368.
457	1922.	196.79	-534.6	13715.	1204.
463	2107.	203.55	-518.7	12574.	1054.
470	2340.		-501.1		892.6
476	2555.		-486.6		762.4
482	2785.		-472.6		632.4
488	3032.		-459.2		471.4

COMPOUND CONSTANTS:

MP   144.607 K   TC   499.983 K   VC   357.58 ML/MOL   ZC   0.2690
NBP   331.128 K   PC   3.127 MPA   DC   0.2410 G/ML   OM   0.2480
MU   0.0200 DEBYE   RG   3.5209 ANGSTROM

VAPOR PRESSURE CORRELATION CONSTANTS, K AND KPA:

WAGNER	VAPRES-2	WAGNER	WAGNER
259. - 500. K	259. - 349. K	259. - 414. K	312. - 500. K
48 POINTS	34 POINTS	41 POINTS	29 POINTS
RMSD = 0.9750	RMSD = 0.0280	RMSD = 0.1950	RMSD = 0.9580
-0.69444536E+01	0.17883502E+04	-0.70365771E+01	-0.68713395E+01
0.63150875E+00	-0.39675911E+05	0.86408056E+00	0.37874781E+00
-0.95001787E+00	0.10531941E+01	-0.13602497E+01	-0.11652854E-01
-0.95223554E+01	-0.55523406E-03	-0.85026957E+01	-0.16545931E+02
	-0.33636750E+03		

LIQUID DENSITY CORRELATION CONSTANTS, K AND G/ML:

FRANCIS1	FRANCIS1	FRANCIS1	FRANCIS2
273. - 490. K	273. - 344. K	273. - 464. K	453. - 500. K
42 POINTS	18 POINTS	38 POINTS	9 POINTS
RMSD = 0.00047	RMSD = 0.00011	RMSD = 0.00018	RMSD = 0.00264
0.92520744E+00	0.91661257E+00	0.92362469E+00	0.23479375E-03
0.75286860E-03	0.77414466E-03	0.70620747E-03	0.30182006E+01
0.11021470E+02	0.59999990E+01	0.14992791E+02	
0.55055786E+03	0.50667334E+03	0.56589917E+03	

SECOND VIRIAL COEFFICIENT CORRELATION CONSTANTS, ML/MOL:

PITZER-CURL: TC=499.983   PC=3.127   VC=357.58   OM=0.2480

LITERATURE DOCUMENTS FOR CORRELATED VAPOR PRESSURE VALUES:

   1485   3571   4463   5017   6553   8958   9918

LITERATURE DOCUMENTS FOR CORRELATED LIQUID DENSITY VALUES:

   1488   1771   6541   6553   6656 10808

LITERATURE DOCUMENTS REPORTING VIRIAL COEFFICIENT DATA:

   4463   6458

T,K	VAPOR PRESSURE, KPA	SATURATED LIQUID VOLUME, ML/MOL	SECOND VIRIAL COEFFICIENT, ML/MOL	HEAT OF VAPORIZATION, J/MOL	SATURATED VAPOR VOLUME, ML/MOL
259	4.752		-2448.		450706.
264	6.294		-2304.		346425.
269	8.225		-2174.		269730.
275	11.15	127.07	-2036.	30711.	202918.
280	14.20	127.92	-1932.	30350.	162010.
286	18.70	128.96	-1819.	29933.	125303.
291	23.27	129.85	-1734.	29596.	102189.
297	29.90	130.94	-1641.	29204.	80913.
302	36.50	131.87	-1570.	28885.	67188.
308	45.89	133.01	-1491.	28511.	54271.
313	55.09	133.99	-1431.	28205.	45762.
319	67.98	135.19	-1364.	27842.	37599.
324	80.44	136.22	-1312.	27542.	32122.
330	97.66	137.49	-1254.	27182.	26778.
335	114.1	138.58	-1209.	26882.	23135.
341	136.6	139.92	-1158.	26520.	19527.
346	157.8	141.08	-1118.	26214.	17033.
352	186.5	142.51	-1073.	25840.	14530.
357	213.4	143.75	-1038.	25521.	12778.
363	249.5	145.29	-998.0	25128.	10999.
368	283.0	146.61	-966.5	24789.	9739.
374	327.6	148.27	-930.7	24368.	8447.
379	368.7	149.71	-902.3	24002.	7522.
384	413.6	151.21	-875.2	23620.	6714.
390	472.8	153.08	-844.2	23138.	5873.
395	526.8	154.72	-819.7	22715.	5263.
401	597.7	156.80	-791.5	22175.	4623.
406	662.1	158.62	-769.1	21696.	4155.
412	746.0	160.93	-743.4	21080.	3659.
417	821.9	162.99	-722.8	20525.	3292.
423	920.4	165.62	-699.2	20485.	2994.
428	1009.	167.97	-680.4	19942.	2709.
434	1124.	171.02	-658.6	19252.	2403.
439	1226.	173.78	-641.3	18641.	2176.
445	1359.	177.40	-621.2	17859.	1930.
450	1477.	180.74	-605.1	17163.	1746.
456	1628.	185.19	-586.5	16265.	1546.
461	1763.	189.37	-571.6	15458.	1395.
467	1936.	195.10	-554.3	14404.	1229.
472	2090.	200.62	-540.4	13440.	1102.
478	2286.	208.44	-524.4	12151.	960.9
483	2459.	216.29	-511.4	10933.	850.1
489	2681.	227.97	-496.4	9215.	721.9
494	2877.		-484.3		613.2
499	3085.		-472.6		479.2

COMPOUND CONSTANTS:

MP 182.586 K	TC 540.166 K	VC 426.00 ML/MOL	ZC 0.2595
NBP 371.568 K	PC 2.736 MPA	DC 0.2352 G/ML	OM 0.3497
MU 0.0000 DEBYE	RG 4.2665 ANGSTROM		

VAPOR PRESSURE CORRELATION CONSTANTS, K AND KPA:

WAGNER	VAPRES-2	RIEDEL	WAGNER
185. - 540. K	185. - 391. K	185. - 454. K	352. - 540. K
197 POINTS	169 POINTS	178 POINTS	96 POINTS
RMSD = 0.8132	RMSD = 0.0350	RMSD = 0.1469	RMSD = 0.7983
-0.81490460E+01	-0.10961611E+03	0.62808752E+02	-0.81333050E+01
0.25064250E+01	-0.33844196E+04	-0.68986277E+01	0.25210802E+01
-0.52636083E+01	-0.12056836E+00	-0.64742355E+04	-0.56551732E+01
0.34239499E-02	0.71921948E-04	0.22179356E-16	0.40351146E+01
	0.26735293E+02		

LIQUID DENSITY CORRELATION CONSTANTS, K AND G/ML:

FRANCIS1	FRANCIS1	FRANCIS1	RACKETT
183. - 530. K	183. - 384. K	183. - 510. K	493. - 540. K
147 POINTS	123 POINTS	143 POINTS	16 POINTS
RMSD = 0.00032	RMSD = 0.00023	RMSD = 0.00029	RMSD = 0.00146
0.93400484E+00	0.92942894E+00	0.93450126E+00	0.25979145E+00
0.72764489E-03	0.75804070E-03	0.71781618E-03	
0.10929075E+02	0.59999990E+01	0.12174254E+02	
0.58927686E+03	0.54864941E+03	0.59546582E+03	

SECOND VIRIAL COEFFICIENT CORRELATION CONSTANTS, ML/MOL:

TSONOPOULOS: TC=540.166 PC=2.736 VC=426.00 OM=0.3497 A= 0.000000 B=0.0000

LITERATURE DOCUMENTS FOR CORRELATED VAPOR PRESSURE VALUES:

```
    4     90    102    175    236    262    543    555    556    695    727    960   1043
 1290   1485   1486   1529   1648   1711   1788   2097   2902   2983   3571   3646   4821
 4842   4891   5068   6894   7342   7459   7862   7879   7924   7929   8405   9176  10290
10413  10917  11316  11493  15728  17434  17992  18206  18913  19435  21583
```

LITERATURE DOCUMENTS FOR CORRELATED LIQUID DENSITY VALUES:

```
  555    922    985   1294   1487   1490   1524   1769   1773   1832   2288   2475   2824
 2902   2986   3062   4216   4893   5068   5070   5759   7929   8580   8902   8973   9050
10592  10917  11234  12856  13803  13832  13838  14487  15650  17517  18913  20436  20784
21304  21388  21432  21804  22448  40176  40683
```

LITERATURE DOCUMENTS REPORTING VIRIAL COEFFICIENT DATA:

```
 1415   1803   1938  11493  40535
```

T,K	VAPOR PRESSURE, KPA	SATURATED LIQUID VOLUME, ML/MOL	SECOND VIRIAL COEFFICIENT, ML/MOL	HEAT OF VAPORIZATION, J/MOL	SATURATED VAPOR VOLUME, ML/MOL
183		129.47	-40610.		
191	0.0006628	130.54	-30368.	42806.	0.2396E+10
199	0.001949	131.64	-23196.	42377.	0.8491E+09
207	0.005217	132.76	-18078.	41946.	0.3299E+09
215	0.01286	133.90	-14361.	41509.	0.1390E+09
223	0.02945	135.06	-11614.	41068.	0.6295E+08
231	0.06317	136.25	-9552.	40626.	0.3040E+08
239	0.1277	137.47	-7979.	40182.	0.1556E+08
247	0.2450	138.72	-6761.	39738.	8380839.
256	0.4818	140.16	-5699.	39239.	4417923.
264	0.8397	141.47	-4958.	38798.	2614004.
272	1.408	142.81	-4359.	38288.	1603024.
280	2.280	144.19	-3868.	37821.	1018222.
288	3.576	145.61	-3463.	37347.	666994.
296	5.448	147.06	-3123.	36867.	449323.
304	8.081	148.56	-2836.	36378.	310497.
312	11.70	150.10	-2590.	35880.	219600.
321	17.27	151.89	-2355.	35308.	152525.
329	23.89	153.54	-2174.	34787.	112582.
337	32.44	155.24	-2016.	34255.	84559.
345	43.29	157.00	-1876.	33709.	64528.
353	56.87	158.83	-1752.	33151.	49962.
361	73.62	160.72	-1641.	32578.	39198.
369	94.02	162.70	-1541.	31992.	31127.
377	118.6	164.75	-1451.	31392.	24991.
385	147.9	166.90	-1369.	30777.	20267.
394	187.1	169.45	-1286.	30067.	16185.
402	228.3	171.84	-1218.	29420.	13368.
410	276.1	174.35	-1156.	28756.	11122.
418	331.1	177.02	-1099.	28075.	9315.
426	393.9	179.87	-1046.	27373.	7846.
434	465.4	182.91	-996.9	26649.	6643.
442	546.3	186.19	-951.3	25900.	5649.
450	637.3	189.74	-908.7	25119.	4821.
459	753.0	194.14	-864.1	24195.	4048.
467	868.6	198.49	-827.2	23324.	3472.
475	997.4	203.36	-792.5	22393.	2981.
483	1141.	208.88	-759.8	21385.	2559.
491	1299.	215.25	-729.1	20276.	2193.
499	1475.	222.75	-700.1	19029.	1872.
507	1670.	231.81	-672.7	17590.	1587.
515	1885.	243.14	-646.8		1489.
523	2124.	257.89	-622.2		1267.
532	2426.		-596.0		1035.
540	2729.		-574.0		774.9

COMPOUND CONSTANTS:

MP   154.907 K   TC   530.363 K   VC   421.02 ML/MOL   ZC   0.2609
NBP 363.199 K   PC   2.733 MPA   DC   0.2380 G/ML   OM   0.3281
MU   0.0000 DEBYE   RG   4.2779 ANGSTROM

VAPOR PRESSURE CORRELATION CONSTANTS, K AND KPA:

WAGNER	RPM2	WAGNER	WAGNER
273. - 530. K	273. - 364. K	273. - 429. K	349. - 530. K
39 POINTS	24 POINTS	29 POINTS	19 POINTS
RMSD = 1.0500	RMSD = 0.0035	RMSD = 0.2705	RMSD = 1.1160
-0.76073514E+01	0.29007369E+02	-0.71617184E+01	-0.77273067E+01
0.15209925E+01	-0.57792531E+04	0.39732846E+00	0.19310394E+01
-0.39731702E+01	-0.32552413E-01	-0.20104113E+01	-0.53321599E+01
-0.53795115E+00	0.25365548E-04	-0.53209640E+01	0.58754306E+01

LIQUID DENSITY CORRELATION CONSTANTS, K AND G/ML:

FRANCIS1	FRANCIS1	FRANCIS1	FRANCIS2
273. - 518. K	273. - 383. K	273. - 491. K	490. - 530. K
18 POINTS	9 POINTS	16 POINTS	4 POINTS
RMSD = 0.00148	RMSD = 0.00060	RMSD = 0.00154	RMSD = 0.00011
0.94757944E+00	0.90810364E+00	0.94974345E+00	0.41892329E-03
0.84017613E-03	0.61401236E-03	0.84732007E-03	0.26539601E+01
0.59999990E+01	0.10599940E+02	0.59999990E+01	
0.56205054E+03	0.50920801E+03	0.56374414E+03	

SECOND VIRIAL COEFFICIENT CORRELATION CONSTANTS, ML/MOL:

TSONOPOULOS: TC=530.363 PC=2.733 VC=421.02 OM=0.3281 A= 0.000000 B=0.0000

LITERATURE DOCUMENTS FOR CORRELATED VAPOR PRESSURE VALUES:

   1486   1494   9050

LITERATURE DOCUMENTS FOR CORRELATED LIQUID DENSITY VALUES:

   1211   1490   9050 10808

LITERATURE DOCUMENTS REPORTING VIRIAL COEFFICIENT DATA:

T,K	VAPOR PRESSURE, KPA	SATURATED LIQUID VOLUME, ML/MOL	SECOND VIRIAL COEFFICIENT, ML/MOL	HEAT OF VAPORIZATION, J/MOL	SATURATED VAPOR VOLUME, ML/MOL
273	2.327	143.67	-3793.	36125.	971462.
278	3.100	144.62	-3530.	35863.	742182.
284	4.304	145.77	-3253.	35546.	545415.
290	5.881	146.95	-3010.	35224.	406987.
296	7.917	148.16	-2796.	34898.	308042.
302	10.51	149.38	-2607.	34566.	236260.
308	13.77	150.64	-2439.	34229.	183455.
313	17.09	151.70	-2313.	33943.	149903.
319	21.92	153.01	-2175.	33593.	118809.
325	27.79	154.35	-2051.	33236.	95123.
331	34.89	155.71	-1938.	32871.	76882.
337	43.39	157.11	-1836.	32497.	62689.
343	53.46	158.55	-1742.	32114.	51538.
349	65.32	160.02	-1656.	31722.	42698.
354	76.72	161.27	-1590.	31386.	36702.
360	92.41	162.82	-1516.	30975.	30797.
366	110.5	164.41	-1448.	30551.	26005.
372	131.3	166.05	-1385.	30116.	22087.
378	154.9	167.73	-1326.	29669.	18863.
384	181.7	169.48	-1271.	29207.	16190.
389	206.7	170.97	-1228.	28812.	14306.
395	240.0	172.83	-1179.	28324.	12381.
401	277.3	174.75	-1134.	27819.	10758.
407	318.8	176.75	-1091.	27296.	9382.
413	364.8	178.83	-1051.	26753.	8209.
419	415.7	181.00	-1012.	26188.	7203.
425	471.7	183.28	-976.5	25598.	6337.
430	522.6	185.26	-948.0	25084.	5705.
436	588.9	187.76	-915.4	24438.	5037.
442	661.4	190.41	-884.5	23753.	4452.
448	740.4	193.23	-855.1	23023.	3939.
454	826.2	196.24	-827.2	22798.	3567.
460	919.2	199.49	-800.6	22067.	3172.
466	1020.	203.01	-775.2	21298.	2822.
471	1110.	206.20	-754.9	20624.	2560.
477	1226.	210.39	-731.6	19769.	2276.
483	1351.	215.09	-709.3	18856.	2022.
489	1486.	220.42	-688.0	17872.	1793.
495	1630.	226.60	-667.5	16801.	1585.
501	1786.	233.92	-647.9	15619.	1395.
506	1924.	241.22	-632.1	14523.	1248.
512	2101.	252.06	-613.9	13027.	1083.
518	2292.	266.38	-596.5	11233.	925.0
524	2497.		-579.6		765.6
530	2719.		-563.4		553.7

COMPOUND CONSTANTS:

MP 153.754 K	TC 535.255 K	VC 404.04 ML/MOL	ZC 0.2554
NBP 364.995 K	PC 2.813 MPA	DC 0.2480 G/ML	OM 0.3217
MU 0.0000 DEBYE	RG 4.1454 ANGSTROM		

VAPOR PRESSURE CORRELATION CONSTANTS, K AND KPA:

WAGNER	WAGNER	VAPRES-2	WAGNER
283. - 535. K	283. - 366. K	283. - 441. K	351. - 535. K
37 POINTS	25 POINTS	29 POINTS	21 POINTS
RMSD = 3.7650	RMSD = 0.0053	RMSD = 0.5891	RMSD = 1.7360
-0.74902375E+01	-0.74380988E+01	0.75138807E+03	-0.79187850E+01
0.13293349E+01	0.10746784E+01	-0.21328469E+05	0.27842218E+01
-0.37269937E+01	-0.27562361E+01	0.35604721E+00	-0.88597556E+01
-0.13298973E+00	-0.43827551E+01	-0.15925702E-03	0.37064253E+02
		-0.13509964E+03	

LIQUID DENSITY CORRELATION CONSTANTS, K AND G/ML:

FRANCIS1	FRANCIS1	FRANCIS1	FRANCIS2
288. - 526. K	288. - 304. K	288. - 492. K	491. - 535. K
12 POINTS	5 POINTS	10 POINTS	3 POINTS
RMSD = 0.00047	RMSD = 0.00014	RMSD = 0.00030	RMSD = 0.00088
0.91412258E+00	0.90447927E+00	0.91931915E+00	0.45353312E-03
0.64804824E-03	0.62613213E-03	0.71462942E-03	0.26248192E+01
0.10686495E+02	0.60008020E+01	0.59999990E+01	
0.57942847E+03	0.46962524E+03	0.55489111E+03	

SECOND VIRIAL COEFFICIENT CORRELATION CONSTANTS, ML/MOL:

TSONOPOULOS: TC=535.255 PC=2.813 VC=404.04 OM=0.3217 A= 0.000000 B=0.0000

LITERATURE DOCUMENTS FOR CORRELATED VAPOR PRESSURE VALUES:

  1486 3571 9050

LITERATURE DOCUMENTS FOR CORRELATED LIQUID DENSITY VALUES:

  1490 9050 9714

LITERATURE DOCUMENTS REPORTING VIRIAL COEFFICIENT DATA:

T,K	VAPOR PRESSURE, KPA	SATURATED LIQUID VOLUME, ML/MOL	SECOND VIRIAL COEFFICIENT, ML/MOL	HEAT OF VAPORIZATION, J/MOL	SATURATED VAPOR VOLUME, ML/MOL
283	3.813		-3318.		613803.
288	4.963	145.05	-3105.	35384.	479383.
294	6.716	146.04	-2879.	35088.	361047.
300	8.963	147.04	-2680.	34785.	275589.
305	11.28	147.90	-2531.	34528.	222163.
311	14.71	148.94	-2370.	34213.	173396.
317	18.95	150.01	-2226.	33890.	136833.
323	24.14	151.11	-2096.	33559.	109096.
328	29.31	152.04	-1998.	33276.	90997.
334	36.66	153.18	-1889.	32929.	73805.
340	45.44	154.36	-1791.	32571.	60372.
346	55.82	155.57	-1701.	32204.	49777.
351	65.85	156.60	-1631.	31889.	42625.
357	79.71	157.87	-1554.	31501.	35612.
363	95.79	159.18	-1483.	31102.	29949.
369	114.3	160.53	-1417.	30690.	25340.
374	131.8	161.69	-1366.	30336.	22142.
380	155.4	163.12	-1308.	29900.	18925.
386	182.2	164.61	-1254.	29450.	16257.
391	207.1	165.89	-1212.	29063.	14373.
397	240.4	167.49	-1164.	28584.	12448.
403	277.6	169.15	-1120.	28088.	10823.
409	319.0	170.89	-1078.	27573.	9445.
414	356.9	172.40	-1045.	27129.	8452.
420	406.8	174.29	-1007.	26576.	7419.
426	461.8	176.29	-971.2	25999.	6528.
432	522.2	178.40	-937.4	25393.	5758.
437	576.9	180.26	-910.7	24865.	5193.
443	648.1	182.62	-880.1	24198.	4595.
449	725.6	185.14	-851.0	23489.	4069.
455	809.9	187.87	-823.3	23247.	3682.
460	885.5	190.30	-801.3	22653.	3341.
466	983.0	193.47	-776.0	21907.	2975.
472	1088.	196.95	-751.8	21122.	2650.
477	1183.	200.14	-732.5	20433.	2407.
483	1303.	204.37	-710.2	19558.	2142.
489	1433.	209.17	-688.9	18623.	1905.
495	1573.	214.68	-668.5	17613.	1690.
500	1697.	219.96	-652.2	16704.	1527.
506	1856.	227.39	-633.3	15509.	1345.
512	2027.	236.42	-615.1	14168.	1178.
518	2209.	247.74	-597.7	12622.	1021.
523	2371.	259.72	-583.7	11100.	895.3
529	2579.		-567.5		741.9
535	2803.		-551.8		533.5

COMPOUND CONSTANTS:

MP   154.589 K   TC   540.636 K   VC   415.78 ML/MOL   ZC   0.2674
NBP   366.618 K   PC   2.891 MPA   DC   0.2410 G/ML   OM   0.3103
MU   0.0000 DEBYE   RG

VAPOR PRESSURE CORRELATION CONSTANTS, K AND KPA:

WAGNER	WAGNER	WAGNER	WAGNER
294. - 541. K	294. - 386. K	294. - 458. K	348. - 541. K
49 POINTS	30 POINTS	38 POINTS	32 POINTS
RMSD = 0.9503	RMSD = 0.0072	RMSD = 0.2508	RMSD = 1.1630
-0.74745360E+01	-0.74239479E+01	-0.74550781E+01	-0.74787884E+01
0.13435354E+01	0.12306687E+01	0.12902576E+01	0.13614511E+01
-0.32677881E+01	-0.31279391E+01	-0.31528388E+01	-0.33709390E+01
-0.27546337E+01	-0.29230028E+01	-0.31428313E+01	-0.13709105E+01

LIQUID DENSITY CORRELATION CONSTANTS, K AND G/ML:

FRANCIS1	FRANCIS1	FRANCIS1	FRANCIS2
163. - 530. K	163. - 364. K	163. - 502. K	501. - 541. K
36 POINTS	24 POINTS	33 POINTS	5 POINTS
RMSD = 0.00142	RMSD = 0.00029	RMSD = 0.00156	RMSD = 0.00176
0.96000803E+00	0.94956046E+00	0.96314865E+00	0.29523711E-03
0.75765885E-03	0.76951482E-03	0.74117235E-03	0.28637951E+01
0.12048149E+02	0.60056467E+01	0.15000000E+02	
0.59581030E+03	0.53145801E+03	0.61258276E+03	

SECOND VIRIAL COEFFICIENT CORRELATION CONSTANTS, ML/MOL:

TSONOPOULOS: TC=540.636 PC=2.891 VC=415.78 OM=0.3103 A= 0.000000 B=0.0000

LITERATURE DOCUMENTS FOR CORRELATED VAPOR PRESSURE VALUES:

    1486 6893 9050

LITERATURE DOCUMENTS FOR CORRELATED LIQUID DENSITY VALUES:

    1490 2986 9050

LITERATURE DOCUMENTS REPORTING VIRIAL COEFFICIENT DATA:

T,K	VAPOR PRESSURE, KPA	SATURATED LIQUID VOLUME, ML/MOL	SECOND VIRIAL COEFFICIENT, ML/MOL	HEAT OF VAPORIZATION, J/MOL	SATURATED VAPOR VOLUME, ML/MOL
163		123.91	-80548.		
171		124.93	-56922.		
180		126.10	-39642.		
188		127.16	-29448.		
197		128.38	-21634.		
205		129.48	-16813.		
214		130.76	-12961.		
223		132.07	-10225.		
231		133.26	-8432.		
240		134.63	-6916.		
248		135.87	-5887.		
257		137.32	-4987.		
266		138.80	-4288.		
274		140.15	-3789.		
283		141.71	-3334.		
291		143.15	-3001.		
300	8.457	144.81	-2689.	35064.	292233.
308	12.20	146.33	-2455.	34591.	207360.
317	17.95	148.09	-2231.	34057.	144529.
326	25.75	149.92	-2040.	33518.	103169.
334	34.80	151.61	-1893.	33033.	77851.
343	47.88	153.57	-1747.	32477.	57765.
351	62.54	155.39	-1634.	31973.	44967.
360	83.08	157.51	-1520.	31392.	34439.
369	108.6	159.73	-1419.	30793.	26759.
377	136.0	161.80	-1338.	30242.	21616.
386	173.1	164.23	-1256.	29600.	17190.
394	212.1	166.51	-1190.	29007.	14147.
403	263.7	169.21	-1122.	28309.	11464.
411	317.1	171.75	-1066.	27659.	9575.
420	386.7	174.79	-1009.	26887.	7874.
429	467.1	178.06	-956.3	26066.	6515.
437	548.7	181.18	-913.0	25286.	5528.
446	652.8	184.99	-867.8	24338.	4612.
454	757.2	188.68	-830.3	23419.	3933.
463	889.0	193.26	-791.1	22840.	3368.
472	1037.	198.39	-754.5	21740.	2838.
480	1184.	203.54	-724.0	20683.	2438.
489	1368.	210.20	-691.8	19385.	2054.
497	1548.	217.12	-664.9	18110.	1758.
506	1772.	226.46	-636.4	16499.	1469.
514	1992.	236.72	-612.4	14850.	1241.
523	2265.	251.59	-586.9	12632.	1009.
532	2568.		-562.9		788.4
540	2866.		-542.6		547.9

COMPOUND CONSTANTS:

MP 149.430 K	TC 520.500 K	VC 415.78 ML/MOL	ZC 0.2664
NBP 352.345 K	PC 2.773 MPA	DC 0.2410 G/ML	OM 0.2864
MU 0.0000 DEBYE	RG 4.0001 ANGSTROM		

VAPOR PRESSURE CORRELATION CONSTANTS, K AND KPA:

WAGNER	WAGNER	WAGNER	WAGNER
285. - 521. K	285. - 354. K	285. - 436. K	332. - 521. K
47 POINTS	36 POINTS	40 POINTS	27 POINTS
RMSD = 1.3500	RMSD = 0.0057	RMSD = 0.5432	RMSD = 0.6940
-0.73327672E+01	-0.75574598E+01	-0.71034130E+01	-0.74958750E+01
0.13059728E+01	0.17957034E+01	0.71518622E+00	0.18738954E+01
-0.32267503E+01	-0.37674483E+01	-0.21078123E+01	-0.52522640E+01
-0.14532542E+01	-0.12038030E+01	-0.47517828E+01	0.12756068E+02

LIQUID DENSITY CORRELATION CONSTANTS, K AND G/ML:

FRANCIS1	FRANCIS1	FRANCIS1	FRANCIS2
153. - 504. K	153. - 354. K	153. - 491. K	490. - 521. K
32 POINTS	26 POINTS	31 POINTS	3 POINTS
RMSD = 0.00080	RMSD = 0.00012	RMSD = 0.00070	RMSD = 0.00001
0.92736036E+00	0.92925358E+00	0.92650038E+00	0.38267387E-03
0.76346612E-03	0.81630168E-03	0.77673863E-03	0.27197151E+01
0.77306700E+01	0.59999990E+01	0.62852316E+01	
0.55655396E+03	0.66112524E+03	0.54695728E+03	

SECOND VIRIAL COEFFICIENT CORRELATION CONSTANTS, ML/MOL:

TSONOPOULOS: TC=520.500 PC=2.773 VC=415.78 OM=0.2864 A= 0.000000 B=0.0000

LITERATURE DOCUMENTS FOR CORRELATED VAPOR PRESSURE VALUES:

    1485   1486   5045   9050

LITERATURE DOCUMENTS FOR CORRELATED LIQUID DENSITY VALUES:

    1490   2986   9050   9714

LITERATURE DOCUMENTS REPORTING VIRIAL COEFFICIENT DATA:

T,K	VAPOR PRESSURE, KPA	SATURATED LIQUID VOLUME, ML/MOL	SECOND VIRIAL COEFFICIENT, ML/MOL	HEAT OF VAPORIZATION, J/MOL	SATURATED VAPOR VOLUME, ML/MOL
153		126.62	-91621.		
161		127.66	-63125.		
169		128.73	-44673.		
178		129.96	-31214.		
186		131.07	-23288.		
194		132.21	-17780.		
203		133.51	-13470.		
211		134.70	-10755.		
219		135.91	-8752.		
228		137.31	-7088.		
236		138.58	-5977.		
244		139.88	-5113.		
253		141.38	-4357.		
261		142.75	-3825.		
269		144.16	-3393.		
278		145.78	-2997.		
286	7.964	147.26	-2707.	32964.	295850.
294	11.63	148.79	-2462.	32545.	207614.
303	17.33	150.56	-2229.	32066.	143122.
311	24.14	152.18	-2052.	31630.	105037.
320	34.24	154.07	-1881.	31127.	75777.
328	45.85	155.81	-1748.	30667.	57678.
336	60.40	157.61	-1631.	30193.	44556.
345	80.92	159.71	-1514.	29640.	33863.
353	103.4	161.65	-1421.	29131.	26877.
361	130.5	163.67	-1338.	28601.	21567.
370	167.3	166.05	-1254.	27979.	17035.
378	206.2	168.27	-1186.	27401.	13944.
386	251.7	170.61	-1124.	26797.	11505.
395	311.6	173.38	-1060.	26081.	9345.
403	373.3	176.00	-1008.	25408.	7818.
411	443.8	178.79	-959.9	24695.	6575.
420	534.6	182.16	-910.0	23837.	5439.
428	626.3	185.40	-868.9	23013.	4611.
436	729.2	188.93	-830.5	22117.	3917.
445	859.6	193.32	-790.3	21526.	3339.
453	989.3	197.69	-757.0	20577.	2857.
462	1152.	203.30	-721.9	19421.	2399.
470	1313.	209.10	-692.7	18300.	2052.
478	1491.	215.95	-665.1	17068.	1751.
487	1712.	225.48	-635.9	15506.	1457.
495	1928.	236.41	-611.5	13898.	1226.
503	2166.	251.15	-588.3	11976.	1016.
512	2462.		-563.7		790.7
520	2754.		-543.0		542.5

COMPOUND CONSTANTS:

MP	138.140 K	TC	537.355 K	VC	406.00 ML/MOL	ZC	0.2645

MP   138.140 K   TC   537.355 K   VC   406.00 ML/MOL   ZC   0.2645
NBP   362.931 K   PC   2.911 MPA   DC   0.2468 G/ML   OM   0.2942
MU   0.0000 DEBYE   RG   3.9210 ANGSTROM

VAPOR PRESSURE CORRELATION CONSTANTS, K AND KPA:

WAGNER	WAGNER	WAGNER	WAGNER
208. - 537. K	208. - 364. K	208. - 442. K	349. - 537. K
47 POINTS	36 POINTS	39 POINTS	19 POINTS
RMSD = 4.4330	RMSD = 0.0093	RMSD = 1.5310	RMSD = 2.1460
-0.72873414E+01	-0.78557566E+01	-0.70757103E+01	-0.78426308E+01
0.11421019E+01	0.23683966E+01	0.66945129E+00	0.30588272E+01
-0.31875718E+01	-0.45636660E+01	-0.26079962E+01	-0.99426853E+01
-0.14019464E+01	0.67420641E-01	-0.21046442E+01	0.44842000E+02

LIQUID DENSITY CORRELATION CONSTANTS, K AND G/ML:

FRANCIS1	FRANCIS1	FRANCIS1	FRANCIS2
293. - 527. K	293. - 374. K	293. - 504. K	493. - 537. K
22 POINTS	8 POINTS	20 POINTS	4 POINTS
RMSD = 0.00069	RMSD = 0.00020	RMSD = 0.00057	RMSD = 0.00270
0.94559073E+00	0.95265055E+00	0.94530666E+00	0.25522811E-03
0.68847346E-03	0.81440387E-03	0.68845483E-03	0.29466599E+01
0.15000000E+02	0.60294313E+01	0.14924578E+02	
0.60152295E+03	0.61480811E+03	0.60132983E+03	

SECOND VIRIAL COEFFICIENT CORRELATION CONSTANTS, ML/MOL:

TSONOPOULOS: TC=537.355 PC=2.911 VC=406.00 OM=0.2942 A= 0.000000 B=0.0000

LITERATURE DOCUMENTS FOR CORRELATED VAPOR PRESSURE VALUES:

   1486   5718   9050

LITERATURE DOCUMENTS FOR CORRELATED LIQUID DENSITY VALUES:

   1490   4094

LITERATURE DOCUMENTS REPORTING VIRIAL COEFFICIENT DATA:

T,K	VAPOR PRESSURE, KPA	SATURATED LIQUID VOLUME, ML/MOL	SECOND VIRIAL COEFFICIENT, ML/MOL	HEAT OF VAPORIZATION, J/MOL	SATURATED VAPOR VOLUME, ML/MOL
208	0.01448		-14109.		0.1194E+09
215	0.02999		-11600.		0.5959E+08
222	0.05896		-9671.		0.3129E+08
230	0.1205		-7983.		0.1586E+08
237	0.2152		-6838.		9148616.
245	0.3985		-5808.		5105862.
252	0.6583		-5091.		3177793.
260	1.124		-4429.		1918411.
267	1.742		-3956.		1270635.
275	2.782		-3509.		818217.
282	4.087		-3184.		570488.
290	6.176		-2869.		387532.
297	8.676	144.83	-2635.	34282.	281975.
305	12.51	146.28	-2406.	33863.	200267.
312	16.93	147.58	-2231.	33489.	151001.
320	23.46	149.11	-2058.	33054.	111318.
327	30.74	150.50	-1924.	32664.	86479.
335	41.19	152.13	-1788.	32207.	65778.
342	52.52	153.61	-1683.	31796.	52401.
350	68.38	155.36	-1575.	31311.	40921.
357	85.16	156.94	-1489.	30873.	33295.
365	108.1	158.83	-1401.	30355.	26583.
372	132.0	160.55	-1331.	29885.	22017.
380	164.0	162.60	-1257.	29327.	17908.
387	196.7	164.48	-1198.	28819.	15054.
395	240.0	166.72	-1136.	28215.	12435.
402	283.4	168.79	-1086.	27662.	10582.
410	340.2	171.29	-1033.	26999.	8852.
417	396.5	173.60	-989.9	26390.	7607.
425	469.1	176.41	-944.0	25656.	6426.
432	540.4	179.03	-906.5	24974.	5563.
440	631.6	182.25	-866.4	24142.	4732.
447	720.3	185.30	-833.4	23359.	4115.
455	832.6	189.08	-798.0	22872.	3583.
462	941.2	192.70	-768.9	22052.	3135.
470	1078.	197.27	-737.5	21055.	2694.
477	1209.	201.73	-711.5	20121.	2360.
484	1352.	206.72	-686.9	19120.	2066.
492	1530.	213.24	-660.1	17877.	1771.
499	1700.	219.87	-638.0	16682.	1543.
507	1911.	228.86	-613.9	15157.	1310.
514	2113.	238.39	-593.9	13634.	1126.
522	2363.	252.03	-572.1	11575.	931.1
529	2600.		-553.9		764.8
537	2897.		-534.0		522.7

COMPOUND CONSTANTS:

MP   153.971 K   TC   519.790 K   VC   417.51 ML/MOL   ZC   0.2644
NBP   353.649 K   PC   2.737 MPA   DC   0.2400 G/ML   OM   0.3016
MU   0.0000 DEBYE   RG   3.9634 ANGSTROM

    VAPOR PRESSURE CORRELATION CONSTANTS, K AND KPA:

    WAGNER   VAPRES-2   WAGNER   WAGNER
    286. - 520. K   286. - 374. K   286. - 437. K   333. - 520. K
    35 POINTS   22 POINTS   27 POINTS   24 POINTS
    RMSD = 2.8180   RMSD = 0.0295   RMSD = 1.2710   RMSD = 1.7700
    -0.74447060E+01   0.11153549E+04   -0.69091093E+01   -0.77071180E+01
    0.15048825E+01   -0.27145403E+05   0.81670173E-01   0.24358652E+01
    -0.41665638E+01   0.62679051E+00   -0.12669044E+01   -0.76983321E+01
    0.24631765E+01   -0.32284039E-03   -0.68976948E+01   0.30006951E+02
                         -0.20708970E+03

    LIQUID DENSITY CORRELATION CONSTANTS, K AND G/ML:

    FRANCIS1   FRANCIS1   FRANCIS1   FRANCIS2
    273. - 505. K   273. - 374. K   273. - 484. K   473. - 520. K
    22 POINTS   11 POINTS   20 POINTS   4 POINTS
    RMSD = 0.00060   RMSD = 0.00051   RMSD = 0.00061   RMSD = 0.00006
    0.94119030E+00   0.95662403E+00   0.94036901E+00   0.38311861E-03
    0.83045685E-03   0.91487728E-03   0.81856456E-03   0.27388759E+01
    0.63011560E+01   0.60251541E+01   0.71560965E+01
    0.55334473E+03   0.69965552E+03   0.55896411E+03

    SECOND VIRIAL COEFFICIENT CORRELATION CONSTANTS, ML/MOL:

    TSONOPOULOS: TC=519.790 PC=2.737 VC=417.51 OM=0.3016 A= 0.000000 B=0.0000

LITERATURE DOCUMENTS FOR CORRELATED VAPOR PRESSURE VALUES:

    1486   1541   5045   9050

LITERATURE DOCUMENTS FOR CORRELATED LIQUID DENSITY VALUES:

    71   1490   4094   9050

LITERATURE DOCUMENTS REPORTING VIRIAL COEFFICIENT DATA:

T,K	VAPOR PRESSURE, KPA	SATURATED LIQUID VOLUME, ML/MOL	SECOND VIRIAL COEFFICIENT, ML/MOL	HEAT OF VAPORIZATION, J/MOL	SATURATED VAPOR VOLUME, ML/MOL
273		144.80	-3301.		
278		145.76	-3083.		
284		146.94	-2852.		
289	8.569	147.93	-2682.	33098.	277694.
295	11.36	149.15	-2499.	32843.	213303.
301	14.88	150.40	-2336.	32575.	165823.
306	18.46	151.46	-2214.	32340.	135600.
312	23.65	152.75	-2082.	32046.	107551.
317	28.85	153.86	-1982.	31790.	89343.
323	36.26	155.21	-1872.	31470.	72133.
329	45.15	156.60	-1772.	31134.	58753.
334	53.83	157.78	-1696.	30844.	49834.
340	65.94	159.23	-1611.	30482.	41193.
345	77.61	160.47	-1546.	30169.	35345.
351	93.69	161.99	-1474.	29779.	29599.
357	112.3	163.56	-1407.	29374.	24947.
362	129.9	164.91	-1355.	29025.	21734.
368	153.7	166.56	-1296.	28591.	18515.
373	176.0	167.99	-1251.	28217.	16265.
379	206.0	169.75	-1200.	27754.	13986.
385	239.7	171.57	-1152.	27273.	12081.
390	270.9	173.14	-1114.	26859.	10728.
396	312.2	175.09	-1071.	26345.	9335.
402	358.1	177.13	-1031.	25810.	8152.
407	400.1	178.89	-999.7	25348.	7299.
413	455.2	181.10	-963.7	24772.	6409.
418	505.3	183.02	-935.2	24271.	5762.
424	570.7	185.44	-902.7	23642.	5080.
430	642.1	188.00	-871.9	22979.	4485.
435	706.6	190.25	-847.4	22394.	4047.
441	790.1	193.12	-819.4	22132.	3654.
446	865.1	195.67	-797.1	21549.	3310.
452	962.0	198.95	-771.5	20818.	2943.
458	1067.	202.53	-747.0	20050.	2617.
463	1160.	205.78	-727.5	19377.	2373.
469	1281.	210.07	-705.1	18524.	2109.
474	1388.	214.05	-687.1	17769.	1910.
480	1525.	219.43	-666.5	16799.	1692.
486	1673.	225.67	-646.6	15742.	1495.
491	1805.	231.75	-630.7	14778.	1343.
497	1973.	240.51	-612.4	13489.	1174.
502	2123.	249.52	-597.6	12265.	1041.
508	2315.		-580.6		888.0
514	2521.		-564.2		731.8
519	2706.		-551.0		565.7

133

COMPOUND CONSTANTS:

MP   138.680 K   TC   536.405 K   VC   414.06 ML/MOL   ZC   0.2735
NBP 359.209 K   PC   2.946 MPA   DC   0.2420 G/ML   OM   0.2674
MU   0.0000 DEBYE   RG   3.7952 ANGSTROM

VAPOR PRESSURE CORRELATION CONSTANTS, K AND KPA:

WAGNER	VAPRES-2	WAGNER	WAGNER
286. - 536. K	286. - 375. K	286. - 448. K	345. - 536. K
56 POINTS	38 POINTS	44 POINTS	31 POINTS
RMSD = 1.2420	RMSD = 0.0084	RMSD = 0.1687	RMSD = 1.2000
-0.73843013E+01	-0.32467975E+03	-0.72947961E+01	-0.74382579E+01
0.15974612E+01	0.17791402E+04	0.13644877E+01	0.17869852E+01
-0.32926858E+01	-0.23581729E+00	-0.28473096E+01	-0.39977803E+01
-0.12582043E+01	0.12974410E-03	-0.25253046E+01	0.40373980E+01
	0.66675423E+02		

LIQUID DENSITY CORRELATION CONSTANTS, K AND G/ML:

FRANCIS1	FRANCIS1	FRANCIS1	FRANCIS2
293. - 524. K	293. - 362. K	293. - 498. K	490. - 536. K
14 POINTS	4 POINTS	11 POINTS	6 POINTS
RMSD = 0.00092	RMSD = 0.00031	RMSD = 0.00071	RMSD = 0.00024
0.98495698E+00	0.94720399E+00	0.98808336E+00	0.31299158E-03
0.92053902E-03	0.75665279E-03	0.93054445E-03	0.27888601E+01
0.59999990E+01	0.59999990E+01	0.59999990E+01	
0.57281689E+03	0.48241528E+03	0.57497070E+03	

SECOND VIRIAL COEFFICIENT CORRELATION CONSTANTS, ML/MOL:

TSONOPOULOS: TC=536.405 PC=2.946 VC=414.06 OM=0.2674 A= 0.000000 B=0.0000

LITERATURE DOCUMENTS FOR CORRELATED VAPOR PRESSURE VALUES:

    1485   1486   9050

LITERATURE DOCUMENTS FOR CORRELATED LIQUID DENSITY VALUES:

    1490   9050

LITERATURE DOCUMENTS REPORTING VIRIAL COEFFICIENT DATA:

T,K	VAPOR PRESSURE, KPA	SATURATED LIQUID VOLUME, ML/MOL	SECOND VIRIAL COEFFICIENT, ML/MOL	HEAT OF VAPORIZATION, J/MOL	SATURATED VAPOR VOLUME, ML/MOL
286	6.208		-2846.		380154.
291	7.922		-2675.		302713.
297	10.48	145.26	-2492.	33017.	233105.
303	13.69	146.54	-2330.	32711.	181720.
308	16.94	147.62	-2208.	32453.	148966.
314	21.64	148.95	-2076.	32139.	118517.
320	27.36	150.31	-1958.	31820.	95243.
325	33.01	151.46	-1867.	31550.	79951.
331	40.98	152.88	-1768.	31220.	65331.
337	50.43	154.33	-1678.	30884.	53825.
342	59.56	155.56	-1609.	30598.	46072.
348	72.20	157.08	-1532.	30248.	38481.
354	86.85	158.63	-1461.	29889.	32360.
359	100.8	159.96	-1406.	29584.	28145.
365	119.7	161.59	-1345.	29208.	23934.
371	141.2	163.27	-1288.	28823.	20467.
377	165.7	165.00	-1235.	28427.	17592.
382	188.4	166.49	-1194.	28088.	15565.
388	218.8	168.32	-1147.	27670.	13493.
394	252.7	170.21	-1103.	27238.	11745.
399	283.9	171.83	-1069.	26868.	10493.
405	325.2	173.85	-1030.	26409.	9196.
411	370.8	175.94	-992.8	25932.	8085.
416	412.3	177.74	-963.6	25521.	7278.
422	466.6	179.99	-930.4	25008.	6431.
428	526.2	182.34	-898.9	24472.	5696.
433	580.0	184.38	-873.9	24005.	5155.
439	649.8	186.94	-845.3	23416.	4580.
445	725.8	189.64	-818.1	22792.	4074.
451	808.1	192.50	-792.3	22124.	3626.
456	882.0	195.01	-771.6	22012.	3360.
462	977.1	198.22	-747.9	21347.	3004.
468	1080.	201.66	-725.3	20648.	2687.
473	1171.	204.73	-707.2	20036.	2449.
479	1289.	208.73	-686.3	19260.	2191.
485	1415.	213.12	-666.3	18432.	1958.
490	1527.	217.15	-650.3	17697.	1781.
496	1670.	222.54	-631.8	16748.	1587.
502	1824.	228.71	-614.0	15712.	1409.
507	1961.	234.62	-599.7	14765.	1272.
513	2135.	242.97	-583.2	13498.	1118.
519	2322.	253.22	-567.3	12037.	973.1
525	2522.		-552.0		832.8
530	2700.		-539.7		713.1
536	2930.		-525.4		522.1

COMPOUND CONSTANTS:

MP	248.496 K	TC	531.123 K	VC	397.63 ML/MOL	ZC	0.2659
NBP	354.023 K	PC	2.953 MPA	DC	0.2520 G/ML	OM	0.2500
MU	0.0000 DEBYE	RG	3.6960 ANGSTROM			

VAPOR PRESSURE CORRELATION CONSTANTS, K AND KPA:

WAGNER	RIEDEL	VAPRES-2	WAGNER
285. - 531. K	285. - 373. K	285. - 447. K	336. - 531. K
57 POINTS	39 POINTS	46 POINTS	38 POINTS
RMSD = 1.3160	RMSD = 0.0112	RMSD = 0.1224	RMSD = 0.4534
-0.72719872E+01	0.54279445E+02	0.30168165E+03	-0.74006645E+01
0.15669030E+01	-0.57866035E+01	-0.11057104E+05	0.20046013E+01
-0.33382194E+01	-0.55813064E+04	0.12289000E+00	-0.48820759E+01
-0.26690417E+00	0.34508435E-16	-0.50737084E-04	0.10776775E+02
		-0.51620175E+02	

LIQUID DENSITY CORRELATION CONSTANTS, K AND G/ML:

FRANCIS1	FRANCIS1	FRANCIS1	FRANCIS2
273. - 514. K	273. - 374. K	273. - 491. K	481. - 531. K
24 POINTS	13 POINTS	23 POINTS	5 POINTS
RMSD = 0.00306	RMSD = 0.00038	RMSD = 0.00306	RMSD = 0.00053
0.92777890E+00	0.94173783E+00	0.92313874E+00	0.37089892E-03
0.63783978E-03	0.80834609E-03	0.62018586E-03	0.27422596E+01
0.15000000E+02	0.59999990E+01	0.15000000E+02	
0.59222827E+03	0.69965552E+03	0.58847363E+03	

SECOND VIRIAL COEFFICIENT CORRELATION CONSTANTS, ML/MOL:

TSONOPOULOS: TC=531.123 PC=2.953 VC=397.63 OM=0.2500 A= 0.000000 B=0.0000

LITERATURE DOCUMENTS FOR CORRELATED VAPOR PRESSURE VALUES:

   1486 6656 6893 9050

LITERATURE DOCUMENTS FOR CORRELATED LIQUID DENSITY VALUES:

   1487 1490 8639 9050 9706 10808

LITERATURE DOCUMENTS REPORTING VIRIAL COEFFICIENT DATA:

T,K	VAPOR PRESSURE, KPA	SATURATED LIQUID VOLUME, ML/MOL	SECOND VIRIAL COEFFICIENT, ML/MOL	HEAT OF VAPORIZATION, J/MOL	SATURATED VAPOR VOLUME, ML/MOL
273		141.80	-3134.		
278		142.59	-2928.		
284		143.56	-2711.		
290	9.447	144.56	-2520.	32275.	252686.
296	12.42	145.57	-2351.	31998.	195824.
302	16.12	146.61	-2201.	31715.	153561.
308	20.68	147.67	-2068.	31425.	121752.
314	26.23	148.76	-1947.	31128.	97528.
319	31.74	149.69	-1856.	30874.	81664.
325	39.54	150.83	-1757.	30563.	66542.
331	48.80	152.00	-1666.	30243.	54679.
337	59.71	153.20	-1583.	29914.	45287.
343	72.47	154.44	-1507.	29576.	37785.
349	87.28	155.72	-1438.	29228.	31742.
355	104.4	157.03	-1373.	28869.	26837.
361	123.9	158.39	-1313.	28500.	22826.
366	142.3	159.55	-1267.	28184.	20030.
372	167.1	161.00	-1215.	27795.	17206.
378	195.0	162.49	-1166.	27392.	14852.
384	226.4	164.05	-1121.	26977.	12878.
390	261.4	165.67	-1078.	26549.	11212.
396	300.4	167.35	-1038.	26105.	9800.
402	343.6	169.11	-1000.	25645.	8595.
407	383.1	170.65	-970.5	25249.	7724.
413	434.8	172.57	-936.6	24755.	6812.
419	491.5	174.59	-904.5	24241.	6024.
425	553.5	176.73	-874.2	23703.	5338.
431	621.3	178.99	-845.4	23137.	4739.
437	694.9	181.39	-818.0	22540.	4214.
443	774.9	183.96	-791.9	21904.	3749.
449	861.4	186.72	-767.1	21663.	3403.
454	938.9	189.18	-747.4	21130.	3099.
460	1039.	192.35	-724.6	20462.	2772.
466	1146.	195.82	-702.9	19758.	2481.
472	1261.	199.63	-682.1	19013.	2219.
478	1385.	203.84	-662.2	18222.	1985.
484	1518.	208.55	-643.1	17375.	1773.
490	1661.	213.88	-624.8	16461.	1580.
495	1788.	218.89	-610.1	15637.	1433.
501	1949.	225.78	-593.1	14557.	1269.
507	2122.	233.90	-576.8	13347.	1117.
513	2307.	243.66	-561.1	11960.	974.1
519	2505.		-546.0		836.4
525	2718.		-531.4		695.6
531	2948.		-517.4		494.7

137

COMPOUND CONSTANTS:

```
MP   216.375 K        TC  568.841 K        VC  492.00 ML/MOL    ZC  0.2588
NBP  398.823 K        PC    2.488 MPA      DC  0.2322 G/ML      OM  0.3979
MU     0.0000 DEBYE   RG    4.6804 ANGSTROM
```

VAPOR PRESSURE CORRELATION CONSTANTS, K AND KPA:

WAGNER	VAPRES-2	RIEDEL	WAGNER
216. - 569. K	216. - 400. K	216. - 484. K	384. - 569. K
87 POINTS	54 POINTS	69 POINTS	44 POINTS
RMSD = 2.1670	RMSD = 0.2820	RMSD = 0.4319	RMSD = 0.4388
-0.81621949E+01	-0.10870845E+04	0.69033523E+02	-0.80052875E+01
0.21052126E+01	0.15527746E+05	-0.77048846E+01	0.17207793E+01
-0.54163890E+01	-0.69846314E+00	-0.73155167E+04	-0.51529172E+01
-0.15830507E+00	0.36523232E-03	0.17543260E-16	0.66244302E+01
	0.21261327E+03		

LIQUID DENSITY CORRELATION CONSTANTS, K AND G/ML:

FRANCIS1	FRANCIS1	FRANCIS1	FRANCIS2
223. - 554. K	223. - 419. K	223. - 539. K	528. - 569. K
234 POINTS	203 POINTS	228 POINTS	10 POINTS
RMSD = 0.00040	RMSD = 0.00041	RMSD = 0.00039	RMSD = 0.00080
0.94450295E+00	0.93760902E+00	0.94463199E+00	0.34038144E-03
0.69548492E-03	0.72470121E-03	0.69144624E-03	0.27593496E+01
0.12603612E+02	0.60558453E+01	0.13132452E+02	
0.62403296E+03	0.56036475E+03	0.62641626E+03	

SECOND VIRIAL COEFFICIENT CORRELATION CONSTANTS, ML/MOL:

TSONOPOULOS: TC=568.841 PC=2.488 VC=492.00 OM=0.3979 A= 0.000000 B=0.0000

LITERATURE DOCUMENTS FOR CORRELATED VAPOR PRESSURE VALUES:

```
   707  1146  1485  1540  2097  2981  4816  4844  8647  9050 10719 11265 17992
18208 40865
```

LITERATURE DOCUMENTS FOR CORRELATED LIQUID DENSITY VALUES:

```
   694  1487  1540  1769  2835  2959  2981  5572  8149  8580  8902  9050  9709
10430 10592 10719 13714 13803 14407 14487 15688 16166 16589 17615 18355 18474
18913 19531 20436 21304 21313 21607 21835 21870 40895 41226
```

LITERATURE DOCUMENTS REPORTING VIRIAL COEFFICIENT DATA:

```
1938  2409 17612
```

T,K	VAPOR PRESSURE, KPA	SATURATED LIQUID VOLUME, ML/MOL	SECOND VIRIAL COEFFICIENT, ML/MOL	HEAT OF VAPORIZATION, J/MOL	SATURATED VAPOR VOLUME, ML/MOL
216	0.001963				
224	0.004956	150.86	-19432.	46336.	0.3758E+09
232	0.01164	152.10	-15760.	45855.	0.1658E+09
240	0.02560	153.38	-12979.	45371.	0.7796E+08
248	0.05309	154.68	-10843.	44887.	0.3884E+08
256	0.1044	156.01	-9179.	44402.	0.2038E+08
264	0.1959	157.37	-7867.	43918.	0.1121E+08
272	0.3517	158.76	-6819.	43436.	6430816.
280	0.6070	160.18	-5971.	42955.	3835590.
288	1.010	161.64	-5278.	42478.	2369677.
296	1.628	163.13	-4705.	41897.	1507959.
304	2.545	164.67	-4226.	41385.	989681.
312	3.869	166.24	-3822.	40866.	667035.
320	5.737	167.86	-3478.	40338.	460639.
328	8.309	169.53	-3182.	39801.	325264.
336	11.78	171.24	-2927.	39252.	234402.
344	16.37	173.01	-2703.	38692.	172107.
352	22.34	174.83	-2507.	38119.	128550.
360	29.97	176.72	-2334.	37533.	97536.
368	39.58	178.67	-2179.	36931.	75079.
376	51.53	180.69	-2041.	36315.	58561.
384	66.18	182.79	-1917.	35681.	46231.
392	83.95	184.97	-1804.	35032.	36907.
400	105.3	187.25	-1702.	34367.	29764.
408	130.6	189.62	-1609.	33686.	24228.
416	160.4	192.11	-1524.	32987.	19890.
424	195.2	194.73	-1446.	32271.	16456.
432	235.5	197.49	-1374.	31536.	13710.
440	281.8	200.40	-1307.	30782.	11495.
448	334.8	203.50	-1245.	30007.	9693.
456	394.9	206.81	-1188.	29209.	8215.
464	462.9	210.35	-1134.	28386.	6993.
472	539.5	214.18	-1084.	27532.	5975.
480	625.2	218.34	-1038.	26644.	5121.
488	720.9	222.91	-993.7	25714.	4400.
496	827.3	227.96	-952.4	24733.	3787.
504	945.3	233.61	-913.7	23689.	3262.
512	1076.	240.02	-877.1	22565.	2809.
520	1220.	247.41	-842.5	21336.	2414.
528	1379.	256.10	-809.9	19966.	2068.
536	1555.	266.55	-779.0	18399.	1759.
544	1748.	279.51	-749.6	16536.	1479.
552	1961.	296.22	-721.8		1448.
560	2196.		-695.3		1220.
568	2458.		-670.1		959.3

COMPOUND CONSTANTS:

MP 164.124 K	TC 559.639 K	VC 487.75 ML/MOL	ZC 0.2604
NBP 390.783 K	PC 2.484 MPA	DC 0.2342 G/ML	OM 0.3777
MU 0.0000 DEBYE	RG 4.7401 ANGSTROM		

VAPOR PRESSURE CORRELATION CONSTANTS, K AND KPA:

WAGNER	VAPRES-2	RPM2	WAGNER
233. - 560. K	233. - 392. K	233. - 474. K	373. - 560. K
65 POINTS	51 POINTS	56 POINTS	22 POINTS
RMSD = 0.7142	RMSD = 0.0241	RMSD = 0.2040	RMSD = 0.9684
-0.78689374E+01	-0.22695519E+03	0.29126567E+02	-0.78929384E+01
0.15801504E+01	-0.12702072E+04	-0.63137460E+04	0.17117887E+01
-0.43377549E+01	-0.18539148E+00	-0.29427080E-01	-0.51505980E+01
-0.12214105E+01	0.10267690E-03	0.20620811E-04	0.69676253E+01
	0.48858015E+02		

LIQUID DENSITY CORRELATION CONSTANTS, K AND G/ML:

FRANCIS1	FRANCIS1	FRANCIS1	FRANCIS2
273. - 544. K	273. - 324. K	273. - 524. K	513. - 560. K
15 POINTS	8 POINTS	13 POINTS	5 POINTS
RMSD = 0.00061	RMSD = 0.00041	RMSD = 0.00033	RMSD = 0.00123
0.94253016E+00	0.94858950E+00	0.94593406E+00	0.36758921E-03
0.71759755E-03	0.78828074E-03	0.76430640E-03	0.27236619E+01
0.10625784E+02	0.59999990E+01	0.69231071E+01	
0.60694360E+03	0.60588574E+03	0.58774756E+03	

SECOND VIRIAL COEFFICIENT CORRELATION CONSTANTS, ML/MOL:

TSONOPOULOS: TC=559.639 PC=2.484 VC=487.75 OM=0.3777 A= 0.000000 B=0.0000

LITERATURE DOCUMENTS FOR CORRELATED VAPOR PRESSURE VALUES:

   1146   1485   5718   7333   9050

LITERATURE DOCUMENTS FOR CORRELATED LIQUID DENSITY VALUES:

   1487   1769   8149   9050 12709

LITERATURE DOCUMENTS REPORTING VIRIAL COEFFICIENT DATA:

T,K	VAPOR PRESSURE, KPA	SATURATED LIQUID VOLUME, ML/MOL	SECOND VIRIAL COEFFICIENT, ML/MOL	HEAT OF VAPORIZATION, J/MOL	SATURATED VAPOR VOLUME, ML/MOL
233	0.02450		-13305.		0.7904E+08
240	0.04710		-11287.		0.4235E+08
247	0.08671		-9680.		0.2367E+08
255	0.1660		-8224.		0.1276E+08
262	0.2821		-7203.		7714651.
270	0.4969		-6257.		4511070.
277	0.7901	160.54	-5578.	40909.	2909257.
284	1.223	161.84	-5009.	40485.	1926301.
292	1.953	163.36	-4466.	40000.	1238543.
299	2.871	164.72	-4065.	39575.	861714.
307	4.347	166.32	-3674.	39086.	583524.
314	6.119	167.76	-3381.	38655.	423216.
322	8.851	169.44	-3091.	38157.	299358.
329	12.01	170.95	-2870.	37715.	224858.
336	16.05	172.50	-2674.	37268.	171354.
344	21.97	174.33	-2476.	36747.	127673.
351	28.51	175.97	-2323.	36283.	99998.
359	37.82	177.91	-2166.	35740.	76686.
366	47.86	179.67	-2043.	35254.	61474.
374	61.82	181.74	-1915.	34685.	48309.
381	76.52	183.62	-1814.	34174.	39495.
388	93.86	185.57	-1722.	33648.	32552.
396	117.3	187.88	-1625.	33030.	26338.
403	141.3	189.99	-1548.	32472.	22043.
411	173.3	192.51	-1466.	31813.	18123.
418	205.6	194.82	-1400.	31215.	15362.
425	242.4	197.24	-1339.	30597.	13089.
433	290.3	200.15	-1273.	29860.	10961.
440	337.9	202.85	-1220.	29186.	9426.
448	399.2	206.12	-1164.	28378.	7968.
455	459.5	209.18	-1117.	27631.	6899.
463	536.6	212.93	-1068.	26724.	5869.
470	611.7	216.48	-1027.	25872.	5103.
477	694.5	220.32	-988.6	24952.	4439.
485	799.2	225.15	-947.2	24528.	3892.
492	900.2	229.83	-912.9	23595.	3407.
500	1027.	235.83	-876.0	22453.	2926.
507	1149.	241.81	-845.3	21375.	2558.
515	1302.	249.74	-812.2	20032.	2190.
522	1449.	257.94	-784.6	18734.	1905.
529	1608.	267.77	-758.3	17284.	1648.
537	1806.	281.87	-729.8	15371.	1382.
544	1996.	297.98	-706.0	13360.	1167.
552	2233.		-680.0		926.5
559	2462.		-658.3		662.5

COMPOUND CONSTANTS:

MP 152.653 K	TC 563.670 K	VC 463.78 ML/MOL	ZC 0.2519
NBP 392.073 K	PC 2.546 MPA	DC 0.2463 G/ML	OM 0.3701
MU 0.0000 DEBYE	RG 4.5932 ANGSTROM		

VAPOR PRESSURE CORRELATION CONSTANTS, K AND KPA:

WAGNER	RPM2	VAPRES-2	WAGNER
238. - 564. K	238. - 399. K	238. - 474. K	373. - 564. K
72 POINTS	55 POINTS	60 POINTS	26 POINTS
RMSD = 0.4152	RMSD = 0.0117	RMSD = 0.0626	RMSD = 0.6663
-0.78265140E+01	0.30914794E+02	0.87964731E+02	-0.78314080E+01
0.15217083E+01	-0.65304167E+04	-0.76488535E+04	0.15419131E+01
-0.40245536E+01	-0.34666001E-01	0.40038406E-03	-0.41220181E+01
-0.22330712E+01	0.25706587E-04	0.70782369E-05	-0.13552276E+01
		-0.10898977E+02	

LIQUID DENSITY CORRELATION CONSTANTS, K AND G/ML:

FRANCIS1	FRANCIS1	FRANCIS1	FRANCIS2
273. - 553. K	273. - 399. K	273. - 534. K	523. - 564. K
24 POINTS	12 POINTS	22 POINTS	5 POINTS
RMSD = 0.00058	RMSD = 0.00032	RMSD = 0.00052	RMSD = 0.00037
0.94242412E+00	0.93919325E+00	0.94123191E+00	0.35590123E-03
0.70508267E-03	0.68480754E-03	0.68544992E-03	0.27501950E+01
0.94656153E+01	0.98918819E+01	0.11196142E+02	
0.60735596E+03	0.59400049E+03	0.61521460E+03	

SECOND VIRIAL COEFFICIENT CORRELATION CONSTANTS, ML/MOL:

TSONOPOULOS: TC=563.670 PC=2.546 VC=463.78 OM=0.3701 A= 0.000000 B=0.0000

LITERATURE DOCUMENTS FOR CORRELATED VAPOR PRESSURE VALUES:

1146   1485   5718   7333   9050 10804

LITERATURE DOCUMENTS FOR CORRELATED LIQUID DENSITY VALUES:

1487   1769   9050 14373

LITERATURE DOCUMENTS REPORTING VIRIAL COEFFICIENT DATA:

T,K	VAPOR PRESSURE, KPA	SATURATED LIQUID VOLUME, ML/MOL	SECOND VIRIAL COEFFICIENT, ML/MOL	HEAT OF VAPORIZATION, J/MOL	SATURATED VAPOR VOLUME, ML/MOL
238	0.03617		-11920.		0.5470E+08
245	0.06786		-10180.		0.3001E+08
252	0.1222		-8786.		0.1714E+08
260	0.2283		-7514.		9459570.
267	0.3805		-6616.		5827509.
275	0.6566	158.64	-5779.	41302.	3476720.
282	1.026	159.87	-5175.	40845.	2279120.
289	1.563	161.13	-4666.	40393.	1532273.
297	2.456	162.60	-4177.	39880.	1001298.
304	3.561	163.92	-3815.	39434.	705893.
312	5.313	165.46	-3460.	38924.	484742.
319	7.392	166.84	-3193.	38478.	355589.
326	10.11	168.25	-2958.	38031.	265140.
334	14.18	169.91	-2724.	37516.	193096.
341	18.77	171.40	-2543.	37062.	148497.
349	25.43	173.15	-2361.	36536.	111714.
356	32.72	174.73	-2218.	36069.	88183.
363	41.62	176.35	-2089.	35595.	70362.
371	54.06	178.26	-1957.	35042.	55026.
378	67.24	179.99	-1852.	34547.	44812.
386	85.27	182.04	-1743.	33969.	35805.
393	104.0	183.90	-1656.	33449.	29666.
400	125.8	185.82	-1576.	32915.	24755.
408	154.9	188.10	-1492.	32286.	20290.
415	184.4	190.19	-1424.	31717.	17158.
423	223.2	192.69	-1352.	31044.	14260.
430	262.1	194.98	-1294.	30432.	12192.
437	306.0	197.38	-1240.	29795.	10469.
445	362.7	200.28	-1182.	29034.	8837.
452	418.6	202.97	-1134.	28333.	7647.
460	490.1	206.25	-1084.	27486.	6502.
467	560.1	209.32	-1042.	26696.	5655.
474	637.3	212.63	-1003.	25849.	4925.
482	735.1	216.73	-960.7	25399.	4304.
489	829.6	220.66	-925.9	24525.	3770.
497	948.6	225.65	-888.3	23461.	3242.
504	1063.	230.54	-857.2	22463.	2842.
511	1188.	236.07	-827.6	21392.	2488.
519	1344.	243.38	-795.5	20056.	2133.
526	1493.	250.94	-768.8	18765.	1857.
534	1679.	261.47	-739.9	17105.	1575.
541	1857.	273.06	-715.7	15427.	1351.
548	2050.	288.05	-692.7	13435.	1142.
556	2291.		-667.5		908.5
563	2523.		-646.5		652.3

143

COMPOUND CONSTANTS:

MP 152.197 K	TC 561.740 K	VC 475.96 ML/MOL	ZC 0.2590
NBP 390.868 K	PC 2.542 MPA	DC 0.2400 G/ML	OM 0.3708
MU 0.0000 DEBYE	RG 4.5581 ANGSTROM		

VAPOR PRESSURE CORRELATION CONSTANTS, K AND KPA:

WAGNER	WAGNER	VAPRES-2	WAGNER
252. - 562. K	252. - 392. K	252. - 474. K	376. - 562. K
43 POINTS	27 POINTS	31 POINTS	20 POINTS
RMSD = 1.6770	RMSD = 0.0091	RMSD = 0.1601	RMSD = 1.4760
-0.77373260E+01	-0.76107100E+01	0.21591734E+03	-0.78886882E+01
0.12760601E+01	0.96623825E+00	-0.10593078E+05	0.17926392E+01
-0.35175317E+01	-0.30295202E+01	0.62078163E-01	-0.53046087E+01
-0.41078726E+01	-0.50146999E+01	-0.19561906E-04	0.80125113E+01
		-0.34426932E+02	

LIQUID DENSITY CORRELATION CONSTANTS, K AND G/ML:

FRANCIS1	FRANCIS1	FRANCIS1	FRANCIS2
273. - 534. K	273. - 324. K	273. - 524. K	513. - 562. K
11 POINTS	4 POINTS	10 POINTS	4 POINTS
RMSD = 0.00048	RMSD = 0.00035	RMSD = 0.00025	RMSD = 0.00086
0.93752861E+00	0.91970289E+00	0.93835384E+00	0.38042494E-03
0.64010872E-03	0.62917708E-03	0.67403866E-03	0.27143385E+01
0.15000000E+02	0.60230255E+01	0.11454981E+02	
0.62223120E+03	0.48911304E+03	0.60772266E+03	

SECOND VIRIAL COEFFICIENT CORRELATION CONSTANTS, ML/MOL:

TSONOPOULOS: TC=561.740 PC=2.542 VC=475.96 OM=0.3708 A= 0.000000 B=0.0000

LITERATURE DOCUMENTS FOR CORRELATED VAPOR PRESSURE VALUES:

1146   1485   9050

LITERATURE DOCUMENTS FOR CORRELATED LIQUID DENSITY VALUES:

1487   1769   9050

LITERATURE DOCUMENTS REPORTING VIRIAL COEFFICIENT DATA:

T,K	VAPOR PRESSURE, KPA	SATURATED LIQUID VOLUME, ML/MOL	SECOND VIRIAL COEFFICIENT, ML/MOL	HEAT OF VAPORIZATION, J/MOL	SATURATED VAPOR VOLUME, ML/MOL
252	0.1254		-8628.		0.1671E+08
259	0.2185		-7524.		9846507.
266	0.3675		-6621.		6011044.
273	0.5981	158.69	-5876.	41680.	3788953.
280	0.9447	159.88	-5254.	41153.	2459070.
287	1.452	161.10	-4731.	40637.	1639120.
294	2.175	162.34	-4287.	40132.	1119578.
301	3.184	163.62	-3907.	39635.	782013.
308	4.563	164.92	-3580.	39147.	557553.
315	6.413	166.26	-3296.	38665.	405084.
322	8.849	167.63	-3048.	38187.	299454.
329	12.01	169.03	-2830.	37712.	224926.
336	16.04	170.48	-2637.	37239.	171447.
343	21.13	171.97	-2465.	36766.	132463.
350	27.45	173.50	-2311.	36292.	103630.
357	35.23	175.08	-2173.	35814.	82011.
364	44.70	176.71	-2048.	35332.	65596.
371	56.09	178.39	-1934.	34844.	52984.
378	69.69	180.13	-1831.	34348.	43186.
385	85.77	181.94	-1736.	33842.	35495.
392	104.6	183.81	-1650.	33325.	29399.
399	126.6	185.76	-1570.	32796.	24523.
406	152.0	187.79	-1496.	32252.	20589.
413	181.3	189.92	-1427.	31691.	17390.
420	214.6	192.14	-1364.	31111.	14768.
427	252.5	194.48	-1305.	30509.	12604.
435	301.9	197.30	-1242.	29791.	10573.
442	350.9	199.92	-1190.	29132.	9104.
449	405.7	202.70	-1142.	28440.	7866.
456	466.7	205.67	-1097.	27709.	6815.
463	534.5	208.85	-1055.	26932.	5918.
470	609.5	212.27	-1014.	26100.	5148.
477	692.2	215.97	-976.6	25201.	4480.
484	783.2	220.01	-940.8	24891.	4001.
491	882.8	224.45	-906.8	23975.	3503.
498	991.9	229.37	-874.6	22999.	3068.
505	1111.	234.89	-844.0	21954.	2685.
512	1240.	241.14	-814.9	20824.	2348.
519	1381.	248.32	-787.3	19594.	2048.
526	1534.	256.72	-760.9	18238.	1780.
533	1700.	266.74	-735.7	16720.	1538.
540	1880.		-711.6		1317.
547	2075.		-688.6		1109.
554	2286.		-666.6		905.6
561	2516.		-645.6		653.4

COMPOUND CONSTANTS:

MP	TC 565.490 K	VC 455.10 ML/MOL	ZC 0.2524
NBP 391.692 K	PC 2.608 MPA	DC 0.2510 G/ML	OM 0.3609
MU 0.0000 DEBYE	RG		

VAPOR PRESSURE CORRELATION CONSTANTS, K AND KPA:

WAGNER	WAGNER	WAGNER	WAGNER
253. - 566. K	253. - 393. K	253. - 474. K	377. - 566. K
42 POINTS	26 POINTS	31 POINTS	20 POINTS
RMSD = 1.1020	RMSD = 0.0090	RMSD = 0.1219	RMSD = 0.6742
-0.76925373E+01	-0.76271693E+01	-0.75997596E+01	-0.77944077E+01
0.12649149E+01	0.11006079E+01	0.10410928E+01	0.16150052E+01
-0.33357926E+01	-0.30634440E+01	-0.29944624E+01	-0.45840161E+01
-0.46552181E+01	-0.51798705E+01	-0.52615188E+01	0.43775456E+01

LIQUID DENSITY CORRELATION CONSTANTS, K AND G/ML:

FRANCIS1	FRANCIS1	FRANCIS1	FRANCIS2
273. - 554. K	273. - 399. K	273. - 534. K	523. - 566. K
18 POINTS	8 POINTS	16 POINTS	5 POINTS
RMSD = 0.00078	RMSD = 0.00015	RMSD = 0.00057	RMSD = 0.00216
0.94684452E+00	0.94977880E+00	0.94474781E+00	0.35107381E-03
0.71059261E-03	0.73694251E-03	0.66948263E-03	0.27964448E+01
0.78373184E+01	0.65273972E+01	0.11616555E+02	
0.60126196E+03	0.60901318E+03	0.62016089E+03	

SECOND VIRIAL COEFFICIENT CORRELATION CONSTANTS, ML/MOL:

TSONOPOULOS: TC=565.490 PC=2.608 VC=455.10 OM=0.3609 A= 0.000000 B=0.0000

LITERATURE DOCUMENTS FOR CORRELATED VAPOR PRESSURE VALUES:

1146   1485   9050

LITERATURE DOCUMENTS FOR CORRELATED LIQUID DENSITY VALUES:

1487   1769   8148   9050

LITERATURE DOCUMENTS REPORTING VIRIAL COEFFICIENT DATA:

T,K	VAPOR PRESSURE, KPA	SATURATED LIQUID VOLUME, ML/MOL	SECOND VIRIAL COEFFICIENT, ML/MOL	HEAT OF VAPORIZATION, J/MOL	SATURATED VAPOR VOLUME, ML/MOL
253	0.1316		-8440.		0.1598E+08
260	0.2287		-7365.		9443904.
267	0.3835		-6484.		5781773.
274	0.6224	156.87	-5757.	41728.	3654721.
281	0.9801	158.06	-5150.	41182.	2378512.
288	1.502	159.28	-4639.	40650.	1589699.
295	2.244	160.52	-4205.	40130.	1088664.
302	3.277	161.79	-3835.	39622.	762344.
309	4.685	163.08	-3515.	39124.	544855.
316	6.568	164.40	-3237.	38633.	396785.
324	9.452	165.95	-2963.	38081.	282022.
331	12.77	167.34	-2753.	37602.	212746.
338	16.99	168.76	-2567.	37127.	162817.
345	22.28	170.21	-2402.	36653.	126272.
352	28.85	171.70	-2253.	36180.	99136.
359	36.90	173.24	-2120.	35704.	78716.
366	46.66	174.81	-1999.	35225.	63156.
373	58.38	176.43	-1890.	34741.	51162.
380	72.32	178.10	-1790.	34251.	41815.
387	88.78	179.81	-1698.	33752.	34456.
395	111.1	181.85	-1603.	33169.	27873.
402	133.9	183.69	-1526.	32647.	23325.
409	160.3	185.60	-1455.	32110.	19643.
416	190.5	187.59	-1390.	31557.	16637.
423	225.0	189.66	-1329.	30986.	14166.
430	264.0	191.81	-1272.	30394.	12120.
437	308.1	194.07	-1219.	29777.	10414.
444	357.5	196.44	-1169.	29133.	8982.
451	412.8	198.93	-1122.	28457.	7773.
458	474.3	201.58	-1078.	27743.	6746.
466	552.8	204.80	-1031.	26873.	5752.
473	629.3	207.83	-992.2	26055.	5012.
480	713.5	211.10	-955.6	25171.	4370.
487	806.0	214.64	-921.0	24871.	3908.
494	907.3	218.52	-888.2	23973.	3427.
501	1018.	222.81	-857.0	23016.	3006.
508	1139.	227.63	-827.4	21992.	2636.
515	1270.	233.11	-799.2	20888.	2309.
522	1412.	239.46	-772.4	19687.	2018.
529	1567.	246.99	-746.8	18364.	1757.
537	1760.	257.68	-719.0	16657.	1489.
544	1943.	269.78	-695.8	14925.	1276.
551	2142.	286.02	-673.6	12848.	1076.
558	2357.		-652.4		878.4
565	2591.		-632.0		625.1

COMPOUND CONSTANTS:

MP   151.972 K   TC   549.868 K   VC   477.95 ML/MOL   ZC   0.2644
NBP 379.986 K   PC   2.527 MPA   DC   0.2390 G/ML   OM   0.3363
MU   0.0000 DEBYE   RG   4.4956 ANGSTROM

VAPOR PRESSURE CORRELATION CONSTANTS, K AND KPA:

RPM2	VAPRES-2	WAGNER	WAGNER
243. - 550. K	243. - 380. K	243. - 464. K	365. - 550. K
38 POINTS	23 POINTS	29 POINTS	18 POINTS
RMSD = 0.6529	RMSD = 0.0088	RMSD = 0.1484	RMSD = 0.6065
0.27554687E+02	-0.32429492E+03	-0.77383832E+01	-0.74755221E+01
-0.58871392E+04	0.11222349E+04	0.17509109E+01	0.98425339E+00
-0.26718545E-01	-0.23780691E+00	-0.44154525E+01	-0.22240690E+01
0.18780640E-04	0.12861024E-03	-0.48584799E+00	-0.17682860E+02
	0.66960311E+02		

LIQUID DENSITY CORRELATION CONSTANTS, K AND G/ML:

FRANCIS1	FRANCIS1	FRANCIS1	FRANCIS2
273. - 539. K	273. - 394. K	273. - 506. K	505. - 550. K
17 POINTS	5 POINTS	14 POINTS	5 POINTS
RMSD = 0.00113	RMSD = 0.00007	RMSD = 0.00098	RMSD = 0.00092
0.94219083E+00	0.92412806E+00	0.94245440E+00	0.35842327E-03
0.70201652E-03	0.69563673E-03	0.68437890E-03	0.27244778E+01
0.12935048E+02	0.59999990E+01	0.14999969E+02	
0.60595776E+03	0.53219702E+03	0.61428906E+03	

SECOND VIRIAL COEFFICIENT CORRELATION CONSTANTS, ML/MOL:

TSONOPOULOS: TC=549.868 PC=2.527 VC=477.95 OM=0.3363 A= 0.000000 B=0.0000

LITERATURE DOCUMENTS FOR CORRELATED VAPOR PRESSURE VALUES:

   1146   1485   9050

LITERATURE DOCUMENTS FOR CORRELATED LIQUID DENSITY VALUES:

   1487   9050

LITERATURE DOCUMENTS REPORTING VIRIAL COEFFICIENT DATA:

T,K	VAPOR PRESSURE, KPA	SATURATED LIQUID VOLUME, ML/MOL	SECOND VIRIAL COEFFICIENT, ML/MOL	HEAT OF VAPORIZATION, J/MOL	SATURATED VAPOR VOLUME, ML/MOL
243	0.1280		-8639.		0.1578E+08
249	0.2066		-7639.		0.1001E+08
256	0.3496		-6679.		6082424.
263	0.5725		-5893.		3813906.
270	0.9099		-5244.		2461952.
277	1.407	161.25	-4702.	38426.	1632165.
284	2.121	162.57	-4245.	38032.	1108873.
291	3.124	163.93	-3857.	37635.	770515.
298	4.504	165.31	-3525.	37232.	546622.
305	6.363	166.74	-3238.	36824.	395271.
312	8.826	168.19	-2989.	36409.	290911.
319	12.03	169.69	-2770.	35987.	217618.
326	16.15	171.23	-2578.	35558.	165257.
333	21.35	172.81	-2407.	35120.	127250.
340	27.84	174.43	-2254.	34674.	99251.
347	35.83	176.11	-2118.	34218.	78337.
354	45.58	177.83	-1995.	33752.	62512.
361	57.34	179.62	-1883.	33275.	50393.
368	71.38	181.46	-1782.	32786.	41005.
375	87.99	183.38	-1689.	32285.	33657.
382	107.5	185.36	-1604.	31772.	27846.
389	130.2	187.42	-1526.	31244.	23209.
396	156.4	189.57	-1454.	30702.	19474.
403	186.6	191.81	-1387.	30144.	16442.
410	221.0	194.15	-1325.	29567.	13960.
417	260.1	196.61	-1267.	28971.	11913.
424	304.2	199.20	-1214.	28352.	10213.
431	353.7	201.93	-1163.	27708.	8791.
438	409.1	204.83	-1116.	27034.	7593.
445	470.8	207.91	-1072.	26326.	6578.
452	539.3	211.20	-1030.	25575.	5712.
459	615.0	214.74	-991.0	24773.	4968.
466	698.4	218.57	-954.0	23908.	4323.
473	790.1	222.74	-919.0	23593.	3859.
480	890.7	227.31	-885.8	22714.	3378.
487	1001.	232.38	-854.3	21775.	2958.
494	1121.	238.05	-824.5	20767.	2589.
501	1251.	244.48	-796.1	19675.	2262.
508	1393.	251.87	-769.1	18479.	1972.
515	1547.	260.54	-743.3	17152.	1713.
521	1690.	269.29	-722.2	15881.	1510.
528	1869.	281.63	-698.6	14189.	1292.
535	2063.	297.23	-676.0	12172.	1088.
542	2272.		-654.4		888.1
549	2498.		-633.8		642.1

COMPOUND CONSTANTS:

MP	TC 563.490 K	VC 468.16 ML/MOL	ZC 0.2626
NBP 388.757 K	PC 2.628 MPA	DC 0.2440 G/ML	OM 0.3463
MU 0.0000 DEBYE	RG 4.4084 ANGSTROM		

VAPOR PRESSURE CORRELATION CONSTANTS, K AND KPA:

WAGNER	RPM2	VAPRES-2	WAGNER
250. - 564. K	250. - 399. K	250. - 473. K	374. - 564. K
38 POINTS	24 POINTS	29 POINTS	18 POINTS
RMSD = 0.6205	RMSD = 0.0132	RMSD = 0.1875	RMSD = 1.1050
-0.75280849E+01	0.32560266E+02	0.13259271E+03	-0.74552180E+01
0.10371372E+01	-0.66205849E+04	-0.85984366E+04	0.79427291E+00
-0.29155868E+01	-0.40265587E-01	0.20587523E-01	-0.21093918E+01
-0.45109803E+01	0.31376729E-04	-0.73059245E-06	-0.96322589E+01
		-0.19076109E+02	

LIQUID DENSITY CORRELATION CONSTANTS, K AND G/ML:

FRANCIS1	FRANCIS1	FRANCIS1	FRANCIS2
273. - 553. K	273. - 399. K	273. - 524. K	523. - 564. K
15 POINTS	5 POINTS	12 POINTS	5 POINTS
RMSD = 0.00084	RMSD = 0.00007	RMSD = 0.00079	RMSD = 0.00124
0.95815301E+00	0.94404763E+00	0.95945835E+00	0.39121808E-03
0.74322382E-03	0.70726709E-03	0.76440047E-03	0.27079994E+01
0.84206161E+01	0.60317430E+01	0.66612835E+01	
0.60175488E+03	0.54824390E+03	0.59109399E+03	

SECOND VIRIAL COEFFICIENT CORRELATION CONSTANTS, ML/MOL:

TSONOPOULOS: TC=563.490 PC=2.628 VC=468.16 OM=0.3463 A= 0.000000 B=0.0000

LITERATURE DOCUMENTS FOR CORRELATED VAPOR PRESSURE VALUES:

   1146   1485   9050

LITERATURE DOCUMENTS FOR CORRELATED LIQUID DENSITY VALUES:

   1487   9050

LITERATURE DOCUMENTS REPORTING VIRIAL COEFFICIENT DATA:

T,K	VAPOR PRESSURE, KPA	SATURATED LIQUID VOLUME, ML/MOL	SECOND VIRIAL COEFFICIENT, ML/MOL	HEAT OF VAPORIZATION, J/MOL	SATURATED VAPOR VOLUME, ML/MOL
250	0.1315		-8494.		0.1580E+08
257	0.2285		-7393.		9345153.
264	0.3830		-6496.		5724599.
271	0.6213		-5756.		3620706.
278	0.9782	157.44	-5140.	40279.	2357740.
285	1.498	158.71	-4624.	39779.	1576694.
292	2.239	160.00	-4187.	39292.	1080324.
299	3.268	161.32	-3813.	38815.	756871.
307	4.907	162.86	-3450.	38281.	516666.
314	6.860	164.24	-3177.	37822.	377363.
321	9.421	165.65	-2939.	37368.	280337.
328	12.73	167.10	-2730.	36919.	211535.
335	16.93	168.58	-2544.	36472.	161929.
342	22.21	170.10	-2379.	36026.	125607.
349	28.75	171.66	-2232.	35579.	98629.
356	36.78	173.26	-2099.	35129.	78321.
364	48.06	175.15	-1963.	34611.	60947.
371	60.05	176.85	-1856.	34150.	49439.
378	74.31	178.60	-1758.	33682.	40456.
385	91.12	180.41	-1668.	33204.	33373.
392	110.8	182.28	-1586.	32715.	27736.
399	133.6	184.22	-1510.	32213.	23209.
406	160.0	186.23	-1440.	31695.	19543.
413	190.2	188.31	-1375.	31160.	16552.
421	230.0	190.81	-1307.	30524.	13778.
428	269.7	193.09	-1251.	29943.	11795.
435	314.5	195.48	-1199.	29336.	10140.
442	364.8	197.99	-1150.	28700.	8750.
449	421.0	200.64	-1104.	28030.	7576.
456	483.4	203.45	-1061.	27321.	6577.
463	552.7	206.43	-1021.	26565.	5723.
470	629.2	209.62	-982.5	25756.	4987.
478	726.2	213.57	-941.4	24748.	4265.
485	819.8	217.35	-907.4	24438.	3816.
492	922.4	221.48	-875.2	23534.	3347.
499	1034.	226.05	-844.7	22571.	2937.
506	1156.	231.17	-815.7	21539.	2576.
513	1289.	236.99	-788.1	20425.	2257.
520	1432.	243.73	-761.8	19214.	1974.
527	1588.	251.70	-736.7	17882.	1719.
535	1781.	262.97	-709.4	16165.	1459.
542	1965.	275.65	-686.6	14426.	1251.
549	2164.	292.52	-664.9	12349.	1056.
556	2378.		-644.0		862.7
563	2611.		-624.0		616.0

COMPOUND CONSTANTS:

MP 135.638 K	TC 553.518 K	VC 472.02 ML/MOL	ZC 0.2621
NBP 382.565 K	PC 2.556 MPA	DC 0.2420 G/ML	OM 0.3424
MU 0.0000 DEBYE	RG 4.3463 ANGSTROM		

VAPOR PRESSURE CORRELATION CONSTANTS, K AND KPA:

WAGNER	RPM2	VAPRES-2	WAGNER
246. - 554. K	246. - 398. K	246. - 463. K	368. - 554. K
38 POINTS	25 POINTS	30 POINTS	18 POINTS
RMSD = 1.1840	RMSD = 0.0443	RMSD = 0.2404	RMSD = 0.6812
-0.74908882E+01	0.35681957E+02	0.26693486E+03	-0.75867527E+01
0.99684504E+00	-0.68575425E+04	-0.11425054E+05	0.12953986E+01
-0.28443144E+01	-0.50223975E-01	0.90534654E-01	-0.36356518E+01
-0.52852114E+01	0.41511622E-04	-0.32742495E-04	-0.31810516E+01
		-0.44106263E+02	

LIQUID DENSITY CORRELATION CONSTANTS, K AND G/ML:

FRANCIS1	FRANCIS1	FRANCIS1	FRANCIS2
273. - 543. K	273. - 398. K	273. - 514. K	513. - 554. K
16 POINTS	6 POINTS	13 POINTS	5 POINTS
RMSD = 0.00153	RMSD = 0.00034	RMSD = 0.00165	RMSD = 0.00020
0.97131544E+00	0.92772371E+00	0.97254336E+00	0.31871675E-03
0.85852854E-03	0.66252123E-03	0.86256396E-03	0.27783403E+01
0.59999990E+01	0.67036257E+01	0.59999990E+01	
0.58808838E+03	0.49378076E+03	0.58918115E+03	

SECOND VIRIAL COEFFICIENT CORRELATION CONSTANTS, ML/MOL:

TSONOPOULOS: TC=553.518 PC=2.556 VC=472.02 OM=0.3424 A= 0.000000 B=0.0000

LITERATURE DOCUMENTS FOR CORRELATED VAPOR PRESSURE VALUES:

1146 1485 9050

LITERATURE DOCUMENTS FOR CORRELATED LIQUID DENSITY VALUES:

1487 9050 10757

LITERATURE DOCUMENTS REPORTING VIRIAL COEFFICIENT DATA:

T,K	VAPOR PRESSURE, KPA	SATURATED LIQUID VOLUME, ML/MOL	SECOND VIRIAL COEFFICIENT, ML/MOL	HEAT OF VAPORIZATION, J/MOL	SATURATED VAPOR VOLUME, ML/MOL
246	0.1305		-8444.		0.1566E+08
252	0.2124		-7483.		9859100.
259	0.3616		-6557.		5948025.
266	0.5947		-5797.		3712890.
273	0.9475	159.12	-5167.	39811.	2390545.
280	1.466	160.56	-4640.	39273.	1583122.
287	2.210	162.03	-4195.	38751.	1075668.
294	3.250	163.53	-3817.	38243.	748193.
301	4.676	165.07	-3492.	37747.	531673.
308	6.591	166.65	-3210.	37261.	385282.
315	9.118	168.26	-2965.	36783.	284251.
322	12.40	169.90	-2751.	36312.	213191.
329	16.59	171.59	-2561.	35846.	162328.
336	21.87	173.33	-2392.	35382.	125325.
343	28.43	175.10	-2242.	34920.	97999.
350	36.51	176.93	-2107.	34457.	77536.
357	46.33	178.80	-1985.	33992.	62011.
364	58.16	180.73	-1875.	33521.	50090.
371	72.26	182.72	-1774.	33045.	40832.
378	88.94	184.77	-1682.	32559.	33567.
385	108.5	186.88	-1598.	32064.	27809.
392	131.3	189.07	-1521.	31555.	23204.
399	157.6	191.33	-1449.	31032.	19487.
406	187.8	193.68	-1383.	30492.	16464.
413	222.3	196.11	-1321.	29932.	13986.
420	261.5	198.65	-1264.	29349.	11939.
427	305.8	201.29	-1210.	28740.	10237.
434	355.6	204.06	-1160.	28102.	8813.
441	411.2	206.97	-1114.	27429.	7612.
448	473.3	210.03	-1070.	26716.	6593.
455	542.2	213.27	-1028.	25956.	5724.
462	618.4	216.72	-989.0	25140.	4978.
469	702.4	220.40	-952.1	24255.	4332.
476	794.7	224.36	-917.2	23927.	3866.
483	895.8	228.67	-884.2	23022.	3384.
490	1006.	233.38	-852.9	22058.	2963.
497	1127.	238.60	-823.1	21023.	2594.
504	1258.	244.47	-794.8	19907.	2268.
511	1400.	251.18	-767.9	18693.	1978.
518	1555.	259.03	-742.2	17358.	1719.
525	1722.	268.47	-717.7	15866.	1486.
532	1903.	280.25	-694.3	14159.	1272.
539	2098.	295.67	-672.0	12133.	1072.
546	2309.		-650.6		874.4
553	2538.		-630.1		623.9

COMPOUND CONSTANTS:

MP   181.978 K   TC   550.058 K   VC   481.98 ML/MOL   ZC   0.2621
NBP   382.292 K   PC   2.487 MPA   DC   0.2370 G/ML   OM   0.3571
MU   0.0000 DEBYE   RG   4.5932 ANGSTROM

VAPOR PRESSURE CORRELATION CONSTANTS, K AND KPA:

WAGNER	VAPRES-2	WAGNER	WAGNER
246. - 551. K	246. - 394. K	246. - 458. K	368. - 551. K
40 POINTS	26 POINTS	31 POINTS	19 POINTS
RMSD = 1.5030	RMSD = 0.0352	RMSD = 0.3232	RMSD = 1.0930
-0.76856644E+01	0.54357209E+03	-0.75794721E+01	-0.78524758E+01
0.12587401E+01	-0.16974349E+05	0.10019177E+01	0.18483243E+01
-0.31121753E+01	0.25656193E+00	-0.27184048E+01	-0.55450813E+01
-0.55106588E+01	-0.11974684E-03	-0.62179308E+01	0.19632149E+02
	-0.96723052E+02		

LIQUID DENSITY CORRELATION CONSTANTS, K AND G/ML:

FRANCIS1	FRANCIS1	FRANCIS1	FRANCIS2
273. - 534. K	273. - 394. K	273. - 514. K	513. - 551. K
82 POINTS	59 POINTS	78 POINTS	8 POINTS
RMSD = 0.00033	RMSD = 0.00032	RMSD = 0.00030	RMSD = 0.00036
0.94145274E+00	0.93550581E+00	0.94146502E+00	0.33679185E-03
0.69995713E-03	0.72744605E-03	0.69329864E-03	0.27698136E+01
0.13339068E+02	0.75824556E+01	0.14101482E+02	
0.60877368E+03	0.56117236E+03	0.61187915E+03	

SECOND VIRIAL COEFFICIENT CORRELATION CONSTANTS, ML/MOL:

TSONOPOULOS: TC=550.058 PC=2.487 VC=481.98 OM=0.3571 A= 0.000000 B=0.0000

LITERATURE DOCUMENTS FOR CORRELATED VAPOR PRESSURE VALUES:

    1146   1485   6553   9050

LITERATURE DOCUMENTS FOR CORRELATED LIQUID DENSITY VALUES:

    1487   5839   5866   6553   8149   9050 10719 13803 14407

LITERATURE DOCUMENTS REPORTING VIRIAL COEFFICIENT DATA:

T,K	VAPOR PRESSURE, KPA	SATURATED LIQUID VOLUME, ML/MOL	SECOND VIRIAL COEFFICIENT, ML/MOL	HEAT OF VAPORIZATION, J/MOL	SATURATED VAPOR VOLUME, ML/MOL
246	0.1286		-8570.		0.1589E+08
252	0.2099		-7597.		9972191.
259	0.3587		-6660.		5996829.
266	0.5915		-5890.		3733026.
273	0.9445	160.74	-5251.	40000.	2397863.
280	1.465	162.05	-4716.	39440.	1584805.
287	2.211	163.39	-4265.	38897.	1075000.
294	3.257	164.77	-3880.	38368.	746672.
301	4.691	166.17	-3549.	37852.	529965.
308	6.618	167.61	-3263.	37348.	383668.
315	9.161	169.09	-3014.	36853.	282831.
322	12.46	170.61	-2796.	36366.	211985.
329	16.68	172.17	-2602.	35884.	161321.
336	22.00	173.77	-2431.	35407.	124493.
343	28.62	175.42	-2278.	34931.	97314.
349	35.49	176.88	-2159.	34524.	79539.
356	45.12	178.62	-2033.	34047.	63502.
363	56.72	180.43	-1919.	33566.	51214.
370	70.58	182.30	-1815.	33081.	41688.
377	86.98	184.24	-1720.	32588.	34226.
384	106.2	186.25	-1633.	32086.	28320.
391	128.7	188.34	-1553.	31572.	23602.
398	154.6	190.52	-1479.	31045.	19801.
405	184.5	192.79	-1411.	30502.	16712.
412	218.6	195.17	-1347.	29939.	14182.
419	257.4	197.67	-1288.	29356.	12095.
426	301.2	200.30	-1233.	28746.	10360.
433	350.5	203.08	-1182.	28108.	8909.
440	405.7	206.03	-1134.	27435.	7688.
447	467.3	209.17	-1088.	26722.	6652.
453	525.6	212.03	-1052.	26073.	5886.
460	600.3	215.60	-1011.	25264.	5111.
467	682.8	219.47	-972.9	24385.	4441.
474	773.6	223.67	-936.8	24070.	3958.
481	873.4	228.29	-902.6	23170.	3459.
488	982.6	233.40	-870.1	22207.	3024.
495	1102.	239.11	-839.4	21172.	2642.
502	1232.	245.58	-810.1	20050.	2305.
509	1374.	253.01	-782.3	18822.	2006.
516	1528.	261.70	-755.8	17463.	1739.
523	1695.	272.07	-730.6	15932.	1497.
530	1876.	284.76	-706.4	14167.	1276.
537	2073.		-683.4		1067.
544	2286.		-661.3		858.1
550	2485.		-643.2		599.8

COMPOUND CONSTANTS:

MP   147.047 K   TC 562.020 K   VC 442.75 ML/MOL   ZC 0.2514
NBP 385.120 K   PC   2.653 MPA   DC 0.2580 G/ML   OM 0.3203
MU   0.0000 DEBYE   RG 4.3197 ANGSTROM

VAPOR PRESSURE CORRELATION CONSTANTS, K AND KPA:

WAGNER   VAPRES-2   VAPRES-2   WAGNER
247. - 562. K   247. - 399. K   247. - 473. K   370. - 562. K
39 POINTS   25 POINTS   30 POINTS   19 POINTS
RMSD = 1.2390   RMSD = 0.0160   RMSD = 0.2951   RMSD = 0.7773
-0.74512355E+01   0.47079423E+03   0.39891317E+03   -0.75747958E+01
 0.10628308E+01   -0.15969697E+05   -0.14466518E+05   0.14430739E+01
-0.23549408E+01   0.19386892E+00   0.15401634E+00   -0.32995283E+01
-0.70229723E+01   -0.79279454E-04   -0.60010600E-04   -0.50218757E+01
                    -0.81902889E+02   -0.68387073E+02

LIQUID DENSITY CORRELATION CONSTANTS, K AND G/ML:

FRANCIS1   FRANCIS1   FRANCIS1   FRANCIS2
273. - 544. K   273. - 399. K   273. - 524. K   513. - 562. K
16 POINTS   4 POINTS   14 POINTS   6 POINTS
RMSD = 0.00149   RMSD = 0.00039   RMSD = 0.00151   RMSD = 0.00063
0.92694610E+00   0.94234377E+00   0.92911780E+00   0.39190702E-03
0.63577341E-03   0.74763875E-03   0.66863350E-03   0.27293650E+01
0.94204235E+01   0.59999990E+01   0.67525425E+01
0.60227856E+03   0.74718604E+03   0.58728467E+03

SECOND VIRIAL COEFFICIENT CORRELATION CONSTANTS, ML/MOL:

TSONOPOULOS: TC=562.020 PC=2.653 VC=442.75 OM=0.3203 A= 0.000000 B=0.0000

LITERATURE DOCUMENTS FOR CORRELATED VAPOR PRESSURE VALUES:

1146   1485   9050

LITERATURE DOCUMENTS FOR CORRELATED LIQUID DENSITY VALUES:

1487   9050

LITERATURE DOCUMENTS REPORTING VIRIAL COEFFICIENT DATA:

T,K	VAPOR PRESSURE, KPA	SATURATED LIQUID VOLUME, ML/MOL	SECOND VIRIAL COEFFICIENT, ML/MOL	HEAT OF VAPORIZATION, J/MOL	SATURATED VAPOR VOLUME, ML/MOL
247	0.1295		-8402.		0.1585E+08
254	0.2283		-7299.		9244307.
261	0.3870		-6403.		5601246.
268	0.6333		-5666.		3512961.
275	1.004	157.92	-5054.	39917.	2273183.
282	1.545	159.04	-4542.	39319.	1513255.
289	2.315	160.18	-4110.	38744.	1033682.
297	3.570	161.51	-3694.	38112.	688084.
304	5.092	162.70	-3384.	37579.	493013.
311	7.118	163.92	-3116.	37062.	360119.
318	9.770	165.16	-2882.	36561.	267723.
325	13.18	166.43	-2677.	36071.	202267.
332	17.51	167.73	-2495.	35593.	155088.
340	23.81	169.26	-2312.	35056.	116392.
347	30.70	170.64	-2170.	34593.	91746.
354	39.12	172.05	-2042.	34135.	73138.
361	49.28	173.50	-1927.	33678.	58911.
368	61.44	174.99	-1822.	33222.	47906.
375	75.85	176.53	-1726.	32763.	39299.
383	95.45	178.35	-1627.	32233.	31648.
390	115.7	180.00	-1548.	31763.	26393.
397	139.1	181.71	-1475.	31283.	22158.
404	166.0	183.48	-1407.	30792.	18718.
411	196.7	185.32	-1344.	30286.	15901.
418	231.7	187.24	-1286.	29764.	13577.
425	271.3	189.25	-1232.	29222.	11647.
433	322.7	191.66	-1174.	28574.	9824.
440	373.4	193.89	-1127.	27978.	8498.
447	430.0	196.24	-1082.	27350.	7374.
454	492.9	198.74	-1041.	26684.	6416.
461	562.5	201.40	-1001.	25974.	5594.
468	639.3	204.26	-964.4	25211.	4885.
476	736.5	207.80	-924.6	24260.	4187.
483	830.4	211.19	-891.7	23963.	3749.
490	933.1	214.91	-860.5	23107.	3294.
497	1045.	219.05	-830.8	22192.	2895.
504	1167.	223.69	-802.7	21210.	2543.
511	1299.	228.98	-775.8	20150.	2232.
519	1464.	236.08	-746.7	18820.	1917.
526	1621.	243.55	-722.5	17529.	1672.
533	1791.	252.67	-699.3	16083.	1450.
540	1974.	264.20	-677.2	14426.	1246.
547	2172.		-656.0		1055.
554	2386.		-635.8		866.2
561	2618.		-616.3		638.2

COMPOUND CONSTANTS:

MP   TC   568.851 K   VC   466.25 ML/MOL   ZC   0.2654
NBP 390.909 K   PC   2.692 MPA   DC   0.2450 G/ML   OM   0.3388
MU   0.0000 DEBYE   RG   4.4000 ANGSTROM

      VAPOR PRESSURE CORRELATION CONSTANTS, K AND KPA:

      WAGNER   WAGNER   VAPRES-2   WAGNER
      251. - 569. K   251. - 399. K   251. - 473. K   376. - 569. K
       38 POINTS   24 POINTS   29 POINTS   19 POINTS
      RMSD = 0.6744   RMSD = 0.0348   RMSD = 0.1554   RMSD = 0.6093
      -0.76231951E+01   -0.80512682E+01   0.31056873E+02   -0.75770514E+01
       0.13726361E+01   0.23561971E+01   -0.63104061E+04   0.12415247E+01
      -0.34659090E+01   -0.47975889E+01   -0.29414513E-01   -0.32545226E+01
      -0.28493962E+01   -0.82465686E+00   0.21253432E-04   -0.12870822E+01
                                              -0.34256123E+00

      LIQUID DENSITY CORRELATION CONSTANTS, K AND G/ML:

      FRANCIS1   FRANCIS1   FRANCIS1   FRANCIS2
      273. - 553. K   273. - 399. K   273. - 534. K   523. - 569. K
       19 POINTS   7 POINTS   17 POINTS   5 POINTS
      RMSD = 0.00144   RMSD = 0.00033   RMSD = 0.00133   RMSD = 0.00135
      0.95756406E+00   0.94204307E+00   0.95570278E+00   0.40641835E-03
      0.65948674E-03   0.59607788E-03   0.65288274E-03   0.27045561E+01
      0.15000000E+02   0.14808465E+02   0.15000000E+02
      0.63137256E+03   0.60130566E+03   0.62956299E+03

      SECOND VIRIAL COEFFICIENT CORRELATION CONSTANTS, ML/MOL:

      TSONOPOULOS: TC=568.851 PC=2.692 VC=466.24 OM=0.3388 A= 0.000000 B=0.0000

LITERATURE DOCUMENTS FOR CORRELATED VAPOR PRESSURE VALUES:

    1146   1485   9050

LITERATURE DOCUMENTS FOR CORRELATED LIQUID DENSITY VALUES:

    1487   8639   9050 14407

LITERATURE DOCUMENTS REPORTING VIRIAL COEFFICIENT DATA:

T,K	VAPOR PRESSURE, KPA	SATURATED LIQUID VOLUME, ML/MOL	SECOND VIRIAL COEFFICIENT, ML/MOL	HEAT OF VAPORIZATION, J/MOL	SATURATED VAPOR VOLUME, ML/MOL
251	0.1323		-8495.		0.1577E+08
258	0.2280		-7394.		9401528.
265	0.3797		-6495.		5796483.
272	0.6127		-5754.		3685308.
279	0.9606	156.27	-5138.	40225.	2409805.
287	1.555	157.62	-4554.	39712.	1530225.
294	2.309	158.83	-4126.	39271.	1054550.
301	3.354	160.06	-3761.	38833.	742457.
308	4.773	161.33	-3446.	38400.	533111.
316	6.980	162.81	-3138.	37907.	373256.
323	9.555	164.15	-2904.	37477.	278133.
330	12.87	165.51	-2698.	37047.	210429.
337	17.08	166.91	-2516.	36615.	161456.
344	22.36	168.36	-2354.	36180.	125499.
352	29.95	170.05	-2190.	35678.	95487.
359	38.18	171.58	-2061.	35234.	76069.
366	48.13	173.16	-1945.	34783.	61218.
373	60.06	174.79	-1839.	34323.	49730.
381	76.44	176.72	-1730.	33787.	39633.
388	93.50	178.47	-1643.	33307.	32774.
395	113.4	180.29	-1563.	32815.	27301.
402	136.5	182.17	-1489.	32309.	22896.
409	163.1	184.13	-1420.	31788.	19322.
417	198.1	186.47	-1348.	31173.	16026.
424	233.4	188.62	-1290.	30615.	13683.
431	273.2	190.87	-1236.	30035.	11737.
438	318.0	193.24	-1185.	29432.	10111.
446	375.8	196.10	-1130.	28710.	8565.
453	432.6	198.76	-1086.	28045.	7434.
460	495.7	201.58	-1044.	27344.	6470.
467	565.5	204.59	-1005.	26601.	5643.
474	642.5	207.82	-967.7	25807.	4930.
482	739.9	211.82	-927.7	24824.	4228.
489	833.8	215.64	-894.7	24523.	3789.
496	936.5	219.82	-863.4	23646.	3332.
503	1048.	224.42	-833.7	22712.	2930.
511	1189.	230.32	-801.4	21566.	2530.
518	1322.	236.19	-774.7	20481.	2221.
525	1468.	242.88	-749.2	19301.	1946.
532	1625.	250.63	-724.9	18003.	1699.
539	1795.	259.78	-701.6	16556.	1476.
547	2006.	272.60	-676.3	14650.	1242.
554	2207.		-655.2		1052.
561	2425.		-634.9		862.8
568	2662.		-615.5		630.2

COMPOUND CONSTANTS:

MP 158.198 K	TC 567.091 K	VC 442.75 ML/MOL	ZC 0.2535
NBP 388.789 K	PC 2.700 MPA	DC 0.2580 G/ML	OM 0.3294
MU 0.0000 DEBYE	RG		

VAPOR PRESSURE CORRELATION CONSTANTS, K AND KPA:

WAGNER	VAPRES-2	RPM2	WAGNER
249. - 567. K	249. - 399. K	249. - 473. K	374. - 567. K
51 POINTS	26 POINTS	31 POINTS	30 POINTS
RMSD = 1.1900	RMSD = 0.0295	RMSD = 0.1132	RMSD = 1.5500
-0.76476740E+01	-0.17904121E+03	0.28057695E+02	-0.76544443E+01
0.15623556E+01	-0.19037819E+04	-0.60781787E+04	0.15923820E+01
-0.38309478E+01	-0.14989366E+00	-0.27640726E-01	-0.39776560E+01
-0.16329622E+01	0.82267921E-04	0.19458809E-04	-0.90765203E+00
	0.39308475E+02		

LIQUID DENSITY CORRELATION CONSTANTS, K AND G/ML:

FRANCIS1	FRANCIS1	FRANCIS1	FRANCIS2
273. - 554. K	273. - 399. K	273. - 534. K	523. - 567. K
19 POINTS	6 POINTS	16 POINTS	5 POINTS
RMSD = 0.00109	RMSD = 0.00063	RMSD = 0.00111	RMSD = 0.00059
0.94740307E+00	0.95343709E+00	0.94708109E+00	0.35708146E-03
0.64998236E-03	0.75533590E-03	0.62578241E-03	0.27673493E+01
0.12308627E+02	0.59999990E+01	0.14993712E+02	
0.62068359E+03	0.74651074E+03	0.63140918E+03	

SECOND VIRIAL COEFFICIENT CORRELATION CONSTANTS, ML/MOL:

TSONOPOULOS: TC=567.091 PC=2.700 VC=442.75 OM=0.3294 A= 0.000000 B=0.0000

LITERATURE DOCUMENTS FOR CORRELATED VAPOR PRESSURE VALUES:

1146   1485   9050

LITERATURE DOCUMENTS FOR CORRELATED LIQUID DENSITY VALUES:

1487   9050

LITERATURE DOCUMENTS REPORTING VIRIAL COEFFICIENT DATA:

T,K	VAPOR PRESSURE, KPA	SATURATED LIQUID VOLUME, ML/MOL	SECOND VIRIAL COEFFICIENT, ML/MOL	HEAT OF VAPORIZATION, J/MOL	SATURATED VAPOR VOLUME, ML/MOL
249	0.1314		-8513.		0.1575E+08
256	0.2263		-7398.		9397095.
263	0.3768		-6490.		5797045.
270	0.6080		-5743.		3686358.
277	0.9535	156.15	-5123.	39689.	2410217.
285	1.544	157.45	-4536.	39223.	1529860.
292	2.295	158.62	-4107.	38816.	1053731.
299	3.336	159.81	-3741.	38410.	741393.
306	4.752	161.02	-3427.	38004.	531959.
314	6.957	162.45	-3118.	37537.	372120.
321	9.532	163.73	-2885.	37126.	277069.
328	12.85	165.04	-2680.	36711.	209462.
335	17.07	166.38	-2499.	36292.	160596.
342	22.37	167.75	-2337.	35867.	124744.
350	29.98	169.37	-2174.	35373.	94845.
357	38.24	170.83	-2046.	34933.	75516.
364	48.24	172.33	-1930.	34484.	60743.
371	60.22	173.88	-1825.	34026.	49324.
379	76.69	175.70	-1717.	33489.	39296.
386	93.83	177.36	-1630.	33007.	32487.
393	113.8	179.07	-1551.	32512.	27058.
400	137.0	180.85	-1477.	32004.	22690.
408	167.8	182.96	-1400.	31404.	18699.
415	198.9	184.89	-1338.	30862.	15884.
422	234.3	186.91	-1280.	30302.	13563.
429	274.2	189.01	-1226.	29722.	11637.
436	319.1	191.22	-1176.	29120.	10027.
444	377.0	193.89	-1122.	28401.	8498.
451	433.9	196.37	-1078.	27740.	7379.
458	497.1	198.99	-1037.	27046.	6424.
465	566.9	201.78	-997.7	26313.	5606.
473	655.4	205.22	-955.8	25416.	4807.
480	741.1	208.48	-921.3	24569.	4206.
487	834.9	212.01	-888.6	24271.	3770.
494	937.3	215.87	-857.5	23415.	3317.
502	1066.	220.78	-823.9	22373.	2866.
509	1189.	225.59	-796.1	21396.	2522.
516	1322.	231.03	-769.6	20344.	2216.
523	1467.	237.26	-744.3	19203.	1943.
530	1624.	244.51	-720.2	17950.	1698.
538	1819.	254.50	-694.0	16338.	1445.
545	2005.	265.32	-672.1	14709.	1244.
552	2206.	279.01	-651.1	12774.	1054.
559	2424.		-631.1		866.3
567	2697.		-609.1		574.1

COMPOUND CONSTANTS:

MP   182.308 K   TC   576.583 K   VC   455.10 ML/MOL   ZC   0.2665
NBP   391.419 K   PC   2.807 MPA   DC   0.2510 G/ML   OM   0.3039
MU   0.0000 DEBYE   RG

VAPOR PRESSURE CORRELATION CONSTANTS, K AND KPA:

WAGNER   RPM2   VAPRES-2   WAGNER
249. - 577. K   249. - 399. K   249. - 489. K   376. - 577. K
  40 POINTS   24 POINTS   30 POINTS   20 POINTS
RMSD = 0.9367   RMSD = 0.0124   RMSD = 0.0531   RMSD = 0.9309
-0.75113723E+01   0.29174706E+02   0.10005138E+03   -0.75731433E+01
 0.14527376E+01   -0.61595949E+04   -0.75699410E+04   0.16647955E+01
-0.31424787E+01   -0.32053436E-01   0.10611915E-01   -0.39272445E+01
-0.25696733E+01   0.24323223E-04   0.20317592E-05   0.37250452E+01
                                            -0.13494371E+02

LIQUID DENSITY CORRELATION CONSTANTS, K AND G/ML:

FRANCIS1   FRANCIS1   FRANCIS1   FRANCIS2
273. - 563. K   273. - 399. K   273. - 544. K   534. - 577. K
  17 POINTS   4 POINTS   15 POINTS   5 POINTS
RMSD = 0.00092   RMSD = 0.00006   RMSD = 0.00093   RMSD = 0.00074
0.95571357E+00   0.94342220E+00   0.95564467E+00   0.42841277E-03
0.64278254E-03   0.65775961E-03   0.64935256E-03   0.26615392E+01
0.13634657E+02   0.59999990E+01   0.12819571E+02
0.63217554E+03   0.54834229E+03   0.62856323E+03

SECOND VIRIAL COEFFICIENT CORRELATION CONSTANTS, ML/MOL:

TSONOPOULOS: TC=576.583 PC=2.807 VC=455.10 OM=0.3039 A= 0.000000 B=0.0000

LITERATURE DOCUMENTS FOR CORRELATED VAPOR PRESSURE VALUES:

    1146   1485   9050

LITERATURE DOCUMENTS FOR CORRELATED LIQUID DENSITY VALUES:

    1487   9050

LITERATURE DOCUMENTS REPORTING VIRIAL COEFFICIENT DATA:

T,K	VAPOR PRESSURE, KPA	SATURATED LIQUID VOLUME, ML/MOL	SECOND VIRIAL COEFFICIENT, ML/MOL	HEAT OF VAPORIZATION, J/MOL	SATURATED VAPOR VOLUME, ML/MOL
249	0.1305		-8644.		0.1585E+08
256	0.2237		-7498.		9507530.
263	0.3705		-6567.		5894982.
271	0.6351		-5705.		3542078.
278	0.9874	154.67	-5087.	39201.	2335854.
286	1.584	155.95	-4504.	38734.	1496401.
293	2.337	157.09	-4077.	38330.	1038518.
301	3.548	158.42	-3666.	37874.	701691.
308	5.006	159.61	-3361.	37478.	508165.
316	7.256	161.01	-3061.	37028.	358988.
323	9.863	162.26	-2834.	36634.	269429.
330	13.20	163.54	-2634.	36240.	205201.
338	18.10	165.05	-2434.	35788.	152826.
345	23.51	166.40	-2280.	35390.	119675.
353	31.24	167.99	-2123.	34930.	91772.
360	39.58	169.43	-2000.	34521.	73575.
368	51.19	171.12	-1874.	34047.	57834.
375	63.44	172.64	-1775.	33623.	47300.
383	80.18	174.45	-1672.	33129.	37969.
390	97.50	176.08	-1590.	32686.	31583.
397	117.6	177.78	-1514.	32231.	26455.
405	144.4	179.79	-1434.	31696.	21777.
412	171.6	181.62	-1370.	31213.	18483.
420	207.3	183.80	-1303.	30642.	15426.
427	242.9	185.80	-1248.	30123.	13239.
435	289.2	188.20	-1190.	29507.	11176.
442	334.9	190.41	-1142.	28944.	9679.
450	393.6	193.08	-1092.	28269.	8247.
457	451.2	195.56	-1050.	27648.	7192.
464	514.8	198.18	-1011.	26993.	6289.
472	595.6	201.39	-968.8	26197.	5408.
479	673.8	204.42	-934.2	25450.	4747.
487	772.4	208.16	-896.8	24526.	4094.
494	867.2	211.73	-866.0	24237.	3681.
502	986.0	216.22	-832.6	23289.	3194.
509	1100.	220.59	-804.9	22403.	2822.
517	1242.	226.19	-774.8	21315.	2448.
524	1377.	231.77	-749.9	20285.	2159.
531	1523.	238.16	-726.0	19166.	1901.
539	1705.	246.76	-700.1	17752.	1636.
546	1877.	255.81	-678.4	16359.	1428.
554	2091.	268.65	-654.8	14524.	1209.
561	2293.	283.06	-635.0	12601.	1029.
569	2545.		-613.4		826.2
576	2786.		-595.4		597.8

COMPOUND CONSTANTS:

MP	160.891 K	TC	563.500 K	VC	435.99 ML/MOL	ZC	0.2539
NBP	382.996 K	PC	2.729 MPA	DC	0.2620 G/ML	OM	0.2973
MU	0.0000 DEBYE	RG	4.1618 ANGSTRON				

VAPOR PRESSURE CORRELATION CONSTANTS, K AND KPA:

WAGNER	RPM2	VAPRES-2	WAGNER
244. - 564. K	244. - 399. K	244. - 473. K	368. - 564. K
37 POINTS	24 POINTS	28 POINTS	18 POINTS
RMSD = 0.8746	RMSD = 0.0101	RMSD = 0.0608	RMSD = 0.6252
-0.74785182E+01	0.30208248E+02	0.98942076E+02	-0.75622229E+01
0.14853469E+01	-0.61642577E+04	-0.75076413E+04	0.17759710E+01
-0.33020191E+01	-0.35236264E-01	0.72114338E-02	-0.43612069E+01
-0.29070043E+01	0.27270980E-04	0.46303456E-05	0.48661198E+01
		-0.13140887E+02	

LIQUID DENSITY CORRELATION CONSTANTS, K AND G/ML:

FRANCIS1	FRANCIS1	FRANCIS1	FRANCIS2
273. - 553. K	273. - 399. K	273. - 534. K	523. - 564. K
16 POINTS	4 POINTS	14 POINTS	6 POINTS
RMSD = 0.00140	RMSD = 0.00050	RMSD = 0.00091	RMSD = 0.00200
0.94544941E+00	0.95151567E+00	0.94268590E+00	0.34137770E-03
0.69735921E-03	0.76306285E-03	0.66734827E-03	0.27826123E+01
0.78367720E+01	0.59999990E+01	0.10055699E+02	
0.59971118E+03	0.74784229E+03	0.60985107E+03	

SECOND VIRIAL COEFFICIENT CORRELATION CONSTANTS, ML/MOL:

TSONOPOULOS: TC=563.500 PC=2.729 VC=435.99 OM=0.2973 A= 0.000000 B=0.0000

LITERATURE DOCUMENTS FOR CORRELATED VAPOR PRESSURE VALUES:

1146 1485 9050

LITERATURE DOCUMENTS FOR CORRELATED LIQUID DENSITY VALUES:

1487 9050

LITERATURE DOCUMENTS REPORTING VIRIAL COEFFICIENT DATA:

T,K	VAPOR PRESSURE, KPA	SATURATED LIQUID VOLUME, ML/MOL	SECOND VIRIAL COEFFICIENT, ML/MOL	HEAT OF VAPORIZATION, J/MOL	SATURATED VAPOR VOLUME, ML/MOL
244	0.1314		-8457.		0.1543E+08
251	0.2285		-7324.		9126154.
258	0.3832		-6405.		5591002.
265	0.6220		-5653.		3536540.
273	1.043	156.25	-4952.	38533.	2171166.
280	1.592	157.41	-4447.	38092.	1457413.
287	2.372	158.60	-4020.	37657.	1002127.
294	3.452	159.81	-3658.	37228.	704431.
302	5.167	161.22	-3307.	36743.	482615.
309	7.204	162.49	-3044.	36321.	353567.
316	9.867	163.78	-2815.	35901.	263438.
323	13.29	165.10	-2614.	35482.	199359.
331	18.35	166.65	-2413.	35002.	147534.
338	23.96	168.04	-2258.	34580.	114972.
345	30.89	169.46	-2119.	34154.	90684.
352	39.35	170.92	-1995.	33725.	72326.
360	51.18	172.63	-1867.	33227.	56556.
367	63.70	174.17	-1767.	32784.	46067.
374	78.52	175.76	-1675.	32333.	37851.
381	95.92	177.40	-1591.	31873.	31350.
389	119.3	179.33	-1503.	31335.	25505.
396	143.3	181.08	-1433.	30851.	21446.
403	170.7	182.89	-1368.	30354.	18145.
411	206.9	185.05	-1300.	29767.	15095.
418	243.1	187.01	-1244.	29236.	12920.
425	283.9	189.05	-1192.	28685.	11110.
432	329.7	191.19	-1144.	28113.	9594.
440	388.7	193.76	-1092.	27430.	8150.
447	446.5	196.13	-1050.	26802.	7091.
454	510.6	198.64	-1010.	26141.	6185.
461	581.3	201.30	-972.6	25441.	5408.
469	670.9	204.55	-932.2	24584.	4646.
476	757.4	207.63	-899.0	23772.	4072.
483	852.0	210.96	-867.5	23476.	3654.
490	955.2	214.59	-837.6	22655.	3219.
498	1084.	219.20	-805.3	21654.	2787.
505	1208.	223.74	-778.4	20715.	2456.
512	1342.	228.89	-752.9	19704.	2162.
519	1487.	234.84	-728.5	18607.	1899.
527	1667.	242.96	-702.0	17218.	1630.
534	1839.	251.72	-679.9	15848.	1418.
541	2024.	262.78	-658.7	14276.	1223.
548	2224.	277.43	-638.5	12402.	1039.
556	2473.		-616.4		831.6
563	2711.		-597.9		596.5

COMPOUND CONSTANTS:

MP	165.801 K	TC	543.957 K	VC	470.79 ML/MOL	ZC	0.2673
NBP	372.396 K	PC	2.568 MPA	DC	0.2426 G/ML	OM	0.3036
MU	0.0000 DEBYE	RG	4.1714 ANGSTROM				

    VAPOR PRESSURE CORRELATION CONSTANTS, K AND KPA:

WAGNER	VAPRES-2	RIEDEL	WAGNER
193. - 544. K	193. - 392. K	193. - 454. K	353. - 544. K
100 POINTS	72 POINTS	78 POINTS	45 POINTS
RMSD = 1.7900	RMSD = 0.0603	RMSD = 0.3198	RMSD = 0.5493
-0.76501166E+01	-0.19677213E+03	0.55246443E+02	-0.75078877E+01
0.18899385E+01	-0.13826558E+04	-0.58432393E+01	0.15128556E+01
-0.42070574E+01	-0.17079505E+00	-0.59886585E+04	-0.36302332E+01
-0.22322060E+00	0.98661708E-04	0.17711720E-16	0.85819232E+00
	0.43078676E+02		

    LIQUID DENSITY CORRELATION CONSTANTS, K AND G/ML:

FRANCIS1	FRANCIS1	FRANCIS1	RACKETT
173. - 534. K	173. - 373. K	173. - 514. K	508. - 544. K
106 POINTS	86 POINTS	101 POINTS	10 POINTS
RMSD = 0.00044	RMSD = 0.00027	RMSD = 0.00040	RMSD = 0.00097
0.93769777E+00	0.93451047E+00	0.93788779E+00	0.26731919E+00
0.71540871E-03	0.75369887E-03	0.71046059E-03	
0.10661614E+02	0.59999990E+01	0.11250001E+02	
0.59147583E+03	0.57136255E+03	0.59428955E+03	

    SECOND VIRIAL COEFFICIENT CORRELATION CONSTANTS, ML/MOL:

    TSONOPOULOS: TC=543.957 PC=2.568 VC=470.79 OM=0.3036 A= 0.000000 B=0.0000

LITERATURE DOCUMENTS FOR CORRELATED VAPOR PRESSURE VALUES:

    10   122   123   620   1146   1485   1520   1716   2982   2983   3571   4576   6802
6894   9050   9352 10335 15307 16244

LITERATURE DOCUMENTS FOR CORRELATED LIQUID DENSITY VALUES:

    71   336   776   986   1487   1520   2982   2986   5063   8460   8545   9008   9050
9053   9352 10258 16589 18474 19532

LITERATURE DOCUMENTS REPORTING VIRIAL COEFFICIENT DATA:

T,K	VAPOR PRESSURE, KPA	SATURATED LIQUID VOLUME, ML/MOL	SECOND VIRIAL COEFFICIENT, ML/MOL	HEAT OF VAPORIZATION, J/MOL	SATURATED VAPOR VOLUME, ML/MOL
173		144.88	-60970.		
181		146.03	-44356.		
189		147.20	-33009.		
198	0.002750	148.55	-24294.	40136.	0.5987E+09
206	0.007057	149.78	-18907.	39761.	0.2427E+09
215	0.01855	151.18	-14594.	39335.	0.9633E+08
223	0.04071	152.46	-11820.	38953.	0.4553E+08
232	0.09159	153.94	-9515.	38520.	0.2105E+08
240	0.1777	155.27	-7977.	38132.	0.1122E+08
248	0.3283	156.64	-6784.	37741.	6274177.
257	0.6214	158.22	-5741.	37297.	3432708.
265	1.051	159.66	-5012.	36898.	2091832.
274	1.818	161.32	-4355.	36443.	1248531.
282	2.861	162.84	-3882.	36032.	815674.
291	4.600	164.59	-3445.	35562.	522468.
299	6.822	166.19	-3122.	35134.	361247.
307	9.876	167.84	-2847.	34698.	255590.
316	14.58	169.75	-2585.	34195.	177564.
324	20.18	171.50	-2384.	33736.	131073.
333	28.45	173.54	-2189.	33204.	95090.
341	37.91	175.42	-2038.	32716.	72691.
350	51.41	177.61	-1887.	32151.	54655.
358	66.37	179.63	-1769.	31632.	43004.
366	84.56	181.73	-1663.	31097.	34238.
375	109.5	184.19	-1555.	30474.	26834.
383	136.0	186.48	-1469.	29901.	21833.
392	171.6	189.18	-1382.	29233.	17493.
400	208.8	191.71	-1311.	28616.	14486.
409	257.7	194.71	-1237.	27893.	11815.
417	308.0	197.54	-1178.	27222.	9922.
425	365.3	200.54	-1122.	26521.	8378.
434	438.9	204.16	-1065.	25689.	6964.
442	513.3	207.63	-1018.	24904.	5931.
451	607.8	211.87	-968.0	23955.	4968.
459	702.2	216.00	-926.9	23039.	4250.
468	821.0	221.14	-883.8	22495.	3656.
476	938.9	226.27	-847.9	21543.	3149.
484	1069.	232.06	-814.1	20521.	2712.
493	1232.	239.61	-778.4	19267.	2290.
501	1392.	247.55	-748.5	18035.	1965.
510	1591.	258.45	-716.7	16468.	1644.
518	1786.	270.68	-690.1	14851.	1391.
527	2029.	288.96	-661.7	12636.	1131.
535	2268.		-637.8		909.0
543	2534.		-615.1		642.7

167

COMPOUND CONSTANTS:

MP   172.472 K   TC 573.562 K   VC 455.10 ML/MOL   ZC 0.2691
NBP 387.886 K   PC   2.820 MPA   DC 0.2510 G/ML   OM 0.2888
MU   0.0000 DEBYE   RG   4.0859 ANGSTROM

   VAPOR PRESSURE CORRELATION CONSTANTS, K AND KPA:

   WAGNER   WAGNER   VAPRES-2   WAGNER
   247. - 574. K   247. - 399. K   247. - 474. K   373. - 574. K
    40 POINTS   26 POINTS   30 POINTS   19 POINTS
   RMSD = 2.6000   RMSD = 0.0293   RMSD = 0.0980   RMSD = 0.9756
   -0.73288884E+01   -0.64624298E+01   0.41329535E+03   -0.76138681E+01
    0.11378442E+01   -0.86502826E+00   -0.14515593E+05   0.20467893E+01
   -0.23616826E+01   0.37747266E+00   0.16845565E+00   -0.49529832E+01
   -0.53041746E+01   -0.94126343E+01   -0.68626471E-04   0.58427280E+01
                                             -0.71513387E+02

   LIQUID DENSITY CORRELATION CONSTANTS, K AND G/ML:

   FRANCIS1   FRANCIS1   FRANCIS1   FRANCIS2
   273. - 554. K   273. - 399. K   273. - 543. K   533. - 574. K
    15 POINTS   4 POINTS   14 POINTS   4 POINTS
   RMSD = 0.00089   RMSD = 0.00008   RMSD = 0.00092   RMSD = 0.00001
   0.96466982E+00   0.93906927E+00   0.96479589E+00   0.38570227E-03
   0.69855968E-03   0.63822744E-03   0.70147263E-03   0.27228978E+01
   0.11221236E+02   0.59999990E+01   0.10924476E+02
   0.62384644E+03   0.52510718E+03   0.62253662E+03

   SECOND VIRIAL COEFFICIENT CORRELATION CONSTANTS, ML/MOL:

   TSONOPOULOS: TC=573.562 PC=2.820 VC=455.10 OM=0.2888 A= 0.000000 B=0.0000

LITERATURE DOCUMENTS FOR CORRELATED VAPOR PRESSURE VALUES:

   1146 1485 9050

LITERATURE DOCUMENTS FOR CORRELATED LIQUID DENSITY VALUES:

   1487 9050

LITERATURE DOCUMENTS REPORTING VIRIAL COEFFICIENT DATA:

T,K	VAPOR PRESSURE, KPA	SATURATED LIQUID VOLUME, ML/MOL	SECOND VIRIAL COEFFICIENT, ML/MOL	HEAT OF VAPORIZATION, J/MOL	SATURATED VAPOR VOLUME, ML/MOL
247	0.1298		-8425.		0.1581E+08
254	0.2258		-7304.		9346182.
261	0.3785		-6394.		5726842.
269	0.6557		-5553.		3405155.
276	1.027	154.44	-4951.	39198.	2229983.
284	1.657	155.78	-4383.	38613.	1420218.
291	2.454	156.97	-3969.	38118.	982079.
298	3.551	158.19	-3615.	37637.	694198.
306	5.280	159.62	-3272.	37103.	478568.
313	7.321	160.90	-3015.	36648.	352448.
321	10.41	162.39	-2761.	36139.	253573.
328	13.93	163.74	-2568.	35703.	193221.
336	19.07	165.31	-2374.	35211.	144048.
343	24.76	166.72	-2224.	34786.	112931.
350	31.73	168.17	-2090.	34363.	89573.
358	41.54	169.88	-1953.	33881.	69638.
365	51.99	171.42	-1845.	33458.	56461.
373	66.38	173.23	-1734.	32972.	44918.
380	81.39	174.87	-1646.	32542.	37098.
388	101.7	176.80	-1554.	32042.	30091.
395	122.5	178.55	-1480.	31596.	25242.
402	146.5	180.37	-1413.	31140.	21309.
410	178.1	182.52	-1341.	30604.	17685.
417	210.0	184.49	-1283.	30119.	15110.
425	251.5	186.83	-1222.	29545.	12697.
432	292.8	188.98	-1172.	29021.	10955.
439	339.0	191.23	-1126.	28475.	9490.
447	398.4	193.93	-1076.	27819.	8088.
454	456.5	196.43	-1035.	27213.	7055.
462	530.6	199.47	-991.5	26477.	6054.
469	602.5	202.30	-955.5	25789.	5307.
477	693.4	205.78	-916.8	24941.	4573.
484	781.2	209.06	-884.8	24134.	4017.
491	876.9	212.61	-854.5	23837.	3611.
499	996.9	217.05	-821.7	22891.	3133.
506	1112.	221.35	-794.4	22004.	2768.
514	1255.	226.85	-764.9	20913.	2401.
521	1391.	232.31	-740.4	19879.	2118.
529	1560.	239.51	-713.7	18587.	1830.
536	1721.	246.92	-691.5	17337.	1605.
543	1894.	255.73	-670.2	15944.	1401.
551	2108.	268.23	-647.0	14113.	1186.
558	2311.		-627.5		1011.
566	2561.		-606.3		811.8
573	2800.		-588.5		588.3

COMPOUND CONSTANTS:

```
MP    163.954 K        TC   566.411 K       VC   460.60 ML/MOL    ZC  0.2670
NBP   386.624 K        PC     2.730 MPA      DC   0.2480 G/ML      OM  0.3156
MU      0.0000 DEBYE   RG     4.2052 ANGSTROM
```

VAPOR PRESSURE CORRELATION CONSTANTS, K AND KPA:

WAGNER	VAPRES-2	VAPRES-2	WAGNER
222. - 566. K	222. - 407. K	222. - 474. K	368. - 566. K
84 POINTS	65 POINTS	75 POINTS	35 POINTS
RMSD = 0.6559	RMSD = 0.0063	RMSD = 0.1450	RMSD = 0.5738
-0.75421426E+01	0.52944723E+02	0.52780172E+02	-0.74849401E+01
0.14438733E+01	-0.64971191E+04	-0.64956579E+04	0.12996129E+01
-0.34414414E+01	-0.13923648E-01	-0.14114575E-01	-0.32680774E+01
-0.19478247E+01	0.13309801E-04	0.13454230E-04	-0.46061774E+00
	-0.47214727E+01	-0.46857144E+01	

LIQUID DENSITY CORRELATION CONSTANTS, K AND G/ML:

FRANCIS1	FRANCIS1	FRANCIS1	FRANCIS2
273. - 554. K	273. - 399. K	273. - 535. K	523. - 566. K
19 POINTS	6 POINTS	17 POINTS	5 POINTS
RMSD = 0.00158	RMSD = 0.00021	RMSD = 0.00160	RMSD = 0.00058
0.94922262E+00	0.95287704E+00	0.94935948E+00	0.41784573E-03
0.63072750E-03	0.74076955E-03	0.63104532E-03	0.26838198E+01
0.14998841E+02	0.77576332E+01	0.14999187E+02	
0.62533325E+03	0.74656714E+03	0.62526270E+03	

SECOND VIRIAL COEFFICIENT CORRELATION CONSTANTS, ML/MOL:

TSONOPOULOS: TC=566.411 PC=2.730 VC=460.60 OM=0.3156 A= 0.000000 B=0.0000

LITERATURE DOCUMENTS FOR CORRELATED VAPOR PRESSURE VALUES:

1146   1485   5718   7333   9050

LITERATURE DOCUMENTS FOR CORRELATED LIQUID DENSITY VALUES:

1487   8639   9050

LITERATURE DOCUMENTS REPORTING VIRIAL COEFFICIENT DATA:

T,K	VAPOR PRESSURE, KPA	SATURATED LIQUID VOLUME, ML/MOL	SECOND VIRIAL COEFFICIENT, ML/MOL	HEAT OF VAPORIZATION, J/MOL	SATURATED VAPOR VOLUME, ML/MOL
222	0.01402		-15423.		0.1316E+09
229	0.02806		-12852.		0.6783E+08
237	0.05852		-10590.		0.3366E+08
245	0.1154		-8858.		0.1765E+08
253	0.2163		-7513.		9715883.
261	0.3877		-6453.		5590609.
268	0.6246		-5703.		3561725.
276	1.040	156.01	-5002.	38902.	2201063.
284	1.674	157.31	-4430.	38448.	1405706.
292	2.614	158.64	-3957.	37998.	924937.
300	3.966	160.01	-3563.	37549.	625313.
308	5.866	161.41	-3230.	37101.	433297.
315	8.103	162.67	-2980.	36707.	320196.
323	11.49	164.15	-2732.	36255.	231036.
331	15.96	165.67	-2518.	35798.	169885.
339	21.77	167.24	-2331.	35335.	127097.
347	29.20	168.86	-2166.	34865.	96600.
355	38.55	170.54	-2021.	34385.	74490.
362	48.58	172.05	-1907.	33957.	59988.
370	62.48	173.84	-1789.	33457.	47376.
378	79.37	175.69	-1683.	32943.	37838.
386	99.65	177.63	-1588.	32413.	30532.
394	123.8	179.64	-1501.	31867.	24868.
402	152.2	181.75	-1421.	31303.	20429.
409	181.0	183.67	-1357.	30792.	17311.
417	218.9	185.98	-1290.	30186.	14421.
425	262.6	188.41	-1227.	29555.	12091.
433	312.6	190.98	-1170.	28896.	10196.
441	369.5	193.71	-1116.	28203.	8641.
448	425.4	196.25	-1072.	27566.	7504.
456	496.9	199.34	-1026.	26798.	6408.
464	577.1	202.68	-982.0	25977.	5489.
472	666.6	206.30	-941.0	25094.	4711.
480	766.1	210.26	-902.6	24131.	4047.
488	876.5	214.63	-866.4	23717.	3567.
495	982.5	218.86	-836.5	22853.	3141.
503	1115.	224.26	-804.1	21798.	2717.
511	1261.	230.43	-773.5	20656.	2348.
519	1421.	237.57	-744.6	19407.	2025.
527	1596.	246.03	-717.2	18024.	1740.
535	1788.	256.27	-691.1	16464.	1486.
542	1970.	267.28	-669.4	14901.	1284.
550	2196.	283.29	-645.7	12778.	1070.
558	2443.		-623.2		860.7
566	2715.		-601.7		593.3

COMPOUND CONSTANTS:

MP   373.780 K   TC   567.871 K   VC   460.60 ML/MOL   ZC   0.2797
NBP   379.362 K   PC   2.867 MPA   DC   0.2480 G/ML   OM
MU   0.0000 DEBYE   RG   3.8146 ANGSTROM

VAPOR PRESSURE CORRELATION CONSTANTS, K AND KPA:

WAGNER	WAGNER	WAGNER	WAGNER
374. - 383. K	374. - 383. K	374. - 383. K	374. - 383. K
6 POINTS	6 POINTS	6 POINTS	6 POINTS
RMSD = 0.0282	RMSD = 0.0282	RMSD = 0.0282	RMSD = 0.0282
0.13934861E+03	0.13934861E+03	0.13934861E+03	0.13934861E+03
-0.37384362E+03	-0.37384362E+03	-0.37384362E+03	-0.37384362E+03
0.72379530E+03	0.72379530E+03	0.72379530E+03	0.72379530E+03
-0.25904075E+04	-0.25904075E+04	-0.25904075E+04	-0.25904075E+04

LIQUID DENSITY CORRELATION CONSTANTS, K AND G/ML:

FRANCIS1	FRANCIS1	FRANCIS1
373. - 376. K	373. - 376. K	373. - 376. K
6 POINTS	6 POINTS	6 POINTS
RMSD = 0.00004	RMSD = 0.00004	RMSD = 0.00004
0.96011889E+00	0.96011889E+00	0.96011889E+00
0.72281645E-03	0.72281645E-03	0.72281645E-03
0.65472984E+01	0.65472984E+01	0.65472984E+01
0.57143213E+03	0.57143213E+03	0.57143213E+03

SECOND VIRIAL COEFFICIENT CORRELATION CONSTANTS, ML/MOL:

LITERATURE DOCUMENTS FOR CORRELATED VAPOR PRESSURE VALUES:

  6660

LITERATURE DOCUMENTS FOR CORRELATED LIQUID DENSITY VALUES:

  17545

LITERATURE DOCUMENTS REPORTING VIRIAL COEFFICIENT DATA:

T,K	VAPOR PRESSURE, KPA	SATURATED LIQUID VOLUME, ML/MOL	SECOND VIRIAL COEFFICIENT, ML/MOL	HEAT OF VAPORIZATION, J/MOL	SATURATED VAPOR VOLUME, ML/MOL
373		173.73			
374	87.16	173.97		32270.	34466.
375	89.70	174.20		31997.	33555.
376	92.28	174.44		31787.	32678.
377	94.90				31833.
378	97.58				31016.
379	100.3				30225.
380	103.1				29456.
381	106.0				28708.
382	108.9				27979.
383	112.0				27266.

COMPOUND CONSTANTS:

MP   219.645 K   TC   594.635 K   VC   548.01 ML/MOL   ZC
NBP   423.968 K   PC   2.288 MPA   DC   0.2340 G/ML   OM 0.4445
MU   0.0000 DEBYE   RG   5.1263 ANGSTROM

VAPOR PRESSURE CORRELATION CONSTANTS, K AND KPA:

WAGNER	VAPRES-2	VAPRES-2	WAGNER
343. - 425. K	223. - 425. K	223. - 425. K	408. - 425. K
36 POINTS	42 POINTS	42 POINTS	10 POINTS
RMSD = 0.0032	RMSD = 0.1339	RMSD = 0.1339	RMSD = 0.0022
-0.81449877E+01	-0.16476676E+04	-0.16476676E+04	-0.48851517E+01
0.14299870E+01	0.27512569E+05	0.27512569E+05	-0.74335982E+01
-0.43379960E+01	-0.96917185E+00	-0.96917185E+00	0.16188666E+02
-0.36949966E+01	0.47629229E-03	0.47629229E-03	-0.10753002E+03
	0.31616334E+03	0.31616334E+03	

LIQUID DENSITY CORRELATION CONSTANTS, K AND G/ML:

FRANCIS1	FRANCIS1	FRANCIS1	FRANCIS1
258. - 511. K	258. - 428. K	258. - 511. K	423. - 511. K
30 POINTS	26 POINTS	30 POINTS	6 POINTS
RMSD = 0.00056	RMSD = 0.00016	RMSD = 0.00056	RMSD = 0.00020
0.95642257E+00	0.94283748E+00	0.95642257E+00	0.10008402E+01
0.67875860E-03	0.69457525E-03	0.67875860E-03	0.84680831E-03
0.15000000E+02	0.59999990E+01	0.15000000E+02	0.91097889E+01
0.67176563E+03	0.57194946E+03	0.67176563E+03	0.69150537E+03

SECOND VIRIAL COEFFICIENT CORRELATION CONSTANTS, ML/MOL:

LITERATURE DOCUMENTS FOR CORRELATED VAPOR PRESSURE VALUES:

    1485 1486 4816

LITERATURE DOCUMENTS FOR CORRELATED LIQUID DENSITY VALUES:

    1211 1490 6516 8580 8902 9709 17433 21304 21607 40895

LITERATURE DOCUMENTS REPORTING VIRIAL COEFFICIENT DATA:

T,K	VAPOR PRESSURE, KPA	SATURATED LIQUID VOLUME, ML/MOL	SECOND VIRIAL COEFFICIENT, ML/MOL	HEAT OF VAPORIZATION, J/MOL	SATURATED VAPOR VOLUME, ML/MOL
258		172.15			
263		173.04			
269		174.12			
275		175.22			
281		176.35			
286		177.29			
292		178.45			
298		179.63			
304		180.82			
309		181.84			
315		183.08			
321		184.34			
327		185.62			
332		186.71			
338		188.04			
344	6.555	189.40		43250.	433882.
350	8.499	190.79		42825.	340010.
355	10.47	191.97		42471.	279697.
361	13.31	193.41		42047.	223249.
367	16.76	194.89		41621.	179830.
373	20.92	196.40		41194.	146101.
378	24.99	197.69		40836.	123638.
384	30.71	199.26		40403.	101898.
390	37.45	200.88		39964.	84581.
396	45.33	202.54		39520.	70678.
401	52.87	203.96		39145.	61148.
407	63.20	205.70		38687.	51671.
413	75.08	207.49		38221.	43906.
419	88.67	209.34		37745.	37502.
424	101.4	210.92		37341.	33008.
430		212.87			
436		214.89			
442		216.97			
447		218.76			
453		220.99			
459		223.30			
465		225.70			
470		227.78			
476		230.38			
482		233.10			
488		235.96			
493		238.45			
499		241.58			
505		244.90			
511		248.43			

COMPOUND CONSTANTS:

MP   170.220 K   TC 577.923 K   VC 535.01 ML/MOL   ZC
NBP   408.374 K   PC   2.300 MPA   DC 0.2397 G/ML   OM
MU   0.0000 DEBYE   RG

   VAPOR PRESSURE CORRELATION CONSTANTS, K AND KPA:

   LIQUID DENSITY CORRELATION CONSTANTS, K AND G/ML:

   FRANCIS1   FRANCIS1   FRANCIS1
   286. - 360. K   286. - 360. K   286. - 360. K
    6 POINTS   6 POINTS   6 POINTS
   RMSD = 0.00013   RMSD = 0.00013   RMSD = 0.00013
   0.94024700E+00   0.94024700E+00   0.94024700E+00
   0.71316352E-03   0.71316352E-03   0.71316352E-03
   0.62458658E+01   0.62458658E+01   0.62458658E+01
   0.58554761E+03   0.58554761E+03   0.58554761E+03

   SECOND VIRIAL COEFFICIENT CORRELATION CONSTANTS, ML/MOL:

LITERATURE DOCUMENTS FOR CORRELATED VAPOR PRESSURE VALUES:

LITERATURE DOCUMENTS FOR CORRELATED LIQUID DENSITY VALUES:

   14407

LITERATURE DOCUMENTS REPORTING VIRIAL COEFFICIENT DATA:

T,K	VAPOR PRESSURE, KPA	SATURATED LIQUID VOLUME, ML/MOL	SECOND VIRIAL COEFFICIENT, ML/MOL	HEAT OF VAPORIZATION, J/MOL	SATURATED VAPOR VOLUME, ML/MOL
286		179.27			
287		179.47			
289		179.86			
291		180.26			
292		180.46			
294		180.86			
296		181.26			
297		181.46			
299		181.87			
301		182.28			
302		182.48			
304		182.89			
306		183.31			
307		183.52			
309		183.94			
311		184.36			
312		184.57			
314		184.99			
316		185.42			
317		185.63			
319		186.06			
321		186.50			
322		186.72			
324		187.15			
326		187.59			
328		188.04			
329		188.26			
331		188.71			
333		189.16			
334		189.39			
336		189.84			
338		190.30			
339		190.53			
341		191.00			
343		191.46			
344		191.70			
346		192.17			
348		192.64			
349		192.88			
351		193.36			
353		193.85			
354		194.09			
356		194.58			
358		195.07			
359		195.32			

COMPOUND CONSTANTS:

MP		TC 588.024 K		VC 498.01 ML/MOL	ZC
NBP 406.759 K		PC 2.493 MPA		DC 0.2575 G/ML	OM 0.3358
MU 0.0000 DEBYE		RG 4.6093 ANGSTROM			

VAPOR PRESSURE CORRELATION CONSTANTS, K AND KPA:

WAGNER	WAGNER	WAGNER
238. - 407. K	238. - 407. K	238. - 407. K
15 POINTS	15 POINTS	15 POINTS
RMSD = 0.0026	RMSD = 0.0026	RMSD = 0.0026
-0.75116625E+01	-0.75116625E+01	-0.75116625E+01
0.12341142E+01	0.12341142E+01	0.12341142E+01
-0.37384358E+01	-0.37384358E+01	-0.37384358E+01
-0.20227091E+01	-0.20227091E+01	-0.20227091E+01

LIQUID DENSITY CORRELATION CONSTANTS, K AND G/ML:

FRANCIS1	FRANCIS1	FRANCIS1
293. - 299. K	293. - 299. K	293. - 299. K
4 POINTS	4 POINTS	4 POINTS
RMSD = 0.00004	RMSD = 0.00004	RMSD = 0.00004
0.94979548E+00	0.94979548E+00	0.94979548E+00
0.67463052E-03	0.67463052E-03	0.67463052E-03
0.59999990E+01	0.59999990E+01	0.59999990E+01
0.55903052E+03	0.55903052E+03	0.55903052E+03

SECOND VIRIAL COEFFICIENT CORRELATION CONSTANTS, ML/MOL:

LITERATURE DOCUMENTS FOR CORRELATED VAPOR PRESSURE VALUES:

   3269   5718

LITERATURE DOCUMENTS FOR CORRELATED LIQUID DENSITY VALUES:

   3269

LITERATURE DOCUMENTS REPORTING VIRIAL COEFFICIENT DATA:

T,K	VAPOR PRESSURE, KPA	SATURATED LIQUID VOLUME, ML/MOL	SECOND VIRIAL COEFFICIENT, ML/MOL	HEAT OF VAPORIZATION, J/MOL	SATURATED VAPOR VOLUME, ML/MOL
238	0.01839				0.1076E+09
241	0.02440				0.8212E+08
245	0.03513				0.5798E+08
249	0.04990				0.4149E+08
253	0.06995				0.3007E+08
257	0.09685				0.2206E+08
261	0.1325				0.1637E+08
264	0.1664				0.1319E+08
268	0.2234				9973123.
272	0.2967				7618751.
276	0.3903				5876083.
280	0.5087				4573469.
284	0.6571				3590618.
287	0.7917				3011113.
291	1.008				2397856.
295	1.273	176.16		41495.	1924080.
299	1.596	176.90		41257.	1555163.
303	1.986				1265724.
307	2.456				1036995.
310	2.866				896735.
314	3.504				742723.
318	4.257				618767.
322	5.142				518385.
326	6.176				436615.
330	7.379				369627.
334	8.771				314450.
337	9.952				279403.
341	11.73				239617.
345	13.76				206390.
349	16.07				178509.
353	18.69				155010.
357	21.65				135117.
360	24.11				122185.
364	27.74				107176.
368	31.81				94329.
372	36.33				83292.
376	41.36				73774.
380	46.93				65539.
383	51.49				60084.
387	58.11				53640.
391	65.39				48013.
395	73.36				43086.
399	82.09				38758.
403	91.61				34946.
406	99.30				32382.

COMPOUND CONSTANTS:

MP 153.153 K	TC 573.622 K	VC 507.01 ML/MOL	ZC
NBP 399.700 K	PC 2.381 MPA	DC 0.2530 G/ML	OM 0.3498
MU 0.0000 DEBYE	RG 4.5391 ANGSTROM		

VAPOR PRESSURE CORRELATION CONSTANTS, K AND KPA:

WAGNER	WAGNER	WAGNER
238. - 400. K	238. - 400. K	238. - 400. K
15 POINTS	15 POINTS	15 POINTS
RMSD = 0.0007	RMSD = 0.0007	RMSD = 0.0007
-0.77064292E+01	-0.77064292E+01	-0.77064292E+01
0.14344532E+01	0.14344532E+01	0.14344532E+01
-0.36231602E+01	-0.36231602E+01	-0.36231602E+01
-0.21350762E+01	-0.21350762E+01	-0.21350762E+01

LIQUID DENSITY CORRELATION CONSTANTS, K AND G/ML:

FRANCIS1	FRANCIS1	FRANCIS1
293. - 299. K	293. - 299. K	293. - 299. K
4 POINTS	4 POINTS	4 POINTS
RMSD = 0.00004	RMSD = 0.00004	RMSD = 0.00004
0.93413663E+00	0.93413663E+00	0.93413663E+00
0.66882488E-03	0.66882488E-03	0.66882488E-03
0.59999990E+01	0.59999990E+01	0.59999990E+01
0.55901392E+03	0.55901392E+03	0.55901392E+03

SECOND VIRIAL COEFFICIENT CORRELATION CONSTANTS, ML/MOL:

LITERATURE DOCUMENTS FOR CORRELATED VAPOR PRESSURE VALUES:

5718 21602

LITERATURE DOCUMENTS FOR CORRELATED LIQUID DENSITY VALUES:

21602

LITERATURE DOCUMENTS REPORTING VIRIAL COEFFICIENT DATA:

T,K	VAPOR PRESSURE, KPA	SATURATED LIQUID VOLUME, ML/MOL	SECOND VIRIAL COEFFICIENT, ML/MOL	HEAT OF VAPORIZATION, J/MOL	SATURATED VAPOR VOLUME, ML/MOL
238	0.03019				0.6555E+08
241	0.03964				0.5055E+08
245	0.05631				0.3617E+08
249	0.07894				0.2622E+08
252	0.1009				0.2077E+08
256	0.1384				0.1538E+08
260	0.1876				0.1152E+08
263	0.2341				9338354.
267	0.3115				7122752.
271	0.4105				5486191.
274	0.5017				4538355.
278	0.6502				3551882.
282	0.8355				2803637.
285	1.003				2360442.
289	1.270				1889381.
293	1.596	179.23		40047.	1523659.
296	1.885	179.79		39874.	1302719.
300	2.340				1063495.
304	2.884				873912.
307	3.359				757368.
311	4.095				629091.
315	4.961				525566.
318	5.707				460935.
322	6.846				388753.
326	8.169				329549.
330	9.698				280724.
333	10.99				249678.
337	12.94				214398.
341	15.16				184897.
344	17.03				165916.
348	19.80				144104.
352	22.93				125637.
355	25.54				113630.
359	29.37				99692.
363	33.66				87760.
366	37.20				79927.
370	42.38				70753.
374	48.12				62819.
377	52.81				57567.
381	59.63				51365.
385	67.14				45952.
388	73.24				42341.
392	82.04				38044.
396	91.66				34264.
399	99.44				31724.

COMPOUND CONSTANTS:

MP   167.387 K   TC   568.021 K   VC   519.01 ML/MOL   ZC
NBP   397.243 K   PC   2.330 MPA   DC   0.2471 G/ML   OM 0.3573
MU   0.0000 DEBYE   RG

VAPOR PRESSURE CORRELATION CONSTANTS, K AND KPA:

WAGNER	WAGNER	WAGNER	WAGNER
238. - 399. K	238. - 399. K	238. - 399. K	382. - 399. K
30 POINTS	30 POINTS	30 POINTS	5 POINTS
RMSD = 0.0107	RMSD = 0.0107	RMSD = 0.0107	RMSD = 0.0018
-0.75553182E+01	-0.75553182E+01	-0.75553182E+01	-0.14342507E+02
0.10714017E+01	0.10714017E+01	0.10714017E+01	0.19033412E+02
-0.34996917E+01	-0.34996917E+01	-0.34996917E+01	-0.41898268E+02
-0.32726540E+01	-0.32726540E+01	-0.32726540E+01	0.16303228E+03

LIQUID DENSITY CORRELATION CONSTANTS, K AND G/ML:

FRANCIS1	FRANCIS1	FRANCIS1
293. - 304. K	293. - 304. K	293. - 304. K
5 POINTS	5 POINTS	5 POINTS
RMSD = 0.00002	RMSD = 0.00002	RMSD = 0.00002
0.91377491E+00	0.91377491E+00	0.91377491E+00
0.57944143E-03	0.57944143E-03	0.57944143E-03
0.60153685E+01	0.60153685E+01	0.60153685E+01
0.45707104E+03	0.45707104E+03	0.45707104E+03

SECOND VIRIAL COEFFICIENT CORRELATION CONSTANTS, ML/MOL:

LITERATURE DOCUMENTS FOR CORRELATED VAPOR PRESSURE VALUES:

  1486 5718

LITERATURE DOCUMENTS FOR CORRELATED LIQUID DENSITY VALUES:

  1490

LITERATURE DOCUMENTS REPORTING VIRIAL COEFFICIENT DATA:

T,K	VAPOR PRESSURE, KPA	SATURATED LIQUID VOLUME, ML/MOL	SECOND VIRIAL COEFFICIENT, ML/MOL	HEAT OF VAPORIZATION, J/MOL	SATURATED VAPOR VOLUME, ML/MOL
238	0.02933				0.6746E+08
241	0.03877				0.5169E+08
245	0.05550				0.3670E+08
248	0.07199				0.2864E+08
252	0.1006				0.2082E+08
256	0.1389				0.1532E+08
259	0.1755				0.1226E+08
263	0.2374				9209265.
267	0.3175				6989604.
270	0.3921				5721870.
274	0.5151				4419360.
278	0.6703				3445284.
281	0.8118				2875133.
285	1.040				2275774.
289	1.321				1816035.
292	1.573				1541103.
296	1.971	181.95		40322.	1246215.
300	2.451	182.79		40063.	1014930.
303	2.874	183.44		39870.	873983.
307	3.533				720103.
311	4.314				597019.
314	4.991				520752.
318	6.030				436170.
322	7.243				367330.
325	8.281				324040.
329	9.855				275358.
333	11.67				235119.
336	13.20				209481.
340	15.50				180289.
344	18.12				155826.
347	20.31				140056.
351	23.56				121899.
355	27.22				106495.
358	30.25				96461.
362	34.72				84792.
366	39.70				74783.
369	43.80				68201.
373	49.79				60478.
377	56.41				53787.
380	61.81				49350.
384	69.66				44101.
388	78.26				39513.
391	85.25				36447.
395	95.32				32794.
398	103.5				30342.

COMPOUND CONSTANTS:

MP   159.780 K   TC   581.624 K   VC   500.01 ML/MOL   ZC
NBP   403.810 K   PC   2.432 MPA   DC   0.2565 G/ML   OM   0.3421
MU   0.0000 DEBYE   RG

VAPOR PRESSURE CORRELATION CONSTANTS, K AND KPA:

WAGNER	WAGNER	WAGNER	WAGNER
323. - 405. K	323. - 405. K	323. - 405. K	388. - 405. K
17 POINTS	17 POINTS	17 POINTS	5 POINTS
RMSD = 0.0035	RMSD = 0.0035	RMSD = 0.0035	RMSD = 0.0034
-0.79010053E+01	-0.79010053E+01	-0.79010053E+01	0.30515439E+02
0.20148395E+01	0.20148395E+01	0.20148395E+01	-0.99473562E+02
-0.46191445E+01	-0.46191445E+01	-0.46191445E+01	0.21097985E+03
0.45384011E+00	0.45384011E+00	0.45384011E+00	-0.91782222E+03

LIQUID DENSITY CORRELATION CONSTANTS, K AND G/ML:

FRANCIS1	FRANCIS1	FRANCIS1
293. - 304. K	293. - 304. K	293. - 304. K
4 POINTS	4 POINTS	4 POINTS
RMSD = 0.00005	RMSD = 0.00005	RMSD = 0.00005
0.94604588E+00	0.94604588E+00	0.94604588E+00
0.67987316E-03	0.67987316E-03	0.67987316E-03
0.60890713E+01	0.60890713E+01	0.60890713E+01
0.55895679E+03	0.55895679E+03	0.55895679E+03

SECOND VIRIAL COEFFICIENT CORRELATION CONSTANTS, ML/MOL:

LITERATURE DOCUMENTS FOR CORRELATED VAPOR PRESSURE VALUES:

  1486

LITERATURE DOCUMENTS FOR CORRELATED LIQUID DENSITY VALUES:

  1490

LITERATURE DOCUMENTS REPORTING VIRIAL COEFFICIENT DATA:

T,K	VAPOR PRESSURE, KPA	SATURATED LIQUID VOLUME, ML/MOL	SECOND VIRIAL COEFFICIENT, ML/MOL	HEAT OF VAPORIZATION, J/MOL	SATURATED VAPOR VOLUME, ML/MOL
293		177.16			
295		177.54			
298		178.11			
300		178.49			
303		179.07			
323	6.178				432392.
326	7.062				381556.
328	7.708				351567.
331	8.769				311638.
333	9.541				287988.
336	10.81				256377.
338	11.72				237582.
341	13.22				212366.
343	14.30				197317.
346	16.06				177057.
348	17.33				164923.
351	19.38				148532.
354	21.64				134051.
356	23.25				125333.
359	25.86				113496.
361	27.72				106347.
364	30.72				96612.
366	32.86				90714.
369	36.30				82660.
371	38.74				77767.
374	42.66				71067.
376	45.44				66985.
379	49.88				61381.
382	54.66				56334.
384	58.05				53247.
387	63.43				48992.
389	67.24				46382.
392	73.28				42777.
394	77.54				40562.
397	84.29				37493.
399	89.04				35603.
402	96.56				32980.
404	101.8				31361.

COMPOUND CONSTANTS:

MP   240.060 K   TC 610.026 K   VC 473.01 ML/MOL   ZC
NBP 419.336 K   PC   2.675 MPA   DC 0.2711 G/ML   OM
MU   0.0000 DEBYE   RG

   VAPOR PRESSURE CORRELATION CONSTANTS, K AND KPA:

   WAGNER   WAGNER   WAGNER   WAGNER
   336. - 421. K   336. - 421. K   336. - 421. K   403. - 421. K
   17 POINTS   17 POINTS   17 POINTS   5 POINTS
   RMSD = 0.0043   RMSD = 0.0043   RMSD = 0.0043   RMSD = 0.0041
   -0.80140519E+01   -0.80140519E+01   -0.80140519E+01   -0.12112738E+02
    0.22125913E+01   0.22125913E+01   0.22125913E+01   0.12899263E+02
   -0.43129564E+01   -0.43129564E+01   -0.43129564E+01   -0.26126979E+02
   -0.52715447E-02   -0.52715447E-02   -0.52715447E-02   0.85570504E+02

   LIQUID DENSITY CORRELATION CONSTANTS, K AND G/ML:

   FRANCIS1   FRANCIS1   FRANCIS1
   273. - 304. K   273. - 304. K   273. - 304. K
    6 POINTS   6 POINTS   6 POINTS
   RMSD = 0.00014   RMSD = 0.00014   RMSD = 0.00014
   0.96241248E+00   0.96241248E+00   0.96241248E+00
   0.53525832E-03   0.53525832E-03   0.53525832E-03
   0.14322139E+02   0.14322139E+02   0.14322139E+02
   0.56838794E+03   0.56838794E+03   0.56838794E+03

   SECOND VIRIAL COEFFICIENT CORRELATION CONSTANTS, ML/MOL:

LITERATURE DOCUMENTS FOR CORRELATED VAPOR PRESSURE VALUES:

   1486

LITERATURE DOCUMENTS FOR CORRELATED LIQUID DENSITY VALUES:

   1490 2523

LITERATURE DOCUMENTS REPORTING VIRIAL COEFFICIENT DATA:

T,K	VAPOR PRESSURE, KPA	SATURATED LIQUID VOLUME, ML/MOL	SECOND VIRIAL COEFFICIENT, ML/MOL	HEAT OF VAPORIZATION, J/MOL	SATURATED VAPOR VOLUME, ML/MOL
273		167.04			
276		167.50			
279		167.97			
283		168.59			
286		169.07			
289		169.55			
293		170.20			
296		170.69			
299		171.19			
303		171.87			
336	6.394				435078.
340	7.575				371383.
343	8.576				330770.
346	9.684				295309.
350	11.35				254792.
353	12.74				228696.
357	14.82				198681.
360	16.55				179223.
363	18.45				162000.
367	21.25				142015.
370	23.57				128948.
373	26.09				117298.
377	29.80				103673.
380	32.84				94693.
383	36.14				86635.
387	40.94				77140.
390	44.86				70838.
394	50.55				63376.
397	55.19				58399.
400	60.16				53888.
404	67.33				48514.
407	73.14				44908.
410	79.34				41622.
414	88.24				37684.
417	95.42				35026.
420	103.0				32592.

COMPOUND CONSTANTS:

MP   263.255 K   TC   607.626 K   VC   478.01 ML/MOL   ZC
NBP 413.440 K   PC   2.741 MPA   DC   0.2683 G/ML   OM
MU   0.0000 DEBYE   RG   4.1556 ANGSTROM

VAPOR PRESSURE CORRELATION CONSTANTS, K AND KPA:

WAGNER	WAGNER	WAGNER	WAGNER
330. - 416. K	330. - 416. K	330. - 416. K	395. - 416. K
36 POINTS	36 POINTS	36 POINTS	10 POINTS
RMSD = 0.0024	RMSD = 0.0024	RMSD = 0.0024	RMSD = 0.0010
-0.74616377E+01	-0.74616377E+01	-0.74616377E+01	-0.73783732E+01
0.13504465E+01	0.13504465E+01	0.13504465E+01	0.11474022E+01
-0.30581863E+01	-0.30581863E+01	-0.30581863E+01	-0.27248199E+01
-0.31853870E+01	-0.31853870E+01	-0.31853870E+01	-0.39450341E+01

LIQUID DENSITY CORRELATION CONSTANTS, K AND G/ML:

FRANCIS1	FRANCIS1	FRANCIS1
293. - 304. K	293. - 304. K	293. - 304. K
4 POINTS	4 POINTS	4 POINTS
RMSD = 0.00003	RMSD = 0.00003	RMSD = 0.00003
0.97175360E+00	0.97175360E+00	0.97175360E+00
0.65948558E-03	0.65948558E-03	0.65948558E-03
0.59999990E+01	0.59999990E+01	0.59999990E+01
0.56838794E+03	0.56838794E+03	0.56838794E+03

SECOND VIRIAL COEFFICIENT CORRELATION CONSTANTS, ML/MOL:

LITERATURE DOCUMENTS FOR CORRELATED VAPOR PRESSURE VALUES:

    1486   7333

LITERATURE DOCUMENTS FOR CORRELATED LIQUID DENSITY VALUES:

    1490

LITERATURE DOCUMENTS REPORTING VIRIAL COEFFICIENT DATA:

T,K	VAPOR PRESSURE, KPA	SATURATED LIQUID VOLUME, ML/MOL	SECOND VIRIAL COEFFICIENT, ML/MOL	HEAT OF VAPORIZATION, J/MOL	SATURATED VAPOR VOLUME, ML/MOL
293		169.49			
295		169.82			
298		170.32			
301		170.83			
304		171.33			
332	6.708				409696.
334	7.310				378112.
337	8.296				335981.
340	9.390				299312.
343	10.60				267307.
346	11.94				239299.
348	12.90				222566.
351	14.47				200012.
354	16.19				180141.
357	18.08				162590.
360	20.14				147053.
362	21.61				137683.
365	23.99				124938.
368	26.58				113589.
371	29.39				103461.
374	32.43				94404.
376	34.59				88897.
379	38.05				81347.
382	41.79				74560.
385	45.81				68448.
388	50.14				62934.
390	53.20				59557.
393	58.06				54894.
396	63.27				50668.
399	68.85				46832.
402	74.80				43344.
404	78.99				41194.
407	85.62				38207.
410	92.68				35480.
413	100.2				32987.
415	105.4				31443.

COMPOUND CONSTANTS:

MP 152.062 K	TC 592.725 K	VC 490.01 ML/MOL	ZC
NBP 406.181 K	PC 2.602 MPA	DC 0.2617 G/ML	OM
MU 0.0000 DEBYE	RG		

VAPOR PRESSURE CORRELATION CONSTANTS, K AND KPA:

WAGNER	WAGNER	WAGNER	WAGNER
325. - 408. K	325. - 408. K	325. - 408. K	390. - 408. K
19 POINTS	19 POINTS	19 POINTS	6 POINTS
RMSD = 0.0032	RMSD = 0.0032	RMSD = 0.0032	RMSD = 0.0033
-0.76154042E+01	-0.76154042E+01	-0.76154042E+01	-0.98436573E+01
0.16458501E+01	0.16458501E+01	0.16458501E+01	0.73988513E+01
-0.37401798E+01	-0.37401798E+01	-0.37401798E+01	-0.15149244E+02
-0.15003250E+01	-0.15003250E+01	-0.15003250E+01	0.40884576E+02

LIQUID DENSITY CORRELATION CONSTANTS, K AND G/ML:

FRANCIS1	FRANCIS1	FRANCIS1
293. - 304. K	293. - 304. K	293. - 304. K
4 POINTS	4 POINTS	4 POINTS
RMSD = 0.00002	RMSD = 0.00002	RMSD = 0.00002
0.95690703E+00	0.95690703E+00	0.95690703E+00
0.66914340E-03	0.66914340E-03	0.66914340E-03
0.60015259E+01	0.60015259E+01	0.60015259E+01
0.56838794E+03	0.56838794E+03	0.56838794E+03

SECOND VIRIAL COEFFICIENT CORRELATION CONSTANTS, ML/MOL:

LITERATURE DOCUMENTS FOR CORRELATED VAPOR PRESSURE VALUES:

  1486 12551

LITERATURE DOCUMENTS FOR CORRELATED LIQUID DENSITY VALUES:

  1490

LITERATURE DOCUMENTS REPORTING VIRIAL COEFFICIENT DATA:

T,K	VAPOR PRESSURE, KPA	SATURATED LIQUID VOLUME, ML/MOL	SECOND VIRIAL COEFFICIENT, ML/MOL	HEAT OF VAPORIZATION, J/MOL	SATURATED VAPOR VOLUME, ML/MOL
293		173.54			
295		173.89			
298		174.43			
300		174.78			
303		175.32			
326	6.648				405892.
329	7.577				359229.
332	8.611				318787.
334	9.364				294816.
337	10.59				262751.
340	11.95				234753.
342	12.94				218055.
345	14.54				195589.
347	15.70				182143.
350	17.57				163988.
353	19.62				147961.
355	21.10				138316.
358	23.47				125227.
360	25.18				117326.
363	27.92				106570.
366	30.89				96982.
368	33.01				91166.
371	36.42				83213.
374	40.09				76084.
376	42.70				71741.
379	46.88				65777.
381	49.83				62133.
384	54.54				57117.
387	59.60				52585.
389	63.17				49805.
392	68.85				45964.
394	72.85				43602.
397	79.19				40331.
400	85.96				37355.
402	90.72				35519.
405	98.24				32966.
407	103.5				31389.

COMPOUND CONSTANTS:

MP   206.643 K   TC   574.722 K   VC 504.01 ML/MOL   ZC
NBP   395.442 K   PC   2.485 MPA   DC 0.2545 G/ML   OM 0.3118
MU   0.0000 DEBYE   RG

VAPOR PRESSURE CORRELATION CONSTANTS, K AND KPA:

WAGNER	WAGNER	WAGNER	WAGNER
315. - 423. K	315. - 423. K	315. - 423. K	380. - 423. K
40 POINTS	40 POINTS	40 POINTS	14 POINTS
RMSD = 0.0119	RMSD = 0.0119	RMSD = 0.0119	RMSD = 0.0039
-0.75974807E+01	-0.75974807E+01	-0.75974807E+01	-0.71183484E+01
0.15994698E+01	0.15994698E+01	0.15994698E+01	0.33345822E+00
-0.35699444E+01	-0.35699444E+01	-0.35699444E+01	-0.89569449E+00
-0.20233706E+01	-0.20233706E+01	-0.20233706E+01	-0.13023011E+02

LIQUID DENSITY CORRELATION CONSTANTS, K AND G/ML:

FRANCIS1	FRANCIS1	FRANCIS1
293. - 304. K	293. - 304. K	293. - 304. K
4 POINTS	4 POINTS	4 POINTS
RMSD = 0.00005	RMSD = 0.00005	RMSD = 0.00005
0.94732237E+00	0.94732237E+00	0.94732237E+00
0.70295599E-03	0.70295599E-03	0.70295599E-03
0.59999990E+01	0.59999990E+01	0.59999990E+01
0.56838794E+03	0.56838794E+03	0.56838794E+03

SECOND VIRIAL COEFFICIENT CORRELATION CONSTANTS, ML/MOL:

LITERATURE DOCUMENTS FOR CORRELATED VAPOR PRESSURE VALUES:

   1486   6893   7333

LITERATURE DOCUMENTS FOR CORRELATED LIQUID DENSITY VALUES:

   1490

LITERATURE DOCUMENTS REPORTING VIRIAL COEFFICIENT DATA:

T,K	VAPOR PRESSURE, KPA	SATURATED LIQUID VOLUME, ML/MOL	SECOND VIRIAL COEFFICIENT, ML/MOL	HEAT OF VAPORIZATION, J/MOL	SATURATED VAPOR VOLUME, ML/MOL
293		178.24			
295		178.63			
298		179.22			
301		179.81			
304		180.41			
316	6.381				409516.
319	7.301				361086.
322	8.329				319295.
325	9.473				283121.
328	10.74				251716.
331	12.15				224373.
334	13.71				200500.
337	15.43				179599.
340	17.32				161253.
343	19.39				145108.
346	21.66				130865.
349	24.15				118269.
352	26.85				· 107104.
355	29.80				97185.
357	31.91				91185.
360	35.29				83001.
363	38.95				75688.
366	42.92				69141.
369	47.19				63266.
372	51.80				57986.
375	56.76				53230.
378	62.08				48940.
381	67.80				45062.
384	73.91				41551.
387	80.46				38366.
390	87.45				35473.
393	94.90				32841.
396	102.8				30442.
399	111.3				28253.
402	120.3				26251.
405	129.8				24420.
408	139.9				22740.
411	150.6				21199.
414	161.9				19782.
417	173.9				18478.
420	186.5				17276.
422	195.3				16527.

COMPOUND CONSTANTS:

MP 171.047 K	TC 607.626 K	VC 481.01 ML/MOL	ZC
NBP 414.717 K	PC 2.716 MPA	DC 0.2666 G/ML	OM
MU 0.0000 DEBYE	RG		

    VAPOR PRESSURE CORRELATION CONSTANTS, K AND KPA:

WAGNER	WAGNER	WAGNER	WAGNER
332. - 416. K	332. - 416. K	332. - 416. K	399. - 416. K
18 POINTS	18 POINTS	18 POINTS	7 POINTS
RMSD = 0.0031	RMSD = 0.0031	RMSD = 0.0031	RMSD = 0.0031
-0.76532929E+01	-0.76532929E+01	-0.76532929E+01	-0.85238824E+01
0.17201713E+01	0.17201713E+01	0.17201713E+01	0.39749910E+01
-0.37935769E+01	-0.37935769E+01	-0.37935769E+01	-0.83048107E+01
-0.10600143E+01	-0.10600143E+01	-0.10600143E+01	0.15936261E+02

    LIQUID DENSITY CORRELATION CONSTANTS, K AND G/ML:

FRANCIS1	FRANCIS1	FRANCIS1
293. - 304. K	293. - 304. K	293. - 304. K
4 POINTS	4 POINTS	4 POINTS
RMSD = 0.00003	RMSD = 0.00003	RMSD = 0.00003
0.96633476E+00	0.96633476E+00	0.96633476E+00
0.64407452E-03	0.64407452E-03	0.64407452E-03
0.60532598E+01	0.60532598E+01	0.60532598E+01
0.55873145E+03	0.55873145E+03	0.55873145E+03

    SECOND VIRIAL COEFFICIENT CORRELATION CONSTANTS, ML/MOL:

LITERATURE DOCUMENTS FOR CORRELATED VAPOR PRESSURE VALUES:

    1486 12551

LITERATURE DOCUMENTS FOR CORRELATED LIQUID DENSITY VALUES:

    1490

LITERATURE DOCUMENTS REPORTING VIRIAL COEFFICIENT DATA:

T,K	VAPOR PRESSURE, KPA	SATURATED LIQUID VOLUME, ML/MOL	SECOND VIRIAL COEFFICIENT, ML/MOL	HEAT OF VAPORIZATION, J/MOL	SATURATED VAPOR VOLUME, ML/MOL
293		169.91			
295		170.24			
298		170.74			
301		171.24			
304		171.75			
332	6.360				432199.
334	6.934				398711.
337	7.875				354049.
340	8.919				315188.
343	10.08				281284.
346	11.36				251625.
348	12.28				233913.
351	13.78				210048.
354	15.44				189033.
357	17.25				170483.
360	19.23				154070.
362	20.65				144177.
365	22.94				130728.
368	25.43				118759.
371	28.14				108086.
374	31.08				98548.
376	33.17				92751.
379	36.52				84808.
382	40.14				77674.
385	44.03				71254.
388	48.23				65467.
390	51.20				61924.
393	55.92				57036.
396	60.98				52609.
399	66.40				48593.
402	72.20				44944.
404	76.28				42696.
407	82.73				39576.
410	89.61				36729.
413	96.93				34128.
415	102.1				32519.

COMPOUND CONSTANTS:

MP 145.730 K	TC 398.260 K	VC 162.80 ML/MOL	ZC 0.2741
NBP 240.370 K	PC 5.575 MPA	DC 0.2585 G/ML	OM 0.1298
MU 0.0000 DEBYE	RG		

VAPOR PRESSURE CORRELATION CONSTANTS, K AND KPA:

WAGNER	VAPRES-2	RPM2	WAGNER
146. - 398. K	146. - 241. K	146. - 338. K	225. - 398. K
44 POINTS	17 POINTS	32 POINTS	32 POINTS
RMSD = 5.5580	RMSD = 0.1470	RMSD = 1.6720	RMSD = 1.6390
-0.67469899E+01	-0.18565639E+04	0.26109717E+02	-0.65889930E+01
0.16449332E+01	0.21878349E+05	-0.35296614E+04	0.12482239E+01
-0.22498072E+01	-0.20601408E+01	-0.38661875E-01	-0.17917701E+01
-0.26444557E+01	0.17495788E-02	0.43196655E-04	-0.15080535E+01
	0.39478404E+03		

LIQUID DENSITY CORRELATION CONSTANTS, K AND G/ML:

FRANCIS1	FRANCIS1	FRANCIS1	FRANCIS1
193. - 388. K	193. - 243. K	193. - 338. K	223. - 388. K
29 POINTS	9 POINTS	19 POINTS	26 POINTS
RMSD = 0.00213	RMSD = 0.00097	RMSD = 0.00212	RMSD = 0.00194
0.92811120E+00	0.95654356E+00	0.91856074E+00	0.92207634E+00
0.71606180E-03	0.10180303E-02	0.82407543E-03	0.69484650E-03
0.14999308E+02	0.66000118E+01	0.59999990E+01	0.15000000E+02
0.44909351E+03	0.45592383E+03	0.40447095E+03	0.44842554E+03

SECOND VIRIAL COEFFICIENT CORRELATION CONSTANTS, ML/MOL:

PITZER-CURL: TC=398.250   PC=5.575   VC=162.80   OM=0.1298

LITERATURE DOCUMENTS FOR CORRELATED VAPOR PRESSURE VALUES:

   1899   1922   2344 21292

LITERATURE DOCUMENTS FOR CORRELATED LIQUID DENSITY VALUES:

   452   1894   2344

LITERATURE DOCUMENTS REPORTING VIRIAL COEFFICIENT DATA:

   1939   2209   4462

T,K	VAPOR PRESSURE, KPA	SATURATED LIQUID VOLUME, ML/MOL	SECOND VIRIAL COEFFICIENT, ML/MOL	HEAT OF VAPORIZATION, J/MOL	SATURATED VAPOR VOLUME, ML/MOL
146	0.06115		-3268.		0.1985E+08
151	0.1202		-2759.		0.1044E+08
157	0.2530		-2297.		5157253.
163	0.4993		-1950.		2712078.
168	0.8423		-1725.		1656527.
174	1.505		-1509.		959473.
180	2.570		-1337.		580918.
186	4.213		-1199.		365862.
191	6.184		-1103.		255693.
197	9.508	57.84	-1005.	22159.	171264.
203	14.19	58.30	-922.7	21868.	118049.
208	19.39	58.69	-863.4	21628.	88304.
214	27.59	59.18	-801.2	21341.	63680.
220	38.38	59.68	-746.9	21055.	46906.
226	52.30	60.20	-699.2	20769.	35218.
231	66.72	60.64	-663.5	20530.	28106.
237	87.98	61.19	-624.8	20240.	21753.
243	114.2	61.76	-590.0	19947.	17080.
249	146.1	62.35	-558.5	19649.	13587.
254	177.6	62.87	-534.3	19396.	11330.
260	222.0	63.51	-507.6	19087.	9202.
266	274.3	64.17	-483.1	18770.	7547.
271	324.6	64.75	-464.2	18499.	6442.
277	393.7	65.48	-442.9	18165.	5367.
283	473.3	66.25	-423.2	17819.	4504.
289	564.3	67.06	-404.9	17460.	3805.
294	649.5	67.78	-390.5	17150.	3321.
300	763.8	68.68	-374.3	16969.	2869.
306	892.3	69.65	-359.1	16595.	2465.
312	1036.	70.69	-344.8	16206.	2128.
317	1168.	71.62	-333.5	15869.	1888.
323	1342.	72.82	-320.6	15446.	1640.
329	1534.	74.13	-308.5	15000.	1429.
334	1709.	75.33	-298.9	14609.	1276.
340	1938.	76.91	-287.9	14111.	1116.
346	2189.	78.67	-277.5	13578.	976.9
352	2463.	80.68	-267.6	13001.	855.2
357	2710.	82.57	-259.7	12479.	765.0
363	3031.	85.19	-250.6	11792.	668.0
369	3380.	88.29	-242.0	11021.	581.2
375	3759.	92.05	-233.8	10134.	502.5
380	4100.	95.87	-227.2	9271.	441.6
386	4541.	101.65	-219.6	8010.	371.9
392	5023.		-212.4		300.9
397	5459.		-206.5		119.6

COMPOUND CONSTANTS:

MP	182.450 K	TC	460.000 K	VC	210.00 ML/MOL	ZC	0.2740
NBP	285.660 K	PC	4.985 MPA	DC	0.2672 G/ML	OM	
MU	0.0000 DEBYE	RG					

VAPOR PRESSURE CORRELATION CONSTANTS, K AND KPA:

WAGNER	WAGNER	WAGNER
199. - 286. K	199. - 286. K	199. - 286. K
23 POINTS	23 POINTS	23 POINTS
RMSD = 0.0746	RMSD = 0.0746	RMSD = 0.0746
-0.69494863E+01	-0.69494863E+01	-0.69494863E+01
0.13782052E+01	0.13782052E+01	0.13782052E+01
-0.18686267E+01	-0.18686267E+01	-0.18686267E+01
-0.20948481E+01	-0.20948481E+01	-0.20948481E+01

LIQUID DENSITY CORRELATION CONSTANTS, K AND G/ML:

FRANCIS1	FRANCIS1	FRANCIS1
195. - 273. K	195. - 273. K	195. - 273. K
4 POINTS	4 POINTS	4 POINTS
RMSD = 0.00028	RMSD = 0.00028	RMSD = 0.00028
0.10467691E+01	0.10467691E+01	0.10467691E+01
0.11028063E-02	0.11028063E-02	0.11028063E-02
0.82401743E+01	0.82401743E+01	0.82401743E+01
0.46870190E+03	0.46870190E+03	0.46870190E+03

SECOND VIRIAL COEFFICIENT CORRELATION CONSTANTS, ML/MOL:

KREGLEWSKI: A= -835.0 B= 0.00 V*= 78.00 R1=0.7000 R2=1.1300 R3=5.1300

LITERATURE DOCUMENTS FOR CORRELATED VAPOR PRESSURE VALUES:

2817   3036

LITERATURE DOCUMENTS FOR CORRELATED LIQUID DENSITY VALUES:

2649   6310

LITERATURE DOCUMENTS REPORTING VIRIAL COEFFICIENT DATA:

3269

T,K	VAPOR PRESSURE, KPA	SATURATED LIQUID VOLUME, ML/MOL	SECOND VIRIAL COEFFICIENT, ML/MOL	HEAT OF VAPORIZATION, J/MOL	SATURATED VAPOR VOLUME, ML/MOL
195		69.99	-2202.		
197		70.20	-2126.		
199	0.8274	70.42	-2054.	27502.	1997628.
201	0.9762	70.63	-1985.	27416.	1710019.
203	1.147	70.85	-1921.	27331.	1469139.
205	1.344	71.07	-1860.	27246.	1266625.
207	1.568	71.29	-1802.	27162.	1095739.
209	1.824	71.51	-1746.	27078.	951027.
211	2.114	71.74	-1694.	26995.	828055.
213	2.443	71.96	-1644.	26913.	723206.
215	2.815	72.19	-1597.	26831.	633518.
217	3.233	72.42	-1552.	26749.	556557.
219	3.702	72.65	-1509.	26668.	490315.
221	4.228	72.89	-1468.	26587.	433129.
223	4.815	73.12	-1428.	26507.	383618.
226	5.824	73.48	-1372.	26387.	321281.
228	6.590	73.72	-1337.	26307.	286331.
230	7.438	73.96	-1304.	26228.	255786.
232	8.376	74.21	-1271.	26149.	229022.
234	9.410	74.45	-1240.	26070.	205513.
236	10.55	74.70	-1211.	25991.	184815.
238	11.80	74.95	-1182.	25912.	166549.
240	13.17	75.20	-1155.	25834.	150393.
242	14.66	75.46	-1128.	25756.	136072.
244	16.30	75.72	-1103.	25677.	123350.
246	18.08	75.98	-1078.	25599.	112025.
248	20.02	76.24	-1055.	25520.	101924.
250	22.13	76.50	-1032.	25442.	92895.
252	24.41	76.77	-1010.	25363.	84811.
254	26.88	77.04	-988.9	25285.	77558.
257	30.97	77.45	-958.6	25166.	68029.
259	33.96	77.73	-939.2	25087.	62457.
261	37.18	78.01	-920.5	25008.	57426.
263	40.65	78.29	-902.3	24928.	52878.
265	44.37	78.57	-884.8	24848.	48759.
267	48.36	78.86	-867.8	24768.	45023.
269	52.63	79.15	-851.4	24687.	41628.
271	57.20	79.45	-835.4	24606.	38540.
273	62.08	79.74	-820.0	24525.	35726.
275	67.28		-805.0		33158.
277	72.83		-790.5		30811.
279	78.73		-776.4		28664.
281	85.01		-762.7		26697.
283	91.68		-749.4		24893.
285	98.75		-736.4		23235.

COMPOUND CONSTANTS:

MP	95.850 K	TC		VC		ZC
NBP	273.880 K	PC		DC		OM
MU	0.0000 DEBYE	RG				

VAPOR PRESSURE CORRELATION CONSTANTS, K AND KPA:

VAPRES-2	VAPRES-2	VAPRES-2
177. - 278. K	177. - 278. K	177. - 278. K
10 POINTS	10 POINTS	10 POINTS
RMSD = 0.1060	RMSD = 0.1060	RMSD = 0.1060
0.22849232E+03	0.22849232E+03	0.22849232E+03
-0.83612107E+04	-0.83612107E+04	-0.83612107E+04
0.55904066E-01	0.55904066E-01	0.55904066E-01
0.16424031E-04	0.16424031E-04	0.16424031E-04
-0.37421438E+02	-0.37421438E+02	-0.37421438E+02

LIQUID DENSITY CORRELATION CONSTANTS, K AND G/ML:

SECOND VIRIAL COEFFICIENT CORRELATION CONSTANTS, ML/MOL:

LITERATURE DOCUMENTS FOR CORRELATED VAPOR PRESSURE VALUES:

   1146

LITERATURE DOCUMENTS FOR CORRELATED LIQUID DENSITY VALUES:

LITERATURE DOCUMENTS REPORTING VIRIAL COEFFICIENT DATA:

T,K	VAPOR PRESSURE, KPA	SATURATED LIQUID VOLUME, ML/MOL	SECOND VIRIAL COEFFICIENT, ML/MOL	HEAT OF VAPORIZATION, J/MOL	SATURATED VAPOR VOLUME, ML/MOL
177	0.1306				0.1127E+08
179	0.1645				9046158.
181	0.2058				7311702.
183	0.2558				5948095.
186	0.3503				4414963.
188	0.4287				3645939.
190	0.5218				3027738.
193	0.6934				2314308.
195	0.8327				1947071.
197	0.9951				1646070.
199	1.183				1398100.
202	1.522				1103555.
204	1.790				947552.
206	2.097				816909.
209	2.638				658694.
211	3.060				573280.
213	3.537				500710.
216	4.366				411310.
218	5.004				362226.
220	5.716				319987.
222	6.510				283517.
225	7.870				237719.
227	8.899				212096.
229	10.04				189729.
232	11.96				161274.
234	13.40				145148.
236	14.99				130934.
238	16.72				118371.
241	19.62				102153.
243	21.76				92828.
245	24.10				84518.
248	27.99				73680.
250	30.85				67388.
252	33.94				61737.
255	39.05				54301.
257	42.78				49944.
259	46.81				46006.
261	51.13				42438.
264	58.23				37696.
266	63.39				34889.
268	68.92				32332.
271	77.94				28908.
273	84.48				26867.
275	91.46				24998.
277	98.91				23284.

COMPOUND CONSTANTS:

MP 179.310 K	TC 511.890 K	VC 259.76 ML/MOL	ZC 0.2760
NBP 322.390 K	PC 4.517 MPA	DC 0.2700 G/ML	OM 0.1928
MU 0.0000 DEBYE	RG		

VAPOR PRESSURE CORRELATION CONSTANTS, K AND KPA:

WAGNER	WAGNER	VAPRES-2	WAGNER
226. - 511. K	226. - 323. K	226. - 433. K	304. - 511. K
43 POINTS	32 POINTS	37 POINTS	22 POINTS
RMSD = 0.5060	RMSD = 0.0180	RMSD = 0.1680	RMSD = 0.9870
-0.69057861E+01	-0.74921960E+01	0.63506861E+02	-0.69444545E+01
0.13558178E+01	0.26324305E+01	-0.52772684E+04	0.14821527E+01
-0.27011259E+01	-0.41996590E+01	0.33801443E-02	-0.30917509E+01
-0.11151872E+01	0.73362365E+00	0.37644458E-05	0.91592674E+00
		-0.76179560E+01	

LIQUID DENSITY CORRELATION CONSTANTS, K AND G/ML:

RACKETT	FRANCIS1	RACKETT	FRANCIS2
193. - 508. K	193. - 333. K	193. - 473. K	473. - 511. K
30 POINTS	19 POINTS	26 POINTS	5 POINTS
RMSD = 0.00400	RMSD = 0.00078	RMSD = 0.00427	RMSD = 0.00018
0.27492412E+00	0.95868111E+00	0.27481123E+00	0.52476745E-03
	0.50999993E-03		0.26648675E+01
	0.10681012E+02		
	0.46123022E+03		

SECOND VIRIAL COEFFICIENT CORRELATION CONSTANTS, ML/MOL:

NOTHNAGEL ET AL.: TC=511.890   NBP=322.390   B=205.0   D=0.278

LITERATURE DOCUMENTS FOR CORRELATED VAPOR PRESSURE VALUES:

  1485   1496   1503   1613   1809   3491   3773   4896

LITERATURE DOCUMENTS FOR CORRELATED LIQUID DENSITY VALUES:

  1487   1496   1559   3491   4670   9923 10808 21388

LITERATURE DOCUMENTS REPORTING VIRIAL COEFFICIENT DATA:

  1503

T,K	VAPOR PRESSURE, KPA	SATURATED LIQUID VOLUME, ML/MOL	SECOND VIRIAL COEFFICIENT, ML/MOL	HEAT OF VAPORIZATION, J/MOL	SATURATED VAPOR VOLUME, ML/MOL
193		83.85	-3548.		
200		84.45	-3167.		
207		85.06	-2847.		
214		85.69	-2576.		
221		86.34	-2343.		
229	1.041	87.09	-2116.	31548.	1827729.
236	1.698	87.77	-1945.	31255.	1153414.
243	2.684	88.47	-1796.	30963.	750929.
250	4.119	89.19	-1665.	30670.	502917.
258	6.506	90.03	-1533.	30333.	328197.
265	9.455	90.79	-1431.	30036.	231601.
272	13.44	91.57	-1340.	29735.	166917.
279	18.72	92.38	-1258.	29429.	122636.
286	25.60	93.21	-1184.	29118.	91702.
294	35.83	94.19	-1108.	28754.	67088.
301	47.30	95.08	-1048.	28427.	51843.
308	61.53	96.00	-993.8	28090.	40602.
315	78.97	96.95	-943.9	27744.	32191.
323	103.5	98.08	-891.8	27335.	25032.
330	129.4	99.10	-850.1	26963.	20308.
337	160.3	100.17	-811.6	26579.	16632.
344	196.4	101.28	-776.1	26180.	13738.
351	238.6	102.44	-743.1	25766.	11437.
359	294.8	103.82	-708.2	25271.	9358.
366	351.8	105.09	-680.0	24818.	7905.
373	416.8	106.42	-653.6	24344.	6717.
380	490.3	107.82	-628.9	23848.	5738.
388	585.7	109.51	-602.7	23249.	4819.
395	679.9	111.07	-581.2	22694.	4155.
402	784.8	112.73	-561.0	22104.	3594.
409	901.2	114.49	-542.0	21474.	3117.
417	1049.	116.65	-521.6	20695.	2655.
424	1193.	118.68	-504.9	19950.	2309.
431	1350.	120.87	-489.0	19750.	2069.
438	1522.	123.25	-474.0	19038.	1815.
445	1710.	125.84	-459.7	18282.	1594.
453	1945.	129.13	-444.4	17358.	1375.
460	2169.	132.36	-431.6	16485.	1206.
467	2412.	136.02	-419.5	15539.	1057.
474	2676.	140.22	-408.0	14504.	923.5
482	3002.	145.98	-395.4	13177.	786.7
489	3313.	152.25	-385.0	11841.	677.8
496	3648.	160.48	-375.0	10258.	575.9
503	4011.	172.67	-365.5	8231.	476.3
510	4404.		-356.4		362.4

COMPOUND CONSTANTS:

MP	166.150 K	TC		VC		ZC	
NBP	312.117 K	PC		DC		OM	
MU	0.0000 DEBYE	RG					

VAPOR PRESSURE CORRELATION CONSTANTS, K AND KPA:

RIEDEL	RIEDEL	RIEDEL	RIEDEL
277. - 344. K	277. - 344. K	277. - 344. K	277. - 344. K
14 POINTS	14 POINTS	14 POINTS	14 POINTS
RMSD = 0.0080	RMSD = 0.0080	RMSD = 0.0080	RMSD = 0.0080
0.47400538E+02	0.47400538E+02	0.47400538E+02	0.47400538E+02
-0.47845484E+01	-0.47845484E+01	-0.47845484E+01	-0.47845484E+01
-0.47861538E+04	-0.47861538E+04	-0.47861538E+04	-0.47861538E+04
0.34503672E-16	0.34503672E-16	0.34503672E-16	0.34503672E-16

LIQUID DENSITY CORRELATION CONSTANTS, K AND G/ML:

SECOND VIRIAL COEFFICIENT CORRELATION CONSTANTS, ML/MOL:

LITERATURE DOCUMENTS FOR CORRELATED VAPOR PRESSURE VALUES:

6463

LITERATURE DOCUMENTS FOR CORRELATED LIQUID DENSITY VALUES:

LITERATURE DOCUMENTS REPORTING VIRIAL COEFFICIENT DATA:

T,K	VAPOR PRESSURE, KPA	SATURATED LIQUID VOLUME, ML/MOL	SECOND VIRIAL COEFFICIENT, ML/MOL	HEAT OF VAPORIZATION, J/MOL	SATURATED VAPOR VOLUME, ML/MOL
277	25.26				91177.
278	26.43				87457.
280	28.90				80555.
281	30.20				77353.
283	32.96				71400.
284	34.40				68633.
286	37.46				63480.
287	39.07				61082.
289	42.45				56608.
290	44.22				54522.
292	47.96				50626.
293	49.91				48806.
295	54.02				45402.
296	56.18				43809.
298	60.69				40826.
299	63.05				39429.
301	67.99				36807.
302	70.58				35577.
304	75.98				33267.
305	78.80				32182.
307	84.69				30141.
308	87.76				29180.
310	94.16				27372.
312	100.9				25701.
313	104.5				24914.
315	111.8				23428.
316	115.6				22726.
318	123.5				21401.
319	127.7				20775.
321	136.2				19590.
322	140.7				19030.
324	149.9				17968.
325	154.7				17466.
327	164.6				16513.
328	169.8				16061.
330	180.5				15203.
331	186.0				14796.
333	197.4				14023.
334	203.4				13656.
336	215.6				12957.
337	221.9				12625.
339	235.0				11993.
340	241.8				11692.
342	255.8				11118.
343	263.0				10845.

COMPOUND CONSTANTS:

MP	279.701 K	TC	553.640 K	VC	317.07 ML/MOL	ZC	0.2807
NBP	353.871 K	PC	4.075 MPA	DC	0.2654 G/ML	OM	0.2095
MU	0.0000 DEBYE	RG	3.2605 ANGSTROM				

VAPOR PRESSURE CORRELATION CONSTANTS, K AND KPA:

WAGNER	WAGNER	VAPRES-2	WAGNER
279. - 554. K	279. - 374. K	279. - 470. K	333. - 554. K
256 POINTS	212 POINTS	234 POINTS	93 POINTS
RMSD = 0.9690	RMSD = 0.0560	RMSD = 0.3500	RMSD = 1.4930
-0.69650246E+01	-0.69647641E+01	-0.26766952E+02	-0.68713138E+01
0.13085228E+01	0.13517692E+01	-0.40555090E+04	0.10128423E+01
-0.26892309E+01	-0.29164844E+01	-0.51401305E-01	-0.18311178E+01
-0.27127151E+01	-0.18325037E+01	0.30375449E-04	-0.69946415E+01
		0.97515448E+01	

LIQUID DENSITY CORRELATION CONSTANTS, K AND G/ML:

FRANCIS1	FRANCIS1	FRANCIS1	FRANCIS2
279. - 544. K	279. - 374. K	279. - 524. K	508. - 554. K
170 POINTS	148 POINTS	167 POINTS	9 POINTS
RMSD = 0.00082	RMSD = 0.00017	RMSD = 0.00073	RMSD = 0.00096
0.10600863E+01	0.10512676E+01	0.10605993E+01	0.42173067E-03
0.82413526E-03	0.83394744E-03	0.80384128E-03	0.28374343E+01
0.12462436E+02	0.77268763E+01	0.15000000E+02	
0.60657153E+03	0.56643433E+03	0.61739478E+03	

SECOND VIRIAL COEFFICIENT CORRELATION CONSTANTS, ML/MOL:

PITZER-CURL: TC=553.640   PC=4.075   VC=317.07   OM=0.2095

LITERATURE DOCUMENTS FOR CORRELATED VAPOR PRESSURE VALUES:

```
   39    102    124    193    194    236    243    269    304    608    691    960   1095
 1173   1485   1548   1688   1698   4076   4114   4139   4328   4604   4939   4946   4972
 5636   5638   5640   5936   6554   6740   6745   6764   7815   7828   7869   7873   7927
 8628   8990   9623   9943  10241  10335  10841  11023  11101  11128  11160  11161  11166
11248  11250  11336  15380  16059  17270  19662  20129  20442  20654  20895  20949  21202
21432  21537  21718  21852  40032  40349  40772  40796  40900
```


LITERATURE DOCUMENTS FOR CORRELATED LIQUID DENSITY VALUES:

```
   71     76    228    583    691    719    774    922   1291   1489   1688   2573   4216
 4856   4886   4983   5203   5364   5882   5992   6298   6351   6554   7867   8899   9053
 9923  10255  10894  10919  11475  12507  12559  15650  16363  17056  17129  17986  20090
20895  20954  21010  21121  21213  21804  40349
```


LITERATURE DOCUMENTS REPORTING VIRIAL COEFFICIENT DATA:

```
  905    932   1257   1803   2402   2412   5283  10754  10972  15380  40253
```

T,K	VAPOR PRESSURE, KPA	SATURATED LIQUID VOLUME, ML/MOL	SECOND VIRIAL COEFFICIENT, ML/MOL	HEAT OF VAPORIZATION, J/MOL	SATURATED VAPOR VOLUME, ML/MOL
279	5.114	106.25			
285	6.966	107.01	-1988.		
291	9.344	107.79	-1865.		
297	12.36	108.59	-1755.		
304	16.85	109.53	-1642.		
310	21.68	110.36	-1555.		
316	27.58	111.21	-1476.		
322	34.72	112.08	-1404.		
329	44.85	113.11	-1327.		
335	55.30	114.02	-1268.		
341	67.60	114.95	-1212.		
347	81.96	115.91	-1161.		
354	101.6	117.05	-1106.		
360	121.3	118.06	-1063.		
366	143.7	119.10	-1022.		
372	169.2	120.16	-983.2		
379	203.3	121.45	-941.4		
385	236.4	122.59	-907.7		
391	273.5	123.76	-876.0		
397	314.9	124.97	-846.0		
404	368.9	126.44	-812.9		
410	420.5	127.75	-786.2		
416	477.3	129.11	-760.7		
422	539.6	130.53	-736.5		
429	619.8	132.26	-709.7		
435	695.2	133.81	-687.9		
441	777.2	135.44	-667.0		
447	866.1	137.16	-647.0		
454	979.1	139.28	-624.8		
460	1084.	141.22	-606.6		
466	1197.	143.28	-589.2		
472	1319.	145.48	-572.4		
479	1472.	148.27	-553.7		
485	1614.	150.86	-538.3		
491	1765.	153.69	-523.5		
497	1926.	156.80	-509.2		
504	2128.	160.85	-493.2		
510	2314.	164.79	-480.0		
516	2511.	169.26	-467.3		
522	2721.	174.42	-455.0		
529	2983.	181.59	-441.1		
535	3223.	189.11	-429.7		
541	3479.	198.41	-418.7		
547	3751.		-408.0		

COMPOUND CONSTANTS:

MP 130.682 K	TC 532.794 K	VC 326.43 ML/MOL	ZC 0.2789
NBP 344.956 K	PC 3.785 MPA	DC 0.2578 G/ML	OM 0.2347
MU 0.0000 DEBYE	RG		

VAPOR PRESSURE CORRELATION CONSTANTS, K AND KPA:

WAGNER	WAGNER	VAPRES-2	WAGNER
256. - 533. K	256. - 346. K	256. - 451. K	325. - 533. K
26 POINTS	22 POINTS	24 POINTS	15 POINTS
RMSD = 0.5608	RMSD = 0.0190	RMSD = 0.1160	RMSD = 1.1990
-0.72196762E+01	-0.85308294E+01	-0.52398368E+03	-0.64998140E+01
0.16658141E+01	0.47369789E+01	0.66777362E+04	-0.58501121E+00
-0.34038796E+01	-0.78487776E+01	-0.32346867E+00	0.28585933E+01
0.29353652E+00	0.82841924E+01	0.16200486E-03	-0.28995136E+02
		0.10294454E+03	

LIQUID DENSITY CORRELATION CONSTANTS, K AND G/ML:

FRANCIS1	FRANCIS1	FRANCIS1	RACKETT
273. - 494. K	273. - 354. K	273. - 494. K	413. - 494. K
21 POINTS	14 POINTS	21 POINTS	5 POINTS
RMSD = 0.00033	RMSD = 0.00032	RMSD = 0.00033	RMSD = 0.00028
0.10230303E+01	0.10163136E+01	0.10230303E+01	0.27052985E+00
0.76325936E-03	0.82500046E-03	0.76325936E-03	
0.15000000E+02	0.60239277E+01	0.15000000E+02	
0.59167261E+03	0.52947632E+03	0.59167261E+03	

SECOND VIRIAL COEFFICIENT CORRELATION CONSTANTS, ML/MOL:

NOTHNAGEL ET AL.: TC=532.794   NBP=344.956   B=246.0   D=0.290

LITERATURE DOCUMENTS FOR CORRELATED VAPOR PRESSURE VALUES:

    3   914   1485   1496   5101

LITERATURE DOCUMENTS FOR CORRELATED LIQUID DENSITY VALUES:

    1489   1496   4670   9923 10808

LITERATURE DOCUMENTS REPORTING VIRIAL COEFFICIENT DATA:

    1492

T,K	VAPOR PRESSURE, KPA	SATURATED LIQUID VOLUME, ML/MOL	SECOND VIRIAL COEFFICIENT, ML/MOL	HEAT OF VAPORIZATION, J/MOL	SATURATED VAPOR VOLUME, ML/MOL
256	2.116		-2388.		1003564.
262	3.021		-2233.		718823.
268	4.237		-2093.		523849.
274	5.843	109.77	-1966.	32433.	387939.
281	8.335	110.70	-1834.	32154.	278446.
287	11.13	111.51	-1731.	31908.	212706.
293	14.66	112.34	-1637.	31654.	164578.
300	19.89	113.33	-1538.	31349.	123855.
306	25.51	114.20	-1461.	31079.	98238.
312	32.37	115.09	-1389.	30802.	78724.
318	40.65	116.00	-1323.	30515.	63693.
325	52.38	117.10	-1253.	30170.	50307.
331	64.45	118.06	-1197.	29864.	41467.
337	78.64	119.05	-1145.	29548.	34446.
344	98.19	120.25	-1089.	29167.	27995.
350	117.8	121.30	-1044.	28830.	23607.
356	140.4	122.39	-1003.	28482.	20031.
363	170.8	123.70	-957.7	28062.	16658.
369	200.6	124.86	-921.6	27690.	14306.
375	234.4	126.07	-887.7	27305.	12347.
381	272.2	127.32	-855.8	26909.	10706.
388	322.1	128.84	-820.9	26429.	9114.
394	370.1	130.19	-792.7	26002.	7972.
400	423.2	131.61	-766.2	25560.	6998.
407	492.2	133.34	-737.0	25023.	6036.
413	557.7	134.91	-713.4	24542.	5334.
419	629.5	136.55	-691.1	24041.	4725.
425	707.9	138.28	-669.9	23516.	4195.
432	808.2	140.42	-646.4	22869.	3659.
438	902.3	142.38	-627.4	22280.	3259.
444	1004.	144.47	-609.2	21652.	2905.
451	1134.	147.09	-589.1	21209.	2581.
457	1254.	149.53	-572.8	20579.	2312.
463	1384.	152.17	-557.1	19918.	2072.
470	1547.	155.56	-539.8	19102.	1824.
476	1699.	158.78	-525.5	18357.	1635.
482	1861.	162.36	-511.9	17564.	1464.
488	2035.	166.37	-498.9	16712.	1309.
495	2253.		-484.4		1145.
501	2454.		-472.5		1017.
507	2669.		-461.0		899.0
514	2939.		-448.2		769.8
520	3189.		-437.7		663.6
526	3456.		-427.6		555.8
532	3744.		-417.8		419.1

COMPOUND CONSTANTS:

MP   265.150 K   TC   604.376 K   VC   353.01 ML/MOL   ZC   0.2712
NBP   391.957 K   PC   3.860 MPA   DC   0.2781 G/ML   OM   0.2413
MU   0.0000 DEBYE   RG

VAPOR PRESSURE CORRELATION CONSTANTS, K AND KPA:

WAGNER	RPM2	WAGNER	WAGNER
283. - 604. K	283. - 412. K	283. - 509. K	372. - 604. K
47 POINTS	30 POINTS	36 POINTS	27 POINTS
RMSD = 5.7400	RMSD = 0.0131	RMSD = 1.0790	RMSD = 3.4120
-0.69104748E+01	0.26004005E+02	-0.66151482E+01	-0.73041851E+01
0.81524706E+00	-0.58250918E+04	0.10255187E+00	0.21553283E+01
-0.22015760E+01	-0.22661648E-01	-0.11009371E+01	-0.67796252E+01
-0.29498700E+01	0.15350346E-04	-0.50483598E+01	0.26586734E+02

LIQUID DENSITY CORRELATION CONSTANTS, K AND G/ML:

FRANCIS1	FRANCIS1	FRANCIS1	FRANCIS2
289. - 581. K	289. - 388. K	289. - 566. K	565. - 604. K
17 POINTS	12 POINTS	16 POINTS	2 POINTS
RMSD = 0.00028	RMSD = 0.00017	RMSD = 0.00021	RMSD = 0.00000
0.10667591E+01	0.10601254E+01	0.10665588E+01	0.23122069E-03
0.74581406E-03	0.77824504E-03	0.75924769E-03	0.32338185E+01
0.14053654E+02	0.62128267E+01	0.12150435E+02	
0.67081763E+03	0.58985620E+03	0.66171582E+03	

SECOND VIRIAL COEFFICIENT CORRELATION CONSTANTS, ML/MOL:

TSONOPOULOS: TC=604.376 PC=3.860 VC=353.01 OM=0.2413 A= 0.000000 B=0.0000

LITERATURE DOCUMENTS FOR CORRELATED VAPOR PRESSURE VALUES:

    6550   8098 10271 10940 40284

LITERATURE DOCUMENTS FOR CORRELATED LIQUID DENSITY VALUES:

    4216 10940 40284

LITERATURE DOCUMENTS REPORTING VIRIAL COEFFICIENT DATA:

T,K	VAPOR PRESSURE, KPA	SATURATED LIQUID VOLUME, ML/MOL	SECOND VIRIAL COEFFICIENT, ML/MOL	HEAT OF VAPORIZATION, J/MOL	SATURATED VAPOR VOLUME, ML/MOL
283	1.267		-3934.		1853244.
290	1.891	120.69	-3540.	38735.	1271728.
297	2.759	121.57	-3206.	38360.	891691.
304	3.945	122.47	-2922.	37990.	637740.
312	5.800	123.52	-2646.	37574.	444579.
319	7.976	124.46	-2439.	37213.	330077.
326	10.79	125.42	-2259.	36854.	248862.
334	14.97	126.54	-2079.	36445.	183360.
341	19.65	127.55	-1942.	36088.	142333.
348	25.45	128.57	-1820.	35729.	111824.
356	33.72	129.77	-1696.	35316.	86042.
363	42.63	130.85	-1600.	34951.	69167.
370	53.33	131.96	-1512.	34581.	56132.
377	66.08	133.09	-1433.	34205.	45958.
385	83.49	134.42	-1351.	33766.	36936.
392	101.5	135.62	-1286.	33374.	30756.
399	122.5	136.85	-1226.	32972.	25789.
407	150.5	138.30	-1163.	32499.	21256.
414	178.8	139.62	-1112.	32073.	18063.
421	211.1	140.97	-1065.	31633.	15436.
429	253.3	142.57	-1015.	31112.	12979.
436	295.3	144.03	-975.0	30639.	11208.
443	342.4	145.53	-937.1	30148.	9721.
451	402.9	147.33	-896.7	29563.	8302.
458	462.2	148.96	-863.6	29027.	7259.
465	527.8	150.67	-832.4	28466.	6367.
472	600.2	152.45	-803.0	27879.	5601.
480	691.7	154.60	-771.4	27170.	4852.
487	780.0	156.58	-745.2	26513.	4289.
494	876.3	158.66	-720.5	25815.	3798.
502	996.8	161.20	-693.7	24959.	3309.
509	1112.	163.57	-671.4	24641.	2992.
516	1237.	166.10	-650.3	23871.	2665.
524	1391.	169.22	-627.3	22940.	2338.
531	1538.	172.19	-608.2	22076.	2085.
538	1696.	175.43	-589.9	21160.	1860.
545	1865.	178.98	-572.4	20186.	1658.
553	2074.	183.51	-553.3	18987.	1451.
560	2270.	188.00	-537.3	17851.	1289.
567	2479.	193.09	-522.0	16615.	1140.
575	2736.	199.88	-505.3	15046.	985.4
582	2977.		-491.2		859.5
589	3234.		-477.7		739.4
597	3548.		-462.9		601.3
604	3843.		-450.4		439.5

COMPOUND CONSTANTS:

MP 146.550 K	TC 572.192 K	VC 368.57 ML/MOL	ZC 0.2689
NBP 374.072 K	PC 3.471 MPA	DC 0.2664 G/ML	OM 0.2353
MU 0.0000 DEBYE	RG 3.7467 ANGSTROM		

VAPOR PRESSURE CORRELATION CONSTANTS, K AND KPA:

WAGNER	VAPRES-2	WAGNER	WAGNER
273. - 572. K	273. - 394. K	273. - 484. K	354. - 572. K
820 POINTS	66 POINTS	74 POINTS	35 POINTS
RMSD = 0.4336	RMSD = 0.0417	RMSD = 0.2080	RMSD = 0.3695
-0.71813996E+01	-0.63713260E+03	-0.72195540E+01	-0.71429287E+01
0.15261477E+01	0.82679317E+04	0.16184132E+01	0.14092506E+01
-0.31980777E+01	-0.40253969E+00	-0.33420076E+01	-0.28918723E+01
-0.10504100E+01	0.20782900E-03	-0.76425340E+00	-0.23957087E+01
	0.12509957E+03		

LIQUID DENSITY CORRELATION CONSTANTS, K AND G/ML:

FRANCIS1	FRANCIS1	FRANCIS1	RACKETT
178. - 558. K	178. - 394. K	178. - 534. K	523. - 572. K
80 POINTS	58 POINTS	76 POINTS	10 POINTS
RMSD = 0.00072	RMSD = 0.00024	RMSD = 0.00061	RMSD = 0.00084
0.10266705E+01	0.10284042E+01	0.10258675E+01	0.26891850E+00
0.76240487E-03	0.82871551E-03	0.78169489E-03	
0.10901861E+02	0.60020304E+01	0.84361782E+01	
0.61840088E+03	0.66301611E+03	0.60383008E+03	

SECOND VIRIAL COEFFICIENT CORRELATION CONSTANTS, ML/MOL:

TSONOPOULOS: TC=572.192 PC=3.471 VC=368.57 OM=0.2353 A= 0.000000 B=0.0000

LITERATURE DOCUMENTS FOR CORRELATED VAPOR PRESSURE VALUES:

   692   757   974 1348 1485   1496 3571 4870 5063 7926 7929 9132 15130
17569

LITERATURE DOCUMENTS FOR CORRELATED LIQUID DENSITY VALUES:

   119 1487 1496 1534 2573 4094 5882 6298 7929 9008 10264 10808 14380
15276 17415 17974 19687 20129 20315 21388 21562 21804

LITERATURE DOCUMENTS REPORTING VIRIAL COEFFICIENT DATA:

T,K	VAPOR PRESSURE, KPA	SATURATED LIQUID VOLUME, ML/MOL	SECOND VIRIAL COEFFICIENT, ML/MOL	HEAT OF VAPORIZATION, J/MOL	SATURATED VAPOR VOLUME, ML/MOL
178		113.35	-46080.		
186		114.22	-33782.		
195		115.21	-24443.		
204		116.22	-18151.		
213		117.26	-13816.		
222		118.31	-10764.		
231		119.39	-8573.		
240		120.49	-6967.		
249		121.62	-5769.		
258		122.78	-4859.		
267		123.96	-4156.		
276	1.939	125.17	-3604.	36210.	1179794.
285	3.188	126.42	-3163.	35787.	740013.
294	5.061	127.70	-2807.	35361.	480178.
303	7.781	129.01	-2514.	34931.	321230.
312	11.62	130.36	-2271.	34494.	220886.
321	16.91	131.75	-2066.	34050.	155704.
330	24.03	133.18	-1892.	33596.	112246.
339	33.41	134.65	-1742.	33132.	82578.
348	45.54	136.18	-1612.	32654.	61879.
357	60.96	137.75	-1498.	32163.	47148.
366	80.24	139.38	-1398.	31655.	36471.
375	104.0	141.08	-1309.	31131.	28602.
384	133.0	142.84	-1229.	30588.	22711.
393	167.8	144.68	-1157.	30025.	18238.
401	204.3	146.37	-1099.	29506.	15136.
410	252.2	148.37	-1039.	28899.	12381.
419	308.3	150.47	-984.9	28267.	10211.
428	373.2	152.68	-935.0	27605.	8484.
437	447.9	155.02	-889.0	26911.	7096.
446	533.2	157.50	-846.6	26179.	5968.
455	630.0	160.16	-807.2	25403.	5044.
464	739.3	163.02	-770.7	24573.	4278.
473	862.0	166.12	-736.6	23679.	3639.
482	999.1	169.50	-704.9	23148.	3156.
491	1152.	173.24	-675.1	22201.	2704.
500	1321.	177.43	-647.2	21188.	2319.
509	1508.	182.18	-620.9	20095.	1990.
518	1715.	187.68	-596.2	18905.	1705.
527	1942.	194.20	-572.9	17590.	1456.
536	2192.	202.15	-550.8	16109.	1237.
545	2465.	212.24	-530.0	14396.	1041.
554	2766.	225.71	-510.2	12325.	861.3
563	3097.		-491.4		687.6
572	3463.		-473.6		457.4

COMPOUND CONSTANTS:

MP   134.703 K   TC   569.522 K   VC   375.00 ML/MOL   ZC 0.2690
NBP   376.657 K   PC   3.397 MPA   DC   0.2618 G/ML   OM 0.2755
MU   0.0000 DEBYE   RG

VAPOR PRESSURE CORRELATION CONSTANTS, K AND KPA:

WAGNER	VAPRES-2	RPM2	WAGNER
301. - 570. K	301. - 394. K	301. - 484. K	362. - 570. K
37 POINTS	20 POINTS	29 POINTS	24 POINTS
RMSD = 1.6400	RMSD = 0.0526	RMSD = 0.4680	RMSD = 1.2030
-0.69580436E+01	0.14337649E+04	0.30944919E+02	-0.67734857E+01
0.26028008E+00	-0.34891439E+05	-0.62118845E+04	-0.32838049E+00
-0.68029999E-01	0.77665542E+00	-0.36638337E-01	0.16503460E+01
-0.12660621E+02	-0.38458001E-03	0.27926000E-04	-0.19697509E+02
	-0.26545190E+03		

LIQUID DENSITY CORRELATION CONSTANTS, K AND G/ML:

FRANCIS1	FRANCIS1	FRANCIS1	FRANCIS2
273. - 559. K	273. - 374. K	273. - 534. K	533. - 570. K
22 POINTS	11 POINTS	19 POINTS	7 POINTS
RMSD = 0.00071	RMSD = 0.00024	RMSD = 0.00060	RMSD = 0.00036
0.10371084E+01	0.10077581E+01	0.10366831E+01	0.51435072E-03
0.78770984E-03	0.69810590E-03	0.76965429E-03	0.25388590E+01
0.13071368E+02	0.87517738E+01	0.15000000E+02	
0.62260278E+03	0.53177808E+03	0.62986621E+03	

SECOND VIRIAL COEFFICIENT CORRELATION CONSTANTS, ML/MOL:

TSONOPOULOS: TC=569.522 PC=3.397 VC=375.00 OM=0.2755 A= 0.000000 B=0.0000

LITERATURE DOCUMENTS FOR CORRELATED VAPOR PRESSURE VALUES:

    1486 1496

LITERATURE DOCUMENTS FOR CORRELATED LIQUID DENSITY VALUES:

    1487 1490 1496 9707

LITERATURE DOCUMENTS REPORTING VIRIAL COEFFICIENT DATA:

T,K	VAPOR PRESSURE, KPA	SATURATED LIQUID VOLUME, ML/MOL	SECOND VIRIAL COEFFICIENT, ML/MOL	HEAT OF VAPORIZATION, J/MOL	SATURATED VAPOR VOLUME, ML/MOL
273		125.13	-4054.		
279		126.00	-3694.		
286		127.02	-3336.		
293		128.07	-3033.		
299		128.99	-2808.		
306	7.746	130.09	-2579.	36410.	325855.
313	10.67	131.21	-2380.	35820.	241452.
320	14.44	132.36	-2207.	35262.	182046.
326	18.46	133.36	-2075.	34808.	144718.
333	24.25	134.57	-1938.	34303.	112223.
340	31.40	135.80	-1815.	33822.	88181.
347	40.14	137.07	-1706.	33362.	70127.
353	49.08	138.18	-1621.	32982.	58131.
360	61.43	139.52	-1531.	32552.	47143.
367	76.11	140.90	-1449.	32133.	38587.
374	93.41	142.32	-1374.	31722.	31852.
380	110.6	143.57	-1315.	31374.	27190.
387	133.6	145.08	-1252.	30970.	22756.
394	160.3	146.64	-1193.	30564.	19170.
401	190.8	148.26	-1139.	30152.	16246.
407	220.5	149.70	-1096.	29792.	14159.
414	259.5	151.45	-1049.	29361.	12116.
421	303.6	153.26	-1005.	28914.	10416.
428	353.3	155.16	-964.0	28447.	8991.
434	400.8	156.87	-931.0	28027.	7950.
441	462.1	158.94	-894.7	27511.	6908.
448	530.3	161.14	-860.6	26961.	6019.
455	606.1	163.45	-828.5	26371.	5259.
461	677.3	165.55	-802.4	25829.	4691.
468	768.3	168.16	-773.6	25148.	4112.
475	868.2	170.94	-746.3	24405.	3608.
481	961.5	173.50	-724.0	24191.	3288.
488	1080.	176.73	-699.3	23409.	2899.
495	1208.	180.24	-675.8	22567.	2557.
502	1348.	184.12	-653.5	21661.	2256.
508	1478.	187.78	-635.2	20827.	2026.
515	1640.	192.54	-614.7	19778.	1786.
522	1815.	197.96	-595.2	18635.	1572.
529	2004.	204.23	-576.6	17381.	1379.
535	2177.	210.49	-561.3	16200.	1229.
542	2393.	219.17	-544.1	14668.	1068.
549	2624.	229.91	-527.7	12916.	919.1
556	2871.	243.71	-512.0	10840.	778.0
562	3096.		-499.0		657.0
569	3375.		-484.4		477.1

215

COMPOUND CONSTANTS:

MP	203.389 K	TC	VC	ZC
NBP	360.991 K	PC	DC	OM
MU	0.0000 DEBYE	RG		

VAPOR PRESSURE CORRELATION CONSTANTS, K AND KPA:

RIEDEL	RIEDEL	RIEDEL	RIEDEL
288. - 362. K	288. - 362. K	288. - 362. K	347. - 362. K
17 POINTS	17 POINTS	17 POINTS	7 POINTS
RMSD = 0.0056	RMSD = 0.0056	RMSD = 0.0056	RMSD = 0.0014
0.52036332E+02	0.52036332E+02	0.52036332E+02	-0.79698775E+02
-0.53909362E+01	-0.53909362E+01	-0.53909362E+01	0.14253274E+02
-0.56710693E+04	-0.56710693E+04	-0.56710693E+04	0.30924979E+03
0.17219653E-16	0.17219653E-16	0.17219653E-16	-0.21464333E-15

LIQUID DENSITY CORRELATION CONSTANTS, K AND G/ML:

FRANCIS1	FRANCIS1	FRANCIS1
293. - 304. K	293. - 304. K	293. - 304. K
4 POINTS	4 POINTS	4 POINTS
RMSD = 0.00005	RMSD = 0.00005	RMSD = 0.00005
0.10201178E+01	0.10201178E+01	0.10201178E+01
0.83202799E-03	0.83202799E-03	0.83202799E-03
0.59999990E+01	0.59999990E+01	0.59999990E+01
0.56838794E+03	0.56838794E+03	0.56838794E+03

SECOND VIRIAL COEFFICIENT CORRELATION CONSTANTS, ML/MOL:

KREGLEWSKI: A=-1193.0 B= 0.00 V*=136.00 R1=0.7000 R2=1.1300 R3=3.4000

LITERATURE DOCUMENTS FOR CORRELATED VAPOR PRESSURE VALUES:

    1486

LITERATURE DOCUMENTS FOR CORRELATED LIQUID DENSITY VALUES:

    1490   7040

LITERATURE DOCUMENTS REPORTING VIRIAL COEFFICIENT DATA:

T,K	VAPOR PRESSURE, KPA	SATURATED LIQUID VOLUME, ML/MOL	SECOND VIRIAL COEFFICIENT, ML/MOL	HEAT OF VAPORIZATION, J/MOL	SATURATED VAPOR VOLUME, ML/MOL
288	6.211		-2776.		382751.
289	6.527		-2740.		365383.
291	7.200		-2670.		333358.
293	7.930	130.13	-2603.	33871.	304597.
294	8.317	130.29	-2570.	33821.	291321.
296	9.139	130.60	-2506.	33719.	266768.
298	10.03	130.92	-2445.	33616.	244630.
299	10.50	131.08	-2415.	33565.	234382.
301	11.49	131.40	-2357.	33461.	215376.
303	12.56	131.72	-2302.	33357.	198176.
304	13.13	131.89	-2274.	33305.	190191.
306	14.33		-2222.		175343.
308	15.61		-2171.		161858.
309	16.29		-2146.		155582.
311	17.71		-2097.		143882.
313	19.23		-2051.		133220.
314	20.04		-2028.		128246.
316	21.72		-1983.		118950.
318	23.52		-1941.		110452.
319	24.46		-1920.		106478.
321	26.44		-1879.		99034.
323	28.54		-1839.		92209.
324	29.65		-1820.		89009.
326	31.95		-1782.		83004.
328	34.40		-1746.		77483.
330	37.00		-1710.		72400.
331	38.36		-1693.		70010.
333	41.19		-1659.		65512.
335	44.19		-1627.		61360.
336	45.75		-1611.		59404.
338	49.01		-1579.		55715.
340	52.45		-1549.		52302.
341	54.24		-1534.		50691.
343	57.96		-1505.		47647.
345	61.89		-1477.		44823.
346	63.93		-1463.		43488.
348	68.16		-1436.		40961.
350	72.62		-1410.		38611.
351	74.93		-1397.		37498.
353	79.72		-1372.		35388.
355	84.75		-1347.		33422.
356	87.36		-1335.		32490.
358	92.76		-1312.		30718.
360	98.42		-1289.		29064.
361	101.3		-1277.		28278.

217

COMPOUND CONSTANTS:

MP	219.330 K	TC		VC	ZC
NBP	372.682 K	PC		DC	OM
MU	0.0000 DEBYE	RG			

VAPOR PRESSURE CORRELATION CONSTANTS, K AND KPA:

RIEDEL	RIEDEL	RIEDEL	RIEDEL
298. - 374. K	298. - 374. K	298. - 374. K	358. - 374. K
20 POINTS	20 POINTS	20 POINTS	7 POINTS
RMSD = 0.0043	RMSD = 0.0043	RMSD = 0.0043	RMSD = 0.0027
0.57023684E+02	0.57023684E+02	0.57023684E+02	-0.78714133E+02
-0.60982835E+01	-0.60982835E+01	-0.60982835E+01	0.14057670E+02
-0.60981710E+04	-0.60981710E+04	-0.60981710E+04	0.21450614E+03
0.23825252E-16	0.23825252E-16	0.23825252E-16	-0.17717157E-15

LIQUID DENSITY CORRELATION CONSTANTS, K AND G/ML:

FRANCIS1	FRANCIS1	FRANCIS1
273. - 304. K	273. - 304. K	273. - 304. K
6 POINTS	6 POINTS	6 POINTS
RMSD = 0.00003	RMSD = 0.00003	RMSD = 0.00003
0.10330324E+01	0.10330324E+01	0.10330324E+01
0.78392983E-03	0.78392983E-03	0.78392983E-03
0.83470335E+01	0.83470335E+01	0.83470335E+01
0.56583545E+03	0.56583545E+03	0.56583545E+03

SECOND VIRIAL COEFFICIENT CORRELATION CONSTANTS, ML/MOL:

KREGLEWSKI: A=-1227.0 B= 0.00 V*=134.00 R1=0.7000 R2=1.1300 R3=3.4100

LITERATURE DOCUMENTS FOR CORRELATED VAPOR PRESSURE VALUES:

 1486

LITERATURE DOCUMENTS FOR CORRELATED LIQUID DENSITY VALUES:

 1490 9708

LITERATURE DOCUMENTS REPORTING VIRIAL COEFFICIENT DATA:

T,K	VAPOR PRESSURE, KPA	SATURATED LIQUID VOLUME, ML/MOL	SECOND VIRIAL COEFFICIENT, ML/MOL	HEAT OF VAPORIZATION, J/MOL	SATURATED VAPOR VOLUME, ML/MOL
273		124.21	-3776.		
275		124.49	-3665.		
277		124.76	-3559.		
279		125.05	-3457.		
282		125.47	-3313.		
284		125.76	-3222.		
286		126.04	-3134.		
289		126.48	-3010.		
291		126.77	-2931.		
293		127.06	-2855.		
295		127.36	-2782.		
298	6.260	127.81	-2677.	35585.	393087.
300	6.893	128.11	-2611.	35479.	359223.
302	7.579	128.41	-2547.	35372.	328751.
305	8.713		-2456.		288581.
307	9.544		-2398.		265021.
309	10.44		-2342.		243709.
312	11.91		-2262.		215435.
314	12.99		-2210.		198747.
316	14.14		-2161.		183579.
318	15.38		-2113.		169772.
321	17.40		-2044.		151318.
323	18.86		-2000.		140344.
325	20.42		-1958.		130311.
328	22.96		-1896.		116829.
330	24.79		-1857.		108769.
332	26.74		-1819.		101369.
334	28.81		-1782.		94569.
337	32.16		-1729.		85371.
339	34.55		-1695.		79837.
341	37.09		-1662.		74732.
344	41.18		-1614.		67796.
346	44.10		-1584.		63603.
348	47.19		-1554.		59722.
351	52.13		-1511.		54426.
353	55.65		-1483.		51211.
355	59.36		-1456.		48225.
357	63.26		-1430.		45448.
360	69.48		-1391.		41640.
362	73.89		-1367.		39316.
364	78.52		-1343.		37150.
367	85.89		-1308.		34166.
369	91.10		-1286.		32340.
371	96.55		-1264.		30631.
373	102.2		-1243.		29032.

COMPOUND CONSTANTS:

MP   155.586 K   TC   VC   ZC
NBP   365.020 K   PC   DC   OM
MU   0.0000 DEBYE   RG

VAPOR PRESSURE CORRELATION CONSTANTS, K AND KPA:

VAPRES-2	VAPRES-2	VAPRES-2	RIEDEL
299. - 366. K	299. - 366. K	299. - 366. K	351. - 366. K
17 POINTS	17 POINTS	17 POINTS	6 POINTS
RMSD = 0.0041	RMSD = 0.0041	RMSD = 0.0041	RMSD = 0.0027
0.57933791E+03	0.57933791E+03	0.57933791E+03	0.46636088E+02
-0.17234214E+05	-0.17234214E+05	-0.17234214E+05	-0.45688168E+01
0.27789722E+00	0.27789722E+00	0.27789722E+00	-0.55043578E+04
-0.12889474E-03	-0.12889474E-03	-0.12889474E-03	0.74672626E-17
-0.10369055E+03	-0.10369055E+03	-0.10369055E+03	

LIQUID DENSITY CORRELATION CONSTANTS, K AND G/ML:

FRANCIS1	FRANCIS1	FRANCIS1
273. - 304. K	273. - 304. K	273. - 304. K
7 POINTS	7 POINTS	7 POINTS
RMSD = 0.00005	RMSD = 0.00005	RMSD = 0.00005
0.10127735E+01	0.10127735E+01	0.10127735E+01
0.81745931E-03	0.81745931E-03	0.81745931E-03
0.59999990E+01	0.59999990E+01	0.59999990E+01
0.56838794E+03	0.56838794E+03	0.56838794E+03

SECOND VIRIAL COEFFICIENT CORRELATION CONSTANTS, ML/MOL:

KREGLEWSKI: A=-1209.0 B= 0.00 V*=137.00 R1=0.7000 R2=1.1300 R3=3.3900

LITERATURE DOCUMENTS FOR CORRELATED VAPOR PRESSURE VALUES:

   1486

LITERATURE DOCUMENTS FOR CORRELATED LIQUID DENSITY VALUES:

   1490 9708

LITERATURE DOCUMENTS REPORTING VIRIAL COEFFICIENT DATA:

T,K	VAPOR PRESSURE, KPA	SATURATED LIQUID VOLUME, ML/MOL	SECOND VIRIAL COEFFICIENT, ML/MOL	HEAT OF VAPORIZATION, J/MOL	SATURATED VAPOR VOLUME, ML/MOL
273		127.63	-3627.		
275		127.93	-3522.		
277		128.23	-3421.		
279		128.52	-3325.		
281		128.82	-3232.		
283		129.13	-3144.		
285		129.43	-3059.		
287		129.73	-2978.		
289		130.04	-2899.		
292		130.51	-2788.		
294		130.82	-2717.		
296		131.13	-2649.		
298		131.45	-2583.		
300	9.308	131.76	-2520.	34386.	265445.
302	10.21	132.08	-2459.	34258.	243561.
304	11.17	132.41	-2401.	34131.	223798.
306	12.21		-2344.		205923.
308	13.33		-2290.		189730.
311	15.17		-2211.		168207.
313	16.51		-2162.		155477.
315	17.93		-2113.		143886.
317	19.46		-2067.		133318.
319	21.10		-2022.		123670.
321	22.84		-1978.		114850.
323	24.70		-1936.		106776.
325	26.67		-1895.		99377.
327	28.78		-1856.		92588.
330	32.18		-1799.		83423.
332	34.62		-1763.		77919.
334	37.22		-1727.		72849.
336	39.96		-1693.		68172.
338	42.87		-1660.		63855.
340	45.94		-1628.		59864.
342	49.18		-1596.		56173.
344	52.61		-1566.		52754.
346	56.22		-1536.		49585.
349	62.00		-1494.		45254.
351	66.12		-1466.		42622.
353	70.44		-1439.		40174.
355	74.98		-1413.		37896.
357	79.75		-1388.		35773.
359	84.76		-1363.		33795.
361	90.01		-1339.		31948.
363	95.51		-1316.		30224.
365	101.3		-1293.		28613.

COMPOUND CONSTANTS:

MP	139.451 K	TC		VC		ZC
NBP	363.922 K	PC		DC		OM
MU	0.0000 DEBYE	RG				

VAPOR PRESSURE CORRELATION CONSTANTS, K AND KPA:

RIEDEL	RIEDEL	RIEDEL	RIEDEL
291. - 365. K	291. - 365. K	291. - 365. K	350. - 365. K
22 POINTS	22 POINTS	22 POINTS	7 POINTS
RMSD = 0.0084	RMSD = 0.0084	RMSD = 0.0084	RMSD = 0.0064
0.52932467E+02	0.52932467E+02	0.52932467E+02	0.51331793E+02
-0.55149996E+01	-0.55149996E+01	-0.55149996E+01	-0.52701969E+01
-0.57633514E+04	-0.57633514E+04	-0.57633514E+04	-0.57016134E+04
0.19041882E-16	0.19041882E-16	0.19041882E-16	0.13651339E-16

LIQUID DENSITY CORRELATION CONSTANTS, K AND G/ML:

FRANCIS1	FRANCIS1	FRANCIS1
293. - 304. K	293. - 304. K	293. - 304. K
6 POINTS	6 POINTS	6 POINTS
RMSD = 0.00014	RMSD = 0.00014	RMSD = 0.00014
0.99886858E+00	0.99886858E+00	0.99886858E+00
0.78431028E-03	0.78431028E-03	0.78431028E-03
0.66081905E+01	0.66081905E+01	0.66081905E+01
0.56379565E+03	0.56379565E+03	0.56379565E+03

SECOND VIRIAL COEFFICIENT CORRELATION CONSTANTS, ML/MOL:

KREGLEWSKI: A=-1206.0 B= 0.00 V*=137.00 R1=0.7000 R2=1.1300 R3=3.3290

LITERATURE DOCUMENTS FOR CORRELATED VAPOR PRESSURE VALUES:

1486 · 1492

LITERATURE DOCUMENTS FOR CORRELATED LIQUID DENSITY VALUES:

1490   7040   8335   8397

LITERATURE DOCUMENTS REPORTING VIRIAL COEFFICIENT DATA:

1492

T,K	VAPOR PRESSURE, KPA	SATURATED LIQUID VOLUME, ML/MOL	SECOND VIRIAL COEFFICIENT, ML/MOL	HEAT OF VAPORIZATION, J/MOL	SATURATED VAPOR VOLUME, ML/MOL
291	6.362		-2775.		377516.
292	6.682		-2739.		360556.
294	7.364	132.01	-2669.	34324.	329259.
296	8.102	132.32	-2602.	34222.	301120.
297	8.494	132.48	-2569.	34171.	288122.
299	9.325	132.79	-2506.	34068.	264064.
301	10.22	133.11	-2445.	33964.	242352.
302	10.70	133.27	-2415.	33913.	232293.
304	11.70	133.59	-2357.	33809.	213625.
306	12.78		-2302.		196715.
307	13.35		-2275.		188860.
309	14.56		-2222.		174243.
311	15.85		-2171.		160957.
312	16.53		-2146.		154770.
314	17.96		-2098.		143227.
316	19.49		-2051.		132701.
317	20.30		-2028.		127787.
319	21.99		-1984.		118598.
321	23.79		-1941.		110191.
322	24.74		-1920.		106258.
324	26.72		-1880.		98886.
326	28.84		-1840.		92122.
327	29.94		-1821.		88950.
329	32.25		-1783.		82993.
331	34.71		-1747.		77511.
333	37.31		-1711.		72462.
334	38.66		-1694.		70088.
336	41.50		-1660.		65615.
338	44.50		-1627.		61484.
339	46.06		-1611.		59538.
341	49.32		-1580.		55864.
343	52.75		-1550.		52463.
344	54.54		-1535.		50858.
346	58.26		-1506.		47822.
348	62.18		-1477.		45005.
349	64.22		-1464.		43672.
351	68.45		-1437.		41149.
353	72.89		-1410.		38802.
354	75.20		-1397.		37690.
356	79.98		-1372.		35580.
358	85.00		-1348.		33614.
359	87.60		-1335.		32681.
361	92.99		-1312.		30908.
363	98.63		-1289.		29251.
364	101.6		-1278.		28464.

COMPOUND CONSTANTS:

MP   139.241 K   TC   VC   ZC
NBP   364.872 K   PC   DC   OM
MU   0.0000 DEBYE   RG

    VAPOR PRESSURE CORRELATION CONSTANTS, K AND KPA:

    RIEDEL   RIEDEL   RIEDEL   RIEDEL
    299. - 366. K   299. - 366. K   299. - 366. K   351. - 366. K
      18 POINTS   18 POINTS   18 POINTS   7 POINTS
    RMSD = 0.0026   RMSD = 0.0026   RMSD = 0.0026   RMSD = 0.0018
    0.56715991E+02   0.56715991E+02   0.56715991E+02   0.18469197E+03
    -0.60805653E+01   -0.60805653E+01   -0.60805653E+01   -0.25135801E+02
    -0.59443072E+04   -0.59443072E+04   -0.59443072E+04   -0.11803914E+05
    0.28142866E-16   0.28142866E-16   0.28142866E-16   0.24029932E-15

    LIQUID DENSITY CORRELATION CONSTANTS, K AND G/ML:

    FRANCIS1   FRANCIS1   FRANCIS1
    273. - 374. K   273. - 374. K   273. - 374. K
      6 POINTS   6 POINTS   6 POINTS
    RMSD = 0.00029   RMSD = 0.00029   RMSD = 0.00029
    0.97295642E+00   0.97295642E+00   0.97295642E+00
    0.66608237E-03   0.66608237E-03   0.66608237E-03
    0.59999990E+01   0.59999990E+01   0.59999990E+01
    0.49688330E+03   0.49688330E+03   0.49688330E+03

    SECOND VIRIAL COEFFICIENT CORRELATION CONSTANTS, ML/MOL:

    KREGLEWSKI: A=-1213.0 B= 0.00 V*=138.00 R1=0.7000 R2=1.1300 R3=3.3800

LITERATURE DOCUMENTS FOR CORRELATED VAPOR PRESSURE VALUES:

    1486

LITERATURE DOCUMENTS FOR CORRELATED LIQUID DENSITY VALUES:

    1490 8397 10808

LITERATURE DOCUMENTS REPORTING VIRIAL COEFFICIENT DATA:

T,K	VAPOR PRESSURE, KPA	SATURATED LIQUID VOLUME, ML/MOL	SECOND VIRIAL COEFFICIENT, ML/MOL	HEAT OF VAPORIZATION, J/MOL	SATURATED VAPOR VOLUME, ML/MOL
273		128.47	-3698.		
275		128.73	-3591.		
277		129.00	-3488.		
279		129.27	-3389.		
282		129.67	-3249.		
284		129.95	-3160.		
286		130.22	-3075.		
289		130.64	-2954.		
291		130.92	-2877.		
293		131.20	-2803.		
295		131.49	-2732.		
298		131.92	-2630.		
300	9.393	132.21	-2566.	34209.	262970.
302	10.29	132.51	-2504.	34101.	241389.
305	11.78	132.95	-2415.	33937.	212833.
307	12.87	133.25	-2358.	33828.	196020.
309	14.03	133.55	-2303.	33719.	180768.
312	15.94	134.01	-2225.	33554.	160461.
314	17.33	134.32	-2175.	33443.	148433.
316	18.82	134.63	-2126.	33332.	137469.
318	20.40	134.95	-2080.	33221.	127462.
321	22.99	135.43	-2012.	33054.	114042.
323	24.86	135.75	-1969.	32942.	106035.
325	26.84	136.08	-1928.	32830.	98695.
328	30.06	136.57	-1868.	32661.	88801.
330	32.38	136.91	-1829.	32548.	82867.
332	34.84	137.24	-1792.	32435.	77406.
334	37.44	137.59	-1756.	32322.	72374.
337	41.63	138.11	-1704.	32151.	65549.
339	44.63	138.46	-1671.	32037.	61431.
341	47.80	138.81	-1638.	31923.	57623.
344	52.89	139.36	-1591.	31751.	52434.
346	56.52	139.72	-1561.	31637.	49289.
348	60.34	140.10	-1532.	31522.	46370.
351	66.45	140.66	-1490.	31349.	42377.
353	70.78	141.05	-1462.	31234.	39947.
355	75.34	141.44	-1436.	31119.	37685.
357	80.12	141.83	-1410.	31004.	35577.
360	87.75	142.44	-1373.	30830.	32679.
362	93.14	142.85	-1349.	30715.	30906.
364	98.78	143.27	-1325.	30600.	29250.
367		143.91	-1291.		
369		144.34	-1269.		
371		144.78	-1248.		
373		145.23	-1227.		

COMPOUND CONSTANTS:

MP		TC		VC		ZC
NBP		PC		DC		OM
MU 0.5800 DEBYE		RG				

VAPOR PRESSURE CORRELATION CONSTANTS, K AND KPA:

RPM2	RPM2	RPM2
300. - 364. K	300. - 364. K	300. - 364. K
6 POINTS	6 POINTS	6 POINTS
RMSD = 0.0370	RMSD = 0.0370	RMSD = 0.0370
0.74239391E+01	0.74239391E+01	0.74239391E+01
-0.39296611E+04	-0.39296611E+04	-0.39296611E+04
0.37078075E-01	0.37078075E-01	0.37078075E-01
-0.44859542E-04	-0.44859542E-04	-0.44859542E-04

LIQUID DENSITY CORRELATION CONSTANTS, K AND G/ML:

SECOND VIRIAL COEFFICIENT CORRELATION CONSTANTS, ML/MOL:

LITERATURE DOCUMENTS FOR CORRELATED VAPOR PRESSURE VALUES:

   13773

LITERATURE DOCUMENTS FOR CORRELATED LIQUID DENSITY VALUES:

LITERATURE DOCUMENTS REPORTING VIRIAL COEFFICIENT DATA:

T,K	VAPOR PRESSURE, KPA	SATURATED LIQUID VOLUME, ML/MOL	SECOND VIRIAL COEFFICIENT, ML/MOL	HEAT OF VAPORIZATION, J/MOL	SATURATED VAPOR VOLUME, ML/MOL
300	4.101				608283.
301	4.327				578438.
302	4.563				550260.
304	5.070				498498.
305	5.342				474729.
307	5.922				430994.
308	6.233				410879.
310	6.895				373810.
311	7.248				356733.
313	8.002				325216.
314	8.403				310674.
315	8.822				296881.
317	9.712				271375.
318	10.19				259583.
320	11.19				237744.
321	11.72				227633.
323	12.86				208882.
324	13.46				200188.
326	14.73				184042.
327	15.40				176545.
329	16.82				162605.
330	17.57				156123.
331	18.35				149945.
333	20.00				138436.
334	20.87				133076.
336	22.70				123077.
337	23.66				118415.
339	25.69				109706.
340	26.76				105640.
342	29.00				98037.
343	30.18				94483.
345	32.66				87829.
346	33.96				84714.
347	35.30				81732.
349	38.11				76141.
350	39.58				73520.
352	42.66				68600.
353	44.27				66290.
355	47.64				61951.
356	49.40				59912.
358	53.08				56077.
359	55.00				54273.
361	59.00				50875.
362	61.08				49275.
363	63.22				47738.

COMPOUND CONSTANTS:

MP		TC		VC		ZC
NBP	389.919 K	PC		DC		OM
MU		RG				

VAPOR PRESSURE CORRELATION CONSTANTS, K AND KPA:

RPM2	RPM2	RPM2	RIEDEL
333. - 390. K	333. - 390. K	333. - 390. K	376. - 390. K
14 POINTS	14 POINTS	14 POINTS	7 POINTS
RMSD = 0.0057	RMSD = 0.0057	RMSD = 0.0057	RMSD = 0.0038
0.28797246E+02	0.28797246E+02	0.28797246E+02	-0.44746069E+03
-0.61706990E+04	-0.61706990E+04	-0.61706990E+04	0.68344225E+02
-0.29979844E-01	-0.29979844E-01	-0.29979844E-01	0.18027046E+05
0.21945006E-04	0.21945006E-04	0.21945006E-04	-0.53814815E-15

LIQUID DENSITY CORRELATION CONSTANTS, K AND G/ML:

FRANCIS1	FRANCIS1	FRANCIS1
293. - 339. K	293. - 339. K	293. - 339. K
4 POINTS	4 POINTS	4 POINTS
RMSD = 0.00046	RMSD = 0.00046	RMSD = 0.00046
0.11150045E+01	0.11150045E+01	0.11150045E+01
0.80701662E-03	0.80701662E-03	0.80701662E-03
0.59999990E+01	0.59999990E+01	0.59999990E+01
0.63523047E+03	0.63523047E+03	0.63523047E+03

SECOND VIRIAL COEFFICIENT CORRELATION CONSTANTS, ML/MOL:

LITERATURE DOCUMENTS FOR CORRELATED VAPOR PRESSURE VALUES:

18172

LITERATURE DOCUMENTS FOR CORRELATED LIQUID DENSITY VALUES:

7697

LITERATURE DOCUMENTS REPORTING VIRIAL COEFFICIENT DATA:

T,K	VAPOR PRESSURE, KPA	SATURATED LIQUID VOLUME, ML/MOL	SECOND VIRIAL COEFFICIENT, ML/MOL	HEAT OF VAPORIZATION, J/MOL	SATURATED VAPOR VOLUME, ML/MOL
293		111.70			
295		111.92			
297		112.14			
299		112.37			
301		112.60			
304		112.94			
306		113.17			
308		113.40			
310		113.63			
312		113.86			
315		114.21			
317		114.45			
319		114.68			
321		114.92			
323		115.16			
326		115.52			
328		115.76			
330		116.01			
332		116.25			
334	15.75	116.50		37071.	176303.
337	17.74	116.87		36936.	157976.
339	19.17	117.12		36847.	147028.
341	20.70				136983.
343	22.32				127757.
345	24.05				119273.
348	26.84				107791.
350	28.85				100880.
352	30.97				94500.
354	33.22				88606.
356	35.60				83155.
359	39.42				75725.
361	42.14				71222.
363	45.02				67043.
365	48.05				63161.
367	51.24				59552.
370	56.34				54603.
372	59.96				51584.
374	63.76				48768.
376	67.75				46140.
378	71.94				43686.
381	78.60				40301.
383	83.31				38224.
385	88.24				36278.
387	93.39				34455.
389	98.78				32744.

COMPOUND CONSTANTS:

MP		TC		VC		ZC
NBP	366.257 K	PC		DC		OM
MU		RG				

VAPOR PRESSURE CORRELATION CONSTANTS, K AND KPA:

LIQUID DENSITY CORRELATION CONSTANTS, K AND G/ML:

FRANCIS1	FRANCIS1	FRANCIS1
288. - 343. K	288. - 343. K	288. - 343. K
6 POINTS	6 POINTS	6 POINTS
RMSD = 0.00013	RMSD = 0.00013	RMSD = 0.00013
0.10752735E+01	0.10752735E+01	0.10752735E+01
0.81602135E-03	0.81602135E-03	0.81602135E-03
0.60293818E+01	0.60293818E+01	0.60293818E+01
0.51213379E+03	0.51213379E+03	0.51213379E+03

SECOND VIRIAL COEFFICIENT CORRELATION CONSTANTS, ML/MOL:

LITERATURE DOCUMENTS FOR CORRELATED VAPOR PRESSURE VALUES:

LITERATURE DOCUMENTS FOR CORRELATED LIQUID DENSITY VALUES:

   7697

LITERATURE DOCUMENTS REPORTING VIRIAL COEFFICIENT DATA:

T,K	VAPOR PRESSURE, KPA	SATURATED LIQUID VOLUME, ML/MOL	SECOND VIRIAL COEFFICIENT, ML/MOL	HEAT OF VAPORIZATION, J/MOL	SATURATED VAPOR VOLUME, ML/MOL
288		118.24			
289		118.38			
290		118.51			
291		118.65			
293		118.93			
294		119.07			
295		119.20			
296		119.34			
298		119.63			
299		119.77			
300		119.91			
301		120.05			
303		120.34			
304		120.48			
305		120.63			
306		120.77			
308		121.06			
309		121.21			
310		121.36			
311		121.50			
313		121.80			
314		121.95			
315		122.10			
316		122.25			
318		122.56			
319		122.71			
320		122.86			
321		123.02			
323		123.33			
324		123.48			
325		123.64			
326		123.80			
328		124.11			
329		124.27			
330		124.43			
331		124.59			
333		124.92			
334		125.08			
335		125.25			
336		125.41			
338		125.74			
339		125.91			
340		126.08			
341		126.25			
343		126.59			

COMPOUND CONSTANTS:

MP	TC	VC	ZC
NBP	PC	DC	OM
MU	RG		

VAPOR PRESSURE CORRELATION CONSTANTS, K AND KPA:

RIEDEL	RIEDEL	RIEDEL
302. - 338. K	302. - 338. K	302. - 338. K
8 POINTS	8 POINTS	8 POINTS
RMSD = 0.1111	RMSD = 0.1111	RMSD = 0.1111
-0.65182136E+02	-0.65182136E+02	-0.65182136E+02
0.12662607E+02	0.12662607E+02	0.12662607E+02
-0.14899795E+04	-0.14899795E+04	-0.14899795E+04
-0.45872331E-15	-0.45872331E-15	-0.45872331E-15

LIQUID DENSITY CORRELATION CONSTANTS, K AND G/ML:

SECOND VIRIAL COEFFICIENT CORRELATION CONSTANTS, ML/MOL:

LITERATURE DOCUMENTS FOR CORRELATED VAPOR PRESSURE VALUES:

13773

LITERATURE DOCUMENTS FOR CORRELATED LIQUID DENSITY VALUES:

LITERATURE DOCUMENTS REPORTING VIRIAL COEFFICIENT DATA:

T,K	VAPOR PRESSURE, KPA	SATURATED LIQUID VOLUME, ML/MOL	SECOND VIRIAL COEFFICIENT, ML/MOL	HEAT OF VAPORIZATION, J/MOL	SATURATED VAPOR VOLUME, ML/MOL
302	6.328				396778.
303	6.661				378234.
304	7.008				360682.
305	7.370				344065.
306	7.749				328329.
307	8.144				313421.
308	8.556				299295.
309	8.986				285906.
310	9.434				273210.
311	9.901				261170.
312	10.39				249746.
313	10.89				238906.
314	11.42				228616.
315	11.97				218845.
316	12.54				209565.
317	13.13				200749.
318	13.74				192370.
319	14.38				184405.
320	15.05				176832.
321	15.73				169629.
322	16.45				162777.
323	17.19				156255.
324	17.95				150048.
325	18.75				144137.
326	19.57				138508.
327	20.42				133146.
328	21.30				128036.
329	22.21				123166.
330	23.15				118523.
331	24.12				114095.
332	25.12				109871.
333	26.16				105841.
334	27.23				101995.
335	28.33				98324.
336	29.46				94819.
337	30.63				91472.
338	31.84				88274.

COMPOUND CONSTANTS:

MP		TC		VC		ZC
NBP	377.698 K	PC		DC		OM
MU		RG				

VAPOR PRESSURE CORRELATION CONSTANTS, K AND KPA:

VAPRES-2	VAPRES-2	VAPRES-2	RIEDEL
323. - 378. K	323. - 378. K	323. - 378. K	358. - 378. K
15 POINTS	15 POINTS	15 POINTS	8 POINTS
RMSD = 0.0067	RMSD = 0.0067	RMSD = 0.0067	RMSD = 0.0063
0.97151009E+03	0.97151009E+03	0.97151009E+03	0.76761710E+02
-0.26016144E+05	-0.26016144E+05	-0.26016144E+05	-0.89681286E+01
0.48781509E+00	0.48781509E+00	0.48781509E+00	-0.71920082E+04
-0.22956716E-03	-0.22956716E-03	-0.22956716E-03	0.39978161E-16
-0.17686077E+03	-0.17686077E+03	-0.17686077E+03	

LIQUID DENSITY CORRELATION CONSTANTS, K AND G/ML:

SECOND VIRIAL COEFFICIENT CORRELATION CONSTANTS, ML/MOL:

LITERATURE DOCUMENTS FOR CORRELATED VAPOR PRESSURE VALUES:

   18172

LITERATURE DOCUMENTS FOR CORRELATED LIQUID DENSITY VALUES:

LITERATURE DOCUMENTS REPORTING VIRIAL COEFFICIENT DATA:

T,K	VAPOR PRESSURE, KPA	SATURATED LIQUID VOLUME, ML/MOL	SECOND VIRIAL COEFFICIENT, ML/MOL	HEAT OF VAPORIZATION, J/MOL	SATURATED VAPOR VOLUME, ML/MOL
323	15.46				173672.
324	16.11				167183.
325	16.79				160983.
326	17.48				155059.
328	18.94				143980.
329	19.71				138800.
330	20.50				133843.
331	21.32				129099.
333	23.03				120205.
334	23.93				116037.
335	24.86				112043.
336	25.82				108214.
338	27.82				101022.
339	28.87				97644.
340	29.95				94402.
341	31.06				91290.
343	33.38				85432.
344	34.60				82675.
345	35.84				80026.
346	37.13				77479.
348	39.81				72677.
349	41.21				70412.
350	42.65				68234.
351	44.13				66137.
353	47.21				62175.
354	48.81				60304.
355	50.45				58501.
356	52.14				56765.
358	55.66				53478.
359	57.49				51922.
360	59.36				50422.
361	61.29				48976.
363	65.28				46233.
364	67.35				44934.
365	69.48				43679.
366	71.66				42468.
368	76.17				40168.
369	78.51				39077.
370	80.91				38022.
371	83.36				37003.
373	88.44				35065.
374	91.07				34144.
375	93.76				33253.
376	96.52				32391.
378	102.2				30750.

COMPOUND CONSTANTS:

MP   TC   VC   ZC
NBP   PC   DC   OM
MU   RG

    VAPOR PRESSURE CORRELATION CONSTANTS, K AND KPA:

    RPM2   RPM2   RPM2
    302. - 373. K   302. - 373. K   302. - 373. K
      9 POINTS   9 POINTS   9 POINTS
    RMSD = 0.1638   RMSD = 0.1638   RMSD = 0.1638
    0.10429877E+03   0.10429877E+03   0.10429877E+03
   -0.14485093E+05   -0.14485093E+05   -0.14485093E+05
   -0.25680415E+00   -0.25680415E+00   -0.25680415E+00
    0.24891715E-03   0.24891715E-03   0.24891715E-03

    LIQUID DENSITY CORRELATION CONSTANTS, K AND G/ML:

    SECOND VIRIAL COEFFICIENT CORRELATION CONSTANTS, ML/MOL:

LITERATURE DOCUMENTS FOR CORRELATED VAPOR PRESSURE VALUES:

    13773

LITERATURE DOCUMENTS FOR CORRELATED LIQUID DENSITY VALUES:

LITERATURE DOCUMENTS REPORTING VIRIAL COEFFICIENT DATA:

T,K	VAPOR PRESSURE, KPA	SATURATED LIQUID VOLUME, ML/MOL	SECOND VIRIAL COEFFICIENT, ML/MOL	HEAT OF VAPORIZATION, J/MOL	SATURATED VAPOR VOLUME, ML/MOL
302	4.403				570290.
303	4.638				543138.
305	5.139				493433.
306	5.405				470678.
308	5.971				428909.
310	6.582				391602.
311	6.906				374447.
313	7.591				342822.
314	7.954				328242.
316	8.720				301299.
318	9.544				277023.
319	9.979				265786.
321	10.90				244942.
322	11.38				235272.
324	12.40				217295.
326	13.49				200969.
327	14.06				193368.
329	15.27				179189.
331	16.55				166253.
332	17.23				160210.
334	18.65				148902.
335	19.40				143609.
337	20.96				133687.
339	22.63				124575.
340	23.50				120298.
342	25.33				112256.
343	26.29				108474.
345	28.30				101353.
347	30.44				94775.
348	31.56				91674.
350	33.91				85819.
352	36.41				80392.
353	37.71				77827.
355	40.45				72974.
356	41.88				70678.
358	44.88				66325.
360	48.07				62273.
361	49.73				60351.
363	53.23				56702.
364	55.06				54969.
366	58.89				51675.
368	62.96				48596.
369	65.10				47131.
371	69.56				44342.
372	71.91				43014.

COMPOUND CONSTANTS:

MP 286.644 K	TC 647.200 K	VC 410.00 ML/MOL	ZC 0.2710
NBP 424.314 K	PC 3.557 MPA	DC 0.2737 G/ML	OM 0.2521
MU 0.0000 DEBYE	RG		

VAPOR PRESSURE CORRELATION CONSTANTS, K AND KPA:

WAGNER	WAGNER	WAGNER	WAGNER
288. - 468. K	288. - 439. K	288. - 468. K	410. - 468. K
33 POINTS	29 POINTS	33 POINTS	12 POINTS
RMSD = 0.0141	RMSD = 0.0132	RMSD = 0.0141	RMSD = 0.0168
-0.70157006E+01	-0.70154123E+01	-0.70157006E+01	-0.75144699E+01
0.98300712E+00	0.98222295E+00	0.98300712E+00	0.22886360E+01
-0.26915749E+01	-0.26901213E+01	-0.26915749E+01	-0.53727928E+01
-0.33143474E+01	-0.33173140E+01	-0.33143474E+01	0.70703568E+01

LIQUID DENSITY CORRELATION CONSTANTS, K AND G/ML:

FRANCIS1	FRANCIS1	FRANCIS1	FRANCIS2
288. - 602. K	288. - 402. K	288. - 602. K	601. - 647. K
23 POINTS	22 POINTS	23 POINTS	2 POINTS
RMSD = 0.00022	RMSD = 0.00023	RMSD = 0.00022	RMSD = 0.00000
0.10765619E+01	0.10765047E+01	0.10765619E+01	0.18249828E-03
0.75872545E-03	0.75918762E-03	0.75872545E-03	0.33653870E+01
0.66627989E+01	0.65734262E+01	0.66627989E+01	
0.66469531E+03	0.66369946E+03	0.66469531E+03	

SECOND VIRIAL COEFFICIENT CORRELATION CONSTANTS, ML/MOL:

TSONOPOULOS: TC=647.200 PC=3.557 VC=410.00 OM=0.2521 A= 0.000000 B=0.0000

LITERATURE DOCUMENTS FOR CORRELATED VAPOR PRESSURE VALUES:

   1613   6550   8098 10271 10552 40284

LITERATURE DOCUMENTS FOR CORRELATED LIQUID DENSITY VALUES:

   1613   2840   5640 10219 10552 10940 17936 40284

LITERATURE DOCUMENTS REPORTING VIRIAL COEFFICIENT DATA:

T,K	VAPOR PRESSURE, KPA	SATURATED LIQUID VOLUME, ML/MOL	SECOND VIRIAL COEFFICIENT, ML/MOL	HEAT OF VAPORIZATION, J/MOL	SATURATED VAPOR VOLUME, ML/MOL
288	0.4018	133.53	-5920.	44350.	5953835.
295	0.6224	134.43	-5280.	43890.	3935192.
302	0.9410	135.35	-4743.	43440.	2663555.
309	1.391	136.28	-4289.	42997.	1842789.
316	2.013	137.23	-3901.	42563.	1300991.
323	2.859	138.20	-3568.	42135.	935817.
330	3.987	139.18	-3280.	41713.	684884.
337	5.469	140.17	-3030.	41296.	509330.
345	7.700	141.33	-2781.	40824.	369721.
352	10.23	142.37	-2592.	40415.	283470.
359	13.41	143.43	-2423.	40008.	220109.
366	17.37	144.50	-2273.	39602.	172934.
373	22.23	145.60	-2139.	39196.	137366.
380	28.14	146.72	-2017.	38789.	110231.
387	35.26	147.86	-1908.	38380.	89300.
395	45.11	149.19	-1795.	37910.	70962.
402	55.43	150.38	-1705.	37494.	58546.
409	67.54	151.60	-1623.	37073.	48671.
416	81.65	152.85	-1548.	36646.	40750.
423	97.99	154.13	-1479.	36212.	34346.
430	116.8	155.44	-1414.	35770.	29127.
437	138.3	156.78	-1355.	35319.	24845.
444	162.7	158.16	-1299.	34858.	21306.
452	194.6	159.78	-1240.	34318.	17984.
459	226.2	161.25	-1192.	33833.	15583.
466	261.6	162.76	-1147.	33334.	13560.
473		164.31	-1105.		
480		165.92	-1065.		
487		167.59	-1028.		
494		169.32	-992.5		
502		171.39	-954.4		
509		173.28	-922.9		
516		175.27	-893.0		
523		177.35	-864.7		
530		179.55	-837.6		
537		181.89	-811.9		
544		184.38	-787.3		
552		187.45	-760.6		
559		190.39	-738.3		
566		193.60	-717.0		
573		197.16	-696.6		
580		201.16	-677.0		
587		205.73	-658.2		
594		211.07	-640.1		
601		217.48	-622.8		

COMPOUND CONSTANTS:

MP	161.853 K	TC		VC	ZC
NBP	404.946 K	PC		DC	OM
MU	0.0000 DEBYE	RG			

VAPOR PRESSURE CORRELATION CONSTANTS, K AND KPA:

WAGNER	VAPRES-2	VAPRES-2	WAGNER
258. - 406. K	258. - 406. K	258. - 406. K	389. - 406. K
23 POINTS	23 POINTS	23 POINTS	4 POINTS
RMSD = 0.0620	RMSD = 0.0368	RMSD = 0.0368	RMSD = 0.0000
-0.75763628E+01	-0.58250789E+03	-0.58250789E+03	-0.36525218E+01
0.24720050E+01	0.65203998E+04	0.65203998E+04	-0.71032053E+01
-0.54489582E+01	-0.36740363E+00	-0.36740363E+00	0.10693235E+02
0.15366023E+01	0.18561680E-03	0.18561680E-03	-0.43451841E+02
	0.11482243E+03	0.11482243E+03	

LIQUID DENSITY CORRELATION CONSTANTS, K AND G/ML:

FRANCIS1	FRANCIS1	FRANCIS1
273. - 374. K	273. - 374. K	273. - 374. K
8 POINTS	8 POINTS	8 POINTS
RMSD = 0.00045	RMSD = 0.00045	RMSD = 0.00045
0.10324488E+01	0.10324488E+01	0.10324488E+01
0.74969605E-03	0.74969605E-03	0.74969605E-03
0.82515640E+01	0.82515640E+01	0.82515640E+01
0.62375073E+03	0.62375073E+03	0.62375073E+03

SECOND VIRIAL COEFFICIENT CORRELATION CONSTANTS, ML/MOL:

LITERATURE DOCUMENTS FOR CORRELATED VAPOR PRESSURE VALUES:

1146 1485

LITERATURE DOCUMENTS FOR CORRELATED LIQUID DENSITY VALUES:

1487 10808

LITERATURE DOCUMENTS REPORTING VIRIAL COEFFICIENT DATA:

T,K	VAPOR PRESSURE, KPA	SATURATED LIQUID VOLUME, ML/MOL	SECOND VIRIAL COEFFICIENT, ML/MOL	HEAT OF VAPORIZATION, J/MOL	SATURATED VAPOR VOLUME, ML/MOL
258	0.1268				0.1692E+08
261	0.1585				0.1369E+08
264	0.1973				0.1113E+08
268	0.2619				8507785.
271	0.3221				6995951.
274	0.3941	139.67		41524.	5779929.
278	0.5122	140.24		41419.	4512896.
281	0.6200	140.67		41330.	3768047.
284	0.7473	141.11		41233.	3159828.
288	0.9520	141.69		41092.	2515244.
291	1.136	142.13		40977.	2129851.
294	1.350	142.58		40857.	1810725.
298	1.689	143.18		40686.	1467163.
301	1.989	143.64		40552.	1258515.
305	2.458	144.25		40366.	1031570.
308	2.870	144.72		40221.	892329.
311	3.338	145.18		40072.	774543.
315	4.063	145.81		39869.	644657.
318	4.689	146.29		39714.	563873.
321	5.394	146.77		39556.	494763.
325	6.471	147.42		39343.	417581.
328	7.391	147.91		39183.	368971.
331	8.417	148.41		39022.	326952.
335	9.966	149.08		38807.	279473.
338	11.28	149.58		38646.	249222.
342	13.24	150.27		38434.	214778.
345	14.89	150.78		38277.	192666.
348	16.70	151.31		38122.	173247.
352	19.39	152.01		37920.	150910.
355	21.64	152.55		37772.	136425.
358	24.08	153.09		37628.	123597.
362	27.69	153.82		37443.	108702.
365	30.67	154.37		37310.	98953.
368	33.90	154.93		37183.	90253.
372	38.63	155.68		37022.	80062.
375	42.52				73334.
379	48.18				65409.
382	52.81				60147.
385	57.79				55396.
389	65.00				49756.
392	70.88				45983.
395	77.17				42555.
399	86.26				38457.
402	93.63				35697.
405	101.5				33175.

241

COMPOUND CONSTANTS:

MP 239.630 K	TC	VC	ZC
NBP 392.701 K	PC	DC	OM
MU 0.0000 DEBYE	RG 4.0925 ANGSTROM		

VAPOR PRESSURE CORRELATION CONSTANTS, K AND KPA:

VAPRES-2	VAPRES-2	VAPRES-2	RIEDEL
248. - 394. K	248. - 394. K	248. - 394. K	377. - 394. K
21 POINTS	21 POINTS	21 POINTS	4 POINTS
RMSD = 0.0528	RMSD = 0.0528	RMSD = 0.0528	RMSD = 0.0000
-0.50980192E+03	-0.50980192E+03	-0.50980192E+03	0.57741115E+02
0.52456352E+04	0.52456352E+04	0.52456352E+04	-0.61777909E+01
-0.32989349E+00	-0.32989349E+00	-0.32989349E+00	-0.63962731E+04
0.16941921E-03	0.16941921E-03	0.16941921E-03	0.17821076E-16
0.10120195E+03	0.10120195E+03	0.10120195E+03	

LIQUID DENSITY CORRELATION CONSTANTS, K AND G/ML:

FRANCIS1	FRANCIS1	FRANCIS1
273. - 304. K	273. - 304. K	273. - 304. K
11 POINTS	11 POINTS	11 POINTS
RMSD = 0.00040	RMSD = 0.00040	RMSD = 0.00040
0.10118551E+01	0.10118551E+01	0.10118551E+01
0.68553770E-03	0.68553770E-03	0.68553770E-03
0.59999990E+01	0.59999990E+01	0.59999990E+01
0.49598633E+03	0.49598633E+03	0.49598633E+03

SECOND VIRIAL COEFFICIENT CORRELATION CONSTANTS, ML/MOL:

LITERATURE DOCUMENTS FOR CORRELATED VAPOR PRESSURE VALUES:

   1146   1486

LITERATURE DOCUMENTS FOR CORRELATED LIQUID DENSITY VALUES:

   1490   9734 13929 19948

LITERATURE DOCUMENTS REPORTING VIRIAL COEFFICIENT DATA:

T,K	VAPOR PRESSURE, KPA	SATURATED LIQUID VOLUME, ML/MOL	SECOND VIRIAL COEFFICIENT, ML/MOL	HEAT OF VAPORIZATION, J/MOL	SATURATED VAPOR VOLUME, ML/MOL
248	0.1257				0.1641E+08
251	0.1579				0.1322E+08
254	0.1972				0.1071E+08
257	0.2451				8719271.
261	0.3248				6681446.
264	0.3988				5503536.
267	0.4874				4554909.
271	0.6320				3564973.
274	0.7639	140.80		38966.	2982071.
277	0.9192	141.23		38881.	2505409.
281	1.169	141.81		38755.	1999388.
284	1.392	142.25		38653.	1696299.
287	1.651	142.69		38545.	1444938.
291	2.061	143.30		38392.	1173839.
294	2.423	143.75		38271.	1008889.
297	2.837	144.22		38146.	870304.
301	3.482	144.84		37972.	718642.
304	4.044	145.32		37837.	625019.
307	4.680				545416.
311	5.657				457123.
314	6.496				401889.
317	7.437				354407.
320	8.488				313462.
324	10.08				267319.
327	11.42				237997.
330	12.91				212460.
334	15.15				183360.
337	17.02				164663.
340	19.07				148232.
344	22.12				129316.
347	24.65				117038.
350	27.41				106158.
354	31.47				93512.
357	34.83				85228.
360	38.46				77830.
364	43.76				69156.
367	48.11				63425.
370	52.79				58270.
374	59.60				52177.
377	65.14				48119.
380	71.09				44445.
384	79.68				40070.
387	86.65				37135.
390	94.10				34461.
393	102.0				32021.

COMPOUND CONSTANTS:

MP	223.178 K	TC	VC	ZC
NBP	402.900 K	PC	DC	OM
MU	0.0000 DEBYE	RG 4.0612 ANGSTROM		

VAPOR PRESSURE CORRELATION CONSTANTS, K AND KPA:

RPM2	RPM2	RPM2
257. - 404. K	257. - 404. K	257. - 404. K
24 POINTS	24 POINTS	24 POINTS
RMSD = 0.0237	RMSD = 0.0237	RMSD = 0.0237
0.32943866E+02	0.32943866E+02	0.32943866E+02
-0.68104693E+04	-0.68104693E+04	-0.68104693E+04
-0.41203222E-01	-0.41203222E-01	-0.41203222E-01
0.31903613E-04	0.31903613E-04	0.31903613E-04

LIQUID DENSITY CORRELATION CONSTANTS, K AND G/ML:

FRANCIS1	FRANCIS1	FRANCIS1
273. - 374. K	273. - 374. K	273. - 374. K
5 POINTS	5 POINTS	5 POINTS
RMSD = 0.00029	RMSD = 0.00029	RMSD = 0.00029
0.10216713E+01	0.10216713E+01	0.10216713E+01
0.69492520E-03	0.69492520E-03	0.69492520E-03
0.12436149E+02	0.12436149E+02	0.12436149E+02
0.69965552E+03	0.69965552E+03	0.69965552E+03

SECOND VIRIAL COEFFICIENT CORRELATION CONSTANTS, ML/MOL:

LITERATURE DOCUMENTS FOR CORRELATED VAPOR PRESSURE VALUES:

   1146   1485

LITERATURE DOCUMENTS FOR CORRELATED LIQUID DENSITY VALUES:

   10808 22447

LITERATURE DOCUMENTS REPORTING VIRIAL COEFFICIENT DATA:

T,K	VAPOR PRESSURE, KPA	SATURATED LIQUID VOLUME, ML/MOL	SECOND VIRIAL COEFFICIENT, ML/MOL	HEAT OF VAPORIZATION, J/MOL	SATURATED VAPOR VOLUME, ML/MOL
257	0.1303				0.1640E+08
260	0.1642				0.1316E+08
263	0.2057				0.1063E+08
267	0.2752				8067763.
270	0.3399				6605132.
273	0.4174	139.78		41886.	5437661.
277	0.5444	140.31		41613.	4230320.
280	0.6604	140.71		41411.	3525144.
283	0.7971	141.12		41210.	2951936.
287	1.017	141.67		40946.	2347063.
290	1.214	142.08		40750.	1986684.
293	1.442	142.50		40556.	1688962.
297	1.804	143.06		40301.	1369088.
300	2.123	143.48		40112.	1175104.
303	2.488	143.91		39925.	1012513.
307	3.057	144.48		39681.	834987.
310	3.552	144.92		39500.	725611.
313	4.113	145.36		39322.	632739.
317	4.974	145.94		39088.	529861.
320	5.715	146.39		38917.	465570.
323	6.545	146.84		38748.	410341.
327	7.805	147.44		38527.	348361.
330	8.876	147.90		38365.	309130.
333	10.07	148.36		38206.	275074.
337	11.85	148.99		37999.	236403.
340	13.36	149.46		37847.	211643.
343	15.01	149.93		37699.	189943.
347	17.48	150.57		37507.	165042.
350	19.54	151.05		37367.	148930.
353	21.79	151.54		37230.	134689.
357	25.11	152.20		37054.	118187.
360	27.87	152.70		36926.	107410.
363	30.86	153.20		36802.	97809.
367	35.24	153.88		36642.	86586.
370	38.85	154.39		36527.	79193.
373	42.74	154.91		36416.	72560.
377	48.41				64745.
380	53.05				59556.
383	58.03				54871.
387	65.25				49310.
390	71.12				45592.
393	77.40				42214.
397	86.46				38179.
400	93.78				35462.
403	101.6				32982.

COMPOUND CONSTANTS:

MP   185.000 K   TC   VC   ZC
NBP 396.579 K   PC   DC   OM
MU   0.0000 DEBYE   RG   4.1814 ANGSTROM

      VAPOR PRESSURE CORRELATION CONSTANTS, K AND KPA:

      RPM2   RPM2   RPM2
      252. - 398. K   252. - 398. K   252. - 398. K
       25 POINTS   25 POINTS   25 POINTS
      RMSD = 0.0146   RMSD = 0.0146   RMSD = 0.0146
       0.31019451E+02   0.31019451E+02   0.31019451E+02
      -0.64296154E+04   -0.64296154E+04   -0.64296154E+04
      -0.37142392E-01   -0.37142392E-01   -0.37142392E-01
       0.28875989E-04   0.28875989E-04   0.28875989E-04

      LIQUID DENSITY CORRELATION CONSTANTS, K AND G/ML:

      FRANCIS1   FRANCIS1   FRANCIS1
      273. - 314. K   273. - 314. K   273. - 314. K
       4 POINTS   4 POINTS   4 POINTS
      RMSD = 0.00000   RMSD = 0.00000   RMSD = 0.00000
      0.97665811E+00   0.97665811E+00   0.97665811E+00
      0.56329928E-03   0.56329928E-03   0.56329928E-03
      0.60415678E+01   0.60415678E+01   0.60415678E+01
      0.46317676E+03   0.46317676E+03   0.46317676E+03

      SECOND VIRIAL COEFFICIENT CORRELATION CONSTANTS, ML/MOL:

LITERATURE DOCUMENTS FOR CORRELATED VAPOR PRESSURE VALUES:

   1146   1485

LITERATURE DOCUMENTS FOR CORRELATED LIQUID DENSITY VALUES:

   1487

LITERATURE DOCUMENTS REPORTING VIRIAL COEFFICIENT DATA:

T,K	VAPOR PRESSURE, KPA	SATURATED LIQUID VOLUME, ML/MOL	SECOND VIRIAL COEFFICIENT, ML/MOL	HEAT OF VAPORIZATION, J/MOL	SATURATED VAPOR VOLUME, ML/MOL
252	0.1325				0.1581E+08
255	0.1672				0.1268E+08
258	0.2097				0.1023E+08
261	0.2613				8305755.
265	0.3471				6347939.
268	0.4266				5222855.
271	0.5215				4320309.
275	0.6761	142.11		40089.	3381721.
278	0.8166	142.51		39906.	2830492.
281	0.9815	142.91		39726.	2380395.
285	1.245	143.46		39487.	1903046.
288	1.481	143.88		39311.	1617232.
291	1.753	144.30		39136.	1380151.
295	2.182	144.87		38906.	1124274.
298	2.559	145.31		38736.	968407.
301	2.989	145.76		38567.	837285.
305	3.656	146.37		38346.	693532.
308	4.236	146.83		38183.	604607.
311	4.889	147.31		38022.	528850.
315	5.890				444621.
318	6.748				391792.
321	7.708				346274.
324	8.777				306931.
328	10.39				262473.
331	11.75				234142.
334	13.26				209410.
338	15.51				181152.
341	17.40				162947.
344	19.47				146913.
348	22.53				128407.
351	25.08				116367.
354	27.85				105676.
358	31.93				93224.
361	35.29				85050.
364	38.93				77738.
368	44.25				69152.
371	48.60				63469.
374	53.29				58351.
378	60.10				52295.
381	65.64				48258.
384	71.59				44599.
388	80.17				40240.
391	87.12				37314.
394	94.55				34647.
397	102.5				32214.

COMPOUND CONSTANTS:

MP   197.623 K   TC   VC   ZC
NBP   397.608 K   PC   DC   OM
MU   0.0000 DEBYE   RG   4.0549 ANGSTROM

VAPOR PRESSURE CORRELATION CONSTANTS, K AND KPA:

VAPRES-2	VAPRES-2	VAPRES-2	RPM2
253. - 399. K	253. - 399. K	253. - 399. K	382. - 399. K
23 POINTS	23 POINTS	23 POINTS	4 POINTS
RMSD = 0.0103	RMSD = 0.0103	RMSD = 0.0103	RMSD = 0.0011
-0.10198710E+03	-0.10198710E+03	-0.10198710E+03	0.38372433E+02
-0.35751265E+04	-0.35751265E+04	-0.35751265E+04	-0.74876047E+04
-0.10541433E+00	-0.10541433E+00	-0.10541433E+00	-0.54509198E-01
0.59658585E-04	0.59658585E-04	0.59658585E-04	0.42701797E-04
0.24739791E+02	0.24739791E+02	0.24739791E+02	

LIQUID DENSITY CORRELATION CONSTANTS, K AND G/ML:

FRANCIS1	FRANCIS1	FRANCIS1
273. - 314. K	273. - 314. K	273. - 314. K
5 POINTS	5 POINTS	5 POINTS
RMSD = 0.00094	RMSD = 0.00094	RMSD = 0.00094
0.10294914E+01	0.10294914E+01	0.10294914E+01
0.76570525E-03	0.76570525E-03	0.76570525E-03
0.59999990E+01	0.59999990E+01	0.59999990E+01
0.58713574E+03	0.58713574E+03	0.58713574E+03

SECOND VIRIAL COEFFICIENT CORRELATION CONSTANTS, ML/MOL:

LITERATURE DOCUMENTS FOR CORRELATED VAPOR PRESSURE VALUES:

    1146   1485

LITERATURE DOCUMENTS FOR CORRELATED LIQUID DENSITY VALUES:

    1487   9734

LITERATURE DOCUMENTS REPORTING VIRIAL COEFFICIENT DATA:

T,K	VAPOR PRESSURE, KPA	SATURATED LIQUID VOLUME, ML/MOL	SECOND VIRIAL COEFFICIENT, ML/MOL	HEAT OF VAPORIZATION, J/MOL	SATURATED VAPOR VOLUME, ML/MOL
253	0.1256				0.1674E+08
256	0.1585				0.1343E+08
259	0.1986				0.1084E+08
262	0.2475				8801051.
266	0.3289				6723663.
269	0.4045				5528897.
272	0.4949				4570073.
276	0.6423	140.47		40588.	3572808.
279	0.7766	140.90		40433.	2987184.
282	0.9344	141.35		40277.	2509151.
286	1.187	141.94		40068.	2002463.
289	1.414	142.39		39911.	1699316.
292	1.677	142.84		39753.	1448059.
296	2.091	143.45		39543.	1177172.
299	2.456	143.92		39385.	1012364.
302	2.873	144.38		39227.	873877.
306	3.523	145.01		39017.	722263.
309	4.087	145.48		38860.	628616.
312	4.726	145.96		38704.	548945.
316	5.705				460505.
319	6.547				405129.
322	7.489				357485.
325	8.541				316365.
329	10.13				269979.
332	11.48				240472.
335	12.97				214751.
339	15.20				185415.
342	17.07				166548.
345	19.13				149955.
349	22.18				130836.
352	24.71				118419.
355	27.48				107408.
359	31.55				94604.
362	34.91				86213.
365	38.55				78716.
369	43.88				69925.
372	48.24				64115.
375	52.95				58889.
379	59.78				52714.
382	65.35				48602.
385	71.32				44880.
389	79.96				40451.
392	86.96				37481.
395	94.43				34778.
398	102.4				32312.

COMPOUND CONSTANTS:

MP	183.079 K	TC		VC		ZC	
NBP	393.245 K	PC		DC		OM	
MU	0.0000 DEBYE	RG	4.1462 ANGSTROM				

VAPOR PRESSURE CORRELATION CONSTANTS, K AND KPA:

RIEDEL	RIEDEL	RIEDEL	RIEDEL
250. - 395. K	250. - 395. K	250. - 395. K	378. - 395. K
24 POINTS	24 POINTS	24 POINTS	4 POINTS
RMSD = 0.0159	RMSD = 0.0159	RMSD = 0.0159	RMSD = 0.0000
0.65146648E+02	0.65146648E+02	0.65146648E+02	-0.66255666E+02
-0.72688789E+01	-0.72688789E+01	-0.72688789E+01	0.12095463E+02
-0.67683680E+04	-0.67683680E+04	-0.67683680E+04	-0.38361218E+03
0.29847592E-16	0.29847592E-16	0.29847592E-16	-0.11195687E-15

LIQUID DENSITY CORRELATION CONSTANTS, K AND G/ML:

FRANCIS1	FRANCIS1	FRANCIS1
273. - 314. K	273. - 314. K	273. - 314. K
4 POINTS	4 POINTS	4 POINTS
RMSD = 0.00010	RMSD = 0.00010	RMSD = 0.00010
0.99755430E+00	0.99755430E+00	0.99755430E+00
0.71052113E-03	0.71052113E-03	0.71052113E-03
0.60013008E+01	0.60013008E+01	0.60013008E+01
0.55259399E+03	0.55259399E+03	0.55259399E+03

SECOND VIRIAL COEFFICIENT CORRELATION CONSTANTS, ML/MOL:

LITERATURE DOCUMENTS FOR CORRELATED VAPOR PRESSURE VALUES:

1146   1485

LITERATURE DOCUMENTS FOR CORRELATED LIQUID DENSITY VALUES:

1487   9734

LITERATURE DOCUMENTS REPORTING VIRIAL COEFFICIENT DATA:

T,K	VAPOR PRESSURE, KPA	SATURATED LIQUID VOLUME, ML/MOL	SECOND VIRIAL COEFFICIENT, ML/MOL	HEAT OF VAPORIZATION, J/MOL	SATURATED VAPOR VOLUME, ML/MOL
250	0.1282				0.1622E+08
253	0.1621				0.1298E+08
256	0.2037				0.1045E+08
259	0.2544				8465592.
263	0.3389				6452161.
266	0.4175				5297636.
269	0.5114				4373167.
273	0.6649	143.47		39942.	3413812.
276	0.8047	143.91		39774.	2851597.
279	0.9692	144.35		39607.	2393364.
282	1.162	144.79		39440.	2018040.
286	1.469	145.39		39220.	1618639.
289	1.743	145.84		39055.	1378713.
292	2.059	146.29		38892.	1179177.
296	2.555	146.91		38676.	963207.
299	2.991	147.37		38516.	831286.
302	3.487	147.84		38356.	720065.
306	4.256	148.47		38146.	597838.
309	4.921	148.95		37990.	522053.
312	5.672	149.43		37835.	457371.
315	6.515				401976.
319	7.800				340022.
322	8.896				300942.
325	10.12				267112.
329	11.95				228820.
332	13.51				204380.
335	15.22				183017.
338	17.11				164294.
342	19.91				142804.
345	22.25				128899.
348	24.81				116608.
352	28.59				102365.
355	31.72				93061.
358	35.11				84773.
362	40.09				75085.
365	44.17				68702.
368	48.59				62975.
371	53.34				57827.
375	60.25				51746.
378	65.89				47700.
381	71.93				44039.
385	80.66				39685.
388	87.74				36766.
391	95.31				34111.
394	103.4				31690.

COMPOUND CONSTANTS:

MP	185.755 K	TC	VC	ZC
NBP	397.478 K	PC	DC	OM
MU	0.0000 DEBYE	RG 4.1446 ANGSTROM		

VAPOR PRESSURE CORRELATION CONSTANTS, K AND KPA:

VAPRES-2	VAPRES-2	VAPRES-2	RIEDEL
253. - 399. K	253. - 399. K	253. - 399. K	382. - 399. K
23 POINTS	23 POINTS	23 POINTS	4 POINTS
RMSD = 0.0278	RMSD = 0.0278	RMSD = 0.0278	RMSD = 0.0000
-0.42491680E+03	-0.42491680E+03	-0.42491680E+03	0.15377882E+03
0.32157378E+04	0.32157378E+04	0.32157378E+04	-0.20283502E+02
-0.28621814E+00	-0.28621814E+00	-0.28621814E+00	-0.11213998E+05
0.14894973E-03	0.14894973E-03	0.14894973E-03	0.11461197E-15
0.85491426E+02	0.85491426E+02	0.85491426E+02	

LIQUID DENSITY CORRELATION CONSTANTS, K AND G/ML:

FRANCIS1	FRANCIS1	FRANCIS1
273. - 314. K	273. - 314. K	273. - 314. K
5 POINTS	5 POINTS	5 POINTS
RMSD = 0.00012	RMSD = 0.00012	RMSD = 0.00012
0.10124531E+01	0.10124531E+01	0.10124531E+01
0.67766756E-03	0.67766756E-03	0.67766756E-03
0.64153109E+01	0.64153109E+01	0.64153109E+01
0.50045313E+03	0.50045313E+03	0.50045313E+03

SECOND VIRIAL COEFFICIENT CORRELATION CONSTANTS, ML/MOL:

LITERATURE DOCUMENTS FOR CORRELATED VAPOR PRESSURE VALUES:

   1146   1485

LITERATURE DOCUMENTS FOR CORRELATED LIQUID DENSITY VALUES:

   1487   9734

LITERATURE DOCUMENTS REPORTING VIRIAL COEFFICIENT DATA:

T,K	VAPOR PRESSURE, KPA	SATURATED LIQUID VOLUME, ML/MOL	SECOND VIRIAL COEFFICIENT, ML/MOL	HEAT OF VAPORIZATION, J/MOL	SATURATED VAPOR VOLUME, ML/MOL
253	0.1316				0.1599E+08
256	0.1652				0.1288E+08
259	0.2063				0.1044E+08
262	0.2562				8503738.
266	0.3391				6521911.
269	0.4160				5376322.
272	0.5078				4453523.
276	0.6575	140.82		40241.	3490128.
279	0.7937	141.25		40128.	2922513.
282	0.9539	141.69		40009.	2458059.
286	1.210	142.27		39842.	1964580.
289	1.440	142.72		39711.	1668717.
292	1.706	143.17		39577.	1423127.
296	2.125	143.77		39391.	1157960.
299	2.495	144.24		39248.	996428.
302	2.918	144.70		39102.	860571.
306	3.575	145.34		38903.	711709.
309	4.146	145.82		38751.	619695.
312	4.792	146.31		38598.	541372.
316	5.782				454385.
319	6.632				399895.
322	7.584				353000.
325	8.647				312516.
329	10.25				266834.
332	11.61				237766.
335	13.11				212424.
339	15.36				183512.
342	17.24				164913.
345	19.31				148551.
349	22.37				129693.
352	24.92				117440.
355	27.70				106572.
359	31.78				93928.
362	35.15				85637.
365	38.79				78226.
369	44.12				69530.
372	48.49				63780.
375	53.20				58604.
379	60.04				52482.
382	65.62				48403.
385	71.60				44706.
389	80.25				40303.
392	87.27				37347.
395	94.77				34653.
398	102.8				32194.

COMPOUND CONSTANTS:

MP 236.252 K	TC 587.775 K	VC	ZC
NBP 392.508 K	PC	DC	OM
MU 0.0000 DEBYE	RG 4.1670 ANGSTROM		

VAPOR PRESSURE CORRELATION CONSTANTS, K AND KPA:

RIEDEL	RIEDEL	RIEDEL	RIEDEL
248. - 394. K	248. - 394. K	248. - 394. K	377. - 394. K
25 POINTS	25 POINTS	25 POINTS	4 POINTS
RMSD = 0.0117	RMSD = 0.0117	RMSD = 0.0117	RMSD = 0.0000
0.54038654E+02	0.54038654E+02	0.54038654E+02	0.33067635E+02
-0.56276060E+01	-0.56276060E+01	-0.56276060E+01	-0.25425598E+01
-0.62236580E+04	-0.62236580E+04	-0.62236580E+04	-0.51952085E+04
0.12810981E-16	0.12810981E-16	0.12810981E-16	-0.76003347E-17

LIQUID DENSITY CORRELATION CONSTANTS, K AND G/ML:

FRANCIS1	FRANCIS1	FRANCIS1
273. - 314. K	273. - 314. K	273. - 314. K
5 POINTS	5 POINTS	5 POINTS
RMSD = 0.00031	RMSD = 0.00031	RMSD = 0.00031
0.10056801E+01	0.10056801E+01	0.10056801E+01
0.76057296E-03	0.76057296E-03	0.76057296E-03
0.59999990E+01	0.59999990E+01	0.59999990E+01
0.58713574E+03	0.58713574E+03	0.58713574E+03

SECOND VIRIAL COEFFICIENT CORRELATION CONSTANTS, ML/MOL:

LITERATURE DOCUMENTS FOR CORRELATED VAPOR PRESSURE VALUES:

1146   1485

LITERATURE DOCUMENTS FOR CORRELATED LIQUID DENSITY VALUES:

1487   9734

LITERATURE DOCUMENTS REPORTING VIRIAL COEFFICIENT DATA:

T,K	VAPOR PRESSURE, KPA	SATURATED LIQUID VOLUME, ML/MOL	SECOND VIRIAL COEFFICIENT, ML/MOL	HEAT OF VAPORIZATION, J/MOL	SATURATED VAPOR VOLUME, ML/MOL
248	0.1248				0.1652E+08
251	0.1574				0.1325E+08
254	0.1974				0.1070E+08
257	0.2460				8684562.
261	0.3270				6635357.
264	0.4022				5456999.
267	0.4921				4511375.
271	0.6387				3527783.
274	0.7722	144.21		38998.	2950105.
277	0.9292	144.67		38863.	2478478.
281	1.181	145.29		38684.	1978455.
284	1.406	145.75		38549.	1679209.
287	1.667	146.23		38416.	1431121.
291	2.079	146.86		38238.	1163570.
294	2.443	147.34		38105.	1000740.
297	2.858	147.82		37973.	863880.
301	3.505	148.47		37797.	714006.
304	4.067	148.97		37666.	621409.
307	4.704	149.47		37535.	542616.
311	5.681	150.14		37362.	455134.
314	6.521	150.64		37232.	400348.
317	7.462				353208.
320	8.513				312521.
324	10.10				266621.
327	11.45				237422.
330	12.94				211974.
334	15.18				182949.
337	17.06				164287.
340	19.12				147876.
344	22.18				128974.
347	24.72				116700.
350	27.50				105820.
354	31.59				93174.
357	34.97				84889.
360	38.63				77491.
364	43.98				68821.
367	48.36				63095.
370	53.09				57947.
374	59.95				51868.
377	65.54				47823.
380	71.54				44165.
384	80.19				39814.
387	87.20				36899.
390	94.68				34248.
393	102.7				31832.

COMPOUND CONSTANTS:

MP	155.816 K	TC	VC	ZC
NBP	404.103 K	PC	DC	OM
MU	0.0000 DEBYE	RG		

VAPOR PRESSURE CORRELATION CONSTANTS, K AND KPA:

RIEDEL	RIEDEL	RIEDEL	RIEDEL
324. - 406. K	324. - 406. K	324. - 406. K	389. - 406. K
31 POINTS	31 POINTS	31 POINTS	8 POINTS
RMSD = 0.0070	RMSD = 0.0070	RMSD = 0.0070	RMSD = 0.0097
0.62301463E+02	0.62301463E+02	0.62301463E+02	0.23401226E+02
-0.67614395E+01	-0.67614395E+01	-0.67614395E+01	-0.10560366E+01
-0.69425375E+04	-0.69425375E+04	-0.69425375E+04	-0.49993499E+04
0.17665896E-16	0.17665896E-16	0.17665896E-16	-0.16866000E-16

LIQUID DENSITY CORRELATION CONSTANTS, K AND G/ML:

FRANCIS1	FRANCIS1	FRANCIS1
273. - 374. K	273. - 374. K	273. - 374. K
9 POINTS	9 POINTS	9 POINTS
RMSD = 0.00034	RMSD = 0.00034	RMSD = 0.00034
0.10497246E+01	0.10497246E+01	0.10497246E+01
0.84456243E-03	0.84456243E-03	0.84456243E-03
0.10371231E+02	0.10371231E+02	0.10371231E+02
0.69965552E+03	0.69965552E+03	0.69965552E+03

SECOND VIRIAL COEFFICIENT CORRELATION CONSTANTS, ML/MOL:

LITERATURE DOCUMENTS FOR CORRELATED VAPOR PRESSURE VALUES:

1485 1486

LITERATURE DOCUMENTS FOR CORRELATED LIQUID DENSITY VALUES:

1487 1490 10808

LITERATURE DOCUMENTS REPORTING VIRIAL COEFFICIENT DATA:

T,K	VAPOR PRESSURE, KPA	SATURATED LIQUID VOLUME, ML/MOL	SECOND VIRIAL COEFFICIENT, ML/MOL	HEAT OF VAPORIZATION, J/MOL	SATURATED VAPOR VOLUME, ML/MOL
273		141.18			
276		141.66			
279		142.14			
282		142.63			
285		143.13			
288		143.63			
291		144.13			
294		144.63			
297		145.14			
300		145.66			
303		146.17			
306		146.70			
309		147.22			
312		147.75			
315		148.29			
318		148.83			
321		149.37			
324	6.099	149.92		39825.	441671.
327	6.984	150.48		39677.	389302.
330	7.973	151.04		39530.	344127.
333	9.076	151.60		39383.	305041.
336	10.30	152.17		39238.	271124.
339	11.67	152.74		39094.	241610.
342	13.17	153.32		38951.	215855.
345	14.84	153.91		38810.	193321.
348	16.67	154.50		38670.	173554.
351	18.69	155.09		38531.	156169.
354	20.90	155.69		38394.	140842.
357	23.32	156.30		38258.	127298.
360	25.96	156.91		38123.	115300.
363	28.84	157.53		37991.	104649.
366	31.97	158.15		37860.	95171.
369	35.38	158.78		37731.	86720.
372	39.07	159.42		37603.	79169.
375	43.06				72408.
378	47.37				66342.
381	52.03				60889.
384	57.04				55978.
387	62.42				51548.
390	68.20				47543.
393	74.40				43917.
396	81.04				40628.
399	88.14				37639.
402	95.72				34920.
405	103.8				32442.

COMPOUND CONSTANTS:

MP 161.787 K	TC	VC	ZC
NBP 399.575 K	PC	DC	OM
MU 0.0000 DEBYE	RG		

VAPOR PRESSURE CORRELATION CONSTANTS, K AND KPA:

RPM2	RPM2	RPM2	RPM2
320. - 401. K	320. - 401. K	320. - 401. K	384. - 401. K
31 POINTS	31 POINTS	31 POINTS	8 POINTS
RMSD = 0.0037	RMSD = 0.0037	RMSD = 0.0037	RMSD = 0.0049
0.27897134E+02	0.27897134E+02	0.27897134E+02	0.35474864E+02
-0.61645701E+04	-0.61645701E+04	-0.61645701E+04	-0.71617726E+04
-0.27467215E-01	-0.27467215E-01	-0.27467215E-01	-0.46656693E-01
0.19568036E-04	0.19568036E-04	0.19568036E-04	0.35762083E-04

LIQUID DENSITY CORRELATION CONSTANTS, K AND G/ML:

FRANCIS1	FRANCIS1	FRANCIS1
273. - 314. K	273. - 314. K	273. - 314. K
7 POINTS	7 POINTS	7 POINTS
RMSD = 0.00024	RMSD = 0.00024	RMSD = 0.00024
0.10041599E+01	0.10041599E+01	0.10041599E+01
0.70773158E-03	0.70773158E-03	0.70773158E-03
0.59999990E+01	0.59999990E+01	0.59999990E+01
0.58713574E+03	0.58713574E+03	0.58713574E+03

SECOND VIRIAL COEFFICIENT CORRELATION CONSTANTS, ML/MOL:

LITERATURE DOCUMENTS FOR CORRELATED VAPOR PRESSURE VALUES:

1485 1486

LITERATURE DOCUMENTS FOR CORRELATED LIQUID DENSITY VALUES:

1487 1490 17544

LITERATURE DOCUMENTS REPORTING VIRIAL COEFFICIENT DATA:

T,K	VAPOR PRESSURE, KPA	SATURATED LIQUID VOLUME, ML/MOL	SECOND VIRIAL COEFFICIENT, ML/MOL	HEAT OF VAPORIZATION, J/MOL	SATURATED VAPOR VOLUME, ML/MOL
273		141.71			
275		141.99			
278		142.40			
281		142.82			
284		143.25			
287		143.67			
290		144.10			
293		144.53			
296		144.97			
299		145.41			
302		145.85			
304		146.15			
307		146.60			
310		147.05			
313		147.50			
322	6.938				385903.
325	7.919				341231.
328	9.013				302575.
331	10.23				269026.
334	11.58				239825.
336	12.56				222457.
339	14.15				199146.
342	15.91				178709.
345	17.84				160746.
348	19.97				144917.
351	22.29				130936.
354	24.83				118558.
357	27.59				107572.
360	30.60				97802.
363	33.88				89093.
366	37.42				81314.
368	39.95				76588.
371	43.99				70114.
374	48.36				64300.
377	53.07				59069.
380	58.13				54353.
383	63.57				50094.
386	69.41				46241.
389	75.66				42749.
392	82.35				39579.
395	89.49				36697.
398	97.12				34072.
400	102.5				32451.

COMPOUND CONSTANTS:

MP	129.336 K	TC	VC	ZC
NBP	394.685 K	PC	DC	OM
MU	0.0000 DEBYE	RG		

    VAPOR PRESSURE CORRELATION CONSTANTS, K AND KPA:

VAPRES-2	VAPRES-2	VAPRES-2	RIEDEL
238. - 436. K	238. - 414. K	238. - 436. K	375. - 436. K
76 POINTS	68 POINTS	76 POINTS	25 POINTS
RMSD = 0.0080	RMSD = 0.0064	RMSD = 0.0080	RMSD = 0.0071
0.50243234E+02	0.45036733E+02	0.50243234E+02	0.54923488E+02
-0.65679589E+04	-0.64658225E+04	-0.65679589E+04	-0.57317169E+01
-0.15220164E-01	-0.18412694E-01	-0.15220164E-01	-0.63506278E+04
0.13542833E-04	0.15232992E-04	0.13542833E-04	0.13208418E-16
-0.41963995E+01	-0.32020205E+01	-0.41963995E+01	

    LIQUID DENSITY CORRELATION CONSTANTS, K AND G/ML:

FRANCIS1	FRANCIS1	FRANCIS1
293. - 304. K	293. - 304. K	293. - 304. K
4 POINTS	4 POINTS	4 POINTS
RMSD = 0.00004	RMSD = 0.00004	RMSD = 0.00004
0.10243349E+01	0.10243349E+01	0.10243349E+01
0.75602671E-03	0.75602671E-03	0.75602671E-03
0.59999990E+01	0.59999990E+01	0.59999990E+01
0.56838794E+03	0.56838794E+03	0.56838794E+03

    SECOND VIRIAL COEFFICIENT CORRELATION CONSTANTS, ML/MOL:

LITERATURE DOCUMENTS FOR CORRELATED VAPOR PRESSURE VALUES:

    1486   5718   7333

LITERATURE DOCUMENTS FOR CORRELATED LIQUID DENSITY VALUES:

    1490

LITERATURE DOCUMENTS REPORTING VIRIAL COEFFICIENT DATA:

T,K	VAPOR PRESSURE, KPA	SATURATED LIQUID VOLUME, ML/MOL	SECOND VIRIAL COEFFICIENT, ML/MOL	HEAT OF VAPORIZATION, J/MOL	SATURATED VAPOR VOLUME, ML/MOL
238	0.04191				0.4722E+08
242	0.05954				0.3379E+08
247	0.09068				0.2265E+08
251	0.1252				0.1667E+08
256	0.1843				0.1155E+08
260	0.2479				8720277.
265	0.3540				6224879.
269	0.4654				4805740.
274	0.6467				3522928.
278	0.8329				2775083.
283	1.129				2083414.
287	1.428				1671100.
292	1.894				1281796.
296	2.355	144.14		39029.	1045015.
301	3.062	144.92		38776.	817227.
305	3.751				676140.
310	4.790				538090.
314	5.787				451162.
319	7.271				364784.
323	8.675				309571.
328	10.74				253929.
332	12.67				217873.
337	15.47				181068.
341	18.07				156918.
346	21.80				131975.
350	25.21				115420.
355	30.08				98135.
359	34.49				86542.
364	40.72				74317.
368	46.33				66038.
373	54.19				57227.
377	61.21				51207.
382	70.98				44747.
386	79.64				40296.
391	91.62				35482.
395	102.2				32141.
400	116.7				28500.
404	129.4				25955.
409	146.8				23164.
413	162.0				21200.
418	182.6				19032.
422	200.5				17497.
427	224.8				15794.
431	245.8				14581.
436	274.1				13227.

COMPOUND CONSTANTS:

MP   167.217 K   TC   VC   ZC
NBP 401.211 K   PC   DC   OM
MU   0.0000 DEBYE   RG

VAPOR PRESSURE CORRELATION CONSTANTS, K AND KPA:

RIEDEL	RIEDEL	RIEDEL	RIEDEL
238. - 403. K	238. - 403. K	238. - 403. K	386. - 403. K
27 POINTS	27 POINTS	27 POINTS	4 POINTS
RMSD = 0.0233	RMSD = 0.0233	RMSD = 0.0233	RMSD = 0.0000
0.60540001E+02	0.60540001E+02	0.60540001E+02	0.31014078E+03
-0.65295727E+01	-0.65295727E+01	-0.65295727E+01	-0.43199987E+02
-0.67611834E+04	-0.67611834E+04	-0.67611834E+04	-0.19107452E+05
0.17029508E-16	0.17029508E-16	0.17029508E-16	0.25504341E-15

LIQUID DENSITY CORRELATION CONSTANTS, K AND G/ML:

FRANCIS1	FRANCIS1	FRANCIS1
273. - 304. K	273. - 304. K	273. - 304. K
5 POINTS	5 POINTS	5 POINTS
RMSD = 0.00008	RMSD = 0.00008	RMSD = 0.00008
0.10251179E+01	0.10251179E+01	0.10251179E+01
0.74424618E-03	0.74424618E-03	0.74424618E-03
0.59999990E+01	0.59999990E+01	0.59999990E+01
0.56838794E+03	0.56838794E+03	0.56838794E+03

SECOND VIRIAL COEFFICIENT CORRELATION CONSTANTS, ML/MOL:

LITERATURE DOCUMENTS FOR CORRELATED VAPOR PRESSURE VALUES:

   1486 5718

LITERATURE DOCUMENTS FOR CORRELATED LIQUID DENSITY VALUES:

   1490 9708

LITERATURE DOCUMENTS REPORTING VIRIAL COEFFICIENT DATA:

T,K	VAPOR PRESSURE, KPA	SATURATED LIQUID VOLUME, ML/MOL	SECOND VIRIAL COEFFICIENT, ML/MOL	HEAT OF VAPORIZATION, J/MOL	SATURATED VAPOR VOLUME, ML/MOL
238	0.02741				0.7219E+08
241	0.03598				0.5569E+08
245	0.05111				0.3986E+08
249	0.07165				0.2889E+08
253	0.09924				0.2120E+08
256	0.1257				0.1693E+08
260	0.1706				0.1267E+08
264	0.2292				9578542.
268	0.3046				7315259.
271	0.3747				6013850.
275	0.4897	140.27		41385.	4668675.
279	0.6344	140.84		41179.	3656301.
283	0.8149	141.42		40973.	2887512.
286	0.9779	141.86		40819.	2431763.
290	1.238	142.45		40615.	1947039.
294	1.556	143.05		40412.	1570486.
298	1.942	143.65		40210.	1275734.
301	2.282	144.11		40059.	1096445.
305	2.814				901084.
309	3.448				745181.
313	4.198				619960.
316	4.847				542106.
320	5.841				455499.
324	7.001				384778.
328	8.347				326707.
331	9.493				289919.
335	11.22				248259.
339	13.20				213562.
343	15.46				184524.
346	17.35				165828.
350	20.16				144327.
354	23.34				126111.
358	26.91				110612.
361	29.87				100491.
365	34.22				88696.
369	39.06				78552.
373	44.43				69796.
376	48.84				64008.
380	55.25				57183.
384	62.31				51236.
388	70.07				46037.
391	76.38				42563.
395	85.47				38424.
399	95.40				34776.
403	106.2				31551.

COMPOUND CONSTANTS:

MP   123.137 K   TC   VC   ZC
NBP 394.533 K   PC   DC   OM
MU   0.0000 DEBYE   RG

VAPOR PRESSURE CORRELATION CONSTANTS, K AND KPA:

LIQUID DENSITY CORRELATION CONSTANTS, K AND G/ML:

FRANCIS1   FRANCIS1   FRANCIS1
273. - 294. K   273. - 294. K   273. - 294. K
   4 POINTS   4 POINTS   4 POINTS
RMSD = 0.00004   RMSD = 0.00004   RMSD = 0.00004
0.10080519E+01   0.10080519E+01   0.10080519E+01
0.73543005E-03   0.73543005E-03   0.73543005E-03
0.59999990E+01   0.59999990E+01   0.59999990E+01
0.54964111E+03   0.54964111E+03   0.54964111E+03

SECOND VIRIAL COEFFICIENT CORRELATION CONSTANTS, ML/MOL:

LITERATURE DOCUMENTS FOR CORRELATED VAPOR PRESSURE VALUES:

LITERATURE DOCUMENTS FOR CORRELATED LIQUID DENSITY VALUES:

   9708

LITERATURE DOCUMENTS REPORTING VIRIAL COEFFICIENT DATA:

T,K	VAPOR PRESSURE, KPA	SATURATED LIQUID VOLUME, ML/MOL	SECOND VIRIAL COEFFICIENT, ML/MOL	HEAT OF VAPORIZATION, J/MOL	SATURATED VAPOR VOLUME, ML/MOL
273		142.84			
274		142.99			
275		143.14			
276		143.29			
277		143.44			
278		143.59			
279		143.74			
280		143.89			
281		144.04			
282		144.19			
283		144.34			
284		144.49			
285		144.65			
286		144.80			
287		144.95			
288		145.11			
289		145.26			
290		145.42			
291		145.57			
292		145.73			
293		145.89			
294		146.04			

COMPOUND CONSTANTS:

MP	251.523 K	TC		VC		ZC	
NBP	386.884 K	PC		DC		OM	
MU	0.0000 DEBYE	RG					

    VAPOR PRESSURE CORRELATION CONSTANTS, K AND KPA:

RIEDEL	RIEDEL	RIEDEL	RIEDEL
309. - 388. K	309. - 388. K	309. - 388. K	372. - 388. K
16 POINTS	16 POINTS	16 POINTS	4 POINTS
RMSD = 0.0054	RMSD = 0.0054	RMSD = 0.0054	RMSD = 0.0000
0.56002904E+02	0.56002904E+02	0.56002904E+02	0.47759229E+02
-0.59273330E+01	-0.59273330E+01	-0.59273330E+01	-0.47005156E+01
-0.62397430E+04	-0.62397430E+04	-0.62397430E+04	-0.58620699E+04
0.17703397E-16	0.17703397E-16	0.17703397E-16	0.51816651E-17

    LIQUID DENSITY CORRELATION CONSTANTS, K AND G/ML:

FRANCIS1	FRANCIS1	FRANCIS1
277. - 304. K	277. - 304. K	277. - 304. K
8 POINTS	8 POINTS	8 POINTS
RMSD = 0.00043	RMSD = 0.00043	RMSD = 0.00043
0.10054121E+01	0.10054121E+01	0.10054121E+01
0.72277058E-03	0.72277058E-03	0.72277058E-03
0.59999990E+01	0.59999990E+01	0.59999990E+01
0.56838794E+03	0.56838794E+03	0.56838794E+03

    SECOND VIRIAL COEFFICIENT CORRELATION CONSTANTS, ML/MOL:

LITERATURE DOCUMENTS FOR CORRELATED VAPOR PRESSURE VALUES:

    1486

LITERATURE DOCUMENTS FOR CORRELATED LIQUID DENSITY VALUES:

    1490   7291

LITERATURE DOCUMENTS REPORTING VIRIAL COEFFICIENT DATA:

T,K	VAPOR PRESSURE, KPA	SATURATED LIQUID VOLUME, ML/MOL	SECOND VIRIAL COEFFICIENT, ML/MOL	HEAT OF VAPORIZATION, J/MOL	SATURATED VAPOR VOLUME, ML/MOL
277		143.02			
279		143.31			
282		143.75			
284		144.04			
287		144.48			
289		144.78			
292		145.23			
294		145.53			
297		145.99			
299		146.30			
302		146.76			
304		147.07			
309	6.306				407395.
312	7.238				358392.
314	7.921				329597.
317	9.045				291396.
319	9.865				268850.
322	11.21				238812.
324	12.19				221010.
327	13.79				197198.
329	14.95				183028.
332	16.83				164004.
335	18.91				147316.
337	20.40				137327.
340	22.83				123837.
342	24.57				115734.
345	27.38				104755.
347	29.40				98137.
350	32.64				89142.
352	34.96				83703.
355	38.69				76287.
357	41.35				71790.
360	45.60				65639.
362	48.63				61898.
365	53.46				56768.
367	56.89				53638.
370	62.36				49335.
372	66.23				46704.
375	72.38				43076.
377	76.73				40851.
380	83.63				37777.
382	88.50				35888.
385	96.21				33270.
387	101.6				31657.

COMPOUND CONSTANTS:

MP	130.697 K	TC		VC	ZC
NBP	378.045 K	PC		DC	OM
MU	0.0000 DEBYE	RG			

VAPOR PRESSURE CORRELATION CONSTANTS, K AND KPA:

RIEDEL	RIEDEL	RIEDEL	RIEDEL
302. - 379. K	302. - 379. K	302. - 379. K	363. - 379. K
16 POINTS	16 POINTS	16 POINTS	4 POINTS
RMSD = 0.0041	RMSD = 0.0041	RMSD = 0.0041	RMSD = 0.0000
0.58473669E+02	0.58473669E+02	0.58473669E+02	0.14790015E+03
-0.63245795E+01	-0.63245795E+01	-0.63245795E+01	-0.19558391E+02
-0.61981659E+04	-0.61981659E+04	-0.61981659E+04	-0.10439595E+05
0.26186454E-16	0.26186454E-16	0.26186454E-16	0.14119884E-15

LIQUID DENSITY CORRELATION CONSTANTS, K AND G/ML:

FRANCIS1	FRANCIS1	FRANCIS1
293. - 304. K	293. - 304. K	293. - 304. K
4 POINTS	4 POINTS	4 POINTS
RMSD = 0.00000	RMSD = 0.00000	RMSD = 0.00000
0.99988145E+00	0.99988145E+00	0.99988145E+00
0.77619916E-03	0.77619916E-03	0.77619916E-03
0.66159124E+01	0.66159124E+01	0.66159124E+01
0.56772217E+03	0.56772217E+03	0.56772217E+03

SECOND VIRIAL COEFFICIENT CORRELATION CONSTANTS, ML/MOL:

LITERATURE DOCUMENTS FOR CORRELATED VAPOR PRESSURE VALUES:

   1486

LITERATURE DOCUMENTS FOR CORRELATED LIQUID DENSITY VALUES:

   1490

LITERATURE DOCUMENTS REPORTING VIRIAL COEFFICIENT DATA:

T,K	VAPOR PRESSURE, KPA	SATURATED LIQUID VOLUME, ML/MOL	SECOND VIRIAL COEFFICIENT, ML/MOL	HEAT OF VAPORIZATION, J/MOL	SATURATED VAPOR VOLUME, ML/MOL
293		149.94			
294		150.12			
296		150.47			
298		150.82			
300		151.17			
302	6.384	151.52		35939.	393334.
304	7.014	151.88		35846.	360387.
306	7.694				330670.
308	8.429				303825.
310	9.220				279540.
312	10.07				257538.
314	10.99				237576.
316	11.97				219442.
318	13.03				202944.
320	14.16				187916.
322	15.37				174209.
324	16.66				161691.
326	18.04				150243.
328	19.51				139763.
330	21.08				130157.
332	22.75				121340.
334	24.52				113240.
335	25.45				109438.
337	27.39				102289.
339	29.45				95703.
341	31.63				89629.
343	33.94				84020.
345	36.39				78836.
347	38.97				74040.
349	41.69				69598.
351	44.57				65480.
353	47.60				61658.
355	50.79				58109.
357	54.16				54809.
359	57.69				51739.
361	61.41				48879.
363	65.31				46213.
365	69.40				43726.
367	73.70				41403.
369	78.20				39233.
371	82.92				37202.
373	87.85				35302.
375	93.01				33521.
377	98.41				31852.
378	101.2				31056.

COMPOUND CONSTANTS:

MP	140.592 K	TC	VC	ZC
NBP	389.887 K	PC	DC	OM
MU	0.0000 DEBYE	RG		

VAPOR PRESSURE CORRELATION CONSTANTS, K AND KPA:

RIEDEL	RIEDEL	RIEDEL	RIEDEL
312. - 391. K	312. - 391. K	312. - 391. K	375. - 391. K
17 POINTS	17 POINTS	17 POINTS	4 POINTS
RMSD = 0.0069	RMSD = 0.0069	RMSD = 0.0069	RMSD = 0.0000
0.54090940E+02	0.54090940E+02	0.54090940E+02	-0.20199781E+03
-0.56289785E+01	-0.56289785E+01	-0.56289785E+01	0.32139535E+02
-0.62153957E+04	-0.62153957E+04	-0.62153957E+04	0.61714147E+04
0.14385617E-16	0.14385617E-16	0.14385617E-16	-0.27115014E-15

LIQUID DENSITY CORRELATION CONSTANTS, K AND G/ML:

FRANCIS1	FRANCIS1	FRANCIS1
293. - 304. K	293. - 304. K	293. - 304. K
4 POINTS	4 POINTS	4 POINTS
RMSD = 0.00003	RMSD = 0.00003	RMSD = 0.00003
0.10091028E+01	0.10091028E+01	0.10091028E+01
0.76368381E-03	0.76368381E-03	0.76368381E-03
0.59999990E+01	0.59999990E+01	0.59999990E+01
0.56838794E+03	0.56838794E+03	0.56838794E+03

SECOND VIRIAL COEFFICIENT CORRELATION CONSTANTS, ML/MOL:

LITERATURE DOCUMENTS FOR CORRELATED VAPOR PRESSURE VALUES:

1486

LITERATURE DOCUMENTS FOR CORRELATED LIQUID DENSITY VALUES:

1490

LITERATURE DOCUMENTS REPORTING VIRIAL COEFFICIENT DATA:

T,K	VAPOR PRESSURE, KPA	SATURATED LIQUID VOLUME, ML/MOL	SECOND VIRIAL COEFFICIENT, ML/MOL	HEAT OF VAPORIZATION, J/MOL	SATURATED VAPOR VOLUME, ML/MOL
293		146.96			
295		147.29			
297		147.62			
299		147.94			
301		148.28			
304		148.78			
313	6.697				388570.
315	7.334				357127.
317	8.020				328657.
319	8.758				302844.
321	9.552				279408.
324	10.85				248180.
326	11.80				229667.
328	12.82				212784.
330	13.90				197367.
333	15.67				176680.
335	16.95				164331.
337	18.31				153008.
339	19.76				142615.
341	21.31				133063.
344	23.81				120146.
346	25.60				112377.
348	27.50				105210.
350	29.52				98592.
353	32.76				89591.
355	35.08				84146.
357	37.52				79101.
359	40.11				74422.
362	44.25				68024.
364	47.19				64133.
366	50.29				60513.
368	53.55				57142.
370	56.97				54000.
373	62.43				49679.
375	66.29				47035.
377	70.34				44564.
379	74.58				42253.
382	81.31				39059.
384	86.06				37097.
386	91.03				35256.
388	96.22				33529.
390	101.6				31906.

COMPOUND CONSTANTS:

MP	142.363 K	TC	VC	ZC
NBP	382.442 K	PC	DC	OM
MU	0.0000 DEBYE	RG		

VAPOR PRESSURE CORRELATION CONSTANTS, K AND KPA:

RPM2	RPM2	RPM2	RPM2
306. - 384. K	306. - 384. K	306. - 384. K	368. - 384. K
17 POINTS	17 POINTS	17 POINTS	4 POINTS
RMSD = 0.0035	RMSD = 0.0035	RMSD = 0.0035	RMSD = 0.0004
0.27454750E+02	0.27454750E+02	0.27454750E+02	0.23632538E+01
-0.58267020E+04	-0.58267020E+04	-0.58267020E+04	-0.27183013E+04
-0.27675298E-01	-0.27675298E-01	-0.27675298E-01	0.39816873E-01
0.20396892E-04	0.20396892E-04	0.20396892E-04	-0.40098107E-04

LIQUID DENSITY CORRELATION CONSTANTS, K AND G/ML:

FRANCIS1	FRANCIS1	FRANCIS1
293. - 304. K	293. - 304. K	293. - 304. K
4 POINTS	4 POINTS	4 POINTS
RMSD = 0.00005	RMSD = 0.00005	RMSD = 0.00005
0.99639934E+00	0.99639934E+00	0.99639934E+00
0.77560195E-03	0.77560195E-03	0.77560195E-03
0.59999990E+01	0.59999990E+01	0.59999990E+01
0.56838794E+03	0.56838794E+03	0.56838794E+03

SECOND VIRIAL COEFFICIENT CORRELATION CONSTANTS, ML/MOL:

LITERATURE DOCUMENTS FOR CORRELATED VAPOR PRESSURE VALUES:

1486

LITERATURE DOCUMENTS FOR CORRELATED LIQUID DENSITY VALUES:

1490

LITERATURE DOCUMENTS REPORTING VIRIAL COEFFICIENT DATA:

T,K	VAPOR PRESSURE, KPA	SATURATED LIQUID VOLUME, ML/MOL	SECOND VIRIAL COEFFICIENT, ML/MOL	HEAT OF VAPORIZATION, J/MOL	SATURATED VAPOR VOLUME, ML/MOL
293		150.15			
295		150.49			
297		150.84			
299		151.19			
301		151.54			
303		151.89			
307	6.695				381268.
309	7.345				349804.
311	8.046				321378.
313	8.802				295660.
315	9.616				272358.
317	10.49				251216.
319	11.43				232007.
321	12.44				214531.
324	14.09				191185.
326	15.29				177311.
328	16.57				164631.
330	17.93				153030.
332	19.38				142402.
334	20.93				132655.
336	22.58				123705.
338	24.34				115478.
340	26.20				107906.
342	28.17				100931.
344	30.27				94499.
346	32.48				88560.
348	34.83				83071.
350	37.31				77994.
352	39.93				73292.
355	44.14				66874.
357	47.13				62978.
359	50.29				59358.
361	53.61				55992.
363	57.10				52858.
365	60.77				49939.
367	64.62				47218.
369	68.67				44678.
371	72.91				42306.
373	77.36				40089.
375	82.02				38014.
377	86.90				36072.
379	92.00				34253.
381	97.33				32547.
383	102.9				30946.

COMPOUND CONSTANTS:

MP	224.178 K	TC	VC	ZC
NBP	410.824 K	PC	DC	OM
MU		RG		

VAPOR PRESSURE CORRELATION CONSTANTS, K AND KPA:

RPM2	RPM2	RPM2
297. - 410. K	297. - 410. K	297. - 410. K
6 POINTS	6 POINTS	6 POINTS
RMSD = 0.0001	RMSD = 0.0001	RMSD = 0.0001
0.17367166E+02	0.17367166E+02	0.17367166E+02
-0.53195008E+04	-0.53195008E+04	-0.53195008E+04
0.76136227E-02	0.76136227E-02	0.76136227E-02
-0.17350255E-04	-0.17350255E-04	-0.17350255E-04

LIQUID DENSITY CORRELATION CONSTANTS, K AND G/ML:

FRANCIS1	FRANCIS1	FRANCIS1
273. - 334. K	273. - 334. K	273. - 334. K
4 POINTS	4 POINTS	4 POINTS
RMSD = 0.00003	RMSD = 0.00003	RMSD = 0.00003
0.11197662E+01	0.11197662E+01	0.11197662E+01
0.78876526E-03	0.78876526E-03	0.78876526E-03
0.63271646E+01	0.63271646E+01	0.63271646E+01
0.62461719E+03	0.62461719E+03	0.62461719E+03

SECOND VIRIAL COEFFICIENT CORRELATION CONSTANTS, ML/MOL:

LITERATURE DOCUMENTS FOR CORRELATED VAPOR PRESSURE VALUES:

8900 13729

LITERATURE DOCUMENTS FOR CORRELATED LIQUID DENSITY VALUES:

8900

LITERATURE DOCUMENTS REPORTING VIRIAL COEFFICIENT DATA:

T,K	VAPOR PRESSURE, KPA	SATURATED LIQUID VOLUME, ML/MOL	SECOND VIRIAL COEFFICIENT, ML/MOL	HEAT OF VAPORIZATION, J/MOL	SATURATED VAPOR VOLUME, ML/MOL
273		124.32			
276		124.67			
279		125.03			
282		125.39			
285		125.75			
288		126.11			
291		126.48			
294		126.85			
297	1.206	127.22		42251.	2047829.
301	1.513	127.72		42092.	1653987.
304	1.786	128.10		41969.	1415317.
307	2.100	128.48		41842.	1215483.
310	2.460	128.87		41712.	1047553.
313	2.873	129.25		41577.	905929.
316	3.342	129.64		41439.	786075.
319	3.876	130.04		41297.	684302.
322	4.480	130.43		41151.	597600.
325	5.162	130.83		41000.	523500.
329	6.205	131.37		40794.	440850.
332	7.098	131.77		40634.	388869.
335	8.097				343996.
338	9.210				305148.
341	10.45				271422.
344	11.82				242064.
347	13.33				216442.
350	15.00				194021.
353	16.83				174354.
357	19.56				151765.
360	21.83				137141.
363	24.30				124211.
366	26.99				112753.
369	29.91				102577.
372	33.07				93521.
375	36.49				85444.
378	40.18				78226.
381	44.14				71762.
385	49.89				64164.
388	54.56				59132.
391	59.54				54598.
394	64.86				50505.
397	70.53				46803.
400	76.54				43451.
403	82.92				40409.
406	89.67				37645.
409	96.80				35130.

COMPOUND CONSTANTS:

MP   243.667 K   TC   VC   ZC
NBP 405.077 K   PC   DC   OM
MU   RG

VAPOR PRESSURE CORRELATION CONSTANTS, K AND KPA:

RPM2   RPM2   RPM2
297. - 406. K   297. - 406. K   297. - 406. K
   6 POINTS   6 POINTS   6 POINTS
RMSD = 0.0001   RMSD = 0.0001   RMSD = 0.0001
0.32821375E+02   0.32821375E+02   0.32821375E+02
-0.67810410E+04   -0.67810410E+04   -0.67810410E+04
-0.46278956E-01   -0.46278956E-01   -0.46278956E-01
0.44388637E-04   0.44388637E-04   0.44388637E-04

LIQUID DENSITY CORRELATION CONSTANTS, K AND G/ML:

SECOND VIRIAL COEFFICIENT CORRELATION CONSTANTS, ML/MOL:

LITERATURE DOCUMENTS FOR CORRELATED VAPOR PRESSURE VALUES:

   7287 13729

LITERATURE DOCUMENTS FOR CORRELATED LIQUID DENSITY VALUES:

LITERATURE DOCUMENTS REPORTING VIRIAL COEFFICIENT DATA:

T,K	VAPOR PRESSURE, KPA	SATURATED LIQUID VOLUME, ML/MOL	SECOND VIRIAL COEFFICIENT, ML/MOL	HEAT OF VAPORIZATION, J/MOL	SATURATED VAPOR VOLUME, ML/MOL
297	1.174				2103814.
299	1.314				1891535.
301	1.469				1703451.
304	1.731				1460129.
306	1.927				1320084.
309	2.258				1137966.
311	2.504				1032605.
314	2.917				894916.
316	3.224				814868.
319	3.737				709771.
321	4.116				648388.
324	4.747				567441.
326	5.213				519956.
329	5.985				457074.
331	6.552				420033.
334	7.490				370786.
336	8.177				341663.
339	9.308				302798.
341	10.14				279728.
344	11.49				248831.
346	12.48				230425.
349	14.11				205692.
351	15.29				190909.
353	16.55				177342.
356	18.61				159029.
358	20.11				148034.
361	22.54				133149.
363	24.30				124187.
366	27.17				112018.
368	29.23				104670.
371	32.58				94666.
373	35.00				88609.
376	38.91				80342.
378	41.72				75323.
381	46.27				68457.
383	49.54				64279.
386	54.81				58549.
388	58.60				55054.
391	64.69				50251.
393	69.06				47315.
396	76.09				43271.
398	81.12				40794.
401	89.20				37376.
403	94.98				35278.
405	101.1				33313.

COMPOUND CONSTANTS:

MP	TC	VC	ZC
NBP	PC	DC	OM
MU	RG		

    VAPOR PRESSURE CORRELATION CONSTANTS, K AND KPA:

RIEDEL	RIEDEL	RIEDEL
297. - 343. K	297. - 343. K	297. - 343. K
10 POINTS	10 POINTS	10 POINTS
RMSD = 0.0000	RMSD = 0.0000	RMSD = 0.0000
0.44034984E+02	0.44034984E+02	0.44034984E+02
-0.40617751E+01	-0.40617751E+01	-0.40617751E+01
-0.61663100E+04	-0.61663100E+04	-0.61663100E+04
0.20851812E-18	0.20851812E-18	0.20851812E-18

    LIQUID DENSITY CORRELATION CONSTANTS, K AND G/ML:

    SECOND VIRIAL COEFFICIENT CORRELATION CONSTANTS, ML/MOL:

LITERATURE DOCUMENTS FOR CORRELATED VAPOR PRESSURE VALUES:

  13729

LITERATURE DOCUMENTS FOR CORRELATED LIQUID DENSITY VALUES:

LITERATURE DOCUMENTS REPORTING VIRIAL COEFFICIENT DATA:

T,K	VAPOR PRESSURE, KPA	SATURATED LIQUID VOLUME, ML/MOL	SECOND VIRIAL COEFFICIENT, ML/MOL	HEAT OF VAPORIZATION, J/MOL	SATURATED VAPOR VOLUME, ML/MOL
297	1.158				2132905.
298	1.224				2023494.
299	1.294				1920479.
300	1.368				1823449.
301	1.445				1732011.
302	1.526				1645812.
303	1.610				1564518.
304	1.699				1487815.
305	1.792				1415417.
306	1.889				1347053.
307	1.990				1282474.
308	2.097				1221446.
309	2.208				1163751.
310	2.324				1109186.
311	2.445				1057561.
312	2.572				1008700.
313	2.704				962437.
314	2.842				918618.
315	2.986				877098.
316	3.136				837742.
317	3.293				800424.
318	3.456				765026.
319	3.626				731438.
321	3.988				669280.
322	4.180				640523.
323	4.380				613198.
324	4.587				587225.
325	4.804				562530.
326	5.028				539040.
327	5.262				516691.
328	5.505				495419.
329	5.757				475167.
330	6.019				455879.
331	6.290				437504.
332	6.572				419993.
333	6.865				403300.
334	7.169				387382.
335	7.483				372199.
336	7.810				357712.
337	8.148				343885.
338	8.498				330684.
339	8.861				318078.
340	9.237				306036.
341	9.626				294528.
342	10.03				283530.

COMPOUND CONSTANTS:

MP	TC	VC	ZC
NBP	PC	DC	OM
MU	RG		

VAPOR PRESSURE CORRELATION CONSTANTS, K AND KPA:

RIEDEL	RIEDEL	RIEDEL
296. - 332. K	296. - 332. K	296. - 332. K
8 POINTS	8 POINTS	8 POINTS
RMSD = 0.0000	RMSD = 0.0000	RMSD = 0.0000
0.17889989E+02	0.17889989E+02	0.17889989E+02
-0.35656521E-01	-0.35656521E-01	-0.35656521E-01
-0.52567736E+04	-0.52567736E+04	-0.52567736E+04
-0.33192683E-18	-0.33192683E-18	-0.33192683E-18

LIQUID DENSITY CORRELATION CONSTANTS, K AND G/ML:

SECOND VIRIAL COEFFICIENT CORRELATION CONSTANTS, ML/MOL:

LITERATURE DOCUMENTS FOR CORRELATED VAPOR PRESSURE VALUES:

13729

LITERATURE DOCUMENTS FOR CORRELATED LIQUID DENSITY VALUES:

LITERATURE DOCUMENTS REPORTING VIRIAL COEFFICIENT DATA:

T,K	VAPOR PRESSURE, KPA	SATURATED LIQUID VOLUME, ML/MOL	SECOND VIRIAL COEFFICIENT, ML/MOL	HEAT OF VAPORIZATION, J/MOL	SATURATED VAPOR VOLUME, ML/MOL
296	0.9301				2646126.
297	0.9873				2501269.
298	1.048				2365264.
299	1.111				2237516.
300	1.178				2117478.
301	1.248				2004636.
302	1.323				1898518.
303	1.401				1798683.
304	1.483				1704723.
305	1.569				1616259.
306	1.660				1532936.
307	1.755				1454426.
308	1.855				1380424.
309	1.960				1310644.
310	2.071				1244822.
311	2.186				1182710.
312	2.308				1124078.
313	2.435				1068712.
314	2.569				1016410.
315	2.708				966987.
316	2.855				920266.
317	3.008				876086.
318	3.169				834293.
319	3.337				794746.
320	3.513				757311.
321	3.697				721864.
322	3.890				688287.
323	4.091				656473.
324	4.301				626319.
325	4.521				597728.
326	4.750				570612.
327	4.990				544886.
328	5.240				520471.
329	5.501				497294.
330	5.773				475285.
331	6.057				454378.
332	6.353				434513.

COMPOUND CONSTANTS:

MP   178.272 K   TC   VC   ZC
NBP 429.897 K   PC   DC   OM
MU   0.0000 DEBYE   RG

   VAPOR PRESSURE CORRELATION CONSTANTS, K AND KPA:

   RPM2   RPM2   RPM2   RPM2
   345. - 431. K   345. - 431. K   345. - 431. K   414. - 431. K
    18 POINTS   18 POINTS   18 POINTS   5 POINTS
   RMSD = 0.0040   RMSD = 0.0040   RMSD = 0.0040   RMSD = 0.0051
   0.28113438E+02   0.28113438E+02   0.28113438E+02   0.87086659E+02
  -0.67615631E+04   -0.67615631E+04   -0.67615631E+04   -0.15047380E+05
  -0.24830113E-01   -0.24830113E-01   -0.24830113E-01   -0.16472493E+00
   0.15732895E-04   0.15732895E-04   0.15732895E-04   0.12633800E-03

   LIQUID DENSITY CORRELATION CONSTANTS, K AND G/ML:

   FRANCIS1   FRANCIS1   FRANCIS1
   273. - 314. K   273. - 314. K   273. - 314. K
    5 POINTS   5 POINTS   5 POINTS
   RMSD = 0.00006   RMSD = 0.00006   RMSD = 0.00006
   0.10208874E+01   0.10208874E+01   0.10208874E+01
   0.70593762E-03   0.70593762E-03   0.70593762E-03
   0.59999990E+01   0.59999990E+01   0.59999990E+01
   0.58713574E+03   0.58713574E+03   0.58713574E+03

   SECOND VIRIAL COEFFICIENT CORRELATION CONSTANTS, ML/MOL:

LITERATURE DOCUMENTS FOR CORRELATED VAPOR PRESSURE VALUES:

   1486

LITERATURE DOCUMENTS FOR CORRELATED LIQUID DENSITY VALUES:

   1487 1490

LITERATURE DOCUMENTS REPORTING VIRIAL COEFFICIENT DATA:

T,K	VAPOR PRESSURE, KPA	SATURATED LIQUID VOLUME, ML/MOL	SECOND VIRIAL COEFFICIENT, ML/MOL	HEAT OF VAPORIZATION, J/MOL	SATURATED VAPOR VOLUME, ML/MOL
273		156.03			
276		156.48			
280		157.08			
283		157.53			
287		158.14			
290		158.60			
294		159.22			
298		159.85			
301		160.32			
305		160.96			
308		161.44			
312		162.09			
348	7.015				412442.
351	7.946				367275.
355	9.345				315853.
359	10.94				272776.
362	12.28				245022.
366	14.28				213088.
369	15.95				192380.
373	18.42				168406.
377	21.19				147943.
380	23.48				134548.
384	26.85				118900.
387	29.63				108600.
391	33.69				96503.
395	38.18				86012.
398	41.86				79051.
402	47.20				70811.
405	51.55				65318.
409	57.85				58786.
413	64.74				53044.
416	70.32				49188.
420	78.34				44573.
423	84.82				41462.
427	94.11				37723.
430	101.6				35193.

COMPOUND CONSTANTS:

MP 189.792 K	TC	VC	ZC
NBP 427.735 K	PC	DC	OM
MU 0.0000 DEBYE	RG		

VAPOR PRESSURE CORRELATION CONSTANTS, K AND KPA:

RPM2	RPM2	RPM2	RPM2
343. - 429. K	343. - 429. K	343. - 429. K	411. - 429. K
18 POINTS	18 POINTS	18 POINTS	5 POINTS
RMSD = 0.0057	RMSD = 0.0057	RMSD = 0.0057	RMSD = 0.0016
0.30725290E+02	0.30725290E+02	0.30725290E+02	0.37684357E+02
-0.70332114E+04	-0.70332114E+04	-0.70332114E+04	-0.80276389E+04
-0.32019784E-01	-0.32019784E-01	-0.32019784E-01	-0.48226905E-01
0.22037966E-04	0.22037966E-04	0.22037966E-04	0.34598789E-04

LIQUID DENSITY CORRELATION CONSTANTS, K AND G/ML:

FRANCIS1	FRANCIS1	FRANCIS1
273. - 314. K	273. - 314. K	273. - 314. K
5 POINTS	5 POINTS	5 POINTS
RMSD = 0.00013	RMSD = 0.00013	RMSD = 0.00013
0.10052309E+01	0.10052309E+01	0.10052309E+01
0.57608844E-03	0.57608844E-03	0.57608844E-03
0.59999990E+01	0.59999990E+01	0.59999990E+01
0.46832153E+03	0.46832153E+03	0.46832153E+03

SECOND VIRIAL COEFFICIENT CORRELATION CONSTANTS, ML/MOL:

LITERATURE DOCUMENTS FOR CORRELATED VAPOR PRESSURE VALUES:

   1486

LITERATURE DOCUMENTS FOR CORRELATED LIQUID DENSITY VALUES:

   1487   1490

LITERATURE DOCUMENTS REPORTING VIRIAL COEFFICIENT DATA:

T,K	VAPOR PRESSURE, KPA	SATURATED LIQUID VOLUME, ML/MOL	SECOND VIRIAL COEFFICIENT, ML/MOL	HEAT OF VAPORIZATION, J/MOL	SATURATED VAPOR VOLUME, ML/MOL
273		154.47			
276		154.89			
280		155.46			
283		155.89			
287		156.47			
290		156.92			
294		157.52			
297		157.98			
301		158.60			
304		159.08			
308		159.72			
311		160.22			
343	6.236				457355.
347	7.384				390700.
350	8.357				348207.
354	9.818				299801.
358	11.48				259222.
361	12.88				233059.
365	14.95				202934.
368	16.68				183384.
372	19.24				160731.
375	21.36				145939.
379	24.49				128698.
382	27.06				117373.
386	30.83				104101.
389	33.93				95336.
393	38.44				85009.
397	43.42				76021.
400	47.49				70038.
404	53.37				62935.
407	58.16				58184.
411	65.07				52520.
414	70.66				48714.
418	78.71				44157.
421	85.20				41083.
425	94.51				37387.
428	102.0				34884.

COMPOUND CONSTANTS:

MP		TC		VC		ZC
NBP	421.616 K	PC		DC		OM
MU	0.0000 DEBYE	RG				

VAPOR PRESSURE CORRELATION CONSTANTS, K AND KPA:

RPM2	RPM2	RPM2	RPM2
348. - 465. K	348. - 436. K	348. - 465. K	407. - 465. K
21 POINTS	17 POINTS	21 POINTS	9 POINTS
RMSD = 0.0047	RMSD = 0.0014	RMSD = 0.0047	RMSD = 0.0013
0.27727358E+02	0.28061560E+02	0.27727358E+02	0.26630677E+02
-0.65409407E+04	-0.65840787E+04	-0.65409407E+04	-0.63845189E+04
-0.24830645E-01	-0.25691158E-01	-0.24830645E-01	-0.22271217E-01
0.16167595E-04	0.16903949E-04	0.16167595E-04	0.14179324E-04

LIQUID DENSITY CORRELATION CONSTANTS, K AND G/ML:

FRANCIS1	FRANCIS1	FRANCIS1
293. - 299. K	293. - 299. K	293. - 299. K
4 POINTS	4 POINTS	4 POINTS
RMSD = 0.00007	RMSD = 0.00007	RMSD = 0.00007
0.10244875E+01	0.10244875E+01	0.10244875E+01
0.73983613E-03	0.73983613E-03	0.73983613E-03
0.59999990E+01	0.59999990E+01	0.59999990E+01
0.55903052E+03	0.55903052E+03	0.55903052E+03

SECOND VIRIAL COEFFICIENT CORRELATION CONSTANTS, ML/MOL:

LITERATURE DOCUMENTS FOR CORRELATED VAPOR PRESSURE VALUES:

5718

LITERATURE DOCUMENTS FOR CORRELATED LIQUID DENSITY VALUES:

3269

LITERATURE DOCUMENTS REPORTING VIRIAL COEFFICIENT DATA:

T,K	VAPOR PRESSURE, KPA	SATURATED LIQUID VOLUME, ML/MOL	SECOND VIRIAL COEFFICIENT, ML/MOL	HEAT OF VAPORIZATION, J/MOL	SATURATED VAPOR VOLUME, ML/MOL
293		160.78			
296		161.29			
351	10.68				273218.
355	12.49				236345.
359	14.54				205275.
363	16.86				178983.
367	19.48				156643.
371	22.42				137586.
375	25.71				121266.
378	28.43				110553.
382	32.41				98003.
386	36.83				87150.
390	41.71				77733.
394	47.11				69535.
398	53.05				62377.
402	59.57				56107.
406	66.71				50598.
410	74.52				45746.
414	83.03				41459.
418	92.28				37662.
421	99.74				35096.
425	110.4				32006.
429	121.9				29250.
433	134.4				26786.
437	147.8				24579.
441	162.3				22596.
445	177.8				20811.
449	194.4				19202.
453	212.2				17747.
457	231.3				16430.
461	251.6				15235.
464	267.7				14411.

COMPOUND CONSTANTS:

MP   207.433 K   TC   VC   ZC
NBP 409.790 K   PC   DC   OM
MU   0.0000 DEBYE   RG

VAPOR PRESSURE CORRELATION CONSTANTS, K AND KPA:

RPM2	RPM2	RPM2	RPM2
327. - 411. K	327. - 411. K	327. - 411. K	394. - 411. K
17 POINTS	17 POINTS	17 POINTS	· 5 POINTS
RMSD = 0.0056	RMSD = 0.0056	RMSD = 0.0056	RMSD = 0.0031
0.27758987E+02	0.27758987E+02	0.27758987E+02	0.62369564E+02
-0.62763432E+04	-0.62763432E+04	-0.62763432E+04	-0.10936934E+05
-0.26679519E-01	-0.26679519E-01	-0.26679519E-01	-0.11233541E+00
0.18510071E-04	0.18510071E-04	0.18510071E-04	0.89156104E-04

LIQUID DENSITY CORRELATION CONSTANTS, K AND G/ML:

FRANCIS1	FRANCIS1	FRANCIS1
273. - 314. K	273. - 314. K	273. - 314. K
5 POINTS	5 POINTS	5 POINTS
RMSD = 0.00001	RMSD = 0.00001	RMSD = 0.00001
0.99150646E+00	0.99150646E+00	0.99150646E+00
0.61482145E-03	0.61482145E-03	0.61482145E-03
0.66303387E+01	0.66303387E+01	0.66303387E+01
0.49739160E+03	0.49739160E+03	0.49739160E+03

SECOND VIRIAL COEFFICIENT CORRELATION CONSTANTS, ML/MOL:

LITERATURE DOCUMENTS FOR CORRELATED VAPOR PRESSURE VALUES:

   1486

LITERATURE DOCUMENTS FOR CORRELATED LIQUID DENSITY VALUES:

   1487   1490

LITERATURE DOCUMENTS REPORTING VIRIAL COEFFICIENT DATA:

T,K	VAPOR PRESSURE, KPA	SATURATED LIQUID VOLUME, ML/MOL	SECOND VIRIAL COEFFICIENT, ML/MOL	HEAT OF VAPORIZATION, J/MOL	SATURATED VAPOR VOLUME, ML/MOL
273		158.97			
276		159.42			
279		159.88			
282		160.34			
285		160.80			
288		161.27			
291		161.75			
294		162.23			
298		162.88			
301		163.38			
304		163.88			
307		164.39			
310		164.91			
313		165.43			
329	6.739				405899.
332	7.668				360010.
335	8.700				320164.
338	9.844				285466.
341	11.11				255170.
345	13.01				220549.
348	14.60				198245.
351	16.34				178602.
354	18.25				161261.
357	20.34				145916.
360	22.62				132306.
363	25.11				120207.
367	28.76				106102.
370	31.77				96833.
373	35.03				88536.
376	38.55				81094.
379	42.35				74407.
382	46.44				68385.
385	50.85				62953.
389	57.23				56515.
392	62.42				52216.
395	67.97				48315.
398	73.91				44770.
401	80.26				41544.
404	87.02				38602.
407	94.22				35916.
410	101.9				33461.

COMPOUND CONSTANTS:

MP	229.955 K	TC		VC		ZC
NBP	411.690 K	PC		DC		OM
MU	0.0000 DEBYE	RG				

VAPOR PRESSURE CORRELATION CONSTANTS, K AND KPA:

RPM2	RPM2	RPM2
318. - 412. K	318. - 412. K	318. - 412. K
6 POINTS	6 POINTS	6 POINTS
RMSD = 0.0378	RMSD = 0.0378	RMSD = 0.0378
0.10416216E+02	0.10416216E+02	0.10416216E+02
-0.40197681E+04	-0.40197681E+04	-0.40197681E+04
0.17011262E-01	0.17011262E-01	0.17011262E-01
-0.17919661E-04	-0.17919661E-04	-0.17919661E-04

LIQUID DENSITY CORRELATION CONSTANTS, K AND G/ML:

FRANCIS1	FRANCIS1	FRANCIS1
293. - 299. K	293. - 299. K	293. - 299. K
4 POINTS	4 POINTS	4 POINTS
RMSD = 0.00004	RMSD = 0.00004	RMSD = 0.00004
0.98412693E+00	0.98412693E+00	0.98412693E+00
0.65460918E-03	0.65460918E-03	0.65460918E-03
0.59999990E+01	0.59999990E+01	0.59999990E+01
0.55903052E+03	0.55903052E+03	0.55903052E+03

SECOND VIRIAL COEFFICIENT CORRELATION CONSTANTS, ML/MOL:

LITERATURE DOCUMENTS FOR CORRELATED VAPOR PRESSURE VALUES:

7362   9132

LITERATURE DOCUMENTS FOR CORRELATED LIQUID DENSITY VALUES:

3269

LITERATURE DOCUMENTS REPORTING VIRIAL COEFFICIENT DATA:

T,K	VAPOR PRESSURE, KPA	SATURATED LIQUID VOLUME, ML/MOL	SECOND VIRIAL COEFFICIENT, ML/MOL	HEAT OF VAPORIZATION, J/MOL	SATURATED VAPOR VOLUME, ML/MOL
293		164.00			
295		164.31			
298		164.79			
320	4.319				615995.
322	4.721				567042.
325	5.385				501809.
328	6.127				445124.
330	6.668				411462.
333	7.557				366367.
336	8.545				326933.
338	9.263				303394.
341	10.43				271703.
344	11.73				243826.
347	13.16				219253.
349	14.19				204488.
352	15.86				184482.
355	17.70				166752.
357	19.02				156052.
360	21.15				141491.
363	23.48				128520.
366	26.02				116945.
368	27.84				109919.
371	30.75				100307.
374	33.91				91688.
376	36.17				86437.
379	39.77				79225.
382	43.67				72730.
384	46.43				68758.
387	50.84				63285.
390	55.59				58335.
393	60.68				53852.
395	64.27				51098.
398	69.98				47286.
401	76.09				43819.
403	80.39				41683.
406	87.19				38717.
409	94.44				36010.
411	99.52				34336.

COMPOUND CONSTANTS:

MP	165.180 K	TC		VC		ZC
NBP	429.773 K	PC		DC		OM
MU	0.0000 DEBYE	RG				

VAPOR PRESSURE CORRELATION CONSTANTS, K AND KPA:

LIQUID DENSITY CORRELATION CONSTANTS, K AND G/ML:

FRANCIS1	FRANCIS1	FRANCIS1
288. - 314. K	288. - 314. K	288. - 314. K
5 POINTS	5 POINTS	5 POINTS
RMSD = 0.00012	RMSD = 0.00012	RMSD = 0.00012
0.10180206E+01	0.10180206E+01	0.10180206E+01
0.72628655E-03	0.72628655E-03	0.72628655E-03
0.59999990E+01	0.59999990E+01	0.59999990E+01
0.58713574E+03	0.58713574E+03	0.58713574E+03

SECOND VIRIAL COEFFICIENT CORRELATION CONSTANTS, ML/MOL:

LITERATURE DOCUMENTS FOR CORRELATED VAPOR PRESSURE VALUES:

LITERATURE DOCUMENTS FOR CORRELATED LIQUID DENSITY VALUES:

9707 10808 17544

LITERATURE DOCUMENTS REPORTING VIRIAL COEFFICIENT DATA:

T,K	VAPOR PRESSURE, KPA	SATURATED LIQUID VOLUME, ML/MOL	SECOND VIRIAL COEFFICIENT, ML/MOL	HEAT OF VAPORIZATION, J/MOL	SATURATED VAPOR VOLUME, ML/MOL
288		160.04			
289		160.20			
290		160.37			
291		160.53			
292		160.69			
293		160.85			
294		161.02			
295		161.18			
296		161.34			
297		161.51			
298		161.67			
299		161.84			
300		162.01			
301		162.17			
302		162.34			
303		162.51			
304		162.67			
305		162.84			
306		163.01			
307		163.18			
308		163.35			
309		163.52			
310		163.69			
311		163.86			
312		164.03			
313		164.20			
314		164.38			

COMPOUND CONSTANTS:

MP   236.472 K   TC   VC   ZC
NBP   441.031 K   PC   DC   OM
MU   RG

VAPOR PRESSURE CORRELATION CONSTANTS, K AND KPA:

VAPRES-2	VAPRES-2	VAPRES-2	RPM2
260. - 466. K	260. - 466. K	260. - 466. K	424. - 466. K
59 POINTS	59 POINTS	59 POINTS	14 POINTS
RMSD = 0.0064	RMSD = 0.0064	RMSD = 0.0064	RMSD = 0.0108
0.68705049E+02	0.68705049E+02	0.68705049E+02	0.30515983E+02
-0.79067252E+04	-0.79067252E+04	-0.79067252E+04	-0.72559464E+04
-0.69429738E-02	-0.69429738E-02	-0.69429738E-02	-0.29955445E-01
0.83610041E-05	0.83610041E-05	0.83610041E-05	0.19360988E-04
-0.73447673E+01	-0.73447673E+01	-0.73447673E+01	

LIQUID DENSITY CORRELATION CONSTANTS, K AND G/ML:

FRANCIS1	FRANCIS1	FRANCIS1
288. - 389. K	288. - 389. K	288. - 389. K
9 POINTS	9 POINTS	9 POINTS
RMSD = 0.00031	RMSD = 0.00031	RMSD = 0.00031
0.11240416E+01	0.11240416E+01	0.11240416E+01
0.76977769E-03	0.76977769E-03	0.76977769E-03
0.59999990E+01	0.59999990E+01	0.59999990E+01
0.72779004E+03	0.72779004E+03	0.72779004E+03

SECOND VIRIAL COEFFICIENT CORRELATION CONSTANTS, ML/MOL:

LITERATURE DOCUMENTS FOR CORRELATED VAPOR PRESSURE VALUES:

6668   7333

LITERATURE DOCUMENTS FOR CORRELATED LIQUID DENSITY VALUES:

6668 15029

LITERATURE DOCUMENTS REPORTING VIRIAL COEFFICIENT DATA:

T,K	VAPOR PRESSURE, KPA	SATURATED LIQUID VOLUME, ML/MOL	SECOND VIRIAL COEFFICIENT, ML/MOL	HEAT OF VAPORIZATION, J/MOL	SATURATED VAPOR VOLUME, ML/MOL
260	0.02266				0.9541E+08
264	0.03178				0.6907E+08
269	0.04772				0.4687E+08
274	0.07041				0.3235E+08
278	0.09499				0.2433E+08
283	0.1362				0.1728E+08
288	0.1924	139.78		46685.	0.1244E+08
292	0.2512	140.29		46447.	9666405.
297	0.3462	140.93		46152.	7131864.
302	0.4714	141.57		45861.	5327063.
306	0.5981	142.09		45630.	4253751.
311	0.7973	142.75		45345.	3243183.
316	1.051	143.41		45063.	2498975.
320	1.302	143.95		44840.	2043335.
325	1.686	144.62		44564.	1602526.
330	2.163	145.30		44292.	1268335.
334	2.623	145.86		44078.	1058515.
339	3.313	146.55		43813.	850685.
344	4.151	147.26		43552.	689094.
348	4.943	147.82		43346.	585395.
353	6.108	148.54		43092.	480544.
358	7.494	149.27		42842.	397204.
362	8.783	149.85		42645.	342688.
367	10.65	150.59		42403.	286563.
372	12.83	151.34		42165.	241080.
377	15.37	152.10		41931.	203986.
381	17.68	152.71		41747.	179174.
386	20.96	153.49		41521.	153085.
391	24.73				131458.
395	28.12				116784.
400	32.88				101146.
405	38.27				87992.
409	43.07				78955.
414	49.74				69207.
419	57.20				60902.
423	63.79				55132.
428	72.86				48841.
433	82.92				43418.
437	91.72				39613.
442	103.7				35424.
447	117.0				31776.
451	128.5				29192.
456	144.0				26324.
461	161.0				23802.
465	175.7				22002.

COMPOUND CONSTANTS:

MP	213.742 K	TC	VC	ZC
NBP	434.265 K	PC	DC	OM
MU		RG		

VAPOR PRESSURE CORRELATION CONSTANTS, K AND KPA:

VAPRES-2	VAPRES-2	VAPRES-2	RPM2
260. - 481. K	260. - 451. K	260. - 481. K	415. - 481. K
62 POINTS	56 POINTS	62 POINTS	19 POINTS
RMSD = 0.0068	RMSD = 0.0060	RMSD = 0.0068	RMSD = 0.0079
0.62417032E+02	0.66831071E+02	0.62417032E+02	0.28281734E+02
-0.75649568E+04	-0.76575448E+04	-0.75649568E+04	-0.68004553E+04
-0.87760059E-02	-0.63222813E-02	-0.87760059E-02	-0.25501617E-01
0.90660868E-05	0.78698592E-05	0.90660868E-05	0.16283033E-04
-0.63021618E+01	-0.71321124E+01	-0.63021618E+01	

LIQUID DENSITY CORRELATION CONSTANTS, K AND G/ML:

FRANCIS1	FRANCIS1	FRANCIS1
288. - 389. K	288. - 389. K	288. - 389. K
6 POINTS	6 POINTS	6 POINTS
RMSD = 0.00022	RMSD = 0.00022	RMSD = 0.00022
0.11045008E+01	0.11045008E+01	0.11045008E+01
0.77809603E-03	0.77809603E-03	0.77809603E-03
0.59999990E+01	0.59999990E+01	0.59999990E+01
0.72779004E+03	0.72779004E+03	0.72779004E+03

SECOND VIRIAL COEFFICIENT CORRELATION CONSTANTS, ML/MOL:

LITERATURE DOCUMENTS FOR CORRELATED VAPOR PRESSURE VALUES:

6668   7333

LITERATURE DOCUMENTS FOR CORRELATED LIQUID DENSITY VALUES:

6668 15029

LITERATURE DOCUMENTS REPORTING VIRIAL COEFFICIENT DATA:

T,K	VAPOR PRESSURE, KPA	SATURATED LIQUID VOLUME, ML/MOL	SECOND VIRIAL COEFFICIENT, ML/MOL	HEAT OF VAPORIZATION, J/MOL	SATURATED VAPOR VOLUME, ML/MOL
260	0.03364				0.6427E+08
265	0.05062				0.4352E+08
270	0.07486				0.2999E+08
275	0.1089				0.2100E+08
280	0.1559				0.1493E+08
285	0.2200				0.1077E+08
290	0.3060	143.59		45242.	7879243.
295	0.4202	144.26		44960.	5837387.
300	0.5698	144.95		44680.	4377691.
305	0.7636	145.63		44404.	3321082.
310	1.012	146.33		44131.	2547159.
315	1.327	147.04		43861.	1973906.
320	1.722	147.75		43595.	1544747.
325	2.215	148.47		43331.	1220201.
330	2.822	149.19		43072.	972400.
335	3.564	149.93		42815.	781457.
340	4.466	150.67		42563.	633042.
345	5.551	151.42		42314.	516723.
350	6.850	152.18		42069.	424836.
355	8.392	152.95		41827.	351700.
360	10.21	153.73		41590.	293069.
365	12.35	154.52		41356.	245743.
370	14.84	155.31		41127.	207290.
375	17.73	156.12		40901.	175850.
380	21.06	156.93		40680.	149990.
385	24.89	157.76		40463.	128597.
390	29.27				110802.
395	34.24				95921.
400	39.87				83414.
405	46.22				72852.
410	53.36				63891.
415	61.34				56254.
420	70.24				49717.
425	80.13				44099.
430	91.08				39251.
435	103.2				35053.
440	116.5				31403.
445	131.1				28219.
450	147.1				25432.
455	164.6				22984.
460	183.6				20828.
465	204.3				18923.
470	226.7				17234.
475	251.0				15733.
480	277.2				14396.

COMPOUND CONSTANTS:

MP   TC   VC   ZC
NBP   PC   DC   OM
MU   RG

VAPOR PRESSURE CORRELATION CONSTANTS, K AND KPA:

RPM2   RPM2   RPM2
294. - 335. K   294. - 335. K   294. - 335. K
  9 POINTS   9 POINTS   9 POINTS
RMSD = 0.0000   RMSD = 0.0000   RMSD = 0.0000
 0.19437768E+02   0.19437768E+02   0.19437768E+02
-0.60986630E+04   -0.60986630E+04   -0.60986630E+04
-0.99443944E-03   -0.99443944E-03   -0.99443944E-03
 0.93111343E-06   0.93111343E-06   0.93111343E-06

LIQUID DENSITY CORRELATION CONSTANTS, K AND G/ML:

SECOND VIRIAL COEFFICIENT CORRELATION CONSTANTS, ML/MOL:

LITERATURE DOCUMENTS FOR CORRELATED VAPOR PRESSURE VALUES:

  13729

LITERATURE DOCUMENTS FOR CORRELATED LIQUID DENSITY VALUES:

LITERATURE DOCUMENTS REPORTING VIRIAL COEFFICIENT DATA:

T,K	VAPOR PRESSURE, KPA	SATURATED LIQUID VOLUME, ML/MOL	SECOND VIRIAL COEFFICIENT, ML/MOL	HEAT OF VAPORIZATION, J/MOL	SATURATED VAPOR VOLUME, ML/MOL
294	0.2192				0.1115E+08
295	0.2350				0.1044E+08
296	0.2519				9768771.
297	0.2699				9148894.
298	0.2890				8572235.
299	0.3094				8035515.
300	0.3310				7535742.
301	0.3540				7070152.
302	0.3784				6636210.
303	0.4043				6231583.
304	0.4318				5854117.
305	0.4609				5501836.
306	0.4918				5172909.
307	0.5246				4865659.
308	0.5593				4578531.
309	0.5961				4310091.
310	0.6350				4059019.
311	0.6762				3824090.
312	0.7197				3604175.
313	0.7658				3398231.
314	0.8145				3205290.
315	0.8660				3024460.
316	0.9203				2854910.
317	0.9777				2695876.
318	1.038				2546647.
319	1.102				2406561.
320	1.169				2275010.
321	1.241				2151423.
322	1.315				2035280.
323	1.394				1926086.
324	1.477				1823390.
325	1.565				1726770.
326	1.657				1635831.
327	1.754				1550211.
328	1.856				1469567.
329	1.963				1393585.
330	2.076				1321969.
331	2.194				1254446.
332	2.318				1190759.
333	2.449				1130669.
334	2.586				1073955.
335	2.730				1020410.

COMPOUND CONSTANTS:

MP		TC		VC		ZC
NBP	425.550 K	PC		DC		OM
MU		RG				

VAPOR PRESSURE CORRELATION CONSTANTS, K AND KPA:

LIQUID DENSITY CORRELATION CONSTANTS, K AND G/ML:

FRANCIS1	FRANCIS1	FRANCIS1
303. - 334. K	303. - 334. K	303. - 334. K
4 POINTS	4 POINTS	4 POINTS
RMSD = 0.00007	RMSD = 0.00007	RMSD = 0.00007
0.11089907E+01	0.11089907E+01	0.11089907E+01
0.80265361E-03	0.80265361E-03	0.80265361E-03
0.59999990E+01	0.59999990E+01	0.59999990E+01
0.62465552E+03	0.62465552E+03	0.62465552E+03

SECOND VIRIAL COEFFICIENT CORRELATION CONSTANTS, ML/MOL:

LITERATURE DOCUMENTS FOR CORRELATED VAPOR PRESSURE VALUES:

LITERATURE DOCUMENTS FOR CORRELATED LIQUID DENSITY VALUES:

18182

LITERATURE DOCUMENTS REPORTING VIRIAL COEFFICIENT DATA:

T,K	VAPOR PRESSURE, KPA	SATURATED LIQUID VOLUME, ML/MOL	SECOND VIRIAL COEFFICIENT, ML/MOL	HEAT OF VAPORIZATION, J/MOL	SATURATED VAPOR VOLUME, ML/MOL
303		146.64			
304		146.79			
305		146.94			
306		147.09			
307		147.24			
308		147.39			
309		147.54			
310		147.69			
311		147.85			
312		148.00			
313		148.15			
314		148.30			
315		148.46			
316		148.61			
317		148.76			
318		148.92			
319		149.07			
320		149.23			
321		149.38			
322		149.54			
323		149.70			
324		149.85			
325		150.01			
326		150.17			
327		150.33			
328		150.49			
329		150.64			
330		150.80			
331		150.96			
332		151.12			
333		151.28			
334		151.45			

COMPOUND CONSTANTS:

MP	TC	VC	ZC
NBP	PC	DC	OM
MU	RG		

VAPOR PRESSURE CORRELATION CONSTANTS, K AND KPA:

RPM2	RPM2	RPM2
298. - 393. K	298. - 393. K	298. - 393. K
15 POINTS	15 POINTS	15 POINTS
RMSD = 0.0040	RMSD = 0.0040	RMSD = 0.0040
0.29487743E+02	0.29487743E+02	0.29487743E+02
-0.63041848E+04	-0.63041848E+04	-0.63041848E+04
-0.31981435E-01	-0.31981435E-01	-0.31981435E-01
0.23379954E-04	0.23379954E-04	0.23379954E-04

LIQUID DENSITY CORRELATION CONSTANTS, K AND G/ML:

SECOND VIRIAL COEFFICIENT CORRELATION CONSTANTS, ML/MOL:

LITERATURE DOCUMENTS FOR CORRELATED VAPOR PRESSURE VALUES:

17569 40700

LITERATURE DOCUMENTS FOR CORRELATED LIQUID DENSITY VALUES:

LITERATURE DOCUMENTS REPORTING VIRIAL COEFFICIENT DATA:

T,K	VAPOR PRESSURE, KPA	SATURATED LIQUID VOLUME, ML/MOL	SECOND VIRIAL COEFFICIENT, ML/MOL	HEAT OF VAPORIZATION, J/MOL	SATURATED VAPOR VOLUME, ML/MOL
298	2.407				1029221.
300	2.674				932821.
302	2.965				846835.
304	3.283				770006.
306	3.628				701244.
308	4.004				639601.
310	4.412				584251.
313	5.088				511473.
315	5.586				468896.
317	6.123				430460.
319	6.703				395710.
321	7.327				364250.
323	7.999				335726.
326	9.102				297780.
328	9.905				275321.
330	10.77				254863.
332	11.69				236204.
334	12.67				219163.
336	13.72				203583.
339	15.43				182641.
341	16.67				170119.
343	17.98				158623.
345	19.37				148055.
347	20.86				138331.
349	22.43				129374.
351	24.10				121113.
354	26.78				109895.
356	28.70				103118.
358	30.74				96846.
360	32.88				91036.
362	35.14				85648.
364	37.53				80647.
367	41.35				73802.
369	44.06				69635.
371	46.91				65755.
373	49.91				62139.
375	53.06				58766.
377	56.36				55618.
380	61.61				51278.
382	65.33				48617.
384	69.22				46127.
386	73.28				43793.
388	77.54				41606.
390	81.98				39553.
392	86.62				37625.

COMPOUND CONSTANTS:

MP	TC	VC	ZC
NBP	PC	DC	OM
MU	RG		

VAPOR PRESSURE CORRELATION CONSTANTS, K AND KPA:

RPM2	RPM2	RPM2
345. - 411. K	345. - 411. K	345. - 411. K
14 POINTS	14 POINTS	14 POINTS
RMSD = 0.0073	RMSD = 0.0073	RMSD = 0.0073
0.27640322E+02	0.27640322E+02	0.27640322E+02
-0.64314479E+04	-0.64314479E+04	-0.64314479E+04
-0.25055191E-01	-0.25055191E-01	-0.25055191E-01
0.16471733E-04	0.16471733E-04	0.16471733E-04

LIQUID DENSITY CORRELATION CONSTANTS, K AND G/ML:

SECOND VIRIAL COEFFICIENT CORRELATION CONSTANTS, ML/MOL:

LITERATURE DOCUMENTS FOR CORRELATED VAPOR PRESSURE VALUES:

17569

LITERATURE DOCUMENTS FOR CORRELATED LIQUID DENSITY VALUES:

LITERATURE DOCUMENTS REPORTING VIRIAL COEFFICIENT DATA:

T,K	VAPOR PRESSURE, KPA	SATURATED LIQUID VOLUME, ML/MOL	SECOND VIRIAL COEFFICIENT, ML/MOL	HEAT OF VAPORIZATION, J/MOL	SATURATED VAPOR VOLUME, ML/MOL
345	10.12				283322.
346	10.54				272945.
348	11.41				253523.
349	11.87				244434.
351	12.83				227399.
352	13.34				219416.
354	14.40				204438.
355	14.95				197410.
357	16.11				184207.
358	16.72				178006.
360	17.99				166341.
361	18.66				160855.
363	20.05				150525.
364	20.78				145662.
366	22.29				136493.
367	23.09				132172.
369	24.74				124017.
370	25.60				120169.
372	27.40				112901.
373	28.33				109467.
375	30.28				102974.
376	31.29				99905.
378	33.40				94094.
379	34.50				91343.
381	36.78				86132.
382	37.96				83663.
384	40.42				78981.
385	41.70				76761.
387	44.35				72546.
388	45.73				70546.
390	48.58				66745.
391	50.06				64939.
393	53.13				61506.
394	54.71				59873.
396	58.00				56766.
397	59.70				55287.
399	63.23				52471.
400	65.05				51129.
402	68.81				48571.
403	70.76				47352.
405	74.79				45026.
406	76.86				43917.
408	81.16				41798.
409	83.37				40787.
411	87.95				38854.

COMPOUND CONSTANTS:

MP   103.986 K   TC   282.343 K   VC   130.99 ML/MOL   ZC   0.2813
NBP   169.410 K   PC   5.040 MPA   DC   0.2141 G/ML   OM   0.0870
MU   0.0000 DEBYE   RG   1.5382 ANGSTROM

     VAPOR PRESSURE CORRELATION CONSTANTS, K AND MPA:

     GOODWIN
    -0.90172866E+01
     0.82095798E+01
     0.43154241E+01
    -0.16925860E+01
    -0.19764956E+00
     0.34465011E+01

     LIQUID DENSITY CORRELATION CONSTANTS, K AND MOL/DM^3:

     MCCARTY1
     104. - 282. K
     0.23342967E+02
    -0.47904706E+01
     0.15138135E-01
    -0.40345608E+00
     0.50868392E+01
    -0.24671199E+02
     0.98003092E+01
    -0.21684652E+01

     SECOND VIRIAL COEFFICIENT CORRELATION CONSTANTS, ML/MOL:

     HAYDEN-O'CONNELL: TC=283.100   PC=5.117   RG=1.5380   MU=0.000   ETA=0.000

LITERATURE DOCUMENTS FOR CORRELATED VAPOR PRESSURE VALUES:

   42045

LITERATURE DOCUMENTS FOR CORRELATED LIQUID DENSITY VALUES:

   42045

LITERATURE DOCUMENTS REPORTING VIRIAL COEFFICIENT DATA:

T,K	VAPOR PRESSURE, KPA	SATURATED LIQUID VOLUME, ML/MOL	SECOND VIRIAL COEFFICIENT, ML/MOL	HEAT OF VAPORIZATION, J/MOL	SATURATED VAPOR VOLUME, ML/MOL
104	0.1216	42.84	-1837.	15949.	7109464.
108	0.2401	43.17	-1578.	15803.	3738674.
112	0.4490	43.51	-1374.	15657.	2072690.
116	0.8000	43.86	-1210.	15510.	1204425.
120	1.365	44.22	-1078.	15364.	729908.
124	2.240	44.58	-968.9	15218.	459292.
128	3.549	44.95	-878.2	15073.	298927.
132	5.449	45.33	-801.8	14928.	200549.
136	8.131	45.72	-736.8	14783.	138279.
140	11.82	46.12	-681.0	14639.	97727.
144	16.79	46.53	-632.6	14494.	70628.
148	23.34	46.95	-590.3	14349.	52088.
152	31.82	47.38	-553.0	14202.	39126.
156	42.60	47.82	-520.0	14053.	29886.
160	56.11	48.28	-490.4	13902.	23177.
164	72.81	48.75	-463.9	13748.	18225.
168	93.18	49.24	-439.9	13591.	14513.
172	117.7	49.75	-418.1	13430.	11692.
176	147.0	50.27	-398.2	13265.	9519.
180	181.6	50.81	-379.9	13094.	7825.
184	222.1	51.37	-363.1	12918.	6489.
188	269.1	51.96	-347.6	12736.	5425.
192	323.2	52.56	-333.2	12548.	4569.
197	401.8	53.36	-316.6	12302.	3722.
201	474.3	54.03	-304.3	12096.	3180.
205	556.0	54.73	-292.8	11882.	2732.
209	647.7	55.47	-282.0	11658.	2358.
213	750.0	56.25	-271.9	11424.	2045.
217	863.7	57.07	-262.3	11179.	1781.
221	989.4	57.94	-253.3	10921.	1556.
225	1128.	58.87	-244.7	10650.	1364.
229	1280.	59.86	-236.6	10363.	1199.
233	1446.	60.92	-228.9	10061.	1056.
237	1627.	62.07	-221.6	9740.	931.7
241	1824.	63.32	-214.6	9397.	823.1
245	2038.	64.68	-208.0	9032.	727.7
249	2269.	66.19	-201.6	8638.	643.4
253	2519.	67.88	-195.5	8213.	568.5
257	2788.	69.79	-189.7	7749.	501.5
261	3078.	72.00	-184.2	7238.	441.0
265	3390.	74.64	-178.8	6666.	386.1
269	3725.	77.89	-173.7	6012.	335.4
273	4085.	82.18	-168.8	5232.	287.6
277	4472.	88.50	-164.0	4223.	240.2
281	4891.	100.32	-159.5	2568.	184.0

COMPOUND CONSTANTS:

MP 87.890 K	TC 365.570 K	VC 188.37 ML/MOL	ZC 0.2891
NBP 225.438 K	PC 4.665 MPA	DC 0.2234 G/ML	OM 0.1410
MU 0.3500 DEBYE	RG 2.2283 ANGSTROM		

    VAPOR PRESSURE CORRELATION CONSTANTS, K AND MPA:

    WAGNER-3
    88. - 365. K
 -0.65535175E+01
  0.95764593E+00
 -0.47470264E+01
  0.19314209E+01

    LIQUID DENSITY CORRELATION CONSTANTS, K AND MOL/DM^3:

    ANGUS-2
    88. - 365. K
 -0.67446675E+01
  0.10417066E+03
 -0.36109921E+03
  0.57039923E+03
 -0.43953921E+03
  0.13550818E+03
 -0.13231188E+01

    SECOND VIRIAL COEFFICIENT CORRELATION CONSTANTS, ML/MOL:

    HAYDEN-O'CONNELL: TC=365.100 PC=4.600 RG=2.2280 MU=0.370 ETA=0.000

LITERATURE DOCUMENTS FOR CORRELATED VAPOR PRESSURE VALUES:

  23680

LITERATURE DOCUMENTS FOR CORRELATED LIQUID DENSITY VALUES:

  23680

LITERATURE DOCUMENTS REPORTING VIRIAL COEFFICIENT DATA:

T,K	VAPOR PRESSURE, KPA	SATURATED LIQUID VOLUME, ML/MOL	SECOND VIRIAL COEFFICIENT, ML/MOL	HEAT OF VAPORIZATION, J/MOL	SATURATED VAPOR VOLUME, ML/MOL
88	0.994E-06	54.74	-37744.	23829.	0.7363E+12
94	0.788E-05	55.25	-23893.	23625.	0.9923E+11
100	0.0000479	55.77	-15979.	23415.	0.1735E+11
106	0.0002343	56.29	-11197.	23201.	0.3761E+10
113	0.001186	56.91	-7777.	22946.	0.7919E+09
119	0.004041	57.44	-5905.	22725.	0.2448E+09
125	0.01211	57.99	-4618.	22502.	0.8584E+08
132	0.03792	58.64	-3580.	22240.	0.2894E+08
138	0.09112	59.20	-2948.	22015.	0.1259E+08
144	0.2019	59.77	-2475.	21789.	5928378.
150	0.4167	60.36	-2113.	21563.	2991671.
157	0.8970	61.05	-1789.	21298.	1453869.
163	1.631	61.66	-1573.	21071.	829419.
169	2.828	62.29	-1398.	20842.	495669.
176	5.092	63.04	-1232.	20572.	286309.
182	8.091	63.69	-1116.	20338.	186029.
188	12.43	64.37	-1018.	20100.	124820.
195	19.75	65.18	-920.9	19816.	81243.
201	28.51	65.90	-850.3	19567.	57798.
207	40.18	66.64	-788.8	19312.	42063.
213	55.39	67.41	-734.8	19050.	31243.
220	78.56	68.35	-679.6	18734.	22601.
226	103.9	69.18	-637.8	18455.	17429.
232	135.3	70.04	-600.3	18167.	13640.
239	180.6	71.10	-561.1	17818.	10418.
245	227.9	72.06	-530.8	17508.	8376.
251	284.2	73.06	-503.2	17186.	6806.
257	350.3	74.11	-477.9	16852.	5583.
264	441.3	75.42	-450.9	16445.	4478.
270	532.5	76.61	-429.7	16080.	3737.
276	636.9	77.88	-410.0	15698.	3140.
283	777.1	79.47	-388.9	15228.	2581.
289	914.3	80.94	-372.0	14801.	2193.
295	1068.	82.53	-356.3	14349.	1872.
302	1272.	84.56	-339.1	13784.	1562.
308	1467.	86.48	-325.4	13262.	1342.
314	1684.	88.60	-312.5	12700.	1154.
320	1923.	90.97	-300.3	12089.	993.8
327	2233.	94.13	-286.9	11301.	833.9
333	2526.	97.30	-276.1	10546.	715.8
339	2848.	101.05	-265.8	9697.	611.9
346	3260.	106.57	-254.5	8541.	504.6
352	3649.	112.97	-245.3	7336.	421.2
358	4073.	122.81	-236.6	5761.	341.0
364	4536.	147.15	-228.3	2975.	248.3

COMPOUND CONSTANTS:

MP   87.800 K   TC   419.600 K   VC   240.00 ML/MOL   ZC   0.2770
NBP   266.890 K   PC   4.023 MPA   DC   0.2338 G/ML   OM   0.1914
MU   0.3300 DEBYE   RG   2.7458 ANGSTROM

VAPOR PRESSURE CORRELATION CONSTANTS, K AND KPA:

RIEDEL	VAPRES-2	RPM2	WAGNER
213. - 419. K	213. - 278. K	213. - 378. K	248. - 419. K
57 POINTS	31 POINTS	46 POINTS	47 POINTS
RMSD = 1.6950	RMSD = 0.1480	RMSD = 1.1660	RMSD = 0.9680
0.36102690E+02	0.33601945E+04	0.22452690E+02	-0.69575743E+01
-0.32206489E+01	-0.60003262E+05	-0.35474538E+04	0.14900911E+01
-0.36040370E+04	0.25480394E+01	-0.23021277E-01	-0.29554733E+01
0.43559060E-16	-0.16372876E-02	0.22527694E-04	0.27338687E+01
	-0.66122822E+03		

LIQUID DENSITY CORRELATION CONSTANTS, K AND G/ML:

FRANCIS1	FRANCIS1	FRANCIS1	FRANCIS2
195. - 389. K	195. - 283. K	195. - 378. K	371. - 416. K
59 POINTS	41 POINTS	57 POINTS	10 POINTS
RMSD = 0.00098	RMSD = 0.00028	RMSD = 0.00094	RMSD = 0.00071
0.92767817E+00	0.92271107E+00	0.92572629E+00	0.51670481E-03
0.88149426E-03	0.10170920E-02	0.87117031E-03	0.25835753E+01
0.14999969E+02	0.59999990E+01	0.15000000E+02	
0.48893799E+03	0.49720728E+03	0.48656494E+03	

SECOND VIRIAL COEFFICIENT CORRELATION CONSTANTS, ML/MOL:

TSONOPOULOS: TC=419.600 PC=4.023 VC=240.00 OM=0.1914 A= 0.000000 B=0.0000

LITERATURE DOCUMENTS FOR CORRELATED VAPOR PRESSURE VALUES:

    391   392   393   589   624   2650   2751   5215   5719   9566

LITERATURE DOCUMENTS FOR CORRELATED LIQUID DENSITY VALUES:

    375   392   393   420   477   589   1911   2098   2777   3016   4127   9566

LITERATURE DOCUMENTS REPORTING VIRIAL COEFFICIENT DATA:

    1898   1900   1901   2650

T,K	VAPOR PRESSURE, KPA	SATURATED LIQUID VOLUME, ML/MOL	SECOND VIRIAL COEFFICIENT, ML/MOL	HEAT OF VAPORIZATION, J/MOL	SATURATED VAPOR VOLUME, ML/MOL
195		79.61	-2438.		
200		80.21	-2193.		
205		80.83	-1986.		
210		81.46	-1811.		
215	7.745	82.10	-1661.	24070.	229124.
220	10.53	82.76	-1532.	23902.	172090.
225	14.11	83.43	-1419.	23729.	131127.
230	18.64	84.12	-1321.	23549.	101258.
235	24.30	84.82	-1234.	23363.	79168.
240	31.28	85.55	-1157.	23170.	62615.
245	39.81	86.29	-1088.	22969.	50056.
250	50.12	87.05	-1026.	22760.	40417.
256	65.21	87.99	-960.2	22480.	31625.
261	80.36	88.80	-910.7	22246.	26032.
266	98.16	89.64	-865.6	22002.	21598.
271	118.9	90.50	-824.3	21749.	18053.
276	143.0	91.39	-786.3	21487.	15193.
281	170.6	92.31	-751.3	21214.	12869.
286	202.2	93.25	-718.9	20932.	10965.
291	238.0	94.24	-688.8	20640.	9396.
296	278.5	95.26	-660.9	20338.	8092.
301	324.1	96.32	-634.8	20026.	7003.
306	375.0	97.42	-610.3	19705.	6088.
312	443.8	98.80	-583.0	19305.	5174.
317	507.9	100.01	-561.7	18960.	4537.
322	578.7	101.28	-541.7	18605.	3992.
327	656.5	102.61	-522.8	18238.	3523.
332	741.9	104.01	-504.9	17859.	3119.
337	835.1	105.49	-487.9	17467.	2768.
342	936.7	107.05	-471.9	17061.	2461.
347	1047.	108.71	-456.6	16640.	2194.
352	1167.	110.48	-442.1	16201.	1958.
357	1296.	112.37	-428.3	15743.	1750.
362	1436.	114.41	-415.1	15262.	1565.
368	1619.	117.07	-400.1	14649.	1370.
373	1784.	119.50	-388.2	14103.	1226.
378	1961.	122.17	-376.7	13518.	1096.
383	2151.	125.11	-365.8	12885.	979.1
388	2355.	128.38	-355.4	12191.	872.2
393	2574.		-345.3		774.0
398	2808.		-335.7		682.8
403	3059.		-326.4		596.8
408	3328.		-317.5		513.3
413	3615.		-308.9		427.1
418	3924.		-300.6		

COMPOUND CONSTANTS:

MP 134.230 K	TC 435.580 K	VC 234.00 ML/MOL	ZC 0.2720
NBP 276.870 K	PC 4.205 MPA	DC 0.2398 G/ML	OM 0.2018
MU 0.3300 DEBYE	RG 2.7765 ANGSTROM		

VAPOR PRESSURE CORRELATION CONSTANTS, K AND KPA:

WAGNER	VAPRES-2	WAGNER	WAGNER
195. - 303. K	195. - 296. K	195. - 303. K	195. - 303. K
35 POINTS	33 POINTS	35 POINTS	35 POINTS
RMSD = 0.0684	RMSD = 0.0500	RMSD = 0.0680	RMSD = 0.0680
-0.68273780E+01	0.58630977E+02	-0.68273780E+01	-0.68273780E+01
0.10280750E+01	-0.47061182E+04	0.10280750E+01	0.10280750E+01
-0.20518604E+01	-0.13624228E-01	-0.20518604E+01	-0.20518604E+01
-0.33169767E+01	0.23248888E-04	-0.33169767E+01	-0.33169767E+01
	-0.62283037E+01		

LIQUID DENSITY CORRELATION CONSTANTS, K AND G/ML:

FRANCIS1	FRANCIS1	FRANCIS1	FRANCIS1
213. - 343. K	213. - 293. K	213. - 343. K	213. - 343. K
18 POINTS	13 POINTS	18 POINTS	18 POINTS
RMSD = 0.00032	RMSD = 0.00031	RMSD = 0.00032	RMSD = 0.00032
0.93834585E+00	0.93047488E+00	0.93834585E+00	0.93834585E+00
0.91353524E-03	0.91759511E-03	0.91353524E-03	0.91353524E-03
0.96594048E+01	0.60115328E+01	0.96594048E+01	0.96594048E+01
0.49157373E+03	0.44377148E+03	0.49157373E+03	0.49157373E+03

SECOND VIRIAL COEFFICIENT CORRELATION CONSTANTS, ML/MOL:

TSONOPOULOS: TC=435.550 PC=4.205 VC=234.00 OM=0.2018 A= 0.000000 B=0.0000

LITERATURE DOCUMENTS FOR CORRELATED VAPOR PRESSURE VALUES:

    624   625   2649   3026

LITERATURE DOCUMENTS FOR CORRELATED LIQUID DENSITY VALUES:

    375   420   2026   9566

LITERATURE DOCUMENTS REPORTING VIRIAL COEFFICIENT DATA:

    1898

T,K	VAPOR PRESSURE, KPA	SATURATED LIQUID VOLUME, ML/MOL	SECOND VIRIAL COEFFICIENT, ML/MOL	HEAT OF VAPORIZATION, J/MOL	SATURATED VAPOR VOLUME, ML/MOL
195	0.8887		-2928.		1821396.
198	1.150		-2731.		1429187.
201	1.474		-2554.		1131458.
205	2.025		-2346.		839545.
208	2.545		-2208.		677429.
211	3.173		-2083.		550726.
215	4.212	79.36	-1934.	26468.	422484.
218	5.166	79.71	-1834.	26307.	349026.
221	6.294	80.07	-1742.	26148.	290171.
225	8.112	80.55	-1632.	25938.	228964.
228	9.745	80.91	-1557.	25782.	192965.
231	11.64	81.29	-1488.	25627.	163503.
235	14.63	81.79	-1404.	25422.	132141.
238	17.26	82.17	-1346.	25269.	113273.
242	21.36	82.68	-1276.	25066.	92895.
245	24.93	83.08	-1227.	24913.	80463.
248	28.97	83.48	-1182.	24761.	69982.
252	35.15	84.02	-1126.	24558.	58454.
255	40.46	84.43	-1087.	24405.	51295.
258	46.38	84.84	-1050.	24252.	45172.
262	55.34	85.41	-1005.	24013.	38275.
265	62.92	85.84	-972.9	23853.	33958.
268	71.31	86.28	-942.8	23691.	30217.
272	83.83	86.87	-905.2	23473.	25980.
275	94.31	87.32	-878.7	23307.	23271.
279	109.8	87.94	-845.5	23084.	20175.
282	122.8	88.41	-821.9	22915.	18178.
285	136.8	88.89	-799.5	22743.	16418.
289	157.5	89.53	-771.3	22511.	14384.
292	174.5	90.03	-751.2	22334.	13058.
295	192.9	90.54	-732.0	22154.	11878.
299	219.7	91.23	-707.7	21911.	10501.
302	241.6	91.75	-690.4	21725.	9594.
305		92.29	-673.8		
309		93.02	-652.6		
312		93.58	-637.5		
316		94.35	-618.2		
319		94.94	-604.4		
322		95.55	-591.0		
326		96.37	-574.0		
329		97.01	-561.7		
332		97.66	-549.9		
336		98.55	-534.7		
339		99.24	-523.8		
342		99.95	-513.2		

313

COMPOUND CONSTANTS:

MP   167.620 K   TC   428.630 K   VC   238.00 ML/MOL   ZC   0.2740
NBP   274.030 K   PC   4.104 MPA   DC   0.2357 G/ML   OM   0.2186
MU   0.0000 DEBYE   RG   2.7123 ANGSTROM

VAPOR PRESSURE CORRELATION CONSTANTS, K AND KPA:

WAGNER	VAPRES-2	VAPRES-2	WAGNER
193. - 378. K	193. - 294. K	193. - 378. K	259. - 378. K
40 POINTS	35 POINTS	40 POINTS	21 POINTS
RMSD = 0.7620	RMSD = 0.0790	RMSD = 0.3580	RMSD = 0.1140
-0.72976526E+01	0.33662138E+03	0.17829927E+03	-0.74657692E+01
0.17545824E+01	-0.92002656E+04	-0.66493106E+04	0.22277579E+01
-0.22729805E+01	0.20777586E+00	0.77212462E-01	-0.33645734E+01
-0.36040243E+01	-0.12241726E-03	-0.33549111E-04	0.60859077E+00
	-0.61670908E+02	-0.29939184E+02	

LIQUID DENSITY CORRELATION CONSTANTS, K AND G/ML:

FRANCIS1	FRANCIS1	FRANCIS1	RACKETT
213. - 378. K	213. - 293. K	213. - 378. K	257. - 411. K
21 POINTS	14 POINTS	21 POINTS	17 POINTS
RMSD = 0.00019	RMSD = 0.00017	RMSD = 0.00019	RMSD = 0.00088
0.91228151E+00	0.90795279E+00	0.91228151E+00	0.27236106E+00
0.86129270E-03	0.89781685E-03	0.86129270E-03	
0.10017007E+02	0.60161753E+01	0.10017007E+02	
0.47377881E+03	0.44205469E+03	0.47377881E+03	

SECOND VIRIAL COEFFICIENT CORRELATION CONSTANTS, ML/MOL:

TSONOPOULOS: TC=428.610 PC=4.104 VC=238.00 OM=0.2186 A= 0.000000 B=0.0000

LITERATURE DOCUMENTS FOR CORRELATED VAPOR PRESSURE VALUES:

    624   625   2653   4450   9566

LITERATURE DOCUMENTS FOR CORRELATED LIQUID DENSITY VALUES:

    375   420   2815   9566

LITERATURE DOCUMENTS REPORTING VIRIAL COEFFICIENT DATA:

    1898

T,K	VAPOR PRESSURE, KPA	SATURATED LIQUID VOLUME, ML/MOL	SECOND VIRIAL COEFFICIENT, ML/MOL	HEAT OF VAPORIZATION, J/MOL	SATURATED VAPOR VOLUME, ML/MOL
193	0.9404		-2987.		1703358.
197	1.324		-2720.		1234627.
201	1.833		-2490.		909382.
205	2.498		-2289.		679896.
209	3.358		-2115.		515433.
214	4.768	81.38	-1926.	25807.	371271.
218	6.224	81.87	-1795.	25582.	289420.
222	8.032	82.35	-1678.	25362.	228123.
226	10.25	82.85	-1574.	25146.	181674.
230	12.96	83.36	-1481.	24935.	146086.
235	17.13	84.00	-1377.	24676.	112651.
239	21.21	84.53	-1304.	24473.	92368.
243	26.04	85.06	-1236.	24272.	76329.
247	31.72	85.61	-1175.	24073.	63538.
251	38.36	86.17	-1119.	23876.	53253.
256	48.18	86.88	-1056.	23630.	43092.
260	57.40	87.46	-1009.	23435.	36623.
264	67.96	88.06	-966.3	23239.	31300.
268	80.01	88.67	-926.6	23042.	26891.
272	93.67	89.29	-889.6	22844.	23217.
277	113.3	90.09	-846.9	22595.	19449.
281	131.1	90.74	-815.3	22392.	16962.
285	151.1	91.42	-785.6	22187.	14852.
289	173.4	92.11	-757.8	21978.	13055.
293	198.1	92.81	-731.6	21764.	11517.
298	232.7	93.73	-701.0	21491.	9892.
302	263.6	94.48	-678.0	21266.	8790.
306	297.6	95.25	-656.3	21035.	7834.
310	334.7	96.05	-635.7	20796.	7000.
314	375.4	96.88	-616.2	20550.	6272.
319	431.3	97.95	-593.1	20231.	5485.
323	480.3	98.84	-575.6	19964.	4940.
327	533.5	99.76	-559.0	19687.	4457.
331	590.9	100.72	-543.1	19398.	4030.
335	652.9	101.72	-527.9	19096.	3649.
340	737.0	103.03	-509.9	18699.	3230.
344	809.9	104.13	-496.1	18364.	2935.
348	887.9	105.28	-483.0	18011.	2669.
352	971.4	106.50	-470.3	17638.	2430.
356	1061.	107.77	-458.2	17243.	2213.
361	1180.	109.47	-443.7	17020.	2005.
365	1283.	110.92	-432.5	16615.	1833.
369	1392.	112.47	-421.8	16191.	1677.
373	1507.	114.13	-411.5	15748.	1535.
377	1630.	115.91	-401.6	15284.	1406.

COMPOUND CONSTANTS:

MP   132.790 K   TC   417.910 K   VC 239.00 ML/MOL   ZC 0.2750
NBP   266.250 K   PC   4.000 MPA   DC   0.2348 G/ML   OM 0.1984
MU   0.5000 DEBYE   RG   2.8281 ANGSTROM

VAPOR PRESSURE CORRELATION CONSTANTS, K AND KPA:

WAGNER	RPM2	WAGNER	WAGNER
194. - 380. K	194. - 286. K	194. - 373. K	248. - 380. K
47 POINTS	39 POINTS	46 POINTS	37 POINTS
RMSD = 3.5300	RMSD = 0.4190	RMSD = 1.4930	RMSD = 4.8920
-0.68629943E+01	0.19426746E+02	-0.67844451E+01	-0.66660189E+01
0.10471164E+01	-0.34067471E+04	0.85983873E+00	0.48822668E+00
-0.13966360E+01	-0.56147197E-02	-0.11140230E+01	-0.23577594E+00
-0.58525339E+01	-0.72551865E-05	-0.63781754E+01	-0.76858473E+01

LIQUID DENSITY CORRELATION CONSTANTS, K AND G/ML:

FRANCIS1	FRANCIS1	FRANCIS1	FRANCIS1
203. - 353. K	203. - 285. K	203. - 353. K	203. - 353. K
33 POINTS	26 POINTS	33 POINTS	33 POINTS
RMSD = 0.00022	RMSD = 0.00019	RMSD = 0.00022	RMSD = 0.00022
0.91364640E+00	0.91236049E+00	0.91364640E+00	0.91364640E+00
0.85729244E-03	0.90000150E-03	0.85729244E-03	0.85729244E-03
0.12100804E+02	0.89868736E+01	0.12100804E+02	0.12100804E+02
0.47367725E+03	0.46298730E+03	0.47367725E+03	0.47367725E+03

SECOND VIRIAL COEFFICIENT CORRELATION CONSTANTS, ML/MOL:

TSONOPOULOS: TC=417.890 PC=4.000 VC=239.00 OM=0.1984 A= 0.000000 B=0.0000

LITERATURE DOCUMENTS FOR CORRELATED VAPOR PRESSURE VALUES:

   395   429   587   624   625   1339   2814   4287   9566

LITERATURE DOCUMENTS FOR CORRELATED LIQUID DENSITY VALUES:

   375   420   1911   2027

LITERATURE DOCUMENTS REPORTING VIRIAL COEFFICIENT DATA:

   1898   2657

T,K	VAPOR PRESSURE, KPA	SATURATED LIQUID VOLUME, ML/MOL	SECOND VIRIAL COEFFICIENT, ML/MOL	HEAT OF VAPORIZATION, J/MOL	SATURATED VAPOR VOLUME, ML/MOL
194	1.644		-2492.		978400.
198	2.280		-2285.		719712.
202	3.111		-2105.		537746.
206	4.182	81.10	-1948.	25329.	407647.
210	5.542	81.58	-1810.	25067.	313204.
215	7.741	82.20	-1660.	24749.	229267.
219	9.976	82.71	-1555.	24503.	180951.
223	12.72	83.22	-1462.	24263.	144343.
227	16.04	83.74	-1377.	24029.	116287.
232	21.15	84.41	-1284.	23743.	89909.
236	26.12	84.96	-1217.	23519.	73890.
240	31.99	85.51	-1156.	23298.	61207.
244	38.86	86.08	-1100.	23081.	51077.
248	46.87	86.66	-1049.	22867.	42920.
253	58.65	87.41	-991.0	22602.	34844.
257	69.67	88.01	-948.6	22391.	29691.
261	82.25	88.64	-909.3	22181.	25442.
265	96.54	89.27	-872.8	21971.	21915.
270	117.0	90.09	-830.7	21676.	18284.
274	135.7	90.77	-799.5	21460.	15913.
278	156.7	91.46	-770.3	21242.	13909.
282	180.0	92.17	-742.9	21022.	12205.
286	206.0	92.89	-717.2	20799.	10751.
291	242.3	93.83	-687.1	20517.	9219.
295	274.8	94.61	-664.5	20286.	8182.
299	310.4	95.41	-643.2	20051.	7284.
303	349.5	96.24	-622.9	19811.	6503.
308	403.4	97.31	-599.1	19504.	5664.
312	450.9	98.20	-581.1	19251.	5086.
316	502.4	99.13	-563.9	18992.	4577.
320	558.3	100.09	-547.6	18725.	4128.
325	634.6	101.34	-528.2	18379.	3638.
329	701.0	102.39	-513.5	18093.	3295.
333	772.5	103.49	-499.4	17797.	2989.
337	849.3	104.64	-485.9	17490.	2715.
341	931.6	105.84	-473.0	17172.	2470.
346	1043.	107.43	-457.6	16756.	2198.
350	1138.	108.79	-445.8	16408.	2005.
354	1241.		-434.4		1830.
358	1349.		-423.5		1671.
363	1495.		-410.5		1493.
367	1620.		-400.4		1365.
371	1752.		-390.8		1247.
375	1893.		-381.5		1140.
379	2041.		-372.4		1041.

COMPOUND CONSTANTS:

MP   107.930 K   TC   464.780 K   VC   296.00 ML/MOL   ZC 0.2700
NBP   303.108 K   PC   3.526 MPA   DC   0.2369 G/ML   OM 0.2329
MU   0.4000 DEBYE   RG   3.1956 ANGSTROM

VAPOR PRESSURE CORRELATION CONSTANTS, K AND KPA:

WAGNER	WAGNER	VAPRES-2	WAGNER
193. - 464. K	193. - 319. K	193. - 393. K	286. - 464. K
36 POINTS	25 POINTS	32 POINTS	28 POINTS
RMSD = 0.7610	RMSD = 0.0630	RMSD = 0.2060	RMSD = 0.5720
-0.69428383E+01	-0.69311318E+01	0.14655033E+03	-0.68834273E+01
0.92146976E+00	0.90377335E+00	-0.68417358E+04	0.74725093E+00
-0.19183481E+01	-0.19254081E+01	0.43016185E-01	-0.15108061E+01
-0.46905014E+01	-0.46249121E+01	-0.11508119E-04	-0.58159547E+01
		-0.22985547E+02	

LIQUID DENSITY CORRELATION CONSTANTS, K AND G/ML:

RACKETT	RACKETT	RACKETT	FRANCIS2
168. - 448. K	168. - 448. K	168. - 448. K	423. - 464. K
12 POINTS	12 POINTS	12 POINTS	2 POINTS
RMSD = 0.00235	RMSD = 0.00235	RMSD = 0.00235	RMSD = 0.00000
0.27034569E+00	0.27034569E+00	0.27034569E+00	0.45115583E-03
			0.26785225E+01

SECOND VIRIAL COEFFICIENT CORRELATION CONSTANTS, ML/MOL:

PITZER-CURL: TC=464.780 PC=3.526 VC=296.00 OM=0.2329

LITERATURE DOCUMENTS FOR CORRELATED VAPOR PRESSURE VALUES:

1146 1991 5048 6284 21583

LITERATURE DOCUMENTS FOR CORRELATED LIQUID DENSITY VALUES:

3491 5048 13422

LITERATURE DOCUMENTS REPORTING VIRIAL COEFFICIENT DATA:

1901 6284

T,K	VAPOR PRESSURE, KPA	SATURATED LIQUID VOLUME, ML/MOL	SECOND VIRIAL COEFFICIENT, ML/MOL	HEAT OF VAPORIZATION, J/MOL	SATURATED VAPOR VOLUME, ML/MOL
168		92.44	-9837.		
174		93.06	-8023.		
181		93.80	-6459.		
188		94.56	-5309.		
194	0.1512	95.23	-4557.	31903.	0.1066E+08
201	0.2995	96.03	-3875.	31373.	5576696.
208	0.5607	96.85	-3346.	30865.	3080820.
215	0.9988	97.69	-2930.	30378.	1786822.
221	1.581	98.44	-2639.	29976.	1159605.
228	2.603	99.33	-2360.	29524.	725988.
235	4.133	100.25	-2129.	29087.	470660.
241	5.985	101.06	-1963.	28725.	332828.
248	8.970	102.03	-1796.	28313.	228061.
255	13.09	103.04	-1654.	27910.	160309.
262	18.64	104.08	-1531.	27515.	115299.
268	24.80	105.00	-1438.	27181.	88372.
275	33.97	106.12	-1342.	26794.	65938.
282	45.67	107.27	-1256.	26407.	50046.
289	60.39	108.47	-1180.	26018.	38575.
295	75.78	109.54	-1121.	25681.	31206.
302	97.46	110.83	-1058.	25282.	24658.
309	123.7	112.18	-1001.	24874.	19711.
315	150.3	113.38	-955.7	24516.	16406.
322	186.8	114.85	-907.2	24084.	13362.
329	229.6	116.39	-862.5	23637.	10977.
336	279.6	118.01	-821.1	23170.	9088.
342	328.7	119.46	-788.1	22752.	7773.
349	394.0	121.25	-752.1	22240.	6515.
356	468.5	123.15	-718.5	21698.	5491.
363	553.2	125.17	-687.1	21120.	4650.
369	634.5	127.01	-661.7	20593.	4044.
376	740.2	129.30	-633.8	19932.	3447.
383	858.5	131.77	-607.6	19213.	2944.
389	970.5	134.05	-586.3	18538.	2573.
396	1115.	136.93	-562.8	18348.	2277.
403	1274.	140.10	-540.5	17563.	1964.
410	1450.	143.62	-519.4	16716.	1694.
416	1614.	146.98	-502.2	15934.	1492.
423	1822.	151.42	-483.1	14942.	1285.
430	2051.	156.59	-464.9	13847.	1102.
437	2300.	162.78	-447.6	12618.	940.8
443	2531.	169.30	-433.5	11422.	815.2
450	2822.		-417.7		679.6
457	3139.		-402.5		548.7
463	3434.		-390.1		421.6

319

COMPOUND CONSTANTS:

MP   121.780 K   TC   474.940 K   VC   292.00 ML/MOL   ZC   0.2730
NBP   310.084 K   PC   3.695 MPA   DC   0.2402 G/ML   OM   0.2538
MU   RG   3.2763 ANGSTROM

VAPOR PRESSURE CORRELATION CONSTANTS, K AND KPA:

WAGNER	WAGNER	WAGNER	WAGNER.
275. - 474. K	275. - 326. K	275. - 342. K	295. - 474. K
19 POINTS	11 POINTS	14 POINTS	15 POINTS
RMSD = 0.8700	RMSD = 0.0030	RMSD = 0.0040	RMSD = 1.0820
-0.79849933E+01	-0.74956420E+01	-0.74243288E+01	-0.80056584E+01
0.31889545E+01	0.19603473E+01	0.17860531E+01	0.32767313E+01
-0.56697503E+01	-0.34616974E+01	-0.31705744E+01	-0.60470033E+01
0.50266633E+01	-0.13312473E+01	-0.20759779E+01	0.77682905E+01

LIQUID DENSITY CORRELATION CONSTANTS, K AND G/ML:

FRANCIS1	FRANCIS1	FRANCIS1	FRANCIS1
135. - 448. K	135. - 448. K	135. - 448. K	135. - 448. K
14 POINTS	14 POINTS	14 POINTS	14 POINTS
RMSD = 0.00189	RMSD = 0.00189	RMSD = 0.00189	RMSD = 0.00189
0.95686620E+00	0.95686620E+00	0.95686620E+00	0.95686620E+00
0.90765092E-03	0.90765092E-03	0.90765092E-03	0.90765092E-03
0.75719318E+01	0.75719318E+01	0.75719318E+01	0.75719318E+01
0.51535083E+03	0.51535083E+03	0.51535083E+03	0.51535083E+03

SECOND VIRIAL COEFFICIENT CORRELATION CONSTANTS, ML/MOL:

PITZER-CURL: TC=474.940   PC=3.695   VC=292.00   OM=0.2538

LITERATURE DOCUMENTS FOR CORRELATED VAPOR PRESSURE VALUES:

3965   6627

LITERATURE DOCUMENTS FOR CORRELATED LIQUID DENSITY VALUES:

10493 16162

LITERATURE DOCUMENTS REPORTING VIRIAL COEFFICIENT DATA:

T,K	VAPOR PRESSURE, KPA	SATURATED LIQUID VOLUME, ML/MOL	SECOND VIRIAL COEFFICIENT, ML/MOL	HEAT OF VAPORIZATION, J/MOL	SATURATED VAPOR VOLUME, ML/MOL
135		86.11	-52310.		
142		86.83	-36237.		
150		87.67	-24637.		
158		88.52	-17338.		
165		89.29	-13098.		
173		90.18	-9790.		
181		91.09	-7536.		
188		91.91	-6129.		
196		92.86	-4955.		
204		93.84	-4098.		
212		94.84	-3457.		
219		95.74	-3023.		
227		96.79	-2631.		
235		97.88	-2320.		
242		98.85	-2098.		
250		99.99	-1887.		
258		101.16	-1712.		
265		102.22	-1582.		
273		103.47	-1453.		
281	33.02	104.75	-1343.	27772.	69394.
289	46.14	106.09	-1246.	27378.	50801.
296	60.79	107.29	-1171.	27007.	39278.
304	81.83	108.72	-1094.	26554.	29754.
312	108.2	110.20	-1025.	26072.	22899.
319	136.3	111.54	-971.1	25628.	18428.
327	175.1	113.14	-914.5	25096.	14554.
335	221.7	114.82	-863.1	24540.	11632.
343	277.2	116.58	-816.1	23961.	9396.
350	333.7	118.19	-778.1	23435.	7856.
358	408.5	120.13	-737.9	22812.	6453.
366	495.1	122.19	-700.8	22163.	5340.
373	581.4	124.10	-670.5	21573.	4548.
381	693.2	126.44	-638.2	20866.	3803.
389	820.2	128.96	-608.0	20116.	3192.
396	944.9	131.34	-583.3	19414.	2744.
404	1104.	134.31	-556.7	19179.	2384.
412	1283.	137.61	-531.7	18392.	2024.
420	1484.	141.33	-508.2	17549.	1719.
427	1678.	145.03	-488.8	16751.	1489.
435	1925.	149.92	-467.8	15745.	1260.
443	2201.	155.81	-447.9	14600.	1059.
450	2470.		-431.4		903.1
458	2811.		-413.5		740.0
466	3197.		-396.5		584.7
473	3579.		-382.3		431.8

COMPOUND CONSTANTS:

MP 132.930 K	TC 475.000 K	VC 292.00 ML/MOL	ZC 0.2700
NBP 309.490 K	PC 3.648 MPA	DC 0.2402 G/ML	OM 0.2395
MU	RG 3.2826 ANGSTROM		

VAPOR PRESSURE CORRELATION CONSTANTS, K AND KPA:

VAPRES-2	VAPRES-2	VAPRES-2	WAGNER
215. - 341. K	215. - 341. K	215. - 341. K	294. - 341. K
16 POINTS	16 POINTS	16 POINTS	10 POINTS
RMSD = 0.0140	RMSD = 0.0140	RMSD = 0.0140	RMSD = 0.0040
-0.32074290E+03	-0.32074290E+03	-0.32074290E+03	-0.68809532E+01
0.24300953E+04	0.24300953E+04	0.24300953E+04	0.75037911E+00
-0.23520154E+00	-0.23520154E+00	-0.23520154E+00	-0.20064827E+01
0.13290244E-03	0.13290244E-03	0.13290244E-03	-0.37740272E+01
0.65837274E+02	0.65837274E+02	0.65837274E+02	

LIQUID DENSITY CORRELATION CONSTANTS, K AND G/ML:

RACKETT	FRANCIS1	FRANCIS1	FRANCIS1
203. - 353. K	203. - 323. K	203. - 353. K	293. - 353. K
16 POINTS	13 POINTS	16 POINTS	7 POINTS
RMSD = 0.00022	RMSD = 0.00017	RMSD = 0.00016	RMSD = 0.00014
0.27027376E+00	0.91531509E+00	0.92154640E+00	0.92486346E+00
	0.77700499E-03	0.74444618E-03	0.81071514E-03
	0.67697983E+01	0.11659627E+02	0.73751087E+01
	0.46477808E+03	0.50413208E+03	0.48226050E+03

SECOND VIRIAL COEFFICIENT CORRELATION CONSTANTS, ML/MOL:

PITZER-CURL: TC=475.000   PC=3.648   VC=292.00   OM=0.2395

LITERATURE DOCUMENTS FOR CORRELATED VAPOR PRESSURE VALUES:

3491   6627

LITERATURE DOCUMENTS FOR CORRELATED LIQUID DENSITY VALUES:

8581

LITERATURE DOCUMENTS REPORTING VIRIAL COEFFICIENT DATA:

T,K	VAPOR PRESSURE, KPA	SATURATED LIQUID VOLUME, ML/MOL	SECOND VIRIAL COEFFICIENT, ML/MOL	HEAT OF VAPORIZATION, J/MOL	SATURATED VAPOR VOLUME, ML/MOL
203		95.85	-4103.		
206		96.20	-3841.		
209		96.54	-3606.		
213		97.01	-3328.		
216	0.8174	97.37	-3142.		2193935.
220	1.097	97.85	-2921.		1664033.
223	1.359	98.22	-2772.		1361201.
226	1.675	98.59	-2636.		1119467.
230	2.193	99.09	-2472.		869561.
233	2.668	99.47	-2361.		723698.
237	3.439	99.98	-2225.		570818.
240	4.136	100.37	-2133.		480364.
243	4.950	100.77	-2047.		406072.
247	6.247	101.30	-1942.		326784.
250	7.400	101.71	-1870.		279022.
254	9.212	102.26	-1781.		227448.
257	10.81	102.68	-1719.		196020.
260	12.62	103.10	-1661.		169573.
264	15.44	103.68	-1589.		140572.
267	17.88	104.12	-1538.		122627.
271	21.61	104.71	-1475.		102750.
274	24.82	105.16	-1431.		90330.
277	28.41	105.62	-1389.		79661.
281	33.84	106.25	-1337.		67687.
284	38.44	106.72	-1300.		60107.
288	45.34	107.37	-1253.		51527.
291	51.15	107.86	-1220.		46050.
295	59.80	108.54	-1178.		39800.
298	67.03	109.05	-1148.		35779.
301	74.93	109.57	-1120.		32241.
305	86.59	110.28	-1083.		28161.
308	96.23	110.83	-1057.		25508.
312	110.4	111.57	-1025.		22428.
315	122.0	112.14	-1001.		20412.
318	134.6	112.72	-978.3		18613.
322	152.9	113.52	-949.3		16504.
325	167.8	114.13	-928.5		15112.
329	189.5	114.96	-901.9		13471.
332	207.1	115.61	-882.7		12380.
335	225.9	116.27	-864.1		11395.
339	253.0	117.17	-840.4		10224.
342		117.87	-823.2		
346		118.82	-801.1		
349		119.57	-785.2		
352		120.33	-769.7		

COMPOUND CONSTANTS:

MP   135.580 K   TC   465.000 K   VC   292.00 ML/MOL   ZC   0.2650
NBP 304.303 K   PC   3.505 MPA   DC   0.2402 G/ML   OM   0.2424
MU   0.5000 DEBYE   RG   3.2239 ANGSTROM

 VAPOR PRESSURE CORRELATION CONSTANTS, K AND KPA:

 WAGNER   WAGNER   VAPRES-2   WAGNER
 274. - 447. K   274. - 320. K   274. - 382. K   289. - 447. K
  22 POINTS   11 POINTS   17 POINTS   19 POINTS
 RMSD = 1.0350   RMSD = 0.0080   RMSD = 0.0880   RMSD = 0.8810
 -0.68845239E+01   -0.71634882E+01   -0.15228899E+03   -0.68487350E+01
  0.63907169E+00   0.13942272E+01   -0.77578693E+03   0.51449606E+00
 -0.13571691E+01   -0.30070567E+01   -0.13009367E+00   -0.88485838E+00
 -0.78719810E+01   -0.14735442E+01   0.75920404E-04   -0.11631168E+02
                                            0.33580565E+02

 LIQUID DENSITY CORRELATION CONSTANTS, K AND G/ML:

 RACKETT   RACKETT   RACKETT   RACKETT
 293. - 298. K   293. - 298. K   293. - 298. K   293. - 298. K
  1 POINTS   1 POINTS   1 POINTS   1 POINTS
 RMSD = 0.00000   RMSD = 0.00000   RMSD = 0.00000   RMSD = 0.00000
 0.26271476E+00   0.26271476E+00   0.26271476E+00   0.26271476E+00

 SECOND VIRIAL COEFFICIENT CORRELATION CONSTANTS, ML/MOL:

 PITZER-CURL: TC=465.000 PC=3.505 VC=292.00 OM=0.2424

LITERATURE DOCUMENTS FOR CORRELATED VAPOR PRESSURE VALUES:

 6284   7091 15272

LITERATURE DOCUMENTS FOR CORRELATED LIQUID DENSITY VALUES:

 9852

LITERATURE DOCUMENTS REPORTING VIRIAL COEFFICIENT DATA:

 6284

T,K	VAPOR PRESSURE, KPA	SATURATED LIQUID VOLUME, ML/MOL	SECOND VIRIAL COEFFICIENT, ML/MOL	HEAT OF VAPORIZATION, J/MOL	SATURATED VAPOR VOLUME, ML/MOL
274	30.71		-1381.		72773.
277	35.06		-1342.		64328.
281	41.60		-1292.		54844.
285	49.07		-1246.		47009.
289	57.58		-1202.		40496.
293	67.20	107.81	-1161.	26159.	35051.
297	78.05	108.54	-1122.	25916.	30474.
301	90.23		-1085.		26606.
305	103.8		-1051.		23321.
309	119.0		-1018.		20518.
313	135.8		-986.5		18116.
317	154.4		-956.8		16048.
321	175.0		-928.6		14260.
325	197.5		-901.7		12708.
329	222.3		-876.0		11357.
332	242.4		-857.4		10456.
336	271.2		-833.7		9385.
340	302.6		-810.9		8443.
344	336.8		-789.1		7613.
348	373.7		-768.2		6877.
352	413.7		-748.0		6224.
356	456.8		-728.7		5643.
360	503.2		-710.1		5124.
364	553.1		-692.2		4659.
368	606.6		-674.9		4242.
372	663.9		-658.3		3866.
376	725.1		-642.3		3526.
380	790.4		-626.8		3219.
384	860.1		-611.8		2939.
388	934.2		-597.3		2685.
391	992.9		-586.8		2584.
395	1075.		-573.1		2371.
399	1163.		-559.9		2176.
403	1255.		-547.0		1998.
407	1353.		-534.6		1835.
411	1456.		-522.6		1685.
415	1565.		-510.9		1547.
419	1680.		-499.6		1420.
423	1801.		-488.6		1302.
427	1929.		-477.9		1193.
431	2063.		-467.5		1091.
435	2204.		-457.4		996.4
439	2352.		-447.6		907.5
443	2507.		-438.0		823.6
446	2628.		-431.0		763.5

COMPOUND CONSTANTS:

MP 104.720 K	TC 464.800 K	VC 300.00 ML/MOL	ZC 0.2670
NBP 293.200 K	PC 3.435 MPA	DC 0.2338 G/ML	OM 0.0908
MU	RG 3.2437 ANGSTROM		

VAPOR PRESSURE CORRELATION CONSTANTS, K AND KPA:

WAGNER	WAGNER	WAGNER	WAGNER
273. - 324. K	273. - 324. K	273. - 324. K	273. - 324. K
23 POINTS	23 POINTS	23 POINTS	23 POINTS
RMSD = 0.1440	RMSD = 0.1440	RMSD = 0.1440	RMSD = 0.1440
-0.35428088E+01	-0.35428088E+01	-0.35428088E+01	-0.35428088E+01
-0.48042457E+01	-0.48042457E+01	-0.48042457E+01	-0.48042457E+01
0.37875967E+01	0.37875967E+01	0.37875967E+01	0.37875967E+01
-0.11253769E+02	-0.11253769E+02	-0.11253769E+02	-0.11253769E+02

LIQUID DENSITY CORRELATION CONSTANTS, K AND G/ML:

RACKETT	RACKETT	RACKETT	RACKETT
288. - 298. K	288. - 298. K	288. - 298. K	288. - 298. K
2 POINTS	2 POINTS	2 POINTS	2 POINTS
RMSD = 0.00000	RMSD = 0.00000	RMSD = 0.00000	RMSD = 0.00000
0.26789528E+00	0.26789528E+00	0.26789528E+00	0.26789528E+00

SECOND VIRIAL COEFFICIENT CORRELATION CONSTANTS, ML/MOL:

TSONOPOULOS: TC=464.800 PC=3.435 VC=300.00 OM=0.0908 A= 0.000000 B=0.0000

LITERATURE DOCUMENTS FOR CORRELATED VAPOR PRESSURE VALUES:

6627 15272 15485

LITERATURE DOCUMENTS FOR CORRELATED LIQUID DENSITY VALUES:

7343 9851

LITERATURE DOCUMENTS REPORTING VIRIAL COEFFICIENT DATA:

7812

T,K	VAPOR PRESSURE, KPA	SATURATED LIQUID VOLUME, ML/MOL	SECOND VIRIAL COEFFICIENT, ML/MOL	HEAT OF VAPORIZATION, J/MOL	SATURATED VAPOR VOLUME, ML/MOL
273	46.67		-1203.		47405.
274	48.64		-1192.		45615.
275	50.68		-1181.		43906.
276	52.78		-1171.		42275.
277	54.95		-1160.		40716.
278	57.20		-1150.		39227.
279	59.51		-1140.		37804.
281	64.37		-1120.		35141.
282	66.91		-1110.		33895.
283	69.53		-1101.		32702.
284	72.23		-1092.		31560.
285	75.01		-1082.		30467.
286	77.88		-1073.		29419.
288	83.88	110.95	-1056.	24292.	27451.
289	87.01	111.13	-1047.	24242.	26527.
290	90.23	111.31	-1039.	24192.	25641.
291	93.54	111.49	-1030.	24142.	24791.
292	96.95	111.68	-1022.	24091.	23974.
293	100.5	111.86	-1014.	24040.	23190.
295	107.8	112.23	-998.0	23938.	21714.
296	111.6	112.42	-990.2	23887.	21018.
297	115.5	112.61	-982.5	23835.	20350.
298	119.5	112.80	-975.0	23783.	19707.
299	123.6		-967.5		19088.
300	127.9		-960.1		18493.
301	132.2		-952.9		17921.
303	141.3		-938.7		16839.
304	146.0		-931.7		16327.
305	150.8		-924.8		15835.
306	155.7		-918.0		15360.
307	160.8		-911.3		14903.
308	166.0		-904.7		14461.
310	176.8		-891.7		13626.
311	182.4		-885.3		13230.
312	188.1		-879.1		12847.
313	194.0		-872.8		12478.
314	200.0		-866.7		12122.
315	206.1		-860.7		11778.
317	218.8		-848.8		11125.
318	225.4		-842.9		10815.
319	232.2		-837.1		10515.
320	239.0		-831.4		10225.
321	246.1		-825.8		9945.
322	253.3		-820.2		9673.
323	260.7		-814.7		9411.

COMPOUND CONSTANTS:

MP 139.400 K	TC 470.390 K	VC 292.00 ML/MOL	ZC 0.2510
NBP 311.590 K	PC 3.360 MPA	DC 0.2402 G/ML	OM 0.2713
MU	RG 3.2301 ANGSTROM		

VAPOR PRESSURE CORRELATION CONSTANTS, K AND KPA:

WAGNER	RIEDEL	VAPRES-2	WAGNER'
194. - 470. K	194. - 329. K	194. - 387. K	292. - 470. K
37 POINTS	26 POINTS	31 POINTS	21 POINTS
RMSD = 2.4790	RMSD = 0.1000	RMSD = 0.3050	RMSD = 2.4110
-0.72173104E+01	0.49783715E+02	0.98577350E+02	-0.74298398E+01
0.11812457E+01	-0.51883934E+01	-0.56984671E+04	0.18467675E+01
-0.29400964E+01	-0.48157873E+04	0.29915376E-01	-0.47668584E+01
-0.45253103E+00	0.87600282E-16	-0.11728980E-04	0.70861541E+01
		-0.14604127E+02	

LIQUID DENSITY CORRELATION CONSTANTS, K AND G/ML:

RACKETT	FRANCIS1	FRANCIS1	FRANCIS1
203. - 353. K	203. - 323. K	203. - 353. K	293. - 353. K
27 POINTS	24 POINTS	27 POINTS	13 POINTS
RMSD = 0.00283	RMSD = 0.00050	RMSD = 0.00126	RMSD = 0.00144
0.25834344E+00	0.91272086E+00	0.92053634E+00	0.99335533E+00
	0.70468942E-03	0.60519902E-03	0.91177155E-03
	0.59999990E+01	0.14996624E+02	0.14997362E+02
	0.42910938E+03	0.47746851E+03	0.53125830E+03

SECOND VIRIAL COEFFICIENT CORRELATION CONSTANTS, ML/MOL:

TSONOPOULOS: TC=470.390 PC=3.360 VC=292.00 OM=0.2713 A= 0.000000 B=0.0000

LITERATURE DOCUMENTS FOR CORRELATED VAPOR PRESSURE VALUES:

624   5113   6284   7091

LITERATURE DOCUMENTS FOR CORRELATED LIQUID DENSITY VALUES:

8581   9851   9920

LITERATURE DOCUMENTS REPORTING VIRIAL COEFFICIENT DATA:

6284

T,K	VAPOR PRESSURE, KPA	SATURATED LIQUID VOLUME, ML/MOL	SECOND VIRIAL COEFFICIENT, ML/MOL	HEAT OF VAPORIZATION, J/MOL	SATURATED VAPOR VOLUME, ML/MOL
194	0.09404		-7044.		0.1714E+08
200	0.1689		-5977.		9842123.
206	0.2919	93.86	-5132.	31108.	5862893.
212	0.4874	94.57	-4455.	30877.	3612134.
219	0.8518	95.42	-3826.	30606.	2133736.
225	1.333	96.16	-3392.	30371.	1400435.
231	2.031	96.93	-3032.	30134.	942741.
237	3.021	97.71	-2730.	29893.	649561.
244	4.669	98.66	-2438.	29607.	432075.
250	6.633	99.49	-2227.	29357.	311141.
256	9.248	100.34	-2045.	29101.	228093.
262	12.67	101.22	-1888.	28840.	169990.
269	17.93	102.28	-1729.	28525.	122960.
275	23.77	103.22	-1611.	28248.	94570.
281	31.06	104.19	-1506.	27962.	73674.
288	41.78	105.36	-1397.	27616.	55883.
294	53.17	106.40	-1315.	27310.	44619.
300	66.93	107.48	-1241.	26992.	35985.
306	83.38	108.59	-1173.	26662.	29292.
313	106.4	109.95	-1102.	26263.	23291.
319	130.0	111.15	-1047.	25906.	19300.
325	157.4	112.41	-996.4	25535.	16109.
331	189.0	113.72	-949.6	25150.	13536.
338	232.0	115.32	-899.6	24680.	11135.
344	274.4	116.76	-860.0	24258.	9476.
350	322.6	118.27	-823.2	23819.	8105.
357	386.5		-783.4		6794.
363	448.6		-751.6		5866.
369	517.9		-721.9		5083.
375	594.9		-694.0		4418.
382	695.2		-663.4		3763.
388	790.7		-638.9		3287.
394	895.7		-615.7		2874.
400	1011.		-593.8		2568.
407	1158.		-569.6		2210.
413	1297.		-550.0		1944.
419	1448.		-531.4		1711.
426	1640.		-510.9		1472.
432	1820.		-494.2		1291.
438	2014.		-478.2		1129.
444	2223.		-463.0		982.5
451	2489.		-446.1		826.8
457	2735.		-432.2		702.5
463	3001.		-419.0		580.5
469	3290.		-406.3		438.8

329

COMPOUND CONSTANTS:

MP	133.318 K	TC	504.030 K	VC	354.00 ML/MOL	ZC	0.2650

MP   133.318 K   TC   504.030 K   VC   354.00 ML/MOL   ZC   0.2650
NBP   336.628 K   PC   3.140 MPA   DC   0.2377 G/ML   OM   0.2785
MU   0.4000 DEBYE   RG   3.6472 ANGSTROM

VAPOR PRESSURE CORRELATION CONSTANTS, K AND KPA:

WAGNER	WAGNER	WAGNER	WAGNER
283. - 354. K	283. - 354. K	283. - 354. K	317. - 354. K
20 POINTS	20 POINTS	20 POINTS	12 POINTS
RMSD = 0.0290	RMSD = 0.0290	RMSD = 0.0290	RMSD = 0.0220
-0.60228102E+01	-0.60228102E+01	-0.60228102E+01	-0.36859246E+01
-0.18093123E+01	-0.18093123E+01	-0.18093123E+01	-0.77652358E+01
0.19840906E+01	0.19840906E+01	0.19840906E+01	0.13293974E+02
-0.14231793E+02	-0.14231793E+02	-0.14231793E+02	-0.51977241E+02

LIQUID DENSITY CORRELATION CONSTANTS, K AND G/ML:

FRANCIS1	FRANCIS1	FRANCIS1
283. - 334. K	283. - 334. K	283. - 334. K
11 POINTS	11 POINTS	11 POINTS
RMSD = 0.00023	RMSD = 0.00023	RMSD = 0.00023
0.95143622E+00	0.95143622E+00	0.95143622E+00
0.82382443E-03	0.82382443E-03	0.82382443E-03
0.12157918E+02	0.12157918E+02	0.12157918E+02
0.62463721E+03	0.62463721E+03	0.62463721E+03

SECOND VIRIAL COEFFICIENT CORRELATION CONSTANTS, ML/MOL:

PITZER-CURL: TC=504.030   PC=3.140   VC=354.00   OM=0.2785

LITERATURE DOCUMENTS FOR CORRELATED VAPOR PRESSURE VALUES:

   1631   1981   1985   1991   5311 10561

LITERATURE DOCUMENTS FOR CORRELATED LIQUID DENSITY VALUES:

   1985   1991 10246 21562 21601

LITERATURE DOCUMENTS REPORTING VIRIAL COEFFICIENT DATA:

   1901

T,K	VAPOR PRESSURE, KPA	SATURATED LIQUID VOLUME, ML/MOL	SECOND VIRIAL COEFFICIENT, ML/MOL	HEAT OF VAPORIZATION, J/MOL	SATURATED VAPOR VOLUME, ML/MOL
283	12.55	123.28	-2005.	31592.	185414.
284	13.17	123.44	-1985.	31521.	177320.
286	14.47	123.78	-1944.	31382.	162370.
287	15.16	123.95	-1925.	31313.	155466.
289	16.62	124.29	-1887.	31177.	142687.
291	18.19	124.63	-1850.	31044.	131151.
292	19.02	124.81	-1832.	30978.	125806.
294	20.77	125.15	-1797.	30849.	115882.
295	21.69	125.33	-1780.	30785.	111275.
297	23.63	125.68	-1747.	30658.	102707.
299	25.72	126.03	-1715.	30534.	94921.
300	26.81	126.21	-1699.	30472.	91297.
302	29.12	126.57	-1668.	30350.	84535.
303	30.33	126.75	-1653.	30290.	81382.
305	32.87	127.11	-1624.	30170.	75491.
307	35.58	127.47	-1596.	30052.	70107.
308	37.00	127.65	-1582.	29994.	67590.
310	39.98	128.02	-1555.	29878.	62875.
312	43.15	128.39	-1529.	29763.	58551.
313	44.81	128.58	-1516.	29706.	56524.
315	48.27	128.95	-1491.	29592.	52719.
316	50.09	129.14	-1478.	29536.	50933.
318	53.88	129.52	-1454.	29424.	47575.
320	57.89	129.90	-1431.	29312.	44480.
321	59.99	130.09	-1419.	29257.	43024.
323	64.36	130.48	-1397.	29146.	40282.
324	66.64	130.67	-1386.	29091.	38989.
326	71.39	131.06	-1364.	28981.	36552.
328	76.40	131.45	-1343.	28870.	34296.
329	79.01	131.65	-1333.	28815.	33231.
331	84.45	132.05	-1313.	28705.	31218.
333	90.18	132.45	-1293.	28595.	29350.
334	93.15	132.65	-1283.	28539.	28467.
336	99.34		-1264.		26795.
337	102.6		-1255.		26003.
339	109.2		-1236.		24502.
341	116.2		-1219.		23104.
342	119.9		-1210.		22441.
344	127.4		-1192.		21182.
345	131.3		-1184.		20584.
347	139.4		-1167.		19448.
349	147.9		-1151.		18386.
350	152.3		-1143.		17882.
352	161.4		-1127.		16921.
353	166.2		-1119.		16464.

COMPOUND CONSTANTS:

MP	131.998 K	TC		VC		ZC
NBP	342.026 K	PC		DC		OM
MU		RG				

VAPOR PRESSURE CORRELATION CONSTANTS, K AND KPA:

RPM2	RPM2	RPM2	RPM2
297. - 343. K	297. - 343. K	297. - 343. K	297. - 343. K
9 POINTS	9 POINTS	9 POINTS	9 POINTS
RMSD = 0.0460	RMSD = 0.0460	RMSD = 0.0460	RMSD = 0.0460
0.14025340E+03	0.14025340E+03	0.14025340E+03	0.14025340E+03
-0.17422945E+05	-0.17422945E+05	-0.17422945E+05	-0.17422945E+05
-0.37854387E+00	-0.37854387E+00	-0.37854387E+00	-0.37854387E+00
0.38277050E-03	0.38277050E-03	0.38277050E-03	0.38277050E-03

LIQUID DENSITY CORRELATION CONSTANTS, K AND G/ML:

FRANCIS1	FRANCIS1	FRANCIS1
293. - 304. K	293. - 304. K	293. - 304. K
4 POINTS	4 POINTS	4 POINTS
RMSD = 0.00000	RMSD = 0.00000	RMSD = 0.00000
0.93853176E+00	0.93853176E+00	0.93853176E+00
0.71244803E-03	0.71244803E-03	0.71244803E-03
0.82087164E+01	0.82087164E+01	0.82087164E+01
0.48638013E+03	0.48638013E+03	0.48638013E+03

SECOND VIRIAL COEFFICIENT CORRELATION CONSTANTS, ML/MOL:

LITERATURE DOCUMENTS FOR CORRELATED VAPOR PRESSURE VALUES:

6669

LITERATURE DOCUMENTS FOR CORRELATED LIQUID DENSITY VALUES:

6669

LITERATURE DOCUMENTS REPORTING VIRIAL COEFFICIENT DATA:

T,K	VAPOR PRESSURE, KPA	SATURATED LIQUID VOLUME, ML/MOL	SECOND VIRIAL COEFFICIENT, ML/MOL	HEAT OF VAPORIZATION, J/MOL	SATURATED VAPOR VOLUME, ML/MOL
293		122.45			
294		122.61			
295		122.78			
296		122.95			
297	18.66	123.12		33955.	132312.
298	19.54	123.29		33771.	126781.
299	20.45	123.46		33592.	121547.
300	21.39	123.63		33418.	116591.
302	23.37	123.98		33086.	107440.
303	24.41	124.15		32928.	103214.
304	25.48	124.33		32775.	99202.
305	26.58				95389.
306	27.73				91765.
307	28.90				88317.
308	30.12				85035.
310	32.65				78930.
311	33.98				76090.
312	35.35				73379.
313	36.76				70792.
314	38.21				68320.
315	39.71				65958.
316	41.25				63699.
317	42.83				61537.
319	46.14				57486.
320	47.86				55587.
321	49.64				53766.
322	51.47				52019.
323	53.35				50343.
324	55.28				48733.
325	57.27				47186.
327	61.41				44270.
328	63.58				42895.
329	65.80				41571.
330	68.09				40296.
331	70.44				39068.
332	72.86				37885.
333	75.35				36743.
335	80.55				34580.
336	83.26				33555.
337	86.04				32565.
338	88.91				31608.
339	91.86				30683.
340	94.90				29789.
341	98.02				28924.
342	101.2				28087.

333

COMPOUND CONSTANTS:

MP 140.171 K	TC	VC	ZC
NBP 341.022 K	PC	DC	OM
MU	RG 3.6964 ANGSTROM		

VAPOR PRESSURE CORRELATION CONSTANTS, K AND KPA:

RPM2	RPM2	RPM2	RPM2
292. - 342. K	292. - 342. K	292. - 342. K	292. - 342. K
13 POINTS	13 POINTS	13 POINTS	13 POINTS
RMSD = 0.0090	RMSD = 0.0090	RMSD = 0.0090	RMSD = 0.0090
0.40340507E+02	0.40340507E+02	0.40340507E+02	0.40340507E+02
-0.66130548E+04	-0.66130548E+04	-0.66130548E+04	-0.66130548E+04
-0.70621493E-01	-0.70621493E-01	-0.70621493E-01	-0.70621493E-01
0.66667683E-04	0.66667683E-04	0.66667683E-04	0.66667683E-04

LIQUID DENSITY CORRELATION CONSTANTS, K AND G/ML:

FRANCIS1	FRANCIS1	FRANCIS1
293. - 309. K	293. - 309. K	293. - 309. K
5 POINTS	5 POINTS	5 POINTS
RMSD = 0.00020	RMSD = 0.00020	RMSD = 0.00020
0.95827532E+00	0.95827532E+00	0.95827532E+00
0.87121152E-03	0.87121152E-03	0.87121152E-03
0.70313587E+01	0.70313587E+01	0.70313587E+01
0.57776392E+03	0.57776392E+03	0.57776392E+03

SECOND VIRIAL COEFFICIENT CORRELATION CONSTANTS, ML/MOL:

LITERATURE DOCUMENTS FOR CORRELATED VAPOR PRESSURE VALUES:

6669

LITERATURE DOCUMENTS FOR CORRELATED LIQUID DENSITY VALUES:

2369   6669 10246

LITERATURE DOCUMENTS REPORTING VIRIAL COEFFICIENT DATA:

T,K	VAPOR PRESSURE, KPA	SATURATED LIQUID VOLUME, ML/MOL	SECOND VIRIAL COEFFICIENT, ML/MOL	HEAT OF VAPORIZATION, J/MOL	SATURATED VAPOR VOLUME, ML/MOL
292	15.74				154286.
293	16.47	124.07		32433.	147902.
294	17.23	124.25		32374.	141837.
295	18.03	124.42		32316.	136071.
296	18.85	124.60		32258.	130588.
297	19.70	124.78		32201.	125372.
298	20.58	124.96		32144.	120408.
299	21.49	125.14		32089.	115681.
301	23.41	125.49		31980.	106888.
302	24.43	125.68		31927.	102799.
303	25.47	125.86		31874.	98899.
304	26.56	126.04		31823.	95178.
305	27.68	126.22		31772.	91628.
306	28.83	126.40		31722.	88238.
307	30.03	126.59		31672.	85001.
309	32.54	126.96		31576.	78952.
310	33.86				76126.
311	35.22				73423.
312	36.62				70837.
313	38.07				68362.
314	39.56				65993.
315	41.10				63723.
316	42.69				61549.
318	46.01				57468.
319	47.74				55553.
320	49.53				53716.
321	51.37				51953.
322	53.27				50261.
323	55.22				48636.
324	57.22				47075.
326	61.41				44135.
327	63.60				42750.
328	65.84				41418.
329	68.15				40137.
330	70.52				38905.
331	72.96				37719.
332	75.47				36577.
334	80.68				34418.
335	83.40				33398.
336	86.18				32415.
337	89.04				31467.
338	91.98				30553.
339	94.99				29672.
340	98.08				28821.
341	101.3				28001.

COMPOUND CONSTANTS:

MP	135.321 K	TC	VC	ZC
NBP	339.592 K	PC	DC	OM
MU	0.3400 DEBYE	RG		

VAPOR PRESSURE CORRELATION CONSTANTS, K AND KPA:

RPM2	RPM2	RPM2	RPM2
286. - 341. K	286. - 341. K	286. - 341. K	286. - 341. K
14 POINTS	14 POINTS	14 POINTS	14 POINTS
RMSD = 0.0070	RMSD = 0.0070	RMSD = 0.0070	RMSD = 0.0070
0.26616978E+02	0.26616978E+02	0.26616978E+02	0.26616978E+02
-0.51231194E+04	-0.51231194E+04	-0.51231194E+04	-0.51231194E+04
-0.27887308E-01	-0.27887308E-01	-0.27887308E-01	-0.27887308E-01
0.22179250E-04	0.22179250E-04	0.22179250E-04	0.22179250E-04

LIQUID DENSITY CORRELATION CONSTANTS, K AND G/ML:

FRANCIS1	FRANCIS1	FRANCIS1
293. - 304. K	293. - 304. K	293. - 304. K
5 POINTS	5 POINTS	5 POINTS
RMSD = 0.00016	RMSD = 0.00016	RMSD = 0.00016
0.95523560E+00	0.95523560E+00	0.95523560E+00
0.84537710E-03	0.84537710E-03	0.84537710E-03
0.75951376E+01	0.75951376E+01	0.75951376E+01
0.56838794E+03	0.56838794E+03	0.56838794E+03

SECOND VIRIAL COEFFICIENT CORRELATION CONSTANTS, ML/MOL:

LITERATURE DOCUMENTS FOR CORRELATED VAPOR PRESSURE VALUES:

6669

LITERATURE DOCUMENTS FOR CORRELATED LIQUID DENSITY VALUES:

6669 9854

LITERATURE DOCUMENTS REPORTING VIRIAL COEFFICIENT DATA:

T,K	VAPOR PRESSURE, KPA	SATURATED LIQUID VOLUME, ML/MOL	SECOND VIRIAL COEFFICIENT, ML/MOL	HEAT OF VAPORIZATION, J/MOL	SATURATED VAPOR VOLUME, ML/MOL
286	12.71				187093.
287	13.32				179085.
288	13.96				171484.
289	14.63				164267.
291	16.03				150898.
292	16.78				144705.
293	17.55	123.77		31939.	138816.
294	18.35	123.95		31897.	133213.
296	20.04	124.29		31813.	122804.
297	20.93	124.47		31772.	117969.
298	21.86	124.64		31730.	113363.
299	22.81	124.82		31689.	108972.
301	24.83	125.17		31607.	100793.
302	25.89	125.35		31566.	96983.
303	26.99	125.53		31526.	93348.
304	28.12	125.70		31485.	89876.
306	30.51				83392.
307	31.76				80365.
308	33.06				77470.
309	34.39				74702.
311	37.19				69520.
312	38.66				67094.
313	40.18				64771.
314	41.74				62546.
316	45.01				58371.
317	46.72				56412.
318	48.48				54533.
319	50.30				52731.
321	54.09				49342.
322	56.07				47749.
323	58.11				46218.
324	60.20				44748.
326	64.57				41978.
327	66.85				40672.
328	69.19				39417.
329	71.59				38209.
331	76.60				35929.
332	79.20				34853.
333	81.88				33816.
334	84.62				32818.
336	90.32				30929.
337	93.29				30035.
338	96.33				29174.
339	99.44				28343.
341	105.9				26769.

COMPOUND CONSTANTS:

MP 159.730 K	TC	VC	ZC
NBP 340.229 K	PC	DC	OM
MU 0.0000 DEBYE	RG		

VAPOR PRESSURE CORRELATION CONSTANTS, K AND KPA:

VAPRES-2	VAPRES-2	VAPRES-2	VAPRES-2
291. - 341. K	291. - 341. K	291. - 341. K	291. - 341. K
13 POINTS	13 POINTS	13 POINTS	13 POINTS
RMSD = 0.0060	RMSD = 0.0060	RMSD = 0.0060	RMSD = 0.0060
0.26081451E+03	0.26081451E+03	0.26081451E+03	0.26081451E+03
-0.98343508E+04	-0.98343508E+04	-0.98343508E+04	-0.98343508E+04
0.11373691E+00	0.11373691E+00	0.11373691E+00	0.11373691E+00
-0.52775345E-04	-0.52775345E-04	-0.52775345E-04	-0.52775345E-04
-0.44579009E+02	-0.44579009E+02	-0.44579009E+02	-0.44579009E+02

LIQUID DENSITY CORRELATION CONSTANTS, K AND G/ML:

FRANCIS1	FRANCIS1	FRANCIS1
293. - 309. K	293. - 309. K	293. - 309. K
7 POINTS	7 POINTS	7 POINTS
RMSD = 0.00032	RMSD = 0.00032	RMSD = 0.00032
0.93628305E+00	0.93628305E+00	0.93628305E+00
0.76421746E-03	0.76421746E-03	0.76421746E-03
0.60243330E+01	0.60243330E+01	0.60243330E+01
0.46609448E+03	0.46609448E+03	0.46609448E+03

SECOND VIRIAL COEFFICIENT CORRELATION CONSTANTS, ML/MOL:

LITERATURE DOCUMENTS FOR CORRELATED VAPOR PRESSURE VALUES:

6669

LITERATURE DOCUMENTS FOR CORRELATED LIQUID DENSITY VALUES:

2369   6669   9854 10246

LITERATURE DOCUMENTS REPORTING VIRIAL COEFFICIENT DATA:

T,K	VAPOR PRESSURE, KPA	SATURATED LIQUID VOLUME, ML/MOL	SECOND VIRIAL COEFFICIENT, ML/MOL	HEAT OF VAPORIZATION, J/MOL	SATURATED VAPOR VOLUME, ML/MOL
291	15.44				156739.
292	16.16				150240.
293	16.91	124.21		32248.	144063.
294	17.69	124.39		32204.	138189.
295	18.50	124.57		32161.	132602.
296	19.34	124.75		32118.	127284.
297	20.20	124.93		32076.	122222.
298	21.10	125.11		32033.	117401.
300	23.00	125.47		31949.	108431.
301	24.00	125.66		31908.	104257.
302	25.04	125.84		31866.	100277.
303	26.11	126.03		31825.	96479.
304	27.22	126.22		31785.	92855.
305	28.37	126.41		31744.	89395.
306	29.55	126.60		31704.	86090.
308	32.04	126.98		31624.	79916.
309	33.35	127.17		31584.	77031.
310	34.70				74273.
311	36.10				71634.
312	37.54				69108.
313	39.02				66691.
314	40.55				64377.
315	42.13				62160.
317	45.44				58000.
318	47.17				56049.
319	48.96				54177.
320	50.79				52382.
321	52.68				50660.
322	54.63				49006.
323	56.63				47419.
325	60.82				44431.
326	63.00				43024.
327	65.24				41672.
328	67.55				40373.
329	69.92				39123.
330	72.36				37921.
331	74.86				36764.
333	80.07				34579.
334	82.78				33548.
335	85.56				32554.
336	88.41				31597.
337	91.34				30675.
338	94.35				29786.
339	97.43				28929.
340	100.6				28103.

COMPOUND CONSTANTS:

MP	137.419 K	TC	VC	ZC
NBP	335.253 K	PC	DC	OM
MU		RG		

VAPOR PRESSURE CORRELATION CONSTANTS, K AND KPA:

RPM2	RPM2	RPM2	RPM2
283. - 336. K	283. - 336. K	283. - 336. K	283. - 336. K
13 POINTS	13 POINTS	13 POINTS	13 POINTS
RMSD = 0.0040	RMSD = 0.0040	RMSD = 0.0040	RMSD = 0.0040
0.28368185E+02	0.28368185E+02	0.28368185E+02	0.28368185E+02
-0.52324472E+04	-0.52324472E+04	-0.52324472E+04	-0.52324472E+04
-0.33893060E-01	-0.33893060E-01	-0.33893060E-01	-0.33893060E-01
0.28652199E-04	0.28652199E-04	0.28652199E-04	0.28652199E-04

LIQUID DENSITY CORRELATION CONSTANTS, K AND G/ML:

FRANCIS1	FRANCIS1	FRANCIS1
293. - 304. K	293. - 304. K	293. - 304. K
4 POINTS	4 POINTS	4 POINTS
RMSD = 0.00000	RMSD = 0.00000	RMSD = 0.00000
0.94801825E+00	0.94801825E+00	0.94801825E+00
0.81138685E-03	0.81138685E-03	0.81138685E-03
0.61974878E+01	0.61974878E+01	0.61974878E+01
0.49768018E+03	0.49768018E+03	0.49768018E+03

SECOND VIRIAL COEFFICIENT CORRELATION CONSTANTS, ML/MOL:

LITERATURE DOCUMENTS FOR CORRELATED VAPOR PRESSURE VALUES:

1981   6669

LITERATURE DOCUMENTS FOR CORRELATED LIQUID DENSITY VALUES:

6669

LITERATURE DOCUMENTS REPORTING VIRIAL COEFFICIENT DATA:

T,K	VAPOR PRESSURE, KPA	SATURATED LIQUID VOLUME, ML/MOL	SECOND VIRIAL COEFFICIENT, ML/MOL	HEAT OF VAPORIZATION, J/MOL	SATURATED VAPOR VOLUME, ML/MOL
283	13.22				177945.
284	13.87				170297.
285	14.53				163041.
286	15.23				156155.
287	15.95				149615.
289	17.48				137499.
290	18.28				131887.
291	19.12				126549.
292	19.99				121470.
293	20.89	123.77		31264.	116637.
295	22.78	124.12		31177.	107650.
296	23.78	124.29		31133.	103472.
297	24.82	124.47		31090.	99490.
298	25.89	124.65		31048.	95693.
299	27.00	124.83		31005.	92071.
301	29.33	125.19		30921.	85315.
302	30.56	125.37		30879.	82165.
303	31.83	125.55		30838.	79155.
304	33.14	125.73		30797.	76279.
305	34.49				73530.
307	37.32				68387.
308	38.81				65981.
309	40.35				63679.
310	41.93				61474.
311	43.56				59362.
313	46.97				55401.
314	48.76				53542.
315	50.60				51760.
316	52.49				50051.
317	54.44				48411.
319	58.52				45325.
320	60.64				43874.
321	62.83				42480.
322	65.08				41140.
323	67.39				39853.
325	72.20				37424.
326	74.71				36279.
327	77.29				35178.
328	79.93				34117.
329	82.65				33097.
331	88.30				31167.
332	91.24				30255.
333	94.25				29376.
334	97.34				28529.
335	100.5				27712.

COMPOUND CONSTANTS:

MP	120.139 K	TC		VC		ZC
NBP	327.310 K	PC		DC		OM
MU		RG				

VAPOR PRESSURE CORRELATION CONSTANTS, K AND KPA:

RPM2	RPM2	RPM2	RPM2
287. - 328. K	287. - 328. K	287. - 328. K	287. - 328. K
11 POINTS	11 POINTS	11 POINTS	11 POINTS
RMSD = 0.0210	RMSD = 0.0210	RMSD = 0.0210	RMSD = 0.0210
0.58546799E+02	0.58546799E+02	0.58546799E+02	0.58546799E+02
-0.81424212E+04	-0.81424212E+04	-0.81424212E+04	-0.81424212E+04
-0.13356994E+00	-0.13356994E+00	-0.13356994E+00	-0.13356994E+00
0.13690697E-03	0.13690697E-03	0.13690697E-03	0.13690697E-03

LIQUID DENSITY CORRELATION CONSTANTS, K AND G/ML:

FRANCIS1	FRANCIS1	FRANCIS1
293. - 304. K	293. - 304. K	293. - 304. K
4 POINTS	4 POINTS	4 POINTS
RMSD = 0.00007	RMSD = 0.00007	RMSD = 0.00007
0.91645706E+00	0.91645706E+00	0.91645706E+00
0.72684302E-03	0.72684302E-03	0.72684302E-03
0.60483646E+01	0.60483646E+01	0.60483646E+01
0.46179395E+03	0.46179395E+03	0.46179395E+03

SECOND VIRIAL COEFFICIENT CORRELATION CONSTANTS, ML/MOL:

LITERATURE DOCUMENTS FOR CORRELATED VAPOR PRESSURE VALUES:

6669

LITERATURE DOCUMENTS FOR CORRELATED LIQUID DENSITY VALUES:

6669

LITERATURE DOCUMENTS REPORTING VIRIAL COEFFICIENT DATA:

T,K	VAPOR PRESSURE, KPA	SATURATED LIQUID VOLUME, ML/MOL	SECOND VIRIAL COEFFICIENT, ML/MOL	HEAT OF VAPORIZATION, J/MOL	SATURATED VAPOR VOLUME, ML/MOL
287	22.61				105547.
288	23.62				101390.
289	24.66				97436.
290	25.74				93671.
291	26.86				90087.
292	28.01				86672.
293	29.20	126.05		29580.	83418.
294	30.44	126.23		29514.	80315.
295	31.71	126.41		29451.	77354.
296	33.02	126.59		29389.	74529.
297	34.38	126.77		29329.	71832.
298	35.78	126.95		29271.	69255.
299	37.22	127.14		29214.	66793.
300	38.71	127.32		29160.	64440.
301	40.24	127.50		29107.	62189.
302	41.82	127.69		29056.	60035.
303	43.45	127.88		29007.	57974.
304	45.13	128.07		28960.	56001.
305	46.87				54110.
306	48.65				52299.
307	50.48				50562.
308	52.37				48897.
309	54.32				47300.
310	56.32				45767.
311	58.38				44295.
312	60.49				42882.
313	62.67				41524.
314	64.91				40219.
315	67.22				38965.
316	69.58				37758.
317	72.02				36598.
318	74.52				35481.
319	77.09				34406.
320	79.73				33371.
321	82.44				32374.
322	85.23				31413.
323	88.09				30487.
324	91.03				29594.
325	94.04				28733.
326	97.14				27902.
327	100.3				27101.
328	103.6				26327.

COMPOUND CONSTANTS:

MP	119.509 K	TC		VC		ZC	
NBP	327.004 K	PC		DC		OM	
MU		RG					

VAPOR PRESSURE CORRELATION CONSTANTS, K AND KPA:

VAPRES-2	VAPRES-2	VAPRES-2	VAPRES-2
283. - 328. K	283. - 328. K	283. - 328. K	283. - 328. K
14 POINTS	14 POINTS	14 POINTS	14 POINTS
RMSD = 0.0090	RMSD = 0.0090	RMSD = 0.0090	RMSD = 0.0090
-0.22156417E+03	-0.22156417E+03	-0.22156417E+03	-0.22156417E+03
0.21492581E+03	0.21492581E+03	0.21492581E+03	0.21492581E+03
-0.17511663E+00	-0.17511663E+00	-0.17511663E+00	-0.17511663E+00
0.10019415E-03	0.10019415E-03	0.10019415E-03	0.10019415E-03
0.46990746E+02	0.46990746E+02	0.46990746E+02	0.46990746E+02

LIQUID DENSITY CORRELATION CONSTANTS, K AND G/ML:

FRANCIS1	FRANCIS1	FRANCIS1
293. - 304. K	293. - 304. K	293. - 304. K
5 POINTS	5 POINTS	5 POINTS
RMSD = 0.00003	RMSD = 0.00003	RMSD = 0.00003
0.91519970E+00	0.91519970E+00	0.91519970E+00
0.72382623E-03	0.72382623E-03	0.72382623E-03
0.65300446E+01	0.65300446E+01	0.65300446E+01
0.45945288E+03	0.45945288E+03	0.45945288E+03

SECOND VIRIAL COEFFICIENT CORRELATION CONSTANTS, ML/MOL:

LITERATURE DOCUMENTS FOR CORRELATED VAPOR PRESSURE VALUES:

1981 6669

LITERATURE DOCUMENTS FOR CORRELATED LIQUID DENSITY VALUES:

6669 21601

LITERATURE DOCUMENTS REPORTING VIRIAL COEFFICIENT DATA:

T,K	VAPOR PRESSURE, KPA	SATURATED LIQUID VOLUME, ML/MOL	SECOND VIRIAL COEFFICIENT, ML/MOL	HEAT OF VAPORIZATION, J/MOL	SATURATED VAPOR VOLUME, ML/MOL
283	19.02				123711.
284	19.89				118712.
285	20.79				113956.
286	21.73				109428.
287	22.70				105116.
288	23.71				101008.
289	24.75				97093.
290	25.83				93361.
291	26.94				89802.
292	28.10				86407.
293	29.29	126.77		29557.	83167.
294	30.53	126.95		29521.	80074.
295	31.80	127.14		29486.	77120.
296	33.12	127.32		29451.	74299.
297	34.49	127.51		29416.	71602.
298	35.90	127.70		29381.	69025.
299	37.35	127.89		29345.	66560.
300	38.85	128.08		29310.	64203.
301	40.40	128.27		29275.	61947.
302	42.00	128.46		29240.	59788.
303	43.65	128.66		29205.	57721.
304	45.35	128.85		29170.	55741.
305	47.10				53844.
306	48.90				52026.
307	50.76				50284.
308	52.68				48613.
309	54.65				47010.
310	56.68				45473.
311	58.77				43997.
312	60.92				42580.
313	63.14				41219.
314	65.41				39912.
315	67.75				38657.
316	70.16				37450.
317	72.63				36290.
318	75.17				35174.
319	77.78				34101.
320	80.46				33068.
321	83.21				32074.
322	86.04				31118.
323	88.94				30197.
324	91.91				29309.
325	94.96				28455.
326	98.10				27631.
327	101.3				26837.

345

COMPOUND CONSTANTS:

MP	138.070 K	TC	VC	ZC
NBP	340.447 K	PC	DC	OM
MU		RG		

VAPOR PRESSURE CORRELATION CONSTANTS, K AND KPA:

RIEDEL	RIEDEL	RIEDEL	RIEDEL
291. - 341. K	291. - 341. K	291. - 341. K	291. - 341. K
13 POINTS	13 POINTS	13 POINTS	13 POINTS
RMSD = 0.0130	RMSD = 0.0130	RMSD = 0.0130	RMSD = 0.0130
0.56774757E+02	0.56774757E+02	0.56774757E+02	0.56774757E+02
-0.61179013E+01	-0.61179013E+01	-0.61179013E+01	-0.61179013E+01
-0.56351337E+04	-0.56351337E+04	-0.56351337E+04	-0.56351337E+04
0.41538943E-16	0.41538943E-16	0.41538943E-16	0.41538943E-16

LIQUID DENSITY CORRELATION CONSTANTS, K AND G/ML:

FRANCIS1	FRANCIS1	FRANCIS1
293. - 304. K	293. - 304. K	293. - 304. K
4 POINTS	4 POINTS	4 POINTS
RMSD = 0.00002	RMSD = 0.00002	RMSD = 0.00002
0.88569254E+00	0.88569254E+00	0.88569254E+00
0.51131030E-03	0.51131030E-03	0.51131030E-03
0.60000572E+01	0.60000572E+01	0.60000572E+01
0.41487866E+03	0.41487866E+03	0.41487866E+03

SECOND VIRIAL COEFFICIENT CORRELATION CONSTANTS, ML/MOL:

LITERATURE DOCUMENTS FOR CORRELATED VAPOR PRESSURE VALUES:

6669

LITERATURE DOCUMENTS FOR CORRELATED LIQUID DENSITY VALUES:

6669

LITERATURE DOCUMENTS REPORTING VIRIAL COEFFICIENT DATA:

T,K	VAPOR PRESSURE, KPA	SATURATED LIQUID VOLUME, ML/MOL	SECOND VIRIAL COEFFICIENT, ML/MOL	HEAT OF VAPORIZATION, J/MOL	SATURATED VAPOR VOLUME, ML/MOL
291	15.28				158362.
292	16.00				151784.
293	16.74	122.57		32306.	145532.
294	17.51	122.73		32263.	139587.
295	18.31	122.90		32220.	133932.
296	19.14	123.07		32178.	128551.
297	20.01	123.23		32136.	123428.
298	20.90	123.41		32094.	118550.
300	22.78	123.75		32010.	109474.
301	23.78	123.93		31968.	105252.
302	24.81	124.11		31927.	101226.
303	25.87	124.29		31885.	97385.
304	26.97	124.47		31844.	93720.
305	28.11				90221.
306	29.28				86879.
308	31.76				80636.
309	33.06				77720.
310	34.40				74931.
311	35.78				72264.
312	37.21				69712.
313	38.69				67269.
314	40.21				64930.
315	41.78				62691.
317	45.06				58488.
318	46.78				56517.
319	48.55				54626.
320	50.38				52813.
321	52.26				51074.
322	54.19				49404.
323	56.18				47801.
325	60.34				44784.
326	62.51				43364.
327	64.74				41999.
328	67.03				40686.
329	69.38				39424.
330	71.81				38211.
331	74.29				37043.
333	79.47				34838.
334	82.17				33796.
335	84.94				32793.
336	87.78				31827.
337	90.69				30896.
338	93.68				29999.
339	96.75				29134.
340	99.89				28300.

COMPOUND CONSTANTS:

MP	138.300 K	TC		VC		ZC
NBP	340.844 K	PC		DC		OM
MU		RG				

   VAPOR PRESSURE CORRELATION CONSTANTS, K AND KPA:

RPM2	RPM2	RPM2	RPM2
291. - 342. K	291. - 342. K	291. - 342. K	291. - 342. K
13 POINTS	13 POINTS	13 POINTS	13 POINTS
RMSD = 0.0100	RMSD = 0.0100	RMSD = 0.0100	RMSD = 0.0100
0.22109108E+02	0.22109108E+02	0.22109108E+02	0.22109108E+02
-0.46522069E+04	-0.46522069E+04	-0.46522069E+04	-0.46522069E+04
-0.13969393E-01	-0.13969393E-01	-0.13969393E-01	-0.13969393E-01
0.79163393E-05	0.79163393E-05	0.79163393E-05	0.79163393E-05

   LIQUID DENSITY CORRELATION CONSTANTS, K AND G/ML:

FRANCIS1	FRANCIS1	FRANCIS1
293. - 304. K	293. - 304. K	293. - 304. K
4 POINTS	4 POINTS	4 POINTS
RMSD = 0.00006	RMSD = 0.00006	RMSD = 0.00006
0.96257335E+00	0.96257335E+00	0.96257335E+00
0.80646062E-03	0.80646062E-03	0.80646062E-03
0.60296249E+01	0.60296249E+01	0.60296249E+01
0.50478857E+03	0.50478857E+03	0.50478857E+03

   SECOND VIRIAL COEFFICIENT CORRELATION CONSTANTS, ML/MOL:

LITERATURE DOCUMENTS FOR CORRELATED VAPOR PRESSURE VALUES:

   6669

LITERATURE DOCUMENTS FOR CORRELATED LIQUID DENSITY VALUES:

   6669

LITERATURE DOCUMENTS REPORTING VIRIAL COEFFICIENT DATA:

T,K	VAPOR PRESSURE, KPA	SATURATED LIQUID VOLUME, ML/MOL	SECOND VIRIAL COEFFICIENT, ML/MOL	HEAT OF VAPORIZATION, J/MOL	SATURATED VAPOR VOLUME, ML/MOL
291	15.29				158206.
292	16.00				151705.
293	16.74	120.61		31994.	145520.
294	17.51	120.77		31959.	139636.
295	18.30	120.93		31923.	134035.
296	19.12	121.10		31888.	128702.
297	19.98	121.26		31852.	123622.
299	21.78	121.60		31782.	114167.
300	22.72	121.76		31746.	109768.
301	23.71	121.93		31711.	105571.
302	24.72	122.10		31675.	101567.
303	25.77	122.27		31639.	97746.
304	26.86	122.44		31604.	94096.
306	29.15				87281.
307	30.35				84099.
308	31.59				81057.
309	32.88				78147.
310	34.20				75364.
311	35.57				72700.
313	38.43				67710.
314	39.94				65372.
315	41.49				63132.
316	43.08				60985.
317	44.73				58928.
318	46.42				56954.
319	48.17				55062.
321	51.82				51503.
322	53.73				49831.
323	55.69				48224.
324	57.71				46681.
325	59.78				45199.
326	61.92				43775.
328	66.37				41089.
329	68.69				39823.
330	71.07				38605.
331	73.52				37433.
332	76.03				36305.
333	78.61				35219.
335	83.98				33166.
336	86.77				32196.
337	89.63				31261.
338	92.57				30360.
339	95.58				29491.
340	98.66				28653.
341	101.8				27845.

COMPOUND CONSTANTS:

MP   134.693 K   TC   VC   ZC
NBP   343.580 K   PC   DC   OM
MU   RG

VAPOR PRESSURE CORRELATION CONSTANTS, K AND KPA:

RPM2	RPM2	RPM2	RPM2
290. - 345. K	290. - 345. K	290. - 345. K	290. - 345. K
14 POINTS	14 POINTS	14 POINTS	14 POINTS
RMSD = 0.0090	RMSD = 0.0090	RMSD = 0.0090	RMSD = 0.0090
0.26081103E+02	0.26081103E+02	0.26081103E+02	0.26081103E+02
-0.51511567E+04	-0.51511567E+04	-0.51511567E+04	-0.51511567E+04
-0.25554522E-01	-0.25554522E-01	-0.25554522E-01	-0.25554522E-01
0.19567312E-04	0.19567312E-04	0.19567312E-04	0.19567312E-04

LIQUID DENSITY CORRELATION CONSTANTS, K AND G/ML:

FRANCIS1	FRANCIS1	FRANCIS1
293. - 304. K	293. - 304. K	293. - 304. K
4 POINTS	4 POINTS	4 POINTS
RMSD = 0.00001	RMSD = 0.00001	RMSD = 0.00001
0.94920295E+00	0.94920295E+00	0.94920295E+00
0.76798303E-03	0.76798303E-03	0.76798303E-03
0.59999990E+01	0.59999990E+01	0.59999990E+01
0.48757422E+03	0.48757422E+03	0.48757422E+03

SECOND VIRIAL COEFFICIENT CORRELATION CONSTANTS, ML/MOL:

LITERATURE DOCUMENTS FOR CORRELATED VAPOR PRESSURE VALUES:

  6669

LITERATURE DOCUMENTS FOR CORRELATED LIQUID DENSITY VALUES:

  6669

LITERATURE DOCUMENTS REPORTING VIRIAL COEFFICIENT DATA:

T,K	VAPOR PRESSURE, KPA	SATURATED LIQUID VOLUME, ML/MOL	SECOND VIRIAL COEFFICIENT, ML/MOL	HEAT OF VAPORIZATION, J/MOL	SATURATED VAPOR VOLUME, ML/MOL
290	12.85				187650.
291	13.47				179678.
292	14.11				172108.
293	14.77	121.38		32749.	164917.
295	16.18	121.71		32666.	151585.
296	16.93	121.87		32624.	145406.
297	17.70	122.04		32583.	139527.
298	18.50	122.20		32542.	133932.
300	20.19	122.54		32459.	123531.
301	21.08	122.70		32419.	118696.
302	22.01	122.87		32378.	114088.
303	22.97	123.04		32337.	109695.
305	24.98				101506.
306	26.04				97690.
307	27.14				94046.
308	28.28				90566.
310	30.66				84064.
311	31.91				81026.
312	33.21				78121.
313	34.54				75342.
315	37.34				70137.
316	38.81				67699.
317	40.32				65364.
318	41.88				63128.
320	45.15				58929.
321	46.86				56958.
322	48.62				55067.
323	50.43				53254.
325	54.22				49841.
326	56.19				48237.
327	58.22				46695.
328	60.32				45214.
330	64.68				42422.
331	66.95				41106.
332	69.29				39840.
333	71.69				38622.
335	76.68				36322.
336	79.28				35236.
337	81.95				34190.
338	84.69				33182.
340	90.39				31276.
341	93.34				30374.
342	96.38				29504.
343	99.49				28665.
345	105.9				27076.

COMPOUND CONSTANTS:

MP	138.710 K	TC		VC		ZC	
NBP	329.526 K	PC		DC		OM	
MU		RG					

VAPOR PRESSURE CORRELATION CONSTANTS, K AND KPA:

RIEDEL	RIEDEL	RIEDEL	RIEDEL
289. - 330. K	289. - 330. K	289. - 330. K	289. - 330. K
11 POINTS	11 POINTS	11 POINTS	11 POINTS
RMSD = 0.0110	RMSD = 0.0110	RMSD = 0.0110	RMSD = 0.0110
0.59821857E+02	0.59821857E+02	0.59821857E+02	0.59821857E+02
-0.66446406E+01	-0.66446406E+01	-0.66446406E+01	-0.66446406E+01
-0.55242878E+04	-0.55242878E+04	-0.55242878E+04	-0.55242878E+04
0.65740830E-16	0.65740830E-16	0.65740830E-16	0.65740830E-16

LIQUID DENSITY CORRELATION CONSTANTS, K AND G/ML:

FRANCIS1	FRANCIS1	FRANCIS1
293. - 304. K	293. - 304. K	293. - 304. K
4 POINTS	4 POINTS	4 POINTS
RMSD = 0.00002	RMSD = 0.00002	RMSD = 0.00002
0.89651567E+00	0.89651567E+00	0.89651567E+00
0.62399684E-03	0.62399684E-03	0.62399684E-03
0.60182428E+01	0.60182428E+01	0.60182428E+01
0.42871606E+03	0.42871606E+03	0.42871606E+03

SECOND VIRIAL COEFFICIENT CORRELATION CONSTANTS, ML/MOL:

LITERATURE DOCUMENTS FOR CORRELATED VAPOR PRESSURE VALUES:

6669

LITERATURE DOCUMENTS FOR CORRELATED LIQUID DENSITY VALUES:

6669

LITERATURE DOCUMENTS REPORTING VIRIAL COEFFICIENT DATA:

T,K	VAPOR PRESSURE, KPA	SATURATED LIQUID VOLUME, ML/MOL	SECOND VIRIAL COEFFICIENT, ML/MOL	HEAT OF VAPORIZATION, J/MOL	SATURATED VAPOR VOLUME, ML/MOL
289	22.06				108941.
290	23.04				104637.
291	24.07				100538.
292	25.12				96633.
293	26.22	125.74		30311.	92911.
294	27.35	125.92		30269.	89362.
295	28.53	126.10		30227.	85978.
296	29.74	126.28		30185.	82748.
297	31.00	126.46		30144.	79666.
298	32.29	126.65		30102.	76723.
299	33.63	126.84		30061.	73912.
300	35.02	127.02		30021.	71227.
301	36.45	127.21		29981.	68660.
302	37.93	127.41		29940.	66206.
303	39.45	127.60		29901.	63858.
304	41.02	127.79		29861.	61613.
305	42.65				59463.
306	44.32				57406.
307	46.05				55435.
308	47.82				53547.
309	49.66				51738.
310	51.55				50004.
311	53.49				48341.
312	55.49				46746.
313	57.55				45216.
314	59.68				43748.
315	61.86				42338.
316	64.11				40984.
317	66.42				39683.
318	68.79				38434.
319	71.24				37233.
320	73.75				36078.
321	76.33				34968.
322	78.98				33899.
323	81.70				32872.
324	84.49				31883.
325	87.36				30930.
326	90.31				30013.
327	93.33				29130.
328	96.44				28279.
329	99.62				27459.
330	102.9				26668.

COMPOUND CONSTANTS:

MP	132.327 K	TC	VC	ZC
NBP	331.752 K	PC	DC	OM
MU		RG		

VAPOR PRESSURE CORRELATION CONSTANTS, K AND KPA:

RIEDEL	RIEDEL	RIEDEL	RIEDEL
292. - 333. K	292. - 333. K	292. - 333. K	292. - 333. K
11 POINTS	11 POINTS	11 POINTS	11 POINTS
RMSD = 0.0110	RMSD = 0.0110	RMSD = 0.0110	RMSD = 0.0110
0.50904933E+02	0.50904933E+02	0.50904933E+02	0.50904933E+02
-0.52738035E+01	-0.52738035E+01	-0.52738035E+01	-0.52738035E+01
-0.52138833E+04	-0.52138833E+04	-0.52138833E+04	-0.52138833E+04
0.30609820E-16	0.30609820E-16	0.30609820E-16	0.30609820E-16

LIQUID DENSITY CORRELATION CONSTANTS, K AND G/ML:

FRANCIS1	FRANCIS1	FRANCIS1
293. - 304. K	293. - 304. K	293. - 304. K
4 POINTS	4 POINTS	4 POINTS
RMSD = 0.00001	RMSD = 0.00001	RMSD = 0.00001
0.90145648E+00	0.90145648E+00	0.90145648E+00
0.64667105E-03	0.64667105E-03	0.64667105E-03
0.60067968E+01	0.60067968E+01	0.60067968E+01
0.43200732E+03	0.43200732E+03	0.43200732E+03

SECOND VIRIAL COEFFICIENT CORRELATION CONSTANTS, ML/MOL:

LITERATURE DOCUMENTS FOR CORRELATED VAPOR PRESSURE VALUES:

6669

LITERATURE DOCUMENTS FOR CORRELATED LIQUID DENSITY VALUES:

6669

LITERATURE DOCUMENTS REPORTING VIRIAL COEFFICIENT DATA:

T,K	VAPOR PRESSURE, KPA	SATURATED LIQUID VOLUME, ML/MOL	SECOND VIRIAL COEFFICIENT, ML/MOL	HEAT OF VAPORIZATION, J/MOL	SATURATED VAPOR VOLUME, ML/MOL
292	22.88				106132.
293	23.89	125.84		30748.	101982.
294	24.94	126.03		30709.	98028.
295	26.02	126.21		30671.	94257.
296	27.15	126.39		30632.	90662.
297	28.31	126.58		30594.	87232.
298	29.51	126.76		30556.	83958.
299	30.75	126.95		30518.	80833.
300	32.04	127.14		30480.	77848.
301	33.37	127.33		30442.	74997.
302	34.74	127.52		30404.	72273.
303	36.16	127.72		30366.	69668.
304	37.63	127.92		30329.	67177.
305	39.14				64795.
306	40.70				62515.
307	42.31				60333.
308	43.97				58244.
309	45.68				56243.
310	47.44				54326.
311	49.26				52488.
312	51.14				50727.
313	53.07				49038.
314	55.06				47418.
315	57.10				45864.
316	59.21				44372.
317	61.38				42940.
318	63.61				41564.
319	65.91				40243.
320	68.27				38974.
321	70.69				37754.
322	73.19				36581.
323	75.75				35453.
324	78.38				34368.
325	81.09				33324.
326	83.87				32320.
327	86.72				31353.
328	89.64				30422.
329	92.65				29525.
330	95.73				28661.
331	98.89				27829.
332	102.1				27026.
333	105.5				26253.

COMPOUND CONSTANTS:

MP	141.612 K	TC		VC		ZC
NBP	337.829 K	PC		DC		OM
MU		RG				

VAPOR PRESSURE CORRELATION CONSTANTS, K AND KPA:

RPM2	RPM2	RPM2	RPM2
289. - 339. K	289. - 339. K	289. - 339. K	289. - 339. K
13 POINTS	13 POINTS	13 POINTS	13 POINTS
RMSD = 0.0140	RMSD = 0.0140	RMSD = 0.0140	RMSD = 0.0140
0.17672819E+02	0.17672819E+02	0.17672819E+02	0.17672819E+02
-0.41741292E+04	-0.41741292E+04	-0.41741292E+04	-0.41741292E+04
0.63065177E-03	0.63065177E-03	0.63065177E-03	0.63065177E-03
-0.79893574E-05	-0.79893574E-05	-0.79893574E-05	-0.79893574E-05

LIQUID DENSITY CORRELATION CONSTANTS, K AND G/ML:

FRANCIS1	FRANCIS1	FRANCIS1
293. - 304. K	293. - 304. K	293. - 304. K
4 POINTS	4 POINTS	4 POINTS
RMSD = 0.00000	RMSD = 0.00000	RMSD = 0.00000
0.94698727E+00	0.94698727E+00	0.94698727E+00
0.75598271E-03	0.75598271E-03	0.75598271E-03
0.65917912E+01	0.65917912E+01	0.65917912E+01
0.47729443E+03	0.47729443E+03	0.47729443E+03

SECOND VIRIAL COEFFICIENT CORRELATION CONSTANTS, ML/MOL:

LITERATURE DOCUMENTS FOR CORRELATED VAPOR PRESSURE VALUES:

6669

LITERATURE DOCUMENTS FOR CORRELATED LIQUID DENSITY VALUES:

6669

LITERATURE DOCUMENTS REPORTING VIRIAL COEFFICIENT DATA:

T,K	VAPOR PRESSURE, KPA	SATURATED LIQUID VOLUME, ML/MOL	SECOND VIRIAL COEFFICIENT, ML/MOL	HEAT OF VAPORIZATION, J/MOL	SATURATED VAPOR VOLUME, ML/MOL
289	15.56				154466.
290	16.29				148060.
291	17.04				141969.
292	17.83				136175.
293	18.64	122.02		31784.	130662.
294	19.49	122.19		31752.	125415.
295	20.37	122.36		31719.	120418.
296	21.28	122.53		31686.	115658.
298	23.20	122.87		31619.	106800.
299	24.21	123.05		31585.	102679.
300	25.26	123.22		31551.	98747.
301	26.34	123.39		31516.	94996.
302	27.47	123.57		31482.	91416.
303	28.63	123.75		31447.	87998.
304	29.83	123.92		31411.	84734.
306	32.35				78635.
307	33.68				75786.
308	35.05				73061.
309	36.46				70456.
310	37.93				67962.
311	39.43				65576.
312	40.99				63291.
313	42.59				61103.
315	45.95				56999.
316	47.71				55074.
317	49.52				53229.
318	51.38				51459.
319	53.30				49761.
320	55.28				48132.
321	57.31				46568.
323	61.56				43626.
324	63.77				42241.
325	66.05				40910.
326	68.39				39632.
327	70.80				38403.
328	73.27				37221.
329	75.81				36084.
331	81.09				33938.
332	83.84				32925.
333	86.66				31950.
334	89.55				31011.
335	92.52				30106.
336	95.56				29235.
337	98.68				28395.
338	101.9				27585.

357

COMPOUND CONSTANTS:

MP   115.872 K   TC   VC   ZC
NBP   328.756 K   PC   DC   OM
MU   RG

VAPOR PRESSURE CORRELATION CONSTANTS, K AND KPA:

RPM2	RPM2	RPM2	RPM2
288. - 330. K	288. - 330. K	288. - 330. K	288. - 330. K
11 POINTS	11 POINTS	11 POINTS	11 POINTS
RMSD = 0.0110	RMSD = 0.0110	RMSD = 0.0110	RMSD = 0.0110
0.23747690E+02	0.23747690E+02	0.23747690E+02	0.23747690E+02
-0.46036601E+04	-0.46036601E+04	-0.46036601E+04	-0.46036601E+04
-0.20743270E-01	-0.20743270E-01	-0.20743270E-01	-0.20743270E-01
0.15668004E-04	0.15668004E-04	0.15668004E-04	0.15668004E-04

LIQUID DENSITY CORRELATION CONSTANTS, K AND G/ML:

FRANCIS1	FRANCIS1	FRANCIS1
293. - 304. K	293. - 304. K	293. - 304. K
7 POINTS	7 POINTS	7 POINTS
RMSD = 0.00006	RMSD = 0.00006	RMSD = 0.00006
0.95093602E+00	0.95093602E+00	0.95093602E+00
0.82232663E-03	0.82232663E-03	0.82232663E-03
0.76598463E+01	0.76598463E+01	0.76598463E+01
0.53344189E+03	0.53344189E+03	0.53344189E+03

SECOND VIRIAL COEFFICIENT CORRELATION CONSTANTS, ML/MOL:

LITERATURE DOCUMENTS FOR CORRELATED VAPOR PRESSURE VALUES:

6669

LITERATURE DOCUMENTS FOR CORRELATED LIQUID DENSITY VALUES:

6656   6669 21809

LITERATURE DOCUMENTS REPORTING VIRIAL COEFFICIENT DATA:

T,K	VAPOR PRESSURE, KPA	SATURATED LIQUID VOLUME, ML/MOL	SECOND VIRIAL COEFFICIENT, ML/MOL	HEAT OF VAPORIZATION, J/MOL	SATURATED VAPOR VOLUME, ML/MOL
288	21.94				109146.
289	22.92				104851.
290	23.93				100760.
291	24.98				96860.
292	26.07				93143.
293	27.19	124.11		29982.	89597.
294	28.35	124.28		29947.	86214.
295	29.56	124.46		29911.	82985.
296	30.80	124.63		29876.	79902.
297	32.09	124.81		29841.	76958.
298	33.42	124.99		29805.	74146.
299	34.79	125.17		29770.	71457.
300	36.21	125.35		29735.	68888.
301	37.67	125.53		29700.	66430.
302	39.19	125.71		29665.	64079.
303	40.75	125.89		29630.	61829.
304	42.36	126.07		29595.	59676.
305	44.02				57614.
306	45.73				55639.
307	47.49				53746.
308	49.31				51933.
309	51.18				50194.
310	53.11				48526.
311	55.10				46927.
312	57.15				45392.
313	59.25				43919.
314	61.42				42505.
315	63.65				41147.
316	65.94				39842.
317	68.30				38588.
318	70.73				37383.
319	73.22				36225.
320	75.78				35111.
321	78.41				34039.
322	81.11				33008.
323	83.88				32015.
324	86.73				31060.
325	89.65				30140.
326	92.65				29254.
327	95.73				28401.
328	98.89				27578.
329	102.1				26785.
330	105.4				26021.

COMPOUND CONSTANTS:

MP 157.958 K	TC	VC	ZC
NBP 314.396 K	PC	DC	OM
MU	RG		

VAPOR PRESSURE CORRELATION CONSTANTS, K AND KPA:

RIEDEL	RIEDEL	RIEDEL	RIEDEL
263. - 315. K	263. - 315. K	263. - 315. K	263. - 315. K
24 POINTS	24 POINTS	24 POINTS	24 POINTS
RMSD = 0.0280	RMSD = 0.0280	RMSD = 0.0280	RMSD = 0.0280
0.46340766E+02	0.46340766E+02	0.46340766E+02	0.46340766E+02
-0.46767584E+01	-0.46767584E+01	-0.46767584E+01	-0.46767584E+01
-0.46703854E+04	-0.46703854E+04	-0.46703854E+04	-0.46703854E+04
0.28055333E-16	0.28055333E-16	0.28055333E-16	0.28055333E-16

LIQUID DENSITY CORRELATION CONSTANTS, K AND G/ML:

FRANCIS1	FRANCIS1	FRANCIS1
293. - 314. K	293. - 314. K	293. - 314. K
9 POINTS	9 POINTS	9 POINTS
RMSD = 0.00008	RMSD = 0.00008	RMSD = 0.00008
0.94462693E+00	0.94462693E+00	0.94462693E+00
0.87856688E-03	0.87856688E-03	0.87856688E-03
0.93365412E+01	0.93365412E+01	0.93365412E+01
0.56677783E+03	0.56677783E+03	0.56677783E+03

SECOND VIRIAL COEFFICIENT CORRELATION CONSTANTS, ML/MOL:

LITERATURE DOCUMENTS FOR CORRELATED VAPOR PRESSURE VALUES:

6669 10403

LITERATURE DOCUMENTS FOR CORRELATED LIQUID DENSITY VALUES:

6656 6669 10403

LITERATURE DOCUMENTS REPORTING VIRIAL COEFFICIENT DATA:

T,K	VAPOR PRESSURE, KPA	SATURATED LIQUID VOLUME, ML/MOL	SECOND VIRIAL COEFFICIENT, ML/MOL	HEAT OF VAPORIZATION, J/MOL	SATURATED VAPOR VOLUME, ML/MOL
263	12.58				173787.
264	13.22				165985.
265	13.89				158599.
266	14.59				151606.
267	15.31				144981.
268	16.07				138701.
270	17.66				127099.
271	18.51				121739.
272	19.39				116650.
273	20.30				111817.
274	21.25				107225.
275	22.23				102859.
277	24.30				94759.
278	25.40				91001.
279	26.53				87422.
280	27.71				84015.
281	28.93				80768.
283	31.49				74722.
284	32.84				71908.
285	34.23				69222.
286	35.67				66659.
287	37.16				64211.
288	38.70				61873.
290	41.93				57504.
291	43.62				55463.
292	45.37				53510.
293	47.17	128.86		27629.	51641.
294	49.03	129.06		27594.	49853.
296	52.92	129.46		27523.	46501.
297	54.96	129.66		27489.	44930.
298	57.06	129.86		27454.	43425.
299	59.22	130.07		27419.	41982.
300	61.44	130.27		27384.	40598.
301	63.73	130.47		27349.	39271.
303	68.51	130.88		27280.	36775.
304	71.00	131.09		27246.	35601.
305	73.56	131.30		27211.	34474.
306	76.19	131.51		27177.	33391.
307	78.90	131.71		27143.	32351.
309	84.54	132.14		27074.	30390.
310	87.47	132.35		27040.	29465.
311	90.49	132.56		27006.	28576.
312	93.58	132.77		26972.	27720.
313	96.76	132.99		26939.	26897.
314	100.0	133.20		26905.	26104.

361

COMPOUND CONSTANTS:

MP	198.903 K	TC	VC	ZC
NBP	346.363 K	PC	DC	OM
MU		RG		

VAPOR PRESSURE CORRELATION CONSTANTS, K AND KPA:

RIEDEL	RIEDEL	RIEDEL	RIEDEL
288. - 348. K	288. - 348. K	288. - 348. K	288. - 348. K
40 POINTS	40 POINTS	40 POINTS	40 POINTS
RMSD = 0.0230	RMSD = 0.0230	RMSD = 0.0230	RMSD = 0.0230
0.45827710E+02	0.45827710E+02	0.45827710E+02	0.45827710E+02
-0.44403138E+01	-0.44403138E+01	-0.44403138E+01	-0.44403138E+01
-0.52818421E+04	-0.52818421E+04	-0.52818421E+04	-0.52818421E+04
0.27386044E-17	0.27386044E-17	0.27386044E-17	0.27386044E-17

LIQUID DENSITY CORRELATION CONSTANTS, K AND G/ML:

FRANCIS1	FRANCIS1	FRANCIS1
293. - 324. K	293. - 324. K	293. - 324. K
10 POINTS	10 POINTS	10 POINTS
RMSD = 0.00016	RMSD = 0.00016	RMSD = 0.00016
0.94494808E+00	0.94494808E+00	0.94494808E+00
0.67919795E-03	0.67919795E-03	0.67919795E-03
0.71775827E+01	0.71775827E+01	0.71775827E+01
0.48310547E+03	0.48310547E+03	0.48310547E+03

SECOND VIRIAL COEFFICIENT CORRELATION CONSTANTS, ML/MOL:

NOTHNAGEL ET AL.: TC=521.150   NBP=346.363   B=258.5   D=0.328

LITERATURE DOCUMENTS FOR CORRELATED VAPOR PRESSURE VALUES:

6450   6669 10403 10749

LITERATURE DOCUMENTS FOR CORRELATED LIQUID DENSITY VALUES:

6669 10403 21809

LITERATURE DOCUMENTS REPORTING VIRIAL COEFFICIENT DATA:

6450

T,K	VAPOR PRESSURE, KPA	SATURATED LIQUID VOLUME, ML/MOL	SECOND VIRIAL COEFFICIENT, ML/MOL	HEAT OF VAPORIZATION, J/MOL	SATURATED VAPOR VOLUME, ML/MOL
288	10.43		-2151.		227518.
289	10.94		-2128.		217514.
290	11.47		-2105.		208026.
292	12.61		-2059.		190477.
293	13.21	118.84	-2037.	32732.	182363.
294	13.83	118.99	-2016.	32683.	174655.
296	15.16	119.29	-1973.	32584.	160366.
297	15.86	119.44	-1953.	32534.	153744.
298	16.58	119.59	-1932.	32484.	147444.
300	18.12	119.89	-1893.	32383.	135741.
301	18.93	120.04	-1873.	32332.	130306.
302	19.77	120.20	-1854.	32280.	125127.
304	21.54	120.51	-1817.	32176.	115488.
305	22.48	120.66	-1798.	32124.	111002.
307	24.44	120.98	-1763.	32018.	102638.
308	25.48	121.14	-1745.	31965.	98740.
309	26.55	121.30	-1728.	31912.	95017.
311	28.80	121.62	-1695.	31803.	88064.
312	29.98	121.78	-1678.	31749.	84816.
313	31.20	121.94	-1662.	31694.	81712.
315	33.76	122.27	-1631.	31583.	75902.
316	35.11	122.44	-1615.	31527.	73184.
317	36.50	122.61	-1600.	31470.	70582.
319	39.40	122.94	-1570.	31356.	65704.
320	40.92	123.11	-1555.	31299.	63418.
322	44.11	123.46	-1527.	31183.	59126.
323	45.78	123.63	-1513.	31125.	57112.
324	47.49	123.81	-1499.	31066.	55181.
326	51.08		-1472.		51549.
327	52.95		-1459.		49842.
328	54.88		-1446.		48202.
330	58.90		-1420.		45115.
331	61.00		-1408.		43662.
332	63.16		-1395.		42265.
334	67.65		-1371.		39630.
335	69.99		-1359.		38388.
337	74.86		-1336.		36043.
338	77.39		-1324.		34935.
339	79.99		-1313.		33869.
341	85.40		-1291.		31853.
342	88.21		-1280.		30900.
343	91.09		-1270.		29982.
345	97.07		-1249.		28243.
346	100.2		-1238.		27420.
347	103.4		-1228.		26626.

COMPOUND CONSTANTS:

MP   154.119 K   TC   537.295 K   VC   ZC
NBP   366.774 K   PC   DC   OM
MU   0.3400 DEBYE   RG   4.0971 ANGSTROM

    VAPOR PRESSURE CORRELATION CONSTANTS, K AND KPA:

    RPM2   RPM2   RPM2
    255. - 368. K   255. - 368. K   255. - 368. K
     31 POINTS   31 POINTS   31 POINTS
    RMSD = 0.0475   RMSD = 0.0475   RMSD = 0.0475
    0.32851183E+02   0.32851183E+02   0.32851183E+02
   -0.62502961E+04   -0.62502961E+04   -0.62502961E+04
   -0.44332128E-01   -0.44332128E-01   -0.44332128E-01
    0.37679095E-04   0.37679095E-04   0.37679095E-04

    LIQUID DENSITY CORRELATION CONSTANTS, K AND G/ML:

    FRANCIS1   FRANCIS1   FRANCIS1
    283. - 304. K   283. - 304. K   283. - 304. K
     4 POINTS   4 POINTS   4 POINTS
    RMSD = 0.00026   RMSD = 0.00026   RMSD = 0.00026
    0.97429627E+00   0.97429627E+00   0.97429627E+00
    0.87047019E-03   0.87047019E-03   0.87047019E-03
    0.59999990E+01   0.59999990E+01   0.59999990E+01
    0.56838794E+03   0.56838794E+03   0.56838794E+03

    SECOND VIRIAL COEFFICIENT CORRELATION CONSTANTS, ML/MOL:

    NOTHNAGEL ET AL.: TC=537.295 NBP=366.774 B=300.0 D=0.344

LITERATURE DOCUMENTS FOR CORRELATED VAPOR PRESSURE VALUES:

    1991   5056 17432 17434

LITERATURE DOCUMENTS FOR CORRELATED LIQUID DENSITY VALUES:

    1991 11064

LITERATURE DOCUMENTS REPORTING VIRIAL COEFFICIENT DATA:

    1901

T,K	VAPOR PRESSURE, KPA	SATURATED LIQUID VOLUME, ML/MOL	SECOND VIRIAL COEFFICIENT, ML/MOL	HEAT OF VAPORIZATION, J/MOL	SATURATED VAPOR VOLUME, ML/MOL
255	0.5979		-4984.		3541253.
257	0.6882		-4836.		3100227.
260	0.8456		-4625.		2551714.
262	0.9671		-4492.		2248046.
265	1.177		-4302.		1867307.
267	1.338		-4182.		1654818.
270	1.615		-4011.		1386370.
272	1.825		-3903.		1235415.
275	2.184		-3749.		1043339.
278	2.601		-3603.		885161.
280	2.915		-3510.		795216.
283	3.445	138.89	-3378.	36458.	679552.
285	3.843	139.27	-3294.	36328.	613357.
288	4.510	139.83	-3173.	36134.	527712.
290	5.007	140.21	-3096.	36005.	478399.
293	5.839	140.78	-2986.	35812.	414227.
296	6.782	141.36	-2882.	35620.	359967.
298	7.479	141.75	-2815.	35493.	328442.
301	8.635	142.33	-2719.	35304.	287065.
303	9.486	142.73	-2658.	35178.	262900.
306	10.89		-2570.		231024.
308	11.92		-2514.		212315.
311	13.61		-2433.		187521.
314	15.49		-2355.		166102.
316	16.87		-2306.		153438.
319	19.11		-2234.		136536.
321	20.73		-2189.		126500.
324	23.38		-2123.		113051.
326	25.29		-2080.		105034.
329	28.40		-2019.		94247.
332	31.81		-1960.		84764.
334	34.26		-1922.		79075.
337	38.23		-1868.		71379.
339	41.07		-1832.		66746.
342	45.65		-1782.		60456.
344	48.92		-1749.		56657.
347	54.19		-1701.		51483.
350	59.90		-1656.		46867.
352	63.97		-1626.		44065.
355	70.48		-1583.		40231.
357	75.11		-1556.		37896.
360	82.51		-1516.		34692.
362	87.76		-1490.		32736.
365	96.13		-1453.		30044.
367	102.1		-1428.		28397.

COMPOUND CONSTANTS:

MP	TC	VC	ZC
NBP 371.593 K	PC	DC	OM
MU	RG		

VAPOR PRESSURE CORRELATION CONSTANTS, K AND KPA:

RIEDEL	RIEDEL	RIEDEL	RIEDEL
314. - 372. K	314. - 372. K	314. - 372. K	357. - 372. K
12 POINTS	12 POINTS	12 POINTS	9 POINTS
RMSD = 0.0073	RMSD = 0.0073	RMSD = 0.0073	RMSD = 0.0081
0.59799445E+02	0.59799445E+02	0.59799445E+02	-0.10390534E+02
-0.64681822E+01	-0.64681822E+01	-0.64681822E+01	0.39445785E+01
-0.63076214E+04	-0.63076214E+04	-0.63076214E+04	-0.30254806E+04
0.26902528E-16	0.26902528E-16	0.26902528E-16	-0.73094134E-16

LIQUID DENSITY CORRELATION CONSTANTS, K AND G/ML:

FRANCIS1	FRANCIS1	FRANCIS1
288. - 299. K	288. - 299. K	288. - 299. K
4 POINTS	4 POINTS	4 POINTS
RMSD = 0.00040	RMSD = 0.00040	RMSD = 0.00040
0.94120413E+00	0.94120413E+00	0.94120413E+00
0.72130188E-03	0.72130188E-03	0.72130188E-03
0.59999990E+01	0.59999990E+01	0.59999990E+01
0.55903076E+03	0.55903076E+03	0.55903076E+03

SECOND VIRIAL COEFFICIENT CORRELATION CONSTANTS, ML/MOL:

LITERATURE DOCUMENTS FOR CORRELATED VAPOR PRESSURE VALUES:

17432

LITERATURE DOCUMENTS FOR CORRELATED LIQUID DENSITY VALUES:

3269 10284

LITERATURE DOCUMENTS REPORTING VIRIAL COEFFICIENT DATA:

T,K	VAPOR PRESSURE, KPA	SATURATED LIQUID VOLUME, ML/MOL	SECOND VIRIAL COEFFICIENT, ML/MOL	HEAT OF VAPORIZATION, J/MOL	SATURATED VAPOR VOLUME, ML/MOL
288		138.03			
289		138.19			
291		138.50			
293		138.82			
295		139.14			
297		139.46			
299		139.78			
314	12.80				204024.
316	13.96				188202.
318	15.21				173827.
320	16.55				160750.
322	17.99				148836.
324	19.53				137970.
326	21.17				128046.
328	22.92				118971.
329	23.84				114726.
331	25.78				106772.
333	27.83				99478.
335	30.02				92780.
337	32.35				86622.
339	34.82				80955.
341	37.44				75734.
343	40.21				70919.
345	43.15				66473.
347	46.26				62364.
349	49.55				58562.
350	51.26				56768.
352	54.83				53379.
354	58.59				50235.
356	62.56				47317.
358	66.73				44605.
360	71.13				42083.
362	75.75				39736.
364	80.60				37548.
366	85.70				35508.
368	91.05				33604.
370	96.66				31825.
371	99.57				30980.

COMPOUND CONSTANTS:

```
MP    163.684 K        TC                      VC                    ZC
NBP   371.096 K        PC                      DC                    OM
MU                     RG    4.1878 ANGSTROM
```

VAPOR PRESSURE CORRELATION CONSTANTS, K AND KPA:

RPM2	RPM2	RPM2	RIEDEL
314. - 372. K	314. - 372. K	314. - 372. K	357. - 372. K
13 POINTS	13 POINTS	13 POINTS	9 POINTS
RMSD = 0.0089	RMSD = 0.0089	RMSD = 0.0089	RMSD = 0.0100
0.29449197E+02	0.29449197E+02	0.29449197E+02	-0.28648182E+02
-0.59875699E+04	-0.59875699E+04	-0.59875699E+04	0.66650531E+01
-0.32850163E-01	-0.32850163E-01	-0.32850163E-01	-0.21873331E+04
0.25375771E-04	0.25375771E-04	0.25375771E-04	-0.10444024E-15

LIQUID DENSITY CORRELATION CONSTANTS, K AND G/ML:

FRANCIS1	FRANCIS1	FRANCIS1
288. - 299. K	288. - 299. K	288. - 299. K
4 POINTS	4 POINTS	4 POINTS
RMSD = 0.00010	RMSD = 0.00010	RMSD = 0.00010
0.90266448E+00	0.90266448E+00	0.90266448E+00
0.51654549E-03	0.51654549E-03	0.51654549E-03
0.60013981E+01	0.60013981E+01	0.60013981E+01
0.41246802E+03	0.41246802E+03	0.41246802E+03

SECOND VIRIAL COEFFICIENT CORRELATION CONSTANTS, ML/MOL:

LITERATURE DOCUMENTS FOR CORRELATED VAPOR PRESSURE VALUES:

17432

LITERATURE DOCUMENTS FOR CORRELATED LIQUID DENSITY VALUES:

3269 10284 20206

LITERATURE DOCUMENTS REPORTING VIRIAL COEFFICIENT DATA:

T,K	VAPOR PRESSURE, KPA	SATURATED LIQUID VOLUME, ML/MOL	SECOND VIRIAL COEFFICIENT, ML/MOL	HEAT OF VAPORIZATION, J/MOL	SATURATED VAPOR VOLUME, ML/MOL
288		139.14			
289		139.32			
291		139.68			
293		140.05			
295		140.43			
297		140.82			
299		141.21			
314	13.03				200313.
316	14.22				184799.
318	15.49				170704.
320	16.85				157881.
322	18.31				146200.
324	19.87				135544.
326	21.54				125812.
328	23.33				116912.
329	24.26				112748.
331	26.22				104947.
333	28.31				97791.
335	30.53				91220.
337	32.90				85179.
339	35.40				79618.
341	38.06				74494.
343	40.88				69767.
345	43.86				65403.
347	47.01				61369.
349	50.35				57635.
350	52.08				55874.
352	55.70				52545.
354	59.51				49457.
356	63.53				46589.
358	67.76				43925.
360	72.22				41446.
362	76.90				39138.
364	81.82				36988.
366	86.99				34982.
368	92.41				33109.
370	98.10				31360.
371	101.0				30529.

369

COMPOUND CONSTANTS:

MP	TC	VC	ZC
NBP 368.777 K	PC	DC	OM
MU	RG		

VAPOR PRESSURE CORRELATION CONSTANTS, K AND KPA:

RIEDEL	RIEDEL	RIEDEL	RPM2
312. - 369. K	312. - 369. K	312. - 369. K	355. - 369. K
13 POINTS	13 POINTS	13 POINTS	9 POINTS
RMSD = 0.0068	RMSD = 0.0068	RMSD = 0.0068	RMSD = 0.0071
0.56039878E+02	0.56039878E+02	0.56039878E+02	0.62270403E+02
-0.59202884E+01	-0.59202884E+01	-0.59202884E+01	-0.98840763E+04
-0.60808465E+04	-0.60808465E+04	-0.60808465E+04	-0.12427223E+00
0.22952427E-16	0.22952427E-16	0.22952427E-16	0.11014290E-03

LIQUID DENSITY CORRELATION CONSTANTS, K AND G/ML:

FRANCIS1	FRANCIS1	FRANCIS1
288. - 299. K	288. - 299. K	288. - 299. K
4 POINTS	4 POINTS	4 POINTS
RMSD = 0.00081	RMSD = 0.00081	RMSD = 0.00081
0.90376961E+00	0.90376961E+00	0.90376961E+00
0.51001902E-03	0.51001902E-03	0.51001902E-03
0.68899994E+01	0.68899994E+01	0.68899994E+01
0.42913818E+03	0.42913818E+03	0.42913818E+03

SECOND VIRIAL COEFFICIENT CORRELATION CONSTANTS, ML/MOL:

LITERATURE DOCUMENTS FOR CORRELATED VAPOR PRESSURE VALUES:

17437

LITERATURE DOCUMENTS FOR CORRELATED LIQUID DENSITY VALUES:

25   3269   6633

LITERATURE DOCUMENTS REPORTING VIRIAL COEFFICIENT DATA:

T,K	VAPOR PRESSURE, KPA	SATURATED LIQUID VOLUME, ML/MOL	SECOND VIRIAL COEFFICIENT, ML/MOL	HEAT OF VAPORIZATION, J/MOL	SATURATED VAPOR VOLUME, ML/MOL
288		138.67			
289		138.84			
291		139.18			
293		139.53			
295		139.88			
297		140.24			
299		140.60			
313	13.66				190478.
315	14.90				175814.
317	16.22				162481.
319	17.64				150344.
321	19.16				139280.
322	19.96				134116.
324	21.64				124464.
326	23.44				115638.
328	25.36				107557.
330	27.40				100150.
332	29.57				93353.
334	31.88				87108.
335	33.09				84176.
337	35.62				78665.
339	38.30				73588.
341	41.15				68905.
343	44.16				64582.
345	47.35				60586.
346	49.01				58702.
348	52.47				55145.
350	56.12				51850.
352	59.98				48794.
354	64.04				45957.
356	68.33				43321.
357	70.55				42073.
359	75.17				39708.
361	80.03				37506.
363	85.13				35453.
365	90.49				33538.
367	96.11				31750.
368	99.02				30901.

COMPOUND CONSTANTS:

MP	136.509 K	TC		VC		ZC
NBP	368.856 K	PC		DC		OM
MU		RG	4.1702 ANGSTROM			

VAPOR PRESSURE CORRELATION CONSTANTS, K AND KPA:

RPM2	RPM2	RPM2	RIEDEL
312. - 369. K	312. - 369. K	312. - 369. K	355. - 369. K
13 POINTS	13 POINTS	13 POINTS	9 POINTS
RMSD = 0.0083	RMSD = 0.0083	RMSD = 0.0083	RMSD = 0.0077
0.28290551E+02	0.28290551E+02	0.28290551E+02	0.15788659E+03
-0.58057272E+04	-0.58057272E+04	-0.58057272E+04	-0.21071332E+02
-0.29952272E-01	-0.29952272E-01	-0.29952272E-01	-0.10769269E+05
0.22900241E-04	0.22900241E-04	0.22900241E-04	0.18594616E-15

LIQUID DENSITY CORRELATION CONSTANTS, K AND G/ML:

FRANCIS1	FRANCIS1	FRANCIS1
288. - 299. K	288. - 299. K	288. - 299. K
4 POINTS	4 POINTS	4 POINTS
RMSD = 0.00020	RMSD = 0.00020	RMSD = 0.00020
0.96057600E+00	0.96057600E+00	0.96057600E+00
0.81057707E-03	0.81057707E-03	0.81057707E-03
0.66354876E+01	0.66354876E+01	0.66354876E+01
0.55903076E+03	0.55903076E+03	0.55903076E+03

SECOND VIRIAL COEFFICIENT CORRELATION CONSTANTS, ML/MOL:

LITERATURE DOCUMENTS FOR CORRELATED VAPOR PRESSURE VALUES:

17432

LITERATURE DOCUMENTS FOR CORRELATED LIQUID DENSITY VALUES:

3269 10284

LITERATURE DOCUMENTS REPORTING VIRIAL COEFFICIENT DATA:

T,K	VAPOR PRESSURE, KPA	SATURATED LIQUID VOLUME, ML/MOL	SECOND VIRIAL COEFFICIENT, ML/MOL	HEAT OF VAPORIZATION, J/MOL	SATURATED VAPOR VOLUME, ML/MOL
288		139.74			
289		139.92			
291		140.28			
293		140.64			
295		141.01			
297		141.38			
299		141.75			
313	13.60				191299.
315	14.83				176546.
317	16.16				163136.
319	17.57				150931.
321	19.09				139808.
322	19.89				134618.
324	21.57				124918.
326	23.36				116049.
328	25.27				107930.
330	27.30				100490.
332	29.47				93663.
334	31.78				87391.
335	32.98				84448.
337	35.51				78914.
339	38.18				73816.
341	41.02				69115.
343	44.03				64775.
345	47.21				60765.
346	48.86				58874.
348	52.32				55304.
350	55.97				51997.
352	59.81				48930.
354	63.87				46082.
356	68.14				43437.
357	70.36				42185.
359	74.97				39812.
361	79.82				37602.
363	84.92				35542.
365	90.27				33620.
367	95.88				31826.
368	98.78				30974.

COMPOUND CONSTANTS:

MP	TC	VC	ZC
NBP 368.554 K	PC	DC	OM
MU	RG		

VAPOR PRESSURE CORRELATION CONSTANTS, K AND KPA:

RIEDEL	RIEDEL	RIEDEL
302. - 369. K	302. - 369. K	302. - 369. K
12 POINTS	12 POINTS	12 POINTS
RMSD = 0.0205	RMSD = 0.0205	RMSD = 0.0205
0.53403690E+02	0.53403690E+02	0.53403690E+02
-0.54949813E+01	-0.54949813E+01	-0.54949813E+01
-0.60166362E+04	-0.60166362E+04	-0.60166362E+04
0.50693729E-17	0.50693729E-17	0.50693729E-17

LIQUID DENSITY CORRELATION CONSTANTS, K AND G/ML:

FRANCIS1	FRANCIS1	FRANCIS1
293. - 304. K	293. - 304. K	293. - 304. K
4 POINTS	4 POINTS	4 POINTS
RMSD = 0.00036	RMSD = 0.00036	RMSD = 0.00036
0.98526138E+00	0.98526138E+00	0.98526138E+00
0.84382435E-03	0.84382435E-03	0.84382435E-03
0.68792238E+01	0.68792238E+01	0.68792238E+01
0.56838794E+03	0.56838794E+03	0.56838794E+03

SECOND VIRIAL COEFFICIENT CORRELATION CONSTANTS, ML/MOL:

LITERATURE DOCUMENTS FOR CORRELATED VAPOR PRESSURE VALUES:

6518

LITERATURE DOCUMENTS FOR CORRELATED LIQUID DENSITY VALUES:

6518

LITERATURE DOCUMENTS REPORTING VIRIAL COEFFICIENT DATA:

T,K	VAPOR PRESSURE, KPA	SATURATED LIQUID VOLUME, ML/MOL	SECOND VIRIAL COEFFICIENT, ML/MOL	HEAT OF VAPORIZATION, J/MOL	SATURATED VAPOR VOLUME, ML/MOL
293		137.70			
294		137.88			
296		138.25			
298		138.61			
299		138.80			
301		139.17			
303	8.618	139.54		36223.	292320.
305	9.470				267793.
306	9.921				256448.
308	10.88				235426.
310	11.91				216423.
311	12.46				207609.
313	13.61				191233.
315	14.85				176378.
317	16.18				162884.
318	16.88				156603.
320	18.36				144891.
322	19.95				134217.
324	21.64				124477.
325	22.53				119928.
327	24.40				111418.
329	26.40				103628.
330	27.44				99981.
332	29.64				93145.
334	31.97				86868.
336	34.45				81100.
337	35.74				78392.
339	38.45				73300.
341	41.33				68607.
343	44.37				64277.
344	45.95				62239.
346	49.26				58395.
348	52.76				54840.
349	54.58				53163.
351	58.37				49995.
353	62.37				47058.
355	66.58				44333.
356	68.77				43043.
358	73.31				40602.
360	78.09				38332.
362	83.10				36218.
363	85.70				35216.
365	91.10				33314.
367	96.75				31539.
368	99.68				30697.

COMPOUND CONSTANTS:

MP		TC		VC		ZC
NBP	366.690 K	PC		DC		OM
MU		RG				

VAPOR PRESSURE CORRELATION CONSTANTS, K AND KPA:

RIEDEL	RIEDEL	RIEDEL	RIEDEL
300. - 368. K	300. - 368. K	300. - 368. K	346. - 368. K
15 POINTS	15 POINTS	15 POINTS	7 POINTS
RMSD = 0.0060	RMSD = 0.0060	RMSD = 0.0060	RMSD = 0.0036
0.59484180E+02	0.59484180E+02	0.59484180E+02	0.83509890E+02
-0.64332478E+01	-0.64332478E+01	-0.64332478E+01	-0.10010252E+02
-0.62123780E+04	-0.62123780E+04	-0.62123780E+04	-0.73132241E+04
0.25134786E-16	0.25134786E-16	0.25134786E-16	0.65000752E-16

LIQUID DENSITY CORRELATION CONSTANTS, K AND G/ML:

FRANCIS1	FRANCIS1	FRANCIS1
293. - 304. K	293. - 304. K	293. - 304. K
4 POINTS	4 POINTS	4 POINTS
RMSD = 0.00024	RMSD = 0.00024	RMSD = 0.00024
0.96293306E+00	0.96293306E+00	0.96293306E+00
0.79076923E-03	0.79076923E-03	0.79076923E-03
0.59999990E+01	0.59999990E+01	0.59999990E+01
0.56838794E+03	0.56838794E+03	0.56838794E+03

SECOND VIRIAL COEFFICIENT CORRELATION CONSTANTS, ML/MOL:

LITERATURE DOCUMENTS FOR CORRELATED VAPOR PRESSURE VALUES:

6518

LITERATURE DOCUMENTS FOR CORRELATED LIQUID DENSITY VALUES:

6518

LITERATURE DOCUMENTS REPORTING VIRIAL COEFFICIENT DATA:

T,K	VAPOR PRESSURE, KPA	SATURATED LIQUID VOLUME, ML/MOL	SECOND VIRIAL COEFFICIENT, ML/MOL	HEAT OF VAPORIZATION, J/MOL	SATURATED VAPOR VOLUME, ML/MOL
293		138.40			
294		138.57			
296		138.91			
298		139.26			
299		139.43			
301	8.572	139.77		35816.	291958.
303	9.420	140.12		35720.	267428.
304	9.870	140.30		35673.	256085.
306	10.82				235072.
308	11.85				216083.
310	12.96				198899.
311	13.54				190921.
313	14.78				176087.
315	16.11				162615.
316	16.80				156345.
318	18.28				144655.
320	19.85				134003.
321	20.68				129033.
323	22.43				119744.
325	24.29				111253.
327	26.27				103481.
328	27.31				99843.
330	29.50				93021.
332	31.82				86759.
333	33.03				83822.
335	35.57				78302.
337	38.27				73220.
339	41.13				68536.
340	42.62				66332.
342	45.73				62180.
344	49.02				58343.
345	50.74				56534.
347	54.31				53118.
349	58.09				49954.
350	60.05				48459.
352	64.14				45633.
354	68.44				43007.
356	72.96				40567.
357	75.31				39412.
359	80.19				37221.
361	85.32				35180.
362	87.98				34212.
364	93.49				32373.
366	99.26				30656.
367	102.3				29841.

COMPOUND CONSTANTS:

MP 149.089 K	TC	VC	ZC
NBP 354.755 K	PC	DC	OM
MU	RG		

VAPOR PRESSURE CORRELATION CONSTANTS, K AND KPA:

VAPRES-2	VAPRES-2	VAPRES-2	RPM2
289. - 356. K	289. - 356. K	289. - 356. K	334. - 356. K
15 POINTS	15 POINTS	15 POINTS	7 POINTS
RMSD = 0.0156	RMSD = 0.0156	RMSD = 0.0156	RMSD = 0.0042
-0.21701146E+04	-0.21701146E+04	-0.21701146E+04	0.26838750E+02
0.39571151E+05	0.39571151E+05	0.39571151E+05	-0.53663789E+04
-0.13095066E+01	-0.13095066E+01	-0.13095066E+01	-0.27293163E-01
0.67857429E-03	0.67857429E-03	0.67857429E-03	0.20571718E-04
0.41597086E+03	0.41597086E+03	0.41597086E+03	

LIQUID DENSITY CORRELATION CONSTANTS, K AND G/ML:

FRANCIS1	FRANCIS1	FRANCIS1
293. - 304. K	293. - 304. K	293. - 304. K
4 POINTS	4 POINTS	4 POINTS
RMSD = 0.00015	RMSD = 0.00015	RMSD = 0.00015
0.94550091E+00	0.94550091E+00	0.94550091E+00
0.77933352E-03	0.77933352E-03	0.77933352E-03
0.60571909E+01	0.60571909E+01	0.60571909E+01
0.55508350E+03	0.55508350E+03	0.55508350E+03

SECOND VIRIAL COEFFICIENT CORRELATION CONSTANTS, ML/MOL:

LITERATURE DOCUMENTS FOR CORRELATED VAPOR PRESSURE VALUES:

6518

LITERATURE DOCUMENTS FOR CORRELATED LIQUID DENSITY VALUES:

6518   6633

LITERATURE DOCUMENTS REPORTING VIRIAL COEFFICIENT DATA:

T,K	VAPOR PRESSURE, KPA	SATURATED LIQUID VOLUME, ML/MOL	SECOND VIRIAL COEFFICIENT, ML/MOL	HEAT OF VAPORIZATION, J/MOL	SATURATED VAPOR VOLUME, ML/MOL
289	8.215				292493.
290	8.620				279723.
292	9.480				256085.
293	9.937	141.47		33453.	245148.
295	10.91	141.83		33417.	224878.
296	11.42	142.01		33396.	215487.
298	12.51	142.36		33348.	198059.
299	13.09	142.54		33322.	189973.
301	14.30	142.91		33265.	174949.
302	14.95	143.09		33234.	167969.
304	16.31	143.45		33169.	154983.
305	17.03				148941.
307	18.54				137686.
308	19.34				132443.
310	21.01				122662.
311	21.89				118100.
313	23.75				109577.
314	24.72				105597.
316	26.77				98151.
317	27.84				94668.
319	30.09				88146.
320	31.27				85091.
322	33.73				79362.
324	36.35				74101.
325	37.72				71631.
327	40.59				66989.
328	42.08				64808.
330	45.20				60702.
331	46.83				58770.
333	50.22				55130.
334	51.99				53415.
336	55.67				50178.
337	57.59				48651.
339	61.59				45767.
340	63.66				44404.
342	67.98				41827.
343	70.23				40608.
345	74.90				38299.
346	77.32				37206.
348	82.36				35132.
349	84.97				34150.
351	90.40				32283.
352	93.21				31398.
354	99.06				29714.
355	102.1				28913.

COMPOUND CONSTANTS:

MP	136.539 K	TC		VC		ZC
NBP	345.660 K	PC		DC		OM
MU		RG				

VAPOR PRESSURE CORRELATION CONSTANTS, K AND KPA:

RPM2	RPM2	RPM2	RIEDEL
290. - 347. K	290. - 347. K	290. - 347. K	326. - 347. K
14 POINTS	14 POINTS	14 POINTS	7 POINTS
RMSD = 0.0066	RMSD = 0.0066	RMSD = 0.0066	RMSD = 0.0041
0.27743792E+02	0.27743792E+02	0.27743792E+02	0.63882248E+02
-0.52541164E+04	-0.52541164E+04	-0.52541164E+04	-0.72346853E+01
-0.32056592E-01	-0.32056592E-01	-0.32056592E-01	-0.59014724E+04
0.26409924E-04	0.26409924E-04	0.26409924E-04	0.58159025E-16

LIQUID DENSITY CORRELATION CONSTANTS, K AND G/ML:

FRANCIS1	FRANCIS1	FRANCIS1
273. - 304. K	273. - 304. K	273. - 304. K
5 POINTS	5 POINTS	5 POINTS
RMSD = 0.00013	RMSD = 0.00013	RMSD = 0.00013
0.94797796E+00	0.94797796E+00	0.94797796E+00
0.83128386E-03	0.83128386E-03	0.83128386E-03
0.59999990E+01	0.59999990E+01	0.59999990E+01
0.56838794E+03	0.56838794E+03	0.56838794E+03

SECOND VIRIAL COEFFICIENT CORRELATION CONSTANTS, ML/MOL:

LITERATURE DOCUMENTS FOR CORRELATED VAPOR PRESSURE VALUES:

   6518

LITERATURE DOCUMENTS FOR CORRELATED LIQUID DENSITY VALUES:

   6518 9865

LITERATURE DOCUMENTS REPORTING VIRIAL COEFFICIENT DATA:

T,K	VAPOR PRESSURE, KPA	SATURATED LIQUID VOLUME, ML/MOL	SECOND VIRIAL COEFFICIENT, ML/MOL	HEAT OF VAPORIZATION, J/MOL	SATURATED VAPOR VOLUME, ML/MOL
273		140.12			
274		140.30			
276		140.67			
278		141.03			
279		141.21			
281		141.58			
283		141.95			
284		142.14			
286		142.51			
288		142.89			
289		143.08			
291	13.41	143.46		31911.	180360.
293	14.68	143.84		31822.	165973.
294	15.34	144.03		31778.	159300.
296	16.75	144.42		31691.	146903.
298	18.26	144.81		31604.	135655.
299	19.06	145.00		31561.	130423.
301	20.74	145.39		31475.	120676.
303	22.53	145.79		31391.	111802.
304	23.48	145.99		31349.	107663.
306	25.46				99933.
308	27.57				92870.
309	28.68				89568.
311	31.01				83384.
313	33.49				77717.
315	36.12				72516.
316	37.49				70076.
318	40.37				65493.
320	43.42				61275.
321	45.01				59291.
323	48.34				55558.
325	51.85				52111.
326	53.69				50487.
328	57.50				47424.
330	61.53				44589.
331	63.63				43251.
333	67.99				40721.
335	72.58				38373.
336	74.97				37263.
338	79.93				35160.
340	85.14				33204.
341	87.84				32278.
343	93.44				30519.
345	99.32				28880.
346	102.4				28102.

COMPOUND CONSTANTS:

MP   TC   VC   ZC
NBP   369.099 K   PC   DC   OM
MU   RG

VAPOR PRESSURE CORRELATION CONSTANTS, K AND KPA:

LIQUID DENSITY CORRELATION CONSTANTS, K AND G/ML:

FRANCIS1   FRANCIS1   FRANCIS1
273. - 304. K   273. - 304. K   273. - 304. K
  5 POINTS   5 POINTS   5 POINTS
RMSD = 0.00017   RMSD = 0.00017   RMSD = 0.00017
0.97579509E+00   0.97579509E+00   0.97579509E+00
0.79638953E-03   0.79638953E-03   0.79638953E-03
0.59999990E+01   0.59999990E+01   0.59999990E+01
0.56838794E+03   0.56838794E+03   0.56838794E+03

SECOND VIRIAL COEFFICIENT CORRELATION CONSTANTS, ML/MOL:

LITERATURE DOCUMENTS FOR CORRELATED VAPOR PRESSURE VALUES:

LITERATURE DOCUMENTS FOR CORRELATED LIQUID DENSITY VALUES:

  2523   3269

LITERATURE DOCUMENTS REPORTING VIRIAL COEFFICIENT DATA:

T,K	VAPOR PRESSURE, KPA	SATURATED LIQUID VOLUME, ML/MOL	SECOND VIRIAL COEFFICIENT, ML/MOL	HEAT OF VAPORIZATION, J/MOL	SATURATED VAPOR VOLUME, ML/MOL
273		133.03			
274		133.19			
275		133.35			
276		133.50			
277		133.66			
278		133.82			
279		133.98			
280		134.14			
281		134.30			
282		134.46			
283		134.62			
284		134.78			
285		134.94			
286		135.10			
287		135.26			
288		135.43			
289		135.59			
290		135.75			
291		135.92			
292		136.08			
293		136.25			
294		136.41			
295		136.58			
296		136.75			
297		136.91			
298		137.08			
299		137.25			
300		137.42			
301		137.59			
302		137.76			
303		137.93			
304		138.10			

COMPOUND CONSTANTS:

MP   145.446 K   TC   VC   ZC
NBP   356.452 K   PC   DC   OM
MU   RG

VAPOR PRESSURE CORRELATION CONSTANTS, K AND KPA:

RPM2	RPM2	RPM2	RIEDEL'
291. - 357. K	291. - 357. K	291. - 357. K	343. - 357. K
14 POINTS	14 POINTS	14 POINTS	6 POINTS
RMSD = 0.0217	RMSD = 0.0217	RMSD = 0.0217	RMSD = 0.0282
0.17532248E+02	0.17532248E+02	0.17532248E+02	0.31228527E+03
-0.44652333E+04	-0.44652333E+04	-0.44652333E+04	-0.44139504E+02
0.26990204E-02	0.26990204E-02	0.26990204E-02	-0.17552073E+05
-0.10617983E-04	-0.10617983E-04	-0.10617983E-04	0.46151550E-15

LIQUID DENSITY CORRELATION CONSTANTS, K AND G/ML:

FRANCIS1	FRANCIS1	FRANCIS1
273. - 304. K	273. - 304. K	273. - 304. K
4 POINTS	4 POINTS	4 POINTS
RMSD = 0.00068	RMSD = 0.00068	RMSD = 0.00068
0.96920675E+00	0.96920675E+00	0.96920675E+00
0.86270901E-03	0.86270901E-03	0.86270901E-03
0.59999990E+01	0.59999990E+01	0.59999990E+01
0.56838794E+03	0.56838794E+03	0.56838794E+03

SECOND VIRIAL COEFFICIENT CORRELATION CONSTANTS, ML/MOL:

LITERATURE DOCUMENTS FOR CORRELATED VAPOR PRESSURE VALUES:

6518

LITERATURE DOCUMENTS FOR CORRELATED LIQUID DENSITY VALUES:

6518   8404

LITERATURE DOCUMENTS REPORTING VIRIAL COEFFICIENT DATA:

T,K	VAPOR PRESSURE, KPA	SATURATED LIQUID VOLUME, ML/MOL	SECOND VIRIAL COEFFICIENT, ML/MOL	HEAT OF VAPORIZATION, J/MOL	SATURATED VAPOR VOLUME, ML/MOL
273		137.64			
274		137.82			
276		138.18			
278		138.54			
280		138.91			
282		139.28			
284		139.65			
286		140.02			
288		140.40			
290		140.78			
292	8.357	141.16		34626.	290501.
293	8.774	141.35		34593.	277646.
295	9.660	141.73		34527.	253895.
297	10.62	142.12		34458.	232507.
299	11.66	142.51		34389.	213221.
301	12.78	142.90		34319.	195803.
303	13.99	143.30		34247.	180051.
305	15.30				165786.
307	16.70				152850.
309	18.21				141104.
311	19.83				130423.
313	21.56				120700.
314	22.47				116166.
316	24.40				107700.
318	26.45				99968.
320	28.64				92899.
322	30.98				86428.
324	33.47				80497.
326	36.11				75056.
328	38.93				70058.
330	41.91				65462.
332	45.08				61231.
334	48.44				57333.
335	50.19				55499.
337	53.84				52043.
339	57.70				48851.
341	61.77				45899.
343	66.07				43167.
345	70.59				40635.
347	75.35				38287.
349	80.36				36108.
351	85.63				34082.
353	91.15				32199.
355	96.95				30446.
356	99.95				29615.

COMPOUND CONSTANTS:

MP 137.680 K	TC	VC	ZC
NBP 353.574 K	PC	DC	OM
MU	RG		

VAPOR PRESSURE CORRELATION CONSTANTS, K AND KPA:

RPM2	RPM2	RPM2	RPM2
290. - 354. K	290. - 354. K	290. - 354. K	333. - 354. K
14 POINTS	14 POINTS	14 POINTS	7 POINTS
RMSD = 0.0068	RMSD = 0.0068	RMSD = 0.0068	RMSD = 0.0026
0.29177255E+02	0.29177255E+02	0.29177255E+02	0.16382953E+02
-0.55602094E+04	-0.55602094E+04	-0.55602094E+04	-0.41118796E+04
-0.35296867E-01	-0.35296867E-01	-0.35296867E-01	0.23515810E-02
0.29171556E-04	0.29171556E-04	0.29171556E-04	-0.77319132E-05

LIQUID DENSITY CORRELATION CONSTANTS, K AND G/ML:

FRANCIS1	FRANCIS1	FRANCIS1
293. - 304. K	293. - 304. K	293. - 304. K
4 POINTS	4 POINTS	4 POINTS
RMSD = 0.00003	RMSD = 0.00003	RMSD = 0.00003
0.95368588E+00	0.95368588E+00	0.95368588E+00
0.71452977E-03	0.71452977E-03	0.71452977E-03
0.12321428E+02	0.12321428E+02	0.12321428E+02
0.56838794E+03	0.56838794E+03	0.56838794E+03

SECOND VIRIAL COEFFICIENT CORRELATION CONSTANTS, ML/MOL:

LITERATURE DOCUMENTS FOR CORRELATED VAPOR PRESSURE VALUES:

6518

LITERATURE DOCUMENTS FOR CORRELATED LIQUID DENSITY VALUES:

6518

LITERATURE DOCUMENTS REPORTING VIRIAL COEFFICIENT DATA:

T,K	VAPOR PRESSURE, KPA	SATURATED LIQUID VOLUME, ML/MOL	SECOND VIRIAL COEFFICIENT, ML/MOL	HEAT OF VAPORIZATION, J/MOL	SATURATED VAPOR VOLUME, ML/MOL
290	9.220				261521.
291	9.669				250240.
292	10.14				239536.
294	11.12	140.53		33169.	219730.
295	11.65	140.70		33122.	210566.
297	12.76	141.06		33027.	193579.
298	13.34	141.24		32980.	185706.
300	14.58	141.60		32887.	171086.
301	15.23	141.78		32841.	164299.
303	16.61	142.15		32750.	151674.
304	17.34	142.33		32705.	145803.
305	18.09				140204.
307	19.67				129770.
308	20.50				124907.
310	22.25				115828.
311	23.17				111591.
313	25.10				103668.
314	26.12				99965.
316	28.24				93031.
317	29.36				89785.
319	31.69				83699.
320	32.91				80846.
321	34.17				78111.
323	36.80				72974.
324	38.18				70562.
326	41.05				66025.
327	42.55				63891.
329	45.69				59874.
330	47.32				57982.
332	50.73				54415.
333	52.50				52734.
335	56.20				49560.
336	58.13				48062.
337	60.10				46619.
339	64.22				43893.
340	66.35				42604.
342	70.80				40164.
343	73.11				39010.
345	77.90				36822.
346	80.39				35786.
348	85.55				33821.
349	88.23				32888.
351	93.78				31119.
352	96.66				30279.
353	99.60				29467.

COMPOUND CONSTANTS:

MP   157.923 K   TC   VC   ZC
NBP 349.883 K   PC   DC   OM
MU   RG

VAPOR PRESSURE CORRELATION CONSTANTS, K AND KPA:

RPM2	RPM2	RPM2	RPM2
288. - 351. K	288. - 351. K	288. - 351. K	337. - 351. K
13 POINTS	13 POINTS	13 POINTS	6 POINTS
RMSD = 0.0078	RMSD = 0.0078	RMSD = 0.0078	RMSD = 0.0051
0.26531518E+02	0.26531518E+02	0.26531518E+02	-0.11196998E+03
-0.52844628E+04	-0.52844628E+04	-0.52844628E+04	0.10549905E+05
-0.26499471E-01	-0.26499471E-01	-0.26499471E-01	0.37729169E+00
0.20111791E-04	0.20111791E-04	0.20111791E-04	-0.37226846E-03

LIQUID DENSITY CORRELATION CONSTANTS, K AND G/ML:

FRANCIS1	FRANCIS1	FRANCIS1
293. - 304. K	293. - 304. K	293. - 304. K
4 POINTS	4 POINTS	4 POINTS
RMSD = 0.00029	RMSD = 0.00029	RMSD = 0.00029
0.94519722E+00	0.94519722E+00	0.94519722E+00
0.79642376E-03	0.79642376E-03	0.79642376E-03
0.60485601E+01	0.60485601E+01	0.60485601E+01
0.55381372E+03	0.55381372E+03	0.55381372E+03

SECOND VIRIAL COEFFICIENT CORRELATION CONSTANTS, ML/MOL:

LITERATURE DOCUMENTS FOR CORRELATED VAPOR PRESSURE VALUES:

   6518

LITERATURE DOCUMENTS FOR CORRELATED LIQUID DENSITY VALUES:

   6518   9863

LITERATURE DOCUMENTS REPORTING VIRIAL COEFFICIENT DATA:

T,K	VAPOR PRESSURE, KPA	SATURATED LIQUID VOLUME, ML/MOL	SECOND VIRIAL COEFFICIENT, ML/MOL	HEAT OF VAPORIZATION, J/MOL	SATURATED VAPOR VOLUME, ML/MOL
288	9.198				260332.
289	9.656				248844.
290	10.13				237955.
292	11.15				217834.
293	11.68	142.58		33412.	208538.
295	12.82	142.95		33324.	191329.
296	13.42	143.13		33280.	183364.
298	14.70	143.50		33193.	168595.
299	15.37	143.69		33150.	161747.
300	16.07	143.88		33107.	155230.
302	17.54	144.25		33021.	143120.
303	18.32	144.44		32978.	137493.
305	19.96				127018.
306	20.83				122143.
308	22.65				113055.
309	23.61				108818.
310	24.60				104772.
312	26.68				97215.
313	27.78				93686.
315	30.08				87083.
316	31.28				83994.
318	33.81				78208.
319	35.13				75498.
320	36.50				72902.
322	39.35				68031.
323	40.85				65745.
325	43.97				61449.
326	45.61				59430.
328	49.02				55633.
329	50.80				53846.
330	52.63				52130.
332	56.45				48896.
333	58.45				47371.
335	62.60				44496.
336	64.76				43139.
338	69.26				40577.
339	71.60				39366.
340	74.00				38201.
342	79.00				35995.
343	81.59				34952.
345	86.98				32977.
346	89.78				32042.
348	95.59				30269.
349	98.60				29428.
350	101.7				28617.

COMPOUND CONSTANTS:

MP		TC		VC		ZC	
NBP	359.512 K	PC		DC		OM	
MU		RG					

VAPOR PRESSURE CORRELATION CONSTANTS, K AND KPA:

RIEDEL	RIEDEL	RIEDEL	RPM2
290. - 360. K	290. - 360. K	290. - 360. K	339. - 360. K
16 POINTS	16 POINTS	16 POINTS	7 POINTS
RMSD = 0.0109	RMSD = 0.0109	RMSD = 0.0109	RMSD = 0.0041
0.53147394E+02	0.53147394E+02	0.53147394E+02	-0.49179331E+01
-0.55095367E+01	-0.55095367E+01	-0.55095367E+01	-0.17775168E+04
-0.57982596E+04	-0.57982596E+04	-0.57982596E+04	0.64400612E-01
0.98372318E-17	0.98372318E-17	0.98372318E-17	-0.67097466E-04

LIQUID DENSITY CORRELATION CONSTANTS, K AND G/ML:

FRANCIS1	FRANCIS1	FRANCIS1
293. - 304. K	293. - 304. K	293. - 304. K
4 POINTS	4 POINTS	4 POINTS
RMSD = 0.00003	RMSD = 0.00003	RMSD = 0.00003
0.97217888E+00	0.97217888E+00	0.97217888E+00
0.82097831E-03	0.82097831E-03	0.82097831E-03
0.60001097E+01	0.60001097E+01	0.60001097E+01
0.55735596E+03	0.55735596E+03	0.55735596E+03

SECOND VIRIAL COEFFICIENT CORRELATION CONSTANTS, ML/MOL:

LITERATURE DOCUMENTS FOR CORRELATED VAPOR PRESSURE VALUES:

6518

LITERATURE DOCUMENTS FOR CORRELATED LIQUID DENSITY VALUES:

6518

LITERATURE DOCUMENTS REPORTING VIRIAL COEFFICIENT DATA:

T,K	VAPOR PRESSURE, KPA	SATURATED LIQUID VOLUME, ML/MOL	SECOND VIRIAL COEFFICIENT, ML/MOL	HEAT OF VAPORIZATION, J/MOL	SATURATED VAPOR VOLUME, ML/MOL
290	6.827				353202.
291	7.176				337181.
293	7.919	138.50		34862.	307645.
294	8.314	138.68		34818.	294031.
296	9.153	139.03		34729.	268889.
297	9.598	139.21		34685.	257279.
299	10.54	139.57		34597.	235800.
301	11.56	139.94		34508.	216425.
302	12.10	140.12		34464.	207452.
304	13.25	140.49		34376.	190806.
305	13.85				183084.
307	15.13				168734.
309	16.50				155714.
310	17.22				149659.
312	18.75				138376.
313	19.55				133121.
315	21.24				123315.
317	23.05				114370.
318	23.99				110192.
320	25.99				102380.
321	27.03				98727.
323	29.23				91885.
324	30.38				88681.
326	32.79				82672.
328	35.35				77154.
329	36.69				74564.
331	39.49				69697.
332	40.95				67410.
334	44.00				63107.
336	47.24				59137.
337	48.93				57268.
339	52.44				53744.
340	54.28				52083.
342	58.09				48947.
344	62.12				46043.
345	64.21				44671.
347	68.57				42077.
348	70.83				40850.
350	75.53				38529.
352	80.47				36370.
353	83.03				35347.
355	88.35				33408.
356	91.11				32489.
358	96.82				30744.
359	99.78				29916.

COMPOUND CONSTANTS:

MP	163.314 K	TC	VC	ZC
NBP	351.035 K	PC	DC	OM
MU		RG		

VAPOR PRESSURE CORRELATION CONSTANTS, K AND KPA:

RIEDEL	RIEDEL	RIEDEL	RIEDEL
288. - 352. K	288. - 352. K	288. - 352. K	331. - 352. K
15 POINTS	15 POINTS	15 POINTS	7 POINTS
RMSD = 0.0065	RMSD = 0.0065	RMSD = 0.0065	RMSD = 0.0020
0.52126745E+02	0.52126745E+02	0.52126745E+02	0.41824499E+02
-0.54382426E+01	-0.54382426E+01	-0.54382426E+01	-0.38928714E+01
-0.55034115E+04	-0.55034115E+04	-0.55034115E+04	-0.50513872E+04
0.22532142E-16	0.22532142E-16	0.22532142E-16	-0.26951599E-18

LIQUID DENSITY CORRELATION CONSTANTS, K AND G/ML:

FRANCIS1	FRANCIS1	FRANCIS1
273. - 324. K	273. - 324. K	273. - 324. K
5 POINTS	5 POINTS	5 POINTS
RMSD = 0.00066	RMSD = 0.00066	RMSD = 0.00066
0.96574610E+00	0.96574610E+00	0.96574610E+00
0.82386518E-03	0.82386518E-03	0.82386518E-03
0.59999990E+01	0.59999990E+01	0.59999990E+01
0.60588574E+03	0.60588574E+03	0.60588574E+03

SECOND VIRIAL COEFFICIENT CORRELATION CONSTANTS, ML/MOL:

LITERATURE DOCUMENTS FOR CORRELATED VAPOR PRESSURE VALUES:

6518

LITERATURE DOCUMENTS FOR CORRELATED LIQUID DENSITY VALUES:

6518 10808

LITERATURE DOCUMENTS REPORTING VIRIAL COEFFICIENT DATA:

T,K	VAPOR PRESSURE, KPA	SATURATED LIQUID VOLUME, ML/MOL	SECOND VIRIAL COEFFICIENT, ML/MOL	HEAT OF VAPORIZATION, J/MOL	SATURATED VAPOR VOLUME, ML/MOL
273		135.84			
274		136.01			
276		136.34			
278		136.67			
280		137.01			
281		137.18			
283		137.52			
285		137.86			
287		138.20			
289	9.792	138.54		32861.	245401.
290	10.26	138.72		32819.	234907.
292	11.27	139.06		32737.	215486.
294	12.35	139.41		32655.	197956.
296	13.51	139.77		32572.	182110.
298	14.77	140.12		32491.	167764.
299	15.43	140.30		32450.	161103.
301	16.83	140.65		32369.	148711.
303	18.33	141.01		32288.	137453.
305	19.93	141.37		32207.	127209.
307	21.65	141.74		32127.	117877.
308	22.56	141.92		32087.	113523.
310	24.46	142.29		32007.	105386.
312	26.48	142.66		31927.	97949.
314	28.64	143.03		31848.	91142.
316	30.94	143.40		31769.	84905.
317	32.15	143.59		31730.	81983.
319	34.67	143.97		31652.	76499.
321	37.35	144.35		31574.	71458.
323	40.19	144.73		31497.	66821.
325	43.20				62548.
326	44.77				60538.
328	48.05				56753.
330	51.52				53256.
332	55.18				50023.
334	59.05				47030.
335	61.06				45617.
337	65.24				42946.
339	69.65				40468.
341	74.29				38165.
343	79.16				36025.
344	81.69				35011.
346	86.94				33089.
348	92.45				31297.
350	98.23				29626.
351	101.2				28833.

COMPOUND CONSTANTS:

MP	171.461 K	TC		VC	ZC
NBP	394.437 K	PC		DC	OM
MU	0.3400 DEBYE	RG	4.5342 ANGSTROM		

VAPOR PRESSURE CORRELATION CONSTANTS, K AND KPA:

VAPRES-2	VAPRES-2	VAPRES-2	RIEDEL
264. - 396. K	264. - 396. K	264. - 396. K	380. - 396. K
26 POINTS	26 POINTS	26 POINTS	5 POINTS
RMSD = 0.2233	RMSD = 0.2233	RMSD = 0.2233	RMSD = 0.0343
0.60614607E+04	0.60614607E+04	0.60614607E+04	-0.10983339E+04
-0.13483127E+06	-0.13483127E+06	-0.13483127E+06	0.16395905E+03
0.32794178E+01	0.32794178E+01	0.32794178E+01	0.50167211E+05
-0.15867412E-02	-0.15867412E-02	-0.15867412E-02	-0.11400396E-14
-0.11311951E+04	-0.11311951E+04	-0.11311951E+04	

LIQUID DENSITY CORRELATION CONSTANTS, K AND G/ML:

FRANCIS1	FRANCIS1	FRANCIS1
283. - 524. K	283. - 399. K	283. - 524. K
19 POINTS	18 POINTS	19 POINTS
RMSD = 0.00029	RMSD = 0.00029	RMSD = 0.00029
0.96701348E+00	0.96791929E+00	0.96701348E+00
0.79335924E-03	0.79749012E-03	0.79335924E-03
0.59999990E+01	0.60334005E+01	0.59999990E+01
0.60459521E+03	0.61147852E+03	0.60459521E+03

SECOND VIRIAL COEFFICIENT CORRELATION CONSTANTS, ML/MOL:

KREGLEWSKI: A=-1308.0 B= 0.00 V*=166.00 R1=0.7590 R2=1.0970 R3=4.5210

LITERATURE DOCUMENTS FOR CORRELATED VAPOR PRESSURE VALUES:

1981 1991 4392 8405 40796 40865

LITERATURE DOCUMENTS FOR CORRELATED LIQUID DENSITY VALUES:

1991 11064 14419 18101 20472 21562

LITERATURE DOCUMENTS REPORTING VIRIAL COEFFICIENT DATA:

1901

T,K	VAPOR PRESSURE, KPA	SATURATED LIQUID VOLUME, ML/MOL	SECOND VIRIAL COEFFICIENT, ML/MOL	HEAT OF VAPORIZATION, J/MOL	SATURATED VAPOR VOLUME, ML/MOL
264	0.2074		-6284.		0.1058E+08
269	0.3212		-5821.		6957262.
275	0.5188		-5332.		4401619.
281	0.8031		-4905.		2904400.
287	1.199	155.76	-4531.	43795.	1985810.
293	1.736	156.88	-4200.	42278.	1399110.
299	2.449	158.01	-3906.	41050.	1011036.
305	3.381	159.17	-3644.	40075.	746405.
311	4.580	160.34	-3410.	39317.	561115.
317	6.107	161.54	-3199.	38742.	428373.
323	8.031	162.76	-3009.	38315.	331380.
328	9.996	163.79	-2864.	38046.	269918.
334	12.87	165.06	-2705.	37795.	212993.
340	16.41	166.34	-2560.	37592.	169617.
346	20.74	167.66	-2427.	37399.	136203.
352	26.00	169.00	-2305.	37183.	110227.
358	32.31	170.37	-2193.	36908.	89874.
364	39.84	171.77	-2090.	36542.	73821.
370	48.71	173.20	-1994.	36051.	61090.
376	59.07	174.66	-1906.	35404.	50946.
382	70.99	176.16	-1823.	34574.	42834.
388	84.53	177.70	-1747.	33534.	36328.
393	97.03	179.01	-1687.	32488.	31895.
399		180.62	-1619.		
405		182.27	-1556.		
411		183.97	-1496.		
417		185.72	-1441.		
423		187.53	-1388.		
429		189.39	-1339.		
435		191.32	-1292.		
441		193.32	-1248.		
447		195.39	-1206.		
453		197.55	-1166.		
458		199.41	-1134.		
464		201.74	-1098.		
470		204.19	-1064.		
476		206.76	-1031.		
482		209.48	******		
488		212.37	-970.2		
494		215.45	-941.7		
500		218.75	-914.5		
506		222.33	-888.4		
512		226.23	-863.5		
518		230.53	-839.5		
523		234.49	-820.3		

COMPOUND CONSTANTS:

MP	172.211 K	TC		VC		ZC
NBP	398.811 K	PC		DC		OM
MU	0.3100 DEBYE	RG				

VAPOR PRESSURE CORRELATION CONSTANTS, K AND KPA:

RIEDEL	RIEDEL	RIEDEL
356. - 400. K	356. - 400. K	356. - 400. K
5 POINTS	5 POINTS	5 POINTS
RMSD = 0.0026	RMSD = 0.0026	RMSD = 0.0026
0.45029305E+02	0.45029305E+02	0.45029305E+02
-0.41739044E+01	-0.41739044E+01	-0.41739044E+01
-0.61366223E+04	-0.61366223E+04	-0.61366223E+04
-0.70318609E-17	-0.70318609E-17	-0.70318609E-17

LIQUID DENSITY CORRELATION CONSTANTS, K AND G/ML:

SECOND VIRIAL COEFFICIENT CORRELATION CONSTANTS, ML/MOL:

LITERATURE DOCUMENTS FOR CORRELATED VAPOR PRESSURE VALUES:

8405

LITERATURE DOCUMENTS FOR CORRELATED LIQUID DENSITY VALUES:

LITERATURE DOCUMENTS REPORTING VIRIAL COEFFICIENT DATA:

T,K	VAPOR PRESSURE, KPA	SATURATED LIQUID VOLUME, ML/MOL	SECOND VIRIAL COEFFICIENT, ML/MOL	HEAT OF VAPORIZATION, J/MOL	SATURATED VAPOR VOLUME, ML/MOL
356	25.94				114095.
357	26.90				110333.
358	27.89				106720.
359	28.91				103249.
360	29.96				99913.
361	31.04				96708.
362	32.15				93626.
363	33.29				90663.
364	34.46				87812.
365	35.67				85071.
366	36.92				82432.
367	38.19				79893.
368	39.51				77448.
369	40.86				75094.
370	42.24				72827.
371	43.66				70644.
372	45.13				68539.
373	46.63				66511.
374	48.17				64557.
375	49.75				62672.
376	51.37				60855.
377	53.04				59101.
378	54.74				57410.
379	56.49				55778.
380	58.29				54203.
381	60.13				52683.
382	62.02				51215.
383	63.95				49797.
384	65.93				48428.
385	67.96				47105.
386	70.03				45826.
387	72.16				44591.
388	74.34				43397.
389	76.57				42242.
390	78.85				41126.
391	81.18				40046.
392	83.57				39001.
393	86.01				37990.
394	88.51				37012.
395	91.06				36065.
396	93.67				35149.
397	96.34				34261.
398	99.07				33402.
399	101.9				32570.
400	104.7				31763.

397

COMPOUND CONSTANTS:

MP	185.380 K	TC	VC	ZC
NBP	398.113 K	PC	DC	OM
MU	0.0000 DEBYE	RG 4.5867 ANGSTROM		

VAPOR PRESSURE CORRELATION CONSTANTS, K AND KPA:

RIEDEL	RIEDEL	RIEDEL
356. - 399. K	356. - 399. K	356. - 399. K
5 POINTS	5 POINTS	5 POINTS
RMSD = 0.0024	RMSD = 0.0024	RMSD = 0.0024
0.96766982E+02	0.96766982E+02	0.96766982E+02
-0.11816176E+02	-0.11816176E+02	-0.11816176E+02
-0.86113038E+04	-0.86113038E+04	-0.86113038E+04
0.55752280E-16	0.55752280E-16	0.55752280E-16

LIQUID DENSITY CORRELATION CONSTANTS, K AND G/ML:

SECOND VIRIAL COEFFICIENT CORRELATION CONSTANTS, ML/MOL:

LITERATURE DOCUMENTS FOR CORRELATED VAPOR PRESSURE VALUES:

8405

LITERATURE DOCUMENTS FOR CORRELATED LIQUID DENSITY VALUES:

LITERATURE DOCUMENTS REPORTING VIRIAL COEFFICIENT DATA:

T,K	VAPOR PRESSURE, KPA	SATURATED LIQUID VOLUME, ML/MOL	SECOND VIRIAL COEFFICIENT, ML/MOL	HEAT OF VAPORIZATION, J/MOL	SATURATED VAPOR VOLUME, ML/MOL
356	26.37				112249.
357	27.35				108526.
358	28.36				104952.
359	29.40				101521.
360	30.47				98227.
361	31.57				95062.
362	32.71				92022.
363	33.87				89100.
364	35.07				86290.
365	36.31				83589.
366	37.57				80991.
367	38.88				78491.
368	40.21				76085.
369	41.59				73769.
370	43.00				71540.
371	44.45				69392.
372	45.94				67324.
373	47.47				65330.
374	49.04				63409.
375	50.65				61557.
376	52.30				59772.
377	54.00				58049.
378	55.74				56388.
379	57.52				54784.
380	59.35				53237.
381	61.22				51743.
382	63.14				50300.
383	65.11				48907.
384	67.13				47561.
385	69.20				46260.
386	71.31				45003.
387	73.48				43788.
388	75.70				42613.
389	77.98				41477.
390	80.31				40378.
391	82.69				39315.
392	85.13				38285.
393	87.63				37289.
394	90.18				36325.
395	92.80				35391.
396	95.47				34487.
397	98.21				33611.
398	101.0				32762.
399	103.9				31940.

COMPOUND CONSTANTS:

MP	163.113 K	TC	VC	ZC
NBP	396.450 K	PC	DC	OM
MU	0.0000 DEBYE	RG		

VAPOR PRESSURE CORRELATION CONSTANTS, K AND KPA:

RIEDEL	RIEDEL	RIEDEL
354. - 397. K	354. - 397. K	354. - 397. K
6 POINTS	6 POINTS	6 POINTS
RMSD = 0.0201	RMSD = 0.0201	RMSD = 0.0201
0.56339430E+02	0.56339430E+02	0.56339430E+02
-0.58555386E+01	-0.58555386E+01	-0.58555386E+01
-0.66291980E+04	-0.66291980E+04	-0.66291980E+04
0.80715815E-17	0.80715815E-17	0.80715815E-17

LIQUID DENSITY CORRELATION CONSTANTS, K AND G/ML:

SECOND VIRIAL COEFFICIENT CORRELATION CONSTANTS, ML/MOL:

LITERATURE DOCUMENTS FOR CORRELATED VAPOR PRESSURE VALUES:

6655 8405

LITERATURE DOCUMENTS FOR CORRELATED LIQUID DENSITY VALUES:

LITERATURE DOCUMENTS REPORTING VIRIAL COEFFICIENT DATA:

T,K	VAPOR PRESSURE, KPA	SATURATED LIQUID VOLUME, ML/MOL	SECOND VIRIAL COEFFICIENT, ML/MOL	HEAT OF VAPORIZATION, J/MOL	SATURATED VAPOR VOLUME, ML/MOL
354	26.07				112890.
355	27.04				109151.
356	28.04				105561.
357	29.07				102113.
358	30.13				98800.
359	31.22				95618.
360	32.34				92559.
361	33.49				89618.
362	34.68				86790.
363	35.90				84070.
364	37.16				81454.
365	38.45				78936.
366	39.77				76512.
367	41.14				74179.
368	42.54				71933.
369	43.97				69769.
370	45.45				67684.
371	46.97				65675.
372	48.53				63739.
373	50.12				61873.
374	51.76				60073.
375	53.45				58337.
376	55.17				56663.
377	56.94				55048.
378	58.76				53489.
379	60.62				51984.
380	62.53				50531.
381	64.48				49128.
382	66.48				47773.
383	68.54				46464.
384	70.64				45199.
385	72.79				43977.
386	74.99				42795.
387	77.25				41653.
388	79.56				40548.
389	81.92				39479.
390	84.34				38446.
391	86.82				37446.
392	89.35				36478.
393	91.94				35541.
394	94.58				34634.
395	97.29				33756.
396	100.1				32906.
397	102.9				32082.

COMPOUND CONSTANTS:

MP	154.134 K	TC	VC	ZC
NBP	395.706 K	PC	DC	OM
MU	0.2600 DEBYE	RG		

VAPOR PRESSURE CORRELATION CONSTANTS, K AND KPA:

RIEDEL	RIEDEL	RIEDEL
353. - 396. K	353. - 396. K	353. - 396. K
5 POINTS	5 POINTS	5 POINTS
RMSD = 0.0040	RMSD = 0.0040	RMSD = 0.0040
0.38097593E+02	0.38097593E+02	0.38097593E+02
-0.31789759E+01	-0.31789759E+01	-0.31789759E+01
-0.57061753E+04	-0.57061753E+04	-0.57061753E+04
-0.12140667E-16	-0.12140667E-16	-0.12140667E-16

LIQUID DENSITY CORRELATION CONSTANTS, K AND G/ML:

SECOND VIRIAL COEFFICIENT CORRELATION CONSTANTS, ML/MOL:

LITERATURE DOCUMENTS FOR CORRELATED VAPOR PRESSURE VALUES:

8405

LITERATURE DOCUMENTS FOR CORRELATED LIQUID DENSITY VALUES:

LITERATURE DOCUMENTS REPORTING VIRIAL COEFFICIENT DATA:

T,K	VAPOR PRESSURE, KPA	SATURATED LIQUID VOLUME, ML/MOL	SECOND VIRIAL COEFFICIENT, ML/MOL	HEAT OF VAPORIZATION, J/MOL	SATURATED VAPOR VOLUME, ML/MOL
353	26.05				112676.
354	27.01				108971.
355	28.00				105411.
356	29.02				101991.
357	30.07				98705.
358	31.15				95545.
359	32.27				92507.
360	33.41				89586.
361	34.59				86776.
362	35.80				84072.
363	37.05				81470.
364	38.33				78965.
365	39.64				76554.
366	40.99				74231.
367	42.38				71994.
368	43.81				69839.
369	45.28				67763.
370	46.78				65761.
371	48.32				63832.
372	49.91				61971.
373	51.54				60177.
374	53.20				58446.
375	54.92				56776.
376	56.67				55164.
377	58.47				53608.
378	60.32				52107.
379	62.21				50657.
380	64.14				49256.
381	66.13				47903.
382	68.16				46596.
383	70.24				45333.
384	72.38				44113.
385	74.56				42932.
386	76.79				41791.
387	79.08				40688.
388	81.42				39621.
389	83.82				38588.
390	86.26				37589.
391	88.77				36622.
392	91.33				35687.
393	93.95				34781.
394	96.62				33904.
395	99.36				33054.
396	102.2				32231.

403

COMPOUND CONSTANTS:

MP   179.367 K   TC   VC   ZC
NBP 395.477 K   PC   DC   OM
MU   0.0000 DEBYE   RG

VAPOR PRESSURE CORRELATION CONSTANTS, K AND KPA:

RIEDEL   RIEDEL   RIEDEL
353. - 396. K   353. - 396. K   353. - 396. K
  5 POINTS   5 POINTS   5 POINTS
RMSD = 0.0010   RMSD = 0.0010   RMSD = 0.0010
 0.49111607E+02   0.49111607E+02   0.49111607E+02
-0.47965360E+01   -0.47965360E+01   -0.47965360E+01
-0.62532766E+04   -0.62532766E+04   -0.62532766E+04
 0.64116961E-18   0.64116961E-18   0.64116961E-18

LIQUID DENSITY CORRELATION CONSTANTS, K AND G/ML:

SECOND VIRIAL COEFFICIENT CORRELATION CONSTANTS, ML/MOL:

LITERATURE DOCUMENTS FOR CORRELATED VAPOR PRESSURE VALUES:

8405

LITERATURE DOCUMENTS FOR CORRELATED LIQUID DENSITY VALUES:

LITERATURE DOCUMENTS REPORTING VIRIAL COEFFICIENT DATA:

T,K	VAPOR PRESSURE, KPA	SATURATED LIQUID VOLUME, ML/MOL	SECOND VIRIAL COEFFICIENT, ML/MOL	HEAT OF VAPORIZATION, J/MOL	SATURATED VAPOR VOLUME, ML/MOL
353	26.04				112730.
354	27.00				108999.
355	28.00				105415.
356	29.03				101973.
357	30.08				98666.
358	31.17				95489.
359	32.29				92434.
360	33.44				89498.
361	34.63				86674.
362	35.85				83958.
363	37.10				81344.
364	38.39				78830.
365	39.72				76409.
366	41.08				74078.
367	42.48				71834.
368	43.92				69672.
369	45.39				67590.
370	46.91				65583.
371	48.46				63649.
372	50.06				61784.
373	51.70				59986.
374	53.38				58252.
375	55.11				56579.
376	56.88				54965.
377	58.69				53408.
378	60.55				51904.
379	62.46				50453.
380	64.41				49051.
381	66.41				47697.
382	68.47				46389.
383	70.57				45126.
384	72.72				43904.
385	74.92				42724.
386	77.18				41583.
387	79.49				40479.
388	81.85				39412.
389	84.27				38380.
390	86.75				37381.
391	89.28				36414.
392	91.87				35478.
393	94.51				34573.
394	97.22				33696.
395	99.99				32847.
396	102.8				32024.

COMPOUND CONSTANTS:

MP		TC		VC		ZC
NBP	395.664 K	PC		DC		OM
MU		RG				

VAPOR PRESSURE CORRELATION CONSTANTS, K AND KPA:

VAPRES-2	VAPRES-2	VAPRES-2
257. - 396. K	257. - 396. K	257. - 396. K
9 POINTS	9 POINTS	9 POINTS
RMSD = 0.0175	RMSD = 0.0175	RMSD = 0.0175
-0.77935987E+03	-0.77935987E+03	-0.77935987E+03
0.97638960E+04	0.97638960E+04	0.97638960E+04
-0.51681062E+00	-0.51681062E+00	-0.51681062E+00
0.27876180E-03	0.27876180E-03	0.27876180E-03
0.15385571E+03	0.15385571E+03	0.15385571E+03

LIQUID DENSITY CORRELATION CONSTANTS, K AND G/ML:

SECOND VIRIAL COEFFICIENT CORRELATION CONSTANTS, ML/MOL:

LITERATURE DOCUMENTS FOR CORRELATED VAPOR PRESSURE VALUES:

1146

LITERATURE DOCUMENTS FOR CORRELATED LIQUID DENSITY VALUES:

LITERATURE DOCUMENTS REPORTING VIRIAL COEFFICIENT DATA:

T,K	VAPOR PRESSURE, KPA	SATURATED LIQUID VOLUME, ML/MOL	SECOND VIRIAL COEFFICIENT, ML/MOL	HEAT OF VAPORIZATION, J/MOL	SATURATED VAPOR VOLUME, ML/MOL
257	0.1327				0.1610E+08
260	0.1669				0.1295E+08
263	0.2088				0.1047E+08
266	0.2597				8516376.
269	0.3214				6959358.
272	0.3957				5715309.
275	0.4848				4716434.
279	0.6308				3677569.
282	0.7643				3067889.
285	0.9218				2570685.
288	1.107				2163404.
291	1.323				1828332.
294	1.576				1551497.
298	1.976				1254183.
301	2.330				1074031.
304	2.738				923186.
307	3.205				796398.
310	3.739				689435.
313	4.346				598870.
317	5.283				498887.
320	6.094				436624.
323	7.006				383302.
326	8.031				337486.
329	9.179				297996.
332	10.46				263853.
335	11.89				234245.
339	14.05				200665.
342	15.87				179192.
345	17.88				160392.
348	20.11				143890.
351	22.56				129366.
354	25.25				116551.
358	29.26				101732.
361	32.60				92070.
364	36.25				83477.
367	40.25				75816.
370	44.60				68971.
373	49.35				62843.
377	56.34				55637.
380	62.12				50863.
383	68.39				46559.
386	75.21				42672.
389	82.60				39155.
392	90.62				35967.
395	99.30				33072.

COMPOUND CONSTANTS:

MP	135.788 K	TC	VC	ZC
NBP	378.596 K	PC	DC	OM
MU		RG		

VAPOR PRESSURE CORRELATION CONSTANTS, K AND KPA:

VAPRES-2	VAPRES-2	VAPRES-2	RIEDEL
305. - 380. K	305. - 380. K	305. - 380. K	364. - 380. K
16 POINTS	16 POINTS	16 POINTS	4 POINTS
RMSD = 0.0191	RMSD = 0.0191	RMSD = 0.0191	RMSD = 0.0000
0.20948636E+04	0.20948636E+04	0.20948636E+04	0.14128070E+03
-0.50227665E+05	-0.50227665E+05	-0.50227665E+05	-0.18481939E+02
0.10961779E+01	0.10961779E+01	0.10961779E+01	-0.10316640E+05
-0.52342534E-03	-0.52342534E-03	-0.52342534E-03	0.10351560E-15
-0.38702469E+03	-0.38702469E+03	-0.38702469E+03	

LIQUID DENSITY CORRELATION CONSTANTS, K AND G/ML:

FRANCIS1	FRANCIS1	FRANCIS1
293. - 304. K	293. - 304. K	293. - 304. K
4 POINTS	4 POINTS	4 POINTS
RMSD = 0.00005	RMSD = 0.00005	RMSD = 0.00005
0.95608085E+00	0.95608085E+00	0.95608085E+00
0.75664301E-03	0.75664301E-03	0.75664301E-03
0.59999990E+01	0.59999990E+01	0.59999990E+01
0.56838794E+03	0.56838794E+03	0.56838794E+03

SECOND VIRIAL COEFFICIENT CORRELATION CONSTANTS, ML/MOL:

LITERATURE DOCUMENTS FOR CORRELATED VAPOR PRESSURE VALUES:

6518

LITERATURE DOCUMENTS FOR CORRELATED LIQUID DENSITY VALUES:

6518

LITERATURE DOCUMENTS REPORTING VIRIAL COEFFICIENT DATA:

T,K	VAPOR PRESSURE, KPA	SATURATED LIQUID VOLUME, ML/MOL	SECOND VIRIAL COEFFICIENT, ML/MOL	HEAT OF VAPORIZATION, J/MOL	SATURATED VAPOR VOLUME, ML/MOL
293		157.47			
294		157.66			
296		158.03			
298		158.40			
300		158.78			
302		159.16			
304		159.54			
306	7.179				354404.
308	7.887				324681.
310	8.652				297921.
312	9.475				273784.
314	10.36				251973.
316	11.31				232230.
318	12.34				214330.
320	13.43				198073.
322	14.61				183286.
324	15.86				169816.
326	17.21				157528.
328	18.64				146301.
330	20.17				136030.
332	21.80				126622.
334	23.54				117993.
336	25.38				110068.
338	27.34				102781.
340	29.42				96074.
342	31.63				89893.
344	33.97				84190.
346	36.45				78924.
348	39.07				74055.
350	41.84				69550.
352	44.77				65376.
354	47.85				61507.
356	51.11				57917.
358	54.53				54583.
360	58.14				51483.
362	61.93				48600.
364	65.91				45915.
366	70.10				43413.
368	74.48				41081.
370	79.07				38905.
372	83.88				36872.
374	88.91				34973.
376	94.17				33198.
378	99.66				31537.
379	102.5				30746.

COMPOUND CONSTANTS:

MP		TC	VC		ZC
NBP	374.042 K	PC	DC		OM
MU		RG			

VAPOR PRESSURE CORRELATION CONSTANTS, K AND KPA:

VAPRES-2	VAPRES-2	VAPRES-2	RIEDEL
302. - 375. K	302. - 375. K	302. - 375. K	360. - 375. K
16 POINTS	16 POINTS	16 POINTS	4 POINTS
RMSD = 0.0093	RMSD = 0.0093	RMSD = 0.0093	RMSD = 0.0000
0.13774385E+04	0.13774385E+04	0.13774385E+04	-0.48403743E+03
-0.34977302E+05	-0.34977302E+05	-0.34977302E+05	0.74233398E+02
0.70181519E+00	0.70181519E+00	0.70181519E+00	0.19056192E+05
-0.33146070E-03	-0.33146070E-03	-0.33146070E-03	-0.75837680E-15
-0.25242237E+03	-0.25242237E+03	-0.25242237E+03	

LIQUID DENSITY CORRELATION CONSTANTS, K AND G/ML:

FRANCIS1	FRANCIS1	FRANCIS1
293. - 304. K	293. - 304. K	293. - 304. K
4 POINTS	4 POINTS	4 POINTS
RMSD = 0.00001	RMSD = 0.00001	RMSD = 0.00001
0.96429086E+00	0.96429086E+00	0.96429086E+00
0.80960826E-03	0.80960826E-03	0.80960826E-03
0.61868601E+01	0.61868601E+01	0.61868601E+01
0.56044971E+03	0.56044971E+03	0.56044971E+03

SECOND VIRIAL COEFFICIENT CORRELATION CONSTANTS, ML/MOL:

LITERATURE DOCUMENTS FOR CORRELATED VAPOR PRESSURE VALUES:

6518

LITERATURE DOCUMENTS FOR CORRELATED LIQUID DENSITY VALUES:

6518

LITERATURE DOCUMENTS REPORTING VIRIAL COEFFICIENT DATA:

T,K	VAPOR PRESSURE, KPA	SATURATED LIQUID VOLUME, ML/MOL	SECOND VIRIAL COEFFICIENT, ML/MOL	HEAT OF VAPORIZATION, J/MOL	SATURATED VAPOR VOLUME ML/MOL
293		159.41			
294		159.61			
296		160.02			
298		160.43			
300		160.84			
302	6.671	161.26		37356.	376408.
304	7.356	161.68		37193.	343631.
306	8.097				314224.
307	8.490				300658.
309	9.323				275581.
311	10.22				252979.
313	11.19				232574.
315	12.23				214120.
317	13.35				197405.
319	14.55				182240.
320	15.19				175186.
322	16.52				162044.
324	17.95				150076.
326	19.48				139162.
328	21.11				129196.
330	22.85				120082.
332	24.70				111737.
333	25.68				107830.
335	27.71				100502.
337	29.88				93771.
339	32.18				87582.
341	34.62				81884.
343	37.21				76633.
345	39.96				71788.
347	42.86				67313.
348	44.37				65205.
350	47.53				61225.
352	50.86				57541.
354	54.38				54126.
356	58.09				50958.
358	61.99				48017.
360	66.10				45283.
361	68.23				43989.
363	72.66				41536.
365	77.32				39252.
367	82.20				37122.
369	87.32				35136.
371	92.68				33282.
373	98.30				31550.
374	101.2				30727.

411

COMPOUND CONSTANTS:

MP		TC		VC		ZC
NBP	382.410 K	PC		DC		OM
MU		RG				

VAPOR PRESSURE CORRELATION CONSTANTS, K AND KPA:

RIEDEL	RIEDEL	RIEDEL	RIEDEL
308. - 383. K	308. - 383. K	308. - 383. K	368. - 383. K
16 POINTS	16 POINTS	16 POINTS	4 POINTS
RMSD = 0.0077	RMSD = 0.0077	RMSD = 0.0077	RMSD = 0.0000
0.55651308E+02	0.55651308E+02	0.55651308E+02	-0.38117361E+03
-0.58519594E+01	-0.58519594E+01	-0.58519594E+01	0.58760954E+02
-0.62264795E+04	-0.62264795E+04	-0.62264795E+04	0.14548590E+05
0.15293617E-16	0.15293617E-16	0.15293617E-16	-0.53533272E-15

LIQUID DENSITY CORRELATION CONSTANTS, K AND G/ML:

FRANCIS1	FRANCIS1	FRANCIS1
293. - 304. K	293. - 304. K	293. - 304. K
4 POINTS	4 POINTS	4 POINTS
RMSD = 0.00005	RMSD = 0.00005	RMSD = 0.00005
0.96596307E+00	0.96596307E+00	0.96596307E+00
0.74880337E-03	0.74880337E-03	0.74880337E-03
0.59999990E+01	0.59999990E+01	0.59999990E+01
0.56838794E+03	0.56838794E+03	0.56838794E+03

SECOND VIRIAL COEFFICIENT CORRELATION CONSTANTS, ML/MOL:

LITERATURE DOCUMENTS FOR CORRELATED VAPOR PRESSURE VALUES:

6518

LITERATURE DOCUMENTS FOR CORRELATED LIQUID DENSITY VALUES:

6518

LITERATURE DOCUMENTS REPORTING VIRIAL COEFFICIENT DATA:

T,K	VAPOR PRESSURE, KPA	SATURATED LIQUID VOLUME, ML/MOL	SECOND VIRIAL COEFFICIENT, ML/MOL	HEAT OF VAPORIZATION, J/MOL	SATURATED VAPOR VOLUME, ML/MOL
293		154.83			
295		155.18			
297		155.54			
299		155.90			
301		156.26			
303		156.62			
309	7.120				360831.
311	7.809				331128.
313	8.553				304283.
315	9.354				279985.
317	10.22				257963.
319	11.15				237975.
321	12.14				219810.
323	13.21				203279.
325	14.36				188217.
327	15.58				174474.
329	16.89				161921.
331	18.29				150440.
333	19.79				139927.
335	21.38				130289.
337	23.07				121444.
340	25.82				109501.
342	27.79				102327.
344	29.88				95718.
346	32.10				89623.
348	34.45				83995.
350	36.93				78793.
352	39.56				73982.
354	42.33				69526.
356	45.26				65395.
358	48.35				61563.
360	51.60				58004.
362	55.03				54696.
364	58.63				51619.
366	62.42				48753.
368	66.40				46082.
370	70.57				43591.
372	74.95				41265.
374	79.54				39092.
376	84.35				37061.
378	89.39				35159.
380	94.66				33379.
382	100.2				31710.

413

COMPOUND CONSTANTS:

MP	179.697 K	TC	VC	ZC
NBP	374.612 K	PC	DC	OM
MU		RG		

VAPOR PRESSURE CORRELATION CONSTANTS, K AND KPA:

VAPRES-2	VAPRES-2	VAPRES-2	RIEDEL
301. - 404. K	301. - 394. K	301. - 404. K	360. - 404. K
20 POINTS	18 POINTS	20 POINTS	8 POINTS
RMSD = 0.2447	RMSD = 0.1234	RMSD = 0.2447	RMSD = 0.2287
0.37458509E+04	0.58309851E+04	0.37458509E+04	0.21460975E+03
-0.83745265E+05	-0.12813615E+06	-0.83745265E+05	-0.29107224E+02
0.20602000E+01	0.32054287E+01	0.20602000E+01	-0.14152633E+05
-0.10160316E-02	-0.15734292E-02	-0.10160316E-02	0.99238530E-16
-0.69978928E+03	-0.10908577E+04	-0.69978928E+03	

LIQUID DENSITY CORRELATION CONSTANTS, K AND G/ML:

FRANCIS1	FRANCIS1	FRANCIS1
293. - 304. K	293. - 304. K	293. - 304. K
4 POINTS	4 POINTS	4 POINTS
RMSD = 0.00007	RMSD = 0.00007	RMSD = 0.00007
0.96026617E+00	0.96026617E+00	0.96026617E+00
0.76243468E-03	0.76243468E-03	0.76243468E-03
0.59999990E+01	0.59999990E+01	0.59999990E+01
0.56838794E+03	0.56838794E+03	0.56838794E+03

SECOND VIRIAL COEFFICIENT CORRELATION CONSTANTS, ML/MOL:

LITERATURE DOCUMENTS FOR CORRELATED VAPOR PRESSURE VALUES:

   6518   6800

LITERATURE DOCUMENTS FOR CORRELATED LIQUID DENSITY VALUES:

   6518

LITERATURE DOCUMENTS REPORTING VIRIAL COEFFICIENT DATA:

T,K	VAPOR PRESSURE, KPA	SATURATED LIQUID VOLUME, ML/MOL	SECOND VIRIAL COEFFICIENT, ML/MOL	HEAT OF VAPORIZATION, J/MOL	SATURATED VAPOR VOLUME, ML/MOL
293		156.92			
295		157.29			
298		157.85			
300		158.23			
303	7.498	158.80		35950.	335978.
305	8.232				308043.
308	9.441				271245.
310	10.32				249663.
313	11.77				221062.
315	12.83				204189.
318	14.55				181708.
320	15.80				168375.
323	17.84				150525.
325	19.32				139889.
328	21.72				125587.
330	23.44				117030.
333	26.25				105477.
335	28.27				98539.
338	31.53				89139.
340	33.87				83475.
343	37.63				75777.
345	40.33				71125.
348	44.66				64784.
351	49.36				59125.
353	52.70				55690.
356	58.05				50991.
358	61.84				48133.
361	67.89				44214.
363	72.16				41825.
366	78.95				38543.
368	83.74				36540.
371	91.31				33783.
373	96.62				32097.
376	105.0				29774.
378	110.9				28352.
381	120.0				26390.
383	126.4				25188.
386	136.4				23527.
388	143.3				22509.
391	154.1				21102.
393	161.5				20238.
396	172.9				19045.
398	180.7				18313.
401	192.7				17301.
403	200.9				16680.

COMPOUND CONSTANTS:

MP 166.837 K	TC	VC	ZC
NBP 378.063 K	PC	DC	OM
MU	RG		

VAPOR PRESSURE CORRELATION CONSTANTS, K AND KPA:

VAPRES-2	VAPRES-2	VAPRES-2	RIEDEL
263. - 379. K	263. - 379. K	263. - 379. K	364. - 379. K
16 POINTS	16 POINTS	16 POINTS	4 POINTS
RMSD = 0.0275	RMSD = 0.0275	RMSD = 0.0275	RMSD = 0.0000
-0.20835103E+04	-0.20835103E+04	-0.20835103E+04	0.22034939E+03
0.38364917E+05	0.38364917E+05	0.38364917E+05	-0.30230151E+02
-0.12129709E+01	-0.12129709E+01	-0.12129709E+01	-0.13985059E+05
0.60731778E-03	0.60731778E-03	0.60731778E-03	0.23222176E-15
0.39737183E+03	0.39737183E+03	0.39737183E+03	

LIQUID DENSITY CORRELATION CONSTANTS, K AND G/ML:

FRANCIS1	FRANCIS1	FRANCIS1
293. - 304. K	293. - 304. K	293. - 304. K
4 POINTS	4 POINTS	4 POINTS
RMSD = 0.00000	RMSD = 0.00000	RMSD = 0.00000
0.96108747E+00	0.96108747E+00	0.96108747E+00
0.71255304E-03	0.71255304E-03	0.71255304E-03
0.66464462E+01	0.66464462E+01	0.66464462E+01
0.50770923E+03	0.50770923E+03	0.50770923E+03

SECOND VIRIAL COEFFICIENT CORRELATION CONSTANTS, ML/MOL:

LITERATURE DOCUMENTS FOR CORRELATED VAPOR PRESSURE VALUES:

4392   6518

LITERATURE DOCUMENTS FOR CORRELATED LIQUID DENSITY VALUES:

6518

LITERATURE DOCUMENTS REPORTING VIRIAL COEFFICIENT DATA:

T,K	VAPOR PRESSURE, KPA	SATURATED LIQUID VOLUME, ML/MOL	SECOND VIRIAL COEFFICIENT, ML/MOL	HEAT OF VAPORIZATION, J/MOL	SATURATED VAPOR VOLUME, ML/MOL
263	0.6552				3337484.
265	0.7423				2968146.
268	0.8931				2495094.
270	1.009				2225917.
273	1.207				1880187.
276	1.441				1592913.
278	1.618				1428635.
281	1.920				1216558.
284	2.272				1039204.
286	2.537				937214.
289	2.986				804803.
291	3.322				728375.
294	3.888	155.75		37277.	628783.
297	4.535	156.31		37243.	544529.
299	5.016	156.68		37210.	495596.
302	5.820	157.25		37144.	431443.
305	6.731				376763.
307	7.403				344807.
310	8.516				302649.
313	9.767				266446.
315	10.68				245154.
318	12.19				216890.
320	13.29				200201.
323	15.09				177963.
326	17.09				158634.
328	18.53				147150.
331	20.89				131755.
334	23.48				118276.
336	25.35				110219.
339	28.37				99354.
342	31.67				89774.
344	34.04				84014.
347	37.86				76201.
349	40.59				71487.
352	44.98				65071.
355	49.73				59351.
357	53.12				55881.
360	58.54				51131.
363	64.40				46869.
365	68.55				44269.
368	75.19				40692.
371	82.34				37463.
373	87.40				35483.
376	95.47				32745.
378	101.2				31060.

417

COMPOUND CONSTANTS:

MP 191.812 K	TC 593.250 K	VC 528.00 ML/MOL	ZC
NBP 420.037 K	PC 2.330 MPA	DC 0.2391 G/ML	OM 0.4174
MU	RG 4.9687 ANGSTROM		

VAPOR PRESSURE CORRELATION CONSTANTS, K AND KPA:

WAGNER	WAGNER	WAGNER	WAGNER
339. - 422. K	339. - 422. K	339. - 422. K	405. - 422. K
19 POINTS	19 POINTS	19 POINTS	7 POINTS
RMSD = 0.0073	RMSD = 0.0073	RMSD = 0.0073	RMSD = 0.0049
-0.80254074E+01	-0.80254074E+01	-0.80254074E+01	-0.60588720E+01
0.15655676E+01	0.15655676E+01	0.15655676E+01	-0.37111274E+01
-0.49747721E+01	-0.49747721E+01	-0.49747721E+01	0.67893158E+01
0.23758496E+00	0.23758496E+00	0.23758496E+00	-0.55417651E+02

LIQUID DENSITY CORRELATION CONSTANTS, K AND G/ML:

FRANCIS1	FRANCIS1	FRANCIS1
293. - 334. K	293. - 334. K	293. - 334. K
8 POINTS	8 POINTS	8 POINTS
RMSD = 0.00006	RMSD = 0.00006	RMSD = 0.00006
0.95968002E+00	0.95968002E+00	0.95968002E+00
0.67370338E-03	0.67370338E-03	0.67370338E-03
0.86231489E+01	0.86231489E+01	0.86231489E+01
0.55489722E+03	0.55489722E+03	0.55489722E+03

SECOND VIRIAL COEFFICIENT CORRELATION CONSTANTS, ML/MOL:

LITERATURE DOCUMENTS FOR CORRELATED VAPOR PRESSURE VALUES:

1991

LITERATURE DOCUMENTS FOR CORRELATED LIQUID DENSITY VALUES:

1991 17433

LITERATURE DOCUMENTS REPORTING VIRIAL COEFFICIENT DATA:

T,K	VAPOR PRESSURE, KPA	SATURATED LIQUID VOLUME, ML/MOL	SECOND VIRIAL COEFFICIENT, ML/MOL	HEAT OF VAPORIZATION, J/MOL	SATURATED VAPOR VOLUME, ML/MOL
293		173.08			
295		173.47			
298		174.04			
301		174.62			
304		175.21			
307		175.81			
310		176.41			
313		177.02			
316		177.63			
319		178.25			
322		178.88			
325		179.52			
328		180.16			
331		180.82			
334		181.48			
339	6.175				454006.
342	7.051				400902.
345	8.028				354945.
348	9.116				315062.
351	10.32				280355.
354	11.66				250073.
357	13.14				223584.
360	14.78				200353.
363	16.57				179931.
366	18.55				161935.
369	20.71				146041.
372	23.07				131970.
375	25.65				119487.
378	28.46				108389.
380	30.47				101673.
383	33.70				92508.
386	37.19				84315.
389	40.98				76978.
392	45.07				70394.
395	49.48				64475.
398	54.23				59144.
401	59.34				54335.
404	64.82				49988.
407	70.70				46054.
410	76.99				42487.
413	83.72				39247.
416	90.90				36300.
419	98.56				33616.
421	103.9				31959.

COMPOUND CONSTANTS:

MP   TC   VC   ZC
NBP   423.973 K   PC   DC   OM
MU   RG

    VAPOR PRESSURE CORRELATION CONSTANTS, K AND KPA:

    RPM2   RPM2   RPM2
    379. - 425. K   379. - 425. K   379. - 425. K
      5 POINTS   5 POINTS   5 POINTS
    RMSD = 0.0045   RMSD = 0.0045   RMSD = 0.0045
    0.23704024E+02   0.23704024E+02   0.23704024E+02
    -0.62121906E+04   -0.62121906E+04   -0.62121906E+04
    -0.12690106E-01   -0.12690106E-01   -0.12690106E-01
    0.52678312E-05   0.52678312E-05   0.52678312E-05

    LIQUID DENSITY CORRELATION CONSTANTS, K AND G/ML:

    FRANCIS1   FRANCIS1   FRANCIS1
    293. - 334. K   293. - 334. K   293. - 334. K
      5 POINTS   5 POINTS   5 POINTS
    RMSD = 0.00001   RMSD = 0.00001   RMSD = 0.00001
    0.97073853E+00   0.97073853E+00   0.97073853E+00
    0.71296655E-03   0.71296655E-03   0.71296655E-03
    0.65637407E+01   0.65637407E+01   0.65637407E+01
    0.57792334E+03   0.57792334E+03   0.57792334E+03

    SECOND VIRIAL COEFFICIENT CORRELATION CONSTANTS, ML/MOL:

LITERATURE DOCUMENTS FOR CORRELATED VAPOR PRESSURE VALUES:

    8405

LITERATURE DOCUMENTS FOR CORRELATED LIQUID DENSITY VALUES:

    17433

LITERATURE DOCUMENTS REPORTING VIRIAL COEFFICIENT DATA:

T,K	VAPOR PRESSURE, KPA	SATURATED LIQUID VOLUME, ML/MOL	SECOND VIRIAL COEFFICIENT, ML/MOL	HEAT OF VAPORIZATION, J/MOL	SATURATED VAPOR VOLUME, ML/MOL
293		170.87			
296		171.43			
299		171.98			
302		172.55			
305		173.11			
308		173.69			
311		174.27			
314		174.85			
317		175.44			
320		176.03			
323		176.63			
326		177.24			
329		177.85			
332		178.47			
380	26.97				117164.
383	29.86				106633.
386	33.01				97228.
389	36.42				88810.
392	40.11				81264.
395	44.09				74485.
398	48.39				68385.
401	53.02				62887.
404	57.99				57922.
407	63.33				53432.
410	69.06				49364.
413	75.18				45674.
416	81.73				42320.
419	88.72				39267.
422	96.17				36485.
425	104.1				33946.

COMPOUND CONSTANTS:

MP		TC		VC		ZC
NBP	423.223 K	PC		DC		OM
MU		RG				

VAPOR PRESSURE CORRELATION CONSTANTS, K AND KPA:

RPM2	RPM2	RPM2
379. - 424. K	379. - 424. K	379. - 424. K
6 POINTS	6 POINTS	6 POINTS
RMSD = 0.0109	RMSD = 0.0109	RMSD = 0.0109
0.22090394E+02	0.22090394E+02	0.22090394E+02
-0.60096262E+04	-0.60096262E+04	-0.60096262E+04
-0.84674268E-02	-0.84674268E-02	-0.84674268E-02
0.17374476E-05	0.17374476E-05	0.17374476E-05

LIQUID DENSITY CORRELATION CONSTANTS, K AND G/ML:

FRANCIS1	FRANCIS1	FRANCIS1
293. - 334. K	293. - 334. K	293. - 334. K
5 POINTS	5 POINTS	5 POINTS
RMSD = 0.00003	RMSD = 0.00003	RMSD = 0.00003
0.97052610E+00	0.97052610E+00	0.97052610E+00
0.72344765E-03	0.72344765E-03	0.72344765E-03
0.85213394E+01	0.85213394E+01	0.85213394E+01
0.62381372E+03	0.62381372E+03	0.62381372E+03

SECOND VIRIAL COEFFICIENT CORRELATION CONSTANTS, ML/MOL:

LITERATURE DOCUMENTS FOR CORRELATED VAPOR PRESSURE VALUES:

8405

LITERATURE DOCUMENTS FOR CORRELATED LIQUID DENSITY VALUES:

17433

LITERATURE DOCUMENTS REPORTING VIRIAL COEFFICIENT DATA:

T,K	VAPOR PRESSURE, KPA	SATURATED LIQUID VOLUME, ML/MOL	SECOND VIRIAL COEFFICIENT, ML/MOL	HEAT OF VAPORIZATION, J/MOL	SATURATED VAPOR VOLUME, ML/MOL
293		172.27			
295		172.65			
298		173.22			
301		173.80			
304		174.38			
307		174.96			
310		175.55			
313		176.15			
316		176.75			
319		177.36			
322		177.97			
325		178.59			
328		179.21			
331		179.84			
334		180.48			
379	26.42				119254.
382	29.29				108423.
385	32.41				98759.
388	35.80				90119.
391	39.46				82379.
394	43.43				75434.
397	47.71				69190.
400	52.32				63568.
403	57.28				58496.
406	62.61				53913.
409	68.33				49765.
412	74.46				46005.
415	81.01				42591.
418	88.01				39487.
421	95.48				36661.
423	100.7				34917.

COMPOUND CONSTANTS:

MP		TC		VC		ZC
NBP	421.159 K	PC		DC		OM
MU		RG				

VAPOR PRESSURE CORRELATION CONSTANTS, K AND KPA:

RPM2	RPM2	RPM2
377. - 422. K	377. - 422. K	377. - 422. K
5 POINTS	5 POINTS	5 POINTS
RMSD = 0.0098	RMSD = 0.0098	RMSD = 0.0098
0.23570266E+02	0.23570266E+02	0.23570266E+02
-0.61354558E+04	-0.61354558E+04	-0.61354558E+04
-0.12652373E-01	-0.12652373E-01	-0.12652373E-01
0.53263389E-05	0.53263389E-05	0.53263389E-05

LIQUID DENSITY CORRELATION CONSTANTS, K AND G/ML:

FRANCIS1	FRANCIS1	FRANCIS1
293. - 334. K	293. - 334. K	293. - 334. K
5 POINTS	5 POINTS	5 POINTS
RMSD = 0.00003	RMSD = 0.00003	RMSD = 0.00003
0.97531253E+00	0.97531253E+00	0.97531253E+00
0.76107774E-03	0.76107774E-03	0.76107774E-03
0.59999990E+01	0.59999990E+01	0.59999990E+01
0.62463721E+03	0.62463721E+03	0.62463721E+03

SECOND VIRIAL COEFFICIENT CORRELATION CONSTANTS, ML/MOL:

LITERATURE DOCUMENTS FOR CORRELATED VAPOR PRESSURE VALUES:

8405

LITERATURE DOCUMENTS FOR CORRELATED LIQUID DENSITY VALUES:

17433

LITERATURE DOCUMENTS REPORTING VIRIAL COEFFICIENT DATA:

T,K	VAPOR PRESSURE, KPA	SATURATED LIQUID VOLUME, ML/MOL	SECOND VIRIAL COEFFICIENT, ML/MOL	HEAT OF VAPORIZATION, J/MOL	SATURATED VAPOR VOLUME, ML/MOL
293		171.94			
295		172.32			
298		172.90			
301		173.48			
304		174.07			
307		174.66			
310		175.26			
313		175.86			
316		176.47			
319		177.08			
322		177.70			
325		178.32			
328		178.95			
331		179.58			
334		180.22			
378	27.59				113924.
380	29.53				106977.
383	32.66				97492.
386	36.06				89009.
389	39.73				81408.
392	43.70				74584.
395	47.98				68447.
398	52.59				62918.
401	57.56				57928.
404	62.88				53417.
407	68.60				49332.
410	74.71				45628.
413	81.25				42263.
416	88.23				39202.
419	95.67				36413.
421	100.9				34691.

COMPOUND CONSTANTS:

MP   TC   VC   ZC
NBP   421.329 K   PC   DC   OM
MU   RG

VAPOR PRESSURE CORRELATION CONSTANTS, K AND KPA:

RPM2   RPM2   RPM2
377. - 422. K   377. - 422. K   377. - 422. K
  6 POINTS   6 POINTS   6 POINTS
RMSD = 0.0029   RMSD = 0.0029   RMSD = 0.0029
 0.41154529E+02   0.41154529E+02   0.41154529E+02
-0.85224735E+04   -0.85224735E+04   -0.85224735E+04
-0.56110323E-01   -0.56110323E-01   -0.56110323E-01
 0.41304527E-04   0.41304527E-04   0.41304527E-04

LIQUID DENSITY CORRELATION CONSTANTS, K AND G/ML:

FRANCIS1   FRANCIS1   FRANCIS1
293. - 334. K   293. - 334. K   293. - 334. K
  5 POINTS   5 POINTS   5 POINTS
RMSD = 0.00005   RMSD = 0.00005   RMSD = 0.00005
0.97189420E+00   0.97189420E+00   0.97189420E+00
0.73267356E-03   0.73267356E-03   0.73267356E-03
0.81711779E+01   0.81711779E+01   0.81711779E+01
0.62463721E+03   0.62463721E+03   0.62463721E+03

SECOND VIRIAL COEFFICIENT CORRELATION CONSTANTS, ML/MOL:

LITERATURE DOCUMENTS FOR CORRELATED VAPOR PRESSURE VALUES:

  8405

LITERATURE DOCUMENTS FOR CORRELATED LIQUID DENSITY VALUES:

  17433

LITERATURE DOCUMENTS REPORTING VIRIAL COEFFICIENT DATA:

426

T,K	VAPOR PRESSURE, KPA	SATURATED LIQUID VOLUME, ML/MOL	SECOND VIRIAL COEFFICIENT, ML/MOL	HEAT OF VAPORIZATION, J/MOL	SATURATED VAPOR VOLUME, ML/MOL
293		172.32			
295		172.70			
298		173.28			
301		173.86			
304		174.44			
307		175.03			
310		175.63			
313		176.23			
316		176.83			
319		177.45			
322		178.06			
325		178.68			
328		179.31			
331		179.95			
334		180.59			
378	27.12				115877.
380	29.06				108716.
383	32.18				98960.
386	35.56				90253.
389	39.22				82466.
392	43.18				75487.
395	47.45				69218.
398	52.05				63578.
401	57.00				58493.
404	62.32				53899.
407	68.03				49743.
410	74.15				45974.
413	80.70				42552.
416	87.70				39439.
419	95.18				36603.
421	100.4				34851.

COMPOUND CONSTANTS:

MP		TC		VC		ZC
NBP	420.570 K	PC		DC		OM
MU		RG				

VAPOR PRESSURE CORRELATION CONSTANTS, K AND KPA:

RPM2	RPM2	RPM2
376. - 421. K	376. - 421. K	376. - 421. K
5 POINTS	5 POINTS	5 POINTS
RMSD = 0.0019	RMSD = 0.0019	RMSD = 0.0019
0.18064368E+02	0.18064368E+02	0.18064368E+02
-0.53914714E+04	-0.53914714E+04	-0.53914714E+04
0.11373727E-02	0.11373727E-02	0.11373727E-02
-0.62469235E-05	-0.62469235E-05	-0.62469235E-05

LIQUID DENSITY CORRELATION CONSTANTS, K AND G/ML:

FRANCIS1	FRANCIS1	FRANCIS1
293. - 334. K	293. - 334. K	293. - 334. K
5 POINTS	5 POINTS	5 POINTS
RMSD = 0.00003	RMSD = 0.00003	RMSD = 0.00003
0.97487438E+00	0.97487438E+00	0.97487438E+00
0.74691861E-03	0.74691861E-03	0.74691861E-03
0.65794945E+01	0.65794945E+01	0.65794945E+01
0.60210132E+03	0.60210132E+03	0.60210132E+03

SECOND VIRIAL COEFFICIENT CORRELATION CONSTANTS, ML/MOL:

LITERATURE DOCUMENTS FOR CORRELATED VAPOR PRESSURE VALUES:

8405

LITERATURE DOCUMENTS FOR CORRELATED LIQUID DENSITY VALUES:

17433

LITERATURE DOCUMENTS REPORTING VIRIAL COEFFICIENT DATA:

T,K	VAPOR PRESSURE, KPA	SATURATED LIQUID VOLUME, ML/MOL	SECOND VIRIAL COEFFICIENT, ML/MOL	HEAT OF VAPORIZATION, J/MOL	SATURATED VAPOR VOLUME, ML/MOL
293		171.82			
295		172.20			
298		172.78			
301		173.36			
304		173.95			
307		174.54			
310		175.14			
313		175.74			
316		176.35			
319		176.97			
322		177.58			
324		178.00			
327		178.63			
330		179.27			
333		179.91			
377	27.23				115115.
380	30.16				104766.
383	33.34				95521.
386	36.78				87247.
388	39.24				82213.
391	43.17				75308.
394	47.41				69098.
397	51.98				63505.
400	56.89				58459.
403	62.17				53898.
406	67.83				49770.
409	73.88				46027.
412	80.36				42628.
415	87.27				39538.
418	94.64				36724.
420	99.81				34987.

COMPOUND CONSTANTS:

MP		TC		VC		ZC
NBP	420.923 K	PC		DC		OM
MU		RG				

VAPOR PRESSURE CORRELATION CONSTANTS, K AND KPA:

RPM2	RPM2	RPM2
377. - 421. K	377. - 421. K	377. - 421. K
6 POINTS	6 POINTS	6 POINTS
RMSD = 0.0059	RMSD = 0.0059	RMSD = 0.0059
0.27259203E+02	0.27259203E+02	0.27259203E+02
-0.66434292E+04	-0.66434292E+04	-0.66434292E+04
-0.21685163E-01	-0.21685163E-01	-0.21685163E-01
0.12811591E-04	0.12811591E-04	0.12811591E-04

LIQUID DENSITY CORRELATION CONSTANTS, K AND G/ML:

FRANCIS1	FRANCIS1	FRANCIS1
293. - 334. K	293. - 334. K	293. - 334. K
5 POINTS	5 POINTS	5 POINTS
RMSD = 0.00004	RMSD = 0.00004	RMSD = 0.00004
0.96685982E+00	0.96685982E+00	0.96685982E+00
0.72294055E-03	0.72294055E-03	0.72294055E-03
0.65207682E+01	0.65207682E+01	0.65207682E+01
0.57984839E+03	0.57984839E+03	0.57984839E+03

SECOND VIRIAL COEFFICIENT CORRELATION CONSTANTS, ML/MOL:

LITERATURE DOCUMENTS FOR CORRELATED VAPOR PRESSURE VALUES:

8405

LITERATURE DOCUMENTS FOR CORRELATED LIQUID DENSITY VALUES:

17433

LITERATURE DOCUMENTS REPORTING VIRIAL COEFFICIENT DATA:

T,K	VAPOR PRESSURE, KPA	SATURATED LIQUID VOLUME, ML/MOL	SECOND VIRIAL COEFFICIENT, ML/MOL	HEAT OF VAPORIZATION, J/MOL	SATURATED VAPOR VOLUME, ML/MOL
293		172.39			
295		172.77			
298		173.34			
301		173.92			
304		174.50			
307		175.09			
310		175.68			
313		176.28			
316		176.88			
319		177.49			
322		178.11			
324		178.52			
327		179.15			
330		179.78			
333		180.42			
377	26.65				117598.
380	29.55				106915.
383	32.70				97387.
386	36.11				88870.
388	38.55				83694.
391	42.44				76601.
394	46.65				70229.
397	51.18				64495.
400	56.06				59325.
403	61.31				54656.
406	66.93				50432.
409	72.97				46605.
412	79.42				43131.
415	86.32				39973.
418	93.68				37097.
420	98.86				35323.

COMPOUND CONSTANTS:

MP	TC	VC	ZC
NBP 275.550 K	PC	DC	OM
MU	RG		

VAPOR PRESSURE CORRELATION CONSTANTS, K AND KPA:

VAPRES-2	VAPRES-2	VAPRES-2
210. - 276. K	210. - 276. K	210. - 276. K
10 POINTS	10 POINTS	10 POINTS
RMSD = 0.0990	RMSD = 0.0990	RMSD = 0.0990
0.20530801E+04	0.20530801E+04	0.20530801E+04
-0.35706387E+05	-0.35706387E+05	-0.35706387E+05
0.17009642E+01	0.17009642E+01	0.17009642E+01
-0.11767959E-02	-0.11767959E-02	-0.11767959E-02
-0.40902726E+03	-0.40902726E+03	-0.40902726E+03

LIQUID DENSITY CORRELATION CONSTANTS, K AND G/ML:

SECOND VIRIAL COEFFICIENT CORRELATION CONSTANTS, ML/MOL:

LITERATURE DOCUMENTS FOR CORRELATED VAPOR PRESSURE VALUES:

2817

LITERATURE DOCUMENTS FOR CORRELATED LIQUID DENSITY VALUES:

LITERATURE DOCUMENTS REPORTING VIRIAL COEFFICIENT DATA:

T,K	VAPOR PRESSURE, KPA	SATURATED LIQUID VOLUME, ML/MOL	SECOND VIRIAL COEFFICIENT, ML/MOL	HEAT OF VAPORIZATION, J/MOL	SATURATED VAPOR VOLUME, ML/MOL
210	3.465				503897.
211	3.710				472899.
213	4.243				417433.
214	4.532				392609.
216	5.160				348012.
217	5.501				327974.
219	6.239				291843.
220	6.638				275550.
222	7.501				246074.
223	7.967				232737.
225	8.971				208533.
226	9.512				197548.
228	10.68				177555.
229	11.30				168456.
231	12.65				151850.
232	13.37				144272.
234	14.92				130408.
235	15.75				124066.
237	17.53				112438.
238	18.48				107107.
240	20.51				97310.
241	21.59				92809.
243	23.90				84522.
244	25.14				80707.
246	27.76				73670.
247	29.16				70425.
249	32.13				64429.
250	33.71				61659.
252	37.06				56533.
253	38.84				54162.
255	42.60				49766.
256	44.60				47729.
258	48.81				43949.
259	51.04				42195.
261	55.74				38935.
262	58.21				37420.
264	63.44				34602.
265	66.18				33291.
267	71.96				30849.
268	75.00				29711.
270	81.36				27590.
271	84.70				26602.
273	91.69				24756.
274	95.34				23894.
276	103.0				22285.

COMPOUND CONSTANTS:

MP	138.070 K	TC	506.110 K	VC		ZC	
NBP	317.382 K	PC	4.783 MPA	DC		OM	
MU	0.9500 DEBYE	RG					

MP   138.070 K   TC   506.110 K   VC   ZC
NBP 317.382 K   PC   4.783 MPA   DC   OM
MU   0.9500 DEBYE   RG

VAPOR PRESSURE CORRELATION CONSTANTS, K AND KPA:

VAPRES-2	VAPRES-2	VAPRES-2	WAGNER
230. - 318. K	230. - 318. K	230. - 318. K	300. - 318. K
14 POINTS	14 POINTS	14 POINTS	8 POINTS
RMSD = 0.0290	RMSD = 0.0290	RMSD = 0.0290	RMSD = 0.0030
-0.29497550E+04	-0.29497550E+04	-0.29497550E+04	-0.88105920E+00
0.49247228E+05	0.49247228E+05	0.49247228E+05	-0.13106005E+02
-0.20944627E+01	-0.20944627E+01	-0.20944627E+01	0.19969060E+02
0.12462301E-02	0.12462301E-02	0.12462301E-02	-0.51910148E+02
0.57957557E+03	0.57957557E+03	0.57957557E+03	

LIQUID DENSITY CORRELATION CONSTANTS, K AND G/ML:

RACKETT	RACKETT	RACKETT	RACKETT
273. - 303. K	273. - 303. K	273. - 303. K	273. - 303. K
7 POINTS	7 POINTS	7 POINTS	7 POINTS
RMSD = 0.00036	RMSD = 0.00036	RMSD = 0.00036	RMSD = 0.00036
0.27492715E+00	0.27492715E+00	0.27492715E+00	0.27492715E+00

SECOND VIRIAL COEFFICIENT CORRELATION CONSTANTS, ML/MOL:

LITERATURE DOCUMENTS FOR CORRELATED VAPOR PRESSURE VALUES:

1991 5065

LITERATURE DOCUMENTS FOR CORRELATED LIQUID DENSITY VALUES:

1991 2523 9923

LITERATURE DOCUMENTS REPORTING VIRIAL COEFFICIENT DATA:

T,K	VAPOR PRESSURE, KPA	SATURATED LIQUID VOLUME, ML/MOL	SECOND VIRIAL COEFFICIENT, ML/MOL	HEAT OF VAPORIZATION, J/MOL	SATURATED VAPOR VOLUME, ML/MOL
230	1.406				1359364.
232	1.608				1198211.
234	1.837				1057712.
236	2.096				935122.
238	2.387				828061.
240	2.713				734468.
242	3.078				·652558.
244	3.487				580789.
246	3.942				517826.
248	4.448				462517.
250	5.010				413864.
252	5.632				371005.
254	6.320				333196.
256	7.077				299790.
258	7.910				270230.
260	8.824				244032.
262	9.825				220776.
264	10.92				200099.
266	12.11				181685.
268	13.41				165259.
270	14.82				150582.
272	16.35				137447.
274	18.00	86.06		29577.	125671.
276	19.79	86.28		29446.	115097.
278	21.72	86.50		29311.	105586.
280	23.79	86.73		29174.	97016.
282	26.02	86.95		29036.	89282.
284	28.41	87.18		28898.	82290.
286	30.97	87.41		28761.	75959.
288	33.71	87.64		28627.	70217.
290	36.64	87.88		28495.	65000.
292	39.77	88.11		28366.	60252.
294	43.11	88.35		28243.	55924.
296	46.66	88.59		28125.	51973.
298	50.44	88.84		28014.	48359.
300	54.45	89.08		27911.	45049.
302	58.72	89.33		27816.	42013.
304	63.24				39222.
306	68.04				36654.
308	73.13				34286.
310	78.53				32100.
312	84.24				30078.
314	90.29				28206.
316	96.70				26469.
318	103.5				24856.

435

COMPOUND CONSTANTS:

MP		TC		VC		ZC
NBP	315.340 K	PC		DC		OM
MU		RG				

VAPOR PRESSURE CORRELATION CONSTANTS, K AND KPA:

RIEDEL	RIEDEL	RIEDEL	RIEDEL
290. - 315. K	290. - 315. K	290. - 315. K	290. - 315. K
6 POINTS	6 POINTS	6 POINTS	6 POINTS
RMSD = 0.1810	RMSD = 0.1810	RMSD = 0.1810	RMSD = 0.1810
-0.14343673E+02	-0.14343673E+02	-0.14343673E+02	-0.14343673E+02
0.49284152E+01	0.49284152E+01	0.49284152E+01	0.49284152E+01
-0.27265302E+04	-0.27265302E+04	-0.27265302E+04	-0.27265302E+04
-0.75995909E-15	-0.75995909E-15	-0.75995909E-15	-0.75995909E-15

LIQUID DENSITY CORRELATION CONSTANTS, K AND G/ML:

SECOND VIRIAL COEFFICIENT CORRELATION CONSTANTS, ML/MOL:

LITERATURE DOCUMENTS FOR CORRELATED VAPOR PRESSURE VALUES:

15712

LITERATURE DOCUMENTS FOR CORRELATED LIQUID DENSITY VALUES:

LITERATURE DOCUMENTS REPORTING VIRIAL COEFFICIENT DATA:

T,K	VAPOR PRESSURE, KPA	SATURATED LIQUID VOLUME, ML/MOL	SECOND VIRIAL COEFFICIENT, ML/MOL	HEAT OF VAPORIZATION, J/MOL	SATURATED VAPOR VOLUME, ML/MOL
290	42.35				56934.
291	44.07				54899.
292	45.84				52959.
293	47.66				51110.
294	49.54				49347.
295	51.46				47665.
296	53.43				46061.
297	55.46				44529.
298	57.53				43067.
299	59.66				41671.
300	61.84				40337.
301	64.07				39063.
302	66.35				37846.
303	68.68				36682.
304	71.06				35569.
305	73.49				34505.
306	75.97				33488.
307	78.51				32514.
308	81.08				31582.
309	83.71				30691.
310	86.38				29837.
311	89.10				29020.
312	91.87				28237.
313	94.68				27488.
314	97.52				26770.
315	100.4				26082.

COMPOUND CONSTANTS:

MP 169.657 K	TC 560.479 K	VC	ZC
NBP 356.121 K	PC	DC	OM
MU 0.6600 DEBYE	RG		

VAPOR PRESSURE CORRELATION CONSTANTS, K AND KPA:

VAPRES-2	VAPRES-2	VAPRES-2	RPM2
176. - 365. K	176. - 365. K	176. - 365. K	336. - 365. K
54 POINTS	54 POINTS	54 POINTS	16 POINTS
RMSD = 0.0610	RMSD = 0.0610	RMSD = 0.0610	RMSD = 0.0510
0.22236583E+03	0.22236583E+03	0.22236583E+03	0.57588261E+02
-0.88644145E+04	-0.88644145E+04	-0.88644145E+04	-0.89534287E+04
0.10781013E+00	0.10781013E+00	0.10781013E+00	-0.11573877E+00
-0.57232494E-04	-0.57232494E-04	-0.57232494E-04	-0.10557013E-03
-0.38124377E+02	-0.38124377E+02	-0.38124377E+02	

LIQUID DENSITY CORRELATION CONSTANTS, K AND G/ML:

FRANCIS1	FRANCIS1	FRANCIS1
243. - 374. K	243. - 374. K	243. - 374. K
26 POINTS	26 POINTS	26 POINTS
RMSD = 0.00027	RMSD = 0.00027	RMSD = 0.00270
0.10903921E+01	0.10903921E+01	0.10903921E+01
0.87289442E-03	0.87289442E-03	0.87289442E-03
0.96937876E+01	0.96937876E+01	0.96937876E+01
0.69965552E+03	0.69965552E+03	0.69965552E+03

SECOND VIRIAL COEFFICIENT CORRELATION CONSTANTS, ML/MOL:

LITERATURE DOCUMENTS FOR CORRELATED VAPOR PRESSURE VALUES:

1991 5065 5641 10569 17966 21583 40724

LITERATURE DOCUMENTS FOR CORRELATED LIQUID DENSITY VALUES:

791 1744 1991 2523 6970 10296 10808 12849 20472 20518 40686

LITERATURE DOCUMENTS REPORTING VIRIAL COEFFICIENT DATA:

T,K	VAPOR PRESSURE, KPA	SATURATED LIQUID VOLUME, ML/MOL	SECOND VIRIAL COEFFICIENT, ML/MOL	HEAT OF VAPORIZATION, J/MOL	SATURATED VAPOR VOLUME, ML/MOL
176	0.0003634				0.4026E+10
180	0.0006704				0.2232E+10
185	0.001379				0.1116E+10
189	0.002376				0.6615E+09
194	0.004518				0.3570E+09
198	0.007349				0.2240E+09
203	0.01308				0.1291E+09
207	0.02025				0.8500E+08
212	0.03403				0.5179E+08
216	0.05051				0.3555E+08
221	0.08082				0.2273E+08
225	0.1157				0.1618E+08
230	0.1773				0.1078E+08
234	0.2458				7915391.
239	0.3631				5472774.
243	0.4895	95.85		35949.	4127792.
248	0.6998	96.36		35704.	2946419.
252	0.9205	96.78		35515.	2276092.
257	1.279	97.31		35286.	1670929.
261	1.646	97.74		35109.	1318327.
266	2.229	98.28		34894.	992164.
270	2.815	98.71		34726.	797602.
275	3.726	99.27		34523.	613687.
279	4.624	99.71		34364.	501657.
284	5.998	100.28		34169.	393685.
288	7.330	100.74		34016.	326671.
293	9.336	101.31		33828.	260951.
297	11.25	101.78		33679.	219468.
302	14.09	102.37		33494.	178145.
306	16.78	102.85		33348.	151663.
311	20.70	103.46		33164.	124905.
315	24.36	103.95		33018.	107521.
320	29.65	104.57		32833.	89728.
324	34.53	105.08		32685.	78023.
329	41.51	105.71		32496.	65903.
333	47.87	106.23		32344.	57839.
338	56.89	106.88		32149.	49399.
342	65.03	107.41		31990.	43725.
347	76.47	108.08		31786.	37728.
351	86.71	108.63		31619.	33658.
356	101.0	109.32		31403.	29318.
360	113.6	109.88		31226.	26347.
365	131.1	110.59		30996.	23153.
369		111.16			
374		111.89			

COMPOUND CONSTANTS:

MP	146.617 K	TC	VC	ZC
NBP	348.633 K	PC	DC	OM
MU		RG		

VAPOR PRESSURE CORRELATION CONSTANTS, K AND KPA:

LIQUID DENSITY CORRELATION CONSTANTS, K AND G/ML:

FRANCIS1	FRANCIS1	FRANCIS1
273. - 304. K	273. - 304. K	273. - 304. K
5 POINTS	5 POINTS	5 POINTS
RMSD = 0.00047	RMSD = 0.00047	RMSD = 0.00047
0.10567980E+01	0.10567980E+01	0.10567980E+01
0.87137194E-03	0.87137194E-03	0.87137194E-03
0.59999990E+01	0.59999990E+01	0.59999990E+01
0.56838794E+03	0.56838794E+03	0.56838794E+03

SECOND VIRIAL COEFFICIENT CORRELATION CONSTANTS, ML/MOL:

LITERATURE DOCUMENTS FOR CORRELATED VAPOR PRESSURE VALUES:

LITERATURE DOCUMENTS FOR CORRELATED LIQUID DENSITY VALUES:

4670   9707 21804

LITERATURE DOCUMENTS REPORTING VIRIAL COEFFICIENT DATA:

T,K	VAPOR PRESSURE, KPA	SATURATED LIQUID VOLUME, ML/MOL	SECOND VIRIAL COEFFICIENT, ML/MOL	HEAT OF VAPORIZATION, J/MOL	SATURATED VAPOR VOLUME, ML/MOL
273		102.86			
274		102.98			
275		103.10			
276		103.23			
277		103.35			
278		103.47			
279		103.59			
280		103.72			
281		103.84			
282		103.97			
283		104.09			
284		104.21			
285		104.34			
286		104.47			
287		104.59			
288		104.72			
289		104.84			
290		104.97			
291		105.10			
292		105.23			
293		105.35			
294		105.48			
295		105.61			
296		105.74			
297		105.87			
298		106.00			
299		106.13			
300		106.26			
301		106.40			
302		106.53			
303		106.66			
304		106.79			

COMPOUND CONSTANTS:

MP   217.181 K   TC   VC   ZC
NBP   387.507 K   PC   DC   OM
MU   RG

VAPOR PRESSURE CORRELATION CONSTANTS, K AND KPA:

RIEDEL   RIEDEL   RIEDEL
251. - 388. K   251. - 388. K   251. - 388. K
  4 POINTS   4 POINTS   4 POINTS
RMSD = 0.0000   RMSD = 0.0000   RMSD = 0.0000
0.74473521E+02   0.74473521E+02   0.74473521E+02
-0.87708770E+01   -0.87708770E+01   -0.87708770E+01
-0.69694036E+04   -0.69694036E+04   -0.69694036E+04
0.11876524E-15   0.11876524E-15   0.11876524E-15

LIQUID DENSITY CORRELATION CONSTANTS, K AND G/ML:

FRANCIS1   FRANCIS1   FRANCIS1
286. - 294. K   286. - 294. K   286. - 294. K
  4 POINTS   4 POINTS   4 POINTS
RMSD = 0.00000   RMSD = 0.00000   RMSD = 0.00000
0.11233511E+01   0.11233511E+01   0.11233511E+01
0.88310870E-03   0.88310870E-03   0.88310870E-03
0.75155640E+01   0.75155640E+01   0.75155640E+01
0.50566919E+03   0.50566919E+03   0.50566919E+03

SECOND VIRIAL COEFFICIENT CORRELATION CONSTANTS, ML/MOL:

LITERATURE DOCUMENTS FOR CORRELATED VAPOR PRESSURE VALUES:

5065 22457

LITERATURE DOCUMENTS FOR CORRELATED LIQUID DENSITY VALUES:

2123 22447

LITERATURE DOCUMENTS REPORTING VIRIAL COEFFICIENT DATA:

442

T,K	VAPOR PRESSURE, KPA	SATURATED LIQUID VOLUME, ML/MOL	SECOND VIRIAL COEFFICIENT, ML/MOL	HEAT OF VAPORIZATION, J/MOL	SATURATED VAPOR VOLUME, ML/MOL
251	0.1779				0.1173E+08
254	0.2230				9469060.
257	0.2778				7691643.
260	0.3440				6284959.
263	0.4234				5164846.
266	0.5182				4267652.
269	0.6309				3544916.
272	0.7641				2959521.
275	0.9209				2482860.
279	1.172				1978850.
282	1.397				1677931.
285	1.658				1428883.
288	1.960	115.25		37914.	1221831.
291	2.307	115.68		37768.	1048939.
294	2.704	116.12		37627.	903961.
297	3.158				781892.
300	3.675				678708.
303	4.262				591153.
307	5.165				494231.
310	5.943				433700.
313	6.817				381726.
316	7.797				336952.
319	8.893				298256.
322	10.11				264709.
325	11.47				235539.
328	12.98				210100.
331	14.65				187853.
335	17.15				162382.
338	19.26				145938.
341	21.57				131431.
344	24.12				118599.
347	26.91				107223.
350	29.97				97113.
353	33.31				88108.
356	36.97				80071.
359	40.96				72881.
363	46.84				64437.
366	51.71				58849.
369	57.01				53817.
372	62.77				49278.
375	69.02				45175.
378	75.81				41459.
381	83.17				38090.
384	91.15				35029.
387	99.79				32243.

COMPOUND CONSTANTS:

MP	152.753 K	TC		VC		ZC
NBP	383.450 K	PC		DC		OM
MU		RG				

VAPOR PRESSURE CORRELATION CONSTANTS, K AND KPA:

VAPRES-2	VAPRES-2	VAPRES-2	RIEDEL
309. - 384. K	309. - 384. K	309. - 384. K	369. - 384. K
17 POINTS	17 POINTS	17 POINTS	6 POINTS
RMSD = 0.0141	RMSD = 0.0141	RMSD = 0.0141	RMSD = 0.0010
0.21789735E+04	0.21789735E+04	0.21789735E+04	-0.21683911E+02
-0.52836511E+05	-0.52836511E+05	-0.52836511E+05	0.56022102E+01
0.11154303E+01	0.11154303E+01	0.11154303E+01	-0.25937426E+04
-0.52167237E-03	-0.52167237E-03	-0.52167237E-03	-0.82497441E-16
-0.40132577E+03	-0.40132577E+03	-0.40132577E+03	

LIQUID DENSITY CORRELATION CONSTANTS, K AND G/ML:

FRANCIS1	FRANCIS1	FRANCIS1
287. - 336. K	287. - 336. K	287. - 336. K
7 POINTS	7 POINTS	7 POINTS
RMSD = 0.00054	RMSD = 0.00054	RMSD = 0.00054
0.10868092E+01	0.10868092E+01	0.10868092E+01
0.87563018E-03	0.87563018E-03	0.87563018E-03
0.59999990E+01	0.59999990E+01	0.59999990E+01
0.62857544E+03	0.62857544E+03	0.62857544E+03

SECOND VIRIAL COEFFICIENT CORRELATION CONSTANTS, ML/MOL:

LITERATURE DOCUMENTS FOR CORRELATED VAPOR PRESSURE VALUES:

6518

LITERATURE DOCUMENTS FOR CORRELATED LIQUID DENSITY VALUES:

6518   9923

LITERATURE DOCUMENTS REPORTING VIRIAL COEFFICIENT DATA:

T,K	VAPOR PRESSURE, KPA	SATURATED LIQUID VOLUME, ML/MOL	SECOND VIRIAL COEFFICIENT, ML/MOL	HEAT OF VAPORIZATION, J/MOL	SATURATED VAPOR VOLUME, ML/MOL
287		117.58			
289		117.85			
291		118.11			
293		118.38			
295		118.66			
298		119.07			
300		119.34			
302		119.62			
304		119.90			
306		120.18			
309	6.700	120.60		37792.	383452.
311	7.363	120.88		37615.	351167.
313	8.079	121.17		37448.	322110.
315	8.851	121.45		37290.	295911.
317	9.681	121.74		37140.	272245.
320	11.05	122.18		36931.	240887.
322	12.04	122.47		36800.	222394.
324	13.10	122.76		36677.	205591.
326	14.24	123.06		36560.	190299.
328	15.46	123.35		36450.	176362.
331	17.45	123.80		36295.	157696.
333	18.89	124.10		36198.	146574.
335	20.42	124.41		36106.	136386.
337	22.05				127044.
339	23.79				118466.
342	26.61				106875.
344	28.63				99908.
346	30.77				93485.
348	33.05				87556.
350	35.45				82079.
353	39.33				74623.
355	42.10				70109.
357	45.03				65924.
359	48.11				62041.
361	51.37				58434.
364	56.57				53495.
366	60.28				50486.
368	64.17				47684.
370	68.25				45072.
372	72.54				42635.
375	79.37				39281.
377	84.20				37228.
379	89.25				35307.
381	94.53				33511.
383	100.0				31829.

COMPOUND CONSTANTS:

MP	150.150 K	TC		VC		ZC
NBP	376.000 K	PC		DC		OM
MU		RG				

VAPOR PRESSURE CORRELATION CONSTANTS, K AND KPA:

RIEDEL	RIEDEL	RIEDEL	RIEDEL
318. - 377. K	318. - 377. K	318. - 377. K	362. - 377. K
12 POINTS	12 POINTS	12 POINTS	9 POINTS
RMSD = 0.0096	RMSD = 0.0096	RMSD = 0.0096	RMSD = 0.0073
0.69450519E+02	0.69450519E+02	0.69450519E+02	0.34945980E+03
-0.79093209E+01	-0.79093209E+01	-0.79093209E+01	-0.49435069E+02
-0.67875324E+04	-0.67875324E+04	-0.67875324E+04	-0.19906045E+05
0.42020401E-16	0.42020401E-16	0.42020401E-16	0.43521474E-15

LIQUID DENSITY CORRELATION CONSTANTS, K AND G/ML:

FRANCIS1	FRANCIS1	FRANCIS1
286. - 299. K	286. - 299. K	286. - 299. K
4 POINTS	4 POINTS	4 POINTS
RMSD = 0.00064	RMSD = 0.00064	RMSD = 0.00064
0.10941839E+01	0.10941839E+01	0.10941839E+01
0.92274440E-03	0.92274440E-03	0.92274440E-03
0.59999990E+01	0.59999990E+01	0.59999990E+01
0.55901440E+03	0.55901440E+03	0.55901440E+03

SECOND VIRIAL COEFFICIENT CORRELATION CONSTANTS, ML/MOL:

LITERATURE DOCUMENTS FOR CORRELATED VAPOR PRESSURE VALUES:

17432

LITERATURE DOCUMENTS FOR CORRELATED LIQUID DENSITY VALUES:

3269 22447

LITERATURE DOCUMENTS REPORTING VIRIAL COEFFICIENT DATA:

T,K	VAPOR PRESSURE, KPA	SATURATED LIQUID VOLUME, ML/MOL	SECOND VIRIAL COEFFICIENT, ML/MOL	HEAT OF VAPORIZATION, J/MOL	SATURATED VAPOR VOLUME, ML/MOL
286		118.98			
288		119.28			
290		119.57			
292		119.87			
294		120.18			
296		120.48			
298		120.79			
319	13.72				193380.
321	14.93				178767.
323	16.23				165463.
325	17.62				153335.
327	19.11				142265.
329	20.70				132147.
331	22.39				122889.
333	24.20				114407.
335	26.12				106626.
337	28.17				99480.
339	30.34				92909.
341	32.64				86860.
343	35.08				81287.
345	37.67				76145.
348	41.84				69158.
350	44.81				64935.
352	47.96				61024.
354	51.28				57400.
356	54.78				54037.
358	58.46				50915.
360	62.34				48012.
362	66.43				45311.
364	70.72				42796.
366	75.23				40451.
368	79.96				38263.
370	84.93				36221.
372	90.14				34312.
374	95.60				32526.
376	101.3				30855.

COMPOUND CONSTANTS:

MP		TC		VC		ZC
NBP	376.578 K	PC		DC		OM
MU		RG				

VAPOR PRESSURE CORRELATION CONSTANTS, K AND KPA:

RIEDEL	RIEDEL	RIEDEL	RIEDEL
331. - 387. K	331. - 387. K	331. - 387. K	369. - 387. K
9 POINTS	9 POINTS	9 POINTS	4 POINTS
RMSD = 0.0046	RMSD = 0.0046	RMSD = 0.0046	RMSD = 0.0000
0.58041891E+02	0.58041891E+02	0.58041891E+02	0.58800704E+02
-0.62321812E+01	-0.62321812E+01	-0.62321812E+01	-0.63392077E+01
-0.62208544E+04	-0.62208544E+04	-0.62208544E+04	-0.62666501E+04
0.20939325E-16	0.20939325E-16	0.20939325E-16	0.20104971E-16

LIQUID DENSITY CORRELATION CONSTANTS, K AND G/ML:

FRANCIS1	FRANCIS1	FRANCIS1
290. - 335. K	290. - 335. K	290. - 335. K
6 POINTS	6 POINTS	6 POINTS
RMSD = 0.00067	RMSD = 0.00067	RMSD = 0.00067
0.10730829E+01	0.10730829E+01	0.10730829E+01
0.84664207E-03	0.84664207E-03	0.84664207E-03
0.59999990E+01	0.59999990E+01	0.59999990E+01
0.62745044E+03	0.62745044E+03	0.62745044E+03

SECOND VIRIAL COEFFICIENT CORRELATION CONSTANTS, ML/MOL:

LITERATURE DOCUMENTS FOR CORRELATED VAPOR PRESSURE VALUES:

10569

LITERATURE DOCUMENTS FOR CORRELATED LIQUID DENSITY VALUES:

9923

LITERATURE DOCUMENTS REPORTING VIRIAL COEFFICIENT DATA:

T,K	VAPOR PRESSURE, KPA	SATURATED LIQUID VOLUME, ML/MOL	SECOND VIRIAL· COEFFICIENT, ML/MOL	HEAT OF VAPORIZATION, J/MOL	SATURATED VAPOR VOLUME, ML/MOL
290		118.76			
292		119.03			
294		119.29			
296		119.56			
298		119.83			
301		120.24			
303		120.51			
305		120.78			
307		121.06			
309		121.33			
312		121.75			
314		122.03			
316		122.31			
318		122.60			
320		122.88			
323		123.31			
325		123.60			
327		123.89			
329		124.18			
331	22.54	124.48		34990.	122077.
334	25.27	124.92		34861.	109899.
336	27.23				102598.
338	29.31				95880.
340	31.52				89691.
342	33.86				83985.
345	37.63				76235.
347	40.32				71554.
349	43.17				67222.
351	46.17				63209.
353	49.34				59487.
356	54.41				54399.
358	58.02				51305.
360	61.81				48426.
362	65.79				45746.
364	69.98				43247.
367	76.65				39811.
369	81.36				37708.
371	86.30				35743.
373	91.47				33904.
375	96.88				32182.
378	105.5				29801.
380	111.5				28336.
382	117.8				26960.
384	124.4				25668.
386	131.2				24453.

COMPOUND CONSTANTS:

MP   TC   VC   ZC
NBP   PC   DC   OM
MU   RG

   VAPOR PRESSURE CORRELATION CONSTANTS, K AND KPA:

   RPM2   RPM2   RPM2
   301. - 351. K   301. - 351. K   301. - 351. K
    12 POINTS   12 POINTS   12 POINTS
   RMSD = 0.0825   RMSD = 0.0825   RMSD = 0.0825
    0.12387190E+03   0.12387190E+03   0.12387190E+03
   -0.16247968E+05   -0.16247968E+05   -0.16247968E+05
   -0.31891321E+00   -0.31891321E+00   -0.31891321E+00
    0.31219897E-03   0.31219897E-03   0.31219897E-03

   LIQUID DENSITY CORRELATION CONSTANTS, K AND G/ML:

   SECOND VIRIAL COEFFICIENT CORRELATION CONSTANTS, ML/MOL:

LITERATURE DOCUMENTS FOR CORRELATED VAPOR PRESSURE VALUES:

   13773

LITERATURE DOCUMENTS FOR CORRELATED LIQUID DENSITY VALUES:

LITERATURE DOCUMENTS REPORTING VIRIAL COEFFICIENT DATA:

T,K	VAPOR PRESSURE, KPA	SATURATED LIQUID VOLUME, ML/MOL	SECOND VIRIAL COEFFICIENT, ML/MOL	HEAT OF VAPORIZATION, J/MOL	SATURATED VAPOR VOLUME, ML/MOL
301	8.887				281603.
302	9.325				269274.
303	9.779				257625.
304	10.25				246612.
305	10.74				236193.
306	11.24				226329.
307	11.76				216985.
308	12.30				208129.
310	13.44				191758.
311	14.04				184189.
312	14.66				176998.
313	15.29				170161.
314	15.95				163658.
315	16.63				157469.
316	17.33				151576.
318	18.80				140607.
319	19.57				135502.
320	20.37				130629.
321	21.19				125977.
322	22.03				121533.
323	22.90				117285.
324	23.79				113223.
325	24.71				109338.
327	26.64				102056.
328	27.65				98643.
329	28.68				95372.
330	29.75				92235.
331	30.84				89225.
332	31.97				86337.
333	33.13				83563.
335	35.56				78336.
336	36.82				75874.
337	38.12				73505.
338	39.46				71226.
339	40.83				69031.
340	42.24				66918.
341	43.70				64882.
343	46.73				61027.
344	48.31				59202.
345	49.94				57442.
346	51.61				55742.
347	53.33				54101.
348	55.10				52516.
349	56.91				50984.
350	58.78				49503.

COMPOUND CONSTANTS:

MP		TC		VC		ZC
NBP	389.238 K	PC		DC		OM
MU		RG				

VAPOR PRESSURE CORRELATION CONSTANTS, K AND KPA:

RPM2	RPM2	RPM2	RIEDEL
333. - 390. K	333. - 390. K	333. - 390. K	376. - 390. K
14 POINTS	14 POINTS	14 POINTS	8 POINTS
RMSD = 0.0053	RMSD = 0.0053	RMSD = 0.0053	RMSD = 0.0042
0.22212276E+02	0.22212276E+02	0.22212276E+02	0.67098719E+02
-0.53871849E+04	-0.53871849E+04	-0.53871849E+04	-0.75107161E+01
-0.11641720E-01	-0.11641720E-01	-0.11641720E-01	-0.69228635E+04
0.51332468E-05	0.51332468E-05	0.51332468E-05	0.28940469E-16

LIQUID DENSITY CORRELATION CONSTANTS, K AND G/ML:

SECOND VIRIAL COEFFICIENT CORRELATION CONSTANTS, ML/MOL:

LITERATURE DOCUMENTS FOR CORRELATED VAPOR PRESSURE VALUES:

17731 18172

LITERATURE DOCUMENTS FOR CORRELATED LIQUID DENSITY VALUES:

LITERATURE DOCUMENTS REPORTING VIRIAL COEFFICIENT DATA:

T,K	VAPOR PRESSURE, KPA	SATURATED LIQUID VOLUME, ML/MOL	SECOND VIRIAL COEFFICIENT, ML/MOL	HEAT OF VAPORIZATION, J/MOL	SATURATED VAPOR VOLUME, ML/MOL
333	15.29				181102.
334	15.92				174485.
335	16.56				168156.
336	17.23				162099.
338	18.64				150752.
339	19.38				145437.
340	20.14				140345.
342	21.74				130790.
343	22.58				126306.
344	23.44				122007.
345	24.33				117883.
347	26.20				110129.
348	27.17				106484.
349	28.18				102984.
351	30.28				96392.
352	31.37				93289.
353	32.50				90306.
355	34.86				84681.
356	36.08				82029.
357	37.35				79477.
358	38.65				77022.
360	41.35				72383.
361	42.76				70192.
362	44.21				68082.
364	47.22				64089.
365	48.79				62201.
366	50.40				60381.
367	52.05				58626.
369	55.48				55300.
370	57.26				53725.
371	59.09				52204.
373	62.88				49320.
374	64.85				47952.
375	66.86				46630.
377	71.05				44120.
378	73.21				42928.
379	75.43				41776.
380	77.70				40662.
382	82.40				38543.
383	84.84				37536.
384	87.33				36561.
386	92.48				34704.
387	95.14				33820.
388	97.86				32964.
389	100.6				32135.

COMPOUND CONSTANTS:

MP   287.944 K   TC   637.227 K   VC   ZC
NBP 424.300 K   PC   3.557 MPA   DC   OM
MU   0.6200 DEBYE   RG

   VAPOR PRESSURE CORRELATION CONSTANTS, K AND KPA:

   RIEDEL   RIEDEL   RIEDEL
   273. - 334. K   273. - 334. K   273. - 334. K
     4 POINTS   4 POINTS   4 POINTS
   RMSD = 0.0000   RMSD = 0.0000   RMSD = 0.0000
    0.35793970E+03   0.35793970E+03   0.35793970E+03
   -0.51889092E+02   -0.51889092E+02   -0.51889092E+02
   -0.18830755E+05   -0.18830755E+05   -0.18830755E+05
    0.12870383E-14   0.12870383E-14   0.12870383E-14

   LIQUID DENSITY CORRELATION CONSTANTS, K AND G/ML:

   SECOND VIRIAL COEFFICIENT CORRELATION CONSTANTS, ML/MOL:

LITERATURE DOCUMENTS FOR CORRELATED VAPOR PRESSURE VALUES:

   5065

LITERATURE DOCUMENTS FOR CORRELATED LIQUID DENSITY VALUES:

LITERATURE DOCUMENTS REPORTING VIRIAL COEFFICIENT DATA:

T,K	VAPOR PRESSURE, KPA	SATURATED LIQUID VOLUME, ML/MOL	SECOND VIRIAL COEFFICIENT, ML/MOL	HEAT OF VAPORIZATION, J/MOL	SATURATED VAPOR VOLUME, ML/MOL
273	0.2070				
274	0.2229				
275	0.2397				
277	0.2766				
278	0.2967				
279	0.3181				
281	0.3645				
282	0.3898				
284	0.4446				
285	0.4743				
286	0.5056				
288	0.5733				4174706.
289	0.6099				3937968.
291	0.6888				3510859.
292	0.7313				3318144.
293	0.7759				3137921.
295	0.8719				2811286.
296	0.9234				2663226.
297	0.9775				2524356.
299	1.094				2271580.
300	1.156				2156503.
302	1.289				1946394.
303	1.360				1850448.
304	1.435				1760026.
306	1.594				1594285.
307	1.679				1518305.
309	1.861				1378653.
310	1.958				1314457.
311	2.060				1253695.
313	2.276				1141623.
314	2.392				1089925.
315	2.512				1040883.
317	2.769				950132.
318	2.906				908133.
320	3.198				830237.
321	3.354				794104.
322	3.516				759706.
324	3.863				695723.
325	4.048				665961.
327	4.442				610491.
328	4.652				584637.
329	4.871				559949.
331	5.340				513822.
332	5.590				492273.
333	5.851				471662.

COMPOUND CONSTANTS:

MP		TC		VC		ZC
NBP	410.156 K	PC		DC		OM
MU		RG				

VAPOR PRESSURE CORRELATION CONSTANTS, K AND KPA:

RPM2	RPM2	RPM2	RIEDEL
332. - 411. K	332. - 411. K	332. - 411. K	395. - 411. K
16 POINTS	16 POINTS	16 POINTS	4 POINTS
RMSD = 0.0075	RMSD = 0.0075	RMSD = 0.0075	RMSD = 0.0000
0.33414865E+02	0.33414865E+02	0.33414865E+02	-0.28122816E+03
-0.71684554E+04	-0.71684554E+04	-0.71684554E+04	0.43581173E+02
-0.39387640E-01	-0.39387640E-01	-0.39387640E-01	0.10223138E+05
0.28746625E-04	0.28746625E-04	0.28746625E-04	-0.27017144E-15

LIQUID DENSITY CORRELATION CONSTANTS, K AND G/ML:

FRANCIS1	FRANCIS1	FRANCIS1
293. - 304. K	293. - 304. K	293. - 304. K
4 POINTS	4 POINTS	4 POINTS
RMSD = 0.00007	RMSD = 0.00007	RMSD = 0.00007
0.10753946E+01	0.10753946E+01	0.10753946E+01
0.78968843E-03	0.78968843E-03	0.78968843E-03
0.59999990E+01	0.59999990E+01	0.59999990E+01
0.56838794E+03	0.56838794E+03	0.56838794E+03

SECOND VIRIAL COEFFICIENT CORRELATION CONSTANTS, ML/MOL:

LITERATURE DOCUMENTS FOR CORRELATED VAPOR PRESSURE VALUES:

6518

LITERATURE DOCUMENTS FOR CORRELATED LIQUID DENSITY VALUES:

6518

LITERATURE DOCUMENTS REPORTING VIRIAL COEFFICIENT DATA:

T,K	VAPOR PRESSURE, KPA	SATURATED LIQUID VOLUME, ML/MOL	SECOND VIRIAL COEFFICIENT, ML/MOL	HEAT OF VAPORIZATION, J/MOL	SATURATED VAPOR VOLUME, ML/MOL
293		134.02			
295		134.31			
298		134.74			
301		135.17			
303		135.46			
333	7.096				390161.
335	7.750				359387.
338	8.824				318465.
341	10.02				282978.
343	10.89				261934.
346	12.31				233769.
349	13.87				209161.
351	15.01				194476.
354	16.85				174707.
357	18.87				157311.
360	21.08				141967.
362	22.67				132739.
365	25.24				120223.
368	28.04				109112.
370	30.04				102396.
373	33.26				93241.
376	36.75				85066.
378	39.24				80101.
381	43.22				73301.
384	47.51				67196.
386	50.56				63471.
389	55.43				58349.
392	60.66				53726.
394	64.37				50894.
397	70.26				46983.
400	76.56				43438.
402	81.01				41257.
405	88.07				38234.
408	95.61				35482.
410	100.9				33783.

COMPOUND CONSTANTS:

MP		TC		VC		ZC
NBP	393.157 K	PC		DC		OM
MU		RG				

VAPOR PRESSURE CORRELATION CONSTANTS, K AND KPA:

LIQUID DENSITY CORRELATION CONSTANTS, K AND G/ML:

FRANCIS1	FRANCIS1	FRANCIS1
277. - 299. K	277. - 299. K	277. - 299. K
4 POINTS	4 POINTS	4 POINTS
RMSD = 0.00007	RMSD = 0.00007	RMSD = 0.00007
0.10521784E+01	0.10521784E+01	0.10521784E+01
0.78680483E-03	0.78680483E-03	0.78680483E-03
0.59999990E+01	0.59999990E+01	0.59999990E+01
0.55901392E+03	0.55901392E+03	0.55901392E+03

SECOND VIRIAL COEFFICIENT CORRELATION CONSTANTS, ML/MOL:

LITERATURE DOCUMENTS FOR CORRELATED VAPOR PRESSURE VALUES:

LITERATURE DOCUMENTS FOR CORRELATED LIQUID DENSITY VALUES:

13929

LITERATURE DOCUMENTS REPORTING VIRIAL COEFFICIENT DATA:

T,K	VAPOR PRESSURE, KPA	SATURATED LIQUID VOLUME, ML/MOL	SECOND VIRIAL COEFFICIENT, ML/MOL	HEAT OF VAPORIZATION, J/MOL	SATURATED VAPOR VOLUME, ML/MOL
277		135.55			
278		135.70			
279		135.84			
280		135.99			
281		136.13			
282		136.28			
283		136.42			
284		136.57			
285		136.72			
286		136.86			
287		137.01			
288		137.16			
289		137.31			
290		137.46			
291		137.61			
292		137.76			
293		137.91			
294		138.06			
295		138.21			
296		138.36			
297		138.51			
298		138.66			
299		138.82			

COMPOUND CONSTANTS:

MP	TC	VC	ZC
NBP 396.158 K	PC	DC	OM
MU	RG		

VAPOR PRESSURE CORRELATION CONSTANTS, K AND KPA:

LIQUID DENSITY CORRELATION CONSTANTS, K AND G/ML:

FRANCIS1	FRANCIS1	FRANCIS1
289. - 336. K	289. - 336. K	289. - 336. K
4 POINTS	4 POINTS	4 POINTS
RMSD = 0.00016	RMSD = 0.00016	RMSD = 0.00016
0.10506010E+01	0.10506010E+01	0.10506010E+01
0.80403336E-03	0.80403336E-03	0.80403336E-03
0.59999990E+01	0.59999990E+01	0.59999990E+01
0.62820044E+03	0.62820044E+03	0.62820044E+03

SECOND VIRIAL COEFFICIENT CORRELATION CONSTANTS, ML/MOL:

LITERATURE DOCUMENTS FOR CORRELATED VAPOR PRESSURE VALUES:

LITERATURE DOCUMENTS FOR CORRELATED LIQUID DENSITY VALUES:

9923

LITERATURE DOCUMENTS REPORTING VIRIAL COEFFICIENT DATA:

T,K	VAPOR PRESSURE, KPA	SATURATED LIQUID VOLUME, ML/MOL	SECOND VIRIAL COEFFICIENT, ML/MOL	HEAT OF VAPORIZATION, J/MOL	SATURATED VAPOR VOLUME, ML/MOL
289		137.65			
290		137.80			
291		137.95			
292		138.10			
293		138.25			
294		138.40			
295		138.54			
296		138.69			
297		138.84			
298		138.99			
299		139.15			
300		139.30			
301		139.45			
302		139.60			
303		139.75			
305		140.06			
306		140.21			
307		140.37			
308		140.52			
309		140.68			
310		140.83			
311		140.99			
312		141.14			
313		141.30			
314		141.46			
315		141.61			
316		141.77			
317		141.93			
318		142.09			
319		142.25			
321		142.56			
322		142.72			
323		142.89			
324		143.05			
325		143.21			
326		143.37			
327		143.53			
328		143.70			
329		143.86			
330		144.02			
331		144.19			
332		144.35			
333		144.52			
334		144.68			
335		144.85			

461

COMPOUND CONSTANTS:

MP	TC	VC	ZC
NBP 397.559 K	PC	DC	OM
MU	RG		

VAPOR PRESSURE CORRELATION CONSTANTS, K AND KPA:

LIQUID DENSITY CORRELATION CONSTANTS, K AND G/ML:

FRANCIS1	FRANCIS1	FRANCIS1
289. - 360. K	289. - 360. K	289. - 360. K
4 POINTS	4 POINTS	4 POINTS
RMSD = 0.00026	RMSD = 0.00026	RMSD = 0.00026
0.10587015E+01	0.10587015E+01	0.10587015E+01
0.83040912E-03	0.83040912E-03	0.83040912E-03
0.59999990E+01	0.59999990E+01	0.59999990E+01
0.67470996E+03	0.67470996E+03	0.67470996E+03

SECOND VIRIAL COEFFICIENT CORRELATION CONSTANTS, ML/MOL:

LITERATURE DOCUMENTS FOR CORRELATED VAPOR PRESSURE VALUES:

LITERATURE DOCUMENTS FOR CORRELATED LIQUID DENSITY VALUES:

9923

LITERATURE DOCUMENTS REPORTING VIRIAL COEFFICIENT DATA:

T,K	VAPOR PRESSURE, KPA	SATURATED LIQUID VOLUME, ML/MOL	SECOND VIRIAL COEFFICIENT, ML/MOL	HEAT OF VAPORIZATION, J/MOL	SATURATED VAPOR VOLUME, ML/MOL
289		137.21			
290		137.36			
292		137.66			
293		137.81			
295		138.11			
297		138.41			
298		138.56			
300		138.87			
301		139.02			
303		139.33			
305		139.63			
306		139.79			
308		140.10			
309		140.26			
311		140.57			
313		140.88			
314		141.04			
316		141.36			
318		141.68			
319		141.84			
321		142.16			
322		142.32			
324		142.64			
326		142.97			
327		143.13			
329		143.46			
330		143.63			
332		143.96			
334		144.29			
335		144.46			
337		144.79			
339		145.13			
340		145.30			
342		145.64			
343		145.81			
345		146.15			
347		146.49			
348		146.67			
350		147.01			
351		147.19			
353		147.54			
355		147.89			
356		148.07			
358		148.42			
359		148.60			

COMPOUND CONSTANTS:

MP	TC	VC	ZC
NBP	PC	DC	OM
MU	RG		

VAPOR PRESSURE CORRELATION CONSTANTS, K AND KPA:

LIQUID DENSITY CORRELATION CONSTANTS, K AND G/ML:

FRANCIS1	FRANCIS1	FRANCIS1
277. - 299. K	277. - 299. K	277. - 299. K
5 POINTS	5 POINTS	5 POINTS
RMSD = 0.00009	RMSD = 0.00009	RMSD = 0.00009
0.99370909E+00	0.99370909E+00	0.99370909E+00
0.52907830E-03	0.52907830E-03	0.52907830E-03
0.89936142E+01	0.89936142E+01	0.89936142E+01
0.45107153E+03	0.45107153E+03	0.45107153E+03

SECOND VIRIAL COEFFICIENT CORRELATION CONSTANTS, ML/MOL:

LITERATURE DOCUMENTS FOR CORRELATED VAPOR PRESSURE VALUES:

LITERATURE DOCUMENTS FOR CORRELATED LIQUID DENSITY VALUES:

7291

LITERATURE DOCUMENTS REPORTING VIRIAL COEFFICIENT DATA:

T,K	VAPOR PRESSURE, KPA	SATURATED LIQUID VOLUME, ML/MOL	SECOND VIRIAL COEFFICIENT, ML/MOL	HEAT OF VAPORIZATION, J/MOL	SATURATED VAPOR VOLUME, ML/MOL
277		138.53			
278		138.67			
279		138.82			
280		138.97			
281		139.11			
282		139.26			
283		139.41			
284		139.56			
285		139.71			
286		139.86			
287		140.02			
288		140.17			
289		140.32			
290		140.48			
291		140.64			
292		140.80			
293		140.96			
294		141.12			
295		141.28			
296		141.44			
297		141.61			
298		141.77			
299		141.94			

COMPOUND CONSTANTS:

MP	TC	VC	ZC
NBP 402.523 K	PC	DC	OM
MU	RG		

VAPOR PRESSURE CORRELATION CONSTANTS, K AND KPA:

RPM2	RPM2	RPM2	RIEDEL
273. - 404. K	273. - 404. K	273. - 404. K	383. - 404. K
27 POINTS	27 POINTS	27 POINTS	5 POINTS
RMSD = 0.0004	RMSD = 0.0004	RMSD = 0.0004	RMSD = 0.0001
0.15947102E+02	0.15947102E+02	0.15947102E+02	0.29118797E+02
-0.41013980E+04	-0.41013980E+04	-0.41013980E+04	-0.22048265E+01
-0.29564664E-02	-0.29564664E-02	-0.29564664E-02	-0.45406496E+04
0.31170528E-06	0.31170528E-06	0.31170528E-06	0.94032437E-18

LIQUID DENSITY CORRELATION CONSTANTS, K AND G/ML:

FRANCIS1	FRANCIS1	FRANCIS1
293. - 344. K	293. - 344. K	293. - 344. K
6 POINTS	6 POINTS	6 POINTS
RMSD = 0.00007	RMSD = 0.00007	RMSD = 0.00007
0.10756521E+01	0.10756521E+01	0.10756521E+01
0.75216731E-03	0.75216731E-03	0.75216731E-03
0.68139811E+01	0.68139811E+01	0.68139811E+01
0.56814331E+03	0.56814331E+03	0.56814331E+03

SECOND VIRIAL COEFFICIENT CORRELATION CONSTANTS, ML/MOL:

LITERATURE DOCUMENTS FOR CORRELATED VAPOR PRESSURE VALUES:

6800

LITERATURE DOCUMENTS FOR CORRELATED LIQUID DENSITY VALUES:

7697

LITERATURE DOCUMENTS REPORTING VIRIAL COEFFICIENT DATA:

T,K	VAPOR PRESSURE, KPA	SATURATED LIQUID VOLUME, ML/MOL	SECOND VIRIAL COEFFICIENT, ML/MOL	HEAT OF VAPORIZATION, J/MOL	SATURATED VAPOR VOLUME, ML/MOL
273	1.150				1973728.
275	1.276				1792360.
278	1.486				1555492.
281	1.725				1354407.
284	1.996				1183107.
287	2.302				1036688.
290	2.646				911126.
293	3.033	130.26		32116.	803109.
296	3.467	130.66		32075.	709901.
299	3.951	131.06		32035.	629231.
302	4.490	131.47		31994.	559211.
305	5.089	131.87		31953.	498265.
308	5.754	132.29		31911.	445072.
311	6.488	132.70		31868.	398524.
314	7.299	133.12		31826.	357686.
317	8.191	133.55		31782.	321768.
320	9.171	133.97		31739.	290101.
323	10.25	134.40		31695.	262116.
326	11.42	134.84		31650.	237328.
329	12.70	135.28		31605.	215323.
332	14.10	135.72		31559.	195747.
335	15.62	136.17		31513.	178294.
338	17.27	136.63		31466.	162702.
341	19.06	137.09		31419.	148746.
344	21.00	137.55		31371.	136229.
347	23.08				124981.
350	25.34				114856.
353	27.76				105725.
356	30.37				97475.
359	33.16				90010.
362	36.16				83242.
365	39.36				77098.
368	42.79				71510.
371	46.44				66421.
374	50.33				61779.
377	54.48				57539.
380	58.88				53659.
383	63.55				50106.
386	68.51				46846.
389	73.76				43851.
392	79.31				41097.
395	85.17				38561.
398	91.36				36222.
401	97.88				34062.
403	102.4				32715.

COMPOUND CONSTANTS:

MP   TC   VC   ZC
NBP   PC   DC   OM
MU   RG

VAPOR PRESSURE CORRELATION CONSTANTS, K AND KPA:

LIQUID DENSITY CORRELATION CONSTANTS, K AND G/ML:

FRANCIS1   FRANCIS1   FRANCIS1
273. - 294. K   273. - 294. K   273. - 294. K
  4 POINTS   4 POINTS   4 POINTS
RMSD = 0.00004   RMSD = 0.00004   RMSD = 0.00004
0.10451231E+01   0.10451231E+01   0.10451231E+01
0.73642260E-03   0.73642260E-03   0.73642260E-03
0.59999990E+01   0.59999990E+01   0.59999990E+01
0.54964111E+03   0.54964111E+03   0.54964111E+03

SECOND VIRIAL COEFFICIENT CORRELATION CONSTANTS, ML/MOL:

LITERATURE DOCUMENTS FOR CORRELATED VAPOR PRESSURE VALUES:

LITERATURE DOCUMENTS FOR CORRELATED LIQUID DENSITY VALUES:

  9717

LITERATURE DOCUMENTS REPORTING VIRIAL COEFFICIENT DATA:

T,K	VAPOR PRESSURE, KPA	SATURATED LIQUID VOLUME, ML/MOL	SECOND VIRIAL COEFFICIENT, ML/MOL	HEAT OF VAPORIZATION, J/MOL	SATURATED VAPOR VOLUME, ML/MOL
273		151.05			
274		151.20			
275		151.35			
276		151.50			
277		151.66			
278		151.81			
279		151.96			
280		152.11			
281		152.26			
282		152.42			
283		152.57			
284		152.72			
285		152.88			
286		153.03			
287		153.19			
288		153.35			
289		153.50			
290		153.66			
291		153.82			
292		153.97			
293		154.13			
294		154.29			

COMPOUND CONSTANTS:

MP   TC   VC   ZC
NBP   PC   DC   OM
MU   RG

VAPOR PRESSURE CORRELATION CONSTANTS, K AND KPA:

LIQUID DENSITY CORRELATION CONSTANTS, K AND G/ML:

FRANCIS1   FRANCIS1   FRANCIS1
273. - 294. K   273. - 294. K   273. - 294. K
 4 POINTS   4 POINTS   4 POINTS
RMSD = 0.00004   RMSD = 0.00004   RMSD = 0.00004
0.10480242E+01   0.10480242E+01   0.10480242E+01
0.73642726E-03   0.73642726E-03   0.73642726E-03
0.59999990E+01   0.59999990E+01   0.59999990E+01
0.54964111E+03   0.54964111E+03   0.54964111E+03

SECOND VIRIAL COEFFICIENT CORRELATION CONSTANTS, ML/MOL:

LITERATURE DOCUMENTS FOR CORRELATED VAPOR PRESSURE VALUES:

LITERATURE DOCUMENTS FOR CORRELATED LIQUID DENSITY VALUES:

   9718

LITERATURE DOCUMENTS REPORTING VIRIAL COEFFICIENT DATA:

T,K	VAPOR PRESSURE, KPA	SATURATED LIQUID VOLUME, ML/MOL	SECOND VIRIAL COEFFICIENT, ML/MOL	HEAT OF VAPORIZATION, J/MOL	SATURATED VAPOR VOLUME, ML/MOL
273		150.52			
274		150.67			
275		150.82			
276		150.97			
277		151.12			
278		151.27			
279		151.42			
280		151.57			
281		151.72			
282		151.88			
283		152.03			
284		152.18			
285		152.34			
286		152.49			
287		152.64			
288		152.80			
289		152.95			
290		153.11			
291		153.26			
292		153.42			
293		153.58			
294		153.74			

COMPOUND CONSTANTS:

MP		TC		VC		ZC
NBP	697.225 K	PC		DC		OM
MU		RG				

VAPOR PRESSURE CORRELATION CONSTANTS, K AND KPA:

VAPRES-2	VAPRES-2	VAPRES-2
280. - 698. K	280. - 698. K	280. - 698. K
15 POINTS	15 POINTS	15 POINTS
RMSD = 0.0025	RMSD = 0.0025	RMSD = 0.0025
-0.47007105E+03	-0.47007105E+03	-0.47007105E+03
0.41256717E+04	0.41256717E+04	0.41256717E+04
-0.28571205E+00	-0.28571205E+00	-0.28571205E+00
0.14014050E-03	0.14014050E-03	0.14014050E-03
0.91620943E+02	0.91620943E+02	0.91620943E+02

LIQUID DENSITY CORRELATION CONSTANTS, K AND G/ML:

SECOND VIRIAL COEFFICIENT CORRELATION CONSTANTS, ML/MOL:

LITERATURE DOCUMENTS FOR CORRELATED VAPOR PRESSURE VALUES:

17731

LITERATURE DOCUMENTS FOR CORRELATED LIQUID DENSITY VALUES:

LITERATURE DOCUMENTS REPORTING VIRIAL COEFFICIENT DATA:

T,K	VAPOR PRESSURE, KPA	SATURATED LIQUID VOLUME, ML/MOL	SECOND VIRIAL COEFFICIENT, ML/MOL	HEAT OF VAPORIZATION, J/MOL	SATURATED VAPOR VOLUME, ML/MOL
280	0.0003084				0.7549E+10
289	0.0005541				0.4336E+10
299	0.001016				0.2447E+10
308	0.001689				0.1517E+10
318	0.002858				0.9250E+09
327	0.004447				0.6113E+09
337	0.007037				0.3982E+09
346	0.01036				0.2778E+09
356	0.01548				0.1912E+09
365	0.02175				0.1396E+09
375	0.03101				0.1005E+09
384	0.04190				0.7620E+08
394	0.05746				0.5701E+08
403	0.07522				0.4454E+08
413	0.09996				0.3435E+08
422	0.1276				0.2751E+08
432	0.1653				0.2173E+08
441	0.2068				0.1773E+08
451	0.2628				0.1427E+08
460	0.3239				0.1181E+08
470	0.4060				9625808.
479	0.4951				8044258.
489	0.6147				6614445.
498	0.7447				5559810.
508	0.9198				4591849.
517	1.111				3868400.
527	1.371				3197061.
536	1.656				2690767.
546	2.047				2217739.
555	2.482				1859305.
565	3.083				1523552.
574	3.761				1268991.
584	4.710				1030891.
593	5.794				850984.
603	7.334				683602.
612	9.119				558027.
622	11.69				442220.
631	14.73				356245.
641	19.18				277901.
650	24.51				220518.
660	32.47				168996.
669	42.18				131864.
679	56.97				99094.
688	75.36				75910.
698	103.9				55843.

COMPOUND CONSTANTS:

MP		TC		VC		ZC
NBP	422.750 K	PC		DC		OM
MU		RG				

VAPOR PRESSURE CORRELATION CONSTANTS, K AND KPA:

LIQUID DENSITY CORRELATION CONSTANTS, K AND G/ML:

FRANCIS1	FRANCIS1	FRANCIS1
303. - 334. K	303. - 334. K	303. - 334. K
4 POINTS	4 POINTS	4 POINTS
RMSD = 0.00000	RMSD = 0.00000	RMSD = 0.00000
0.11784296E+01	0.11784296E+01	0.11784296E+01
0.92196837E-03	0.92196837E-03	0.92196837E-03
0.67763071E+01	0.67763071E+01	0.67763071E+01
0.53259692E+03	0.53259692E+03	0.53259692E+03

SECOND VIRIAL COEFFICIENT CORRELATION CONSTANTS, ML/MOL:

LITERATURE DOCUMENTS FOR CORRELATED VAPOR PRESSURE VALUES:

LITERATURE DOCUMENTS FOR CORRELATED LIQUID DENSITY VALUES:

   18182

LITERATURE DOCUMENTS REPORTING VIRIAL COEFFICIENT DATA:

T,K	VAPOR PRESSURE, KPA	SATURATED LIQUID VOLUME, ML/MOL	SECOND VIRIAL COEFFICIENT, ML/MOL	HEAT OF VAPORIZATION, J/MOL	SATURATED VAPOR VOLUME, ML/MOL
303		140.54			
304		140.71			
305		140.88			
306		141.05			
307		141.23			
308		141.40			
309		141.57			
310		141.75			
311		141.92			
312		142.10			
313		142.27			
314		142.45			
315		142.62			
316		142.80			
317		142.98			
318		143.16			
319		143.34			
320		143.52			
321		143.70			
322		143.88			
323		144.06			
324		144.25			
325		144.43			
326		144.62			
327		144.80			
328		144.99			
329		145.17			
330		145.36			
331		145.55			
332		145.74			
333		145.93			
334		146.12			

COMPOUND CONSTANTS:

MP	TC	VC	ZC
NBP 420.530 K	PC	DC	OM
MU	RG		

VAPOR PRESSURE CORRELATION CONSTANTS, K AND KPA:

LIQUID DENSITY CORRELATION CONSTANTS, K AND G/ML:

FRANCIS1	FRANCIS1	FRANCIS1
288. - 344. K	288. - 344. K	288. - 344. K
7 POINTS	7 POINTS	7 POINTS
RMSD = 0.00019	RMSD = 0.00019	RMSD = 0.00019
0.11553621E+01	0.11553621E+01	0.11553621E+01
0.82278857E-03	0.82278857E-03	0.82278857E-03
0.59999990E+01	0.59999990E+01	0.59999990E+01
0.64340552E+03	0.64340552E+03	0.64340552E+03

SECOND VIRIAL COEFFICIENT CORRELATION CONSTANTS, ML/MOL:

LITERATURE DOCUMENTS FOR CORRELATED VAPOR PRESSURE VALUES:

LITERATURE DOCUMENTS FOR CORRELATED LIQUID DENSITY VALUES:

   7697

LITERATURE DOCUMENTS REPORTING VIRIAL COEFFICIENT DATA:

T,K	VAPOR PRESSURE, KPA	SATURATED LIQUID VOLUME, ML/MOL	SECOND VIRIAL COEFFICIENT, ML/MOL	HEAT OF VAPORIZATION, J/MOL	SATURATED VAPOR VOLUME, ML/MOL
288		133.32			
289		133.45			
290		133.58			
291		133.71			
293		133.97			
294		134.10			
295		134.23			
296		134.36			
298		134.63			
299		134.76			
300		134.89			
301		135.02			
303		135.29			
304		135.42			
305		135.55			
307		135.82			
308		135.96			
309		136.09			
310		136.23			
312		136.50			
313		136.64			
314		136.77			
315		136.91			
317		137.18			
318		137.32			
319		137.46			
321		137.74			
322		137.88			
323		138.02			
324		138.16			
326		138.44			
327		138.58			
328		138.72			
329		138.86			
331		139.14			
332		139.29			
333		139.43			
335		139.72			
336		139.86			
337		140.00			
338		140.15			
340		140.44			
341		140.59			
342		140.73			
343		140.88			

477

COMPOUND CONSTANTS:

MP	TC	VC	ZC
NBP 687.430 K	PC	DC	OM
MU	RG		

VAPOR PRESSURE CORRELATION CONSTANTS, K AND KPA:

LIQUID DENSITY CORRELATION CONSTANTS, K AND G/ML:

FRANCIS1	FRANCIS1	FRANCIS1
303. - 334. K	303. - 334. K	303. - 334. K
4 POINTS	4 POINTS	4 POINTS
RMSD = 0.00051	RMSD = 0.00051	RMSD = 0.00051
0.11712351E+01	0.11712351E+01	0.11712351E+01
0.89628319E-03	0.89628319E-03	0.89628319E-03
0.59999990E+01	0.59999990E+01	0.59999990E+01
0.62465552E+03	0.62465552E+03	0.62465552E+03

SECOND VIRIAL COEFFICIENT CORRELATION CONSTANTS, ML/MOL:

LITERATURE DOCUMENTS FOR CORRELATED VAPOR PRESSURE VALUES:

LITERATURE DOCUMENTS FOR CORRELATED LIQUID DENSITY VALUES:

  18182

LITERATURE DOCUMENTS REPORTING VIRIAL COEFFICIENT DATA:

T,K	VAPOR PRESSURE, KPA	SATURATED LIQUID VOLUME, ML/MOL	SECOND VIRIAL COEFFICIENT, ML/MOL	HEAT OF VAPORIZATION, J/MOL	SATURATED VAPOR VOLUME, ML/MOL
303		136.43			
304		136.58			
305		136.72			
306		136.87			
307		137.02			
308		137.17			
309		137.32			
310		137.47			
311		137.62			
312		137.77			
313		137.92			
314		138.08			
315		138.23			
316		138.38			
317		138.53			
318		138.69			
319		138.84			
320		139.00			
321		139.15			
322		139.31			
323		139.46			
324		139.62			
325		139.77			
326		139.93			
327		140.09			
328		140.24			
329		140.40			
330		140.56			
331		140.72			
332		140.88			
333		141.04			
334		141.20			

COMPOUND CONSTANTS:

MP	136.850 K	TC	393.150 K	VC		ZC	
NBP	238.650 K	PC		DC		OM	
MU	0.2000 DEBYE	RG	1.8980 ANGSTROM				

VAPOR PRESSURE CORRELATION CONSTANTS, K AND KPA:

VAPRES-2	VAPRES-2	VAPRES-2	VAPRES-2
138. - 293. K	138. - 293. K	138. - 293. K	138. - 293. K
10 POINTS	10 POINTS	10 POINTS	10 POINTS
RMSD = 0.0480	RMSD = 0.0480	RMSD = 0.0480	RMSD = 0.0480
0.24255853E+03	0.24255853E+03	0.24255853E+03	0.24255853E+03
-0.60275400E+04	-0.60275400E+04	-0.60275400E+04	-0.60275400E+04
0.22675525E+00	0.22675525E+00	0.22675525E+00	0.22675525E+00
-0.19074479E-03	-0.19074479E-03	-0.19074479E-03	-0.19074479E-03
-0.46746040E+02	-0.46746040E+02	-0.46746040E+02	-0.46746040E+02

LIQUID DENSITY CORRELATION CONSTANTS, K AND G/ML:

FRANCIS1	FRANCIS1	FRANCIS1	FRANCIS1
203. - 243. K	203. - 243. K	203. - 243. K	223. - 243. K
6 POINTS	6 POINTS	6 POINTS	4 POINTS
RMSD = 0.00017	RMSD = 0.00017	RMSD = 0.00017	RMSD = 0.00021
0.95773578E+00	0.95773578E+00	0.95773578E+00	0.95821565E+00
0.10570332E-02	0.10570332E-02	0.10570332E-02	0.10790953E-02
0.92157679E+01	0.92157679E+01	0.92157679E+01	0.81870365E+01
0.45547778E+03	0.45547778E+03	0.45547778E+03	0.45566968E+03

SECOND VIRIAL COEFFICIENT CORRELATION CONSTANTS, ML/MOL:

NOTHNAGEL ET AL.: TC=393.150   NBP=238.650   B=123.5   D=0.330

LITERATURE DOCUMENTS FOR CORRELATED VAPOR PRESSURE VALUES:

624   4422

LITERATURE DOCUMENTS FOR CORRELATED LIQUID DENSITY VALUES:

1894

LITERATURE DOCUMENTS REPORTING VIRIAL COEFFICIENT DATA:

1898   1940

T,K	VAPOR PRESSURE, KPA	SATURATED LIQUID VOLUME, ML/MOL	SECOND VIRIAL COEFFICIENT, ML/MOL	HEAT OF VAPORIZATION, J/MOL	SATURATED VAPOR VOLUME, ML/MOL
138	0.02261		-4039.		0.5074E+08
141	0.03527		-3707.		0.3323E+08
145	0.06178		-3323.		0.1951E+08
148	0.09197		-3073.		0.1338E+08
152	0.1521		-2781.		8304583.
155	0.2177		-2588.		5916700.
159	0.3431		-2361.		3851200.
162	0.4747		-2209.		2835152.
166	0.7175		-2030.		1921479.
169	0.9646		-1909.		1454820.
173	1.407		-1764.		1020871.
176	1.844		-1667.		791921.
180	2.606		-1549.		572720.
183	3.342		-1469.		453761.
187	4.596		-1371.		336925.
190	5.781		-1305.		271933.
194	7.759		-1224.		206651.
197	9.595		-1168.		169539.
201	12.60		-1099.		131514.
204	15.34	56.79	-1052.	22133.	109476.
208	19.76	57.18	-993.7	22026.	86493.
211	23.74	57.48	-953.3	21940.	72942.
215	30.04	57.88	-903.3	21815.	58586.
219	37.67	58.30	-857.3	21679.	47465.
222	44.37	58.61	-825.2	21568.	40753.
226	54.80	59.03	-785.3	21406.	33486.
229	63.85	59.36	-757.3	21275.	29043.
233	77.74	59.80	-722.4	21084.	24174.
236	89.67	60.13	-697.9	20929.	21161.
240	107.8	60.59	-667.2	20705.	17823.
243	123.1	60.93	-645.5	20523.	15736.
247	146.2		-618.4		13400.
250	165.5		-599.2		11926.
254	194.3		-575.0		10260.
257	218.2		-557.9		9200.
261	253.3		-536.4		7992.
264	282.2		-521.1		7216.
268	324.3		-501.7		6327.
271	358.5		-487.9		5751.
275	407.9		-470.5		5087.
278	447.7		-458.0		4655.
282	504.5		-442.3		4152.
285	549.9		-431.0		3823.
289	614.0		-416.7		3439.
292	664.7		-406.4		3187.

COMPOUND CONSTANTS:

MP   136.950 K   TC   444.450 K   VC 219.20 ML/MOL   ZC 0.2670
NBP 284.000 K   PC   4.499 MPA   DC 0.2468 G/ML   OM
MU   0.4010 DEBYE   RG   2.7497 ANGSTROM

VAPOR PRESSURE CORRELATION CONSTANTS, K AND KPA:

WAGNER   VAPRES-2   VAPRES-2
200. - 284. K   200. - 284. K   200. - 284. K
18 POINTS   18 POINTS   18 POINTS
RMSD = 0.0058   RMSD = 0.0040   RMSD = 0.0040
-0.82319433E+01   -0.23147903E+02   -0.23147903E+02
 0.35595151E+01   -0.26461710E+04   -0.26461710E+04
-0.50471402E+01   -0.30048180E-01   -0.30048180E-01
 0.40506790E+01   0.11981467E-04   0.11981467E-04
                      0.79042986E+01   0.79042986E+01

LIQUID DENSITY CORRELATION CONSTANTS, K AND G/ML:

RACKETT   RACKETT   RACKETT
272. - 274. K   272. - 274. K   272. - 274. K
 1 POINTS   1 POINTS   1 POINTS
RMSD = 0.00000   RMSD = 0.00000   RMSD = 0.00000
0.26665661E+00   0.26665661E+00   0.26665661E+00

SECOND VIRIAL COEFFICIENT CORRELATION CONSTANTS, ML/MOL:

LITERATURE DOCUMENTS FOR CORRELATED VAPOR PRESSURE VALUES:

  2737

LITERATURE DOCUMENTS FOR CORRELATED LIQUID DENSITY VALUES:

  4092

LITERATURE DOCUMENTS REPORTING VIRIAL COEFFICIENT DATA:

T,K	VAPOR PRESSURE, KPA	SATURATED LIQUID VOLUME, ML/MOL	SECOND VIRIAL COEFFICIENT, ML/MOL	HEAT OF VAPORIZATION, J/MOL	SATURATED VAPOR VOLUME, ML/MOL
200	0.9708				1711837.
201	1.052				1588049.
203	1.231				1369931.
205	1.437				1185435.
207	1.671				1028875.
209	1.938				895605.
211	2.241				781816.
213	2.584				684373.
215	2.971				600687.
217	3.407				528615.
219	3.896				466375.
220	4.163				438469.
222	4.743				388273.
224	5.390				344642.
226	6.110				306623.
228	6.911				273415.
230	7.799				244338.
232	8.781				218822.
234	9.865				196379.
236	11.06				176597.
238	12.37				159122.
240	13.81				143653.
241	14.58				136589.
243	16.23				123657.
245	18.04				112152.
247	20.00				101898.
249	22.14				92740.
251	24.46				84548.
253	26.98				77206.
255	29.71				70614.
257	32.66				64686.
259	35.85				59346.
261	39.28				54528.
262	41.10				52296.
264	44.93				48154.
266	49.04				44402.
268	53.45				40999.
270	58.18				37907.
272	63.23	79.86		24869.	35094.
274	68.63	80.13		24782.	32531.
276	74.38				30193.
278	80.52				28057.
280	87.04				26103.
282	93.98				24313.
283	97.60				23474.

COMPOUND CONSTANTS:

MP 164.250 K	TC 425.000 K	VC 221.00 ML/MOL	ZC 0.2710
NBP 268.740 K	PC 4.327 MPA	DC 0.2448 G/ML	OM 0.1960
MU 0.0000 DEBYE	RG		

VAPOR PRESSURE CORRELATION CONSTANTS, K AND KPA:

WAGNER	VAPRES-2	WAGNER	WAGNER
164. - 423. K	164. - 283. K	164. - 373. K	253. - 423. K
41 POINTS	24 POINTS	35 POINTS	28 POINTS
RMSD = 3.1130	RMSD = 0.1376	RMSD = 0.3673	RMSD = 3.3190
-0.72490863E+01	0.13397259E+03	-0.72216183E+01	-0.72920422E+01
0.20173329E+01	-0.55076158E+04	0.19538014E+01	0.21613512E+01
-0.30766441E+01	0.67343611E-01	-0.29919855E+01	-0.35780518E+01
-0.14654537E+01	-0.41407055E-04	-0.15788375E+01	0.18467970E+01
	-0.22161244E+02		

LIQUID DENSITY CORRELATION CONSTANTS, K AND G/ML:

FRANCIS1	FRANCIS1	FRANCIS1	RACKETT
194. - 369. K	194. - 289. K	194. - 369. K	253. - 369. K
27 POINTS	16 POINTS	27 POINTS	20 POINTS
RMSD = 0.00026	RMSD = 0.00028	RMSD = 0.00026	RMSD = 0.00063
0.94665784E+00	0.94285101E+00	0.94665784E+00	0.27121101E+00
0.92101190E-03	0.95061515E-03	0.92101190E-03	
0.97117214E+01	0.64130383E+01	0.97117214E+01	
0.46777466E+03	0.44197803E+03	0.46777466E+03	

SECOND VIRIAL COEFFICIENT CORRELATION CONSTANTS, ML/MOL:

TSONOPOULOS: TC=425.000 PC=4.327 VC=221.00 OM=0.1960 A= 0.000000 B=0.0000

LITERATURE DOCUMENTS FOR CORRELATED VAPOR PRESSURE VALUES:

2238 3025 3144 3262 5719 9566

LITERATURE DOCUMENTS FOR CORRELATED LIQUID DENSITY VALUES:

2645 3016 3025 9566

LITERATURE DOCUMENTS REPORTING VIRIAL COEFFICIENT DATA:

2768

T,K	VAPOR PRESSURE, KPA	SATURATED LIQUID VOLUME, ML/MOL	SECOND VIRIAL COEFFICIENT, ML/MOL	HEAT OF VAPORIZATION, J/MOL	SATURATED VAPOR VOLUME, ML/MOL
164	0.06731				
169	0.1220		-5005.		0.1151E+08
175	0.2367		-4142.		6144219.
181	0.4363		-3482.		3445438.
187	0.7689		-2970.		2019177.
193	1.301		-2567.		1231110.
199	2.121	74.38	-2246.	25812.	777844.
205	3.345	75.03	-1986.	25526.	507511.
211	5.120	75.70	-1774.	25241.	340892.
216	7.146	76.28	-1625.	25004.	249671.
222	10.42	76.98	-1473.	24719.	175662.
228	14.84	77.70	-1344.	24433.	126345.
234	20.70	78.44	-1235.	24144.	92715.
240	28.32	79.20	-1140.	23854.	69291.
246	38.06	79.98	-1058.	23559.	52655.
252	50.32	80.79	-985.7	23260.	40626.
258	65.53	81.62	-922.0	22905.	31713.
264	84.16	82.47	-865.3	22583.	25113.
269	102.7	83.21	-822.6	22310.	20861.
275	128.9	84.12	-776.1	21975.	16861.
281	160.0	85.07	-734.0	21632.	13764.
287	196.6	86.05	-695.7	21282.	11337.
293	239.3	87.07	-660.8	20923.	9415.
299	288.7	88.13	-628.7	20556.	7878.
305	345.4	89.25	-599.2	20179.	6638.
311	410.1	90.41	-572.0	19793.	5628.
317	483.6	91.64	-546.7	19396.	4799.
322	551.9	92.71	-527.0	19057.	4219.
328	643.0	94.06	-504.8	18638.	3631.
334	744.9	95.49	-484.1	18206.	3137.
340	858.4	97.02	-464.7	17758.	2721.
346	984.1	98.66	-446.5	17293.	2367.
352	1123.	100.43	-429.4	16807.	2065.
358	1276.	102.35	-413.3	16297.	1805.
364	1444.	104.46	-398.1	15758.	1581.
370	1628.		-383.6		1385.
375	1795.		-372.2		1241.
381	2011.		-359.2		1087.
387	2248.		-346.8		950.8
393	2505.		-335.0		828.4
399	2785.		-323.8		717.4
405	3090.		-313.1		614.6
411	3422.		-302.9		516.0
417	3785.		-293.1		409.6
422	4115.		-285.3		-93.77

COMPOUND CONSTANTS:

MP 135.880 K	TC	VC	ZC
NBP 318.000 K	PC	DC	OM
MU 0.5000 DEBYE	RG		

VAPOR PRESSURE CORRELATION CONSTANTS, K AND KPA:

RIEDEL	RIEDEL	RIEDEL	RIEDEL
213. - 319. K	213. - 319. K	213. - 319. K	301. - 319. K
25 POINTS	25 POINTS	25 POINTS	8 POINTS
RMSD = 0.0150	RMSD = 0.0150	RMSD = 0.0150	RMSD = 0.0030
0.55323672E+02	0.55323672E+02	0.55323672E+02	0.96804450E+02
-0.59615425E+01	-0.59615425E+01	-0.59615425E+01	-0.12280320E+02
-0.52190072E+04	-0.52190072E+04	-0.52190072E+04	-0.68894724E+04
0.55587454E-16	0.55587454E-16	0.55587454E-16	0.23100046E-15

LIQUID DENSITY CORRELATION CONSTANTS, K AND G/ML:

FRANCIS1	FRANCIS1	FRANCIS1	FRANCIS1
293. - 303. K	293. - 303. K	293. - 303. K	293. - 303. K
3 POINTS	3 POINTS	3 POINTS	3 POINTS
RMSD = 0.00002	RMSD = 0.00002	RMSD = 0.00002	RMSD = 0.00002
0.98488998E+00	0.98488998E+00	0.98488998E+00	0.98488998E+00
0.99720000E-03	0.99720000E-03	0.99720000E-03	0.99720000E-03

SECOND VIRIAL COEFFICIENT CORRELATION CONSTANTS, ML/MOL:

LITERATURE DOCUMENTS FOR CORRELATED VAPOR PRESSURE VALUES:

1991 10351

LITERATURE DOCUMENTS FOR CORRELATED LIQUID DENSITY VALUES:

1991

LITERATURE DOCUMENTS REPORTING VIRIAL COEFFICIENT DATA:

T,K	VAPOR PRESSURE, KPA	SATURATED LIQUID VOLUME, ML/MOL	SECOND VIRIAL COEFFICIENT, ML/MOL	HEAT OF VAPORIZATION, J/MOL	SATURATED VAPOR VOLUME, ML/MOL
213	0.3214				5510281.
215	0.3819				4680705.
217	0.4522				3990301.
220	0.5786				3161311.
222	0.6791				2718058.
225	0.8581				2180056.
227	0.9990				1889284.
229	1.159				1642232.
232	1.441				1338178.
234	1.661				1171559.
237	2.043				964613.
239	2.337				850174.
241	2.667				751225.
244	3.236				626865.
246	3.670				557283.
249	4.413				469149.
251	4.975				419452.
253	5.597				375836.
256	6.651				320027.
258	7.442				288241.
261	8.775				247296.
263	9.770				223820.
265	10.86				202950.
268	12.67				175831.
270	14.02				160147.
273	16.25				139648.
275	17.90				127724.
278	20.63				112055.
280	22.63				102891.
282	24.78				94619.
285	28.32				83670.
287	30.90				77221.
290	35.13				68645.
292	38.19				63570.
294	41.47	98.48		29298.	58946.
297	46.80	98.90		29182.	52759.
299	50.66	99.19		29106.	49075.
302	56.91	99.63		28994.	44124.
304	61.40				41164.
306	66.18				38446.
309	73.89				34772.
311	79.41				32564.
314	88.29				29569.
316	94.64				27762.
318	101.3				26091.

COMPOUND CONSTANTS:

MP 132.330 K	TC	VC	ZC
NBP 317.208 K	PC	DC	OM
MU	RG		

VAPOR PRESSURE CORRELATION CONSTANTS, K AND KPA:

RIEDEL	RIEDEL	RIEDEL	RIEDEL
213. - 318. K	213. - 318. K	213. - 318. K	300. - 318. K
23 POINTS	23 POINTS	23 POINTS	8 POINTS
RMSD = 0.0060	RMSD = 0.0060	RMSD = 0.0060	RMSD = 0.0020
0.55162489E+02	0.55162489E+02	0.55162489E+02	0.50435979E+02
-0.59501910E+01	-0.59501910E+01	-0.59501910E+01	-0.52309388E+01
-0.51823650E+04	-0.51823650E+04	-0.51823650E+04	-0.49909076E+04
0.62660662E-16	0.62660662E-16	0.62660662E-16	0.43333152E-16

LIQUID DENSITY CORRELATION CONSTANTS, K AND G/ML:

FRANCIS1	FRANCIS1	FRANCIS1	FRANCIS1
293. - 303. K	293. - 303. K	293. - 303. K	293. - 303. K
3 POINTS	3 POINTS	3 POINTS	3 POINTS
RMSD = 0.00000	RMSD = 0.00000	RMSD = 0.00000	RMSD = 0.00000
0.99008000E+00	0.99008000E+00	0.99008000E+00	0.99008000E+00
0.10202000E-02	0.10202000E-02	0.10202000E-02	0.10202000E-02

SECOND VIRIAL COEFFICIENT CORRELATION CONSTANTS, ML/MOL:

LITERATURE DOCUMENTS FOR CORRELATED VAPOR PRESSURE VALUES:

    1991 10351

LITERATURE DOCUMENTS FOR CORRELATED LIQUID DENSITY VALUES:

    1991

LITERATURE DOCUMENTS REPORTING VIRIAL COEFFICIENT DATA:

T,K	VAPOR PRESSURE, KPA	SATURATED LIQUID VOLUME, ML/MOL	SECOND VIRIAL COEFFICIENT, ML/MOL	HEAT OF VAPORIZATION, J/MOL	SATURATED VAPOR VOLUME, ML/MOL
213	0.3455				5125613.
215	0.4100				4360294.
217	0.4847				3722454.
220	0.6190				2955260.
222	0.7255				2544338.
224	0.8475				2197589.
227	1.064				1774312.
229	1.233				1544235.
232	1.530				1260640.
234	1.761				1104994.
236	2.020				971191.
239	2.471				804182.
241	2.817				711374.
244	3.412				594575.
246	3.865				529135.
248	4.369				471981.
251	5.227				399278.
253	5.874				358109.
255	6.587				321859.
258	7.792				275293.
260	8.693				248667.
263	10.21				214243.
265	11.33				194432.
267	12.56				176770.
270	14.60				153741.
272	16.11				140377.
275	18.61				122854.
277	20.45				112628.
279	22.43				103413.
282	25.70				91241.
284	28.08				84086.
286	30.64				77602.
289	34.83				68979.
291	37.88				63876.
294	42.84	98.70		29086.	57059.
296	46.43	98.99		29013.	53007.
298	50.25	99.29		28939.	49303.
301	56.46	99.73		28832.	44326.
303	60.93	100.03		28761.	41349.
306	68.15				37334.
308	73.33				34923.
310	78.81				32703.
313	87.65				29692.
315	93.96				27875.
317	100.6				26194.

COMPOUND CONSTANTS:

MP	185.710 K	TC	VC	ZC
NBP	315.172 K	PC	DC	OM
MU	0.6800 DEBYE	RG		

VAPOR PRESSURE CORRELATION CONSTANTS, K AND KPA:

RIEDEL	RIEDEL	RIEDEL	RIEDEL
213. - 316. K	213. - 316. K	213. - 316. K	213. - 316. K
22 POINTS	22 POINTS	22 POINTS	22 POINTS
RMSD = 0.0020	RMSD = 0.0020	RMSD = 0.0020	RMSD = 0.0020
0.53850482E+02	0.53850482E+02	0.53850482E+02	0.53850482E+02
-0.57756418E+01	-0.57756418E+01	-0.57756418E+01	-0.57756418E+01
-0.50649954E+04	-0.50649954E+04	-0.50649954E+04	-0.50649954E+04
0.67765942E-16	0.67765942E-16	0.67765942E-16	0.67765942E-16

LIQUID DENSITY CORRELATION CONSTANTS, K AND G/ML:

FRANCIS1	FRANCIS1	FRANCIS1	FRANCIS1
293. - 303. K	293. - 303. K	293. - 303. K	293. - 303. K
3 POINTS	3 POINTS	3 POINTS	3 POINTS
RMSD = 0.00002	RMSD = 0.00002	RMSD = 0.00002	RMSD = 0.00002
0.97245496E+00	0.97245496E+00	0.97245496E+00	0.97245496E+00
0.10112000E-02	0.10112000E-02	0.10112000E-02	0.10112000E-02

SECOND VIRIAL COEFFICIENT CORRELATION CONSTANTS, ML/MOL:

LITERATURE DOCUMENTS FOR CORRELATED VAPOR PRESSURE VALUES:

1991 10351

LITERATURE DOCUMENTS FOR CORRELATED LIQUID DENSITY VALUES:

1991

LITERATURE DOCUMENTS REPORTING VIRIAL COEFFICIENT DATA:

T,K	VAPOR PRESSURE, KPA	SATURATED LIQUID VOLUME, ML/MOL	SECOND VIRIAL COEFFICIENT, ML/MOL	HEAT OF VAPORIZATION, J/MOL	SATURATED VAPOR VOLUME, ML/MOL
213	0.4117				4301398.
215	0.4868				3671853.
217	0.5736				3145354.
220	0.7289				2509449.
222	0.8516				2167428.
224	0.9918				1877898.
227	1.239				1523109.
229	1.432				1329518.
231	1.650				1163829.
234	2.031				958111.
236	2.324				844387.
238	2.652				746078.
241	3.219				622552.
243	3.651				553452.
245	4.130				493172.
248	4.950				416601.
250	5.568				373301.
252	6.250				335215.
255	7.404				286351.
257	8.268				258446.
259	9.214				233713.
262	10.80				201692.
264	11.98				183238.
266	13.26				166769.
269	15.40				145267.
271	16.97				132772.
273	18.67				121550.
276	21.49				106784.
278	23.55				98139.
280	25.77				90327.
283	29.42				79976.
285	32.08				73873.
287	34.92				68329.
290	39.57				60933.
292	42.94				56544.
294	46.53	100.89		28581.	52537.
297	52.36	101.35		28478.	47160.
299	56.57	101.65		28411.	43949.
301	61.03	101.96		28344.	41004.
304	68.26				37030.
306	73.44				34643.
308	78.93				32445.
311	87.76				29463.
313	94.08				27663.
315	100.7				25998.

491

COMPOUND CONSTANTS:

MP	124.862 K	TC		VC		ZC	
NBP	299.108 K	PC		DC		OM	
MU	0.3800 DEBYE	RG					

    VAPOR PRESSURE CORRELATION CONSTANTS, K AND KPA:

VAPRES-2	VAPRES-2	VAPRES-2	RIEDEL
194. - 300. K	194. - 300. K	194. - 300. K	288. - 300. K
24 POINTS	24 POINTS	24 POINTS	8 POINTS
RMSD = 0.0280	RMSD = 0.0280	RMSD = 0.0280	RMSD = 0.0040
-0.65649636E+03	-0.65649636E+03	-0.65649636E+03	-0.89487154E+02
0.70317180E+04	0.70317180E+04	0.70317180E+04	0.16164036E+02
-0.56167407E+00	-0.56167407E+00	-0.56167407E+00	0.74082320E+03
0.37277356E-03	0.37277356E-03	0.37277356E-03	-0.72517770E-15
0.13546454E+03	0.13546454E+03	0.13546454E+03	

    LIQUID DENSITY CORRELATION CONSTANTS, K AND G/ML:

    SECOND VIRIAL COEFFICIENT CORRELATION CONSTANTS, ML/MOL:

LITERATURE DOCUMENTS FOR CORRELATED VAPOR PRESSURE VALUES:

    624   1991 10351

LITERATURE DOCUMENTS FOR CORRELATED LIQUID DENSITY VALUES:

LITERATURE DOCUMENTS REPORTING VIRIAL COEFFICIENT DATA:

T,K	VAPOR PRESSURE, KPA	SATURATED LIQUID VOLUME, ML/MOL	SECOND VIRIAL COEFFICIENT, ML/MOL	HEAT OF VAPORIZATION, J/MOL	SATURATED VAPOR VOLUME, ML/MOL
194	0.2065				7811246.
196	0.2489				6546108.
198	0.2990				5505372.
201	0.3910				4273917.
203	0.4655				3625535.
206	0.6009				2850407.
208	0.7093				2438020.
210	0.8346				2091976.
213	1.059				1672571.
215	1.236				1446323.
218	1.550				1169541.
220	1.795				1018823.
222	2.074				890058.
225	2.561				730570.
227	2.937				642634.
230	3.589				532799.
232	4.089				471729.
234	4.647				418707.
237	5.602				351742.
239	6.327				314092.
242	7.559				266185.
244	8.487				239048.
246	9.507				215145.
249	11.23				184429.
251	12.51				166856.
254	14.65				144123.
256	16.24				131032.
259	18.89				113988.
261	20.84				104110.
263	22.96				95255.
266	26.45				83629.
268	29.00				76835.
271	33.20				67866.
273	36.26				62594.
275	39.55				57818.
278	44.91				51464.
280	48.80				47702.
283	55.14				42673.
285	59.72				39681.
287	64.59				36944.
290	72.50				33259.
292	78.18				31053.
295	87.38				28071.
297	93.98				26276.
299	101.0				24621.

COMPOUND CONSTANTS:

MP   147.500 K   TC   VC   ZC
NBP   321.405 K   PC   DC   OM
MU   RG   3.0552 ANGSTROM

VAPOR PRESSURE CORRELATION CONSTANTS, K AND KPA:

VAPRES-2	VAPRES-2	VAPRES-2	RIEDEL
213. - 322. K	213. - 322. K	213. - 322. K	304. - 322. K
24 POINTS	24 POINTS	24 POINTS	8 POINTS
RMSD = 0.0070	RMSD = 0.0070	RMSD = 0.0070	RMSD = 0.0050
0.28834871E+02	0.28834871E+02	0.28834871E+02	0.78431821E+02
-0.50047063E+04	-0.50047063E+04	-0.50047063E+04	-0.94156226E+01
-0.30414429E-01	-0.30414429E-01	-0.30414429E-01	-0.62945584E+04
0.26215403E-04	0.26215403E-04	0.26215403E-04	0.11303323E-15
-0.27333595E+00	-0.27333595E+00	-0.27333595E+00	

LIQUID DENSITY CORRELATION CONSTANTS, K AND G/ML:

FRANCIS1	FRANCIS1	FRANCIS1	FRANCIS1
293. - 303. K	293. - 303. K	293. - 303. K	293. - 303. K
3 POINTS	3 POINTS	3 POINTS	3 POINTS
RMSD = 0.00005	RMSD = 0.00005	RMSD = 0.00005	RMSD = 0.00005
0.99496394E+00	0.99496394E+00	0.99496394E+00	0.99496394E+00
0.10232000E-02	0.10232000E-02	0.10232000E-02	0.10232000E-02

SECOND VIRIAL COEFFICIENT CORRELATION CONSTANTS, ML/MOL:

LITERATURE DOCUMENTS FOR CORRELATED VAPOR PRESSURE VALUES:

    1991 10351

LITERATURE DOCUMENTS FOR CORRELATED LIQUID DENSITY VALUES:

    1991

LITERATURE DOCUMENTS REPORTING VIRIAL COEFFICIENT DATA:

T,K	VAPOR PRESSURE, KPA	SATURATED LIQUID VOLUME, ML/MOL	SECOND VIRIAL COEFFICIENT, ML/MOL	HEAT OF VAPORIZATION, J/MOL	SATURATED VAPOR VOLUME, ML/MOL
213	0.2427				7295588.
215	0.2899				6165619.
217	0.3450				5230028.
220	0.4447				4113511.
222	0.5243				3520180.
225	0.6671				2804175.
227	0.7801				2419433.
230	0.9805				1950296.
232	1.138				1695588.
235	1.414				1381980.
237	1.628				1210062.
240	2.003				996463.
242	2.291				878308.
245	2.789				730257.
247	3.171				647668.
250	3.825				543355.
252	4.323				484704.
255	5.170				410071.
257	5.810				367796.
260	6.893				313624.
262	7.705				282725.
265	9.072				242869.
267	10.09				219986.
269	11.20				199623.
272	13.06				173131.
274	14.44				157791.
277	16.72				137721.
279	18.41				126034.
282	21.19				110660.
284	23.23				101660.
287	26.58				89761.
289	29.03				82760.
292	33.05				73459.
294	35.97	98.13		30120.	67960.
297	40.73	98.57		30002.	60623.
299	44.18	98.86		29924.	56265.
302	49.80	99.30		29810.	50425.
304	53.84				46943.
307	60.41				42256.
309	65.12				39450.
312	72.75				35659.
314	78.21				33381.
317	87.01				30291.
319	93.30				28428.
321	99.94				26705.

COMPOUND CONSTANTS:

MP	159.530 K	TC	VC	ZC
NBP	313.991 K	PC	DC	OM
MU		RG		

VAPOR PRESSURE CORRELATION CONSTANTS, K AND KPA:

RIEDEL	RIEDEL	RIEDEL	RIEDEL
213. - 319. K	213. - 319. K	213. - 319. K	213. - 319. K
23 POINTS	23 POINTS	23 POINTS	23 POINTS
RMSD = 0.0040	RMSD = 0.0040	RMSD = 0.0040	RMSD = 0.0040
0.53746588E+02	0.53746588E+02	0.53746588E+02	0.53746588E+02
-0.57385643E+01	-0.57385643E+01	-0.57385643E+01	-0.57385643E+01
-0.50845007E+04	-0.50845007E+04	-0.50845007E+04	-0.50845007E+04
0.60950559E-16	0.60950559E-16	0.60950559E-16	0.60950559E-16

LIQUID DENSITY CORRELATION CONSTANTS, K AND G/ML:

SECOND VIRIAL COEFFICIENT CORRELATION CONSTANTS, ML/MOL:

LITERATURE DOCUMENTS FOR CORRELATED VAPOR PRESSURE VALUES:

10351

LITERATURE DOCUMENTS FOR CORRELATED LIQUID DENSITY VALUES:

LITERATURE DOCUMENTS REPORTING VIRIAL COEFFICIENT DATA:

T,K	VAPOR PRESSURE, KPA	SATURATED LIQUID VOLUME, ML/MOL	SECOND VIRIAL COEFFICIENT, ML/MOL	HEAT OF VAPORIZATION, J/MOL	SATURATED VAPOR VOLUME, ML/MOL
213	0.4128				4289892.
215	0.4887				3657779.
217	0.5765				3129725.
220	0.7338				2492804.
222	0.8582				2150704.
225	1.079				1733704.
227	1.252				1507365.
229	1.449				1314427.
232	1.793				1076036.
234	2.059				944884.
237	2.522				781383.
239	2.877				690631.
241	3.274				611937.
244	3.957				512694.
246	4.476				456974.
249	5.361				386166.
251	6.030				346109.
253	6.767				310865.
256	8.013				265632.
258	8.946				239791.
261	10.51				206409.
263	11.68				187214.
265	12.95				170112.
268	15.07				147830.
270	16.64				134909.
273	19.24				117978.
275	21.15				108105.
278	24.31				95100.
280	26.61				87475.
282	29.10				80580.
285	33.17				71433.
287	36.14				66033.
290	40.98				58837.
292	44.49				54569.
294	48.24				50675.
297	54.33				45453.
299	58.72				42338.
302	65.83				38143.
304	70.94				35630.
306	76.36				33319.
309	85.10				30191.
311	91.35				28307.
314	101.4				25747.
316	108.6				24200.
318	116.1				22767.

COMPOUND CONSTANTS:

MP	127.190 K	TC	VC	ZC
NBP	307.207 K	PC	DC	OM
MU	0.2600 DEBYE	RG		

VAPOR PRESSURE CORRELATION CONSTANTS, K AND KPA:

RPM2	RPM2	RPM2	RPM2
216. - 308. K	216. - 308. K	216. - 308. K	216. - 308. K
32 POINTS	32 POINTS	32 POINTS	32 POINTS
RMSD = 0.0240	RMSD = 0.0240	RMSD = 0.0240	RMSD = 0.0240
0.26603797E+02	0.26603797E+02	0.26603797E+02	0.26603797E+02
-0.45680185E+04	-0.45680185E+04	-0.45680185E+04	-0.45680185E+04
-0.32375194E-01	-0.32375194E-01	-0.32375194E-01	-0.32375194E-01
0.29987458E-04	0.29987458E-04	0.29987458E-04	0.29987458E-04

LIQUID DENSITY CORRELATION CONSTANTS, K AND G/ML:

FRANCIS1	FRANCIS1	FRANCIS1	FRANCIS1
173. - 303. K	173. - 303. K	173. - 303. K	173. - 303. K
11 POINTS	11 POINTS	11 POINTS	11 POINTS
RMSD = 0.00015	RMSD = 0.00015	RMSD = 0.00015	RMSD = 0.00015
0.96712947E+00	0.96712947E+00	0.96712947E+00	0.96712947E+00
0.78504952E-03	0.78504952E-03	0.78504952E-03	0.78504952E-03
0.15000000E+02	0.15000000E+02	0.15000000E+02	0.15000000E+02
0.55969409E+03	0.55969409E+03	0.55969409E+03	0.55969409E+03

SECOND VIRIAL COEFFICIENT CORRELATION CONSTANTS, ML/MOL:

LITERATURE DOCUMENTS FOR CORRELATED VAPOR PRESSURE VALUES:

1991 5113 10351 15542 21583

LITERATURE DOCUMENTS FOR CORRELATED LIQUID DENSITY VALUES:

1991 6646

LITERATURE DOCUMENTS REPORTING VIRIAL COEFFICIENT DATA:

T,K	VAPOR PRESSURE, KPA	SATURATED LIQUID VOLUME, ML/MOL	SECOND VIRIAL COEFFICIENT, ML/MOL	HEAT OF VAPORIZATION, J/MOL	SATURATED VAPOR VOLUME, ML/MOL
173		85.95			
176		86.24			
179		86.53			
182		86.83			
185		87.12			
188		87.42			
191		87.72			
194		88.03			
197		88.34			
200		88.65			
203		88.96			
206		89.28			
209		89.60			
212		89.92			
215		90.24			
219	1.098	90.68		30306.	1658609.
222	1.374	91.01		30168.	1343349.
225	1.708	91.35		30031.	1095428.
228	2.108	91.69		29894.	899090.
231	2.587	92.03		29760.	742551.
234	3.154	92.38		29626.	616934.
237	3.822	92.73		29494.	515506.
240	4.607	93.08		29363.	433118.
243	5.523	93.44		29233.	365815.
246	6.587	93.80		29105.	310530.
249	7.816	94.17		28979.	264879.
252	9.231	94.54		28854.	226990.
255	10.85	94.91		28731.	195389.
258	12.70	95.29		28610.	168910.
261	14.80	95.67		28491.	146621.
265	18.04	96.19		28335.	122154.
268	20.83	96.58		28220.	106992.
271	23.96	96.98		28107.	94051.
274	27.46	97.38		27996.	82962.
277	31.37	97.78		27888.	73426.
280	35.71	98.20		27781.	65194.
283	40.52	98.61		27677.	58065.
286	45.84	99.04		27575.	51869.
289	51.71	99.47		27476.	46468.
292	58.16	99.90		27378.	41744.
295	65.23	100.34		27284.	37600.
298	72.97	100.79		27191.	33954.
301	81.42	101.24		27102.	30738.
304	90.62				27892.
307	100.6				25368.

499

COMPOUND CONSTANTS:

MP		TC		VC		ZC	
NBP	352.589 K	PC		DC		OM	
MU		RG					

VAPOR PRESSURE CORRELATION CONSTANTS, K AND KPA:

VAPRES-2	VAPRES-2	VAPRES-2	VAPRES-2
256. - 453. K	256. - 453. K	256. - 453. K	256. - 453. K
5 POINTS	5 POINTS	5 POINTS	5 POINTS
RMSD = 0.0170	RMSD = 0.0170	RMSD = 0.0170	RMSD = 0.0170
0.53908551E+03	0.53908551E+03	0.53908551E+03	0.53908551E+03
-0.16412492E+05	-0.16412492E+05	-0.16412492E+05	-0.16412492E+05
0.24694510E+00	0.24694510E+00	0.24694510E+00	0.24694510E+00
-0.10901125E-03	-0.10901125E-03	-0.10901125E-03	-0.10901125E-03
-0.95721672E+02	-0.95721672E+02	-0.95721672E+02	-0.95721672E+02

LIQUID DENSITY CORRELATION CONSTANTS, K AND G/ML:

FRANCIS1	FRANCIS1	FRANCIS1
293. - 299. K	293. - 299. K	293. - 299. K
4 POINTS	4 POINTS	4 POINTS
RMSD = 0.00000	RMSD = 0.00000	RMSD = 0.00000
0.98466581E+00	0.98466581E+00	0.98466581E+00
0.83079678E-03	0.83079678E-03	0.83079678E-03
0.64182644E+01	0.64182644E+01	0.64182644E+01
0.53788379E+03	0.53788379E+03	0.53788379E+03

SECOND VIRIAL COEFFICIENT CORRELATION CONSTANTS, ML/MOL:

LITERATURE DOCUMENTS FOR CORRELATED VAPOR PRESSURE VALUES:

4457

LITERATURE DOCUMENTS FOR CORRELATED LIQUID DENSITY VALUES:

3269

LITERATURE DOCUMENTS REPORTING VIRIAL COEFFICIENT DATA:

T,K	VAPOR PRESSURE, KPA	SATURATED LIQUID VOLUME, ML/MOL	SECOND VIRIAL COEFFICIENT, ML/MOL	HEAT OF VAPORIZATION, J/MOL	SATURATED VAPOR VOLUME, ML/MOL
256	1.290				1650488.
260	1.681				1286021.
264	2.168				1012595.
269	2.937				761455.
273	3.706				612455.
278	4.896				472119.
282	6.061				386829.
287	7.831				304707.
291	9.536				253729.
296	12.08	115.34		33745.	203680.
300	14.50				172026.
305	18.06				140410.
309	21.40				120081.
314	26.25				99462.
318	30.74				86008.
323	37.21				72176.
327	43.13				63031.
332	51.58				53514.
336	59.25				47147.
341	70.09				40449.
345	79.86				35921.
350	93.54				31109.
354	105.8				27825.
358	119.2				24963.
363	137.9				21882.
367	154.5				19755.
372	177.3				17448.
376	197.3				15843.
381	224.8				14090.
385	248.9				12862.
390	281.6				11513.
394	310.1				10562.
399	348.8				9512.
403	382.2				8767.
408	427.3				7939.
412	466.1				7350.
417	518.1				6692.
421	562.7				6221.
426	622.1				5693.
430	672.8				5314.
435	740.1				4887.
439	797.2				4579.
444	872.6				4230.
448	936.2				3978.
452	1003.				3748.

501

COMPOUND CONSTANTS:

MP		TC		VC		ZC
NBP	346.150 K	PC		DC		OM
MU		RG				

VAPOR PRESSURE CORRELATION CONSTANTS, K AND KPA:

RPM2	RPM2	RPM2	RPM2
299. - 320. K	299. - 320. K	299. - 320. K	299. - 320. K
8 POINTS	8 POINTS	8 POINTS	8 POINTS
RMSD = 0.0140	RMSD = 0.0140	RMSD = 0.0140	RMSD = 0.0140
0.61450812E+01	0.61450812E+01	0.61450812E+01	0.61450812E+01
-0.33060304E+04	-0.33060304E+04	-0.33060304E+04	-0.33060304E+04
0.44333941E-01	0.44333941E-01	0.44333941E-01	0.44333941E-01
-0.61039283E-04	-0.61039283E-04	-0.61039283E-04	-0.61039283E-04

LIQUID DENSITY CORRELATION CONSTANTS, K AND G/ML:

FRANCIS1	FRANCIS1	FRANCIS1
298. - 324. K	298. - 324. K	298. - 324. K
4 POINTS	4 POINTS	4 POINTS
RMSD = 0.00005	RMSD = 0.00005	RMSD = 0.00005
0.97582245E+00	0.97582245E+00	0.97582245E+00
0.79138950E-03	0.79138950E-03	0.79138950E-03
0.83753252E+01	0.83753252E+01	0.83753252E+01
0.51401636E+03	0.51401636E+03	0.51401636E+03

SECOND VIRIAL COEFFICIENT CORRELATION CONSTANTS, ML/MOL:

LITERATURE DOCUMENTS FOR CORRELATED VAPOR PRESSURE VALUES:

5641

LITERATURE DOCUMENTS FOR CORRELATED LIQUID DENSITY VALUES:

5641

LITERATURE DOCUMENTS REPORTING VIRIAL COEFFICIENT DATA:

T,K	VAPOR PRESSURE, KPA	SATURATED LIQUID VOLUME, ML/MOL	SECOND VIRIAL COEFFICIENT, ML/MOL	HEAT OF VAPORIZATION, J/MOL	SATURATED VAPOR VOLUME, ML/MOL
298		117.15			
299	17.94	117.31		33281.	138578.
300	18.76	117.47		33228.	132973.
301	19.61	117.64		33173.	127640.
302	20.49	117.80		33117.	122564.
303	21.40	117.97		33061.	117731.
304	22.34	118.13		33002.	113129.
305	23.32	118.30		32943.	108743.
306	24.33	118.47		32883.	104564.
307	25.38	118.64		32821.	100580.
308	26.46	118.81		32759.	96781.
309	27.58	118.98		32695.	93157.
310	28.73	119.15		32630.	89698.
311	29.93	119.32		32563.	86397.
312	31.16	119.49		32496.	83244.
313	32.44	119.67		32427.	80234.
314	33.75	119.84		32357.	77357.
315	35.10	120.02		32286.	74608.
316	36.50	120.19		32214.	71980.
317	37.94	120.37		32140.	69467.
318	39.43	120.55		32065.	67063.
319	40.95	120.73		31989.	64762.
320	42.53	120.91		31912.	62561.
321		121.09			
322		121.27			
323		121.45			
324		121.64			

COMPOUND CONSTANTS:

MP		TC		VC		ZC
NBP	338.150 K	PC		DC		OM
MU		RG				

VAPOR PRESSURE CORRELATION CONSTANTS, K AND KPA:

RPM2	RPM2	RPM2	RPM2
304. - 323. K	304. - 323. K	304. - 323. K	304. - 323. K
8 POINTS	8 POINTS	8 POINTS	8 POINTS
RMSD = 0.0190	RMSD = 0.0190	RMSD = 0.0190	RMSD = 0.0190
-0.78720694E+02	-0.78720694E+02	-0.78720694E+02	-0.78720694E+02
0.59517541E+04	0.59517541E+04	0.59517541E+04	0.59517541E+04
0.30678344E+00	0.30678344E+00	0.30678344E+00	0.30678344E+00
-0.33250629E-03	-0.33250629E-03	-0.33250629E-03	-0.33250629E-03

LIQUID DENSITY CORRELATION CONSTANTS, K AND G/ML:

FRANCIS1	FRANCIS1	FRANCIS1
299. - 324. K	299. - 324. K	299. - 324. K
4 POINTS	4 POINTS	4 POINTS
RMSD = 0.00013	RMSD = 0.00013	RMSD = 0.00013
0.98024690E+00	0.98024690E+00	0.98024690E+00
0.86870207E-03	0.86870207E-03	0.86870207E-03
0.72943087E+01	0.72943087E+01	0.72943087E+01
0.55671313E+03	0.55671313E+03	0.55671313E+03

SECOND VIRIAL COEFFICIENT CORRELATION CONSTANTS, ML/MOL:

LITERATURE DOCUMENTS FOR CORRELATED VAPOR PRESSURE VALUES:

5641

LITERATURE DOCUMENTS FOR CORRELATED LIQUID DENSITY VALUES:

5641

LITERATURE DOCUMENTS REPORTING VIRIAL COEFFICIENT DATA:

T,K	VAPOR PRESSURE, KPA	SATURATED LIQUID VOLUME, ML/MOL	SECOND VIRIAL COEFFICIENT, ML/MOL	HEAT OF VAPORIZATION, J/MOL	SATURATED VAPOR VOLUME, ML/MOL
299		118.67			
300		118.84			
301		119.01			
302		119.18			
303		119.35			
304	29.69	119.52			85141.
305	30.90	119.69			82064.
306	32.16	119.86			79116.
307	33.46	120.03			76293.
308	34.80	120.21			73588.
309	36.19	120.38			70997.
310	37.62	120.56			68514.
311	39.10	120.73			66135.
312	40.62	120.91			63855.
313	42.20	121.08			61670.
314	43.82	121.26			59575.
315	45.50	121.44			57567.
316	47.22	121.62			55642.
317	48.99	121.80			53797.
318	50.82	121.98			52027.
319	52.70	122.16			50330.
320	54.63	122.34			48702.
321	56.62	122.52			47140.
322	58.66	122.71			45643.
323	60.75	122.89			44206.
324		123.07			

COMPOUND CONSTANTS:

MP 132.457 K	TC	VC	ZC
NBP 332.732 K	PC	DC	OM
MU	RG		

VAPOR PRESSURE CORRELATION CONSTANTS, K AND KPA:

RIEDEL	RIEDEL	RIEDEL	RIEDEL
273. - 333. K	273. - 333. K	273. - 333. K	273. - 333. K
18 POINTS	18 POINTS	18 POINTS	18 POINTS
RMSD = 0.0370	RMSD = 0.0370	RMSD = 0.0370	RMSD = 0.0370
0.79598557E+02	0.79598557E+02	0.79598557E+02	0.79598557E+02
-0.96652022E+01	-0.96652022E+01	-0.96652022E+01	-0.96652022E+01
-0.63288869E+04	-0.63288869E+04	-0.63288869E+04	-0.63288869E+04
0.12516819E-15	0.12516819E-15	0.12516819E-15	0.12516819E-15

LIQUID DENSITY CORRELATION CONSTANTS, K AND G/ML:

FRANCIS1	FRANCIS1	FRANCIS1
273. - 324. K	273. - 324. K	273. - 324. K
6 POINTS	6 POINTS	6 POINTS
RMSD = 0.00087	RMSD = 0.00087	RMSD = 0.00087
0.92219526E+00	0.92219526E+00	0.92219526E+00
0.64979238E-03	0.64979238E-03	0.64979238E-03
0.60433254E+01	0.60433254E+01	0.60433254E+01
0.44870898E+03	0.44870898E+03	0.44870898E+03

SECOND VIRIAL COEFFICIENT CORRELATION CONSTANTS, ML/MOL:

LITERATURE DOCUMENTS FOR CORRELATED VAPOR PRESSURE VALUES:

5641 10749

LITERATURE DOCUMENTS FOR CORRELATED LIQUID DENSITY VALUES:

2369 5641 8367

LITERATURE DOCUMENTS REPORTING VIRIAL COEFFICIENT DATA:

T,K	VAPOR PRESSURE, KPA	SATURATED LIQUID VOLUME, ML/MOL	SECOND VIRIAL COEFFICIENT, ML/MOL	HEAT OF VAPORIZATION, J/MOL	SATURATED VAPOR VOLUME, ML/MOL
273	9.497	115.63		31373.	239009.
274	9.988	115.77		31310.	228090.
275	10.50	115.91		31248.	217768.
277	11.59	116.19		31124.	198769.
278	12.16	116.33		31063.	190024.
279	12.76	116.47		31002.	181743.
281	14.04	116.75		30882.	166456.
282	14.71	116.90		30822.	159401.
283	15.41	117.04		30763.	152707.
285	16.89	117.34		30646.	140319.
286	17.67	117.48		30588.	134586.
287	18.48	117.63		30531.	129137.
289	20.19	117.93		30418.	119028.
290	21.09	118.08		30362.	114337.
292	22.99	118.38		30252.	105619.
293	23.99	118.54		30198.	101567.
294	25.02	118.69		30144.	97705.
296	27.19	119.00		30038.	90509.
297	28.33	119.16		29986.	87157.
298	29.51	119.32		29934.	83957.
300	31.99	119.64		29833.	77981.
301	33.28	119.80		29783.	75191.
302	34.62	119.96		29733.	72524.
304	37.43	120.29		29637.	67532.
305	38.90	120.45		29589.	65196.
307	41.97	120.79		29496.	60819.
308	43.58	120.96		29451.	58767.
309	45.23	121.13		29406.	56801.
311	48.69	121.48		29319.	53109.
312	50.49	121.65		29276.	51375.
313	52.35	121.83		29234.	49712.
315	56.22	122.18		29152.	46582.
316	58.24	122.36		29112.	45109.
317	60.32	122.54		29073.	43694.
319	64.65	122.91		28998.	41027.
320	66.90	123.10		28961.	39770.
322	71.59	123.48		28890.	37397.
323	74.03	123.67		28856.	36278.
324	76.53	123.86		28823.	35200.
326	81.74				33161.
327	84.44				32198.
328	87.22				31269.
330	92.98				29510.
331	95.97				28677.
332	99.03				27873.

COMPOUND CONSTANTS:

MP		TC		VC		ZC
NBP	353.150 K	PC		DC		OM
MU		RG				

VAPOR PRESSURE CORRELATION CONSTANTS, K AND KPA:

RIEDEL	RIEDEL	RIEDEL	RIEDEL
304. - 323. K	304. - 323. K	304. - 323. K	304. - 323. K
7 POINTS	7 POINTS	7 POINTS	7 POINTS
RMSD = 0.0030	RMSD = 0.0030	RMSD = 0.0030	RMSD = 0.0030
0.26728003E+03	0.26728003E+03	0.26728003E+03	0.26728003E+03
-0.38056096E+02	-0.38056096E+02	-0.38056096E+02	-0.38056096E+02
-0.14480764E+05	-0.14480764E+05	-0.14480764E+05	-0.14480764E+05
0.82147822E-15	0.82147822E-15	0.82147822E-15	0.82147822E-15

LIQUID DENSITY CORRELATION CONSTANTS, K AND G/ML:

FRANCIS1	FRANCIS1	FRANCIS1
299. - 324. K	299. - 324. K	299. - 324. K
4 POINTS	4 POINTS	4 POINTS
RMSD = 0.00091	RMSD = 0.00091	RMSD = 0.00091
0.10115967E+01	0.10115967E+01	0.10115967E+01
0.94082765E-03	0.94082765E-03	0.94082765E-03
0.59999990E+01	0.59999990E+01	0.59999990E+01
0.60676831E+03	0.60676831E+03	0.60676831E+03

SECOND VIRIAL COEFFICIENT CORRELATION CONSTANTS, ML/MOL:

LITERATURE DOCUMENTS FOR CORRELATED VAPOR PRESSURE VALUES:

5641

LITERATURE DOCUMENTS FOR CORRELATED LIQUID DENSITY VALUES:

5641

LITERATURE DOCUMENTS REPORTING VIRIAL COEFFICIENT DATA:

T,K	VAPOR PRESSURE, KPA	SATURATED LIQUID VOLUME, ML/MOL	SECOND VIRIAL COEFFICIENT, ML/MOL	HEAT OF VAPORIZATION, J/MOL	SATURATED VAPOR VOLUME, ML/MOL
299		115.57			
300		115.73			
301		115.90			
302		116.06			
303		116.23			
304	15.28	116.39		34018.	165413.
305	15.97	116.56		33930.	158797.
306	16.68	116.72		33845.	152505.
307	17.42	116.89		33766.	146517.
308	18.19	117.06		33691.	140817.
309	18.98	117.23		33621.	135387.
310	19.79	117.40		33555.	130211.
311	20.64	117.57		33495.	125275.
312	21.52	117.74		33439.	120567.
313	22.42	117.91		33389.	116072.
314	23.36	118.08		33343.	111779.
315	24.32	118.25		33303.	107678.
316	25.32	118.42		33268.	103757.
317	26.35	118.60		33239.	100007.
318	27.42	118.77		33215.	96418.
319	28.52	118.94		33196.	92983.
320	29.66	119.12		33183.	89693.
321	30.84	119.29		33176.	86540.
322	32.06	119.47		33175.	83518.
323	33.31	119.65			80620.
324		119.82			

COMPOUND CONSTANTS:

MP	197.178 K	TC		VC		ZC
NBP	341.667 K	PC		DC		OM
MU	0.5200 DEBYE	RG				

VAPOR PRESSURE CORRELATION CONSTANTS, K AND KPA:

RIEDEL	RIEDEL	RIEDEL	RIEDEL
273. - 342. K	273. - 342. K	273. - 342. K	273. - 342. K
9 POINTS	9 POINTS	9 POINTS	9 POINTS
RMSD = 0.0100	RMSD = 0.0100	RMSD = 0.0100	RMSD = 0.0100
0.14754378E+02	0.14754378E+02	0.14754378E+02	0.14754378E+02
0.24852085E+00	0.24852085E+00	0.24852085E+00	0.24852085E+00
-0.39049589E+04	-0.39049589E+04	-0.39049589E+04	-0.39049589E+04
-0.98527752E-16	-0.98527752E-16	-0.98527752E-16	-0.98527752E-16

LIQUID DENSITY CORRELATION CONSTANTS, K AND G/ML:

FRANCIS1	FRANCIS1	FRANCIS1
288. - 299. K	288. - 299. K	288. - 299. K
4 POINTS	4 POINTS	4 POINTS
RMSD = 0.00009	RMSD = 0.00009	RMSD = 0.00009
0.94432020E+00	0.94432020E+00	0.94432020E+00
0.51027420E-03	0.51027420E-03	0.51027420E-03
0.13475048E+02	0.13475048E+02	0.13475048E+02
0.49092456E+03	0.49092456E+03	0.49092456E+03

SECOND VIRIAL COEFFICIENT CORRELATION CONSTANTS, ML/MOL:

LITERATURE DOCUMENTS FOR CORRELATED VAPOR PRESSURE VALUES:

10749

LITERATURE DOCUMENTS FOR CORRELATED LIQUID DENSITY VALUES:

3269 20809

LITERATURE DOCUMENTS REPORTING VIRIAL COEFFICIENT DATA:

T,K	VAPOR PRESSURE, KPA	SATURATED LIQUID VOLUME, ML/MOL	SECOND VIRIAL COEFFICIENT, ML/MOL	HEAT OF VAPORIZATION, J/MOL	SATURATED VAPOR VOLUME, ML/MOL
273	6.073				373784.
274	6.398				356070.
276	7.094				323494.
277	7.465				308517.
279	8.257				280926.
280	8.680				268218.
282	9.579				244768.
283	10.06				233949.
285	11.08				213951.
287	12.18				195935.
288	12.76	112.38		32235.	187601.
290	14.01	112.64		32198.	172154.
291	14.66	112.77		32178.	164997.
293	16.06	113.03		32138.	151713.
294	16.79	113.17		32117.	145548.
296	18.35	113.44		32073.	134088.
298	20.03	113.71		32028.	123686.
299	20.92	113.85		32004.	118848.
301	22.79				109831.
302	23.77				105631.
304	25.85				97794.
305	26.94				94138.
307	29.24				87308.
309	31.69				81067.
310	32.98				78149.
312	35.69				72685.
313	37.11				70128.
315	40.09				65333.
316	41.65				63086.
318	44.91				58868.
320	48.38				54991.
321	50.20				53170.
323	53.98				49746.
324	55.96				48137.
326	60.09				45106.
327	62.24				43679.
329	66.73				40991.
331	71.47				38506.
332	73.94				37334.
334	79.07				35122.
335	81.73				34078.
337	87.27				32105.
338	90.15				31173.
340	96.12				29410.
341	99.22				28576.

COMPOUND CONSTANTS:

MP   TC   VC   ZC
NBP   PC   DC   OM
MU   0.2000 DEBYE   RG

VAPOR PRESSURE CORRELATION CONSTANTS, K AND KPA:

RIEDEL   RIEDEL   RIEDEL   RIEDEL
306. - 323. K   306. - 323. K   306. - 323. K   306. - 323. K
 9 POINTS   9 POINTS   9 POINTS   9 POINTS
RMSD = 0.0070   RMSD = 0.0070   RMSD = 0.0070   RMSD = 0.0070
-0.58146228E+03   -0.58146228E+03   -0.58146228E+03   -0.58146228E+03
0.90812588E+02   0.90812588E+02   0.90812588E+02   0.90812588E+02
0.20324069E+05   0.20324069E+05   0.20324069E+05   0.20324069E+05
-0.23103479E-14   -0.23103479E-14   -0.23103479E-14   -0.23103479E-14

LIQUID DENSITY CORRELATION CONSTANTS, K AND G/ML:

FRANCIS1   FRANCIS1   FRANCIS1
298. - 324. K   298. - 324. K   298. - 324. K
 4 POINTS   4 POINTS   4 POINTS
RMSD = 0.00004   RMSD = 0.00004   RMSD = 0.00004
0.10263624E+01   0.10263624E+01   0.10263624E+01
0.75194030E-03   0.75194030E-03   0.75194030E-03
0.60002947E+01   0.60002947E+01   0.60002947E+01
0.49869482E+03   0.49869482E+03   0.49869482E+03

SECOND VIRIAL COEFFICIENT CORRELATION CONSTANTS, ML/MOL:

LITERATURE DOCUMENTS FOR CORRELATED VAPOR PRESSURE VALUES:

   5641

LITERATURE DOCUMENTS FOR CORRELATED LIQUID DENSITY VALUES:

   5641

LITERATURE DOCUMENTS REPORTING VIRIAL COEFFICIENT DATA:

T,K	VAPOR PRESSURE, KPA	SATURATED LIQUID VOLUME, ML/MOL	SECOND VIRIAL COEFFICIENT, ML/MOL	HEAT OF VAPORIZATION, J/MOL	SATURATED VAPOR VOLUME, ML/MOL
298		103.74			
299		103.86			
300		103.99			
301		104.11			
302		104.23			
303		104.35			
304		104.48			
305		104.60			
306	17.00	104.73			149678.
307	17.73	104.85			143926.
308	18.50	104.98			138418.
309	19.30	105.10			133145.
310	20.12	105.23			128098.
311	20.98	105.36			123268.
312	21.86	105.49			118647.
313	22.78	105.61			114226.
314	23.73	105.74			109999.
315	24.72	105.87			105956.
316	25.74	106.00			102092.
317	26.79	106.13			98398.
318	27.87	106.27			94868.
319	28.99	106.40			91495.
320	30.14	106.53			88273.
321	31.33	106.66			85195.
322	32.55	106.80			82256.
323	33.80	106.93			79449.
324		107.07			

COMPOUND CONSTANTS:

MP	TC	VC	ZC
NBP 378.152 K	PC	DC	OM
MU	RG		

VAPOR PRESSURE CORRELATION CONSTANTS, K AND KPA:

LIQUID DENSITY CORRELATION CONSTANTS, K AND G/ML:

FRANCIS1	FRANCIS1	FRANCIS1
273. - 295. K	273. - 295. K	273. - 295. K
4 POINTS	4 POINTS	4 POINTS
RMSD = 0.00007	RMSD = 0.00007	RMSD = 0.00007
0.98357606E+00	0.98357606E+00	0.98357606E+00
0.77280519E-03	0.77280519E-03	0.77280519E-03
0.59999990E+01	0.59999990E+01	0.59999990E+01
0.55245361E+03	0.55245361E+03	0.55245361E+03

SECOND VIRIAL COEFFICIENT CORRELATION CONSTANTS, ML/MOL:

LITERATURE DOCUMENTS FOR CORRELATED VAPOR PRESSURE VALUES:

LITERATURE DOCUMENTS FOR CORRELATED LIQUID DENSITY VALUES:

   22458

LITERATURE DOCUMENTS REPORTING VIRIAL COEFFICIENT DATA:

T,K	VAPOR PRESSURE, KPA	SATURATED LIQUID VOLUME, ML/MOL	SECOND VIRIAL COEFFICIENT, ML/MOL	HEAT OF VAPORIZATION, J/MOL	SATURATED VAPOR VOLUME, ML/MOL
273		128.04			
274		128.18			
275		128.33			
276		128.47			
277		128.62			
278		128.77			
279		128.91			
280		129.06			
281		129.21			
282		129.36			
283		129.51			
284		129.66			
285		129.81			
286		129.96			
287		130.11			
288		130.26			
289		130.41			
290		130.56			
291		130.71			
292		130.87			
293		131.02			
294		131.18			
295		131.33			

COMPOUND CONSTANTS:

MP	TC	VC	ZC
NBP 378.152 K	PC	DC	OM
MU	RG		

VAPOR PRESSURE CORRELATION CONSTANTS, K AND KPA:

LIQUID DENSITY CORRELATION CONSTANTS, K AND G/ML:

FRANCIS1	FRANCIS1	FRANCIS1
273. - 295. K	273. - 295. K	273. - 295. K
4 POINTS	4 POINTS	4 POINTS
RMSD = 0.00007	RMSD = 0.00007	RMSD = 0.00007
0.98357606E+00	0.98357606E+00	0.98357606E+00
0.77280519E-03	0.77280519E-03	0.77280519E-03
0.59999990E+01	0.59999990E+01	0.59999990E+01
0.55245361E+03	0.55245361E+03	0.55245361E+03

SECOND VIRIAL COEFFICIENT CORRELATION CONSTANTS, ML/MOL:

LITERATURE DOCUMENTS FOR CORRELATED VAPOR PRESSURE VALUES:

LITERATURE DOCUMENTS FOR CORRELATED LIQUID DENSITY VALUES:

  22458

LITERATURE DOCUMENTS REPORTING VIRIAL COEFFICIENT DATA:

T,K	VAPOR PRESSURE, KPA	SATURATED LIQUID VOLUME, ML/MOL	SECOND VIRIAL COEFFICIENT, ML/MOL	HEAT OF VAPORIZATION, J/MOL	SATURATED VAPOR VOLUME, ML/MOL
273		128.04			
274		128.18			
275		128.33			
276		128.47			
277		128.62			
278		128.77			
279		128.91			
280		129.06			
281		129.21			
282		129.36			
283		129.51			
284		129.66			
285		129.81			
286		129.96			
287		130.11			
288		130.26			
289		130.41			
290		130.56			
291		130.71			
292		130.87			
293		131.02			
294		131.18			
295		131.33			

COMPOUND CONSTANTS:

MP   TC   VC   ZC
NBP   366.298 K   PC   DC   OM
MU   RG

VAPOR PRESSURE CORRELATION CONSTANTS, K AND KPA:

LIQUID DENSITY CORRELATION CONSTANTS, K AND G/ML:

FRANCIS1   FRANCIS1   FRANCIS1
273. - 297. K   273. - 297. K   273. - 297. K
  4 POINTS   4 POINTS   4 POINTS
RMSD = 0.00011   RMSD = 0.00011   RMSD = 0.00011
0.95326507E+00   0.95326507E+00   0.95326507E+00
0.66073355E-03   0.66073355E-03   0.66073355E-03
0.59999990E+01   0.59999990E+01   0.59999990E+01
0.55526465E+03   0.55526465E+03   0.55526465E+03

SECOND VIRIAL COEFFICIENT CORRELATION CONSTANTS, ML/MOL:

LITERATURE DOCUMENTS FOR CORRELATED VAPOR PRESSURE VALUES:

LITERATURE DOCUMENTS FOR CORRELATED LIQUID DENSITY VALUES:

   7387   8404

LITERATURE DOCUMENTS REPORTING VIRIAL COEFFICIENT DATA:

T,K	VAPOR PRESSURE, KPA	SATURATED LIQUID VOLUME, ML/MOL	SECOND VIRIAL COEFFICIENT, ML/MOL	HEAT OF VAPORIZATION, J/MOL	SATURATED VAPOR VOLUME, ML/MOL
273		127.95			
274		128.08			
275		128.20			
276		128.33			
277		128.46			
278		128.58			
279		128.71			
280		128.84			
281		128.96			
282		129.09			
283		129.22			
284		129.35			
285		129.48			
286		129.61			
287		129.74			
288		129.87			
289		130.00			
290		130.13			
291		130.26			
292		130.40			
293		130.53			
294		130.66			
295		130.79			
296		130.93			
297		131.06			

COMPOUND CONSTANTS:

MP   TC   VC   ZC
NBP   356.145 K   PC   DC   OM
MU   RG

   VAPOR PRESSURE CORRELATION CONSTANTS, K AND KPA:

   LIQUID DENSITY CORRELATION CONSTANTS, K AND G/ML:

   FRANCIS1   FRANCIS1   FRANCIS1
   273. - 296. K   273. - 296. K   273. - 296. K
     4 POINTS   4 POINTS   4 POINTS
   RMSD = 0.00118   RMSD = 0.00118   RMSD = 0.00118
   0.98972327E+00   0.98972327E+00   0.98972327E+00
   0.79004187E-03   0.79004187E-03   0.79004187E-03
   0.59999990E+01   0.59999990E+01   0.59999990E+01
   0.44550244E+03   0.44550244E+03   0.44550244E+03

   SECOND VIRIAL COEFFICIENT CORRELATION CONSTANTS, ML/MOL:

LITERATURE DOCUMENTS FOR CORRELATED VAPOR PRESSURE VALUES:

LITERATURE DOCUMENTS FOR CORRELATED LIQUID DENSITY VALUES:

   8067   8404

LITERATURE DOCUMENTS REPORTING VIRIAL COEFFICIENT DATA:

T,K	VAPOR PRESSURE, KPA	SATURATED LIQUID VOLUME, ML/MOL	SECOND VIRIAL COEFFICIENT, ML/MOL	HEAT OF VAPORIZATION, J/MOL	SATURATED VAPOR VOLUME, ML/MOL
273		130.09			
274		130.27			
275		130.44			
276		130.62			
277		130.80			
278		130.98			
279		131.16			
280		131.34			
281		131.52			
282		131.70			
283		131.88			
284		132.07			
285		132.25			
286		132.44			
287		132.63			
288		132.82			
289		133.01			
290		133.20			
291		133.39			
292		133.58			
293		133.78			
294		133.97			
295		134.17			
296		134.37			

COMPOUND CONSTANTS:

MP		TC		VC		ZC
NBP	395.158 K	PC		DC		OM
MU		RG				

VAPOR PRESSURE CORRELATION CONSTANTS, K AND KPA:

RIEDEL	RIEDEL	RIEDEL
317. - 396. K	317. - 396. K	317. - 396. K
4 POINTS	4 POINTS	4 POINTS
RMSD = 0.0300	RMSD = 0.0300	RMSD = 0.0300
0.18004810E+02	0.18004810E+02	0.18004810E+02
-0.52896640E+04	-0.52896640E+04	-0.52896640E+04

LIQUID DENSITY CORRELATION CONSTANTS, K AND G/ML:

SECOND VIRIAL COEFFICIENT CORRELATION CONSTANTS, ML/MOL:

LITERATURE DOCUMENTS FOR CORRELATED VAPOR PRESSURE VALUES:

8358

LITERATURE DOCUMENTS FOR CORRELATED LIQUID DENSITY VALUES:

LITERATURE DOCUMENTS REPORTING VIRIAL COEFFICIENT DATA:

T,K	VAPOR PRESSURE, KPA	SATURATED LIQUID VOLUME, ML/MOL	SECOND VIRIAL COEFFICIENT, ML/MOL	HEAT OF VAPORIZATION, J/MOL	SATURATED VAPOR VOLUME, ML/MOL
318	3.938				671420.
320	4.369				608928.
322	4.842				552945.
324	5.359				502727.
326	5.923				457621.
327	6.224				436803.
329	6.868				398315.
331	7.568				363637.
333	8.331				332353.
335	9.159				304099.
336	9.600				291004.
338	10.54				266699.
340	11.55				244682.
342	12.65				224718.
344	13.84				206593.
345	14.48				198162.
347	15.81				182453.
349	17.26				168154.
351	18.81				155125.
353	20.49				143241.
354	21.38				137694.
356	23.25				127322.
358	25.26				117838.
360	27.42				109158.
362	29.74				101206.
363	30.96				97481.
365	33.53				90496.
367	36.29				84082.
369	39.24				78187.
371	42.39				72765.
372	44.05				70218.
374	47.53				65426.
376	51.24				61009.
378	55.20				56934.
380	59.42				53171.
381	61.63				51399.
383	66.27				48055.
385	71.20				44962.
387	76.43				42097.
389	82.00				39443.
390	84.91				38190
392	90.99				35819.
394	97.44				33618.
396	104.3				31574.

COMPOUND CONSTANTS:

MP		TC		VC		ZC
NBP	422.169 K	PC		DC		OM
MU		RG				

VAPOR PRESSURE CORRELATION CONSTANTS, K AND KPA:

LIQUID DENSITY CORRELATION CONSTANTS, K AND G/ML:

FRANCIS1	FRANCIS1	FRANCIS1
273. - 292. K	273. - 292. K	273. - 292. K
4 POINTS	4 POINTS	4 POINTS
RMSD = 0.00004	RMSD = 0.00004	RMSD = 0.00004
0.96429694E+00	0.96429694E+00	0.96429694E+00
0.64810226E-03	0.64810226E-03	0.64810226E-03
0.59999990E+01	0.59999990E+01	0.59999990E+01
0.54589233E+03	0.54589233E+03	0.54589233E+03

SECOND VIRIAL COEFFICIENT CORRELATION CONSTANTS, ML/MOL:

LITERATURE DOCUMENTS FOR CORRELATED VAPOR PRESSURE VALUES:

LITERATURE DOCUMENTS FOR CORRELATED LIQUID DENSITY VALUES:

20809 22458

LITERATURE DOCUMENTS REPORTING VIRIAL COEFFICIENT DATA:

T,K	VAPOR PRESSURE, KPA	SATURATED LIQUID VOLUME, ML/MOL	SECOND VIRIAL COEFFICIENT, ML/MOL	HEAT OF VAPORIZATION, J/MOL	SATURATED VAPOR VOLUME, ML/MOL
273		162.31			
274		162.46			
275		162.62			
276		162.77			
277		162.93			
278		163.08			
279		163.24			
280		163.40			
281		163.56			
282		163.71			
283		163.87			
284		164.03			
285		164.19			
286		164.35			
287		164.51			
288		164.67			
289		164.83			
290		165.00			
291		165.16			
292		165.32			

COMPOUND CONSTANTS:

MP	175.950 K	TC		VC		ZC
NBP	313.370 K	PC		DC		OM
MU	0.7000 DEBYE	RG				

    VAPOR PRESSURE CORRELATION CONSTANTS, K AND KPA:

RPM2	RPM2	RPM2	RPM2
271. - 323. K	271. - 323. K	271. - 323. K	271. - 323. K
22 POINTS	22 POINTS	22 POINTS	22 POINTS
RMSD = 0.1980	RMSD = 0.1980	RMSD = 0.1980	RMSD = 0.1980
-0.76214544E+02	-0.76214544E+02	-0.76214544E+02	-0.76214544E+02
0.60556417E+04	0.60556417E+04	0.60556417E+04	0.60556417E+04
0.29613174E+00	0.29613174E+00	0.29613174E+00	0.29613174E+00
-0.31859190E-03	-0.31859190E-03	-0.31859190E-03	-0.31859190E-03

    LIQUID DENSITY CORRELATION CONSTANTS, K AND G/ML:

FRANCIS1	FRANCIS1	FRANCIS1	FRANCIS1
288. - 298. K	288. - 298. K	288. - 298. K	288. - 298. K
3 POINTS	3 POINTS	3 POINTS	3 POINTS
RMSD = 0.00003	RMSD = 0.00003	RMSD = 0.00003	RMSD = 0.00003
0.12648296E+01	0.12648296E+01	0.12648296E+01	0.12648296E+01
0.15705000E-02	0.15705000E-02	0.15705000E-02	0.15705000E-02

    SECOND VIRIAL COEFFICIENT CORRELATION CONSTANTS, ML/MOL:

LITERATURE DOCUMENTS FOR CORRELATED VAPOR PRESSURE VALUES:

   10665 15880 17943

LITERATURE DOCUMENTS FOR CORRELATED LIQUID DENSITY VALUES:

   10665 14703

LITERATURE DOCUMENTS REPORTING VIRIAL COEFFICIENT DATA:

T,K	VAPOR PRESSURE, KPA	SATURATED LIQUID VOLUME, ML/MOL	SECOND VIRIAL COEFFICIENT, ML/MOL	HEAT OF VAPORIZATION, J/MOL	SATURATED VAPOR VOLUME, ML/MOL
271	19.79				113877.
272	20.61				109707.
273	21.48				105693.
274	22.37				101828.
275	23.31				98108.
276	24.28				94528.
278	26.34				87769.
279	27.43				84580.
280	28.56				81512.
281	29.74				78561.
282	30.96				75723.
283	32.24				72994.
285	34.93				67846.
286	36.35				65419.
287	37.82				63086.
288	39.36	81.35			60844.
289	40.94	81.51			58687.
291	44.30	81.83			54622.
292	46.06	81.99			52707.
293	47.89	82.15			50865.
294	49.79	82.31			49096.
295	51.75	82.47			47395.
296	53.78	82.63			45759.
298	58.06	82.96			42677.
299	60.30				41225.
300	62.63				39829.
301	65.03				38487.
302	67.50				37197.
304	72.70				34766.
305	75.43				33620.
306	78.24				32518.
307	81.14				31459.
308	84.13				30441.
309	87.20				29462.
311	93.64				27615.
312	96.99				26745.
313	100.4				25908.
314	104.0				25102.
315	107.7				24328.
317	115.3				22867.
318	119.2				22178.
319	123.3				21514.
320	127.4				20877.
321	131.7				20263.
322	136.1				19672.

COMPOUND CONSTANTS:

MP	223.480 K	TC	VC	ZC
NBP	353.489 K	PC	DC	OM
MU	0.3800 DEBYE	RG		

VAPOR PRESSURE CORRELATION CONSTANTS, K AND KPA:

VAPRES-2	VAPRES-2	VAPRES-2	VAPRES-2
303. - 364. K	303. - 364. K	303. - 364. K	303. - 364. K
14 POINTS	14 POINTS	14 POINTS	14 POINTS
RMSD = 0.0070	RMSD = 0.0070	RMSD = 0.0070	RMSD = 0.0070
0.59823950E+03	0.59823950E+03	0.59823950E+03	0.59823950E+03
-0.17859406E+05	-0.17859406E+05	-0.17859406E+05	-0.17859406E+05
0.27423111E+00	0.27423111E+00	0.27423111E+00	0.27423111E+00
-0.12102858E-03	-0.12102858E-03	-0.12102858E-03	-0.12102858E-03
-0.10649767E+03	-0.10649767E+03	-0.10649767E+03	-0.10649767E+03

LIQUID DENSITY CORRELATION CONSTANTS, K AND G/ML:

FRANCIS1	FRANCIS1	FRANCIS1
293. - 355. K	293. - 355. K	293. - 355. K
7 POINTS	7 POINTS	7 POINTS
RMSD = 0.00048	RMSD = 0.00048	RMSD = 0.00048
0.11043558E+01	0.11043558E+01	0.11043558E+01
0.76830131E-03	0.76830131E-03	0.76830131E-03
0.59999990E+01	0.59999990E+01	0.59999990E+01
0.45624170E+03	0.45624170E+03	0.45624170E+03

SECOND VIRIAL COEFFICIENT CORRELATION CONSTANTS, ML/MOL:

LITERATURE DOCUMENTS FOR CORRELATED VAPOR PRESSURE VALUES:

5641 10569

LITERATURE DOCUMENTS FOR CORRELATED LIQUID DENSITY VALUES:

5641 9922 17061 40686

LITERATURE DOCUMENTS REPORTING VIRIAL COEFFICIENT DATA:

T,K	VAPOR PRESSURE, KPA	SATURATED LIQUID VOLUME, ML/MOL	SECOND VIRIAL COEFFICIENT, ML/MOL	HEAT OF VAPORIZATION, J/MOL	SATURATED VAPOR VOLUME, ML/MOL
293		95.11			
294		95.22			
296		95.45			
297		95.56			
299		95.79			
301		96.03			
302		96.14			
304	16.82	96.38		33461.	150267.
305	17.57	96.50		33403.	144359.
307	19.14	96.74		33290.	133368.
309	20.82	96.98		33181.	123380.
310	21.71	97.11		33127.	118728.
312	23.57	97.36		33023.	110050.
313	24.55	97.48		32972.	106002.
315	26.61	97.73		32872.	98439.
317	28.80	97.99		32774.	91525.
318	29.95	98.12		32727.	88291.
320	32.35	98.38		32634.	82234.
322	34.91	98.65		32543.	76679.
323	36.25	98.78		32499.	74075.
325	39.06	99.05		32413.	69184.
326	40.52	99.19		32370.	66888.
328	43.58	99.47		32287.	62571.
330	46.83	99.75		32206.	58592.
331	48.52	99.89		32166.	56720.
333	52.05	100.18		32089.	53192.
334	53.89	100.33		32050.	51530.
336	57.73	100.63		31976.	48394.
338	61.78	100.93		31902.	45491.
339	63.88	101.08		31867.	44120.
341	68.27	101.39		31796.	41530.
343	72.89	101.71		31727.	39125.
344	75.29	101.87		31693.	37988.
346	80.28	102.19		31626.	35834.
347	82.87	102.36		31593.	34815.
349	88.25	102.69		31528.	32881.
351	93.90	103.04		31464.	31080.
352	96.83	103.21		31432.	30226.
354	102.9	103.57		31370.	28604.
355	106.0	103.75		31339.	27833.
357	112.6				26369.
359	119.4				25001.
360	122.9				24350.
362	130.2				23111.
363	134.0				22521.

529

COMPOUND CONSTANTS:

MP	TC	VC	ZC
NBP	PC	DC	OM
MU	RG		

VAPOR PRESSURE CORRELATION CONSTANTS, K AND KPA:

RIEDEL	RIEDEL	RIEDEL	RIEDEL
304. - 323. K	304. - 323. K	304. - 323. K	304. - 323. K
7 POINTS	7 POINTS	7 POINTS	7 POINTS
RMSD = 0.0280	RMSD = 0.0280	RMSD = 0.0280	RMSD = 0.0280
0.29841027E+03	0.29841027E+03	0.29841027E+03	0.29841027E+03
-0.42854540E+02	-0.42854540E+02	-0.42854540E+02	-0.42854540E+02
-0.15728768E+05	-0.15728768E+05	-0.15728768E+05	-0.15728768E+05
0.10063391E-14	0.10063391E-14	0.10063391E-14	0.10063391E-14

LIQUID DENSITY CORRELATION CONSTANTS, K AND G/ML:

FRANCIS1	FRANCIS1	FRANCIS1
298. - 324. K	298. - 324. K	298. - 324. K
5 POINTS	5 POINTS	5 POINTS
RMSD = 0.00018	RMSD = 0.00018	RMSD = 0.00018
0.11241980E+01	0.11241980E+01	0.11241980E+01
0.80671208E-03	0.80671208E-03	0.80671208E-03
0.59999990E+01	0.59999990E+01	0.59999990E+01
0.48083350E+03	0.48083350E+03	0.48083350E+03

SECOND VIRIAL COEFFICIENT CORRELATION CONSTANTS, ML/MOL:

LITERATURE DOCUMENTS FOR CORRELATED VAPOR PRESSURE VALUES:

5641

LITERATURE DOCUMENTS FOR CORRELATED LIQUID DENSITY VALUES:

5641 40686

LITERATURE DOCUMENTS REPORTING VIRIAL COEFFICIENT DATA:

T,K	VAPOR PRESSURE, KPA	SATURATED LIQUID VOLUME, ML/MOL	SECOND VIRIAL COEFFICIENT, ML/MOL	HEAT OF VAPORIZATION, J/MOL	SATURATED VAPOR VOLUME, ML/MOL
298		94.16			
299		94.27			
300		94.38			
301		94.49			
302		94.60			
303		94.71			
304	11.76	94.82		34488.	214950.
305	12.30	94.94		34411.	206228.
306	12.85	95.05		34340.	197935.
307	13.43	95.16		34274.	190044.
308	14.03	95.28		34214.	182533.
309	14.65	95.39		34160.	175378.
310	15.29	95.51		34112.	168560.
311	15.96	95.62		34070.	162059.
312	16.64	95.74		34034.	155856.
313	17.36	95.85		34004.	149937.
314	18.09	95.97		33980.	144283.
315	18.86	96.09		33963.	138881.
316	19.65	96.21			133718.
317	20.47	96.33			128779.
318	21.31	96.45			124053.
319	22.19	96.57			119529.
320	23.10	96.69			115196.
321	24.03	96.81			111044.
322	25.01	96.93			107063.
323	26.01	97.05			103245.
324		97.18			

COMPOUND CONSTANTS:

MP   197.950 K   TC   VC   ZC
NBP   388.750 K   PC   DC   OM
MU   RG

VAPOR PRESSURE CORRELATION CONSTANTS, K AND KPA:

VAPRES-2   VAPRES-2   VAPRES-2
273. - 339. K   273. - 339. K   273. - 339. K
 24 POINTS   24 POINTS   24 POINTS
RMSD = 0.0114   RMSD = 0.0114   RMSD = 0.0114
-0.20883711E+04   -0.20883711E+04   -0.20883711E+04
 0.35610518E+05   0.35610518E+05   0.35610518E+05
-0.13313742E+01   -0.13313742E+01   -0.13313742E+01
 0.71902187E-03   0.71902187E-03   0.71902187E-03
 0.40422469E+03   0.40422469E+03   0.40422469E+03

LIQUID DENSITY CORRELATION CONSTANTS, K AND G/ML:

FRANCIS1   FRANCIS1   FRANCIS1
290. - 292. K   290. - 292. K   290. - 292. K
 4 POINTS   4 POINTS   4 POINTS
RMSD = 0.00000   RMSD = 0.00000   RMSD = 0.00000
0.13098488E+01   0.13098488E+01   0.13098488E+01
0.69164764E-03   0.69164764E-03   0.69164764E-03
0.10327146E+02   0.10327146E+02   0.10327146E+02
0.33847021E+03   0.33847021E+03   0.33847021E+03

SECOND VIRIAL COEFFICIENT CORRELATION CONSTANTS, ML/MOL:

LITERATURE DOCUMENTS FOR CORRELATED VAPOR PRESSURE VALUES:

  6550

LITERATURE DOCUMENTS FOR CORRELATED LIQUID DENSITY VALUES:

  20811

LITERATURE DOCUMENTS REPORTING VIRIAL COEFFICIENT DATA:

T,K	VAPOR PRESSURE, KPA	SATURATED LIQUID VOLUME, ML/MOL	SECOND VIRIAL COEFFICIENT, ML/MOL	HEAT OF VAPORIZATION, J/MOL	SATURATED VAPOR VOLUME, ML/MOL
273	0.7263				3125287.
274	0.7742				2942749.
276	0.8783				2612717.
277	0.9349				2463581.
279	1.058				2193425.
280	1.124				2071107.
282	1.268				1849104.
283	1.346				1748392.
285	1.514				1565252.
286	1.605				1482007.
288	1.800				1330345.
289	1.905				1261274.
291	2.131	103.41		39185.	1135196.
292	2.253	104.04		39140.	1077667.
294	2.514				972461.
295	2.653				924363.
297	2.953				836240.
298	3.113				795878.
300	3.456				721790.
301	3.639				687793.
303	4.029				625275.
304	4.237				596534.
306	4.680				543589.
307	4.916				519204.
309	5.418				474203.
310	5.684				453441.
312	6.250				415058.
313	6.550				397318.
315	7.186				364467.
316	7.523				349256.
318	8.236				321043.
319	8.613				307957.
321	9.409				283645.
322	9.830				272351.
324	10.72				251331.
325	11.19				241551.
327	12.17				223319.
328	12.69				214822.
330	13.79				198958.
331	14.37				191553.
333	15.58				177707.
334	16.22				171233.
336	17.56				159111.
337	18.26				153434.
339	19.74				142789.

COMPOUND CONSTANTS:

MP	TC	VC	ZC
NBP	PC	DC	OM
MU	RG		

VAPOR PRESSURE CORRELATION CONSTANTS, K AND KPA:

RIEDEL	RIEDEL	RIEDEL
300. - 353. K	300. - 353. K	300. - 353. K
8 POINTS	8 POINTS	8 POINTS
RMSD = 0.0590	RMSD = 0.0590	RMSD = 0.0590
0.15903842E+03	0.15903842E+03	0.15903842E+03
-0.21610194E+02	-0.21610194E+02	-0.21610194E+02
-0.10128210E+05	-0.10128210E+05	-0.10128210E+05
0.35934393E-15	0.35934393E-15	0.35934393E-15

LIQUID DENSITY CORRELATION CONSTANTS, K AND G/ML:

SECOND VIRIAL COEFFICIENT CORRELATION CONSTANTS, ML/MOL:

LITERATURE DOCUMENTS FOR CORRELATED VAPOR PRESSURE VALUES:

13773

LITERATURE DOCUMENTS FOR CORRELATED LIQUID DENSITY VALUES:

LITERATURE DOCUMENTS REPORTING VIRIAL COEFFICIENT DATA:

T,K	VAPOR PRESSURE, KPA	SATURATED LIQUID VOLUME, ML/MOL	SECOND VIRIAL COEFFICIENT, ML/MOL	HEAT OF VAPORIZATION, J/MOL	SATURATED VAPOR VOLUME, ML/MOL
300	9.775				255172.
301	10.23				244627.
302	10.70				234612.
303	11.19				225098.
304	11.70				216053.
306	12.77				199269.
307	13.33				191479.
308	13.91				184061.
309	14.52				176993.
310	15.14				170257.
312	16.45				157706.
313	17.14				151858.
314	17.85				146275.
315	18.58				140942.
316	19.34				135846.
318	20.93				126316.
319	21.77				121860.
320	22.63				117594.
321	23.51				113511.
322	24.43				109599.
324	26.34				102259.
325	27.35				98814.
326	28.38				95510.
327	29.44				92340.
328	30.54				89296.
330	32.83				83567.
331	34.03				80869.
332	35.26				78276.
333	36.53				75782.
334	37.84				73384.
336	40.57				68855.
337	42.00				66716.
338	43.47				64656.
339	44.97				62671.
340	46.53				60758.
342	49.77				57135.
343	51.46				55420.
344	53.20				53764.
345	54.99				52166.
346	56.83				50624.
348	60.67				47694.
349	62.67				46303.
350	64.73				44959.
351	66.84				43659.
352	69.02				42402.

COMPOUND CONSTANTS:

MP	TC	VC	ZC
NBP	PC	DC	OM
MU	RG		

VAPOR PRESSURE CORRELATION CONSTANTS, K AND KPA:

LIQUID DENSITY CORRELATION CONSTANTS, K AND G/ML:

FRANCIS1	FRANCIS1	FRANCIS1
303. - 334. K	303. - 334. K	303. - 334. K
4 POINTS	4 POINTS	4 POINTS
RMSD = 0.00000	RMSD = 0.00000	RMSD = 0.00000
0.11640186E+01	0.11640186E+01	0.11640186E+01
0.87225647E-03	0.87225647E-03	0.87225647E-03
0.77953701E+01	0.77953701E+01	0.77953701E+01
0.59625098E+03	0.59625098E+03	0.59625098E+03

SECOND VIRIAL COEFFICIENT CORRELATION CONSTANTS, ML/MOL:

LITERATURE DOCUMENTS FOR CORRELATED VAPOR PRESSURE VALUES:

LITERATURE DOCUMENTS FOR CORRELATED LIQUID DENSITY VALUES:

18182

LITERATURE DOCUMENTS REPORTING VIRIAL COEFFICIENT DATA:

T,K	VAPOR PRESSURE, KPA	SATURATED LIQUID VOLUME, ML/MOL	SECOND VIRIAL COEFFICIENT, ML/MOL	HEAT OF VAPORIZATION, J/MOL	SATURATED VAPOR VOLUME, ML/MOL
303		123.90			
304		124.04			
305		124.17			
306		124.31			
307		124.45			
308		124.59			
309		124.73			
310		124.87			
311		125.01			
312		125.15			
313		125.29			
314		125.43			
315		125.57			
316		125.71			
317		125.85			
318		126.00			
319		126.14			
320		126.28			
321		126.43			
322		126.57			
323		126.71			
324		126.86			
325		127.01			
326		127.15			
327		127.30			
328		127.45			
329		127.59			
330		127.74			
331		127.89			
332		128.04			
333		128.19			
334		128.34			

COMPOUND CONSTANTS:

MP   268.480 K   TC   VC   ZC
NBP   PC   DC   OM
MU   0.0690 DEBYE   RG

VAPOR PRESSURE CORRELATION CONSTANTS, K AND KPA:

RIEDEL   RIEDEL   RIEDEL
273. - 349. K   273. - 349. K   273. - 349. K
 15 POINTS   15 POINTS   15 POINTS
RMSD = 0.0035   RMSD = 0.0035   RMSD = 0.0035
 0.11829691E+03   0.11829691E+03   0.11829691E+03
-0.15164103E+02   -0.15164103E+02   -0.15164103E+02
-0.95326084E+04   -0.95326084E+04   -0.95326084E+04
 0.16695269E-15   0.16695269E-15   0.16695269E-15

LIQUID DENSITY CORRELATION CONSTANTS, K AND G/ML:

FRANCIS1   FRANCIS1   FRANCIS1
293. - 354. K   293. - 354. K   293. - 354. K
 10 POINTS   10 POINTS   10 POINTS
RMSD = 0.00074   RMSD = 0.00074   RMSD = 0.00074
0.11940374E+01   0.11940374E+01   0.11940374E+01
0.88054477E-03   0.88054477E-03   0.88054477E-03
0.59999990E+01   0.59999990E+01   0.59999990E+01
0.66214453E+03   0.66214453E+03   0.66214453E+03

SECOND VIRIAL COEFFICIENT CORRELATION CONSTANTS, ML/MOL:

LITERATURE DOCUMENTS FOR CORRELATED VAPOR PRESSURE VALUES:

   6471

LITERATURE DOCUMENTS FOR CORRELATED LIQUID DENSITY VALUES:

   2840 13365

LITERATURE DOCUMENTS REPORTING VIRIAL COEFFICIENT DATA:

T,K	VAPOR PRESSURE, KPA	SATURATED LIQUID VOLUME, ML/MOL	SECOND VIRIAL COEFFICIENT, ML/MOL	HEAT OF VAPORIZATION, J/MOL	SATURATED VAPOR VOLUME, ML/MOL
273	0.1990				0.1141E+08
274	0.2142				0.1064E+08
276	0.2476				9269232.
278	0.2854				8098382.
280	0.3282				7093797.
282	0.3764				6229550.
284	0.4306				5484075.
285	0.4601				5150136.
287	0.5244				4550060.
289	0.5964				4029179.
291	0.6766				3575949.
293	0.7659	113.23		43859.	3180652.
295	0.8651	113.46		43682.	2835087.
296	0.9187	113.58		43594.	2678744.
298	1.034	113.81		43422.	2395109.
300	1.162	114.04		43253.	2145784.
302	1.304	114.27		43087.	1926152.
304	1.459	114.50		42925.	1732279.
306	1.630	114.74		42766.	1560795.
307	1.722	114.85		42689.	1482529.
309	1.918	115.09		42536.	1339347.
311	2.133	115.33		42387.	1212082.
313	2.369	115.56		42242.	1098754.
315	2.625	115.80		42101.	997652.
317	2.905	116.04		41964.	907295.
319	3.209	116.28		41832.	826400.
320	3.372	116.41		41768.	789146.
322	3.716	116.65		41643.	720401.
324	4.090	116.89		41522.	658594.
326	4.495	117.14		41407.	602938.
328	4.934	117.38		41296.	552740.
330	5.408	117.63		41191.	507395.
331	5.658	117.76		41141.	486373.
333	6.189	118.01		41044.	447331.
335	6.762	118.26		40953.	411932.
337	7.378	118.51		40867.	379790.
339	8.040	118.76		40788.	350564.
341	8.752	119.02		40715.	323953.
342	9.127	119.15		40681.	311543.
344	9.919	119.40		40617.	288365.
346	10.77	119.66		40560.	267190.
348	11.68	119.92		40510.	247821.
350		120.18			
352		120.44			
353		120.57			

COMPOUND CONSTANTS:

MP   TC   VC   ZC
NBP   433.250 K   PC   DC   OM
MU   RG

VAPOR PRESSURE CORRELATION CONSTANTS, K AND KPA:

LIQUID DENSITY CORRELATION CONSTANTS, K AND G/ML:

FRANCIS1   FRANCIS1   FRANCIS1
288. - 344. K   288. - 344. K   288. - 344. K
   7 POINTS   7 POINTS   7 POINTS
RMSD = 0.00019   RMSD = 0.00019   RMSD = 0.00019
0.11855917E+01   0.11855917E+01   0.11855917E+01
0.82031661E-03   0.82031661E-03   0.82031661E-03
0.59999990E+01   0.59999990E+01   0.59999990E+01
0.64349878E+03   0.64349878E+03   0.64349878E+03

SECOND VIRIAL COEFFICIENT CORRELATION CONSTANTS, ML/MOL:

LITERATURE DOCUMENTS FOR CORRELATED VAPOR PRESSURE VALUES:

LITERATURE DOCUMENTS FOR CORRELATED LIQUID DENSITY VALUES:

   7697

LITERATURE DOCUMENTS REPORTING VIRIAL COEFFICIENT DATA:

T,K	VAPOR PRESSURE, KPA	SATURATED LIQUID VOLUME, ML/MOL	SECOND VIRIAL COEFFICIENT, ML/MOL	HEAT OF VAPORIZATION, J/MOL	SATURATED VAPOR VOLUME, ML/MOL
288		128.90			
289		129.02			
290		129.14			
291		129.26			
293		129.50			
294		129.62			
295		129.75			
296		129.87			
298		130.11			
299		130.24			
300		130.36			
301		130.48			
303		130.73			
304		130.85			
305		130.98			
307		131.23			
308		131.35			
309		131.48			
310		131.60			
312		131.86			
313		131.98			
314		132.11			
315		132.24			
317		132.49			
318		132.62			
319		132.75			
321		133.01			
322		133.14			
323		133.27			
324		133.40			
326		133.66			
327		133.79			
328		133.92			
329		134.05			
331		134.31			
332		134.45			
333		134.58			
335		134.85			
336		134.98			
337		135.11			
338		135.25			
340		135.52			
341		135.65			
342		135.79			
343		135.93			

COMPOUND CONSTANTS:

MP   170.450 K   TC   402.390 K   VC 164.00 ML/MOL   ZC 0.2760
NBP 249.940 K   PC   5.628 MPA   DC   0.2443 G/ML   OM 0.2121
MU   0.7500 DEBYE   RG   1.8864 ANGSTROM

VAPOR PRESSURE CORRELATION CONSTANTS, K AND KPA:

WAGNER	VAPRES-2	VAPRES-2	WAGNER
194. - 402. K	194. - 252. K	194. - 338. K	234. - 402. K
50 POINTS	26 POINTS	32 POINTS	39 POINTS
RMSD = 2.6640	RMSD = 0.9210	RMSD = 1.6350	RMSD = 3.2400
-0.71148325E+01	-0.10263622E+05	0.92337758E+02	-0.71360401E+01
0.16202941E+01	0.15296368E+06	-0.48446228E+04	0.17040778E+01
-0.30625795E+01	-0.93203882E+01	0.22066558E-01	-0.34803711E+01
-0.20427617E+01	0.68815150E-02	-0.41148474E-05	0.14479666E+01
	0.20929963E+04	-0.13327813E+02	

LIQUID DENSITY CORRELATION CONSTANTS, K AND G/ML:

FRANCIS1	FRANCIS1	FRANCIS1	FRANCIS1
216. - 388. K	216. - 261. K	216. - 328. K	231. - 388. K
27 POINTS	16 POINTS	21 POINTS	23 POINTS
RMSD = 0.00125	RMSD = 0.00086	RMSD = 0.00116	RMSD = 0.00130
0.96181381E+00	0.96141177E+00	0.95734137E+00	0.96372926E+00
0.87452028E-03	0.10504571E-02	0.84978808E-03	0.89671416E-03
0.15000000E+02	0.65321779E+01	0.15000000E+02	0.14102544E+02
0.45851831E+03	0.48835571E+03	0.45359912E+03	0.45626758E+03

SECOND VIRIAL COEFFICIENT CORRELATION CONSTANTS, ML/MOL:

TSONOPOULOS: TC=402.380 PC=5.628 VC=164.00 OM=0.2121 A= 0.000000 B=0.0000

LITERATURE DOCUMENTS FOR CORRELATED VAPOR PRESSURE VALUES:

360   425   578   1915

LITERATURE DOCUMENTS FOR CORRELATED LIQUID DENSITY VALUES:

360   425   578

LITERATURE DOCUMENTS REPORTING VIRIAL COEFFICIENT DATA:

364   410

T,K	VAPOR PRESSURE, KPA	SATURATED LIQUID VOLUME, ML/MOL	SECOND VIRIAL COEFFICIENT, ML/MOL	HEAT OF VAPORIZATION, J/MOL	SATURATED VAPOR VOLUME, ML/MOL
194	3.603		-1497.		446202.
198	4.936		-1379.		332165.
203	7.165		-1252.		234297.
208	10.18		-1144.		168666.
212	13.30		-1069.		131436.
217	18.27	56.43	-986.8	23918.	97750.
222	24.68	56.89	-915.0	23653.	73875.
227	32.81	57.36	-852.0	23387.	56664.
231	40.77	57.74	-807.0	23173.	46284.
236	52.85	58.23	-756.4	22902.	36355.
241	67.64	58.74	-711.2	22629.	28895.
245	81.70	59.15	-678.4	22408.	24235.
250	102.4	59.69	-641.1	22128.	19631.
255	127.1	60.24	-607.2	21843.	16052.
260	156.2	60.81	-576.4	21553.	13241.
264	182.9	61.28	-553.6	21317.	11417.
269	221.2	61.88	-527.3	21016.	9551.
274	265.5	62.51	-503.1	20707.	8046.
279	316.1	63.17	-480.7	20392.	6821.
283	361.7	63.71	-463.9	20133.	6003.
288	425.4	64.41	-444.4	19802.	5142.
293	497.3	65.15	-426.2	19460.	4427.
297	561.0	65.77	-412.4	19179.	3941.
302	649.1	66.57	-396.3	18817.	3420.
307	747.0	67.41	-381.2	18441.	2980.
312	855.6	68.30	-366.9	18050.	2605.
316	950.5	69.05	-356.1	17725.	2344.
321	1080.	70.04	-343.4	17301.	2060.
326	1222.	71.10	-331.3	16807.	1809.
331	1377.	72.23	-319.8	16388.	1600.
335	1511.	73.19	-311.1	16042.	1453.
340	1692.	74.48	-300.7	15593.	1290.
345	1888.	75.89	-290.8	15122.	1147.
349	2057.	77.10	-283.3	14728.	1045.
354	2284.	78.76	-274.2	14210.	930.1
359	2529.	80.59	-265.6	13656.	828.0
364	2794.	82.64	-257.4	13060.	736.6
368	3020.	84.48	-251.0	12544.	670.1
373	3322.	87.06	-243.4	11839.	593.7
378	3647.	90.04	-236.2	11044.	523.7
383	3996.	93.56	-229.2	10125.	458.4
387	4295.	96.86	-223.8	9255.	408.1
392	4693.		-217.4		344.3
397	5123.		-211.2		
401	5493.		-206.4		

COMPOUND CONSTANTS:

MP 147.430 K	TC 463.700 K	VC 220.80 ML/MOL	ZC 0.2700
NBP 281.240 K	PC 4.712 MPA	DC 0.2450 G/ML	OM
MU 0.8100 DEBYE	RG 2.7130 ANGSTROM		

VAPOR PRESSURE CORRELATION CONSTANTS, K AND KPA:

WAGNER	VAPRES-2	VAPRES-2
194. - 283. K	194. - 283. K	194. - 283. K
17 POINTS	17 POINTS	17 POINTS
RMSD = 0.0241	RMSD = 0.0080	RMSD = 0.0080
-0.49018707E+01	-0.30915542E+03	-0.30915542E+03
-0.85046469E+00	0.12551180E+04	0.12551180E+04
-0.31478442E+01	-0.29954593E+00	-0.29954593E+00
0.52666004E+00	0.20775326E-03	0.20775326E-03
	0.66875253E+02	0.66875253E+02

LIQUID DENSITY CORRELATION CONSTANTS, K AND G/ML:

FRANCIS1	FRANCIS1	FRANCIS1
242. - 282. K	242. - 282. K	242. - 282. K
11 POINTS	11 POINTS	11 POINTS
RMSD = 0.00025	RMSD = 0.00025	RMSD = 0.00025
0.96013415E+00	0.96013415E+00	0.96013415E+00
0.87237079E-03	0.87237079E-03	0.87237079E-03
0.80509758E+01	0.80509758E+01	0.80509758E+01
0.45740845E+03	0.45740845E+03	0.45740845E+03

SECOND VIRIAL COEFFICIENT CORRELATION CONSTANTS, ML/MOL:

NOTHNAGEL ET AL.: TC=463.700 NBP=281.240 B=174.0 D=0.321

LITERATURE DOCUMENTS FOR CORRELATED VAPOR PRESSURE VALUES:

2752

LITERATURE DOCUMENTS FOR CORRELATED LIQUID DENSITY VALUES:

425 1911

LITERATURE DOCUMENTS REPORTING VIRIAL COEFFICIENT DATA:

T,K	VAPOR PRESSURE, KPA	SATURATED LIQUID VOLUME, ML/MOL	SECOND VIRIAL COEFFICIENT, ML/MOL	HEAT OF VAPORIZATION, J/MOL	SATURATED VAPOR VOLUME, ML/MOL
194	0.5019		-2860.		3210743.
196	0.6026		-2762.		2701433.
198	0.7207		-2669.		2281678.
200	0.8585		-2580.		1934382.
202	1.019		-2496.		1645949.
204	1.205		-2416.		1405516.
206	1.419		-2340.		1204370.
208	1.666		-2268.		1035501.
210	1.950		-2199.		893242.
212	2.274		-2133.		772998.
214	2.643		-2071.		671031.
216	3.063		-2011.		584286.
218	3.539		-1953.		510262.
220	4.076		-1899.		446901.
222	4.680		-1846.		392507.
224	5.360		-1796.		345676.
226	6.121		-1748.		305243.
228	6.971		-1702.		270237.
230	7.918		-1658.		239848.
232	8.971		-1615.		213399.
234	10.14		-1575.		190319.
236	11.43		-1535.		170129.
238	12.85		-1498.		152425.
240	14.42		-1462.		136862.
242	16.15	76.01	-1427.	26896.	123151.
244	18.04	76.23	-1393.	26785.	111043.
246	20.11	76.46	-1361.	26672.	100328.
248	22.37	76.69	-1330.	26558.	90824.
250	24.83	76.92	-1300.	26443.	82377.
252	27.52	77.15	-1271.	26327.	74854.
254	30.43	77.39	-1243.	26210.	68139.
256	33.59	77.62	-1216.	26093.	62136.
258	37.00	77.86	-1190.	25976.	56757.
260	40.70	78.11	-1164.	25857.	51928.
262	44.68	78.35	-1140.	25739.	47586.
264	48.97	78.60	-1117.	25621.	43675.
266	53.59	78.85	-1094.	25502.	40145.
268	58.55	79.10	-1072.	25383.	36953.
270	63.87	79.35	-1050.	25265.	34064.
272	69.57	79.61	-1030.	25147.	31442.
274	75.67	79.87	-1010.	25029.	29061.
276	82.19	80.14	-990.2	24912.	26894.
278	89.14	80.40	-971.4	24795.	24919.
280	96.56	80.67	-953.2	24679.	23117.
282	104.5	80.95	-935.4	24564.	21469.

COMPOUND CONSTANTS:

MP   240.890 K   TC   488.600 K   VC   220.80 ML/MOL   ZC 0.2760
NBP   300.130 K   PC   5.084 MPA   DC   0.2450 G/ML   OM
MU   RG

    VAPOR PRESSURE CORRELATION CONSTANTS, K AND KPA:

WAGNER   WAGNER   WAGNER
242. - 296. K   242. - 296. K   242. - 296. K
  13 POINTS   13 POINTS   13 POINTS
RMSD = 0.0033   RMSD = 0.0033   RMSD = 0.0033
-0.70098420E+01   -0.70098420E+01   -0.70098420E+01
0.24914136E+01   0.24914136E+01   0.24914136E+01
-0.53816755E+01   -0.53816755E+01   -0.53816755E+01
0.23381762E+01   0.23381762E+01   0.23381762E+01

    LIQUID DENSITY CORRELATION CONSTANTS, K AND G/ML:

FRANCIS1   FRANCIS1   FRANCIS1
273. - 300. K   273. - 300. K   273. - 300. K
  4 POINTS   4 POINTS   4 POINTS
RMSD = 0.00131   RMSD = 0.00131   RMSD = 0.00131
0.10061312E+01   0.10061312E+01   0.10061312E+01
0.96174888E-03   0.96174888E-03   0.96174888E-03
0.83306894E+01   0.83306894E+01   0.83306894E+01
0.56295068E+03   0.56295068E+03   0.56295068E+03

    SECOND VIRIAL COEFFICIENT CORRELATION CONSTANTS, ML/MOL:

    KREGLEWSKI: A= -849.0 B= 15.00 V*= 78.28 R1=0.7000 R2=1.1300 R3=5.1300

LITERATURE DOCUMENTS FOR CORRELATED VAPOR PRESSURE VALUES:

    3024

LITERATURE DOCUMENTS FOR CORRELATED LIQUID DENSITY VALUES:

    948   1911   3024

LITERATURE DOCUMENTS REPORTING VIRIAL COEFFICIENT DATA:

    3269

T,K	VAPOR PRESSURE, KPA	SATURATED LIQUID VOLUME, ML/MOL	SECOND VIRIAL COEFFICIENT, ML/MOL	HEAT OF VAPORIZATION, J/MOL	SATURATED VAPOR VOLUME, ML/MOL
242	6.524		-1412.		306997.
243	6.930		-1394.		290162.
244	7.356		-1376.		274401.
245	7.805		-1359.		259635.
247	8.771		-1325.		232816.
248	9.290		-1308.		220636.
249	9.835		-1292.		209201.
251	11.01		-1261.		188363.
252	11.63		-1246.		178868.
253	12.29		-1231.		169936.
255	13.69		-1202.		153607.
256	14.45		-1188.		146144.
257	15.23		-1175.		139108.
259	16.91		-1148.		126209.
260	17.80		-1135.		120296.
261	18.73		-1123.		114710.
263	20.72		-1098.		104441.
264	21.77		-1086.		99721.
265	22.87		-1074.		95253.
267	25.20		-1051.		87018.
268	26.44		-1040.		83222.
269	27.73		-1029.		79623.
270	29.06		-1018.		76210.
272	31.89		-997.4		69898.
273	33.39	75.67	-987.2	27838.	66979.
274	34.94	75.78	-977.1	27783.	64207.
276	38.22	76.01	-957.5	27672.	59066.
277	39.95	76.12	-948.0	27616.	56683.
278	41.74	76.24	-938.6	27559.	54416.
280	45.52	76.46	-920.2	27445.	50202.
281	47.51	76.58	-911.3	27388.	48245.
282	49.57	76.70	-902.5	27330.	46379.
284	53.90	76.93	-885.3	27213.	42905.
285	56.17	77.05	-876.9	27154.	41287.
286	58.52	77.16	-868.7	27095.	39743.
288	63.46	77.40	-852.6	26976.	36862.
289	66.04	77.52	-844.7	26916.	35517.
290	68.71	77.64	-836.9	26855.	34232.
292	74.31	77.88	-821.7	26733.	31829.
293	77.24	78.00	-814.3	26672.	30705.
294	80.25	78.12	-807.0	26610.	29629.
296	86.56	78.36	-792.7	26486.	27614.
297		78.49	-785.7		
298		78.61	-778.9		
299		78.73	-772.1		

COMPOUND CONSTANTS:

MP	141.242 K	TC	VC	ZC
NBP	344.574 K	PC	DC	OM
MU	0.8800 DEBYE	RG		

VAPOR PRESSURE CORRELATION CONSTANTS, K AND KPA:

RPM2	RPM2	RPM2	RPM2
285. - 345. K	285. - 345. K	285. - 345. K	285. - 345. K
6 POINTS	6 POINTS	6 POINTS	6 POINTS
RMSD = 0.2000	RMSD = 0.2000	RMSD = 0.2000	RMSD = 0.2000
-0.26611036E+03	-0.26611036E+03	-0.26611036E+03	-0.26611036E+03
0.26008734E+05	0.26008734E+05	0.26008734E+05	0.26008734E+05
0.88443716E+00	0.88443716E+00	0.88443716E+00	0.88443716E+00
-0.92230326E-03	-0.92230326E-03	-0.92230326E-03	-0.92230326E-03

LIQUID DENSITY CORRELATION CONSTANTS, K AND G/ML:

FRANCIS1	FRANCIS1	FRANCIS1
293. - 304. K	293. - 304. K	293. - 304. K
4 POINTS	4 POINTS	4 POINTS
RMSD = 0.00169	RMSD = 0.00169	RMSD = 0.00169
0.91896635E+00	0.91896635E+00	0.91896635E+00
0.61777560E-03	0.61777560E-03	0.61777560E-03
0.60407877E+01	0.60407877E+01	0.60407877E+01
0.56161011E+03	0.56161011E+03	0.56161011E+03

SECOND VIRIAL COEFFICIENT CORRELATION CONSTANTS, ML/MOL:

LITERATURE DOCUMENTS FOR CORRELATED VAPOR PRESSURE VALUES:

9580 17428

LITERATURE DOCUMENTS FOR CORRELATED LIQUID DENSITY VALUES:

3269 6878 6886 13373

LITERATURE DOCUMENTS REPORTING VIRIAL COEFFICIENT DATA:

T,K	VAPOR PRESSURE, KPA	SATURATED LIQUID VOLUME, ML/MOL	SECOND VIRIAL COEFFICIENT, ML/MOL	HEAT OF VAPORIZATION, J/MOL	SATURATED VAPOR VOLUME, ML/MOL
285	9.963				237844.
286	10.36				229615.
287	10.77				221583.
289	11.66				206130.
290	12.13				198714.
291	12.63				191508.
293	13.71	114.81			177725.
294	14.28	114.93			171148.
295	14.89	115.04			164779.
297	16.18	115.27			152658.
298	16.87	115.38			146901.
299	17.59	115.49			141342.
301	19.13	115.72			130806.
302	19.96	115.84			125822.
304	21.72	116.07			116398.
305	22.65				111950.
306	23.63				107672.
308	25.71				99608.
309	26.81				95812.
310	27.97				92167.
312	30.41				85312.
313	31.70				82093.
314	33.04				79006.
316	35.89				73211.
317	37.39				70494.
319	40.56				65399.
320	42.22				63013.
321	43.95				60729.
323	47.57				56451.
324	49.47				54450.
325	51.43				52536.
327	55.54				48955.
328	57.68				47282.
329	59.88				45682.
331	64.46				42692.
332	66.85				41295.
334	71.78				38686.
335	74.34				37469.
336	76.95				36306.
338	82.33				34135.
339	85.09				33123.
340	87.91				32157.
342	93.67				30356.
343	96.62				29517.
344	99.60				28717.

COMPOUND CONSTANTS:

MP   170.070 K   TC   VC   ZC
NBP 354.461 K   PC   DC   OM
MU   RG

VAPOR PRESSURE CORRELATION CONSTANTS, K AND KPA:

VAPRES-2   VAPRES-2   VAPRES-2   VAPRES-2
253. - 355. K   253. - 355. K   253. - 355. K   253. - 355. K
 22 POINTS   22 POINTS   22 POINTS   22 POINTS
RMSD = 0.2280   RMSD = 0.2280   RMSD = 0.2280   RMSD = 0.2280
-0.37943659E+04   -0.37943659E+04   -0.37943659E+04   -0.37943659E+04
 0.72397703E+05   0.72397703E+05   0.72397703E+05   0.72397703E+05
-0.23035653E+01   -0.23035653E+01   -0.23035653E+01   -0.23035653E+01
 0.12236068E-02   0.12236068E-02   0.12236068E-02   0.12236068E-02
 0.72522836E+03   0.72522836E+03   0.72522836E+03   0.72522836E+03

LIQUID DENSITY CORRELATION CONSTANTS, K AND G/ML:

FRANCIS1   FRANCIS1   FRANCIS1
293. - 299. K   293. - 299. K   293. - 299. K
 4 POINTS   4 POINTS   4 POINTS
RMSD = 0.00004   RMSD = 0.00004   RMSD = 0.00004
0.99776715E+00   0.99776715E+00   0.99776715E+00
0.83916448E-03   0.83916448E-03   0.83916448E-03
0.66157951E+01   0.66157951E+01   0.66157951E+01
0.52031177E+03   0.52031177E+03   0.52031177E+03

SECOND VIRIAL COEFFICIENT CORRELATION CONSTANTS, ML/MOL:

LITERATURE DOCUMENTS FOR CORRELATED VAPOR PRESSURE VALUES:

  3855   7195 10070

LITERATURE DOCUMENTS FOR CORRELATED LIQUID DENSITY VALUES:

  3269   6886

LITERATURE DOCUMENTS REPORTING VIRIAL COEFFICIENT DATA:

T,K	VAPOR PRESSURE, KPA	SATURATED LIQUID VOLUME, ML/MOL	SECOND VIRIAL COEFFICIENT, ML/MOL	HEAT OF VAPORIZATION, J/MOL	SATURATED VAPOR VOLUME, ML/MOL
253	1.326				1586902.
255	1.468				1444095.
257	1.626				1313918.
259	1.801				1195416.
262	2.100				1037547.
264	2.325				944288.
266	2.573				859668.
269	2.993				747291.
271	3.308				681065.
273	3.655				621052.
276	4.238				541434.
278	4.674				494533.
280	5.150				452027.
283	5.948				395600.
285	6.540				362325.
287	7.185				332132.
290	8.258				291979.
292	9.051				268248.
294	9.910	113.80			246674.
297	11.33	114.26			217904.
299	12.38	114.57			200850.
301	13.51				185306.
303	14.72				171127.
306	16.72				152136.
308	18.18				140825.
310	19.75				130477.
313	22.33				116561.
315	24.20				108240.
317	26.20				100601.
320	29.47				90285.
322	31.84				84090.
324	34.37				78383.
327	38.49				70642.
329	41.46				65973.
331	44.64				61656.
334	49.79				55775.
336	53.51				52211.
338	57.47				48904.
341	63.89				44379.
343	68.51				41625.
345	73.44				39061.
348	81.42				35536.
350	87.17				33382.
352	93.30				31369.
354	99.82				29487.

COMPOUND CONSTANTS:

MP   192.282 K   TC   VC   ZC
NBP   373.146 K   PC   DC   OM
MU   0.8700 DEBYE   RG

VAPOR PRESSURE CORRELATION CONSTANTS, K AND KPA:

RPM2	RPM2	RPM2	VAPRES-2
319. - 374. K	319. - 374. K	319. - 374. K	354. - 374. K
12 POINTS	12 POINTS	12 POINTS	9 POINTS
RMSD = 0.0339	RMSD = 0.0339	RMSD = 0.0339	RMSD = 0.0073
-0.58582705E+02	-0.58582705E+02	-0.58582705E+02	0.56424299E+03
0.40824770E+04	0.40824770E+04	0.40824770E+04	-0.17901399E+05
0.21924483E+00	0.21924483E+00	0.21924483E+00	0.25760388E+00
-0.21222653E-03	-0.21222653E-03	-0.21222653E-03	-0.11584640E-03
			-0.99906545E+02

LIQUID DENSITY CORRELATION CONSTANTS, K AND G/ML:

FRANCIS1	FRANCIS1	FRANCIS1
293. - 334. K	293. - 334. K	293. - 334. K
7 POINTS	7 POINTS	7 POINTS
RMSD = 0.00039	RMSD = 0.00039	RMSD = 0.00039
0.99410743E+00	0.99410743E+00	0.99410743E+00
0.79037575E-03	0.79037575E-03	0.79037575E-03
0.87779760E+01	0.87779760E+01	0.87779760E+01
0.59182300E+03	0.59182300E+03	0.59182300E+03

SECOND VIRIAL COEFFICIENT CORRELATION CONSTANTS, ML/MOL:

LITERATURE DOCUMENTS FOR CORRELATED VAPOR PRESSURE VALUES:

  17432

LITERATURE DOCUMENTS FOR CORRELATED LIQUID DENSITY VALUES:

  948   6655 13373 14634 17428

LITERATURE DOCUMENTS REPORTING VIRIAL COEFFICIENT DATA:

T,K	VAPOR PRESSURE, KPA	SATURATED LIQUID VOLUME, ML/MOL	SECOND VIRIAL COEFFICIENT, ML/MOL	HEAT OF VAPORIZATION, J/MOL	SATURATED VAPOR VOLUME, ML/MOL
293		131.18			
294		131.34			
296		131.66			
298		131.98			
300		132.30			
302		132.63			
304		132.96			
305		133.12			
307		133.45			
309		133.79			
311		134.12			
313		134.46			
315		134.80			
316		134.97			
318		135.32			
320	13.48	135.66			197340.
322	14.70	136.01			182073.
324	16.03	136.37			168102.
326	17.45	136.72			155311.
327	18.21	136.90			149325.
329	19.81	137.26			138113.
331	21.53	137.62			127837.
333	23.38	137.98			118416.
335	25.37				109775.
337	27.51				101845.
339	29.81				94564.
340	31.01				91149.
342	33.56				84739.
344	36.28				78845.
346	39.18				73425.
348	42.28				68438.
350	45.58				63846.
351	47.31				61687.
353	50.93				57627.
355	54.78				53883.
357	58.86				50429.
359	63.18				47242.
361	67.76				44299.
362	70.14				42912.
364	75.10				40298.
366	80.33				37880.
368	85.84				35643.
370	91.63				33572.
372	97.71				31654.
373	100.9				30748.

COMPOUND CONSTANTS:

MP		TC		VC		ZC
NBP	385.134 K	PC		DC		OM
MU		RG				

VAPOR PRESSURE CORRELATION CONSTANTS, K AND KPA:

RIEDEL	RIEDEL	RIEDEL	RIEDEL
329. - 386. K	329. - 386. K	329. - 386. K	365. - 386. K
13 POINTS	13 POINTS	13 POINTS	10 POINTS
RMSD = 0.0075	RMSD = 0.0075	RMSD = 0.0075	RMSD = 0.0079
0.56911642E+02	0.56911642E+02	0.56911642E+02	0.14310337E+03
-0.58673187E+01	-0.58673187E+01	-0.58673187E+01	-0.18606957E+02
-0.67011061E+04	-0.67011061E+04	-0.67011061E+04	-0.10817657E+05
0.11560092E-16	0.11560092E-16	0.11560092E-16	0.11681955E-15

LIQUID DENSITY CORRELATION CONSTANTS, K AND G/ML:

FRANCIS1	FRANCIS1	FRANCIS1
288. - 299. K	288. - 299. K	288. - 299. K
4 POINTS	4 POINTS	4 POINTS
RMSD = 0.00005	RMSD = 0.00005	RMSD = 0.00005
0.10091105E+01	0.10091105E+01	0.10091105E+01
0.81475684E-03	0.81475684E-03	0.81475684E-03
0.59999990E+01	0.59999990E+01	0.59999990E+01
0.55901440E+03	0.55901440E+03	0.55901440E+03

SECOND VIRIAL COEFFICIENT CORRELATION CONSTANTS, ML/MOL:

LITERATURE DOCUMENTS FOR CORRELATED VAPOR PRESSURE VALUES:

17432

LITERATURE DOCUMENTS FOR CORRELATED LIQUID DENSITY VALUES:

3269

LITERATURE DOCUMENTS REPORTING VIRIAL COEFFICIENT DATA:

T,K	VAPOR PRESSURE, KPA	SATURATED LIQUID VOLUME, ML/MOL	SECOND VIRIAL COEFFICIENT, ML/MOL	HEAT OF VAPORIZATION, J/MOL	SATURATED VAPOR VOLUME, ML/MOL
288		127.83			
290		128.14			
292		128.45			
294		128.76			
296		129.07			
299		129.54			
330	13.40				204797.
332	14.62				188790.
334	15.94				174242.
336	17.35				161005.
339	19.67				143320.
341	21.35				132808.
343	23.15				123203.
345	25.07				114415.
348	28.20				102597.
350	30.46				95527.
352	32.87				89034.
354	35.43				83064.
357	39.58				74986.
359	42.56				70125.
361	45.73				65640.
363	49.08				61497.
365	52.63				57667.
368	58.34				52450.
370	62.41				49289.
372	66.72				46357.
374	71.26				43635.
377	78.54				39909.
379	83.72				37640.
381	89.17				35527.
383	94.89				33558.
385	100.9				31721.

COMPOUND CONSTANTS:

MP   142.643 K   TC   VC   ZC
NBP   380.060 K   PC   DC   OM
MU   RG

    VAPOR PRESSURE CORRELATION CONSTANTS, K AND KPA:

RPM2	RPM2	RPM2	RPM2
342. - 381. K	342. - 381. K	342. - 381. K	361. - 381. K
12 POINTS	12 POINTS	12 POINTS	10 POINTS
RMSD = 0.0067	RMSD = 0.0067	RMSD = 0.0067	RMSD = 0.0067
0.36392806E+02	0.36392806E+02	0.36392806E+02	0.68387968E+02
-0.73117225E+04	-0.73117225E+04	-0.73117225E+04	-0.11280542E+05
-0.47203812E-01	-0.47203812E-01	-0.47203812E-01	-0.13316798E+00
0.37412955E-04	0.37412955E-04	0.37412955E-04	0.11439015E-03

    LIQUID DENSITY CORRELATION CONSTANTS, K AND G/ML:

FRANCIS1	FRANCIS1	FRANCIS1
288. - 299. K	288. - 299. K	288. - 299. K
4 POINTS	4 POINTS	4 POINTS
RMSD = 0.00005	RMSD = 0.00005	RMSD = 0.00005
0.99950755E+00	0.99950755E+00	0.99950755E+00
0.81478548E-03	0.81478548E-03	0.81478548E-03
0.59999990E+01	0.59999990E+01	0.59999990E+01
0.55901440E+03	0.55901440E+03	0.55901440E+03

    SECOND VIRIAL COEFFICIENT CORRELATION CONSTANTS, ML/MOL:

LITERATURE DOCUMENTS FOR CORRELATED VAPOR PRESSURE VALUES:

  17432

LITERATURE DOCUMENTS FOR CORRELATED LIQUID DENSITY VALUES:

  3269

LITERATURE DOCUMENTS REPORTING VIRIAL COEFFICIENT DATA:

T,K	VAPOR PRESSURE, KPA	SATURATED LIQUID VOLUME, ML/MOL	SECOND VIRIAL COEFFICIENT, ML/MOL	HEAT OF VAPORIZATION, J/MOL	SATURATED VAPOR VOLUME, ML/MOL
288		129.49			
290		129.80			
292		130.12			
294		130.43			
296		130.75			
298		131.08			
342	25.68				110723.
345	29.00				98925.
347	31.40				91893.
349	33.96				85451.
351	36.69				79544.
353	39.60				74121.
355	42.69				69137.
357	45.98				64551.
359	49.48				60326.
361	53.19				56431.
364	59.18				51142.
366	63.47				47947.
368	68.01				44991.
370	72.81				42253.
372	77.88				39714.
374	83.24				37358.
376	88.89				35170.
378	94.85				33135.
380	101.1				31242.

COMPOUND CONSTANTS:

MP   193.583 K   TC   VC   ZC
NBP   399.420 K   PC   DC   OM
MU   RG

VAPOR PRESSURE CORRELATION CONSTANTS, K AND KPA:

RIEDEL   RIEDEL   RIEDEL   RIEDEL
340. - 400. K   340. - 400. K   340. - 400. K   385. - 400. K
  8 POINTS   8 POINTS   8 POINTS   4 POINTS
RMSD = 0.0034   RMSD = 0.0034   RMSD = 0.0034   RMSD = 0.0000
 0.57256817E+02   0.57256817E+02   0.57256817E+02   0.32280841E+03
-0.59628186E+01   -0.59628186E+01   -0.59628186E+01   -0.44988040E+02
-0.67773272E+04   -0.67773272E+04   -0.67773272E+04   -0.19889474E+05
 0.11514857E-16   0.11514857E-16   0.11514857E-16   0.26715561E-15

LIQUID DENSITY CORRELATION CONSTANTS, K AND G/ML:

FRANCIS1   FRANCIS1   FRANCIS1
293. - 366. K   293. - 366. K   293. - 366. K
 14 POINTS   14 POINTS   14 POINTS
RMSD = 0.00031   RMSD = 0.00031   RMSD = 0.00031
0.10012398E+01   0.10012398E+01   0.10012398E+01
0.81011257E-03   0.81011257E-03   0.81011257E-03
0.65672951E+01   0.65672951E+01   0.65672951E+01
0.68615112E+03   0.68615112E+03   0.68615112E+03

SECOND VIRIAL COEFFICIENT CORRELATION CONSTANTS, ML/MOL:

LITERATURE DOCUMENTS FOR CORRELATED VAPOR PRESSURE VALUES:

  17432

LITERATURE DOCUMENTS FOR CORRELATED LIQUID DENSITY VALUES:

  13373 14634 21510

LITERATURE DOCUMENTS REPORTING VIRIAL COEFFICIENT DATA:

T,K	VAPOR PRESSURE, KPA	SATURATED LIQUID VOLUME, ML/MOL	SECOND VIRIAL COEFFICIENT, ML/MOL	HEAT OF VAPORIZATION, J/MOL	SATURATED VAPOR VOLUME, ML/MOL
293		147.49			
295		147.83			
297		148.16			
300		148.68			
302		149.02			
305		149.54			
307		149.89			
310		150.41			
312		150.77			
314		151.12			
317		151.65			
319		152.01			
322		152.56			
324		152.92			
327		153.47			
329		153.84			
331		154.21			
334		154.77			
336		155.15			
339		155.72			
341	13.81	156.10		39721.	205258.
344	15.61	156.67		39588.	183269.
346	16.91	157.06		39500.	170163.
348	18.29	157.45		39412.	158163.
351	20.55	158.04		39281.	142004.
353	22.18	158.43		39194.	132327.
356	24.82	159.03		39064.	119253.
358	26.72	159.43		38978.	111397.
361	29.79	160.04		38849.	100749.
363	32.00	160.44		38764.	94331.
365	34.33	160.85		38679.	88401.
368	38.09				80329.
370	40.78				75443.
373	45.09				68773.
375	48.17				64724.
378	53.11				59182.
380	56.61				55808.
382	60.31				52666.
385	66.21				48350.
387	70.39				45713.
390	77.06				42082.
392	81.77				39857.
395	89.28				36786.
397	94.58				34901.
399	100.1				33134.

COMPOUND CONSTANTS:

MP	211.582 K	TC	VC	ZC
NBP	411.385 K	PC	DC	OM
MU		RG		

VAPOR PRESSURE CORRELATION CONSTANTS, K AND KPA:

RIEDEL	RIEDEL	RIEDEL	RIEDEL
350. - 412. K	350. - 412. K	350. - 412. K	396. - 412. K
8 POINTS	8 POINTS	8 POINTS	4 POINTS
RMSD = 0.0015	RMSD = 0.0015	RMSD = 0.0015	RMSD = 0.0000
0.56611037E+02	0.56611037E+02	0.56611037E+02	0.15402426E+03
-0.58331596E+01	-0.58331596E+01	-0.58331596E+01	-0.20079118E+02
-0.69628431E+04	-0.69628431E+04	-0.69628431E+04	-0.11911643E+05
0.93960488E-17	0.93960488E-17	0.93960488E-17	0.85800953E-16

LIQUID DENSITY CORRELATION CONSTANTS, K AND G/ML:

SECOND VIRIAL COEFFICIENT CORRELATION CONSTANTS, ML/MOL:

LITERATURE DOCUMENTS FOR CORRELATED VAPOR PRESSURE VALUES:

17432

LITERATURE DOCUMENTS FOR CORRELATED LIQUID DENSITY VALUES:

LITERATURE DOCUMENTS REPORTING VIRIAL COEFFICIENT DATA:

T,K	VAPOR PRESSURE, KPA	SATURATED LIQUID VOLUME, ML/MOL	SECOND VIRIAL COEFFICIENT, ML/MOL	HEAT OF VAPORIZATION, J/MOL	SATURATED VAPOR VOLUME, ML/MOL
350	12.99				224014.
351	13.53				215774.
352	14.08				207891.
354	15.24				193128.
355	15.85				186216.
357	17.13				173256.
358	17.80				167181.
359	18.50				161358.
361	19.95				150423.
362	20.72				145289.
364	22.31				135637.
365	23.15				131100.
366	24.01				126743.
368	25.81				118541.
369	26.75				114680.
371	28.72				107402.
372	29.75				103972.
373	30.81				100673.
375	33.01				94447.
376	34.16				91508.
378	36.56				85957.
379	37.81				83334.
380	39.10				80808.
382	41.78				76028.
383	43.17				73767.
385	46.07				69486.
386	47.58				67459.
388	50.71				63616.
389	52.34				61795.
390	54.01				60037.
392	57.48				56701.
393	59.28				55119.
395	63.02				52112.
396	64.96				50684.
397	66.95				49304.
399	71.07				46679.
400	73.20				45432.
402	77.63				43057.
403	79.92				41927.
404	82.26				40833.
406	87.12				38749.
407	89.63				37756.
409	94.82				35863.
410	97.51				34961.
411	100.3				34086.

COMPOUND CONSTANTS:

MP	169.169 K	TC	VC	ZC
NBP	406.383 K	PC	DC	OM
MU		RG		

VAPOR PRESSURE CORRELATION CONSTANTS, K AND KPA:

RIEDEL	RIEDEL	RIEDEL	RIEDEL
360. - 407. K	360. - 407. K	360. - 407. K	392. - 407. K
8 POINTS	8 POINTS	8 POINTS	4 POINTS
RMSD = 0.0051	RMSD = 0.0051	RMSD = 0.0051	RMSD = 0.0000
0.53694953E+02	0.53694953E+02	0.53694953E+02	-0.80942956E+02
-0.53996824E+01	-0.53996824E+01	-0.53996824E+01	0.14342256E+02
-0.67730859E+04	-0.67730859E+04	-0.67730859E+04	-0.37863678E+02
0.61458848E-17	0.61458848E-17	0.61458848E-17	-0.11183162E-15

LIQUID DENSITY CORRELATION CONSTANTS, K AND G/ML:

SECOND VIRIAL COEFFICIENT CORRELATION CONSTANTS, ML/MOL:

LITERATURE DOCUMENTS FOR CORRELATED VAPOR PRESSURE VALUES:

17432

LITERATURE DOCUMENTS FOR CORRELATED LIQUID DENSITY VALUES:

LITERATURE DOCUMENTS REPORTING VIRIAL COEFFICIENT DATA:

T,K	VAPOR PRESSURE, KPA	SATURATED LIQUID VOLUME, ML/MOL	SECOND VIRIAL COEFFICIENT, ML/MOL	HEAT OF VAPORIZATION, J/MOL	SATURATED VAPOR VOLUME, ML/MOL
360	22.45				133351.
361	23.30				128817.
362	24.18				124467.
363	25.09				120292.
364	26.03				116284.
365	26.99				112435.
366	27.99				108738.
367	29.01				105187.
368	30.06				101774.
369	31.15				98493.
370	32.27				95340.
371	33.42				92307.
372	34.60				89390.
373	35.82				86584.
374	37.07				83883.
376	39.68				78782.
377	41.04				76372.
378	42.44				74052.
379	43.88				71816.
380	45.35				69662.
381	46.87				67586.
382	48.43				65584.
383	50.03				63655.
384	51.67				61794.
385	53.35				59999.
386	55.08				58267.
387	56.85				56596.
388	58.67				54983.
389	60.54				53425.
390	62.45				51922.
392	66.42				49067.
393	68.48				47712.
394	70.60				46403.
395	72.76				45138.
396	74.98				43914.
397	77.25				42732.
398	79.57				41588.
399	81.95				40481.
400	84.39				39411.
401	86.88				38376.
402	89.43				37373.
403	92.04				36403.
404	94.72				35464.
405	97.45				34555.
406	100.2				33675.

COMPOUND CONSTANTS:

MP 170.620 K	TC	VC	ZC
NBP 404.982 K	PC	DC	OM
MU	RG		

VAPOR PRESSURE CORRELATION CONSTANTS, K AND KPA:

RIEDEL	RIEDEL	RIEDEL	RIEDEL
358. - 405. K	358. - 405. K	358. - 405. K	390. - 405. K
8 POINTS	8 POINTS	8 POINTS	4 POINTS
RMSD = 0.0029	RMSD = 0.0029	RMSD = 0.0029	RMSD = 0.0000
0.70657367E+02	0.70657367E+02	0.70657367E+02	0.22358179E+03
-0.79064508E+01	-0.79064508E+01	-0.79064508E+01	-0.30336759E+02
-0.75662872E+04	-0.75662872E+04	-0.75662872E+04	-0.15202401E+05
0.25623847E-16	0.25623847E-16	0.25623847E-16	0.16141647E-15

LIQUID DENSITY CORRELATION CONSTANTS, K AND G/ML:

FRANCIS1	FRANCIS1	FRANCIS1
293. - 304. K	293. - 304. K	293. - 304. K
4 POINTS	4 POINTS	4 POINTS
RMSD = 0.00005	RMSD = 0.00005	RMSD = 0.00005
0.99535042E+00	0.99535042E+00	0.99535042E+00
0.76067448E-03	0.76067448E-03	0.76067448E-03
0.59999990E+01	0.59999990E+01	0.59999990E+01
0.56838794E+03	0.56838794E+03	0.56838794E+03

SECOND VIRIAL COEFFICIENT CORRELATION CONSTANTS, ML/MOL:

LITERATURE DOCUMENTS FOR CORRELATED VAPOR PRESSURE VALUES:

17432

LITERATURE DOCUMENTS FOR CORRELATED LIQUID DENSITY VALUES:

13373

LITERATURE DOCUMENTS REPORTING VIRIAL COEFFICIENT DATA:

T,K	VAPOR PRESSURE, KPA	SATURATED LIQUID VOLUME, ML/MOL	SECOND VIRIAL COEFFICIENT, ML/MOL	HEAT OF VAPORIZATION, J/MOL	SATURATED VAPOR VOLUME, ML/MOL
293		146.80			
295		147.13			
298		147.63			
300		147.96			
303		148.46			
359	22.65				131802.
361	24.40				122996.
364	27.25				111078.
366	29.29				103903.
369	32.58				94161.
371	34.94				88276.
374	38.74				80263.
376	41.46				75408.
379	45.82				68777.
382	50.54				62843.
384	53.90				59231.
387	59.28				54276.
389	63.10				51253.
392	69.21				47094.
394	73.53				44550.
397	80.43				41042.
399	85.30				38890.
402	93.06				35916.
404	98.54				34088.

COMPOUND CONSTANTS:

MP   223.178 K   TC   VC   ZC
NBP   423.870 K   PC   DC   OM
MU   RG

    VAPOR PRESSURE CORRELATION CONSTANTS, K AND KPA:

    LIQUID DENSITY CORRELATION CONSTANTS, K AND G/ML:

FRANCIS1   FRANCIS1   FRANCIS1
293. - 359. K   293. - 359. K   293. - 359. K
  4 POINTS   4 POINTS   4 POINTS
RMSD = 0.00001   RMSD = 0.00001   RMSD = 0.00001
0.97660637E+00   0.97660637E+00   0.97660637E+00
0.65518869E-03   0.65518869E-03   0.65518869E-03
0.60076742E+01   0.60076742E+01   0.60076742E+01
0.51942065E+03   0.51942065E+03   0.51942065E+03

    SECOND VIRIAL COEFFICIENT CORRELATION CONSTANTS, ML/MOL:

LITERATURE DOCUMENTS FOR CORRELATED VAPOR PRESSURE VALUES:

LITERATURE DOCUMENTS FOR CORRELATED LIQUID DENSITY VALUES:

  14634

LITERATURE DOCUMENTS REPORTING VIRIAL COEFFICIENT DATA:

T,K	VAPOR PRESSURE, KPA	SATURATED LIQUID VOLUME, ML/MOL	SECOND VIRIAL COEFFICIENT, ML/MOL	HEAT OF VAPORIZATION, J/MOL	SATURATED VAPOR VOLUME, ML/MOL
293		163.86			
294		164.03			
296		164.37			
297		164.54			
299		164.88			
300		165.05			
302		165.39			
303		165.56			
305		165.91			
306		166.08			
308		166.43			
309		166.61			
311		166.97			
312		167.14			
314		167.50			
315		167.68			
317		168.05			
318		168.23			
320		168.60			
321		168.78			
323		169.15			
324		169.34			
326		169.72			
327		169.91			
329		170.29			
330		170.48			
332		170.87			
333		171.06			
335		171.45			
336		171.65			
338		172.05			
339		172.25			
341		172.65			
342		172.85			
344		173.26			
345		173.47			
347		173.88			
348		174.09			
350		174.51			
351		174.73			
353		175.16			
354		175.37			
356		175.81			
357		176.03			
359		176.47			

COMPOUND CONSTANTS:

MP   236.750 K   TC   VC   ZC
NBP   282.850 K   PC   DC   OM
MU   RG

VAPOR PRESSURE CORRELATION CONSTANTS, K AND KPA:

VAPRES-2	VAPRES-2	VAPRES-2
238. - 283. K	238. - 283. K	238. - 283. K
18 POINTS	18 POINTS	18 POINTS
RMSD = 0.4270	RMSD = 0.4270	RMSD = 0.4270
-0.24583918E+05	-0.24583918E+05	-0.24583918E+05
0.42123871E+06	0.42123871E+06	0.42123871E+06
-0.18424218E+02	-0.18424218E+02	-0.18424218E+02
0.11627729E-01	0.11627729E-01	0.11627729E-01
0.48504320E+04	0.48504320E+04	0.48504320E+04

LIQUID DENSITY CORRELATION CONSTANTS, K AND G/ML:

FRANCIS1	FRANCIS1	FRANCIS1
238. - 273. K	238. - 273. K	238. - 273. K
4 POINTS	4 POINTS	4 POINTS
RMSD = 0.00069	RMSD = 0.00069	RMSD = 0.00069
0.10724297E+01	0.10724297E+01	0.10724297E+01
0.10980645E-02	0.10980645E-02	0.10980645E-02
0.86587925E+01	0.86587925E+01	0.86587925E+01
0.51215552E+03	0.51215552E+03	0.51215552E+03

SECOND VIRIAL COEFFICIENT CORRELATION CONSTANTS, ML/MOL:

LITERATURE DOCUMENTS FOR CORRELATED VAPOR PRESSURE VALUES:

   2299   3020   4372

LITERATURE DOCUMENTS FOR CORRELATED LIQUID DENSITY VALUES:

   2297   4372

LITERATURE DOCUMENTS REPORTING VIRIAL COEFFICIENT DATA:

T,K	VAPOR PRESSURE, KPA	SATURATED LIQUID VOLUME, ML/MOL	SECOND VIRIAL COEFFICIENT, ML/MOL	HEAT OF VAPORIZATION, J/MOL	SATURATED VAPOR VOLUME, ML/MOL
238	12.76	64.22			155047.
239	13.47	64.32			147514.
240	14.22	64.42			140338.
241	15.01	64.52			133509.
242	15.84	64.62			127020.
243	16.72	64.72			120858.
244	17.64	64.83			115014.
245	18.61	64.93			109474.
246	19.62	65.03			104227.
247	20.69	65.14			99260.
248	21.81	65.24			94560.
249	22.97	65.34			90115.
250	24.19	65.45			85912.
251	25.47	65.55			81940.
252	26.80	65.66			78185.
253	28.18	65.76			74638.
254	29.63	65.87			71285.
255	31.13	65.98			68118.
256	32.68	66.08			65124.
257	34.30	66.19			62295.
258	35.98	66.30			59620.
259	37.72	66.41			57092.
260	39.52	66.52			54700.
261	41.38	66.63			52437.
262	43.31	66.74			50296.
263	45.30	66.85			48268.
264	47.36	66.96			46347.
265	49.48	67.07			44526.
266	51.67	67.18			42800.
267	53.93	67.29			41162.
268	56.26	67.40			39607.
269	58.66	67.52			38129.
270	61.13	67.63			36724.
271	63.67	67.74			35388.
272	66.29	67.86			34115.
273	68.99	67.97			32902.
274	71.76				31745.
275	74.62				30640.
276	77.57				29585.
277	80.60				28575.
278	83.72				27608.
279	86.94				26682.
280	90.26				25793.
281	93.68				24939.
282	97.22				24118.

COMPOUND CONSTANTS:

MP   TC   VC   ZC
NBP   278.450 K   PC   DC   OM
MU   RG

VAPOR PRESSURE CORRELATION CONSTANTS, K AND KPA:

RPM2   RPM2   RPM2   RPM2
203. - 313. K   203. - 313. K   203. - 313. K   203. - 313. K
   9 POINTS   9 POINTS   9 POINTS   9 POINTS
RMSD = 0.1850   RMSD = 0.1850   RMSD = 0.1850   RMSD = 0.1850
0.34817871E+01   0.34817871E+01   0.34817871E+01   0.34817871E+01
-0.24016618E+04   -0.24016618E+04   -0.24016618E+04   -0.24016618E+04
0.61523730E-01   0.61523730E-01   0.61523730E-01   0.61523730E-01
-0.94929187E-04   -0.94929187E-04   -0.94929187E-04   -0.94929187E-04

LIQUID DENSITY CORRELATION CONSTANTS, K AND G/ML:

SECOND VIRIAL COEFFICIENT CORRELATION CONSTANTS, ML/MOL:

LITERATURE DOCUMENTS FOR CORRELATED VAPOR PRESSURE VALUES:

  2238   4091   8849

LITERATURE DOCUMENTS FOR CORRELATED LIQUID DENSITY VALUES:

LITERATURE DOCUMENTS REPORTING VIRIAL COEFFICIENT DATA:

T,K	VAPOR PRESSURE, KPA	SATURATED LIQUID VOLUME, ML/MOL	SECOND VIRIAL COEFFICIENT, ML/MOL	HEAT OF VAPORIZATION, J/MOL	SATURATED VAPOR VOLUME, ML/MOL
203	1.256				1343287.
205	1.476				1154786.
208	1.869				925443.
210	2.179				801262.
213	2.729				648843.
215	3.161				565585.
218	3.919				462542.
220	4.508				405786.
223	5.535				334990.
225	6.327				295688.
228	7.697				246298.
230	8.745				218675.
233	10.54				183715.
235	11.91				164024.
238	14.24				138936.
240	16.00				124710.
243	18.98				106469.
245	21.21				96060.
248	24.95				82631.
250	27.74				74922.
253	32.40				64919.
255	35.85				59143.
258	41.57				51608.
260	45.77				47233.
263	52.70				41495.
265	57.76				38146.
268	66.06				33732.
270	72.08				31143.
273	81.90				27715.
275	88.99				25695.
278	100.5				23007.
280	108.7				21416.
283	122.0				19291.
285	131.4				18028.
288	146.6				16334.
290	157.4				15323.
293	174.5				13961.
295	186.6				13146.
298	205.7				12044.
300	219.1				11382.
303	240.3				10484.
305	255.0				9943.
308	278.1				9207.
310	294.2				8762.
313	319.1				8155.

COMPOUND CONSTANTS:

MP		TC		VC		ZC
NBP	356.144 K	PC		DC		OM
MU	0.1000 DEBYE	RG				

VAPOR PRESSURE CORRELATION CONSTANTS, K AND KPA:

RIEDEL	RIEDEL	RIEDEL	RIEDEL
273. - 364. K	273. - 364. K	273. - 364. K	273. - 364. K
8 POINTS	8 POINTS	8 POINTS	8 POINTS
RMSD = 1.9670	RMSD = 1.9670	RMSD = 1.9670	RMSD = 1.9670
0.12001877E+03	0.12001877E+03	0.12001877E+03	0.12001877E+03
-0.15558983E+02	-0.15558983E+02	-0.15558983E+02	-0.15558983E+02
-0.86798037E+04	-0.86798037E+04	-0.86798037E+04	-0.86798037E+04
0.18887842E-15	0.18887842E-15	0.18887842E-15	0.18887842E-15

LIQUID DENSITY CORRELATION CONSTANTS, K AND G/ML:

SECOND VIRIAL COEFFICIENT CORRELATION CONSTANTS, ML/MOL:

LITERATURE DOCUMENTS FOR CORRELATED VAPOR PRESSURE VALUES:

4091

LITERATURE DOCUMENTS FOR CORRELATED LIQUID DENSITY VALUES:

LITERATURE DOCUMENTS REPORTING VIRIAL COEFFICIENT DATA:

T,K	VAPOR PRESSURE, KPA	SATURATED LIQUID VOLUME, ML/MOL	SECOND VIRIAL COEFFICIENT, ML/MOL	HEAT OF VAPORIZATION, J/MOL	SATURATED VAPOR VOLUME, ML/MOL
273	2.787				814300.
275	3.147				726666.
277	3.543				649979.
279	3.981				582705.
281	4.463				523550.
283	4.991				471413.
285	5.571				425356.
287	6.205				384581.
289	6.897				348403.
291	7.651				316237.
293	8.471				287579.
295	9.362				261995.
297	10.33				239111.
299	11.37				218601.
301	12.50				200186.
304	14.36				175966.
306	15.73				161786.
308	17.19				148975.
310	18.76				137381.
312	20.45				126871.
314	22.25				117328.
316	24.18				108649.
318	26.24				100745.
320	28.45				93534.
322	30.79				86947.
324	33.29				80921.
326	35.95				75400.
328	38.77				70334.
330	41.77				65680.
332	44.96				61400.
335	50.09				55601.
337	53.77				52107.
339	57.67				48878.
341	61.78				45891.
343	66.13				43124.
345	70.73				40558.
347	75.57				38176.
349	80.69				35962.
351	86.08				33902.
353	91.77				31983.
355	97.76				30193.
357	104.1				28523.
359	110.7				26962.
361	117.7				25503.
363	125.0				24136.

COMPOUND CONSTANTS:

MP 278.680 K	TC 562.160 K	VC 258.65 ML/MOL	ZC 0.2710
NBP 353.235 K	PC 4.898 MPA	DC 0.3020 G/ML	OM 0.2092
MU 0.0000 DEBYE	RG 3.0037 ANGSTROM		

VAPOR PRESSURE CORRELATION CONSTANTS, K AND KPA:

WAGNER	WAGNER	VAPRES-2	WAGNER
278. - 563. K	278. - 374. K	278. - 478. K	333. - 563. K
352 POINTS	268 POINTS	314 POINTS	176 POINTS
RMSD = 0.8720	RMSD = 0.0400	RMSD = 0.2070	RMSD = 1.2280
-0.69740008E+01	-0.69650565E+01	0.12486728E+03	-0.69755790E+01
0.13185461E+01	0.12975649E+01	-0.75647711E+04	0.13234387E+01
-0.26337465E+01	-0.26030426E+01	0.24669095E-01	-0.26475417E+01
-0.32587657E+01	-0.33168888E+01	-0.40448273E-05	-0.31913135E+01
		-0.18244438E+02	

LIQUID DENSITY CORRELATION CONSTANTS, K AND G/ML:

FRANCIS1	FRANCIS1	FRANCIS1	RACKETT
280. - 553. K	280. - 374. K	280. - 529. K	513. - 563. K
336 POINTS	279 POINTS	331 POINTS	13 POINTS
RMSD = 0.00055	RMSD = 0.00018	RMSD = 0.00045	RMSD = 0.00503
0.11886606E+01	0.11971445E+01	0.11878071E+01	0.27227339E+00
0.89817494E-03	0.96949888E-03	0.92299492E-03	
0.15000000E+02	0.11815362E+02	0.11831548E+02	
0.61569116E+03	0.64129077E+03	0.60160205E+03	

SECOND VIRIAL COEFFICIENT CORRELATION CONSTANTS, ML/MOL:

TSONOPOULOS: TC=562.160 PC=4.898 VC=258.65 OM=0 2092 A= 0.000000 B=0.0000

LITERATURE DOCUMENTS FOR CORRELATED VAPOR PRESSURE VALUES:
```
    10    102    124    162    194    196    212    228    262    295    687    914   1485
  1486   1526   1540   1690   1716   1770   1981   2284   2415   2418   2925   2977   3047
  3118   3569   3571   4184   4233   4455   4519   4818   4821   4908   4939   5101   5173
  5233   6634   6893   6894   6947   7746   7904   7952   8111   8628   8801   9623   9760
 10135  10216  10416  10428  10834  11019  11100  11157  11336  12536  15514  15728  16004
 18044  18270  19427  19840  20232  20654  21533  21537  21718  21855  21969  40032  40360
```

LITERATURE DOCUMENTS FOR CORRELATED LIQUID DENSITY VALUES:
```
    70    138    182    228    929   1291   1489   1490   1540   1549   2281   2283   2573
  2959   2976   3610   3907   4565   4614   4909   5994   7234   7904   8546   8899  10193
 10258  10574  10894  11019  11295  12358  12385  12387  12507  12993  13053  13464  14049
 14116  14299  15650  16363  16488  17144  17937  17974  18837  19433  20064  20065  20066
 20074  20075  20076  20954  21026  21080  21121  21306  21432  21805  21816  40685  40813
 40895
```

LITERATURE DOCUMENTS REPORTING VIRIAL COEFFICIENT DATA:
```
   905    932   1002   1088   1257   1497   1803   2409   2410   2413   2416   2420   2422
  2423   5283  10428  10754  10972  11493
```

T,K	VAPOR PRESSURE, KPA	SATURATED LIQUID VOLUME, ML/MOL	SECOND VIRIAL COEFFICIENT, ML/MOL	HEAT OF VAPORIZATION, J/MOL	SATURATED VAPOR VOLUME, ML/MOL
278	4.611				
284	6.348	87.93	-1989.	34623.	369952.
290	8.602	88.55	-1841.	34276.	278463.
297	12.03	89.29	-1692.	33875.	203509.
303	15.81	89.94	-1580.	33534.	157761.
310	21.39	90.71	-1465.	33140.	119012.
316	27.37	91.38	-1378.	32804.	94603.
323	35.98	92.19	-1287.	32412.	73325.
329	44.99	92.90	-1217.	32076.	59552.
336	57.70	93.75	-1144.	31683.	47241.
342	70.74	94.49	-1088.	31344.	39080.
349	88.77	95.38	-1028.	30945.	31625.
355	107.0	96.17	-981.0	30599.	26579.
362	131.7	97.11	-930.9	30191.	21880.
368	156.3	97.94	-891.4	29836.	18641.
375	189.3	98.94	-848.9	29415.	15574.
381	221.6	99.82	-815.1	29047.	13425.
388	264.5	100.88	-778.6	28608.	11359.
394	306.1	101.83	-749.4	28223.	9891.
401	360.7	102.97	-717.7	27762.	8460.
407	413.0	103.98	-692.2	27356.	7430.
414	481.1	105.21	-664.3	26868.	6414.
420	545.9	106.31	-641.9	26436.	5674.
426	616.9	107.45	-620.6	25990.	5034.
433	708.2	108.85	-597.2	25449.	4392.
439	794.2	110.10	-578.3	24966.	3918.
446	903.9	111.64	-557.4	24377.	3437.
452	1007.	113.04	-540.4	23846.	3078.
459	1137.	114.76	-521.6	23192.	2711.
465	1258.	116.33	-506.2	22596.	2434.
472	1411.	118.29	-489.2	22265.	2185.
478	1553.	120.10	-475.3	21657.	1970.
485	1732.	122.38	-459.8	20913.	1748.
491	1896.	124.51	-447.1	20241.	1577.
498	2103.	127.24	-432.9	19411.	1400.
504	2293.	129.82	-421.3	18654.	1263.
511	2530.	133.20	-408.3	17708.	1119.
517	2748.	136.49	-397.6	16835.	1006.
524	3020.	140.89	-385.6	15724.	887.1
530	3268.	145.30	-375.8	14674.	793.0
537	3578.	151.46	-364.7	13298.	691.2
543	3862.	157.93	-355.6	11942.	609.0
550	4216.	167.51	-345.4	10044.	516.0
556	4541.		-336.9		433.8
562	4888.		-328.8		316.4

COMPOUND CONSTANTS:

MP 178.150 K	TC 594.025 K	VC 298.00 ML/MOL	ZC 0.2544
NBP 383.758 K	PC 4.236 MPA	DC 0.3092 G/ML	OM 0.2607
MU 0.4300 DEBYE	RG 3.4431 ANGSTROM		

VAPOR PRESSURE CORRELATION CONSTANTS, K AND KPA:

WAGNER	RPM2	WAGNER	WAGNER
195. - 594. K	195. - 403. K	195. - 504. K	364. - 594. K
310 POINTS	255 POINTS	278 POINTS	106 POINTS
RMSD = 0.4922	RMSD = 0.0510	RMSD = 0.1681	RMSD = 0.6481
-0.73732812E+01	0.30629588E+02	-0.73620355E+01	-0.73990193E+01
0.15868655E+01	-0.62893763E+04	0.15620913E+01	0.16587799E+01
-0.29506351E+01	-0.35932503E-01	-0.29211777E+01	-0.30544936E+01
-0.27393712E+01	0.28297458E-04	-0.27733571E+01	-0.32741666E+01

LIQUID DENSITY CORRELATION CONSTANTS, K AND G/ML:

FRANCIS1	FRANCIS1	FRANCIS1	FRANCIS2
179. - 553. K	179. - 400. K	179. - 553. K	553. - 594. K
267 POINTS	253 POINTS	267 POINTS	3 POINTS
RMSD = 0.00048	RMSD = 0.00033	RMSD = 0.00048	RMSD = 0.00028
0.11470699E+01	0.11439199E+01	0.11470699E+01	0.90605525E-03
0.83579519E-03	0.87978248E-03	0.83579519E-03	0.22017707E+01
0.11802405E+02	0.60034208E+01	0.11802405E+02	
0.63049780E+03	0.60689600E+03	0.63049780E+03	

SECOND VIRIAL COEFFICIENT CORRELATION CONSTANTS, ML/MOL:

TSONOPOULOS: TC=594.025 PC=4.236 VC=298.00 OM=0.2607 A= 0.000000 B=0.0000

LITERATURE DOCUMENTS FOR CORRELATED VAPOR PRESSURE VALUES:

```
   91   123   162   175   189   224   287   314   764   939  1043  1098  1176
 1266  1348  1485  1486  1521  1548  1636  1732  1740  1978  1981  2850  2925
 2979  3262  3571  3822  4233  4392  4470  4895  5101  5635  6802  6951  7404
 7820  7904  7926  7952  9039  9903 10290 10315 10432 10552 11081 11160 11231
15728 18196 18208 18268 19427 19662 19712 20447 20752 20957 21028 40775 40936
```


LITERATURE DOCUMENTS FOR CORRELATED LIQUID DENSITY VALUES:

```
   33    47    71   131   146   182   197   694   779  1487  1490  1549  2573
 3087  3262  3610  3894  4015  4060  4094  4565  4700  4895  5204  5231  5235
 5364  6408  8717  9008  9068 10193 10258 10264 10315 11100 11193 11246 11295
12089 13016 13053 13364 13380 13838 13877 13945 14001 14416 15062 15281 15290
15291 15650 15872 15904 16006 16466 16837 17453 17521 17827 17936 18314 18837
19746 20015 20315 20636 20700 21432 21804 21810 21816 40228 40329 40425 40816
```


LITERATURE DOCUMENTS REPORTING VIRIAL COEFFICIENT DATA:

```
 1497  2410  2411  7706  8421 11125
```

T,K	VAPOR PRESSURE, KPA	SATURATED LIQUID VOLUME, ML/MOL	SECOND VIRIAL COEFFICIENT, ML/MOL	HEAT OF VAPORIZATION, J/MOL	SATURATED VAPOR VOLUME, ML/MOL
178		94.83	-54030.		
188		95.65	-37328.		
197	0.0006846	96.46	-26894.	44570.	0.2392E+10
206	0.002228	97.29	-19858.	43912.	0.7686E+09
216	0.007234	98.22	-14571.	43198.	0.2483E+09
225	0.01880	99.08	-11288.	42571.	0.9947E+08
235	0.04915	100.06	-8711.	41894.	0.3974E+08
244	0.1078	100.96	-7043.	41300.	0.1881E+08
254	0.2389	101.99	-5679.	40660.	8832613.
263	0.4599	102.93	-4762.	40099.	4749471.
273	0.8966	104.01	-3985.	39492.	2527533.
282	1.558	105.00	-3443.	38959.	1501280.
291	2.600	106.02	-3011.	38438.	927597.
301	4.400	107.18	-2628.	37870.	566187.
310	6.824	108.25	-2348.	37366.	375356.
320	10.74	109.48	-2092.	36813.	245710.
329	15.70	110.62	-1900.	36319.	172342.
339	23.28	111.92	-1720.	35771.	119340.
348	32.43	113.13	-1582.	35276.	87599.
357	44.31	114.38	-1463.	34778.	65497.
367	61.35	115.82	-1347.	34219.	48349.
376	80.82	117.16	-1256.	33707.	37379.
386	107.9	118.71	-1167.	33127.	28534.
395	137.9	120.16	-1096.	32592.	22670.
405	178.4	121.84	-1025.	31981.	17785.
414	222.3	123.42	-967.4	31413.	14447.
424	280.3	125.27	-909.7	30759.	11588.
433	341.9	127.02	-862.4	30146.	9583.
442	413.2	128.86	-819.0	29508.	7982.
452	505.0	131.04	-774.8	28763.	6563.
461	600.2	133.12	-738.2	28055.	5535.
471	721.0	135.62	-700.6	27219.	4605.
480	844.7	138.04	-669.3	26413.	3918.
490	999.9	140.99	-636.9	25443.	3284.
499	1157.	143.91	-609.8	24484.	2807.
509	1353.	147.54	-581.7	23885.	2412.
518	1549.	151.24	-558.0	22869.	2075.
527	1766.	155.49	-535.7	21769.	1786.
537	2034.	161.08	-512.4	20426.	1509.
546	2300.	167.21	-492.7	19077.	1293.
556	2627.		-472.1		1083.
565	2952.		-454.5		915.5
575	3350.		-436.0		745.6
584	3745.		-420.3		599.1
593	4183.		-405.3		423.6

577

COMPOUND CONSTANTS:

MP   178.226 K   TC   617.146 K   VC   374.49 ML/MOL   ZC   0.2634
NBP   409.377 K   PC   3.609 MPA   DC   0.2835 G/ML   OM   0.3035
MU   0.3600 DEBYE   RG   3.8211 ANGSTROM

VAPOR PRESSURE CORRELATION CONSTANTS, K AND KPA:

WAGNER	WAGNER	WAGNER	WAGNER
263. - 617. K	263. - 421. K	263. - 524. K	392. - 617. K
126 POINTS	95 POINTS	108 POINTS	47 POINTS
RMSD = 0.5945	RMSD = 0.0381	RMSD = 0.1841	RMSD = 0.9398
-0.73985246E+01	-0.78511138E+01	-0.74118548E+01	-0.73977454E+01
0.12286416E+01	0.22409377E+01	0.12598652E+01	0.12409198E+01
-0.29794715E+01	-0.42581507E+01	-0.30237051E+01	-0.31320918E+01
-0.32193163E+01	-0.14489009E+01	-0.31478726E+01	-0.14299501E+01

LIQUID DENSITY CORRELATION CONSTANTS, K AND G/ML:

FRANCIS1	FRANCIS1	FRANCIS1
178. - 491. K	178. - 421. K	178. - 491. K
78 POINTS	73 POINTS	78 POINTS
RMSD = 0.00045	RMSD = 0.00044	RMSD = 0.00045
0.11289759E+01	0.11295567E+01	0.11289759E+01
0.82616881E-03	0.83080516E-03	0.82616881E-03
0.59999990E+01	0.59999990E+01	0.59999990E+01
0.60095581E+03	0.61312036E+03	0.60095581E+03

SECOND VIRIAL COEFFICIENT CORRELATION CONSTANTS, ML/MOL:

TSONOPOULOS: TC=617.146 PC=3.609 VC=374.49 OM=0.3035 A= 0.000000 B=0.0000

LITERATURE DOCUMENTS FOR CORRELATED VAPOR PRESSURE VALUES:

   162   175   224   1089   1146   1485   1486   2925   3571   5868   6520   6776 11081
  13074 15532 16662 19427 40733

LITERATURE DOCUMENTS FOR CORRELATED LIQUID DENSITY VALUES:

   182   224   929   1099   1487   1490   2573   4612   5364 10193 10809 12591 12709
  13621 13877 14416 15872 20540 40802

LITERATURE DOCUMENTS REPORTING VIRIAL COEFFICIENT DATA:

T,K	VAPOR PRESSURE, KPA	SATURATED LIQUID VOLUME, ML/MOL	SECOND VIRIAL COEFFICIENT, ML/MOL	HEAT OF VAPORIZATION, J/MOL	SATURATED VAPOR VOLUME, ML/MOL
178		109.71			
187		110.59	-69808.		
197		111.59	-47936.		
207		112.62	-33876.		
217		113.66	-24601.		
227		114.73	-18335.		
237		115.82	-14005.		
247		116.93	-10947.		
257		118.07	-8743.		
267	0.1647	119.23	-7123.	44415.	0.1347E+08
277	0.3374	120.42	-5909.	43733.	6819106.
287	0.6512	121.65	-4985.	43071.	3659417.
297	1.192	122.90	-4267.	42427.	2067989.
307	2.080	124.18	-3702.	41800.	1223522.
317	3.480	125.50	-3250.	41187.	754068.
327	5.607	126.85	-2882.	40585.	481983.
337	8.732	128.25	-2580.	39991.	318272.
347	13.19	129.68	-2328.	39403.	216392.
357	19.38	131.16	-2116.	38816.	151028.
367	27.77	132.69	-1935.	38228.	107918.
377	38.90	134.27	-1779.	37635.	78765.
387	53.37	135.90	-1643.	37034.	58595.
397	71.87	137.60	-1525.	36423.	44347.
407	95.14	139.37	-1420.	35798.	34088.
417	124.0	141.21	-1327.	35155.	26572.
427	159.2	143.14	-1243.	34492.	20975.
437	201.8	145.17	-1168.	33804.	16746.
447	252.8	147.31	-1101.	33089.	13506.
457	313.0	149.59	-1039.	32340.	10992.
467	383.7	152.02	-982.7	31553.	9018.
477	465.8	154.65	-931.1	30722.	7450.
487	560.6	157.52	-883.7	29837.	6192.
497	669.3		-839.9		5172.
507	793.0		-799.4		4335.
517	933.2		-761.7		3643.
527	1091.		-726.7		3133.
537	1268.		-694.0		2653.
547	1466.		-663.4		2248.
557	1687.		-634.8		1903.
567	1932.		-607.9		1607.
577	2202.		-582.6		1350.
587	2502.		-558.7		1123.
597	2832.		-536.2		918.5
607	3198.		-514.8		724.0
616	3560.		-496.6		516.1

COMPOUND CONSTANTS:

MP   247.965 K   TC   630.327 K   VC   369.02 ML/MOL   ZC   0.2633
NBP   417.579 K   PC   3.740 MPA   DC   0.2877 G/ML   OM   0.3123
MU   0.5200 DEBYE RG   3.7889 ANGSTROM

VAPOR PRESSURE CORRELATION CONSTANTS, K AND KPA:

WAGNER   VAPRES-2   RPM2   WAGNER
269. - 630. K   269. - 434. K   269. - 534. K   402. - 630. K
102 POINTS   68 POINTS   81 POINTS   47 POINTS
RMSD = 1.2630   RMSD = 0.0080   RMSD = 0.1900   RMSD = 1.7790
-0.76448287E+01   -0.11817699E+03   0.27147243E+02   -0.75848092E+01
 0.16936422E+01   -0.33203049E+04   -0.64808741E+04   0.15141449E+01
-0.36388493E+01   -0.99587059E-01   -0.22850788E-01   -0.31896226E+01
-0.14019536E+01   0.50416065E-04   0.14530686E-04   -0.32463467E+01
                    0.27101082E+02

LIQUID DENSITY CORRELATION CONSTANTS, K AND G/ML:

FRANCIS1   FRANCIS1   FRANCIS1   FRANCIS2
242. - 614. K   242. - 434. K   242. - 574. K   603. - 630. K
111 POINTS   87 POINTS   109 POINTS   3 POINTS
RMSD = 0.00105   RMSD = 0.00048   RMSD = 0.00067   RMSD = 0.00073
0.11404505E+01   0.11265430E+01   0.11360540E+01   0.59339950E-03
0.75655570E-03   0.69812522E-03   0.74034859E-03   0.26693403E+01
0.15000000E+02   0.14648122E+02   0.15000000E+02
0.68433008E+03   0.64029639E+03   0.67736523E+03

SECOND VIRIAL COEFFICIENT CORRELATION CONSTANTS, ML/MOL:

TSONOPOULOS: TC=630.327 PC=3.740 VC=369.02 OM=0.3123 A= 0.000000 B=0.0000

LITERATURE DOCUMENTS FOR CORRELATED VAPOR PRESSURE VALUES:

   162   1095   1098   1146   1485   1486   1499   1602   2925 15532 18668

LITERATURE DOCUMENTS FOR CORRELATED LIQUID DENSITY VALUES:

    76   146   147   1099   1487   1490   1602   1776   2523   2959   2975   4015   5364
  6838   8069   8967 10193 11040 13621 13832 13877 18668 21441 21816 22720

LITERATURE DOCUMENTS REPORTING VIRIAL COEFFICIENT DATA:

  1497   2410

T,K	VAPOR PRESSURE, KPA	SATURATED LIQUID VOLUME, ML/MOL	SECOND VIRIAL COEFFICIENT, ML/MOL	HEAT OF VAPORIZATION, J/MOL	SATURATED VAPOR VOLUME, ML/MOL
242		114.97			
250		115.80	-11594.		
259		116.77	-9454.		
268		117.75	-7833.		
277	0.2312	118.75	-6586.	44488.	9954852.
286	0.4234	119.77	-5613.	43995.	5610624.
294	0.6992	120.70	-4921.	43559.	3491256.
303	1.185	121.77	-4290.	43073.	2122321.
312	1.937	122.86	-3779.	42589.	1335499.
321	3.066	123.98	-3359.	42106.	867094.
330	4.713	125.13	-3012.	41623.	579193.
339	7.051	126.31	-2720.	41138.	397002.
347	9.881	127.38	-2499.	40705.	289457.
356	14.14	128.62	-2285.	40213.	207048.
365	19.81	129.89	-2100.	39715.	151055.
374	27.23	131.21	-1939.	39209.	112210.
383	36.78	132.56	-1799.	38695.	84740.
392	48.87	133.95	-1675.	38169.	64966.
400	62.13	135.23	-1577.	37692.	51899.
409	80.33	136.72	-1477.	37143.	40799.
418	102.5	138.26	-1388.	36578.	32449.
427	129.3	139.86	-1308.	35998.	26085.
436	161.2	141.52	-1235.	35399.	21175.
444	194.5	143.06	-1175.	34850.	17723.
453	238.0	144.86	-1114.	34211.	14620.
462	288.6	146.75	-1058.	33549.	12150.
471	347.1	148.74	-1006.	32861.	10165.
480	414.3	150.83	-957.8	32142.	8555.
489	490.8	153.04	-913.4	31389.	7238.
497	567.4	155.13	-876.6	30687.	6263.
506	664.1	157.63	-837.9	29854.	5341.
515	772.6	160.32	-801.8	28966.	4570.
524	894.0	163.22	-767.9	28012.	3918.
533	1029.	166.38	-736.2	26974.	3363.
542	1179.	169.86	-706.4	26514.	2958.
550	1326.	173.28	-681.4	25564.	2597.
559	1507.	177.58	-654.7	24421.	2244.
568	1707.	182.49	-629.6	23182.	1938.
577	1926.	188.19	-605.8	21826.	1670.
586	2167.	194.96	-583.3	20320.	1433.
595	2431.	203.24	-561.9	18615.	1223.
603	2687.	212.41	-543.8	16867.	1053.
612	3000.	225.86	-524.4	14506.	875.6
621	3345.		-505.9		702.7
630	3725.		-488.3		476.6

COMPOUND CONSTANTS:

MP   225.306 K   TC   617.046 K   VC 373.72 ML/MOL   ZC 0.2579
NBP 412.267 K   PC   3.541 MPA   DC   0.2841 G/ML   OM 0.3263
MU   0.3300 DEBYE   RG   3.8966 ANGSTROM

   VAPOR PRESSURE CORRELATION CONSTANTS, K AND KPA:

   WAGNER   WAGNER   WAGNER   WAGNER
   264. - 617. K   264. - 421. K   264. - 521. K   392. - 617. K
    90 POINTS   81 POINTS   86 POINTS   20 POINTS
   RMSD = 0.3843   RMSD = 0.0194   RMSD = 0.1471   RMSD = 0.2911
   -0.75949384E+01   -0.73565011E+01   -0.75672626E+01   -0.76312483E+01
    0.13851325E+01   0.85481262E+00   0.13220356E+01   0.14824278E+01
   -0.31689679E+01   -0.25107059E+01   -0.30855605E+01   -0.33377901E+01
   -0.24208452E+01   -0.32979650E+01   -0.25423557E+01   -0.24392934E+01

   LIQUID DENSITY CORRELATION CONSTANTS, K AND G/ML:

   FRANCIS1   FRANCIS1   FRANCIS1   RACKETT
   233. - 584. K   233. - 424. K   233. - 584. K   573. - 617. K
    57 POINTS   53 POINTS   57 POINTS   7 POINTS
   RMSD = 0.00042   RMSD = 0.00036   RMSD = 0.00042   RMSD = 0.00133
   0.11224260E+01   0.11156979E+01   0.11224260E+01   0.25794342E+00
   0.74312417E-03   0.78894501E-03   0.74312417E-03
   0.15000000E+02   0.60660095E+01   0.15000000E+02
   0.66854980E+03   0.59869238E+03   0.66854980E+03

   SECOND VIRIAL COEFFICIENT CORRELATION CONSTANTS, ML/MOL:

   TSONOPOULOS: TC=617.046 PC=3.541 VC=373.72 OM=0.3263 A= 0.000000 B=0.0000

LITERATURE DOCUMENTS FOR CORRELATED VAPOR PRESSURE VALUES:

   162   708   1098   1146   1485   1486   1499   2925   4392   6632 15532 40270 41221

LITERATURE DOCUMENTS FOR CORRELATED LIQUID DENSITY VALUES:

   182   1776   2852   2959   2975   3610   5364   5990   6838   7904   9746 10193 10809
   13621 17885 20540 21441

LITERATURE DOCUMENTS REPORTING VIRIAL COEFFICIENT DATA:

   1497   2410

T,K	VAPOR PRESSURE, KPA	SATURATED LIQUID VOLUME, ML/MOL	SECOND VIRIAL COEFFICIENT, ML/MOL	HEAT OF VAPORIZATION, J/MOL	SATURATED VAPOR VOLUME ML/MOL
233		116.05	-16654.		
241		116.89	-13534.		
250		117.86	-10900.		
259		118.84	-8930.		
267	0.1436	119.74	-7579.	44501.	0.1545E+08
276	0.2751	120.77	-6390.	43949.	8336386.
285	0.5020	121.82	-5461.	43408.	4714518.
294	0.8773	122.89	-4725.	42878.	2781559.
302	1.394	123.87	-4193.	42415.	1796969.
311	2.269	125.00	-3701.	41901.	1135705.
320	3.576	126.15	-3297.	41393.	740654.
328	5.227	127.20	-2996.	40945.	518775.
337	7.807	128.41	-2708.	40445.	356157.
346	11.38	129.66	-2464.	39944.	250355.
355	16.21	130.94	-2256.	39443.	179799.
363	21.82	132.12	-2094.	38994.	136187.
372	29.94	133.48	-1935.	38484.	101332.
381	40.36	134.88	-1795.	37967.	76642.
390	53.54	136.33	-1672.	37440.	58840.
398	67.97	137.66	-1574.	36962.	47056.
407	87.74	139.21	-1475.	36411.	37031.
416	111.8	140.83	-1386.	35845.	29478.
424	137.3	142.31	-1314.	35326.	24280.
433	171.3	144.06	-1240.	34724.	19697.
442	211.4	145.88	-1173.	34099.	16119.
451	258.5	147.80	-1112.	33449.	13296.
459	306.7	149.58	-1061.	32847.	11271.
468	368.9	151.70	-1009.	32139.	9417.
477	440.4	153.94	-960.3	31394.	7912.
486	521.9	156.34	-915.4	30606.	6681.
494	603.6	158.60	-878.2	29864.	5769.
503	706.6	161.34	-839.0	28974.	4906.
512	822.4	164.31	-802.4	28013.	4183.
520	936.9	167.19	-771.8	27085.	3635.
529	1080.	170.73	-739.5	26592.	3177.
538	1238.	174.70	-709.1	25502.	2730.
547	1414.	179.18	-680.5	24327.	2347.
555	1585.	183.71	-656.5	23198.	2051.
564	1797.	189.64	-630.9	21811.	1759.
573	2030.	196.74	-606.7	20270.	1503.
582	2285.	205.50	-583.8	18524.	1277.
590	2533.		-564.4		1095.
599	2837.		-543.6		905.7
608	3171.		-523.9		722.5
616	3496.		-507.1		524.8

COMPOUND CONSTANTS:

MP 286.405 K	TC 616.226 K	VC 378.22 ML/MOL	ZC 0.2592
NBP 411.516 K	PC 3.511 MPA	DC 0.2807 G/ML	OM 0.3215
MU 0.0500 DEBYE	RG 3.7962 ANGSTROM		

VAPOR PRESSURE CORRELATION CONSTANTS, K AND KPA:

WAGNER	RIEDEL	VAPRES-2	WAGNER
287. - 616. K	287. - 428. K	287. - 524. K	391. - 616. K
114 POINTS	86 POINTS	103 POINTS	48 POINTS
RMSD = 0.8205	RMSD = 0.0140	RMSD = 0.1614	RMSD = 0.3430
-0.76547358E+01	0.60475817E+02	0.54819207E+02	-0.75674246E+01
0.15715207E+01	-0.64629700E+01	-0.70361813E+04	0.12906932E+01
-0.33449053E+01	-0.70060504E+04	-0.10991850E-01	-0.24421482E+01
-0.22123689E+01	0.15150764E-16	0.10204513E-04	-0.82109545E+01
		-0.50345590E+01	

LIQUID DENSITY CORRELATION CONSTANTS, K AND G/ML:

FRANCIS1	FRANCIS1	FRANCIS1
273. - 491. K	273. - 421. K	273. - 491. K
89 POINTS	85 POINTS	89 POINTS
RMSD = 0.00029	RMSD = 0.00027	RMSD = 0.00029
0.11177034E+01	0.11135645E+01	0.11177034E+01
0.75861323E-03	0.79222070E-03	0.75861323E-03
0.11861958E+02	0.59999990E+01	0.11861958E+02
0.63896191E+03	0.58869189E+03	0.63896191E+03

SECOND VIRIAL COEFFICIENT CORRELATION CONSTANTS, ML/MOL:

TSONOPOULOS: TC=616.226 PC=3.511 VC=378.22 OM=0.3215 A= 0.000000 B=0.0000

LITERATURE DOCUMENTS FOR CORRELATED VAPOR PRESSURE VALUES:

   162   708 1098 1146 1485 1486 1499 2851 2925 5718 7333 7820 15532
 15843 20447 40204 40292 41188 41221

LITERATURE DOCUMENTS FOR CORRELATED LIQUID DENSITY VALUES:

    47   76   146   147   182 1099 1487 1490 1776 2573 2959 2975 4015
  4612 5364 6838 7904 8069 10193 10809 11040 12195 12709 13276 13352 13877
 14292 14586 17145 19607 21816 22720 40928 41188

LITERATURE DOCUMENTS REPORTING VIRIAL COEFFICIENT DATA:

 1497 2410

T,K	VAPOR PRESSURE, KPA	SATURATED LIQUID VOLUME, ML/MOL	SECOND VIRIAL COEFFICIENT, ML/MOL	HEAT OF VAPORIZATION, J/MOL	SATURATED VAPOR VOLUME, ML/MOL
273		120.89			
280		121.72			
288	0.6423	122.68	-5159.	42937.	3722988.
296	1.041	123.66	-4552.	42472.	2359733.
304	1.637	124.66	-4050.	42013.	1539954.
311	2.378	125.55	-3681.	41615.	1083535.
319	3.559	126.60	-3322.	41165.	741908.
327	5.203	127.66	-3017.	40717.	519561.
335	7.443	128.76	-2757.	40271.	371436.
343	10.44	129.88	-2532.	39825.	270614.
350	13.83	130.89	-2360.	39434.	208077.
358	18.76	132.06	-2187.	38985.	156438.
366	25.06	133.28	-2034.	38532.	119375.
374	32.98	134.52	-1898.	38075.	92347.
382	42.82	135.81	-1778.	37611.	72343.
389	53.26	136.96	-1683.	37199.	58990.
397	67.60	138.33	-1584.	36719.	47193.
405	84.84	139.74	-1495.	36230.	38133.
413	105.4	141.20	-1414.	35728.	31097.
421	129.7	142.72	-1340.	35214.	25574.
428	154.4	144.09	-1280.	34752.	21691.
436	186.9	145.73	-1217.	34209.	18091.
444	224.5	147.45	-1159.	33648.	15189.
452	267.7	149.24	-1106.	33067.	12829.
460	317.0	151.14	-1056.	32464.	10896.
467	365.6	152.88	-1015.	31917.	9484.
475	427.9	154.98	-971.9	31264.	8127.
483	497.9	157.23	-931.2	30581.	6991.
491	576.3	159.64	-893.2	29860.	6035.
499	663.7		-857.4		5225.
506	748.1		-827.9		4615.
514	854.2		-796.1		4010.
522	971.3		-766.0		3486.
530	1100.		-737.6		3106.
538	1242.		-710.8		2716.
545	1376.		-688.4		2416.
553	1544.		-664.1		2112.
561	1726.		-641.0		1844.
569	1925.		-619.0		1606.
577	2142.		-598.1		1393.
584	2346.		-580.5		1224.
592	2599.		-561.4		1046.
600	2874.		-543.1		878.5
608	3173.		-525.6		713.1
616	3501.		-508.8		490.0

COMPOUND CONSTANTS:

MP	242.540 K	TC	VC	ZC
NBP	418.387 K	PC	DC	OM
MU	0.1300 DEBYE	RG		

VAPOR PRESSURE CORRELATION CONSTANTS, K AND KPA:

VAPRES-2	VAPRES-2	VAPRES-2	RIEDEL
246. - 419. K	246. - 419. K	246. - 419. K	403. - 419. K
46 POINTS	46 POINTS	46 POINTS	8 POINTS
RMSD = 0.5948	RMSD = 0.5948	RMSD = 0.5948	RMSD = 0.0489
0.54438280E+04	0.54438280E+04	0.54438280E+04	-0.17955204E+04
-0.12147271E+06	-0.12147271E+06	-0.12147271E+06	0.26447070E+03
0.29499094E+01	0.29499094E+01	0.29499094E+01	0.88199172E+05
-0.14248823E-02	-0.14248823E-02	-0.14248823E-02	-0.13277246E-14
-0.10161135E+04	-0.10161135E+04	-0.10161135E+04	

LIQUID DENSITY CORRELATION CONSTANTS, K AND G/ML:

FRANCIS1	FRANCIS1	FRANCIS1
286. - 419. K	286. - 419. K	286. - 419. K
27 POINTS	27 POINTS	27 POINTS
RMSD = 0.00024	RMSD = 0.00024	RMSD = 0.00024
0.11879330E+01	0.11879330E+01	0.11879330E+01
0.90346299E-03	0.90346299E-03	0.90346299E-03
0.59999990E+01	0.59999990E+01	0.59999990E+01
0.64551636E+03	0.64551636E+03	0.64551636E+03

SECOND VIRIAL COEFFICIENT CORRELATION CONSTANTS, ML/MOL:

LITERATURE DOCUMENTS FOR CORRELATED VAPOR PRESSURE VALUES:

1146   1360   1758   4298   4362   4392   6309   6520   6659 14448 15542 19427 40733

LITERATURE DOCUMENTS FOR CORRELATED LIQUID DENSITY VALUES:

1360 19467 20809 22370

LITERATURE DOCUMENTS REPORTING VIRIAL COEFFICIENT DATA:

T,K	VAPOR PRESSURE, KPA	SATURATED LIQUID VOLUME, ML/MOL	SECOND VIRIAL COEFFICIENT, ML/MOL	HEAT OF VAPORIZATION, J/MOL	SATURATED VAPOR VOLUME, ML/MOL
246	0.01051				0.1947E+09
249	0.01516				0.1366E+09
253	0.02396				0.8778E+08
257	0.03670				0.5822E+08
261	0.05464				0.3972E+08
265	0.07928				0.2779E+08
269	0.1124				0.1989E+08
273	0.1562				0.1453E+08
277	0.2130				0.1081E+08
281	0.2856				8180068.
285	0.3774				6278989.
289	0.4920	114.45		44971.	4883632.
293	0.6338	114.93		44198.	3843455.
297	0.8078	115.42		43543.	3056995.
301	1.020	115.91		42997.	2454649.
304	1.207	116.28		42654.	2093918.
308	1.501	116.78		42278.	1705630.
312	1.854	117.28		41985.	1399235.
316	2.274	117.79		41768.	1155264.
320	2.773	118.30		41617.	959397.
324	3.363	118.82		41523.	800975.
328	4.058	119.34		41476.	671973.
332	4.875	119.87			566286.
336	5.830	120.41			479219.
340	6.943	120.95			407134.
344	8.238	121.50			347182.
348	9.738	122.05			297116.
352	11.47	122.61			255150.
356	13.46	123.17			219854.
360	15.75	123.74			190075.
363	17.67	124.18			170793.
367	20.54	124.76			148529.
371	23.80	125.35			129608.
375	27.47	125.95			113493.
379	31.59	126.55			99738.
383	36.20	127.16			87975.
387	41.31	127.78			77899.
391	46.94	128.41			69254.
395	53.12	129.04			61825.
399	59.84	129.68			55435.
403	67.10	130.34			49933.
407	74.88	131.00			45193.
411	83.13	131.67			41109.
415	91.79	132.34			37590.
418	98.52	132.86			35276.

COMPOUND CONSTANTS:

MP   173.611 K   TC   638.317 K   VC   440.01 ML/MOL   ZC
NBP   432.392 K   PC   3.200 MPA   DC   0.2732 G/ML   OM
MU   0.3600 DEBYE   RG

VAPOR PRESSURE CORRELATION CONSTANTS, K AND KPA:

WAGNER	WAGNER	WAGNER	WAGNER
266. - 434. K	266. - 434. K	266. - 434. K	416. - 434. K
40 POINTS	40 POINTS	40 POINTS	10 POINTS
RMSD = 0.0245	RMSD = 0.0245	RMSD = 0.0245	RMSD = 0.0045
-0.81945655E+01	-0.81945655E+01	-0.81945655E+01	-0.34120454E+01
0.25824838E+01	0.25824838E+01	0.25824838E+01	-0.95641753E+01
-0.49994036E+01	-0.49994036E+01	-0.49994036E+01	0.18001220E+02
-0.70133165E+00	-0.70133165E+00	-0.70133165E+00	-0.79735756E+02

LIQUID DENSITY CORRELATION CONSTANTS, K AND G/ML:

FRANCIS1	FRANCIS1	FRANCIS1
178. - 432. K	178. - 432. K	178. - 432. K
31 POINTS	31 POINTS	31 POINTS
RMSD = 0.00025	RMSD = 0.00025	RMSD = 0.00025
0.11103115E+01	0.11103115E+01	0.11103115E+01
0.78414031E-03	0.78414031E-03	0.78414031E-03
0.59999990E+01	0.59999990E+01	0.59999990E+01
0.62005029E+03	0.62005029E+03	0.62005029E+03

SECOND VIRIAL COEFFICIENT CORRELATION CONSTANTS, ML/MOL:

LITERATURE DOCUMENTS FOR CORRELATED VAPOR PRESSURE VALUES:

   1485   1486   3571   4392

LITERATURE DOCUMENTS FOR CORRELATED LIQUID DENSITY VALUES:

   182   1490   2426   5364   5866   13877

LITERATURE DOCUMENTS REPORTING VIRIAL COEFFICIENT DATA:

T,K	VAPOR PRESSURE, KPA	SATURATED LIQUID VOLUME, ML/MOL	SECOND VIRIAL COEFFICIENT, ML/MOL	HEAT OF VAPORIZATION, J/MOL	SATURATED VAPOR VOLUME, ML/MOL
178		125.57			
183		126.11			
189		126.76			
195		127.42			
201		128.09			
207		128.76			
212		129.33			
218		130.02			
224		130.71			
230		131.42			
236		132.13			
241		132.74			
247		133.47			
253		134.21			
259		134.96			
265		135.72			
271	0.06824	136.49		47862.	0.3302E+08
276	0.1002	137.14		47561.	0.2291E+08
282	0.1554	137.93		47201.	0.1508E+08
288	0.2360	138.73		46842.	0.1014E+08
294	0.3514	139.54		46484.	6954950.
300	0.5133	140.36		46128.	4857456.
305	0.6945	141.06		45832.	3649410.
311	0.9832	141.90		45477.	2628149.
317	1.370	142.76		45124.	1921763.
323	1.882	143.63		44772.	1425486.
329	2.548	144.51		44420.	1071670.
335	3.407	145.41		44069.	815909.
340	4.298	146.17		43776.	656008.
346	5.621	147.10		43425.	510160.
352	7.270	148.04		43073.	400972.
358	9.306	149.00		42720.	318320.
364	11.79	149.97		42367.	255095.
369	14.27	150.80		42071.	213561.
375	17.78	151.80		41714.	173884.
381	21.98	152.83		41356.	142710.
387	26.95	153.88		40995.	118008.
393	32.80	154.95		40631.	98276.
399	39.63	156.04		40264.	82393.
404	46.16	156.96		39956.	71481.
410	55.10	158.10		39582.	60608.
416	65.37	159.26		39205.	51682.
422	77.10	160.45		38824.	44307.
428	90.43	161.67		38438.	38176.
433	102.9				33843.

COMPOUND CONSTANTS:

MP	177.147 K	TC	631.127 K	VC	429.12 ML/MOL	ZC	
NBP	425.563 K	PC	3.209 MPA	DC	0.2801 G/ML	OM	
MU	0.3800 DEBYE	RG	4.1870 ANGSTROM				

VAPOR PRESSURE CORRELATION CONSTANTS, K AND KPA:

WAGNER	WAGNER	WAGNER	WAGNER
264. - 427. K	264. - 427. K	264. - 427. K	410. - 427. K
48 POINTS	48 POINTS	48 POINTS	10 POINTS
RMSD = 0.0132	RMSD = 0.0132	RMSD = 0.0132	RMSD = 0.0049
-0.73149189E+01	-0.73149189E+01	-0.73149189E+01	-0.56302443E+01
0.81706145E+00	0.81706145E+00	0.81706145E+00	-0.35368559E+01
-0.27135490E+01	-0.27135490E+01	-0.27135490E+01	0.59233182E+01
-0.45270910E+01	-0.45270910E+01	-0.45270910E+01	-0.36297026E+02

LIQUID DENSITY CORRELATION CONSTANTS, K AND G/ML:

FRANCIS1	FRANCIS1	FRANCIS1	FRANCIS1
253. - 491. K	253. - 441. K	253. - 491. K	420. - 491. K
26 POINTS	22 POINTS	26 POINTS	6 POINTS
RMSD = 0.00046	RMSD = 0.00049	RMSD = 0.00046	RMSD = 0.00001
0.11156139E+01	0.11173906E+01	0.11156139E+01	0.11207809E+01
0.80124475E-03	0.80996333E-03	0.80124475E-03	0.78714709E-03
0.59999990E+01	0.59999990E+01	0.59999990E+01	0.95286846E+01
0.61362866E+03	0.62711011E+03	0.61362866E+03	0.64744434E+03

SECOND VIRIAL COEFFICIENT CORRELATION CONSTANTS, ML/MOL:

LITERATURE DOCUMENTS FOR CORRELATED VAPOR PRESSURE VALUES:

  1485   1486   3571   4392   7814 19427 40270

LITERATURE DOCUMENTS FOR CORRELATED LIQUID DENSITY VALUES:

  182   929 1487 1490 8654 10193

LITERATURE DOCUMENTS REPORTING VIRIAL COEFFICIENT DATA:

T,K	VAPOR PRESSURE, KPA	SATURATED LIQUID VOLUME, ML/MOL	SECOND VIRIAL COEFFICIENT, ML/MOL	HEAT OF VAPORIZATION, J/MOL	SATURATED VAPOR VOLUME, ML/MOL
253		134.11			
258		134.74			
263		135.39			
269	0.08029	136.17		47750.	0.2785E+08
274	0.1184	136.83		47341.	0.1925E+08
280	0.1843	137.63		46861.	0.1263E+08
285	0.2620	138.31		46471.	9040555.
290	0.3670	139.00		46089.	6567884.
296	0.5397	139.83		45641.	4557637.
301	0.7337	140.54		45277.	3408997.
307	1.043	141.39		44849.	2444350.
312	1.382	142.12		44501.	1875673.
317	1.810	142.85		44159.	1454615.
323	2.468	143.74		43758.	1086344.
328	3.163	144.50		43430.	860589.
334	4.208	145.42		43045.	658317.
339	5.288	146.20		42729.	531410.
344	6.591	146.98		42418.	432336.
350	8.500	147.95		42050.	340819.
355	10.42	148.76		41747.	281691.
361	13.19	149.75		41388.	226064.
366	15.94	150.59		41091.	189485.
371	19.14	151.45		40797.	159759.
377	23.65	152.49		40446.	131132.
382	28.05	153.38		40154.	111880.
388	34.17	154.46		39804.	93076.
393	40.07	155.38		39512.	80257.
399	48.19	156.51		39161.	67576.
404	55.92	157.47		38867.	58823.
409	64.61	158.45		38571.	51407.
415	76.43	159.64		38213.	43949.
420	87.53	160.66		37911.	38719.
426	102.5	161.92		37544.	33406.
431		162.99			
436		164.09			
442		165.44			
447		166.60			
453		168.03			
458		169.27			
463		170.54			
469		172.11			
474		173.48			
480		175.18			
485		176.66			
490		178.20			

COMPOUND CONSTANTS:

MP   192.349 K   TC 653.226 K   VC 429.26 ML/MOL   ZC
NBP   438.330 K   PC   3.141 MPA   DC 0.2800 G/ML   OM
MU   0.5600 DEBYE   RG   4.1296 ANGSTROM

    VAPOR PRESSURE CORRELATION CONSTANTS, K AND KPA:

    WAGNER   WAGNER   WAGNER   WAGNER
    354. - 440. K   354. - 440. K   354. - 440. K   422. - 440. K
    18 POINTS   18 POINTS   18 POINTS   5 POINTS
    RMSD = 0.0171   RMSD = 0.0171   RMSD = 0.0171   RMSD = 0.0027
    -0.74649175E+01   -0.74649175E+01   -0.74649175E+01   -0.11087258E+02
    0.19385796E+01   0.19385796E+01   0.19385796E+01   0.11365171E+02
    -0.62277861E+01   -0.62277861E+01   -0.62277861E+01   -0.25261272E+02
    0.57335924E+01   0.57335924E+01   0.57335924E+01   0.77355994E+02

    LIQUID DENSITY CORRELATION CONSTANTS, K AND G/ML:

    FRANCIS1   FRANCIS1   FRANCIS1
    293. - 436. K   293. - 436. K   293. - 436. K
     4 POINTS   4 POINTS   4 POINTS
    RMSD = 0.00002   RMSD = 0.00002   RMSD = 0.00002
    0.11118250E+01   0.11118250E+01   0.11118250E+01
    0.68589277E-03   0.68589277E-03   0.68589277E-03
    0.72085600E+01   0.72085600E+01   0.72085600E+01
    0.53253345E+03   0.53253345E+03   0.53253345E+03

    SECOND VIRIAL COEFFICIENT CORRELATION CONSTANTS, ML/MOL:

LITERATURE DOCUMENTS FOR CORRELATED VAPOR PRESSURE VALUES:

    1486

LITERATURE DOCUMENTS FOR CORRELATED LIQUID DENSITY VALUES:

    1490 22600

LITERATURE DOCUMENTS REPORTING VIRIAL COEFFICIENT DATA:

T,K	VAPOR PRESSURE, KPA	SATURATED LIQUID VOLUME, ML/MOL	SECOND VIRIAL COEFFICIENT, ML/MOL	HEAT OF VAPORIZATION, J/MOL	SATURATED VAPOR VOLUME, ML/MOL
293		136.47			
296		136.84			
299		137.23			
303		137.74			
306		138.13			
309		138.53			
313		139.06			
316		139.47			
319		139.88			
323		140.43			
326		140.85			
329		141.28			
333		141.86			
336		142.29			
339		142.74			
343		143.34			
346		143.80			
349		144.26			
353		144.89			
356	6.879	145.37		43692.	428608.
359	7.785	145.86		43569.	381752.
363	9.148	146.52		43399.	328284.
366	10.30	147.03		43267.	293899.
369	11.57	147.54		43132.	263672.
373	13.46	148.24		42945.	228877.
376	15.04	148.78		42800.	206309.
379	16.77	149.32		42651.	186328.
383	19.34	150.07		42447.	163142.
386	21.47	150.64		42289.	147984.
389	23.79	151.23		42128.	134474.
393	27.20	152.03		41907.	118680.
396	30.01	152.64		41737.	108278.
399	33.05	153.28		41563.	98950.
403	37.49	154.14		41326.	87970.
406	41.13	154.81		41144.	80688.
409	45.05	155.50		40958.	74122.
413	50.73	156.45		40705.	66341.
416	55.36	157.19		40511.	61149.
419	60.32	157.95		40313.	56441.
423	67.47	159.01		40044.	50830.
426	73.26	159.84		39839.	47063.
429	79.44	160.70		39630.	43631.
433	88.30	161.90		39346.	39518.
436	95.45	162.84		39129.	36741.
439	103.0				34199.

COMPOUND CONSTANTS:

MP   177.625 K   TC   636.227 K   VC   429.26 ML/MOL   ZC
NBP   434.480 K   PC   3.141 MPA   DC   0.2800 G/ML   OM
MU   0.3300 DEBYE   RG   4.2845 ANGSTROM

   VAPOR PRESSURE CORRELATION CONSTANTS, K AND KPA:

   WAGNER   WAGNER   WAGNER   WAGNER
   351. - 436. K   351. - 436. K   351. - 436. K   418. - 436. K
     18 POINTS   18 POINTS   18 POINTS   5 POINTS
   RMSD = 0.0164   RMSD = 0.0164   RMSD = 0.0164   RMSD = 0.0043
   -0.88897002E+01   -0.88897002E+01   -0.88897002E+01   -0.19800710E+02
    0.39933572E+01   0.39933572E+01   0.39933572E+01   0.32396268E+02
   -0.77488142E+01   -0.77488142E+01   -0.77488142E+01   -0.65420388E+02
    0.75519735E+01   0.75519735E+01   0.75519735E+01   0.23098895E+03

   LIQUID DENSITY CORRELATION CONSTANTS, K AND G/ML:

   FRANCIS1   FRANCIS1   FRANCIS1
   293. - 304. K   293. - 304. K   293. - 304. K
     4 POINTS   4 POINTS   4 POINTS
   RMSD = 0.00004   RMSD = 0.00004   RMSD = 0.00004
   0.11052513E+01   0.11052513E+01   0.11052513E+01
   0.74692257E-03   0.74692257E-03   0.74692257E-03
   0.59999990E+01   0.59999990E+01   0.59999990E+01
   0.56838794E+03   0.56838794E+03   0.56838794E+03

   SECOND VIRIAL COEFFICIENT CORRELATION CONSTANTS, ML/MOL:

LITERATURE DOCUMENTS FOR CORRELATED VAPOR PRESSURE VALUES:

   1486

LITERATURE DOCUMENTS FOR CORRELATED LIQUID DENSITY VALUES:

   1490

LITERATURE DOCUMENTS REPORTING VIRIAL COEFFICIENT DATA:

T,K	VAPOR PRESSURE, KPA	SATURATED LIQUID VOLUME, ML/MOL	SECOND VIRIAL COEFFICIENT, ML/MOL	HEAT OF VAPORIZATION, J/MOL	SATURATED VAPOR VOLUME, ML/MOL
293		139.01			
296		139.41			
299		139.82			
302		140.22			
351	6.339				458745.
354	7.194				407558.
358	8.483				349333.
361	9.573				312006.
364	10.78				279276.
367	12.11				250512.
371	14.09				217415.
374	15.75				195954.
377	17.57				176959.
380	19.56				160110.
384	22.50				140520.
387	24.93				127686.
390	27.57				116229.
393	30.44				105981.
397	34.64				93951.
400	38.09				85997.
403	41.81				78840.
406	45.82				72389.
410	51.64				64752.
413	56.38				59658.
416	61.46				55043.
419	66.89				50854.
423	74.73				45855.
426	81.07				42495.
429	87.82				39431.
432	95.01				36632.
436	105.3				33268.

COMPOUND CONSTANTS:

MP 210.832 K	TC 636.227 K	VC 429.26 ML/MOL	ZC
NBP 435.166 K	PC 3.141 MPA	DC 0.2800 G/ML	OM
MU 0.0000 DEBYE	RG 4.1662 ANGSTROM		

VAPOR PRESSURE CORRELATION CONSTANTS, K AND KPA:

WAGNER	WAGNER	WAGNER	WAGNER
351. - 437. K	351. - 437. K	351. - 437. K	419. - 437. K
16 POINTS	16 POINTS	16 POINTS	5 POINTS
RMSD = 0.0086	RMSD = 0.0086	RMSD = 0.0086	RMSD = 0.0001
-0.87786588E+01	-0.87786588E+01	-0.87786588E+01	-0.11299575E+02
0.35075752E+01	0.35075752E+01	0.35075752E+01	0.10135800E+02
-0.63821475E+01	-0.63821475E+01	-0.63821475E+01	-0.20221747E+02
0.37058144E+01	0.37058144E+01	0.37058144E+01	0.59926242E+02

LIQUID DENSITY CORRELATION CONSTANTS, K AND G/ML:

FRANCIS1	FRANCIS1	FRANCIS1
293. - 436. K	293. - 436. K	293. - 436. K
4 POINTS	4 POINTS	4 POINTS
RMSD = 0.00003	RMSD = 0.00003	RMSD = 0.00003
0.11046152E+01	0.11046152E+01	0.11046152E+01
0.75632008E-03	0.75632008E-03	0.75632008E-03
0.77651434E+01	0.77651434E+01	0.77651434E+01
0.64978003E+03	0.64978003E+03	0.64978003E+03

SECOND VIRIAL COEFFICIENT CORRELATION CONSTANTS, ML/MOL:

LITERATURE DOCUMENTS FOR CORRELATED VAPOR PRESSURE VALUES:

1486

LITERATURE DOCUMENTS FOR CORRELATED LIQUID DENSITY VALUES:

1490 5866

LITERATURE DOCUMENTS REPORTING VIRIAL COEFFICIENT DATA:

T,K	VAPOR PRESSURE, KPA	SATURATED LIQUID VOLUME, ML/MOL	SECOND VIRIAL COEFFICIENT, ML/MOL	HEAT OF VAPORIZATION, J/MOL	SATURATED VAPOR VOLUME, ML/MOL
293		139.56			
296		139.96			
299		140.36			
302		140.76			
306		141.31			
309		141.72			
312		142.13			
315		142.55			
319		143.11			
322		143.53			
325		143.96			
328		144.39			
332		144.97			
335		145.41			
338		145.85			
342		146.45			
345		146.90			
348		147.35			
351	6.292	147.81		43360.	462196.
355	7.441	148.43		43174.	395071.
358	8.415	148.90		43033.	352168.
361	9.493	149.37		42889.	314642.
364	10.68	149.84		42743.	281738.
368	12.46	150.48		42544.	243971.
371	13.96	150.97		42393.	219542.
374	15.59	151.46		42240.	197960.
378	18.02	152.11		42032.	172976.
381	20.04	152.61		41873.	156678.
384	22.24	153.12		41713.	142179.
387	24.63	153.63		41550.	129255.
391	28.15	154.32		41330.	114142.
394	31.04	154.84		41162.	104185.
397	34.18	155.36		40993.	95254.
400	37.56	155.89		40821.	87227.
404	42.49	156.61		40589.	77756.
407	46.52	157.16		40413.	71459.
410	50.85	157.71		40235.	65769.
414	57.12	158.45		39994.	59010.
417	62.22	159.01		39812.	54489.
420	67.67	159.58		39627.	50381.
423	73.50	160.16		39440.	46643.
427	81.88	160.94		39189.	42170.
430	88.64	161.54		38998.	39155.
433	95.84	162.14		38806.	36398.
436	103.5	162.75		38612.	33875.

COMPOUND CONSTANTS:

MP	247.790 K	TC	664.466 K	VC	429.01 ML/MOL	ZC	
NBP	449.266 K	PC	3.454 MPA	DC	0.2802 G/ML	OM	
MU	0.5600 DEBYE	RG	4.0996 ANGSTROM				

VAPOR PRESSURE CORRELATION CONSTANTS, K AND KPA:

WAGNER	WAGNER	WAGNER	WAGNER
255. - 451. K	255. - 451. K	255. - 451. K	433. - 451. K
23 POINTS	23 POINTS	23 POINTS	5 POINTS
RMSD = 0.0708	RMSD = 0.0708	RMSD = 0.0708	RMSD = 0.0029
-0.72170230E+01	-0.72170230E+01	-0.72170230E+01	-0.26347492E+02
0.86397293E-01	0.86397293E-01	0.86397293E-01	0.48838074E+02
-0.17310121E+01	-0.17310121E+01	-0.17310121E+01	-0.94927053E+02
-0.40459776E+01	-0.40459776E+01	-0.40459776E+01	0.32096337E+03

LIQUID DENSITY CORRELATION CONSTANTS, K AND G/ML:

FRANCIS1	FRANCIS1	FRANCIS1
273. - 304. K	273. - 304. K	273. - 304. K
4 POINTS	4 POINTS	4 POINTS
RMSD = 0.00051	RMSD = 0.00051	RMSD = 0.00051
0.11377935E+01	0.11377935E+01	0.11377935E+01
0.75626327E-03	0.75626327E-03	0.75626327E-03
0.59999990E+01	0.59999990E+01	0.59999990E+01
0.56838794E+03	0.56838794E+03	0.56838794E+03

SECOND VIRIAL COEFFICIENT CORRELATION CONSTANTS, ML/MOL:

LITERATURE DOCUMENTS FOR CORRELATED VAPOR PRESSURE VALUES:

1486   5041

LITERATURE DOCUMENTS FOR CORRELATED LIQUID DENSITY VALUES:

1490   6629

LITERATURE DOCUMENTS REPORTING VIRIAL COEFFICIENT DATA:

T,K	VAPOR PRESSURE, KPA	SATURATED LIQUID VOLUME, ML/MOL	SECOND VIRIAL COEFFICIENT, ML/MOL	HEAT OF VAPORIZATION, J/MOL	SATURATED VAPOR VOLUME, ML/MOL
255	0.006975				0.3040E+09
259	0.01015				0.2122E+09
263	0.01457				0.1501E+09
268	0.02249				0.9909E+08
272	0.03139				0.7204E+08
277	0.04687	132.41		50040.	0.4914E+08
281	0.06380	132.90		49764.	0.3662E+08
286	0.09247	133.51		49426.	0.2571E+08
290	0.1231	134.01		49161.	0.1959E+08
295	0.1737	134.63		48836.	0.1412E+08
299	0.2266	135.14		48581.	0.1097E+08
303	0.2931	135.65		48330.	8593947.
308	0.3998				6403288.
312	0.5082				5102494.
317	0.6790				3879640.
321	0.8496				3139539.
326	1.114				2431304.
330	1.374				1995171.
335	1.771				1571003.
339	2.156				1305644.
344	2.736				1043726.
348	3.291				877494.
352	3.939				741280.
357	4.899				604232.
361	5.804				515613.
366	7.128				425382.
370	8.363				366354.
375	10.15				305590.
379	11.81				265411.
384	14.19				223632.
388	16.36				195733.
393	19.47				166452.
397	22.29				146721.
401	25.44				129711.
406	29.89				111638.
410	33.89				99314.
415	39.49				86119.
419	44.49				77053.
424	51.47				67277.
428	57.66				60513.
433	66.24				53171.
437	73.81				48058.
442	84.25				42473.
446	93.42				38561.
450	103.4				35074.

COMPOUND CONSTANTS:

MP   229.376 K   TC   649.167 K   VC   429.01 ML/MOL   ZC
NBP   442.532 K   PC   3.232 MPA   DC   0.2802 G/ML   OM 0.3784
MU   0.3000 DEBYE   RG   4.1678 ANGSTROM

     VAPOR PRESSURE CORRELATION CONSTANTS, K AND KPA:

     WAGNER   WAGNER   WAGNER   WAGNER
     255. - 605. K   255. - 444. K   255. - 549. K   426. - 605. K
      31 POINTS   23 POINTS   27 POINTS   13 POINTS
     RMSD = 4.1890   RMSD = 0.0651   RMSD = 2.2820   RMSD = 3.1230
     -0.85061697E+01   -0.72930724E+01   -0.86094402E+01   -0.79343035E+01
      0.29054709E+01   0.16773088E+00   0.31528526E+01   0.91813594E+00
     -0.55096577E+01   -0.19949306E+01   -0.58759829E+01   0.17173491E+01
      0.11087622E+01   -0.35111082E+01   0.16745956E+01   -0.52965876E+02

     LIQUID DENSITY CORRELATION CONSTANTS, K AND G/ML:

     FRANCIS1   FRANCIS1   FRANCIS1
     273. - 369. K   273. - 369. K   273. - 369. K
      29 POINTS   29 POINTS   29 POINTS
     RMSD = 0.00013   RMSD = 0.00013   RMSD = 0.00013
     0.11163378E+01   0.11163378E+01   0.11163378E+01
     0.72690798E-03   0.72690798E-03   0.72690798E-03
     0.98063126E+01   0.98063126E+01   0.98063126E+01
     0.65306201E+03   0.65306201E+03   0.65306201E+03

     SECOND VIRIAL COEFFICIENT CORRELATION CONSTANTS, ML/MOL:

LITERATURE DOCUMENTS FOR CORRELATED VAPOR PRESSURE VALUES:

     1095   1486   5041

LITERATURE DOCUMENTS FOR CORRELATED LIQUID DENSITY VALUES:

     182   1490   3383 13877 20934 40292

LITERATURE DOCUMENTS REPORTING VIRIAL COEFFICIENT DATA:

T,K	VAPOR PRESSURE, KPA	SATURATED LIQUID VOLUME, ML/MOL	SECOND VIRIAL COEFFICIENT, ML/MOL	HEAT OF VAPORIZATION, J/MOL	SATURATED VAPOR VOLUME, ML/MOL
255	0.01040				0.2040E+09
262	0.01934				0.1126E+09
270	0.03763				0.5966E+08
278	0.07015	135.34		48409.	0.3295E+08
286	0.1258	136.32		48048.	0.1890E+08
294	0.2175	137.31		47680.	0.1124E+08
302	0.3640	138.33		47304.	6895849.
310	0.5909	139.37		46922.	4359659.
318	0.9327	140.43		46534.	2832474.
326	1.435	141.51		46138.	1887155.
334	2.154	142.61		45735.	1286890.
342	3.164	143.74		45324.	896613.
350	4.552	144.89		44906.	637238.
358	6.425	146.07		44479.	461310.
366	8.908	147.27		44043.	339697.
374	12.15				254132.
382	16.31				192928.
390	21.59				148471.
398	28.19				115710.
406	36.35				91240.
414	46.32				72731.
422	58.38				58564.
429	70.89				48825.
437	87.69				39986.
445	107.5				33014.
453	130.6				27462.
461	157.5				23002.
469	188.5				19390.
477	223.9				16441.
485	264.4				14015.
493	310.2				12006.
501	361.8				10331.
509	419.7				8925.
517	484.5				7738.
525	556.7				6731.
533	636.8				5870.
541	725.4				5132.
549	823.2				4495.
557	930.9				3942.
565	1049.				3460.
573	1179.				3038.
581	1321.				2667.
589	1477.				2338.
597	1647.				2046.
604	1808.				1815.

COMPOUND CONSTANTS:

MP   228.456 K   TC   637.357 K   VC   429.01 ML/MOL   ZC
NBP   437.876 K   PC   3.127 MPA   DC   0.2802 G/ML   OM
MU   0.0600 DEBYE   RG   4.3408 ANGSTROM

   VAPOR PRESSURE CORRELATION CONSTANTS, K AND KPA:

   WAGNER   WAGNER   WAGNER   WAGNER
   255. - 439. K   255. - 439. K   255. - 439. K   422. - 439. K
    26 POINTS   26 POINTS   26 POINTS   5 POINTS
   RMSD = 0.0189   RMSD = 0.0189   RMSD = 0.0189   RMSD = 0.0010
   -0.80183117E+01   -0.80183117E+01   -0.80183117E+01   -0.87514029E+01
    0.15435029E+01   0.15435029E+01   0.15435029E+01   0.34888729E+01
   -0.37426544E+01   -0.37426544E+01   -0.37426544E+01   -0.79559246E+01
   -0.23219599E+01   -0.23219599E+01   -0.23219599E+01   0.16487034E+02

   LIQUID DENSITY CORRELATION CONSTANTS, K AND G/ML:

   FRANCIS1   FRANCIS1   FRANCIS1
   277. - 438. K   277. - 438. K   277. - 438. K
    36 POINTS   36 POINTS   36 POINTS
   RMSD = 0.00041   RMSD = 0.00041   RMSD = 0.00041
   0.11081476E+01   0.11081476E+01   0.11081476E+01
   0.74045639E-03   0.74045639E-03   0.74045639E-03
   0.92136049E+01   0.92136049E+01   0.92136049E+01
   0.63423853E+03   0.63423853E+03   0.63423853E+03

   SECOND VIRIAL COEFFICIENT CORRELATION CONSTANTS, ML/MOL:

LITERATURE DOCUMENTS FOR CORRELATED VAPOR PRESSURE VALUES:

   1486   1499   5041

LITERATURE DOCUMENTS FOR CORRELATED LIQUID DENSITY VALUES:

   182   1490   5866 10317 10574 13877 17056 18243 40292

LITERATURE DOCUMENTS REPORTING VIRIAL COEFFICIENT DATA:

T,K	VAPOR PRESSURE, KPA	SATURATED LIQUID VOLUME, ML/MOL	SECOND VIRIAL COEFFICIENT, ML/MOL	HEAT OF VAPORIZATION, J/MOL	SATURATED VAPOR VOLUME, ML/MOL
255	0.01142				0.1857E+09
259	0.01649				0.1306E+09
263	0.02351				0.9300E+08
267	0.03310				0.6707E+08
271	0.04604				0.4894E+08
275	0.06332				0.3610E+08
280	0.09290	137.39		48908.	0.2506E+08
284	0.1248	137.91		48646.	0.1892E+08
288	0.1660	138.43		48387.	0.1442E+08
292	0.2188	138.95		48130.	0.1109E+08
296	0.2859	139.48		47876.	8607183.
300	0.3703	140.01		47625.	6734078.
305	0.5059	140.69		47314.	5010368.
309	0.6438	141.24		47068.	3988562.
313	0.8133	141.79		46824.	3197980.
317	1.020	142.35		46583.	2581753.
321	1.271	142.91		46344.	2098006.
326	1.658	143.62		46048.	1633056.
330	2.036	144.20		45813.	1345494.
334	2.487	144.79		45580.	1114927.
338	3.020	145.38		45349.	928952.
342	3.647	145.98		45119.	778080.
346	4.381	146.58		44891.	655010.
351	5.471	147.35		44607.	531773.
355	6.501	147.97		44381.	452446.
359	7.688	148.60		44156.	386658.
363	9.053	149.24		43932.	331842.
367	10.61	149.88		43708.	285962.
372	12.88	150.70		43428.	238734.
376	14.96	151.37		43205.	207505.
380	17.32	152.04		42981.	181010.
384	19.97	152.73		42757.	158442.
388	22.95	153.42		42532.	139149.
392	26.29	154.12		42307.	122595.
397	31.01	155.02		42024.	105095.
401	35.27	155.75		41796.	93219.
405	39.98	156.49		41567.	82917.
409	45.20	157.24		41337.	73953.
413	50.95	158.01		41104.	66130.
418	58.96	158.98		40811.	57707.
422	66.07	159.78		40574.	51886.
426	73.84	160.59		40335.	46760.
430	82.34	161.41		40093.	42234.
434	91.59	162.25		39848.	38227.
438	101.6	163.11		39601.	34670.

COMPOUND CONSTANTS:

MP   TC   VC   ZC
NBP   443.678 K   PC   DC   OM
MU   RG

    VAPOR PRESSURE CORRELATION CONSTANTS, K AND KPA:

    RPM2   RPM2   RPM2
    333. - 445. K   333. - 445. K   333. - 445. K
     4 POINTS   4 POINTS   4 POINTS
    RMSD = 0.0017   RMSD = 0.0017   RMSD = 0.0017
    0.21336727E+03   0.21336727E+03   0.21336727E+03
   -0.31474261E+05   -0.31474261E+05   -0.31474261E+05
   -0.49007389E+00   -0.49007389E+00   -0.49007389E+00
    0.40449795E-03   0.40449795E-03   0.40449795E-03

    LIQUID DENSITY CORRELATION CONSTANTS, K AND G/ML:

    SECOND VIRIAL COEFFICIENT CORRELATION CONSTANTS, ML/MOL:

LITERATURE DOCUMENTS FOR CORRELATED VAPOR PRESSURE VALUES:

  6295   8355

LITERATURE DOCUMENTS FOR CORRELATED LIQUID DENSITY VALUES:

LITERATURE DOCUMENTS REPORTING VIRIAL COEFFICIENT DATA:

T,K	VAPOR PRESSURE, KPA	SATURATED LIQUID VOLUME, ML/MOL	SECOND VIRIAL COEFFICIENT, ML/MOL	HEAT OF VAPORIZATION, J/MOL	SATURATED VAPOR VOLUME, ML/MOL
333	1.665				1663026.
335	1.886				1477104.
338	2.258				1244379.
340	2.536				1114522.
343	3.001				950235.
345	3.345				857593.
348	3.914				739219.
350	4.332				671810.
353	5.018				584875.
355	5.518				534913.
358	6.334				469919.
360	6.925				432245.
363	7.884				382836.
366	8.934				340603.
368	9.688				315812.
371	10.90				282906.
373	11.77				263449.
376	13.17				237444.
378	14.16				221961.
381	15.75				201130.
383	16.88				188648.
386	18.69				171750.
388	19.97				161561.
391	22.01				147688.
394	24.21				135294.
396	25.77				127753.
399	28.26				117395.
401	30.02				111059.
404	32.83				102313.
406	34.82				96937.
409	38.00				89480.
411	40.26				84876.
414	43.87				78463.
416	46.43				74487.
419	50.54				68927.
422	54.99				63806.
424	58.16				60613.
427	63.26				56124.
429	66.90				53316.
432	72.77				49359.
434	76.97				46878.
437	83.77				43374.
439	88.65				41172.
442	96.56				38057.
444	102.3				36097.

COMPOUND CONSTANTS:

MP	TC	VC	ZC
NBP 452.133 K	PC	DC	OM
MU 0.7200 DEBYE	RG		

VAPOR PRESSURE CORRELATION CONSTANTS, K AND KPA:

VAPRES-2	VAPRES-2	VAPRES-2
290. - 453. K	290. - 453. K	290. - 453. K
10 POINTS	10 POINTS	10 POINTS
RMSD = 0.0765	RMSD = 0.0765	RMSD = 0.0765
0.12616464E+04	0.12616464E+04	0.12616464E+04
-0.34633308E+05	-0.34633308E+05	-0.34633308E+05
0.59587129E+00	0.59587129E+00	0.59587129E+00
-0.26021838E-03	-0.26021838E-03	-0.26021838E-03
-0.22843503E+03	-0.22843503E+03	-0.22843503E+03

LIQUID DENSITY CORRELATION CONSTANTS, K AND G/ML:

FRANCIS1	FRANCIS1	FRANCIS1
277. - 299. K	277. - 299. K	277. - 299. K
7 POINTS	7 POINTS	7 POINTS
RMSD = 0.00118	RMSD = 0.00118	RMSD = 0.00118
0.12075224E+01	0.12075224E+01	0.12075224E+01
0.94412523E-03	0.94412523E-03	0.94412523E-03
0.59999990E+01	0.59999990E+01	0.59999990E+01
0.55901392E+03	0.55901392E+03	0.55901392E+03

SECOND VIRIAL COEFFICIENT CORRELATION CONSTANTS, ML/MOL:

LITERATURE DOCUMENTS FOR CORRELATED VAPOR PRESSURE VALUES:

1146

LITERATURE DOCUMENTS FOR CORRELATED LIQUID DENSITY VALUES:

182   8354

LITERATURE DOCUMENTS REPORTING VIRIAL COEFFICIENT DATA:

T,K	VAPOR PRESSURE, KPA	SATURATED LIQUID VOLUME, ML/MOL	SECOND VIRIAL COEFFICIENT, ML/MOL	HEAT OF VAPORIZATION, J/MOL	SATURATED VAPOR VOLUME, ML/MOL
277		127.80			
281		128.36			
285		128.94			
289		129.52			
293	0.1564	130.10		47939.	0.1558E+08
297	0.2036	130.70		47515.	0.1213E+08
301	0.2626				9530149.
305	0.3358				7550694.
309	0.4261				6030003.
313	0.5364				4851572.
317	0.6705				3930863.
321	0.8325				3205938.
325	1.027				2630991.
329	1.260				2171841.
333	1.536				1802771.
337	1.863				1504277.
341	2.248				1261450.
345	2.699				1062814.
349	3.226				899474.
353	3.839				764487.
357	4.550				652404.
361	5.370				558920.
365	6.314				480614.
369	7.397				414755.
373	8.635				359148.
377	10.05				312025.
381	11.65				271948.
385	13.46				237750.
389	15.51				208473.
393	17.82				183331.
397	20.42				161677.
401	23.32				142972.
405	26.56				126772.
409	30.17				112703.
413	34.18				100454.
417	38.62				89765.
421	43.53				80413.
425	48.93				72214.
429	54.87				65009.
433	61.37				58665.
437	68.47				53067.
441	76.20				48118.
445	84.60				43734.
449	93.70				39844.
453	103.5				36385.

607

COMPOUND CONSTANTS:

MP	249.964 K	TC	VC	ZC
NBP	438.576 K	PC	DC	OM
MU	0.7600 DEBYE	RG		

VAPOR PRESSURE CORRELATION CONSTANTS, K AND KPA:

VAPRES-2	VAPRES-2	VAPRES-2
303. - 439. K	303. - 439. K	303. - 439. K
13 POINTS	13 POINTS	13 POINTS
RMSD = 0.0010	RMSD = 0.0010	RMSD = 0.0010
0.21506776E+03	0.21506776E+03	0.21506776E+03
-0.10883936E+05	-0.10883936E+05	-0.10883936E+05
0.72374395E-01	0.72374395E-01	0.72374395E-01
-0.27205378E-04	-0.27205378E-04	-0.27205378E-04
-0.34871454E+02	-0.34871454E+02	-0.34871454E+02

LIQUID DENSITY CORRELATION CONSTANTS, K AND G/ML:

FRANCIS1	FRANCIS1	FRANCIS1
290. - 299. K	290. - 299. K	290. - 299. K
4 POINTS	4 POINTS	4 POINTS
RMSD = 0.00068	RMSD = 0.00068	RMSD = 0.00068
0.12240992E+01	0.12240992E+01	0.12240992E+01
0.95153786E-03	0.95153786E-03	0.95153786E-03
0.86577234E+01	0.86577234E+01	0.86577234E+01
0.53703955E+03	0.53703955E+03	0.53703955E+03

SECOND VIRIAL COEFFICIENT CORRELATION CONSTANTS, ML/MOL:

LITERATURE DOCUMENTS FOR CORRELATED VAPOR PRESSURE VALUES:

1360

LITERATURE DOCUMENTS FOR CORRELATED LIQUID DENSITY VALUES:

3269 20809

LITERATURE DOCUMENTS REPORTING VIRIAL COEFFICIENT DATA:

T,K	VAPOR PRESSURE, KPA	SATURATED LIQUID VOLUME, ML/MOL	SECOND VIRIAL COEFFICIENT, ML/MOL	HEAT OF VAPORIZATION, J/MOL	SATURATED VAPOR VOLUME, ML/MOL
290		129.42			
293		129.89			
296		130.36			
303	0.5131				4909713.
306	0.6119				4158066.
310	0.7690				3351838.
313	0.9087				2863928.
317	1.129				2334962.
320	1.323				2011459.
323	1.545				1738530.
327	1.890				1438483.
330	2.191				1252447.
334	2.655				1046092.
337	3.056				917020.
340	3.507				806100.
344	4.196				681619.
347	4.786				602858.
350	5.444				534531.
354	6.440				457035.
357	7.284				407489.
361	8.553				350909.
364	9.623				314493.
367	10.80				282452.
371	12.56				245533.
374	14.04				221559.
378	16.22				193773.
381	18.04				175627.
384	20.02				159468.
388	22.95				140595.
391	25.36				128175.
394	27.99				117047.
398	31.83				103956.
401	34.99				95283.
405	39.60				85033.
408	43.37				78210.
411	47.44				72038.
415	53.33				64701.
418	58.13				59787.
422	65.07				53921.
425	70.70				49978.
428	76.72				46381.
432	85.39				42065.
435	92.39				39148.
438	99.84				36475.

COMPOUND CONSTANTS:

MP   204.613 K   TC   VC   ZC
NBP   442.979 K   PC   DC   OM
MU   RG

VAPOR PRESSURE CORRELATION CONSTANTS, K AND KPA:

VAPRES-2   VAPRES-2   VAPRES-2
305. - 443. K   305. - 443. K   305. - 443. K
  10 POINTS   10 POINTS   10 POINTS
RMSD = 0.0063   RMSD = 0.0063   RMSD = 0.0063
 0.66149951E+03   0.66149951E+03   0.66149951E+03
-0.22008526E+05   -0.22008526E+05   -0.22008526E+05
 0.27888266E+00   0.27888266E+00   0.27888266E+00
-0.11648342E-03   -0.11648342E-03   -0.11648342E-03
-0.11616922E+03   -0.11616922E+03   -0.11616922E+03

LIQUID DENSITY CORRELATION CONSTANTS, K AND G/ML:

FRANCIS1   FRANCIS1   FRANCIS1
273. - 309. K   273. - 309. K   273. - 309. K
  5 POINTS   5 POINTS   5 POINTS
RMSD = 0.00001   RMSD = 0.00001   RMSD = 0.00001
0.11498384E+01   0.11498384E+01   0.11498384E+01
0.66463370E-03   0.66463370E-03   0.66463370E-03
0.10773781E+02   0.10773781E+02   0.10773781E+02
0.54303760E+03   0.54303760E+03   0.54303760E+03

SECOND VIRIAL COEFFICIENT CORRELATION CONSTANTS, ML/MOL:

LITERATURE DOCUMENTS FOR CORRELATED VAPOR PRESSURE VALUES:

   3263 6662

LITERATURE DOCUMENTS FOR CORRELATED LIQUID DENSITY VALUES:

   3269 6662

LITERATURE DOCUMENTS REPORTING VIRIAL COEFFICIENT DATA:

T,K	VAPOR PRESSURE, KPA	SATURATED LIQUID VOLUME, ML/MOL	SECOND VIRIAL COEFFICIENT, ML/MOL	HEAT OF VAPORIZATION, J/MOL	SATURATED VAPOR VOLUME, ML/MOL
273		127.28			
276		127.61			
280		128.07			
284		128.53			
288		128.99			
292		129.46			
296		129.94			
300		130.42			
303		130.79			
307	0.4341	131.29		48956.	5879588.
311	0.5551				4658218.
315	0.7042				3719315.
319	0.8866				2991617.
323	1.108				2423206.
327	1.376				1975909.
330	1.612				1702601.
334	1.979				1403526.
338	2.415				1163747.
342	2.931				970309.
346	3.537				813329.
350	4.247				685213.
354	5.074				580084.
357	5.780				513533.
361	6.849				438236.
365	8.080				375596.
369	9.491				323245.
373	11.10				279300.
377	12.94				242253.
381	15.02				210896.
385	17.37				184250.
388	19.33				166874.
392	22.22				146654.
396	25.46				129300.
400	29.08				114355.
404	33.11				101443.
408	37.59				90252.
412	42.54				80523.
415	46.59				74056.
419	52.47				66389.
423	58.94				59672.
427	66.02				53773.
431	73.77				48577.
435	82.22				43990.
439	91.41				39931.
442	98.81				37192.

COMPOUND CONSTANTS:

MP   186.840 K   TC   VC   ZC
NBP 444.662 K   PC   DC   OM
MU   RG

    VAPOR PRESSURE CORRELATION CONSTANTS, K AND KPA:

    VAPRES-2   VAPRES-2   VAPRES-2
    314. - 445. K   314. - 445. K   314. - 445. K
     18 POINTS   18 POINTS   18 POINTS
    RMSD = 0.1668   RMSD = 0.1668   RMSD = 0.1668
     0.47207427E+04   0.47207427E+04   0.47207427E+04
    -0.11385034E+06   -0.11385034E+06   -0.11385034E+06
     0.23166926E+01   0.23166926E+01   0.23166926E+01
    -0.10349317E-02   -0.10349317E-02   -0.10349317E-02
    -0.86687353E+03   -0.86687353E+03   -0.86687353E+03

    LIQUID DENSITY CORRELATION CONSTANTS, K AND G/ML:

    FRANCIS1   FRANCIS1   FRANCIS1
    273. - 309. K   273. - 309. K   273. - 309. K
     5 POINTS   5 POINTS   5 POINTS
    RMSD = 0.00002   RMSD = 0.00002   RMSD = 0.00002
    0.11503897E+01   0.11503897E+01   0.11503897E+01
    0.72599528E-03   0.72599528E-03   0.72599528E-03
    0.61867218E+01   0.61867218E+01   0.61867218E+01
    0.53300977E+03   0.53300977E+03   0.53300977E+03

    SECOND VIRIAL COEFFICIENT CORRELATION CONSTANTS, ML/MOL:

LITERATURE DOCUMENTS FOR CORRELATED VAPOR PRESSURE VALUES:

    3269   6662 14448

LITERATURE DOCUMENTS FOR CORRELATED LIQUID DENSITY VALUES:

    3269   6662

LITERATURE DOCUMENTS REPORTING VIRIAL COEFFICIENT DATA:

T,K	VAPOR PRESSURE, KPA	SATURATED LIQUID VOLUME, ML/MOL	SECOND VIRIAL COEFFICIENT, ML/MOL	HEAT OF VAPORIZATION, J/MOL	SATURATED VAPOR VOLUME, ML/MOL
273		127.29			
276		127.63			
280		128.08			
284		128.54			
288		129.01			
292		129.48			
296		129.95			
300		130.43			
304		130.92			
308		131.41			
315	0.6888				3802416.
319	0.8718				3042411.
323	1.093				2456073.
327	1.360				1998937.
331	1.679				1639060.
335	2.058				1353201.
339	2.507				1124253.
343	3.036				939481.
347	3.655				789308.
351	4.379				666459.
355	5.221				565359.
358	5.939				501175.
362	7.027				428341.
366	8.280				367543.
370	9.718				316565.
374	11.36				273644.
378	13.24				237370.
382	15.37				206603.
386	17.79				180422.
390	20.51				158075.
394	23.58				138945.
398	27.01				122526.
401	29.84				111735.
405	33.98				99094.
409	38.57				88174.
413	43.62				78722.
417	49.16				70524.
421	55.21				63401.
425	61.77				57204.
429	68.85				51803.
433	76.45				47092.
437	84.54				42977.
441	93.10				39382.
444	99.81				36986.

COMPOUND CONSTANTS:

MP	239.020 K	TC	VC	ZC
NBP	444.267 K	PC	DC	OM
MU		RG		

VAPOR PRESSURE CORRELATION CONSTANTS, K AND KPA:

VAPRES-2	VAPRES-2	VAPRES-2
304. - 445. K	304. - 445. K	304. - 445. K
15 POINTS	15 POINTS	15 POINTS
RMSD = 0.1292	RMSD = 0.1292	RMSD = 0.1292
-0.83666693E+03	-0.83666693E+03	-0.83666693E+03
0.12724066E+05	0.12724066E+05	0.12724066E+05
-0.44529443E+00	-0.44529443E+00	-0.44529443E+00
0.19905378E-03	0.19905378E-03	0.19905378E-03
0.15930423E+03	0.15930423E+03	0.15930423E+03

LIQUID DENSITY CORRELATION CONSTANTS, K AND G/ML:

FRANCIS1	FRANCIS1	FRANCIS1
273. - 309. K	273. - 309. K	273. - 309. K
5 POINTS	5 POINTS	5 POINTS
RMSD = 0.00001	RMSD = 0.00001	RMSD = 0.00001
0.11677370E+01	0.11677370E+01	0.11677370E+01
0.75875595E-03	0.75875595E-03	0.75875595E-03
0.66336937E+01	0.66336937E+01	0.66336937E+01
0.57182007E+03	0.57182007E+03	0.57182007E+03

SECOND VIRIAL COEFFICIENT CORRELATION CONSTANTS, ML/MOL:

LITERATURE DOCUMENTS FOR CORRELATED VAPOR PRESSURE VALUES:

1360   6662 14448

LITERATURE DOCUMENTS FOR CORRELATED LIQUID DENSITY VALUES:

3269   6662

LITERATURE DOCUMENTS REPORTING VIRIAL COEFFICIENT DATA:

T,K	VAPOR PRESSURE, KPA	SATURATED LIQUID VOLUME, ML/MOL	SECOND VIRIAL COEFFICIENT, ML/MOL	HEAT OF VAPORIZATION, J/MOL	SATURATED VAPOR VOLUME, ML/MOL
273		125.94			
276		126.27			
280		126.73			
284		127.18			
288		127.64			
292		128.11			
296		128.58			
300		129.06			
304	0.3540	129.54		47696.	7140398.
308	0.4522	130.02		47650.	5662550.
312	0.5740				4519380.
315	0.6835				3831882.
319	0.8579				3091582.
323	1.070				2509261.
327	1.327				2048521.
331	1.636				1681894.
335	2.006				1388528.
339	2.446				1152505.
343	2.966				961606.
347	3.578				806406.
351	4.294				679590.
355	5.129				575457.
358	5.842				509519.
362	6.920				434913.
366	8.161				372859.
370	9.583				321015.
374	11.21				277514.
378	13.05				240858.
382	15.14				209844.
386	17.49				183498.
390	20.14				161030.
394	23.10				141798.
398	26.42				125273.
401	29.15				114394.
405	33.14				101616.
409	37.56				90531.
413	42.45				80883.
417	47.85				72459.
421	53.79				65080.
425	60.30				58596.
429	67.45				52883.
433	75.26				47833.
437	83.80				43358.
441	93.11				39380.
444	100.6				36685.

COMPOUND CONSTANTS:

MP	221.779 K	TC		VC		ZC
NBP	451.123 K	PC		DC		OM
MU	0.5300 DEBYE	RG				

VAPOR PRESSURE CORRELATION CONSTANTS, K AND KPA:

RPM2	RPM2	RPM2	RPM2
355. - 483. K	355. - 483. K	355. - 483. K	432. - 483. K
25 POINTS	25 POINTS	25 POINTS	11 POINTS
RMSD = 0.0105	RMSD = 0.0105	RMSD = 0.0105	RMSD = 0.0108
0.28986945E+02	0.28986945E+02	0.28986945E+02	0.24368074E+02
-0.72867078E+04	-0.72867078E+04	-0.72867078E+04	-0.65950197E+04
-0.25184391E-01	-0.25184391E-01	-0.25184391E-01	-0.14915958E-01
0.15453744E-04	0.15453744E-04	0.15453744E-04	0.78536866E-05

LIQUID DENSITY CORRELATION CONSTANTS, K AND G/ML:

FRANCIS1	FRANCIS1	FRANCIS1
277. - 299. K	277. - 299. K	277. - 299. K
6 POINTS	6 POINTS	6 POINTS
RMSD = 0.00012	RMSD = 0.00012	RMSD = 0.00012
0.11978998E+01	0.11978998E+01	0.11978998E+01
0.76206075E-03	0.76206075E-03	0.76206075E-03
0.59999990E+01	0.59999990E+01	0.59999990E+01
0.55901392E+03	0.55901392E+03	0.55901392E+03

SECOND VIRIAL COEFFICIENT CORRELATION CONSTANTS, ML/MOL:

LITERATURE DOCUMENTS FOR CORRELATED VAPOR PRESSURE VALUES:

40278

LITERATURE DOCUMENTS FOR CORRELATED LIQUID DENSITY VALUES:

182

LITERATURE DOCUMENTS REPORTING VIRIAL COEFFICIENT DATA:

T,K	VAPOR PRESSURE, KPA	SATURATED LIQUID VOLUME, ML/MOL	SECOND VIRIAL COEFFICIENT, ML/MOL	HEAT OF VAPORIZATION, J/MOL	SATURATED VAPOR VOLUME, ML/MOL
277		122.40			
281		122.82			
286		123.36			
291		123.91			
295		124.35			
356	4.534				652854.
361	5.610				535032.
365	6.619				458519.
370	8.089				380300.
375	9.823				317398.
379	11.42				275839.
384	13.72				232653.
389	16.39				197309.
394	19.47				168216.
398	22.27				148600.
403	26.21				127820.
408	30.71				110457.
412	34.74				98597.
417	40.37				85876.
422	46.72				75104.
426	52.35				67660.
431	60.14				59587.
436	68.83				52669.
440	76.47				47837.
445	86.96				42545.
450	98.56				37961.
454	108.7				34730.
459	122.5				31159.
464	137.6				28035.
468	150.7				25815.
473	168.5				23341.
478	187.8				21158.
482	204.5				19594.

COMPOUND CONSTANTS:

MP   271.700 K   TC   VC   ZC
NBP   455.757 K   PC   DC   OM
MU   RG

    VAPOR PRESSURE CORRELATION CONSTANTS, K AND KPA:

    RPM2   RPM2   RPM2
    335. - 458. K   335. - 458. K   335. - 458. K
      8 POINTS   8 POINTS   8 POINTS
    RMSD = 0.0501   RMSD = 0.0501   RMSD = 0.0501
    0.51215154E+02   0.51215154E+02   0.51215154E+02
    -0.10474049E+05   -0.10474049E+05   -0.10474049E+05
    -0.77854962E-01   -0.77854962E-01   -0.77854962E-01
    0.57134999E-04   0.57134999E-04   0.57134999E-04

    LIQUID DENSITY CORRELATION CONSTANTS, K AND G/ML:

    FRANCIS1   FRANCIS1   FRANCIS1
    277. - 299. K   277. - 299. K   277. - 299. K
      6 POINTS   6 POINTS   6 POINTS
    RMSD = 0.00017   RMSD = 0.00017   RMSD = 0.00017
    0.12773809E+01   0.12773809E+01   0.12773809E+01
    0.82388707E-03   0.82388707E-03   0.82388707E-03
    0.59999990E+01   0.59999990E+01   0.59999990E+01
    0.55901392E+03   0.55901392E+03   0.55901392E+03

    SECOND VIRIAL COEFFICIENT CORRELATION CONSTANTS, ML/MOL:

LITERATURE DOCUMENTS FOR CORRELATED VAPOR PRESSURE VALUES:

    19516 22499

LITERATURE DOCUMENTS FOR CORRELATED LIQUID DENSITY VALUES:

    182

LITERATURE DOCUMENTS REPORTING VIRIAL COEFFICIENT DATA:

T,K	VAPOR PRESSURE, KPA	SATURATED LIQUID VOLUME, ML/MOL	SECOND VIRIAL COEFFICIENT, ML/MOL	HEAT OF VAPORIZATION, J/MOL	SATURATED VAPOR VOLUME, ML/MOL
277		113.01			
281		113.41			
285		113.81			
289		114.21			
293		114.62			
297		115.04			
338	1.551				1811370.
342	1.907				1490966.
346	2.329				1235135.
351	2.964				984585.
355	3.571				826657.
359	4.277				697915.
363	5.095				592350.
367	6.039				505302.
371	7.122				433135.
375	8.359				372993.
379	9.767				322624.
383	11.36				280236.
388	13.65				236325.
392	15.74				207092.
396	18.08				182136.
400	20.69				160743.
404	23.60				142334.
408	26.83				126431.
412	30.41				112644.
416	34.37				100648.
420	38.72				90177.
425	44.79				78893.
429	50.18				71088.
433	56.07				64205.
437	62.52				58116.
441	69.55				52716.
445	77.22				47914.
449	85.56				43631.
453	94.63				39803.
457	104.5				36372.

KETONES

COMPOUND CONSTANTS:

MP	178.476 K	TC	508.100 K	VC	208.92 ML/MOL	ZC	0.2324
NBP	329.207 K	PC	4.700 MPA	DC	0.2780 G/ML	OM	0.3073
MU	2.8600 DEBYE	RG	2.7404 ANGSTROM				

VAPOR PRESSURE CORRELATION CONSTANTS, K AND KPA:

WAGNER	WAGNER	WAGNER	WAGNER
247. - 509. K	247. - 348. K	247. - 398. K	309. - 509. K
102 POINTS	79 POINTS	93 POINTS	63 POINTS
RMSD = 1.0200	RMSD = 0.0286	RMSD = 0.0960	RMSD = 1.2450
-0.74974490E+01	-0.75475020E+01	-0.76478446E+01	-0.74271578E+01
0.13173750E+01	0.14452612E+01	0.16786170E+01	0.10922862E+01
-0.26754153E+01	-0.29036443E+01	-0.32366434E+01	-0.19873034E+01
-0.26773997E+01	-0.21275823E+01	-0.15276237E+01	-0.65774782E+01

LIQUID DENSITY CORRELATION CONSTANTS, K AND G/ML:

FRANCIS1	FRANCIS1	FRANCIS1	FRANCIS2
178. - 498. K	178. - 330. K	178. - 476. K	460. - 508. K
58 POINTS	34 POINTS	53 POINTS	13 POINTS
RMSD = 0.00096	RMSD = 0.00021	RMSD = 0.00072	RMSD = 0.00261
0.11210213E+01	0.11081915E+01	0.11216850E+01	0.11984767E-02
0.10215563E-02	0.10551438E-02	0.97801606E-03	0.22339842E+01
0.79379501E+01	0.96902567E+00	0.12052504E+02	
0.53905884E+03	0.40312988E+03	0.55959619E+03	

SECOND VIRIAL COEFFICIENT CORRELATION CONSTANTS, ML/MOL:

TSONOPOULOS: TC=508.100 PC=4.700 VC=208.92 OM=0.3073 A=-0.030100 B=0.0000

LITERATURE DOCUMENTS FOR CORRELATED VAPOR PRESSURE VALUES:

   709 1514 2560 2575 5881   7952 10044 10253 10318 11081 16497 41354 41545
   41685

LITERATURE DOCUMENTS FOR CORRELATED LIQUID DENSITY VALUES:

   709   777 1359 2478 5204 9277 10165 11019 15755 40228

LITERATURE DOCUMENTS REPORTING VIRIAL COEFFICIENT DATA:

   1257 1803 1950 2422 2678 5214 5283 5948 8330 8577 9027 10972 20469

T,K	VAPOR PRESSURE, KPA	SATURATED LIQUID VOLUME, ML/MOL	SECOND VIRIAL COEFFICIENT, ML/MOL	HEAT OF VAPORIZATION, J/MOL	SATURATED VAPOR VOLUME, ML/MOL
178		63.32			
185		63.85	-24095.		
193		64.47	-18558.		
200		65.02	-14968.		
208		65.66	-11878.		
215		66.23	-9818.		
223		66.90	-7997.		
230		67.51	-6754.		
238		68.21	-5629.		
245		68.84	-4844.		
253	2.917	69.57	-4120.	33497.	716868.
260	4.481	70.23	-3605.	33103.	478800.
268	7.085	71.01	-3120.	32656.	311357.
275	10.31	71.70	-2770.	32265.	218962.
283	15.41	72.52	-2434.	31818.	150183.
290	21.45	73.25	-2188.	31423.	110173.
298	30.61	74.11	-1949.	30969.	78949.
305	41.04	74.89	-1771.	30565.	59967.
313	56.32	75.81	-1595.	30096.	44554.
320	73.18	76.63	-1463.	29678.	34830.
328	97.17	77.61	-1331.	29188.	26665.
335	123.0	78.50	-1231.	28748.	21348.
343	158.7	79.55	-1129.	28229.	16754.
351	202.3	80.64	-1041.	27691.	13299.
358	247.6	81.63	-971.7	27203.	10958.
366	308.6	82.82	-901.4	26622.	8859.
373	370.9	83.91	-846.3	26089.	7405.
381	453.6	85.21	-789.6	25449.	6076.
388	537.0	86.41	-744.7	24858.	5137.
396	646.0	87.87	-698.3	24139.	4261.
403	754.6	89.22	-661.3	23465.	3632.
411	895.0	90.87	-622.7	22631.	3035.
418	1033.	92.42	-591.7	21831.	2597.
426	1211.	94.34	-559.2	20810.	2173.
433	1384.	96.18	-533.0	21925.	2043.
441	1604.	98.51	-505.3	21019.	1738.
448	1818.	100.80	-482.9	20164.	1509.
456	2087.	103.78	-459.1	19102.	1283.
463	2348.	106.83	-439.7	18084.	1112.
471	2675.	111.00	-419.0	16791.	940.2
478	2989.	115.53	-402.1	15513.	807.7
486	3383.	122.29	-384.0	13812.	671.8
493	3761.	130.50	-369.1	12005.	562.6
501	4236.		-353.1		441.2
508	4693.		-339.9		296.5

COMPOUND CONSTANTS:

MP   186.490 K   TC 536.780 K   VC 267.06 ML/MOL   ZC 0.2517
NBP 352.747 K   PC   4.207 MPA   DC 0.2700 G/ML   OM 0.3220
MU   2.7800 DEBYE   RG 3.1395 ANGSTROM

    VAPOR PRESSURE CORRELATION CONSTANTS, K AND KPA:

    WAGNER   WAGNER   WAGNER   WAGNER   .
    265. - 537. K   265. - 371. K   265. - 448. K   333. - 537. K
     65 POINTS   39 POINTS   49 POINTS   49 POINTS
    RMSD = 3.3370   RMSD = 0.0540   RMSD = 0.1800   RMSD = 2.9410
    -0.76980638E+01   -0.80361159E+01   -0.76396468E+01   -0.77821606E+01
     0.16857248E+01   0.24595576E+01   0.15330646E+01   0.19660930E+01
    -0.35946384E+01   -0.46452494E+01   -0.33071878E+01   -0.45547369E+01
    -0.14844025E+01   0.20832996E+00   -0.22181141E+01   0.51033664E+01

    LIQUID DENSITY CORRELATION CONSTANTS, K AND G/ML:

    FRANCIS1   FRANCIS1   FRANCIS1
    195. - 324. K   195. - 324. K   195. - 324. K
     47 POINTS   47 POINTS   47 POINTS
    RMSD = 0.00017   RMSD = 0.00017   RMSD = 0.00017
    0.11104259E+01   0.11104259E+01   0.11104259E+01
    0.81403158E-03   0.81403158E-03   0.81403158E-03
    0.19753296E+02   0.19753296E+02   0.19753296E+02
    0.58889380E+03   0.58889380E+03   0.58889380E+03

    SECOND VIRIAL COEFFICIENT CORRELATION CONSTANTS, ML/MOL:

    HAYDEN-O'CONNELL: TC=533.000 PC=4.002 RG=2.7000 MU=0.900 ETA=3.139

LITERATURE DOCUMENTS FOR CORRELATED VAPOR PRESSURE VALUES:

    697   1005   2920   3760 10318 10901 11081

LITERATURE DOCUMENTS FOR CORRELATED LIQUID DENSITY VALUES:

    241   1859   2045   2478   2920   3386   4220   6763   7452   9002   9712 10163 11017
  11243 12590 14102 15411 20486 40566 40936 41379

LITERATURE DOCUMENTS REPORTING VIRIAL COEFFICIENT DATA:

  10901 21584

T,K	VAPOR PRESSURE, KPA	SATURATED LIQUID VOLUME, ML/MOL	SECOND VIRIAL COEFFICIENT, ML/MOL	HEAT OF VAPORIZATION, J/MOL	SATURATED VAPOR VOLUME, ML/MOL
195		79.98	-16551.		
202		80.57	-13564.		
210		81.26	-11006.		
218		81.97	-9086.		
226		82.69	-7615.		
233		83.34	-6596.		
241		84.09	-5659.		
249		84.87	-4906.		
257		85.67	-4293.		
264		86.38	-3845.		
272	3.033	87.23	-3414.	36106.	742119.
280	4.785	88.09	-3051.	35663.	483406.
288	7.328	88.98	-2744.	35216.	324001.
296	10.92	89.91	-2480.	34763.	222848.
303	15.17	90.74	-2280.	34361.	163747.
311	21.62	91.72	-2080.	33893.	117477.
319	30.18	92.73	-1905.	33415.	85949.
327	41.31		-1751.		64012.
334	53.60		-1631.		50128.
342	71.07		-1509.		38440.
350	92.83		-1400.		29880.
358	119.6		-1302.		23516.
365	147.6		-1225.		19247.
373	185.8		-1145.		15456.
381	231.2		-1073.		12529.
389	284.8		-1007.		10241.
397	347.4		-946.9		8434.
404	410.4		-898.7		7156.
412	492.8		-848.1		5963.
420	587.1		-801.6		4993.
428	694.5		-758.8		4198.
435	800.1		-724.1		3615.
443	935.2		-687.2		3052.
451	1087.		-653.1		2575.
459	1256.		-621.5		2380.
467	1445.		-592.1		2040.
474	1627.		-568.0		1784.
482	1857.		-542.2		1529.
490	2110.		-518.2		1309.
498	2390.		-495.6		1116.
505	2657.		-477.1		965.4
513	2992.		-457.2		810.5
521	3361.		-438.4		668.3
529	3768.		-420.8		530.4
536	4161.		-406.3		381.6

COMPOUND CONSTANTS:

MP	TC	VC	ZC
NBP 361.853 K	PC	DC	OM
MU 1.4600 DEBYE	RG		

VAPOR PRESSURE CORRELATION CONSTANTS, K AND KPA:

RPM2	RPM2	RPM2
260. - 363. K	260. - 363. K	260. - 363. K
9 POINTS	9 POINTS	9 POINTS
RMSD = 0.3393	RMSD = 0.3393	RMSD = 0.3393
-0.36104822E+02	-0.36104822E+02	-0.36104822E+02
0.79123257E+03	0.79123257E+03	0.79123257E+03
0.17481982E+00	0.17481982E+00	0.17481982E+00
-0.18881169E-03	-0.18881169E-03	-0.18881169E-03

LIQUID DENSITY CORRELATION CONSTANTS, K AND G/ML:

SECOND VIRIAL COEFFICIENT CORRELATION CONSTANTS, ML/MOL:

LITERATURE DOCUMENTS FOR CORRELATED VAPOR PRESSURE VALUES:

   1732 10537 14792

LITERATURE DOCUMENTS FOR CORRELATED LIQUID DENSITY VALUES:

LITERATURE DOCUMENTS REPORTING VIRIAL COEFFICIENT DATA:

T,K	VAPOR PRESSURE, KPA	SATURATED LIQUID VOLUME, ML/MOL	SECOND VIRIAL COEFFICIENT, ML/MOL	HEAT OF VAPORIZATION, J/MOL	SATURATED VAPOR VOLUME, ML/MOL
260	0.6893				3136327.
262	0.7844				2777152.
264	0.8916				2461816.
267	1.078				2058941.
269	1.222				1830253.
271	1.383				1628804.
274	1.662				1370355.
276	1.876				1223018.
278	2.115				1092785.
281	2.526				924990.
283	2.839				828912.
285	3.186				743692.
288	3.780				633415.
290	4.230				569990.
292	4.728				513535.
295	5.572				440158.
297	6.208				397766.
299	6.908				359898.
302	8.088				310461.
304	8.970				281771.
306	9.936				256051.
309	11.56				222324.
311	12.76				202662.
313	14.07				184972.
316	16.25				161672.
318	17.86				148028.
320	19.61				135708.
323	22.49				119410.
325	24.60				109823.
327	26.88				101136.
330	30.62				89593.
332	33.35				82774.
334	36.27				76573.
337	41.02				68299.
339	44.46				63390.
341	48.13				58910.
344	54.06				52908.
346	58.31				49332.
348	62.82				46058.
351	70.06				41654.
353	75.22				39020.
355	80.64				36601.
358	89.30				33334.
360	95.41				31372.
362	101.8				29565.

COMPOUND CONSTANTS:

MP   196.320 K   TC 561.080 K   VC 301.16 ML/MOL   ZC 0.2385
NBP 375.408 K   PC   3.694 MPA   DC 0.2860 G/ML   OM 0.3470
MU   2.7200 DEBYE   RG

VAPOR PRESSURE CORRELATION CONSTANTS, K AND KPA:

WAGNER	WAGNER	WAGNER	WAGNER
282. - 545. K	282. - 395. K	282. - 473. K	355. - 545. K
37 POINTS	25 POINTS	27 POINTS	30 POINTS
RMSD = 2.6930	RMSD = 0.0291	RMSD = 0.2848	RMSD = 2.6860
-0.73516670E+01	-0.76517516E+01	-0.75098802E+01	-0.72551714E+01
0.58891876E+00	0.13209503E+01	0.98487982E+00	0.25769821E+00
-0.21522346E+01	-0.33462984E+01	-0.28407473E+01	-0.10269338E+01
-0.58445133E+01	-0.32001583E+01	-0.41918431E+01	-0.12997027E+02

LIQUID DENSITY CORRELATION CONSTANTS, K AND G/ML:

FRANCIS1	FRANCIS1	FRANCIS1
233. - 354. K	233. - 354. K	233. - 354. K
9 POINTS	9 POINTS	9 POINTS
RMSD = 0.00012	RMSD = 0.00012	RMSD = 0.00012
0.10936251E+01	0.10936251E+01	0.10936251E+01
0.96370699E-03	0.96370699E-03	0.96370699E-03
0.48774409E+00	0.48774409E+00	0.48774409E+00
0.42092505E+03	0.42092505E+03	0.42092505E+03

SECOND VIRIAL COEFFICIENT CORRELATION CONSTANTS, ML/MOL:

TSONOPOULOS: TC=561.080 PC=3.694 VC=301.16 OM=0.3470 A=-0.017600 B=0.0000

LITERATURE DOCUMENTS FOR CORRELATED VAPOR PRESSURE VALUES:

4418 5067 10318 10901

LITERATURE DOCUMENTS FOR CORRELATED LIQUID DENSITY VALUES:

7439 7470 40489

LITERATURE DOCUMENTS REPORTING VIRIAL COEFFICIENT DATA:

10901

T,K	VAPOR PRESSURE, KPA	SATURATED LIQUID VOLUME, ML/MOL	SECOND VIRIAL COEFFICIENT, ML/MOL	HEAT OF VAPORIZATION, J/MOL	SATURATED VAPOR VOLUME, ML/MOL
233		99.40	-12933.		
240		100.20	-10932.		
247		101.00	-9325.		
254		101.82	-8023.		
261		102.66	-6959.		
268		103.51	-6083.		
275		104.37	-5355.		
282	1.910	105.25	-4745.	39611.	1223039.
289	2.873	106.15	-4232.	39092.	832150.
296	4.220	107.07	-3797.	38588.	579356.
303	6.065	108.01	-3424.	38096.	411919.
310	8.543	108.96	-3105.	37614.	298554.
318	12.36	110.08	-2792.	37075.	211149.
325	16.76	111.09	-2556.	36610.	158639.
332	22.38	112.12	-2350.	36149.	120946.
339	29.46	113.19	-2168.	35692.	93458.
346	38.26	114.28	-2008.	35235.	73116.
353	49.09	115.42	-1866.	34777.	57858.
360	62.26		-1739.		46266.
367	78.13		-1625.		37356.
374	97.07		-1522.		30432.
381	119.5		-1430.		24995.
388	145.8		-1346.		20685.
396	181.3		-1260.		16803.
403	217.5		-1191.		14103.
410	259.2		-1129.		11906.
417	306.8		-1071.		10103.
424	360.9		-1018.		8614.
431	422.1		-968.8		7375.
438	490.9		-923.4		6338.
445	567.9		-881.3		5464.
452	653.9		-842.1		4722.
459	749.4		-805.6		4089.
466	855.1		-771.5		3545.
474	989.2		-735.1		3012.
481	1119.		-705.5		2789.
488	1261.		-677.6		2443.
495	1417.		-651.3		2141.
502	1586.		-626.6		1876.
509	1770.		-603.2		1642.
516	1970.		-581.1		1435.
523	2186.		-560.2		1250.
530	2419.		-540.3		1084.
537	2671.		-521.5		932.2
544	2942.		-503.6		791.7

COMPOUND CONSTANTS:

MP 234.203 K	TC 561.460 K	VC 336.46 ML/MOL	ZC 0.2688
NBP 375.109 K	PC 3.729 MPA	DC 0.2560 G/ML	OM 0.3410
MU 2.7200 DEBYE	RG 3.4817 ANGSTROM		

VAPOR PRESSURE CORRELATION CONSTANTS, K AND KPA:

WAGNER	WAGNER	WAGNER	WAGNER
282. - 545. K	282. - 385. K	282. - 473. K	355. - 545. K
40 POINTS	24 POINTS	29 POINTS	31 POINTS
RMSD = 3.7470	RMSD = 0.0036	RMSD = 1.2720	RMSD = 4.0520
-0.70827199E+01	-0.77760626E+01	-0.64924142E+01	-0.74645307E+01
0.21564015E+00	0.17033533E+01	-0.12998031E+01	0.15101506E+01
-0.25715917E+01	-0.41525350E+01	0.20895716E+00	-0.70957752E+01
-0.24178698E+01	-0.13722948E+01	-0.94551795E+01	0.29279142E+02

LIQUID DENSITY CORRELATION CONSTANTS, K AND G/ML:

FRANCIS1	FRANCIS1	FRANCIS1
273. - 348. K	273. - 348. K	273. - 348. K
19 POINTS	19 POINTS	19 POINTS
RMSD = 0.00007	RMSD = 0.00007	RMSD = 0.00007
0.10943232E+01	0.10943232E+01	0.10943232E+01
0.86304802E-03	0.86304802E-03	0.86304802E-03
0.70360441E+01	0.70360441E+01	0.70360441E+01
0.55340259E+03	0.55340259E+03	0.55340259E+03

SECOND VIRIAL COEFFICIENT CORRELATION CONSTANTS, ML/MOL:

TSONOPOULOS: TC=561.460 PC=3.729 VC=336.46 OM=0.3410 A=-0.017700 B=0.0000

LITERATURE DOCUMENTS FOR CORRELATED VAPOR PRESSURE VALUES:

   4033 5067 10318

LITERATURE DOCUMENTS FOR CORRELATED LIQUID DENSITY VALUES:

   1360 5839 7483 9920 11733 13138 15411 17509 40353 40425 40566 41156

LITERATURE DOCUMENTS REPORTING VIRIAL COEFFICIENT DATA:

   4033 21584

T,K	VAPOR PRESSURE, KPA	SATURATED LIQUID VOLUME, ML/MOL	SECOND VIRIAL COEFFICIENT, ML/MOL	HEAT OF VAPORIZATION, J/MOL	SATURATED VAPOR VOLUME, ML/MOL
273		103.32	-5493.		
279		104.04	-4943.		
285	2.288	104.77	-4469.	39025.	1031365.
291	3.213	105.51	-4059.	38692.	748865.
297	4.442	106.26	-3702.	38359.	552241.
303	6.047	107.03	-3390.	38024.	413163.
310	8.516	107.95	-3074.	37632.	299549.
316	11.26	108.76	-2837.	37293.	230474.
322	14.71	109.58	-2627.	36950.	179391.
328	18.99	110.42	-2440.	36603.	141148.
334	24.25	111.28	-2273.	36250.	112185.
340	30.67	112.16	-2123.	35891.	90012.
347	39.84	113.22	-1967.	35463.	70397.
353	49.38		-1847.		57533.
359	60.68		-1739.		47381.
365	74.00		-1640.		39301.
371	89.55		-1550.		32818.
378	110.9		-1455.		26804.
384	132.2		-1380.		22677.
390	156.7		-1311.		19289.
396	184.5		-1248.		16491.
402	216.1		-1190.		14165.
408	251.8		-1135.		12220.
415	299.0		-1077.		10339.
421	344.6		-1031.		8994.
427	395.4		-987.4		7851.
433	451.6		-946.9		6873.
439	513.8		-909.1		6034.
446	594.2		-867.8		5199.
452	670.4		-834.8		4585.
458	753.7		-803.6		4050.
464	844.5		-774.3		3580.
470	943.2		-746.6		3166.
476	1050.		-720.4		2799.
483	1186.		-691.5		2608.
489	1312.		-668.2		2329.
495	1448.		-646.0		2080.
501	1594.		-624.9		1858.
507	1751.		-604.9		1659.
514	1947.		-582.7		1452.
520	2128.		-564.6		1293.
526	2320.		-547.3		1148.
532	2525.		-530.8		1015.
538	2743.		-514.9		891.8
544	2974.		-499.8		776.2

COMPOUND CONSTANTS:

MP   181.178 K   TC   553.400 K   VC   309.83 ML/MOL   ZC   0.2592
NBP   367.482 K   PC   3.850 MPA   DC   0.2780 G/ML   OM   0.3301
MU   2.7700 DEBYE   RG   3.4148 ANGSTROM

   VAPOR PRESSURE CORRELATION CONSTANTS, K AND KPA:

   WAGNER   WAGNER   WAGNER   WAGNER
   276. - 550. K   276. - 377. K   276. - 377. K   350. - 550. K
    34 POINTS   25 POINTS   25 POINTS   25 POINTS
   RMSD = 4.5050   RMSD = 0.0102   RMSD = 0.0102   RMSD = 1.9260
   -0.72049532E+01   -0.79425250E+01   -0.79425250E+01   -0.77016861E+01
   0.35067807E+00   0.20530981E+01   0.20530981E+01   0.20979165E+01
   -0.13709136E+01   -0.37421557E+01   -0.37421557E+01   -0.74639778E+01
   -0.75396462E+01   -0.34919809E+01   -0.34919809E+01   0.32055908E+02

   LIQUID DENSITY CORRELATION CONSTANTS, K AND G/ML:

   FRANCIS1   FRANCIS1   FRANCIS1
   283. - 324. K   283. - 324. K   283. - 324. K
    3 POINTS   3 POINTS   3 POINTS
   RMSD = 0.00000   RMSD = 0.00000   RMSD = 0.00000
   0.10588190E+01   0.10588190E+01   0.10588190E+01
   0.84827400E-03   0.84827400E-03   0.84827400E-03

   SECOND VIRIAL COEFFICIENT CORRELATION CONSTANTS, ML/MOL:

   TSONOPOULOS: TC=553.400 PC=3.850 VC=309.83 OM=0.3301 A=-0.019500 B=0.0000

LITERATURE DOCUMENTS FOR CORRELATED VAPOR PRESSURE VALUES:

   4033 5067 10318

LITERATURE DOCUMENTS FOR CORRELATED LIQUID DENSITY VALUES:

   1091 21601

LITERATURE DOCUMENTS REPORTING VIRIAL COEFFICIENT DATA:

   4033

T,K	VAPOR PRESSURE, KPA	SATURATED LIQUID VOLUME, ML/MOL	SECOND VIRIAL COEFFICIENT, ML/MOL	HEAT OF VAPORIZATION, J/MOL	SATURATED VAPOR VOLUME, ML/MOL
276	1.970		-4724.		1160344.
282	2.815		-4266.		828711.
288	3.948	105.75	-3870.	37624.	602614.
294	5.443	106.41	-3526.	37176.	445535.
300	7.385	107.09	-3225.	36742.	334482.
307	10.35	107.88	-2921.	36249.	243691.
313	13.62	108.57	-2693.	35837.	188386.
319	17.69	109.28	-2492.	35435.	147381.
325	22.72		-2313.		116583.
332	30.00		-2128.		89839.
338	37.65		-1987.		72593.
344	46.82		-1860.		59172.
350	57.70		-1746.		48623.
356	70.53		-1642.		40257.
363	88.27		-1533.		32583.
369	106.2		-1448.		27371.
375	126.8		-1371.		23131.
381	150.5		-1300.		19656.
388	182.4		-1225.		16361.
394	213.8		-1165.		14052.
400	249.2		-1110.		12121.
406	289.1		-1060.		10498.
412	333.7		-1012.		9125.
419	392.4		-961.5		7782.
425	448.7		-921.0		6811.
431	511.0		-883.1		5977.
437	579.7		-847.7		5257.
444	668.6		-809.1		4536.
450	752.7		-778.2		4005.
456	844.7		-749.1		3538.
462	944.9		-721.7		3127.
469	1073.		-691.6		2706.
475	1192.		-667.3		2571.
481	1322.		-644.3		2293.
487	1462.		-622.5		2046.
493	1612.		-601.8		1826.
500	1802.		-578.9		1597.
506	1977.		-560.4		1422.
512	2165.		-542.6		1264.
518	2366.		-525.7		1121.
525	2618.		-507.0		968.1
531	2849.		-491.7		848.0
537	3095.		-477.0		735.1
543	3357.		-463.0		625.6
549	3635.		-449.5		510.9

COMPOUND CONSTANTS:

MP   249.964 K   TC   VC   ZC
NBP 411.855 K   PC   DC   OM
MU   2.8100 DEBYE   RG

VAPOR PRESSURE CORRELATION CONSTANTS, K AND KPA:

RPM2	RPM2	RPM2	RIEDEL
289. - 412. K	289. - 412. K	289. - 412. K	393. - 412. K
29 POINTS	29 POINTS	29 POINTS	6 POINTS
RMSD = 0.1731	RMSD = 0.1731	RMSD = 0.1731	RMSD = 0.0310
-0.48963358E+02	-0.48963358E+02	-0.48963358E+02	0.33460813E+04
0.24378665E+04	0.24378665E+04	0.24378665E+04	-0.48757859E+03
0.19501662E+00	0.19501662E+00	0.19501662E+00	-0.17259015E+06
-0.19254584E-03	-0.19254584E-03	-0.19254584E-03	0.26929717E-14

LIQUID DENSITY CORRELATION CONSTANTS, K AND G/ML:

FRANCIS1	FRANCIS1	FRANCIS1
277. - 374. K	277. - 374. K	277. - 374. K
27 POINTS	27 POINTS	27 POINTS
RMSD = 0.00020	RMSD = 0.00020	RMSD = 0.00020
0.12644968E+01	0.12644968E+01	0.12644968E+01
0.89787156E-03	0.89787156E-03	0.89787156E-03
0.80934544E+01	0.80934544E+01	0.80934544E+01
0.61981030E+03	0.61981030E+03	0.61981030E+03

SECOND VIRIAL COEFFICIENT CORRELATION CONSTANTS, ML/MOL:

LITERATURE DOCUMENTS FOR CORRELATED VAPOR PRESSURE VALUES:

15791 16370 41430

LITERATURE DOCUMENTS FOR CORRELATED LIQUID DENSITY VALUES:

5858 13839

LITERATURE DOCUMENTS REPORTING VIRIAL COEFFICIENT DATA:

T,K	VAPOR PRESSURE, KPA	SATURATED LIQUID VOLUME, ML/MOL	SECOND VIRIAL COEFFICIENT, ML/MOL	HEAT OF VAPORIZATION, J/MOL	SATURATED VAPOR VOLUME, ML/MOL
277		100.91			
280		101.20			
283		101.50			
286		101.80			
289	0.7791	102.10			3084262.
292	0.9168	102.41			2648156.
295	1.077	102.71			2277350.
298	1.263	103.02			1961705.
301	1.478	103.34			1692694.
304	1.728	103.65			1463141.
307	2.015	103.97			1267007.
310	2.345	104.28			1099199.
313	2.724	104.61			955430.
316	3.158	104.93			832079.
319	3.653	105.26			726094.
323	4.421	105.70			607391.
326	5.089	106.03			532568.
329	5.846	106.36			467951.
332	6.699	106.70			412059.
335	7.660	107.04			363636.
338	8.738	107.39			321617.
341	9.945	107.73			285093.
344	11.29	108.08			253295.
347	12.79	108.44			225564.
350	14.45	108.79			201340.
353	16.29	109.15			180144.
356	18.32	109.52			161566.
359	20.55	109.88			145255.
362	22.99	110.25			130911.
365	25.66	110.63			118275.
369	29.58	111.13			103706.
372	32.82	111.51			94247.
375	36.31				85869.
378	40.07				78437.
381	44.10				71834.
384	48.40				65959.
387	52.99				60724.
390	57.85				56053.
393	62.98				51879.
396	68.39				48145.
399	74.05				44801.
402	79.96				41802.
405	86.10				39111.
408	92.45				36694.
411	98.99				34522.

COMPOUND CONSTANTS:

MP 217.381 K	TC 587.000 K	VC	ZC
NBP 400.733 K	PC 3.323 MPA	DC	OM 0.3942
MU 2.6800 DEBYE	RG		

VAPOR PRESSURE CORRELATION CONSTANTS, K AND KPA:

WAGNER	WAGNER	WAGNER	WAGNER
298. - 428. K	298. - 417. K	298. - 428. K	385. - 428. K
32 POINTS	30 POINTS	32 POINTS	13 POINTS
RMSD = 0.0038	RMSD = 0.0029	RMSD = 0.0038	RMSD = 0.0032
-0.81315482E+01	-0.81216982E+01	-0.81315482E+01	-0.82250937E+01
0.18766746E+01	0.18535284E+01	0.18766746E+01	0.21242391E+01
-0.42287461E+01	-0.41948020E+01	-0.42287461E+01	-0.47585480E+01
-0.27338595E+01	-0.27979991E+01	-0.27338595E+01	-0.41166467E+00

LIQUID DENSITY CORRELATION CONSTANTS, K AND G/ML:

FRANCIS1	FRANCIS1	FRANCIS1
283. - 334. K	283. - 334. K	283. - 334. K
7 POINTS	7 POINTS	7 POINTS
RMSD = 0.00019	RMSD = 0.00019	RMSD = 0.00019
0.10629501E+01	0.10629501E+01	0.10629501E+01
0.76617324E-03	0.76617324E-03	0.76617324E-03
0.67085629E+01	0.67085629E+01	0.67085629E+01
0.54074902E+03	0.54074902E+03	0.54074902E+03

SECOND VIRIAL COEFFICIENT CORRELATION CONSTANTS, ML/MOL:

TSONOPOULOS: TC=587.000 PC=3.323 VC= 0.00 OM=0.3942 A=-0.014000 B=0.0000

LITERATURE DOCUMENTS FOR CORRELATED VAPOR PRESSURE VALUES:

10318

LITERATURE DOCUMENTS FOR CORRELATED LIQUID DENSITY VALUES:

1091 7439 9712 13065

LITERATURE DOCUMENTS REPORTING VIRIAL COEFFICIENT DATA:

T,K	VAPOR PRESSURE, KPA	SATURATED LIQUID VOLUME, ML/MOL	SECOND VIRIAL COEFFICIENT, ML/MOL	HEAT OF VAPORIZATION, J/MOL	SATURATED VAPOR VOLUME, ML/MOL
283		122.13	-6681.		
286		122.52	-6343.		
289		122.91	-6029.		
292		123.31	-5737.		
296		123.85	-5379.		
299	1.627	124.25	-5131.	42863.	1522367.
302	1.932	124.66	-4900.	42654.	1294788.
306	2.413	125.21	-4616.	42375.	1049830.
309	2.837	125.63	-4418.	42168.	901037.
312	3.324	126.05	-4233.	41961.	776178.
315	3.880	126.48	-4059.	41754.	670996.
319	4.741	127.06	-3844.	41479.	555526.
322	5.489	127.50	-3693.	41274.	484014.
325	6.334	127.94	-3552.	41068.	423044.
329	7.628	128.54	-3376.	40795.	355207.
332	8.737	129.00	-3252.	40590.	312638.
335	9.979		-3136.		275950.
339	11.86		-2990.		234633.
342	13.46		-2887.		208396.
345	15.23		-2790.		185563.
348	17.18		-2698.		165636.
352	20.11		-2583.		142879.
355	22.57		-2501.		128228.
358	25.27		-2423.		115333.
362	29.27		-2326.		100463.
365	32.59		-2256.		90800.
368	36.22		-2190.		82227.
371	40.17		-2127.		74607.
375	45.96		-2048.		65720.
378	50.74		-1991.		59881.
381	55.91		-1937.		54655.
385	63.44		-1868.		48512.
388	69.61		-1819.		44446.
391	76.25		-1772.		40783.
395	85.88		-1712.		36447.
398	93.71		-1669.		33557.
401	102.1		-1628.		30938.
404	111.1		-1589.		28562.
408	124.0		-1539.		25726.
411	134.4		-1503.		23820.
414	145.5		-1468.		22081.
418	161.5		-1423.		19994.
421	174.3		-1392.		18582.
424	187.9		-1361.		17289.
427	202.3		-1331.		16102.

COMPOUND CONSTANTS:

MP	217.531 K	TC 582.820 K	VC	ZC
NBP	396.656 K	PC 3.320 MPA	DC	OM 0.3794
MU		RG		

VAPOR PRESSURE CORRELATION CONSTANTS, K AND KPA:

WAGNER	WAGNER	WAGNER	WAGNER
298. - 407. K	298. - 407. K	298. - 407. K	379. - 407. K
24 POINTS	24 POINTS	24 POINTS	14 POINTS
RMSD = 0.0036	RMSD = 0.0036	RMSD = 0.0036	RMSD = 0.0021
-0.79647906E+01	-0.79647906E+01	-0.79647906E+01	-0.77461734E+01
0.17198839E+01	0.17198839E+01	0.17198839E+01	0.11668065E+01
-0.42568912E+01	-0.42568912E+01	-0.42568912E+01	-0.32275202E+01
-0.23051546E+01	-0.23051546E+01	-0.23051546E+01	-0.56463010E+01

LIQUID DENSITY CORRELATION CONSTANTS, K AND G/ML:

FRANCIS1	FRANCIS1	FRANCIS1
297. - 324. K	297. - 324. K	297. - 324. K
4 POINTS	4 POINTS	4 POINTS
RMSD = 0.00001	RMSD = 0.00001	RMSD = 0.00001
0.10622683E+01	0.10622683E+01	0.10622683E+01
0.75031724E-03	0.75031724E-03	0.75031724E-03
0.48805723E+01	0.48805723E+01	0.48805723E+01
0.47581445E+03	0.47581445E+03	0.47581445E+03

SECOND VIRIAL COEFFICIENT CORRELATION CONSTANTS, ML/MOL:

TSONOPOULOS: TC=582.820 PC=3.320 VC= 0.00 OM=0.3794 A=-0.014200 B=0.0000

LITERATURE DOCUMENTS FOR CORRELATED VAPOR PRESSURE VALUES:

4033 10318

LITERATURE DOCUMENTS FOR CORRELATED LIQUID DENSITY VALUES:

13065 13138

LITERATURE DOCUMENTS REPORTING VIRIAL COEFFICIENT DATA:

4033

T,K	VAPOR PRESSURE, KPA	SATURATED LIQUID VOLUME, ML/MOL	SECOND VIRIAL COEFFICIENT, ML/MOL	HEAT OF VAPORIZATION, J/MOL	SATURATED VAPOR VOLUME, ML/MOL
297		123.33	-5028.		
299	1.951	123.61	-4875.	42123.	1269274.
302	2.309	124.02	-4658.	41920.	1082580.
304	2.578	124.30	-4521.	41784.	975753.
307	3.032	124.73	-4327.	41581.	837589.
309	3.370	125.02	-4204.	41446.	758080.
312	3.938	125.45	-4030.	41243.	654690.
314	4.360	125.75	-3920.	41108.	594872.
317	5.064	126.20	-3763.	40905.	516687.
319	5.585	126.50	-3664.	40770.	471220.
322	6.450	126.96	-3522.	40568.	411504.
324	7.088	127.27	-3432.	40433.	376609.
327	8.143		-3304.		330566.
329	8.916		-3223.		303537.
332	10.19		-3106.		267717.
334	11.12		-3032.		246597.
337	12.65		-2926.		218492.
339	13.77		-2859.		201853.
342	15.59		-2762.		179621.
344	16.91		-2700.		166408.
347	19.06		-2612.		148687.
349	20.62		-2555.		138115.
352	23.15		-2474.		123886.
354	24.98		-2422.		115367.
357	27.93		-2347.		103861.
359	30.06		-2299.		96950.
362	33.49		-2230.		87584.
364	35.95		-2186.		81940.
367	39.91		-2123.		74269.
369	42.74		-2082.		69631.
372	47.29		-2023.		63309.
374	50.54		-1985.		59476.
377	55.74		-1931.		54236.
379	59.44		-1895.		51049.
382	65.35		-1845.		46682.
384	69.54		-1812.		44019.
387	76.24		-1765.		40360.
389	80.98		-1734.		38123.
392	88.52		-1690.		35042.
394	93.86		-1662.		33154.
397	102.3		-1621.		30546.
399	108.3		-1594.		28945.
402	117.8		-1556.		26728.
404	124.4		-1531.		25364.
407	135.0		-1495.		23471.

COMPOUND CONSTANTS:

MP	189.181 K	TC	571.000 K	VC	ZC
NBP	388.856 K	PC	3.270 MPA	DC	OM 0.3663
MU		RG			

VAPOR PRESSURE CORRELATION CONSTANTS, K AND KPA:

RIEDEL	WAGNER	WAGNER	WAGNER
283. - 567. K	283. - 389. K	283. - 484. K	388. - 567. K
29 POINTS	11 POINTS	15 POINTS	19 POINTS
RMSD = 1.8630	RMSD = 0.0840	RMSD = 1.3250	RMSD = 1.6100
0.60863450E+02	-0.54683310E+01	-0.69749525E+01	-0.77857409E+01
-0.64483400E+01	-0.42671786E+01	-0.78241822E+00	0.11556420E+01
-0.68459575E+04	0.52946853E+01	0.41328603E+00	0.14013432E+01
0.10731339E-16	-0.22805945E+02	-0.14350688E+02	-0.12915196E+03

LIQUID DENSITY CORRELATION CONSTANTS, K AND G/ML:

FRANCIS1	FRANCIS1	FRANCIS1
273. - 374. K	273. - 374. K	273. - 374. K
18 POINTS	18 POINTS	18 POINTS
RMSD = 0.00012	RMSD = 0.00012	RMSD = 0.00012
0.10666428E+01	0.10666428E+01	0.10666428E+01
0.82335575E-03	0.82335575E-03	0.82335575E-03
0.69826603E+01	0.69826603E+01	0.69826603E+01
0.57530029E+03	0.57530029E+03	0.57530029E+03

SECOND VIRIAL COEFFICIENT CORRELATION CONSTANTS, ML/MOL:

TSONOPOULOS: TC=571.000 PC=3.270 VC= 0.00 OM=0.3663 A=-0.015000 B=0.0000

LITERATURE DOCUMENTS FOR CORRELATED VAPOR PRESSURE VALUES:

1091   5067 11773 13065

LITERATURE DOCUMENTS FOR CORRELATED LIQUID DENSITY VALUES:

5115   7470   9138 11733 11773 15411 22902 23180 40228 41379 41407 41622

LITERATURE DOCUMENTS REPORTING VIRIAL COEFFICIENT DATA:

T,K	VAPOR PRESSURE, KPA	SATURATED LIQUID VOLUME, ML/MOL	SECOND VIRIAL COEFFICIENT, ML/MOL	HEAT OF VAPORIZATION, J/MOL	SATURATED VAPOR VOLUME, ML/MOL
273		122.33	-6791.		
279		123.14	-6102.		
286	1.300	124.11	-5418.	41461.	1823138.
293	1.971	125.10	-4842.	41059.	1230916.
299	2.765	125.96	-4417.	40709.	894674.
306	4.022	126.98	-3988.	40294.	628482.
313	5.737	128.03	-3619.	39871.	449995.
319	7.663	128.95	-3343.	39500.	342749.
326	10.57	130.04	-3059.	39059.	253332.
333	14.35	131.16	-2812.	38606.	190137.
339	18.42	132.14	-2623.	38209.	150379.
346	24.32	133.31	-2428.	37734.	115796.
353	31.70	134.52	-2254.	37246.	90280.
359	39.38	135.58	-2121.	36817.	73608.
366	50.18	136.84	-1980.	36304.	58591.
373	63.24	138.15	-1854.	35776.	47108.
379	76.48		-1755.		39367.
386	94.59		-1651.		32188.
393	115.9		-1556.		26532.
399	137.1		-1481.		22617.
406	165.4		-1401.		18896.
413	198.0		-1328.		15890.
419	229.8		-1270.		13762.
426	271.6		-1207.		11695.
433	318.9		-1149.		9989.
440	372.3		-1096.		8570.
446	423.2		-1053.		7539.
453	489.0		-1006.		6513.
460	562.2		-962.1		5644.
466	631.2		-927.0		5001.
473	719.6		-888.4		4349.
480	816.9		-852.3		3785.
486	908.0		-823.1		3360.
493	1024.		-790.9		3109.
500	1150.		-760.5		2731.
506	1268.		-735.9		2445.
513	1417.		-708.6		2147.
520	1579.		-682.8		1883.
526	1729.		-661.7		1680.
533	1919.		-638.4		1466.
540	2125.		-616.2		1272.
546	2315.		-598.0		1118.
553	2556.		-577.8		951.1
560	2817.		-558.5		789.9
566	3058.		-542.6		645.1

COMPOUND CONSTANTS:

MP	TC	VC	ZC
NBP 388.032 K	PC	DC	OM
MU	RG		

VAPOR PRESSURE CORRELATION CONSTANTS, K AND KPA:

RPM2	RPM2	RPM2
300. - 389. K	300. - 389. K	300. - 389. K
12 POINTS	12 POINTS	12 POINTS
RMSD = 0.5270	RMSD = 0.5270	RMSD = 0.5270
0.67406874E+02	0.67406874E+02	0.67406874E+02
-0.11369426E+05	-0.11369426E+05	-0.11369426E+05
-0.12515635E+00	-0.12515635E+00	-0.12515635E+00
0.10012928E-03	0.10012928E-03	0.10012928E-03

LIQUID DENSITY CORRELATION CONSTANTS, K AND G/ML:

SECOND VIRIAL COEFFICIENT CORRELATION CONSTANTS, ML/MOL:

LITERATURE DOCUMENTS FOR CORRELATED VAPOR PRESSURE VALUES:

8181 13065 15776 21544

LITERATURE DOCUMENTS FOR CORRELATED LIQUID DENSITY VALUES:

LITERATURE DOCUMENTS REPORTING VIRIAL COEFFICIENT DATA:

T,K	VAPOR PRESSURE, KPA	SATURATED LIQUID VOLUME, ML/MOL	SECOND VIRIAL COEFFICIENT, ML/MOL	HEAT OF VAPORIZATION, J/MOL	SATURATED VAPOR VOLUME, ML/MOL
300	2.647				942235.
302	2.988				840230.
304	3.365				751111.
306	3.780				673054.
308	4.236				604516.
310	4.736				544191.
312	5.284				490970.
314	5.881				443909.
316	6.532				402201.
318	7.241				365156.
320	8.009				332183.
322	8.842				302773.
324	9.743				276488.
326	10.72				252948.
328	11.76				231826.
330	12.89				212838.
332	14.10				195736.
334	15.40				180306.
336	16.79				166358.
338	18.28				153729.
340	19.87				142275.
342	21.56				131870.
344	23.37				122401.
346	25.29				113771.
348	27.32				105894.
350	29.49				98693.
352	31.78				92100.
354	34.20				86055.
356	36.77				80504.
358	39.48				75401.
360	42.33				70703.
362	45.35				66372.
364	48.52				62373.
366	51.86				58678.
368	55.37				55258.
370	59.06				52089.
372	62.93				49149.
374	66.99				46419.
376	71.25				43880.
378	75.70				41516.
380	80.37				39314.
382	85.24				37259.
384	90.34				35340.
386	95.67				33546.
388	101.2				31867.

COMPOUND CONSTANTS:

MP   221.179 K   TC   567.000 K   VC   ZC
NBP   379.263 K   PC   3.470 MPA   DC   OM 0.3229
MU   2.8100 DEBYE   RG

   VAPOR PRESSURE CORRELATION CONSTANTS, K AND KPA:

   WAGNER   WAGNER   WAGNER   WAGNER
   284. - 406. K   284. - 395. K   284. - 406. K   362. - 406. K
    38 POINTS   36 POINTS   38 POINTS   16 POINTS
   RMSD = 0.0071   RMSD = 0.0069   RMSD = 0.0071   RMSD = 0.0088
   -0.76569055E+01   -0.76403269E+01   -0.76569055E+01   -0.76862355E+01
    0.16102765E+01   0.15718952E+01   0.16102765E+01   0.16766059E+01
   -0.36188184E+01   -0.35647812E+01   -0.36188184E+01   -0.36938294E+01
   -0.28748554E+01   -0.29702606E+01   -0.28748554E+01   -0.30089991E+01

   LIQUID DENSITY CORRELATION CONSTANTS, K AND G/ML:

   FRANCIS1   FRANCIS1   FRANCIS1
   293. - 354. K   293. - 354. K   293. - 354. K
    13 POINTS   13 POINTS   13 POINTS
   RMSD = 0.00003   RMSD = 0.00003   RMSD = 0.00003
   0.10912037E+01   0.10912037E+01   0.10912037E+01
   0.94584492E-03   0.94584492E-03   0.94584492E-03
   0.63908215E+01   0.63908215E+01   0.63908215E+01
   0.12265347E+04   0.12265347E+04   0.12265347E+04

   SECOND VIRIAL COEFFICIENT CORRELATION CONSTANTS, ML/MOL:

   TSONOPOULOS: TC=567.000 PC=3.470 VC= 0.00 OM=0.3229 A=-0.017200 B=0.0000

LITERATURE DOCUMENTS FOR CORRELATED VAPOR PRESSURE VALUES:

   10318 13065

LITERATURE DOCUMENTS FOR CORRELATED LIQUID DENSITY VALUES:

   10483

LITERATURE DOCUMENTS REPORTING VIRIAL COEFFICIENT DATA:

T,K	VAPOR PRESSURE, KPA	SATURATED LIQUID VOLUME, ML/MOL	SECOND VIRIAL COEFFICIENT, ML/MOL	HEAT OF VAPORIZATION, J/MOL	SATURATED VAPOR VOLUME, ML/MOL
284	1.951		-5023.		1205045.
286	2.192		-4860.		1079811.
289	2.601		-4630.		919185.
292	3.073		-4416.		785677.
295	3.615	124.37	-4216.	38473.	674229.
297	4.020	124.67	-4090.	38342.	610158.
300	4.699	125.11	-3911.	38145.	526914.
303	5.471	125.56	-3744.	37949.	456666.
306	6.348	126.01	-3587.	37753.	397161.
308	6.996	126.32	-3488.	37622.	362540.
311	8.070	126.77	-3347.	37427.	317043.
314	9.278	127.23	-3215.	37231.	278132.
317	10.63	127.70	-3090.	37036.	244741.
320	12.15	128.17	-2973.	36841.	215994.
322	13.25	128.48	-2898.	36710.	199045.
325	15.07	128.95	-2792.	36514.	176489.
328	17.08	129.43	-2692.	36318.	156908.
331	19.31	129.91	-2597.	36122.	139862.
333	20.93	130.23	-2536.	35991.	129720.
336	23.55	130.72	-2450.	35793.	116105.
339	26.44	131.21	-2368.	35595.	104165.
342	29.62	131.70	-2290.	35397.	93666.
344	31.90	132.03	-2241.	35264.	87369.
347	35.58	132.53	-2169.	35064.	78849.
350	39.61	133.04	-2102.	34864.	71309.
353	43.99	133.55	-2037.	34662.	64619.
356	48.75		-1976.		58671.
358	52.15		-1937.		55071.
361	57.60		-1880.		50158.
364	63.49		-1826.		45764.
367	69.86		-1774.		41827.
369	74.38		-1741.		39429.
372	81.58		-1693.		36136.
375	89.33		-1648.		33170.
378	97.65		-1604.		30493.
381	106.6		-1562.		28074.
383	112.8		-1535.		26590.
386	122.8		-1496.		24537.
389	133.5		-1458.		22674.
392	144.8		-1422.		20979.
394	152.8		-1399.		19934.
397	165.4		-1366.		18481.
400	178.8		-1333.		17155.
403	193.0		-1302.		15942.
405	203.0		-1282.		15190.

COMPOUND CONSTANTS:

MP	TC	VC	ZC
NBP	PC	DC	OM
MU	RG		

VAPOR PRESSURE CORRELATION CONSTANTS, K AND KPA:

LIQUID DENSITY CORRELATION CONSTANTS, K AND G/ML:

FRANCIS1	FRANCIS1	FRANCIS1
277. - 374. K	277. - 374. K	277. - 374. K
21 POINTS	21 POINTS	21 POINTS
RMSD = 0.00052	RMSD = 0.00052	RMSD = 0.00052
0.12592354E+01	0.12592354E+01	0.12592354E+01
0.73114224E-03	0.73114224E-03	0.73114224E-03
0.24115524E+02	0.24115524E+02	0.24115524E+02
0.65499878E+03	0.65499878E+03	0.65499878E+03

SECOND VIRIAL COEFFICIENT CORRELATION CONSTANTS, ML/MOL:

LITERATURE DOCUMENTS FOR CORRELATED VAPOR PRESSURE VALUES:

LITERATURE DOCUMENTS FOR CORRELATED LIQUID DENSITY VALUES:

13839

LITERATURE DOCUMENTS REPORTING VIRIAL COEFFICIENT DATA:

T,K	VAPOR PRESSURE, KPA	SATURATED LIQUID VOLUME, ML/MOL	SECOND VIRIAL COEFFICIENT, ML/MOL	HEAT OF VAPORIZATION, J/MOL	SATURATED VAPOR VOLUME, ML/MOL
277		114.96			
279		115.17			
281		115.38			
283		115.59			
285		115.80			
288		116.12			
290		116.34			
292		116.56			
294		116.77			
296		116.99			
299		117.33			
301		117.55			
303		117.77			
305		118.00			
307		118.23			
310		118.57			
312		118.80			
314		119.03			
316		119.27			
318		119.50			
321		119.86			
323		120.10			
325		120.34			
327		120.58			
329		120.82			
332		121.19			
334		121.44			
336		121.69			
338		121.94			
340		122.20			
343		122.58			
345		122.84			
347		123.10			
349		123.36			
351		123.63			
354		124.03			
356		124.30			
358		124.57			
360		124.85			
362		125.12			
365		125.54			
367		125.82			
369		126.11			
371		126.40			
373		126.69			

COMPOUND CONSTANTS:

MP 238.171 K	TC 611.500 K	VC	ZC
NBP 424.206 K	PC 3.436 MPA	DC	OM 0.4857
MU 2.6100 DEBYE	RG		

VAPOR PRESSURE CORRELATION CONSTANTS, K AND KPA:

WAGNER	WAGNER	WAGNER	WAGNER
274. - 453. K	274. - 442. K	274. - 453. K	407. - 453. K
46 POINTS	44 POINTS	46 POINTS	11 POINTS
RMSD = 0.0602	RMSD = 0.0396	RMSD = 0.0602	RMSD = 0.0124
-0.99483801E+01	-0.98709446E+01	-0.99483801E+01	-0.11049744E+02
0.47053877E+01	0.45299059E+01	0.47053877E+01	0.76039453E+01
-0.67889499E+01	-0.65596530E+01	-0.67889499E+01	-0.12885185E+02
-0.17144393E+00	-0.50714254E+00	-0.17144393E+00	0.25715023E+02

LIQUID DENSITY CORRELATION CONSTANTS, K AND G/ML:

FRANCIS1	FRANCIS1	FRANCIS1
288. - 429. K	288. - 429. K	288. - 429. K
16 POINTS	16 POINTS	16 POINTS
RMSD = 0.00011	RMSD = 0.00011	RMSD = 0.00011
0.10823164E+01	0.10823164E+01	0.10823164E+01
0.68128319E-03	0.68128319E-03	0.68128319E-03
0.28077332E+02	0.28077332E+02	0.28077332E+02
0.71180640E+03	0.71180640E+03	0.71180640E+03

SECOND VIRIAL COEFFICIENT CORRELATION CONSTANTS, ML/MOL:

TSONOPOULOS: TC=611.500 PC=3.436 VC= 0.00 OM=0.4857 A=-0.012700 B=0.0000

LITERATURE DOCUMENTS FOR CORRELATED VAPOR PRESSURE VALUES:

1499 10318

LITERATURE DOCUMENTS FOR CORRELATED LIQUID DENSITY VALUES:

9712 10483 12038

LITERATURE DOCUMENTS REPORTING VIRIAL COEFFICIENT DATA:

T,K	VAPOR PRESSURE, KPA	SATURATED LIQUID VOLUME, ML/MOL	SECOND VIRIAL COEFFICIENT, ML/MOL	HEAT OF VAPORIZATION, J/MOL	SATURATED VAPOR VOLUME, ML/MOL
274	0.09403		-10657.		0.2422E+08
278	0.1279		-9853.		0.1806E+08
282	0.1723		-9130.		0.1360E+08
286	0.2297		-8479.		0.1034E+08
290	0.3034	139.56	-7892.	47781.	7938165.
294	0.3972	140.14	-7360.	47504.	6147494.
298	0.5153	140.72	-6878.	47228.	4801239.
302	0.6631	141.31	-6440.	46952.	3780286.
306	0.8465	141.91	-6041.	46676.	2999572.
310	1.072	142.51	-5677.	46399.	2397797.
314	1.349	143.12	-5344.	46123.	1930394.
318	1.684	143.75	-5040.	45846.	1564691.
322	2.089	144.37	-4760.	45569.	1276545.
326	2.575	145.01	-4502.	45292.	1047972.
330	3.155	145.66	-4265.	45014.	865479.
335	4.032	146.47	-3994.	44667.	686867.
339	4.874	147.14	-3796.	44389.	574417.
343	5.862	147.81	-3612.	44110.	482884.
347	7.012	148.50	-3441.	43831.	407969.
351	8.347	149.19	-3283.	43551.	346330.
355	9.887	149.90	-3135.	43271.	295358.
359	11.66	150.61	-2997.	42990.	253001.
363	13.68	151.34	-2869.	42708.	217637.
367	16.00	152.08	-2749.	42426.	187977.
371	18.62	152.82	-2636.	42143.	162994.
375	21.59	153.59	-2531.	41860.	141860.
379	24.93	154.36	-2431.	41576.	123910.
383	28.69	155.15	-2338.	41291.	108604.
387	32.90	155.95	-2250.	41005.	95504.
391	37.59	156.76	-2168.	40718.	84252.
396	44.21	157.80	-2071.	40358.	72339.
400	50.16	158.65	-1998.	40070.	64243.
404	56.73	159.51	-1929.	39780.	57211.
408	63.98	160.39	-1864.	39490.	51085.
412	71.95	161.29	-1803.	39198.	45731.
416	80.70	162.20	-1744.	38905.	41038.
420	90.28	163.13	-1688.	38611.	36913.
424	100.7	164.09	-1635.	38316.	33277.
428	112.1	165.06	-1585.	38020.	30063.
432	124.5		-1537.		27216.
436	138.0		-1491.		24687.
440	152.5		-1447.		22434.
444	168.3		-1406.		20424.
448	185.3		-1366.		18625.
452	203.7		-1328.		17013.

649

COMPOUND CONSTANTS:

MP   234.173 K   TC   VC   ZC
NBP   420.569 K   PC   DC   OM
MU   2.8100 DEBYE   RG

VAPOR PRESSURE CORRELATION CONSTANTS, K AND KPA:

LIQUID DENSITY CORRELATION CONSTANTS, K AND G/ML:

FRANCIS1   FRANCIS1   FRANCIS1
288. - 360. K   288. - 360. K   288. - 360. K
  7 POINTS   7 POINTS   7 POINTS
RMSD = 0.00014   RMSD = 0.00014   RMSD = 0.00014
0.10566082E+01   0.10566082E+01   0.10566082E+01
0.76659396E-03   0.76659396E-03   0.76659396E-03
0.29826403E+01   0.29826403E+01   0.29826403E+01
0.51027295E+03   0.51027295E+03   0.51027295E+03

SECOND VIRIAL COEFFICIENT CORRELATION CONSTANTS, ML/MOL:

LITERATURE DOCUMENTS FOR CORRELATED VAPOR PRESSURE VALUES:

LITERATURE DOCUMENTS FOR CORRELATED LIQUID DENSITY VALUES:

    7601 14417 23251

LITERATURE DOCUMENTS REPORTING VIRIAL COEFFICIENT DATA:

T,K	VAPOR PRESSURE, KPA	SATURATED LIQUID VOLUME, ML/MOL	SECOND VIRIAL COEFFICIENT, ML/MOL	HEAT OF VAPORIZATION, J/MOL	SATURATED VAPOR VOLUME, ML/MOL
288		138.84			
289		138.98			
291		139.26			
292		139.41			
294		139.69			
296		139.97			
297		140.12			
299		140.40			
301		140.69			
302		140.84			
304		141.13			
305		141.27			
307		141.57			
309		141.86			
310		142.01			
312		142.31			
314		142.61			
315		142.76			
317		143.06			
319		143.37			
320		143.52			
322		143.83			
323		143.98			
325		144.29			
327		144.60			
328		144.76			
330		145.08			
332		145.39			
333		145.55			
335		145.87			
337		146.20			
338		146.36			
340		146.69			
341		146.85			
343		147.18			
345		147.51			
346		147.68			
348		148.02			
350		148.35			
351		148.52			
353		148.87			
355		149.21			
356		149.39			
358		149.74			
359		149.91			

651

COMPOUND CONSTANTS:

MP 240.669 K	TC	VC	ZC
NBP 417.259 K	PC	DC	OM
MU 2.5000 DEBYE	RG		

VAPOR PRESSURE CORRELATION CONSTANTS, K AND KPA:

RPM2	RPM2	RPM2
273. - 418. K	273. - 418. K	273. - 418. K
7 POINTS	7 POINTS	7 POINTS
RMSD = 0.0132	RMSD = 0.0132	RMSD = 0.0132
0.12619969E+02	0.12619969E+02	0.12619969E+02
-0.44925265E+04	-0.44925265E+04	-0.44925265E+04
0.11623943E-01	0.11623943E-01	0.11623943E-01
-0.11975873E-04	-0.11975873E-04	-0.11975873E-04

LIQUID DENSITY CORRELATION CONSTANTS, K AND G/ML:

FRANCIS1	FRANCIS1	FRANCIS1
253. - 394. K	253. - 394. K	253. - 394. K
10 POINTS	10 POINTS	10 POINTS
RMSD = 0.00041	RMSD = 0.00041	RMSD = 0.00041
0.10744505E+01	0.10744505E+01	0.10744505E+01
0.73973392E-03	0.73973392E-03	0.73973392E-03
0.12342205E+02	0.12342205E+02	0.12342205E+02
0.59130103E+03	0.59130103E+03	0.59130103E+03

SECOND VIRIAL COEFFICIENT CORRELATION CONSTANTS, ML/MOL:

LITERATURE DOCUMENTS FOR CORRELATED VAPOR PRESSURE VALUES:

1091   2314 11733 16695

LITERATURE DOCUMENTS FOR CORRELATED LIQUID DENSITY VALUES:

7470 11733

LITERATURE DOCUMENTS REPORTING VIRIAL COEFFICIENT DATA:

T,K	VAPOR PRESSURE, KPA	SATURATED LIQUID VOLUME, ML/MOL	SECOND VIRIAL COEFFICIENT, ML/MOL	HEAT OF VAPORIZATION, J/MOL	SATURATED VAPOR VOLUME, ML/MOL
253		134.21			
256		134.61			
260		135.15			
264		135.70			
268		136.26			
271		136.68			
275	0.2404	137.24			9512316.
279	0.3099	137.82			7484666.
283	0.3968	138.40			5929407.
286	0.4755	138.84			5000423.
290	0.6019	139.43			4006204.
294	0.7570	140.03			3229265.
298	0.9463	140.64			2618290.
301	1.115	141.10			2245510.
305	1.379	141.72			1838455.
309	1.698	142.35			1513155.
313	2.079	142.99			1251770.
316	2.412	143.48			1089327.
320	2.928	144.13			908827.
324	3.536	144.80			761742.
328	4.252	145.47			641315.
331	4.869	145.98			565262.
335	5.809	146.68			479455.
339	6.903	147.38			408329.
343	8.169	148.10			349126.
346	9.244	148.64			311207.
350	10.86	149.38			267845.
354	12.72	150.13			231357.
358	14.84	150.90			200540.
361	16.62	151.48			180555.
365	19.28	152.27			157426.
369	22.28	153.07			137703.
373	25.67	153.89			120828.
376	28.48	154.52			109762.
380	32.63	155.37			96819.
384	37.28	156.24			85650.
388	42.46	157.13			75983.
391	46.72	157.80			69581.
395	52.95				62021.
399	59.85				55427.
403	67.47				49660.
406	73.69				45807.
410	82.70				41219.
414	92.59				37178.
418	103.4				33610.

COMPOUND CONSTANTS:

MP		TC	VC	ZC
NBP	406.531 K	PC	DC	OM
MU		RG		

VAPOR PRESSURE CORRELATION CONSTANTS, K AND KPA:

RPM2	RPM2	RPM2	RPM2
349. - 451. K	349. - 421. K	349. - 451. K	390. - 451. K
11 POINTS	8 POINTS	11 POINTS	7 POINTS
RMSD = 0.0007	RMSD = 0.0004	RMSD = 0.0007	RMSD = 0.0007
0.16894049E+02	0.16968053E+02	0.16894049E+02	0.16781712E+02
-0.49772008E+04	-0.49868283E+04	-0.49772008E+04	-0.49615208E+04
-0.12070011E-03	-0.30987682E-03	-0.12070011E-03	0.14711739E-03
0.99568801E-07	0.26042134E-06	0.99568801E-07	-0.11291159E-06

LIQUID DENSITY CORRELATION CONSTANTS, K AND G/ML:

SECOND VIRIAL COEFFICIENT CORRELATION CONSTANTS, ML/MOL:

LITERATURE DOCUMENTS FOR CORRELATED VAPOR PRESSURE VALUES:

13799

LITERATURE DOCUMENTS FOR CORRELATED LIQUID DENSITY VALUES:

LITERATURE DOCUMENTS REPORTING VIRIAL COEFFICIENT DATA:

T,K	VAPOR PRESSURE, KPA	SATURATED LIQUID VOLUME, ML/MOL	SECOND VIRIAL COEFFICIENT, ML/MOL	HEAT OF VAPORIZATION, J/MOL	SATURATED VAPOR VOLUME, ML/MOL
349	13.50				214934.
351	14.64				199315.
353	15.87				184995.
355	17.18				171854.
358	19.31				154118.
360	20.86				143474.
362	22.52				133674.
365	25.21				120397.
367	27.15				112398.
369	29.22				105011.
372	32.57				94967.
374	34.98				88893.
376	37.54				83269.
379	41.69				75594.
381	44.66				70937.
383	47.80				66613.
386	52.88				60693.
388	56.51				57090.
390	60.34				53735.
393	66.51				49129.
395	70.91				46315.
397	75.55				43691.
399	80.44				41240.
402	88.28				37861.
404	93.85				35790.
406	99.72				33852.
409	109.1				31173.
411	115.7				29527.
413	122.7				27983.
416	133.8				25843.
418	141.7				24524.
420	150.0				23285.
423	163.1				21563.
425	172.4				20500.
427	182.1				19498.
430	197.5				18103.
432	208.3				17240.
434	219.7				16425.
437	237.7				15288.
439	250.3				14583.
441	263.5				13916.
444	284.3				12983.
446	299.0				12403.
448	314.2				11854.
450	330.1				11334.

COMPOUND CONSTANTS:

MP	204.183 K	TC	VC	ZC
NBP	396.818 K	PC	DC	OM
MU	2.7300 DEBYE	RG		

VAPOR PRESSURE CORRELATION CONSTANTS, K AND KPA:

VAPRES-2	RPM2	VAPRES-2	RIEDEL
293. - 441. K	293. - 411. K	293. - 441. K	380. - 441. K
15 POINTS	12 POINTS	15 POINTS	7 POINTS
RMSD = 0.0265	RMSD = 0.0305	RMSD = 0.0265	RMSD = 0.0004
0.47099357E+03	0.31024959E+02	0.47099357E+03	0.16486399E+02
-0.15989232E+05	-0.65104600E+04	-0.15989232E+05	-0.20851789E-01
0.19827006E+00	-0.38616077E-01	0.19827006E+00	-0.46599068E+04
-0.79336319E-04	0.33808294E-04	-0.79336319E-04	0.11864683E-18
-0.82270805E+02		-0.82270805E+02	

LIQUID DENSITY CORRELATION CONSTANTS, K AND G/ML:

FRANCIS1	FRANCIS1	FRANCIS1
283. - 374. K	283. - 374. K	283. - 374. K
24 POINTS	24 POINTS	24 POINTS
RMSD = 0.00040	RMSD = 0.00040	RMSD = 0.00040
0.10650043E+01	0.10650043E+01	0.10650043E+01
0.79409732E-03	0.79409732E-03	0.79409732E-03
0.79816027E+01	0.79816027E+01	0.79816027E+01
0.59359106E+03	0.59359106E+03	0.59359106E+03

SECOND VIRIAL COEFFICIENT CORRELATION CONSTANTS, ML/MOL:

LITERATURE DOCUMENTS FOR CORRELATED VAPOR PRESSURE VALUES:

   1361   7406 13799

LITERATURE DOCUMENTS FOR CORRELATED LIQUID DENSITY VALUES:

   10483 13155

LITERATURE DOCUMENTS REPORTING VIRIAL COEFFICIENT DATA:

T,K	VAPOR PRESSURE, KPA	SATURATED LIQUID VOLUME, ML/MOL	SECOND VIRIAL COEFFICIENT, ML/MOL	HEAT OF VAPORIZATION, J/MOL	SATURATED VAPOR VOLUME, ML/MOL
283		140.18			
286		140.63			
290		141.25			
293	1.481	141.71		40853.	1645072.
297	1.855	142.33		40629.	1331213.
300	2.186	142.81		40471.	1141009.
304	2.705	143.44		40272.	934418.
308	3.325	144.09		40088.	770096.
311	3.867	144.58		39959.	668754.
315	4.703	145.24		39797.	556862.
318	5.427	145.74		39685.	487167.
322	6.538	146.41		39546.	409494.
326	7.835	147.09		39418.	345936.
329	8.946	147.61		39329.	305789.
333	10.63	148.31		39221.	260454.
336	12.06	148.84		39146.	231588.
340	14.22	149.55		39054.	198745.
344	16.70	150.27		38972.	171249.
347	18.79	150.82		38916.	153543.
351	21.91	151.57		38848.	133179.
354	24.53	152.13		38802.	119981.
358	28.43	152.89		38747.	104708.
361	31.68	153.47		38710.	94749.
365	36.49	154.26		38666.	83158.
369	41.91	155.05		38627.	73210.
372	46.39	155.66		38602.	66666.
376	53.00				58989.
379	58.45				53914.
383	66.44				47931.
387	75.32				42721.
390	82.61				39253.
394	93.23				35136.
397	101.9				32384.
401	114.6				29104.
405	128.5				26214.
408	139.8				24270.
412	156.1				21940.
415	169.4				20367.
419	188.6				18476.
423	209.4				16792.
426	226.3				15650.
430	250.6				14269.
433	270.1				13330.
437	298.0				12192.
440	320.5				11415.

COMPOUND CONSTANTS:

MP	TC	VC	ZC
NBP	PC	DC	OM
MU	RG		

VAPOR PRESSURE CORRELATION CONSTANTS, K AND KPA:

LIQUID DENSITY CORRELATION CONSTANTS, K AND G/ML:

FRANCIS1	FRANCIS1	FRANCIS1
277. - 374. K	277. - 374. K	277. - 374. K
18 POINTS	18 POINTS	18 POINTS
RMSD = 0.00010	RMSD = 0.00010	RMSD = 0.00010
0.12222862E+01	0.12222862E+01	0.12222862E+01
0.81374589E-03	0.81374589E-03	0.81374589E-03
0.84028273E+01	0.84028273E+01	0.84028273E+01
0.60423389E+03	0.60423389E+03	0.60423389E+03

SECOND VIRIAL COEFFICIENT CORRELATION CONSTANTS, ML/MOL:

LITERATURE DOCUMENTS FOR CORRELATED VAPOR PRESSURE VALUES:

LITERATURE DOCUMENTS FOR CORRELATED LIQUID DENSITY VALUES:

13839

LITERATURE DOCUMENTS REPORTING VIRIAL COEFFICIENT DATA:

T,K	VAPOR PRESSURE, KPA	SATURATED LIQUID VOLUME, ML/MOL	SECOND VIRIAL COEFFICIENT, ML/MOL	HEAT OF VAPORIZATION, J/MOL	SATURATED VAPOR VOLUME, ML/MOL
277		131.97			
279		132.21			
281		132.46			
283		132.70			
285		132.95			
288		133.32			
290		133.57			
292		133.82			
294		134.08			
296		134.33			
299		134.71			
301		134.97			
303		135.23			
305		135.49			
307		135.75			
310		136.14			
312		136.40			
314		136.67			
316		136.94			
318		137.20			
321		137.61			
323		137.88			
325		138.16			
327		138.43			
329		138.71			
332		139.13			
334		139.41			
336		139.69			
338		139.97			
340		140.26			
343		140.69			
345		140.98			
347		141.27			
349		141.57			
351		141.86			
354		142.31			
356		142.61			
358		142.91			
360		143.22			
362		143.53			
365		143.99			
367		144.30			
369		144.61			
371		144.93			
373		145.25			

COMPOUND CONSTANTS:

MP	TC	VC	ZC
NBP	PC	DC	OM
MU	RG		

VAPOR PRESSURE CORRELATION CONSTANTS, K AND KPA:

LIQUID DENSITY CORRELATION CONSTANTS, K AND G/ML:

FRANCIS1	FRANCIS1	FRANCIS1
288. - 374. K	288. - 374. K	288. - 374. K
20 POINTS	20 POINTS	20 POINTS
RMSD = 0.00018	RMSD = 0.00018	RMSD = 0.00018
0.12179537E+01	0.12179537E+01	0.12179537E+01
0.80069271E-03	0.80069271E-03	0.80069271E-03
0.10817964E+02	0.10817964E+02	0.10817964E+02
0.63199292E+03	0.63199292E+03	0.63199292E+03

SECOND VIRIAL COEFFICIENT CORRELATION CONSTANTS, ML/MOL:

LITERATURE DOCUMENTS FOR CORRELATED VAPOR PRESSURE VALUES:

LITERATURE DOCUMENTS FOR CORRELATED LIQUID DENSITY VALUES:

182 13846

LITERATURE DOCUMENTS REPORTING VIRIAL COEFFICIENT DATA:

T,K	VAPOR PRESSURE, KPA	SATURATED LIQUID VOLUME, ML/MOL	SECOND VIRIAL COEFFICIENT, ML/MOL	HEAT OF VAPORIZATION, J/MOL	SATURATED VAPOR VOLUME, ML/MOL
288		134.08			
289		134.21			
291		134.46			
293		134.71			
295		134.97			
297		135.22			
299		135.48			
301		135.74			
303		136.00			
305		136.26			
307		136.52			
309		136.78			
311		137.05			
313		137.31			
315		137.58			
317		137.85			
319		138.12			
321		138.39			
323		138.67			
325		138.94			
327		139.22			
329		139.49			
330		139.63			
332		139.92			
334		140.20			
336		140.48			
338		140.77			
340		141.05			
342		141.34			
344		141.63			
346		141.93			
348		142.22			
350		142.52			
352		142.81			
354		143.11			
356		143.41			
358		143.72			
360		144.02			
362		144.33			
364		144.64			
366		144.95			
368		145.27			
370		145.58			
372		145.90			
373		146.06			

COMPOUND CONSTANTS:

MP   TC   VC   ZC
NBP   PC   DC   OM
MU   RG

VAPOR PRESSURE CORRELATION CONSTANTS, K AND KPA:

LIQUID DENSITY CORRELATION CONSTANTS, K AND G/ML:

FRANCIS1   FRANCIS1   FRANCIS1
313. - 374. K   313. - 374. K   313. - 374. K
 13 POINTS   13 POINTS   13 POINTS
RMSD = 0.00031   RMSD = 0.00031   RMSD = 0.00031
0.13543797E+01   0.13543797E+01   0.13543797E+01
0.89812255E-03   0.89812255E-03   0.89812255E-03
0.94656706E+01   0.94656706E+01   0.94656706E+01
0.10463357E+04   0.10463357E+04   0.10463357E+04

SECOND VIRIAL COEFFICIENT CORRELATION CONSTANTS, ML/MOL:

LITERATURE DOCUMENTS FOR CORRELATED VAPOR PRESSURE VALUES:

LITERATURE DOCUMENTS FOR CORRELATED LIQUID DENSITY VALUES:

  13839

LITERATURE DOCUMENTS REPORTING VIRIAL COEFFICIENT DATA:

T,K	VAPOR PRESSURE, KPA	SATURATED LIQUID VOLUME, ML/MOL	SECOND VIRIAL COEFFICIENT, ML/MOL	HEAT OF VAPORIZATION, J/MOL	SATURATED VAPOR VOLUME, ML/MOL
313		134.06			
314		134.18			
315		134.29			
317		134.53			
318		134.64			
319		134.76			
321		134.99			
322		135.11			
324		135.35			
325		135.47			
326		135.58			
328		135.82			
329		135.94			
331		136.18			
332		136.30			
333		136.42			
335		136.66			
336		136.78			
337		136.90			
339		137.14			
340		137.27			
342		137.51			
343		137.63			
344		137.75			
346		138.00			
347		138.12			
349		138.37			
350		138.49			
351		138.62			
353		138.86			
354		138.99			
355		139.11			
357		139.36			
358		139.49			
360		139.74			
361		139.87			
362		139.99			
364		140.25			
365		140.38			
367		140.63			
368		140.76			
369		140.89			
371		141.14			
372		141.27			
373		141.40			

COMPOUND CONSTANTS:

MP   252.862 K   TC   VC   ZC
NBP   446.425 K   PC   DC   OM
MU   2.7200 DEBYE   RG

VAPOR PRESSURE CORRELATION CONSTANTS, K AND KPA:

RPM2	RPM2	RPM2
293. - 447. K	293. - 447. K	293. - 447. K
12 POINTS	12 POINTS	12 POINTS
RMSD = 0.1342	RMSD = 0.1342	RMSD = 0.1342
-0.58281733E+02	-0.58281733E+02	-0.58281733E+02
0.46022852E+04	0.46022852E+04	0.46022852E+04
0.19032262E+00	0.19032262E+00	0.19032262E+00
-0.16244189E-03	-0.16244189E-03	-0.16244189E-03

LIQUID DENSITY CORRELATION CONSTANTS, K AND G/ML:

FRANCIS1	FRANCIS1	FRANCIS1
253. - 434. K	253. - 434. K	253. - 434. K
17 POINTS	17 POINTS	17 POINTS
RMSD = 0.00026	RMSD = 0.00026	RMSD = 0.00026
0.10774565E+01	0.10774565E+01	0.10774565E+01
0.79408265E-03	0.79408265E-03	0.79408265E-03
0.89678802E+01	0.89678802E+01	0.89678802E+01
0.64906519E+03	0.64906519E+03	0.64906519E+03

SECOND VIRIAL COEFFICIENT CORRELATION CONSTANTS, ML/MOL:

LITERATURE DOCUMENTS FOR CORRELATED VAPOR PRESSURE VALUES:

    5842   9712 15791 18439

LITERATURE DOCUMENTS FOR CORRELATED LIQUID DENSITY VALUES:

    5736   5846   7470 19380 40489

LITERATURE DOCUMENTS REPORTING VIRIAL COEFFICIENT DATA:

T,K	VAPOR PRESSURE, KPA	SATURATED LIQUID VOLUME, ML/MOL	SECOND VIRIAL COEFFICIENT, ML/MOL	HEAT OF VAPORIZATION, J/MOL	SATURATED VAPOR VOLUME, ML/MOL
253		150.15			
257		150.75			
261		151.36			
266		152.13			
270		152.75			
275		153.53			
279		154.16			
283		154.80			
288		155.61			
292		156.27			
297	0.5550	157.10			4449572.
301	0.6557	157.77			3816686.
305	0.7749	158.45			3272575.
310	0.9548	159.30			2699622.
314	1.128	160.00			2314560.
319	1.388	160.88			1910310.
323	1.638	161.59			1639272.
327	1.931	162.31			1407638.
332	2.370	163.22			1164955.
336	2.787	163.96			1002443.
341	3.407	164.89			832189.
345	3.994	165.65			718131.
349	4.676	166.42			620622.
354	5.679	167.39			518311.
358	6.620	168.18			449625.
363	7.997	169.18			377416.
367	9.280	169.99			328827.
372	11.14	171.02			277612.
376	12.86	171.86			243051.
380	14.81	172.71			213285.
385	17.61	173.78			181766.
389	20.17	174.66			160389.
394	23.80	175.77			137671.
398	27.08	176.68			122203.
402	30.73	177.60			108774.
407	35.84	178.77			94419.
411	40.40	179.72			84588.
416	46.72	180.93			74036.
420	52.29	181.92			66780.
424	58.34	182.92			60423.
429	66.60	184.20			53560.
433	73.75	185.25			48813.
438	83.39				43669.
442	91.65				40098.
446	100.4				36945.

COMPOUND CONSTANTS:

MP		TC		VC		ZC
NBP	440.863 K	PC		DC		OM
MU		RG				

VAPOR PRESSURE CORRELATION CONSTANTS, K AND KPA:

RPM2	RPM2	RPM2
293. - 441. K	293. - 441. K	293. - 441. K
11 POINTS	11 POINTS	11 POINTS
RMSD = 0.1354	RMSD = 0.1354	RMSD = 0.1354
-0.16377240E+02	-0.16377240E+02	-0.16377240E+02
-0.14125767E+04	-0.14125767E+04	-0.14125767E+04
0.91139663E-01	0.91139663E-01	0.91139663E-01
-0.82220634E-04	-0.82220634E-04	-0.82220634E-04

LIQUID DENSITY CORRELATION CONSTANTS, K AND G/ML:

FRANCIS1	FRANCIS1	FRANCIS1
293. - 360. K	293. - 360. K	293. - 360. K
6 POINTS	6 POINTS	6 POINTS
RMSD = 0.00013	RMSD = 0.00013	RMSD = 0.00013
0.10447025E+01	0.10447025E+01	0.10447025E+01
0.75790938E-03	0.75790938E-03	0.75790938E-03
0.56298885E-01	0.56298885E-01	0.56298885E-01
0.37720679E+03	0.37720679E+03	0.37720679E+03

SECOND VIRIAL COEFFICIENT CORRELATION CONSTANTS, ML/MOL:

LITERATURE DOCUMENTS FOR CORRELATED VAPOR PRESSURE VALUES:

13138 14417 18439

LITERATURE DOCUMENTS FOR CORRELATED LIQUID DENSITY VALUES:

13138 14417

LITERATURE DOCUMENTS REPORTING VIRIAL COEFFICIENT DATA:

T,K	VAPOR PRESSURE, KPA	SATURATED LIQUID VOLUME, ML/MOL	SECOND VIRIAL COEFFICIENT, ML/MOL	HEAT OF VAPORIZATION, J/MOL	SATURATED VAPOR VOLUME, ML/MOL
293	0.2116	155.98			0.1151E+08
296	0.2526	156.42			9743060.
299	0.3008	156.86			8265600.
303	0.3781	157.45			6662220.
306	0.4477	157.90			5682536.
309	0.5289	158.35			4857935.
313	0.6580	158.96			3955318.
316	0.7730	159.41			3399016.
319	0.9060	159.88			2927377.
323	1.116	160.50			2406841.
326	1.301	160.97			2083355.
329	1.514	161.44			1807189.
333	1.846	162.08			1499970.
336	2.137	162.57			1307526.
340	2.588	163.23			1092291.
343	2.981	163.73			956739.
346	3.426	164.24			839715.
350	4.111	164.93			707857.
353	4.702	165.47			624194.
356	5.367	166.03			551514.
360	6.382	166.82			469033.
363	7.249				416323.
366	8.219				370255.
370	9.686				317614.
373	10.93				283740.
377	12.80				244850.
380	14.38				219707.
383	16.12				197517.
387	18.72				171876.
390	20.89				155191.
393	23.28				140385.
397	26.80				123169.
400	29.72				111896.
403	32.90				101840.
407	37.57				90077.
410	41.41				82329.
414	47.01				73228.
417	51.59				67209.
420	56.51				61795.
424	63.63				55401.
427	69.41				51148.
430	75.58				47305.
434	84.42				42743.
437	91.54				39693.
440	99.08				36924.

667

COMPOUND CONSTANTS:

MP	TC	VC	ZC
NBP	PC	DC	OM
MU	RG		

VAPOR PRESSURE CORRELATION CONSTANTS, K AND KPA:

RPM2	RPM2	RPM2
288. - 424. K	288. - 424. K	288. - 424. K
12 POINTS	12 POINTS	12 POINTS
RMSD = 0.1901	RMSD = 0.1901	RMSD = 0.1901
0.51574275E+02	0.51574275E+02	0.51574275E+02
-0.86523422E+04	-0.86523422E+04	-0.86523422E+04
-0.10643159E+00	-0.10643159E+00	-0.10643159E+00
0.10186297E-03	0.10186297E-03	0.10186297E-03

LIQUID DENSITY CORRELATION CONSTANTS, K AND G/ML:

FRANCIS1	FRANCIS1	FRANCIS1
297. - 324. K	297. - 324. K	297. - 324. K
3 POINTS	3 POINTS	3 POINTS
RMSD = 0.00000	RMSD = 0.00000	RMSD = 0.00000
0.10581158E+01	0.10581158E+01	0.10581158E+01
0.81718800E-03	0.81718800E-03	0.81718800E-03

SECOND VIRIAL COEFFICIENT CORRELATION CONSTANTS, ML/MOL:

LITERATURE DOCUMENTS FOR CORRELATED VAPOR PRESSURE VALUES:

18439

LITERATURE DOCUMENTS FOR CORRELATED LIQUID DENSITY VALUES:

13138

LITERATURE DOCUMENTS REPORTING VIRIAL COEFFICIENT DATA:

T,K	VAPOR PRESSURE, KPA	SATURATED LIQUID VOLUME, ML/MOL	SECOND VIRIAL COEFFICIENT, ML/MOL	HEAT OF VAPORIZATION, J/MOL	SATURATED VAPOR VOLUME, ML/MOL
288	0.5107				4688677.
291	0.6038				4007437.
294	0.7105				3440215.
297	0.8327	157.24		38255.	2965685.
300	0.9717	157.71		38029.	2566883.
303	1.130	158.19		37813.	2230245.
306	1.308	158.67		37610.	1944876.
309	1.510	159.15		37418.	1701978.
312	1.736	159.64		37239.	1494416.
315	1.990	160.13		37072.	1316376.
318	2.273	160.62		36918.	1163099.
321	2.589	161.11		36777.	1030677.
325	3.067				881057.
328	3.471				785685.
331	3.918				702389.
334	4.412				629419.
337	4.956				565310.
340	5.556				508826.
343	6.214				458927.
346	6.937				414729.
349	7.728				375482.
352	8.594				340547.
355	9.541				309377.
359	10.94				272874.
362	12.10				248778.
365	13.36				227124.
368	14.74				207625.
371	16.23				190032.
374	17.86				174131.
377	19.62				159732.
380	21.54				146672.
383	23.62				134807.
386	25.88				124010.
389	28.33				114171.
393	31.92				102375.
396	34.87				94410.
399	38.08				87118.
402	41.55				80434.
405	45.32				74301.
408	49.40				68667.
411	53.83				63486.
414	58.62				58717.
417	63.82				54323.
420	69.46				50271.
423	75.58				46531.

COMPOUND CONSTANTS:

MP   TC   VC   ZC
NBP   428.531 K   PC   DC   OM
MU   RG

VAPOR PRESSURE CORRELATION CONSTANTS, K AND KPA:

RPM2   RPM2   RPM2
380. - 481. K   380. - 481. K   380. - 481. K
 11 POINTS   11 POINTS   11 POINTS
RMSD = 0.0006   RMSD = 0.0006   RMSD = 0.0006
 0.16856127E+02   0.16856127E+02   0.16856127E+02
-0.52338853E+04   -0.52338853E+04   -0.52338853E+04
-0.88892330E-04   -0.88892330E-04   -0.88892330E-04
 0.75498105E-07   0.75498105E-07   0.75498105E-07

LIQUID DENSITY CORRELATION CONSTANTS, K AND G/ML:

SECOND VIRIAL COEFFICIENT CORRELATION CONSTANTS, ML/MOL:

LITERATURE DOCUMENTS FOR CORRELATED VAPOR PRESSURE VALUES:

  13799

LITERATURE DOCUMENTS FOR CORRELATED LIQUID DENSITY VALUES:

LITERATURE DOCUMENTS REPORTING VIRIAL COEFFICIENT DATA:

T,K	VAPOR PRESSURE, KPA	SATURATED LIQUID VOLUME, ML/MOL	SECOND VIRIAL COEFFICIENT, ML/MOL	HEAT OF VAPORIZATION, J/MOL	SATURATED VAPOR VOLUME, ML/MOL
380	21.32				148160.
382	22.92				138586.
384	24.61				129725.
386	26.41				121517.
389	29.32				110312.
391	31.41				103511.
393	33.62				97195.
396	37.18				88546.
398	39.74				83279.
400	42.43				78376.
402	45.29				73808.
405	49.87				67528.
407	53.13				63689.
409	56.58				60105.
412	62.10				55163.
414	66.02				52134.
416	70.16				49300.
419	76.77				45381.
421	81.46				42973.
423	86.38				40714.
425	91.56				38595.
428	99.80				35656.
430	105.6				33843.
432	111.8				32139.
435	121.5				29770.
437	128.4				28306.
439	135.6				26927.
441	143.1				25627.
444	155.0				23815.
446	163.4				22691.
448	172.2				21631.
451	186.1				20149.
453	195.9				19228.
455	206.1				18357.
458	222.2				17138.
460	233.5				16379.
462	245.3				15660.
464	257.5				14979.
467	276.9				14023.
469	290.4				13426.
471	304.5				12860.
474	326.7				12064.
476	342.2				11566.
478	358.3				11092.
480	375.0				10642.

COMPOUND CONSTANTS:

MP   TC   VC   ZC
NBP   418.732 K   PC   DC   OM
MU   RG

VAPOR PRESSURE CORRELATION CONSTANTS, K AND KPA:

RPM2   RPM2   RPM2
359. - 461. K   359. - 461. K   359. - 461. K
 11 POINTS   11 POINTS   11 POINTS
RMSD = 0.0005   RMSD = 0.0005   RMSD = 0.0005
 0.16867186E+02   0.16867186E+02   0.16867186E+02
-0.51189412E+04   -0.51189412E+04   -0.51189412E+04
-0.88156709E-04   -0.88156709E-04   -0.88156709E-04
 0.73679066E-07   0.73679066E-07   0.73679066E-07

LIQUID DENSITY CORRELATION CONSTANTS, K AND G/ML:

SECOND VIRIAL COEFFICIENT CORRELATION CONSTANTS, ML/MOL:

LITERATURE DOCUMENTS FOR CORRELATED VAPOR PRESSURE VALUES:

 13799

LITERATURE DOCUMENTS FOR CORRELATED LIQUID DENSITY VALUES:

LITERATURE DOCUMENTS REPORTING VIRIAL COEFFICIENT DATA:

T,K	VAPOR PRESSURE, KPA	SATURATED LIQUID VOLUME, ML/MOL	SECOND VIRIAL COEFFICIENT, ML/MOL	HEAT OF VAPORIZATION, J/MOL	SATURATED VAPOR VOLUME, ML/MOL
359	13.28				224790.
361	14.37				208888.
363	15.54				194273.
365	16.78				180830.
368	18.81				162636.
370	20.28				151686.
372	21.85				141584.
375	24.39				127860.
377	26.21				119570.
379	28.16				111900.
382	31.31				101445.
384	33.57				95108.
386	35.97				89228.
389	39.84				81187.
391	42.61				76297.
393	45.54				71749.
396	50.26				65509.
398	53.63				61703.
400	57.19				58154.
403	62.90				53271.
405	66.97				50283.
407	71.25				47491.
409	75.77				44880.
412	82.99				41275.
414	88.12				39061.
416	93.51				36987.
419	102.1				34114.
421	108.2				32346.
423	114.6				30686.
426	124.8				28381.
428	132.0				26959.
430	139.5				25620.
433	151.5				23758.
435	160.0				22606.
437	168.8				21521.
440	182.9				20007.
442	192.7				19068.
444	203.0				18181.
447	219.4				16942.
449	230.8				16173.
451	242.8				15444.
454	261.7				14425.
456	274.9				13790.
458	288.7				13188.
460	303.1				12618.

COMPOUND CONSTANTS:

MP	244.147 K	TC	VC	ZC
NBP	408.131 K	PC	DC	OM
MU		RG		

VAPOR PRESSURE CORRELATION CONSTANTS, K AND KPA:

RPM2	RPM2	RPM2
287. - 409. K	287. - 409. K	287. - 409. K
8 POINTS	8 POINTS	8 POINTS
RMSD = 0.0596	RMSD = 0.0596	RMSD = 0.0596
0.33729517E+02	0.33729517E+02	0.33729517E+02
-0.82632914E+04	-0.82632914E+04	-0.82632914E+04
-0.30963951E-01	-0.30963951E-01	-0.30963951E-01
0.22649922E-04	0.22649922E-04	0.22649922E-04

LIQUID DENSITY CORRELATION CONSTANTS, K AND G/ML:

FRANCIS1	FRANCIS1	FRANCIS1
293. - 299. K	293. - 299. K	293. - 299. K
2 POINTS	2 POINTS	2 POINTS
RMSD = 0.00000	RMSD = 0.00000	RMSD = 0.00000
0.10557714E+01	0.10557714E+01	0.10557714E+01
0.85020700E-03	0.85020700E-03	0.85020700E-03

SECOND VIRIAL COEFFICIENT CORRELATION CONSTANTS, ML/MOL:

LITERATURE DOCUMENTS FOR CORRELATED VAPOR PRESSURE VALUES:

  1146

LITERATURE DOCUMENTS FOR CORRELATED LIQUID DENSITY VALUES:

  21602

LITERATURE DOCUMENTS REPORTING VIRIAL COEFFICIENT DATA:

T,K	VAPOR PRESSURE, KPA	SATURATED LIQUID VOLUME, ML/MOL	SECOND VIRIAL COEFFICIENT, ML/MOL	HEAT OF VAPORIZATION, J/MOL	SATURATED VAPOR VOLUME, ML/MOL
287	0.1245				0.1917E+08
289	0.1466				0.1639E+08
292	0.1864				0.1302E+08
295	0.2358	159.28		55968.	0.1040E+08
298	0.2966	159.79		55808.	8354751.
300	0.3446				7239100.
303	0.4297				5862930.
306	0.5332				4771418.
309	0.6585				3901361.
311	0.7561				3420014.
314	0.9266				2817397.
317	1.131				2330980.
320	1.374				1936604.
323	1.662				1615473.
325	1.883				1434678.
328	2.264				1204542.
331	2.711				1015095.
334	3.235				858539.
336	3.631				769326.
339	4.307				654448.
342	5.091				558581.
345	5.997				478302.
347	6.678				432055.
350	7.825				371882.
353	9.142				321047.
356	10.65				277967.
359	12.37				241347.
361	13.64				219990.
364	15.77				191872.
367	18.19				167786.
370	20.91				147096.
372	22.92				134925.
375	26.25				118769.
378	29.99				104793.
381	34.18				92673.
384	38.87				82137.
386	42.30				75879.
389	47.92				67492.
392	54.18				60157.
395	61.13				53726.
397	66.18				49879.
400	74.42				44690.
403	83.53				40116.
406	93.57				36075.
408	100.8				33643.

COMPOUND CONSTANTS:

MP 265.654 K	TC	VC	ZC
NBP 468.465 K	PC	DC	OM
MU	RG		

VAPOR PRESSURE CORRELATION CONSTANTS, K AND KPA:

VAPRES-2	VAPRES-2	VAPRES-2
333. - 469. K	333. - 469. K	333. - 469. K
23 POINTS	23 POINTS	23 POINTS
RMSD = 0.0355	RMSD = 0.0355	RMSD = 0.0355
0.25066182E+04	0.25066182E+04	0.25066182E+04
-0.66711165E+05	-0.66711165E+05	-0.66711165E+05
0.11440382E+01	0.11440382E+01	0.11440382E+01
-0.48792265E-03	-0.48792265E-03	-0.48792265E-03
-0.45344764E+03	-0.45344764E+03	-0.45344764E+03

LIQUID DENSITY CORRELATION CONSTANTS, K AND G/ML:

FRANCIS1	FRANCIS1	FRANCIS1
298. - 439. K	298. - 439. K	298. - 439. K
16 POINTS	16 POINTS	16 POINTS
RMSD = 0.00007	RMSD = 0.00007	RMSD = 0.00007
0.10667076E+01	0.10667076E+01	0.10667076E+01
0.69098081E-03	0.69098081E-03	0.69098081E-03
0.18308731E+02	0.18308731E+02	0.18308731E+02
0.71997681E+03	0.71997681E+03	0.71997681E+03

SECOND VIRIAL COEFFICIENT CORRELATION CONSTANTS, ML/MOL:

LITERATURE DOCUMENTS FOR CORRELATED VAPOR PRESSURE VALUES:

5736 12038

LITERATURE DOCUMENTS FOR CORRELATED LIQUID DENSITY VALUES:

9712 12038

LITERATURE DOCUMENTS REPORTING VIRIAL COEFFICIENT DATA:

T,K	VAPOR PRESSURE, KPA	SATURATED LIQUID VOLUME, ML/MOL	SECOND VIRIAL COEFFICIENT, ML/MOL	HEAT OF VAPORIZATION, J/MOL	SATURATED VAPOR VOLUME, ML/MOL
298		174.01			
301		174.52			
305		175.21			
309		175.90			
313		176.60			
317		177.31			
321		178.02			
325		178.74			
329		179.47			
332		180.03			
336	0.6908	180.77		53985.	4044212.
340	0.8662	181.52		53502.	3263477.
344	1.078	182.29		53057.	2652046.
348	1.334	183.06		52649.	2169467.
352	1.639	183.84		52274.	1785789.
356	2.002	184.63		51929.	1478634.
360	2.431	185.43		51611.	1231133.
364	2.937	186.24		51316.	1030474.
367	3.373	186.85		51109.	904725.
371	4.039	187.68		50849.	763799.
375	4.813	188.52		50604.	647794.
379	5.710	189.37		50371.	551825.
383	6.746	190.23		50146.	472055.
387	7.936	191.10		49928.	405453.
391	9.299	191.99		49712.	349607.
395	10.85	192.89		49495.	302591.
399	12.62	193.80		49274.	262856.
402	14.10	194.49		49104.	237070.
406	16.29	195.42		48869.	207210.
410	18.76	196.37		48622.	181727.
414	21.53	197.34		48359.	159911.
418	24.62	198.31		48077.	141175.
422	28.06	199.31		47773.	125038.
426	31.88	200.32		47444.	111099.
430	36.10	201.35		47087.	99026.
434	40.75	202.39		46698.	88543.
437	44.54	203.19		46384.	81583.
441	49.99				73346.
445	55.94				66146.
449	62.39				59840.
453	69.36				54305.
457	76.86				49437.
461	84.90				45148.
465	93.47				41363.
468	100.2				38815.

COMPOUND CONSTANTS:

MP	TC	VC	ZC
NBP	PC	DC	OM
MU	RG		

VAPOR PRESSURE CORRELATION CONSTANTS, K AND KPA:

LIQUID DENSITY CORRELATION CONSTANTS, K AND G/ML:

FRANCIS1	FRANCIS1	FRANCIS1
293. - 359. K	293. - 359. K	293. - 359. K
4 POINTS	4 POINTS	4 POINTS
RMSD = 0.00260	RMSD = 0.00260	RMSD = 0.00260
0.10525827E+01	0.10525827E+01	0.10525827E+01
0.76424866E-03	0.76424866E-03	0.76424866E-03
0.19214630E+02	0.19214630E+02	0.19214630E+02
0.30584688E+04	0.30584688E+04	0.30584688E+04

SECOND VIRIAL COEFFICIENT CORRELATION CONSTANTS, ML/MOL:

LITERATURE DOCUMENTS FOR CORRELATED VAPOR PRESSURE VALUES:

LITERATURE DOCUMENTS FOR CORRELATED LIQUID DENSITY VALUES:

14417

LITERATURE DOCUMENTS REPORTING VIRIAL COEFFICIENT DATA:

T,K	VAPOR PRESSURE, KPA	SATURATED LIQUID VOLUME, ML/MOL	SECOND VIRIAL COEFFICIENT, ML/MOL	HEAT OF VAPORIZATION, J/MOL	SATURATED VAPOR VOLUME, ML/MOL
293		173.10			
294		173.27			
296		173.59			
297		173.75			
299		174.08			
300		174.24			
302		174.57			
303		174.73			
305		175.06			
306		175.23			
308		175.56			
309		175.73			
311		176.06			
312		176.23			
314		176.56			
315		176.73			
317		177.07			
318		177.24			
320		177.58			
321		177.75			
323		178.09			
324		178.26			
326		178.60			
327		178.78			
329		179.12			
330		179.29			
332		179.64			
333		179.82			
335		180.16			
336		180.34			
338		180.69			
339		180.87			
341		181.22			
342		181.40			
344		181.75			
345		181.93			
347		182.29			
348		182.47			
350		182.83			
351		183.01			
353		183.37			
354		183.55			
356		183.92			
357		184.10			
359		184.47			

COMPOUND CONSTANTS:

MP	267.253 K	TC	640.000 K	VC		ZC
NBP	461.590 K	PC	2.320 MPA	DC		OM 0.5138
MU	2.6900 DEBYE	RG			

VAPOR PRESSURE CORRELATION CONSTANTS, K AND KPA:

WAGNER	WAGNER	WAGNER	WAGNER
298. - 486. K	298. - 480. K	298. - 486. K	443. - 486. K
31 POINTS	30 POINTS	31 POINTS	11 POINTS
RMSD = 0.0079	RMSD = 0.0074	RMSD = 0.0079	RMSD = 0.0041
-0.86650883E+01	-0.86610867E+01	-0.86650883E+01	-0.84860074E+01
0.18886549E+01	0.18792310E+01	0.18886549E+01	0.13629362E+01
-0.55050132E+01	-0.54913237E+01	-0.55050132E+01	-0.40025845E+01
-0.29487241E+01	-0.29722768E+01	-0.29487241E+01	-0.13687968E+02

LIQUID DENSITY CORRELATION CONSTANTS, K AND G/ML:

FRANCIS1	FRANCIS1	FRANCIS1
283. - 360. K	283. - 360. K	283. - 360. K
8 POINTS	8 POINTS	8 POINTS
RMSD = 0.00049	RMSD = 0.00049	RMSD = 0.00049
0.10534096E+01	0.10534096E+01	0.10534096E+01
0.79098716E-03	0.79098716E-03	0.79098716E-03
0.49999952E-02	0.49999952E-02	0.49999952E-02
0.88935620E+03	0.88935620E+03	0.88935620E+03

SECOND VIRIAL COEFFICIENT CORRELATION CONSTANTS, ML/MOL:

TSONOPOULOS: TC=640.000 PC=2.320 VC= 0.00 OM=0.5138 A=-0.008290 B=0.0000

LITERATURE DOCUMENTS FOR CORRELATED VAPOR PRESSURE VALUES:

10318

LITERATURE DOCUMENTS FOR CORRELATED LIQUID DENSITY VALUES:

1091 13138 14417

LITERATURE DOCUMENTS REPORTING VIRIAL COEFFICIENT DATA:

T,K	VAPOR PRESSURE, KPA	SATURATED LIQUID VOLUME, ML/MOL	SECOND VIRIAL COEFFICIENT, ML/MOL	HEAT OF VAPORIZATION, J/MOL	SATURATED VAPOR VOLUME, ML/MOL
283		171.47	-16967.		
287		172.12	-15711.		
292		172.95	-14316.		
296		173.62	-13322.		
301	0.09637	174.46	-12211.	55190.	0.2596E+08
306	0.1380	175.31	-11228.	54746.	0.1843E+08
310	0.1820	176.00	-10521.	54393.	0.1415E+08
315	0.2542	176.86	-9725.	53956.	0.1029E+08
319	0.3289	177.56	-9150.	53610.	8053961.
324	0.4490	178.44	-8499.	53180.	5991555.
329	0.6056	179.33	-7915.	52754.	4508759.
333	0.7633	180.05	-7489.	52416.	3619817.
338	1.010	180.96	-7004.	51996.	2776750.
342	1.253	181.69	-6649.	51662.	2262150.
347	1.628	182.61	-6242.	51247.	1765831.
352	2.095	183.54	-5872.	50834.	1390879.
356	2.548	184.30	-5600.	50505.	1156236.
361	3.228		-5286.		924574.
366	4.057		-4999.		745121.
370	4.843		-4786.		630350.
375	6.004		-4539.		514697.
379	7.094		-4355.		439798.
384	8.685		-4142.		363427.
389	10.56		-3944.		302184.
393	12.30		-3797.		261820.
398	14.80		-3624.		219976.
402	17.08		-3495.		192116.
407	20.34		-3343.		162956.
412	24.09		-3201.		138905.
416	27.49		-3094.		122662.
421	32.27		-2969.		105428.
426	37.70		-2851.		91006.
430	42.56		-2762.		81141.
435	49.34		-2657.		70549.
439	55.36		-2577.		63249.
444	63.70		-2482.		55356.
449	73.01		-2393.		48613.
453	81.22		-2325.		43917.
458	92.49		-2245.		38788.
462	102.4		-2183.		35195.
467	115.9		-2110.		31246.
472	130.7		-2040.		27814.
476	143.7		-1987.		25388.
481	161.2		-1924.		22700.
485	176.5		-1875.		20790.

COMPOUND CONSTANTS:

MP 227.137 K	TC	VC	ZC
NBP 441.414 K	PC	DC	OM
MU 2.6600 DEBYE	RG		

VAPOR PRESSURE CORRELATION CONSTANTS, K AND KPA:

VAPRES-2	VAPRES-2	VAPRES-2	RPM2
333. - 454. K	333. - 454. K	333. - 454. K	423. - 454. K
13 POINTS	13 POINTS	13 POINTS	4 POINTS
RMSD = 0.0011	RMSD = 0.0011	RMSD = 0.0011	RMSD = 0.0006
0.40176615E+03	0.40176615E+03	0.40176615E+03	0.29270221E+02
-0.17039623E+05	-0.17039623E+05	-0.17039623E+05	-0.73333600E+04
0.12875484E+00	0.12875484E+00	0.12875484E+00	-0.24539657E-01
-0.44344156E-04	-0.44344156E-04	-0.44344156E-04	0.14337444E-04
-0.66788250E+02	-0.66788250E+02	-0.66788250E+02	

LIQUID DENSITY CORRELATION CONSTANTS, K AND G/ML:

FRANCIS1	FRANCIS1	FRANCIS1
293. - 359. K	293. - 359. K	293. - 359. K
6 POINTS	6 POINTS	6 POINTS
RMSD = 0.00022	RMSD = 0.00022	RMSD = 0.00022
0.10386066E+01	0.10386066E+01	0.10386066E+01
0.75689703E-03	0.75689703E-03	0.75689703E-03
0.34747953E+01	0.34747953E+01	0.34747953E+01
0.60697998E+03	0.60697998E+03	0.60697998E+03

SECOND VIRIAL COEFFICIENT CORRELATION CONSTANTS, ML/MOL:

LITERATURE DOCUMENTS FOR CORRELATED VAPOR PRESSURE VALUES:

5775

LITERATURE DOCUMENTS FOR CORRELATED LIQUID DENSITY VALUES:

5775   7439

LITERATURE DOCUMENTS REPORTING VIRIAL COEFFICIENT DATA:

T,K	VAPOR PRESSURE, KPA	SATURATED LIQUID VOLUME, ML/MOL	SECOND VIRIAL COEFFICIENT, ML/MOL	HEAT OF VAPORIZATION, J/MOL	SATURATED VAPOR VOLUME, ML/MOL
293		176.53			
296		177.05			
300		177.75			
303		178.28			
307		179.00			
311		179.72			
314		180.26			
318		180.99			
322		181.73			
325		182.29			
329		183.04			
333	1.894	183.80		48232.	1462074.
336	2.211	184.38		47971.	1263247.
340	2.705	185.15		47632.	1045175.
344	3.288	185.93		47303.	869875.
347	3.792	186.52		47063.	760806.
351	4.564	187.32		46750.	639388.
355	5.464	188.12		46447.	540163.
358	6.233	188.73		46226.	477557.
362	7.396				406936.
366	8.735				348370.
369	9.866				310955.
373	11.56				268249.
377	13.49				232366.
380	15.10				209186.
384	17.50				182446.
388	20.20				159714.
391	22.44				144881.
395	25.74				127608.
399	29.42				112768.
402	32.45				102997.
406	36.88				91521.
410	41.79				81567.
413	45.81				74961.
417	51.64				67141.
421	58.05				60300.
424	63.26				55725.
428	70.78				50273.
432	79.00				45466.
435	85.64				42230.
439	95.18				38348.
443	105.5				34902.
446	113.9				32568.
450	125.8				29752.
453	135.3				27838.

COMPOUND CONSTANTS:

MP   TC   VC   ZC
NBP   PC   DC   OM
MU   RG

VAPOR PRESSURE CORRELATION CONSTANTS, K AND KPA:

LIQUID DENSITY CORRELATION CONSTANTS, K AND G/ML:

FRANCIS1   FRANCIS1   FRANCIS1
293. - 359. K   293. - 359. K   293. - 359. K
   4 POINTS   4 POINTS   4 POINTS
RMSD = 0.00007   RMSD = 0.00007   RMSD = 0.00007
0.10355082E+01   0.10355082E+01   0.10355082E+01
0.68153860E-03   0.68153860E-03   0.68153860E-03
0.24201708E+01   0.24201708E+01   0.24201708E+01
0.52146802E+03   0.52146802E+03   0.52146802E+03

SECOND VIRIAL COEFFICIENT CORRELATION CONSTANTS, ML/MOL:

LITERATURE DOCUMENTS FOR CORRELATED VAPOR PRESSURE VALUES:

LITERATURE DOCUMENTS FOR CORRELATED LIQUID DENSITY VALUES:

   14417

LITERATURE DOCUMENTS REPORTING VIRIAL COEFFICIENT DATA:

T,K	VAPOR PRESSURE, KPA	SATURATED LIQUID VOLUME, ML/MOL	SECOND VIRIAL COEFFICIENT, ML/MOL	HEAT OF VAPORIZATION, J/MOL	SATURATED VAPOR VOLUME, ML/MOL
293		189.36			
294		189.53			
296		189.87			
297		190.03			
299		190.37			
300		190.54			
302		190.88			
303		191.05			
305		191.40			
306		191.57			
308		191.91			
309		192.09			
311		192.44			
312		192.61			
314		192.96			
315		193.14			
317		193.49			
318		193.67			
320		194.02			
321		194.20			
323		194.56			
324		194.74			
326		195.10			
327		195.28			
329		195.65			
330		195.83			
332		196.20			
333		196.39			
335		196.76			
336		196.94			
338		197.32			
339		197.51			
341		197.88			
342		198.07			
344		198.45			
345		198.65			
347		199.03			
348		199.22			
350		199.61			
351		199.81			
353		200.20			
354		200.40			
356		200.79			
357		200.99			
359		201.39			

COMPOUND CONSTANTS:

MP	285.944 K	TC		VC		ZC
NBP	506.295 K	PC		DC		OM
MU	2.7000 DEBYE	RG				

VAPOR PRESSURE CORRELATION CONSTANTS, K AND KPA:

RPM2	RPM2	RPM2	RPM2
298. - 539. K	298. - 527. K	298. - 539. K	493. - 539. K
35 POINTS	33 POINTS	35 POINTS	10 POINTS
RMSD = 0.0360	RMSD = 0.0280	RMSD = 0.0360	RMSD = 0.0220
0.39808738E+02	0.39858830E+02	0.39808738E+02	0.57404325E+02
-0.10382444E+05	-0.10388870E+05	-0.10382444E+05	-0.13487763E+05
-0.40885275E-01	-0.41013400E-01	-0.40885275E-01	-0.74065949E-01
0.23468422E-04	0.23576007E-04	0.23468422E-04	0.44291422E-04

LIQUID DENSITY CORRELATION CONSTANTS, K AND G/ML:

FRANCIS1	FRANCIS1	FRANCIS1
293. - 434. K	293. - 434. K	293. - 434. K
28 POINTS	28 POINTS	28 POINTS
RMSD = 0.00009	RMSD = 0.00009	RMSD = 0.00009
0.10617561E+01	0.10617561E+01	0.10617561E+01
0.66399807E-03	0.66399807E-03	0.66399807E-03
0.19551086E+02	0.19551086E+02	0.19551086E+02
0.76776074E+03	0.76776074E+03	0.76776074E+03

SECOND VIRIAL COEFFICIENT CORRELATION CONSTANTS, ML/MOL:

LITERATURE DOCUMENTS FOR CORRELATED VAPOR PRESSURE VALUES:

10318 12038

LITERATURE DOCUMENTS FOR CORRELATED LIQUID DENSITY VALUES:

10483 12038

LITERATURE DOCUMENTS REPORTING VIRIAL COEFFICIENT DATA:

T,K	VAPOR PRESSURE, KPA	SATURATED LIQUID VOLUME, ML/MOL	SECOND VIRIAL COEFFICIENT, ML/MOL	HEAT OF VAPORIZATION, J/MOL	SATURATED VAPOR VOLUME, ML/MOL
293		206.16			
298	0.005908	207.10		66463.	0.4194E+09
304	0.01001	208.25		65872.	0.2525E+09
309	0.01523	209.22		65380.	0.1686E+09
315	0.02468	210.39		64791.	0.1061E+09
320	0.03628	211.38		64302.	0.7334E+08
326	0.05649	212.59		63717.	0.4798E+08
332	0.08622	213.82		63135.	0.3201E+08
337	0.1209	214.85		62653.	0.2317E+08
343	0.1785	216.12		62078.	0.1598E+08
348	0.2437	217.18		61602.	0.1187E+08
354	0.3490	218.48		61035.	8432977.
360	0.4923	219.80		60473.	6079449.
365	0.6487	220.92		60010.	4678561.
371	0.8918	222.28		59459.	3458873.
376	1.151	223.43		59005.	2715457.
382	1.547	224.83		58466.	2053733.
388	2.053	226.26		57935.	1571192.
393	2.578	227.47		57498.	1267412.
399	3.355	228.95		56982.	988713.
404	4.147	230.20		56558.	809984.
410	5.300	231.73		56057.	643141.
415	6.458	233.03		55647.	534327.
421	8.119	234.62		55163.	431128.
427	10.12	236.24		54689.	350700.
432	12.09	237.62		54303.	297014.
438	14.87				244975.
443	17.56				209770.
449	21.31				175211.
455	25.69				147283.
460	29.87				128037.
466	35.61				108809.
471	41.04				95414.
477	48.43				81894.
483	56.85				70646.
488	64.72				62691.
494	75.30				54547.
499	85.13				48738.
505	98.24				42740.
511	112.9				37630.
516	126.4				33939.
522	144.3				30084.
527	160.6				27280.
533	182.1				24331.
538	201.8				22172.

COMPOUND CONSTANTS:

MP	287.744 K	TC	VC	ZC
NBP	500.554 K	PC	DC	OM
MU	2.6800 DEBYE	RG		

VAPOR PRESSURE CORRELATION CONSTANTS, K AND KPA:

VAPRES-2	VAPRES-2	VAPRES-2	RPM2
298. - 532. K	298. - 521. K	298. - 532. K	481. - 532. K
32 POINTS	30 POINTS	32 POINTS	11 POINTS
RMSD = 0.0129	RMSD = 0.0136	RMSD = 0.0129	RMSD = 0.0143
0.38789128E+03	0.40188328E+03	0.38789128E+03	0.33956207E+02
-0.19284923E+05	-0.19622726E+05	-0.19284923E+05	-0.92916871E+04
0.10027672E+00	0.10659951E+00	0.10027672E+00	-0.29565440E-01
-0.29008985E-04	-0.31594792E-04	-0.29008985E-04	0.16060508E-04
-0.62369530E+02	-0.64916970E+02	-0.62369530E+02	

LIQUID DENSITY CORRELATION CONSTANTS, K AND G/ML:

FRANCIS1	FRANCIS1	FRANCIS1
293. - 360. K	293. - 360. K	293. - 360. K
5 POINTS	5 POINTS	5 POINTS
RMSD = 0.00013	RMSD = 0.00013	RMSD = 0.00013
0.10162287E+01	0.10162287E+01	0.10162287E+01
0.57980116E-03	0.57980116E-03	0.57980116E-03
0.45092754E+01	0.45092754E+01	0.45092754E+01
0.50070825E+03	0.50070825E+03	0.50070825E+03

SECOND VIRIAL COEFFICIENT CORRELATION CONSTANTS, ML/MOL:

LITERATURE DOCUMENTS FOR CORRELATED VAPOR PRESSURE VALUES:

10318 13138

LITERATURE DOCUMENTS FOR CORRELATED LIQUID DENSITY VALUES:

13138 14417

LITERATURE DOCUMENTS REPORTING VIRIAL COEFFICIENT DATA:

T,K	VAPOR PRESSURE, KPA	SATURATED LIQUID VOLUME, ML/MOL	SECOND VIRIAL COEFFICIENT, ML/MOL	HEAT OF VAPORIZATION, J/MOL	SATURATED VAPOR VOLUME, ML/MOL
293		206.51			
298	0.007892	207.37		67084.	0.3140E+09
303	0.01231	208.25		66343.	0.2047E+09
309	0.02045	209.32		65480.	0.1256E+09
314	0.03063	210.23		64782.	0.8524E+08
320	0.04863	211.35		63970.	0.5471E+08
325	0.07026	212.30		63313.	0.3846E+08
331	0.1072	213.47		62549.	0.2568E+08
336	0.1500	214.47		61932.	0.1862E+08
341	0.2073	215.49		61332.	0.1367E+08
347	0.3008	216.74		60634.	9592530.
352	0.4048	217.82		60071.	7229306.
358	0.5700	219.15		59416.	5222141.
363	0.7495				4027060.
369	1.028				2985617.
374	1.323				2349947.
379	1.689				1865516.
385	2.240				1429343.
390	2.808				1154578.
396	3.649				902266.
401	4.504				740270.
407	5.747				588846.
412	6.992				489942.
417	8.455				410066.
423	10.54				333677.
428	12.59				282669.
434	15.47				233204.
439	18.28				199726.
445	22.18				166848.
450	25.92				144321.
455	30.18				125362.
461	36.01				106433.
466	41.55				93256.
472	49.08				79965.
477	56.16				70621.
483	65.72				61109.
488	74.64				54361.
493	84.50				48508.
499	97.67				42477.
504	109.8				38147.
510	126.0				33652.
515	140.8				30401.
521	160.4				27002.
526	178.3				24527.
531	197.7				22332.

COMPOUND CONSTANTS:

MP	300.641 K	TC	VC	ZC
NBP	541.172 K	PC	DC	OM
MU		RG		

VAPOR PRESSURE CORRELATION CONSTANTS, K AND KPA:

VAPRES-2	VAPRES-2	VAPRES-2
333. - 542. K	333. - 542. K	333. - 542. K
19 POINTS	19 POINTS	19 POINTS
RMSD = 0.0537	RMSD = 0.0537	RMSD = 0.0537
-0.62855974E+03	-0.62855974E+03	-0.62855974E+03
0.57720465E+04	0.57720465E+04	0.57720465E+04
-0.31642310E+00	-0.31642310E+00	-0.31642310E+00
0.12610738E-03	0.12610738E-03	0.12610738E-03
0.12024952E+03	0.12024952E+03	0.12024952E+03

LIQUID DENSITY CORRELATION CONSTANTS, K AND G/ML:

FRANCIS1	FRANCIS1	FRANCIS1
303. - 434. K	303. - 434. K	303. - 434. K
14 POINTS	14 POINTS	14 POINTS
RMSD = 0.00003	RMSD = 0.00003	RMSD = 0.00003
0.10458326E+01	0.10458326E+01	0.10458326E+01
0.69680461E-03	0.69680461E-03	0.69680461E-03
0.47666807E+01	0.47666807E+01	0.47666807E+01
0.67093164E+03	0.67093164E+03	0.67093164E+03

SECOND VIRIAL COEFFICIENT CORRELATION CONSTANTS, ML/MOL:

LITERATURE DOCUMENTS FOR CORRELATED VAPOR PRESSURE VALUES:

10318 12038

LITERATURE DOCUMENTS FOR CORRELATED LIQUID DENSITY VALUES:

12038

LITERATURE DOCUMENTS REPORTING VIRIAL COEFFICIENT DATA:

T,K	VAPOR PRESSURE, KPA	SATURATED LIQUID VOLUME, ML/MOL	SECOND VIRIAL COEFFICIENT, ML/MOL	HEAT OF VAPORIZATION, J/MOL	SATURATED VAPOR VOLUME, ML/MOL
303		241.37			
308		242.45			
313		243.55			
319		244.87			
324		245.99			
330		247.35			
335	0.01773	248.49		70532.	0.1571E+09
341	0.02765	249.88		70172.	0.1025E+09
346	0.03950	251.05		69845.	0.7283E+08
351	0.05578	252.24		69496.	0.5232E+08
357	0.08312	253.68		69049.	0.3571E+08
362	0.1145	254.89		68655.	0.2628E+08
368	0.1659	256.37		68159.	0.1844E+08
373	0.2234	257.62		67729.	0.1388E+08
379	0.3153	259.14		67193.	9995377.
384	0.4158	260.42		66733.	7679204.
389	0.5434	261.72		66262.	5951989.
395	0.7408	263.30		65685.	4433377.
400	0.9503	264.63		65195.	3499591.
406	1.268	266.26		64600.	2662120.
411	1.599	267.64		64099.	2137044.
417	2.092	269.31		63495.	1657395.
422	2.597	270.73		62991.	1351207.
427	3.202	272.17		62488.	1108875.
433	4.081	273.93		61887.	882063.
438	4.963				733795.
444	6.226				592899.
449	7.475				499418.
455	9.242				409327.
460	10.97				348728.
465	12.95				298535.
471	15.71				249254.
476	18.37				215498.
482	22.02				181967.
487	25.51				158736.
493	30.27				135409.
498	34.77				119077.
503	39.80				105069.
509	46.61				90799.
514	52.98				80668.
520	61.54				70255.
525	69.51				62797.
531	80.18				55066.
536	90.06				49484.
541	100.9				44565.

COMPOUND CONSTANTS:

MP   312.389 K   TC   VC   ZC
NBP   567.786 K   PC   DC   OM
MU   2.7300 DEBYE   RG

    VAPOR PRESSURE CORRELATION CONSTANTS, K AND KPA:

RPM2	RPM2	RPM2	RPM2
437. - 601. K	437. - 576. K	437. - 601. K	562. - 601. K
16 POINTS	14 POINTS	16 POINTS	6 POINTS
RMSD = 0.5110	RMSD = 0.2288	RMSD = 0.5110	RMSD = 0.1750
0.50594333E+02	0.39391556E+02	0.50594333E+02	0.42746661E+03
-0.13634600E+05	-0.11775235E+05	-0.13634600E+05	-0.84065918E+05
-0.56890913E-01	-0.34494615E-01	-0.56890913E-01	-0.72805420E+00
0.32071991E-04	0.17216390E-04	0.32071991E-04	0.42989476E-03

    LIQUID DENSITY CORRELATION CONSTANTS, K AND G/ML:

FRANCIS1	FRANCIS1	FRANCIS1
312. - 354. K	312. - 354. K	312. - 354. K
3 POINTS	3 POINTS	3 POINTS
RMSD = 0.00000	RMSD = 0.00000	RMSD = 0.00000
0.10552623E+01	0.10552623E+01	0.10552623E+01
0.76081000E-03	0.76081000E-03	0.76081000E-03

    SECOND VIRIAL COEFFICIENT CORRELATION CONSTANTS, ML/MOL:

LITERATURE DOCUMENTS FOR CORRELATED VAPOR PRESSURE VALUES:

  10318

LITERATURE DOCUMENTS FOR CORRELATED LIQUID DENSITY VALUES:

  14905 19014

LITERATURE DOCUMENTS REPORTING VIRIAL COEFFICIENT DATA:

T,K	VAPOR PRESSURE, KPA	SATURATED LIQUID VOLUME, ML/MOL	SECOND VIRIAL COEFFICIENT, ML/MOL	HEAT OF VAPORIZATION, J/MOL	SATURATED VAPOR VOLUME, ML/MOL
312		276.81			
318		278.36			
325		280.20			
331		281.79			
338		283.67			
344		285.30			
351		287.23			
443	2.479				1485761.
449	3.157				1182690.
456	4.139				915915.
463	5.369				716957.
469	6.655				585902.
476	8.473				467091.
482	10.34				387408.
489	12.95				313942.
495	15.60				263861.
502	19.24				216975.
509	23.55				179715.
515	27.85				153746.
522	33.66				128921.
528	39.41				111397.
535	47.10				94442.
541	54.63				82334.
548	64.64				70489.
555	76.10				60640.
561	87.20				53493.
568	101.8				46395.
574	115.8				41198.
581	134.2				35990.
587	151.8				32144.
594	174.8				28260.
600	196.6				25370.

COMPOUND CONSTANTS:

MP   356.844 K   TC   VC   ZC
NBP   PC   DC   OM
MU   RG

     VAPOR PRESSURE CORRELATION CONSTANTS, K AND KPA:

     LIQUID DENSITY CORRELATION CONSTANTS, K AND G/ML:

FRANCIS1   FRANCIS1   FRANCIS1
363. - 574. K   363. - 574. K   363. - 574. K
  5 POINTS   5 POINTS   5 POINTS
RMSD = 0.00135   RMSD = 0.00135   RMSD = 0.00135
0.10310059E+01   0.10310059E+01   0.10310059E+01
0.64559188E-03   0.64559188E-03   0.64559188E-03
0.19765949E+00   0.19765949E+00   0.19765949E+00
0.75869238E+03   0.75869238E+03   0.75869238E+03

     SECOND VIRIAL COEFFICIENT CORRELATION CONSTANTS, ML/MOL:

LITERATURE DOCUMENTS FOR CORRELATED VAPOR PRESSURE VALUES:

LITERATURE DOCUMENTS FOR CORRELATED LIQUID DENSITY VALUES:

  8125

LITERATURE DOCUMENTS REPORTING VIRIAL COEFFICIENT DATA:

T,K	VAPOR PRESSURE, KPA	SATURATED LIQUID VOLUME, ML/MOL	SECOND VIRIAL COEFFICIENT, ML/MOL	HEAT OF VAPORIZATION, J/MOL	SATURATED VAPOR VOLUME, ML/MOL
363		566.26			
367		568.10			
372		570.43			
377		572.77			
382		575.14			
386		577.04			
391		579.44			
396		581.86			
401		584.30			
406		586.76			
410		588.75			
415		591.24			
420		593.76			
425		596.31			
430		598.87			
434		600.94			
439		603.54			
444		606.17			
449		608.82			
454		611.50			
458		613.65			
463		616.37			
468		619.11			
473		621.88			
478		624.67			
482		626.92			
487		629.76			
492		632.63			
497		635.52			
502		638.43			
506		640.79			
511		643.76			
516		646.75			
521		649.78			
526		652.83			
530		655.30			
535		658.41			
540		661.54			
545		664.71			
550		667.91			
554		670.50			
559		673.76			
564		677.05			
569		680.37			
573		683.06			

COMPOUND CONSTANTS:

MP   222.179 K   TC   VC   ZC
NBP   371.990 K   PC   DC   OM
MU   2.6100 DEBYE   RG   2.7906 ANGSTROM

VAPOR PRESSURE CORRELATION CONSTANTS, K AND KPA:

RPM2	RPM2	RPM2	RPM2
249. - 381. K	249. - 381. K	249. - 381. K	356. - 381. K
16 POINTS	16 POINTS	16 POINTS	5 POINTS
RMSD = 0.0842	RMSD = 0.0842	RMSD = 0.0842	RMSD = 0.0071
0.19664233E+02	0.19664233E+02	0.19664233E+02	-0.41338309E+02
-0.49770013E+04	-0.49770013E+04	-0.49770013E+04	0.26826805E+04
-0.31815092E-02	-0.31815092E-02	-0.31815092E-02	0.15847068E+00
-0.34876835E-05	-0.34876835E-05	-0.34876835E-05	-0.14601138E-03

LIQUID DENSITY CORRELATION CONSTANTS, K AND G/ML:

FRANCIS1	FRANCIS1	FRANCIS1
273. - 363. K	273. - 363. K	273. - 363. K
13 POINTS	13 POINTS	13 POINTS
RMSD = 0.00052	RMSD = 0.00052	RMSD = 0.00052
0.12236357E+01	0.12236357E+01	0.12236357E+01
0.87728817E-03	0.87728817E-03	0.87728817E-03
0.74975948E+01	0.74975948E+01	0.74975948E+01
0.52308423E+03	0.52308423E+03	0.52308423E+03

SECOND VIRIAL COEFFICIENT CORRELATION CONSTANTS, ML/MOL:

LITERATURE DOCUMENTS FOR CORRELATED VAPOR PRESSURE VALUES:

   22762 40284

LITERATURE DOCUMENTS FOR CORRELATED LIQUID DENSITY VALUES:

   3383 6237 40284

LITERATURE DOCUMENTS REPORTING VIRIAL COEFFICIENT DATA:

T,K	VAPOR PRESSURE, KPA	SATURATED LIQUID VOLUME, ML/MOL	SECOND VIRIAL COEFFICIENT, ML/MOL	HEAT OF VAPORIZATION, J/MOL	SATURATED VAPOR VOLUME, ML/MOL
249	0.2639				7844769.
252	0.3299				6351288.
255	0.4100				5170597.
258	0.5069				4231882.
261	0.6233				3481456.
264	0.7626				2878369.
267	0.9284				2391212.
270	1.125				1995752.
273	1.357	73.46		38228.	1673191.
276	1.629	73.69		38144.	1408873.
279	1.947	73.92		38060.	1191310.
282	2.318	74.16		37974.	1011454.
285	2.748	74.40		37886.	862147.
288	3.246	74.64		37798.	737694.
291	3.819	74.88		37707.	633549.
294	4.476	75.13		37615.	546065.
297	5.228	75.37		37522.	472304.
300	6.085	75.62		37427.	409890.
303	7.059	75.88		37331.	356892.
306	8.161	76.13		37233.	311739.
309	9.406	76.39		37134.	273142.
312	10.81	76.65		37033.	240044.
315	12.38	76.92		36930.	211574.
318	14.14	77.18		36826.	187010.
321	16.10	77.46		36720.	165756.
324	18.29	77.73		36612.	147311.
327	20.71	78.01		36503.	131262.
330	23.40	78.29		36392.	117258.
333	26.37	78.57		36279.	105008.
336	29.64	78.86		36164.	94264.
339	33.23	79.15		36048.	84819.
342	37.17	79.45		35929.	76494.
345	41.49	79.75		35809.	69140.
348	46.20	80.06		35687.	62629.
351	51.33	80.37		35564.	56851.
354	56.92	80.68		35438.	51712.
357	62.98	81.00		35310.	47133.
360	69.54	81.33		35180.	43044.
363	76.63	81.66		35048.	39384.
366	84.29				36103.
369	92.53				33156.
372	101.4				30504.
375	110.9				28112.
378	121.1				25952.
381	132.0				23998.

COMPOUND CONSTANTS:

MP   265.454 K   TC   VC   ZC
NBP 399.650 K   PC   DC   OM
MU   RG

     VAPOR PRESSURE CORRELATION CONSTANTS, K AND KPA:

     RPM2   RPM2   RPM2
     343. - 400. K   343. - 400. K   343. - 400. K
      2 POINTS   2 POINTS   2 POINTS
     RMSD = 0.0000   RMSD = 0.0000   RMSD = 0.0000
     0.16936190E+02   0.16936190E+02   0.16936190E+02
     -0.49228300E+04   -0.49228300E+04   -0.49228300E+04

     LIQUID DENSITY CORRELATION CONSTANTS, K AND G/ML:

     FRANCIS1   FRANCIS1   FRANCIS1
     282. - 303. K   282. - 303. K   282. - 303. K
      9 POINTS   9 POINTS   9 POINTS
     RMSD = 0.00014   RMSD = 0.00014   RMSD = 0.00014
     0.14537134E+01   0.14537134E+01   0.14537134E+01
     0.10908672E-02   0.10908672E-02   0.10908672E-02
     0.14505398E+02   0.14505398E+02   0.14505398E+02
     0.65764941E+03   0.65764941E+03   0.65764941E+03

     SECOND VIRIAL COEFFICIENT CORRELATION CONSTANTS, ML/MOL:

LITERATURE DOCUMENTS FOR CORRELATED VAPOR PRESSURE VALUES:

   13997

LITERATURE DOCUMENTS FOR CORRELATED LIQUID DENSITY VALUES:

   13997

LITERATURE DOCUMENTS REPORTING VIRIAL COEFFICIENT DATA:

T,K	VAPOR PRESSURE, KPA	SATURATED LIQUID VOLUME, ML/MOL	SECOND VIRIAL COEFFICIENT, ML/MOL	HEAT OF VAPORIZATION, J/MOL	SATURATED VAPOR VOLUME, ML/MOL
282		75.92			
284		76.08			
287		76.33			
290		76.58			
292		76.74			
295		77.00			
298		77.25			
300		77.42			
303		77.68			
343	13.25				215251.
346	15.00				191727.
349	16.96				171126.
351	18.38				158816.
354	20.69				142223.
357	23.26				127610.
359	25.12				118836.
362	28.14				106956.
365	31.47				96438.
367	33.87				90095.
370	37.76				81471.
373	42.02				73797.
375	45.09				69150.
378	50.04				62806.
381	55.44				57135.
383	59.31				53687.
386	65.55				48964.
389	72.32				44722.
391	77.16				42134.
394	84.92				38576.
397	93.33				35368.
399	99.31				33404.

COMPOUND CONSTANTS:

```
MP     221.879 K        TC                      VC                    ZC
NBP    403.706 K        PC                      DC                    OM
MU      2.9300 DEBYE    RG    3.1662 ANGSTROM
```

VAPOR PRESSURE CORRELATION CONSTANTS, K AND KPA:

```
VAPRES-2            VAPRES-2            VAPRES-2            VAPRES-2
273. - 539. K       273. - 416. K       273. - 539. K       388. - 539. K
 28 POINTS           17 POINTS           28 POINTS           17 POINTS
RMSD = 1.2680       RMSD = 0.0318       RMSD = 1.2680       RMSD = 1.6810
 0.24376810E+03      0.56206527E+03      0.24376810E+03      0.37597781E+03
-0.11456854E+05     -0.18068105E+05     -0.11456854E+05     -0.14596625E+05
 0.75261551E-01      0.25540648E+00      0.75261551E-01      0.13510762E+00
-0.24333219E-04     -0.11377508E-03     -0.24333219E-04     -0.48500018E-04
-0.39526822E+02     -0.99531428E+02     -0.39526822E+02     -0.63633082E+02
```

LIQUID DENSITY CORRELATION CONSTANTS, K AND G/ML:

```
FRANCIS1            FRANCIS1            FRANCIS1
273. - 364. K       273. - 364. K       273. - 364. K
 15 POINTS           15 POINTS           15 POINTS
RMSD = 0.00012      RMSD = 0.00012      RMSD = 0.00012
0.12307463E+01      0.12307463E+01      0.12307463E+01
0.95416536E-03      0.95416536E-03      0.95416536E-03
0.36097822E+01      0.36097822E+01      0.36097822E+01
0.17891855E+04      0.17891855E+04      0.17891855E+04
```

SECOND VIRIAL COEFFICIENT CORRELATION CONSTANTS, ML/MOL:

LITERATURE DOCUMENTS FOR CORRELATED VAPOR PRESSURE VALUES:

   2060   5067   7406 22762 40284

LITERATURE DOCUMENTS FOR CORRELATED LIQUID DENSITY VALUES:

   72   2840   4670   7948 18654 18956 40156

LITERATURE DOCUMENTS REPORTING VIRIAL COEFFICIENT DATA:

T,K	VAPOR PRESSURE, KPA	SATURATED LIQUID VOLUME, ML/MOL	SECOND VIRIAL COEFFICIENT, ML/MOL	HEAT OF VAPORIZATION, J/MOL	SATURATED VAPOR VOLUME, ML/MOL
273	0.3042	86.91		43941.	7462006.
279	0.4603	87.43		43487.	5039939.
285	0.6816	87.95		43053.	3476299.
291	0.9896	88.48		42639.	2444838.
297	1.411	89.02		42245.	1750667.
303	1.976	89.56		41869.	1274717.
309	2.725	90.11		41512.	942682.
315	3.703	90.67		41173.	707274.
321	4.962	91.23		40851.	537835.
327	6.565	91.81		40545.	414149.
333	8.581	92.38		40255.	322661.
339	11.09	92.97		39981.	254149.
345	14.18	93.56		39722.	202244.
351	17.96	94.16		39477.	162489.
357	22.53	94.77		39245.	131728.
363	28.02	95.39		39026.	107695.
369	34.57				88748.
375	42.32				73682.
381	51.42				61604.
387	62.06				51848.
393	74.41				43911.
399	88.68				37408.
405	105.1				32047.
412	127.2				26934.
418	149.0				23331.
424	173.6				20305.
430	201.4				17751.
436	232.6				15584.
442	267.5				13736.
448	306.5				12154.
454	349.7				10794.
460	397.6				9620.
466	450.4				8602.
472	508.6				7716.
478	572.5				6942.
484	642.3				6265.
490	718.6				5669.
496	801.7				5144.
502	892.0				4679.
508	989.7				4267.
514	1095.				3901.
520	1210.				3574.
526	1332.				3282.
532	1464.				3021.
538	1606.				2786.

COMPOUND CONSTANTS:

MP 204.883 K	TC	VC	ZC
NBP 384.870 K	PC	DC	OM
MU 2.8700 DEBYE	RG		

VAPOR PRESSURE CORRELATION CONSTANTS, K AND KPA:

RPM2	RPM2	RPM2
355. - 380. K	355. - 380. K	355. - 380. K
6 POINTS	6 POINTS	6 POINTS
RMSD = 0.0000	RMSD = 0.0000	RMSD = 0.0000
0.16466401E+02	0.16466401E+02	0.16466401E+02
-0.45368554E+04	-0.45368554E+04	-0.45368554E+04
-0.35654411E-03	-0.35654411E-03	-0.35654411E-03
0.31524389E-60	0.31524389E-60	0.31524389E-60

LIQUID DENSITY CORRELATION CONSTANTS, K AND G/ML:

FRANCIS1	FRANCIS1	FRANCIS1
293. - 361. K	293. - 361. K	293. - 361. K
4 POINTS	4 POINTS	4 POINTS
RMSD = 0.00006	RMSD = 0.00006	RMSD = 0.00006
0.11695671E+01	0.11695671E+01	0.11695671E+01
0.90240128E-03	0.90240128E-03	0.90240128E-03
0.82282317E+00	0.82282317E+00	0.82282317E+00
0.43741748E+03	0.43741748E+03	0.43741748E+03

SECOND VIRIAL COEFFICIENT CORRELATION CONSTANTS, ML/MOL:

LITERATURE DOCUMENTS FOR CORRELATED VAPOR PRESSURE VALUES:

13292 15758

LITERATURE DOCUMENTS FOR CORRELATED LIQUID DENSITY VALUES:

14422

LITERATURE DOCUMENTS REPORTING VIRIAL COEFFICIENT DATA:

T,K	VAPOR PRESSURE, KPA	SATURATED LIQUID VOLUME, ML/MOL	SECOND VIRIAL COEFFICIENT, ML/MOL	HEAT OF VAPORIZATION, J/MOL	SATURATED VAPOR VOLUME, ML/MOL
293		93.52			
294		93.62			
296		93.81			
298		94.01			
300		94.21			
302		94.41			
304		94.61			
306		94.82			
308		95.02			
310		95.22			
312		95.43			
314		95.64			
316		95.85			
318		96.06			
320		96.27			
322		96.48			
324		96.69			
326		96.91			
328		97.13			
330		97.35			
332		97.57			
334		97.79			
336		98.01			
338		98.24			
340		98.46			
342		98.69			
344		98.92			
346		99.16			
348		99.39			
350		99.63			
352		99.87			
354		100.11			
356	36.43	100.35		37299.	81241.
358	39.09	100.60		37292.	76138.
360	41.92	100.85		37284.	71409.
362	44.91				67024.
364	48.07				62955.
366	51.43				59175.
368	54.97				55662.
370	58.72				52394.
372	62.67				49351.
374	66.85				46517.
376	71.25				43875.
378	75.90				41410.
379	78.31				40240.

COMPOUND CONSTANTS:

MP   241.969 K   TC   629.000 K   VC   ZC
NBP   428.763 K   PC   3.850 MPA   DC   OM   0.4524
MU   3.0800 DEBYE   RG

VAPOR PRESSURE CORRELATION CONSTANTS, K AND KPA:

WAGNER	WAGNER	WAGNER	WAGNER
302. - 582. K	302. - 434. K	302. - 533. K	413. - 582. K
32 POINTS	23 POINTS	26 POINTS	18 POINTS
RMSD = 6.3310	RMSD = 0.0673	RMSD = 1.2660	RMSD = 5.9960
-0.82635171E+01	-0.10550075E+02	-0.86065609E+01	-0.77777529E+01
0.91380629E+00	0.62701309E+01	0.17456880E+01	-0.64181523E+00
-0.57455934E-01	-0.77788104E+01	-0.13681218E+01	0.44407638E+01
-0.10049146E+02	0.36400539E+01	-0.74119991E+01	-0.30293267E+02

LIQUID DENSITY CORRELATION CONSTANTS, K AND G/ML:

FRANCIS1	FRANCIS1	FRANCIS1
273. - 354. K	273. - 354. K	273. - 354. K
15 POINTS	15 POINTS	15 POINTS
RMSD = 0.00005	RMSD = 0.00005	RMSD = 0.00005
0.12110815E+01	0.12110815E+01	0.12110815E+01
0.84087066E-03	0.84087066E-03	0.84087066E-03
0.62623587E+01	0.62623587E+01	0.62623587E+01
0.64040259E+03	0.64040259E+03	0.64040259E+03

SECOND VIRIAL COEFFICIENT CORRELATION CONSTANTS, ML/MOL:

TSONOPOULOS: TC=629.000 PC=3.850 VC= 0.00 OM=0.4524 A=-0.018700 B=0.0000

LITERATURE DOCUMENTS FOR CORRELATED VAPOR PRESSURE VALUES:

  1095   3023   4670 10697 18421 18956 19429 40158

LITERATURE DOCUMENTS FOR CORRELATED LIQUID DENSITY VALUES:

  1529   2840   2894   4670   6763 40158

LITERATURE DOCUMENTS REPORTING VIRIAL COEFFICIENT DATA:

T,K	VAPOR PRESSURE, KPA	SATURATED LIQUID VOLUME, ML/MOL	SECOND VIRIAL COEFFICIENT, ML/MOL	HEAT OF VAPORIZATION, J/MOL	SATURATED VAPOR VOLUME, ML/MOL
273		101.76	-12456.		
280		102.42	-10815.		
287		103.09	-9454.		
294		103.77	-8316.		
301		104.46	-7360.		
308	0.9472	105.16	-6550.	45282.	2697044.
315	1.401	105.88	-5861.	44709.	1863299.
322	2.029	106.60	-5271.	44168.	1314243.
329	2.881	107.34	-4762.	43656.	944732.
336	4.017	108.09	-4323.	43171.	691030.
343	5.510	108.85	-3940.	42712.	513594.
350	7.443	109.63	-3605.	42276.	387362.
357	9.911		-3312.		296127.
364	13.03		-3052.		229212.
371	16.92		-2823.		179464.
378	21.72		-2618.		142008.
385	27.60		-2436.		113473.
392	34.74		-2272.		91494.
399	43.32		-2125.		74392.
406	53.57		-1992.		60956.
413	65.72		-1871.		50306.
420	80.03		-1762.		41793.
427	96.79		-1662.		34934.
434	116.3		-1571.		29368.
441	138.9		-1487.		24819.
448	164.9		-1410.		21078.
455	194.7		-1339.		17980.
462	228.8		-1274.		15402.
469	267.5		-1213.		13244.
476	311.3		-1157.		11427.
483	360.7		-1104.		9890.
490	416.2		-1055.		8585.
497	478.4		-1010.		7469.
504	547.9		-967.2		6513.
511	625.2		-927.3		5688.
518	710.9		-889.8		4974.
525	805.9		-854.5		4353.
532	910.6		-821.3		3810.
539	1026.		-790.0		3332.
546	1153.		-760.4		3104.
553	1292.		-732.5		2737.
560	1444.		-706.0		2413.
567	1609.		-680.9		2126.
574	1790.		-657.1		1871.
581	1986.		-634.4		1643.

COMPOUND CONSTANTS:

MP		TC	VC	ZC
NBP	453.558 K	PC	DC	OM
MU	3.1000 DEBYE	RG		

VAPOR PRESSURE CORRELATION CONSTANTS, K AND KPA:

RPM2	RPM2	RPM2	RPM2
313. - 465. K	313. - 465. K	313. - 465. K	433. - 465. K
22 POINTS	22 POINTS	22 POINTS	7 POINTS
RMSD = 0.0317	RMSD = 0.0317	RMSD = 0.0317	RMSD = 0.0146
0.30238353E+02	0.30238353E+02	0.30238353E+02	0.43399170E+01
-0.75018223E+04	-0.75018223E+04	-0.75018223E+04	-0.36523985E+04
-0.28252855E-01	-0.28252855E-01	-0.28252855E-01	0.29850045E-01
0.18152509E-04	0.18152509E-04	0.18152509E-04	-0.25314379E-04

LIQUID DENSITY CORRELATION CONSTANTS, K AND G/ML:

FRANCIS1	FRANCIS1	FRANCIS1
297. - 374. K	297. - 374. K	297. - 374. K
7 POINTS	7 POINTS	7 POINTS
RMSD = 0.00008	RMSD = 0.00008	RMSD = 0.00008
0.11986494E+01	0.11986494E+01	0.11986494E+01
0.81006484E-03	0.81006484E-03	0.81006484E-03
0.38449039E+01	0.38449039E+01	0.38449039E+01
0.65876953E+03	0.65876953E+03	0.65876953E+03

SECOND VIRIAL COEFFICIENT CORRELATION CONSTANTS, ML/MOL:

LITERATURE DOCUMENTS FOR CORRELATED VAPOR PRESSURE VALUES:

18654 40284

LITERATURE DOCUMENTS FOR CORRELATED LIQUID DENSITY VALUES:

40284

LITERATURE DOCUMENTS REPORTING VIRIAL COEFFICIENT DATA:

T,K	VAPOR PRESSURE, KPA	SATURATED LIQUID VOLUME, ML/MOL	SECOND VIRIAL COEFFICIENT, ML/MOL	HEAT OF VAPORIZATION, J/MOL	SATURATED VAPOR VOLUME, ML/MOL
297		118.39			
300		118.71			
304		119.13			
308		119.56			
312		119.99			
316	0.5396	120.43		48440.	4868728.
319	0.6416	120.75		48266.	4133812.
323	0.8034	121.19		48036.	3342866.
327	0.9993	121.64		47807.	2720588.
331	1.235	122.09		47581.	2227776.
335	1.518	122.54		47356.	1835014.
338	1.765	122.88		47189.	1592520.
342	2.147	123.34		46968.	1324627.
346	2.597	123.80		46749.	1107637.
350	3.126	124.27		46533.	930912.
354	3.744	124.74		46319.	786217.
358	4.462	125.21		46108.	667143.
361	5.074	125.57		45951.	591590.
365	5.998	126.05		45745.	505929.
369	7.061	126.54		45541.	434508.
373	8.277	127.03		45340.	374695.
377	9.663				324389.
380	10.82				291870.
384	12.55				254341.
388	14.50				222425.
392	16.70				195182.
396	19.16				171842.
400	21.91				151775.
403	24.18				138558.
407	27.51				123028.
411	31.19				109555.
415	35.27				97828.
419	39.77				87592.
422	43.44				80762.
426	48.76				72636.
430	54.59				65487.
434	60.97				59180.
438	67.94				53603.
442	75.53				48658.
445	81.65				45314.
449	90.43				41284.
453	99.94				37687.
457	110.2				34471.
461	121.3				31587.
464	130.3				29619.

COMPOUND CONSTANTS:

MP 259.258 K	TC	VC	ZC
NBP	PC	DC	OM
MU	RG		

VAPOR PRESSURE CORRELATION CONSTANTS, K AND KPA:

LIQUID DENSITY CORRELATION CONSTANTS, K AND G/ML:

FRANCIS1	FRANCIS1	FRANCIS1
292. - 357. K	292. - 357. K	292. - 357. K
7 POINTS	7 POINTS	7 POINTS
RMSD = 0.00145	RMSD = 0.00145	RMSD = 0.00145
0.11601934E+01	0.11601934E+01	0.11601934E+01
0.70832646E-03	0.70832646E-03	0.70832646E-03
0.10340004E+02	0.10340004E+02	0.10340004E+02
0.66066748E+03	0.66066748E+03	0.66066748E+03

SECOND VIRIAL COEFFICIENT CORRELATION CONSTANTS, ML/MOL:

LITERATURE DOCUMENTS FOR CORRELATED VAPOR PRESSURE VALUES:

LITERATURE DOCUMENTS FOR CORRELATED LIQUID DENSITY VALUES:

9862 9923 14219

LITERATURE DOCUMENTS REPORTING VIRIAL COEFFICIENT DATA:

T,K	VAPOR PRESSURE, KPA	SATURATED LIQUID VOLUME, ML/MOL	SECOND VIRIAL COEFFICIENT, ML/MOL	HEAT OF VAPORIZATION, J/MOL	SATURATED VAPOR VOLUME, ML/MOL
292		121.22			
293		121.33			
294		121.43			
296		121.64			
297		121.74			
299		121.95			
300		122.05			
302		122.26			
303		122.37			
305		122.58			
306		122.69			
308		122.90			
309		123.01			
311		123.22			
312		123.33			
314		123.54			
315		123.65			
317		123.87			
318		123.98			
320		124.20			
321		124.31			
323		124.53			
324		124.64			
325		124.75			
327		124.97			
328		125.08			
330		125.31			
331		125.42			
333		125.64			
334		125.76			
336		125.99			
337		126.10			
339		126.33			
340		126.44			
342		126.68			
343		126.79			
345		127.02			
346		127.14			
348		127.38			
349		127.49			
351		127.73			
352		127.85			
354		128.09			
355		128.21			
356		128.33			

709

COMPOUND CONSTANTS:

MP   199.683 K   TC   VC   ZC
NBP   PC   DC   OM
MU   RG

VAPOR PRESSURE CORRELATION CONSTANTS, K AND KPA:

LIQUID DENSITY CORRELATION CONSTANTS, K AND G/ML:

FRANCIS1   FRANCIS1   FRANCIS1
291. - 361. K   291. - 361. K   291. - 361. K
  6 POINTS   6 POINTS   6 POINTS
RMSD = 0.00078   RMSD = 0.00078   RMSD = 0.00078
0.11473150E+01   0.11473150E+01   0.11473150E+01
0.77219424E-03   0.77219424E-03   0.77219424E-03
0.26848648E+02   0.26848648E+02   0.26848648E+02
0.44958086E+04   0.44958086E+04   0.44958086E+04

SECOND VIRIAL COEFFICIENT CORRELATION CONSTANTS, ML/MOL:

LITERATURE DOCUMENTS FOR CORRELATED VAPOR PRESSURE VALUES:

LITERATURE DOCUMENTS FOR CORRELATED LIQUID DENSITY VALUES:

  9923 14219

LITERATURE DOCUMENTS REPORTING VIRIAL COEFFICIENT DATA:

T,K	VAPOR PRESSURE, KPA	SATURATED LIQUID VOLUME, ML/MOL	SECOND VIRIAL COEFFICIENT, ML/MOL	HEAT OF VAPORIZATION, J/MOL	SATURATED VAPOR VOLUME, ML/MOL
291		122.43			
292		122.53			
294		122.74			
295		122.84			
297		123.05			
298		123.16			
300		123.37			
302		123.58			
303		123.68			
305		123.89			
306		124.00			
308		124.21			
310		124.42			
311		124.53			
313		124.75			
314		124.85			
316		125.07			
318		125.28			
319		125.39			
321		125.61			
322		125.72			
324		125.94			
325		126.05			
327		126.27			
329		126.49			
330		126.60			
332		126.82			
333		126.93			
335		127.15			
337		127.38			
338		127.49			
340		127.71			
341		127.83			
343		128.05			
345		128.28			
346		128.39			
348		128.62			
349		128.73			
351		128.96			
353		129.19			
354		129.31			
356		129.54			
357		129.65			
359		129.89			
360		130.00			

COMPOUND CONSTANTS:

MP   232.574 K   TC   VC   ZC
NBP   PC   DC   OM
MU   RG

VAPOR PRESSURE CORRELATION CONSTANTS, K AND KPA:

LIQUID DENSITY CORRELATION CONSTANTS, K AND G/ML:

FRANCIS1   FRANCIS1   FRANCIS1
291. - 360. K   291. - 360. K   291. - 360. K
  8 POINTS   8 POINTS   8 POINTS
RMSD = 0.00377   RMSD = 0.00377   RMSD = 0.00377
0.11351528E+01   0.11351528E+01   0.11351528E+01
0.70543634E-03   0.70543634E-03   0.70543634E-03
0.90721703E+00   0.90721703E+00   0.90721703E+00
0.37699731E+03   0.37699731E+03   0.37699731E+03

SECOND VIRIAL COEFFICIENT CORRELATION CONSTANTS, ML/MOL:

LITERATURE DOCUMENTS FOR CORRELATED VAPOR PRESSURE VALUES:

LITERATURE DOCUMENTS FOR CORRELATED LIQUID DENSITY VALUES:

   1990   9862   9923 10707 14219

LITERATURE DOCUMENTS REPORTING VIRIAL COEFFICIENT DATA:

T,K	VAPOR PRESSURE, KPA	SATURATED LIQUID VOLUME, ML/MOL	SECOND VIRIAL COEFFICIENT, ML/MOL	HEAT OF VAPORIZATION, J/MOL	SATURATED VAPOR VOLUME, ML/MOL
291		122.01			
292		122.13			
294		122.35			
295		122.46			
297		122.69			
298		122.80			
300		123.03			
301		123.15			
303		123.38			
305		123.62			
306		123.74			
308		123.98			
309		124.11			
311		124.36			
312		124.48			
314		124.74			
316		125.00			
317		125.13			
319		125.40			
320		125.54			
322		125.82			
323		125.96			
325		126.26			
327		126.56			
328		126.71			
330		127.03			
331		127.19			
333		127.52			
334		127.69			
336		128.05			
338		128.42			
339		128.62			
341		129.02			
342		129.23			
344		129.68			
345		129.91			
347		130.41			
349		130.96			
350		131.25			
352		131.88			
353		132.23			
355		132.99			
356		133.41			
358		134.36			
359		134.90			

COMPOUND CONSTANTS:

MP	317.350 K	TC		VC		ZC
NBP	474.612 K	PC		DC		OM
MU	2.9600 DEBYE	RG				

VAPOR PRESSURE CORRELATION CONSTANTS, K AND KPA:

RPM2	RPM2	RPM2
394. - 485. K	394. - 485. K	394. - 485. K
13 POINTS	13 POINTS	13 POINTS
RMSD = 0.0392	RMSD = 0.0392	RMSD = 0.0392
0.22435782E+02	0.22435782E+02	0.22435782E+02
-0.67642090E+04	-0.67642090E+04	-0.67642090E+04
-0.84645315E-02	-0.84645315E-02	-0.84645315E-02
0.20066396E-05	0.20066396E-05	0.20066396E-05

LIQUID DENSITY CORRELATION CONSTANTS, K AND G/ML:

FRANCIS1	FRANCIS1	FRANCIS1
329. - 412. K	329. - 412. K	329. - 412. K
8 POINTS	8 POINTS	8 POINTS
RMSD = 0.00007	RMSD = 0.00007	RMSD = 0.00007
0.11752396E+01	0.11752396E+01	0.11752396E+01
0.74465154E-03	0.74465154E-03	0.74465154E-03
0.23469505E+01	0.23469505E+01	0.23469505E+01
0.55955298E+03	0.55955298E+03	0.55955298E+03

SECOND VIRIAL COEFFICIENT CORRELATION CONSTANTS, ML/MOL:

LITERATURE DOCUMENTS FOR CORRELATED VAPOR PRESSURE VALUES:

40284

LITERATURE DOCUMENTS FOR CORRELATED LIQUID DENSITY VALUES:

40284

LITERATURE DOCUMENTS REPORTING VIRIAL COEFFICIENT DATA:

T,K	VAPOR PRESSURE, KPA	SATURATED LIQUID VOLUME, ML/MOL	SECOND VIRIAL COEFFICIENT, ML/MOL	HEAT OF VAPORIZATION, J/MOL	SATURATED VAPOR VOLUME, ML/MOL
329		137.16			
332		137.52			
336		137.99			
339		138.35			
343		138.83			
346		139.20			
350		139.69			
353		140.06			
357		140.56			
360		140.94			
364		141.45			
367		141.83			
371		142.35			
375		142.87			
378		143.27			
382		143.80			
385		144.21			
389		144.75			
392		145.17			
396	10.15	145.72		47255.	324485.
399	11.30	146.15		47132.	293525.
403	13.01	146.72		46968.	257519.
406	14.43	147.15		46843.	233927.
410	16.52	147.73		46676.	206351.
414	18.85				182569.
417	20.78				166867.
421	23.59				148377.
424	25.90				136112.
428	29.26				121607.
431	32.01				111944.
435	36.00				100467.
438	39.25				92789.
442	43.94				83634.
445	47.75				77486.
449	53.23				70128.
453	59.21				63611.
456	64.03				59209.
460	70.94				53910.
463	76.50				50319.
467	84.44				45981.
470	90.81				43032.
474	99.88				39458.
477	107.1				37020.
481	117.4				34058.
484	125.6				32031.

COMPOUND CONSTANTS:

MP   204.933 K   TC   VC   ZC
NBP   456.335 K   PC   DC   OM
MU   RG

VAPOR PRESSURE CORRELATION CONSTANTS, K AND KPA:

RIEDEL	RIEDEL	RIEDEL
332. - 457. K	332. - 457. K	332. - 457. K
4 POINTS	4 POINTS	4 POINTS
RMSD = 0.0000	RMSD = 0.0000	RMSD = 0.0000
0.23473348E+04	0.23473348E+04	0.23473348E+04
-0.34498341E+03	-0.34498341E+03	-0.34498341E+03
-0.11554548E+06	-0.11554548E+06	-0.11554548E+06
0.25357439E-14	0.25357439E-14	0.25357439E-14

LIQUID DENSITY CORRELATION CONSTANTS, K AND G/ML:

FRANCIS1	FRANCIS1	FRANCIS1
273. - 294. K	273. - 294. K	273. - 294. K
3 POINTS	3 POINTS	3 POINTS
RMSD = 0.00000	RMSD = 0.00000	RMSD = 0.00000
0.11457132E+01	0.11457132E+01	0.11457132E+01
0.83233000E-03	0.83233000E-03	0.83233000E-03

SECOND VIRIAL COEFFICIENT CORRELATION CONSTANTS, ML/MOL:

LITERATURE DOCUMENTS FOR CORRELATED VAPOR PRESSURE VALUES:

9717

LITERATURE DOCUMENTS FOR CORRELATED LIQUID DENSITY VALUES:

9717

LITERATURE DOCUMENTS REPORTING VIRIAL COEFFICIENT DATA:

T,K	VAPOR PRESSURE, KPA	SATURATED LIQUID VOLUME, ML/MOL	SECOND VIRIAL COEFFICIENT, ML/MOL	HEAT OF VAPORIZATION, J/MOL	SATURATED VAPOR VOLUME, ML/MOL
273		137.40			
277		137.90			
281		138.40			
285		138.91			
289		139.42			
293		139.93			
335	1.257				2216593.
339	1.597				1765325.
344	2.063				1386212.
348	2.456				1178011.
352	2.854				1025580.
356	3.245				912203.
360	3.621				826697.
364	3.975				761363.
369	4.384				699887.
373	4.684				662134.
377	4.964				631389.
381	5.233				605375.
385	5.498				582219.
390	5.843				554913.
394	6.149				532778.
398	6.499				509210.
402	6.915				483368.
406	7.424				454669.
410	8.062				422824.
415	9.112				378679.
419	10.23				340396.
423	11.71				300290.
427	13.69				259390.
431	16.37				218893.
436	21.23				170733.
440	26.99				135568.
444	35.37				104380.
448	47.91				77747.
452	67.24				55890.
456	98.01				38685.

COMPOUND CONSTANTS:

MP   TC   VC   ZC
NBP   PC   DC   OM
MU   RG

   VAPOR PRESSURE CORRELATION CONSTANTS, K AND KPA:

   RPM2   RPM2   RPM2
   279. - 381. K   279. - 381. K   279. - 381. K
    11 POINTS   11 POINTS   11 POINTS
   RMSD = 0.0000   RMSD = 0.0000   RMSD = 0.0000
   0.17090378E+02   0.17090378E+02   0.17090378E+02
   -0.55896563E+04   -0.55896563E+04   -0.55896563E+04
   -0.45415371E-03   -0.45415371E-03   -0.45415371E-03
   0.39330977E-06   0.39330977E-06   0.39330977E-06

   LIQUID DENSITY CORRELATION CONSTANTS, K AND G/ML:

   SECOND VIRIAL COEFFICIENT CORRELATION CONSTANTS, ML/MOL:

LITERATURE DOCUMENTS FOR CORRELATED VAPOR PRESSURE VALUES:

   13799

LITERATURE DOCUMENTS FOR CORRELATED LIQUID DENSITY VALUES:

LITERATURE DOCUMENTS REPORTING VIRIAL COEFFICIENT DATA:

T,K	VAPOR PRESSURE, KPA	SATURATED LIQUID VOLUME, ML/MOL	SECOND VIRIAL COEFFICIENT, ML/MOL	HEAT OF VAPORIZATION, J/MOL	SATURATED VAPOR VOLUME, ML/MOL
279	0.04782				0.4851E+08
281	0.05512				0.4238E+08
283	0.06341				0.3711E+08
285	0.07281				0.3255E+08
288	0.08925				0.2683E+08
290	0.1020				0.2364E+08
292	0.1163				0.2087E+08
295	0.1412				0.1737E+08
297	0.1604				0.1540E+08
299	0.1818				0.1367E+08
302	0.2188				0.1148E+08
304	0.2470				0.1023E+08
306	0.2784				9137942.
309	0.3322				7732869.
311	0.3731				6931118.
313	0.4183				6221458.
316	0.4953				5304855.
318	0.5533				4778330.
320	0.6173				4309869.
323	0.7256				3701004.
325	0.8068				3349063.
327	0.8960				3034415.
329	0.9937				2752735.
332	1.158				2383887.
334	1.280				2169087.
336	1.414				1975935.
339	1.637				1721639.
341	1.803				1572748.
343	1.983				1438301.
346	2.282				1260392.
348	2.504				1155697.
350	2.743				1060785.
353	3.140				934589.
355	3.432				859970.
357	3.747				792074.
360	4.268				701388.
362	4.648				647526.
364	5.058				598343.
367	5.732				532373.
369	6.223				493024.
371	6.750				456975.
374	7.614				408426.
376	8.241				379352.
378	8.913				352633.
380	9.631				328058.

719

COMPOUND CONSTANTS:

MP	305.050 K	TC	VC	ZC
NBP	493.570 K	PC	DC	OM
MU	2.8500 DEBYE	RG		

VAPOR PRESSURE CORRELATION CONSTANTS, K AND KPA:

RPM2	RPM2	RPM2
333. - 494. K	333. - 494. K	333. - 494. K
18 POINTS	18 POINTS	18 POINTS
RMSD = 0.0007	RMSD = 0.0007	RMSD = 0.0007
0.42065341E+02	0.42065341E+02	0.42065341E+02
-0.96075335E+04	-0.96075335E+04	-0.96075335E+04
-0.56094570E-01	-0.56094570E-01	-0.56094570E-01
0.39838002E-04	0.39838002E-04	0.39838002E-04

LIQUID DENSITY CORRELATION CONSTANTS, K AND G/ML:

SECOND VIRIAL COEFFICIENT CORRELATION CONSTANTS, ML/MOL:

LITERATURE DOCUMENTS FOR CORRELATED VAPOR PRESSURE VALUES:

18654

LITERATURE DOCUMENTS FOR CORRELATED LIQUID DENSITY VALUES:

LITERATURE DOCUMENTS REPORTING VIRIAL COEFFICIENT DATA:

T,K	VAPOR PRESSURE, KPA	SATURATED LIQUID VOLUME, ML/MOL	SECOND VIRIAL COEFFICIENT, ML/MOL	HEAT OF VAPORIZATION, J/MOL	SATURATED VAPOR VOLUME, ML/MOL
333	0.3506				7896214.
336	0.4153				6726659.
340	0.5174				5463946.
343	0.6074				4694830.
347	0.7482				3855869.
351	0.9161				3185732.
354	1.062				2771129.
358	1.287				2312133.
362	1.552				1939419.
365	1.779				1705572.
369	2.126				1443225.
373	2.527				1227018.
376	2.869				1089647.
380	3.384				933678.
384	3.974				803416.
387	4.471				719706.
391	5.213				623630.
395	6.055				542417.
398	6.758				489690.
402	7.799				428579.
406	8.969				376356.
409	9.939				342135.
413	11.37				302118.
417	12.96				267583.
420	14.27				244760.
424	16.18				217857.
428	18.30				194430.
431	20.04				178829.
435	22.56				160304.
439	25.34				144039.
442	27.60				133133.
446	30.88				120095.
450	34.46				108563.
453	37.37				100781.
457	41.56				91421.
461	46.13				83085.
464	49.83				77428.
468	55.13				70585.
472	60.89				64453.
475	65.53				60269.
479	72.17				55183.
483	79.37				50600.
486	85.15				47457.
490	93.40				43619.
493	100.0				40980.

COMPOUND CONSTANTS:

MP	TC	VC	ZC
NBP	PC	DC	OM
MU	RG		

VAPOR PRESSURE CORRELATION CONSTANTS, K AND KPA:

RPM2	RPM2	RPM2
299. - 401. K	299. - 401. K	299. - 401. K
11 POINTS	11 POINTS	11 POINTS
RMSD = 0.0000	RMSD = 0.0000	RMSD = 0.0000
0.17107479E+02	0.17107479E+02	0.17107479E+02
-0.58037434E+04	-0.58037434E+04	-0.58037434E+04
-0.41235569E-03	-0.41235569E-03	-0.41235569E-03
0.35515840E-06	0.35515840E-06	0.35515840E-06

LIQUID DENSITY CORRELATION CONSTANTS, K AND G/ML:

SECOND VIRIAL COEFFICIENT CORRELATION CONSTANTS, ML/MOL:

LITERATURE DOCUMENTS FOR CORRELATED VAPOR PRESSURE VALUES:

  13799

LITERATURE DOCUMENTS FOR CORRELATED LIQUID DENSITY VALUES:

LITERATURE DOCUMENTS REPORTING VIRIAL COEFFICIENT DATA:

T,K	VAPOR PRESSURE, KPA	SATURATED LIQUID VOLUME, ML/MOL	SECOND VIRIAL COEFFICIENT, ML/MOL	HEAT OF VAPORIZATION, J/MOL	SATURATED VAPOR VOLUME, ML/MOL
299	0.09121				0.2726E+08
301	0.1037				0.2413E+08
303	0.1178				0.2139E+08
305	0.1335				0.1900E+08
308	0.1605				0.1595E+08
310	0.1812				0.1422E+08
312	0.2043				0.1270E+08
315	0.2437				0.1075E+08
317	0.2737				9631349.
319	0.3068				8644295.
322	0.3633				7369200.
324	0.4059				6636649.
326	0.4529				5984840.
329	0.5324				5137634.
331	0.5921				4647920.
333	0.6576				4210106.
336	0.7680				3637758.
338	0.8503				3305015.
340	0.9404				3006207.
343	1.091				2613458.
345	1.203				2383893.
347	1.325				2176872.
349	1.458				1989967.
352	1.679				1742703.
354	1.843				1597236.
356	2.020				1465393.
359	2.314				1290147.
361	2.530				1186563.
363	2.763				1092335.
366	3.148				966531.
368	3.431				891839.
370	3.735				823659.
373	4.235				732248.
375	4.600				677748.
377	4.993				627837.
380	5.635				560652.
382	6.103				520438.
384	6.603				483496.
387	7.421				433581.
389	8.014				403591.
391	8.647				375961.
394	9.678				338495.
396	10.42				315905.
398	11.22				295034.
400	12.06				275738.

COMPOUND CONSTANTS:

MP	295.950 K	TC	VC	ZC
NBP		PC	DC	OM
MU	2.7500 DEBYE	RG		

      VAPOR PRESSURE CORRELATION CONSTANTS, K AND KPA:

RPM2	RPM2	RPM2
353. - 424. K	353. - 424. K	353. - 424. K
15 POINTS	15 POINTS	15 POINTS
RMSD = 0.0002	RMSD = 0.0002	RMSD = 0.0002
0.44285192E+02	0.44285192E+02	0.44285192E+02
-0.10504622E+05	-0.10504622E+05	-0.10504622E+05
-0.57468922E-01	-0.57468922E-01	-0.57468922E-01
0.39224351E-04	0.39224351E-04	0.39224351E-04

      LIQUID DENSITY CORRELATION CONSTANTS, K AND G/ML:

      SECOND VIRIAL COEFFICIENT CORRELATION CONSTANTS, ML/MOL:

LITERATURE DOCUMENTS FOR CORRELATED VAPOR PRESSURE VALUES:

  18654

LITERATURE DOCUMENTS FOR CORRELATED LIQUID DENSITY VALUES:

LITERATURE DOCUMENTS REPORTING VIRIAL COEFFICIENT DATA:

T,K	VAPOR PRESSURE, KPA	SATURATED LIQUID VOLUME, ML/MOL	SECOND VIRIAL COEFFICIENT, ML/MOL	HEAT OF VAPORIZATION, J/MOL	SATURATED VAPOR VOLUME, ML/MOL
353	0.4182				7017932.
354	0.4416				6665683.
356	0.4917				6020301.
357	0.5185				5724702.
359	0.5760				5182149.
361	0.6389				4697923.
362	0.6725				4475479.
364	0.7443				4066025.
365	0.7826				3877607.
367	0.8644				3530213.
369	0.9533				3218304.
370	1.001				3074380.
372	1.101				2808312.
373	1.155				2685347.
375	1.269				2457673.
377	1.392				2252121.
378	1.457				2156880.
380	1.596				1980105.
382	1.745				1819984.
383	1.824				1745612.
385	1.992				1607248.
386	2.080				1542892.
388	2.267				1422999.
390	2.468				1313864.
391	2.574				1262989.
393	2.798				1168002.
394	2.915				1123663.
396	3.164				1040777.
398	3.429				964982.
399	3.569				929528.
401	3.863				863114.
403	4.177				802219.
404	4.342				773678.
406	4.688				720108.
407	4.869				694971.
409	5.250				647738.
411	5.655				604253.
412	5.868				583809.
414	6.312				545326.
415	6.545				527214.
417	7.031				493085.
419	7.548				461544.
420	7.818				446674.
422	8.382				418605.
423	8.676				405358.

COMPOUND CONSTANTS:

MP	266.603 K	TC		VC		ZC
NBP		PC		DC		OM
MU	2.8300 DEBYE	RG				

VAPOR PRESSURE CORRELATION CONSTANTS, K AND KPA:

LIQUID DENSITY CORRELATION CONSTANTS, K AND G/ML:

FRANCIS1	FRANCIS1	FRANCIS1
273. - 374. K	273. - 374. K	273. - 374. K
7 POINTS	7 POINTS	7 POINTS
RMSD = 0.00005	RMSD = 0.00005	RMSD = 0.00005
0.11216106E+01	0.11216106E+01	0.11216106E+01
0.75179804E-03	0.75179804E-03	0.75179804E-03
0.11565094E+01	0.11565094E+01	0.11565094E+01
0.58048584E+03	0.58048584E+03	0.58048584E+03

SECOND VIRIAL COEFFICIENT CORRELATION CONSTANTS, ML/MOL:

LITERATURE DOCUMENTS FOR CORRELATED VAPOR PRESSURE VALUES:

LITERATURE DOCUMENTS FOR CORRELATED LIQUID DENSITY VALUES:

   13811

LITERATURE DOCUMENTS REPORTING VIRIAL COEFFICIENT DATA:

T,K	VAPOR PRESSURE, KPA	SATURATED LIQUID VOLUME, ML/MOL	SECOND VIRIAL COEFFICIENT, ML/MOL	HEAT OF VAPORIZATION, J/MOL	SATURATED VAPOR VOLUME, ML/MOL
273		169.02			
275		169.31			
277		169.59			
279		169.88			
282		170.31			
284		170.60			
286		170.88			
289		171.32			
291		171.61			
293		171.90			
295		172.20			
298		172.64			
300		172.94			
302		173.24			
305		173.68			
307		173.99			
309		174.29			
312		174.74			
314		175.05			
316		175.35			
318		175.66			
321		176.12			
323		176.43			
325		176.74			
328		177.21			
330		177.53			
332		177.84			
334		178.16			
337		178.63			
339		178.95			
341		179.28			
344		179.76			
346		180.08			
348		180.41			
351		180.90			
353		181.23			
355		181.56			
357		181.89			
360		182.39			
362		182.73			
364		183.06			
367		183.57			
369		183.91			
371		184.25			
373		184.60			

COMPOUND CONSTANTS:

MP   TC   VC   ZC
NBP   PC   DC   OM
MU   RG

VAPOR PRESSURE CORRELATION CONSTANTS, K AND KPA:

LIQUID DENSITY CORRELATION CONSTANTS, K AND G/ML:

FRANCIS1   FRANCIS1   FRANCIS1
287. - 294. K   287. - 294. K   287. - 294. K
  5 POINTS   5 POINTS   5 POINTS
RMSD = 0.00001   RMSD = 0.00001   RMSD = 0.00001
0.11585274E+01   0.11585274E+01   0.11585274E+01
0.88412291E-03   0.88412291E-03   0.88412291E-03
0.99014626E+01   0.99014626E+01   0.99014626E+01
0.11632017E+04   0.11632017E+04   0.11632017E+04

SECOND VIRIAL COEFFICIENT CORRELATION CONSTANTS, ML/MOL:

LITERATURE DOCUMENTS FOR CORRELATED VAPOR PRESSURE VALUES:

LITERATURE DOCUMENTS FOR CORRELATED LIQUID DENSITY VALUES:

  6837

LITERATURE DOCUMENTS REPORTING VIRIAL COEFFICIENT DATA:

T,K	VAPOR PRESSURE, KPA	SATURATED LIQUID VOLUME, ML/MOL	SECOND VIRIAL COEFFICIENT, ML/MOL	HEAT OF VAPORIZATION, J/MOL	SATURATED VAPOR VOLUME, ML/MOL
287		172.64			
288		172.81			
289		172.99			
290		173.16			
291		173.34			
292		173.51			
293		173.69			
294		173.86			

COMPOUND CONSTANTS:

MP   278.147 K   TC   VC   ZC
NBP   466.413 K   PC   DC   OM
MU   2.9300 DEBYE   RG

VAPOR PRESSURE CORRELATION CONSTANTS, K AND KPA:

RPM2	RPM2	RPM2	RPM2
369. - 521. K	369. - 481. K	369. - 521. K	450. - 521. K
16 POINTS	12 POINTS	16 POINTS	8 POINTS
RMSD = 0.0007	RMSD = 0.0002	RMSD = 0.0007	RMSD = 0.0011
0.16332197E+02	0.16348405E+02	0.16332197E+02	0.16259752E+02
-0.54509453E+04	-0.54532192E+04	-0.54509453E+04	-0.54395626E+04
-0.94791231E-04	-0.13314052E-03	-0.94791231E-04	0.58457462E-04
0.79497020E-07	0.10961990E-06	0.79497020E-07	-0.28264046E-07

LIQUID DENSITY CORRELATION CONSTANTS, K AND G/ML:

SECOND VIRIAL COEFFICIENT CORRELATION CONSTANTS, ML/MOL:

LITERATURE DOCUMENTS FOR CORRELATED VAPOR PRESSURE VALUES:

6619

LITERATURE DOCUMENTS FOR CORRELATED LIQUID DENSITY VALUES:

LITERATURE DOCUMENTS REPORTING VIRIAL COEFFICIENT DATA:

T,K	VAPOR PRESSURE, KPA	SATURATED LIQUID VOLUME, ML/MOL	SECOND VIRIAL COEFFICIENT, ML/MOL	HEAT OF VAPORIZATION, J/MOL	SATURATED VAPOR VOLUME, ML/MOL
369	4.645				660471.
372	5.232				591124.
375	5.882				530034.
379	6.857				459564.
382	7.676				413771.
386	8.898				360681.
389	9.921				326010.
393	11.44				285621.
396	12.71				259120.
400	14.58				228108.
403	16.13				207668.
406	17.83				189332.
410	20.32				167744.
413	22.38				153430.
417	25.40				136510.
420	27.88				125246.
424	31.51				111880.
427	34.49				102949.
431	38.82				92314.
434	42.36				85182.
438	47.51				76658.
441	51.70				70923.
444	56.20				65689.
448	62.70				59405.
451	67.98				55158.
455	75.60				50042.
458	81.76				46573.
462	90.63				42384.
465	97.79				39535.
469	108.1				36084.
472	116.3				33731.
476	128.2				30873.
479	137.7				28919.
482	147.8				27112.
486	162.2				24909.
489	173.8				23397.
493	190.2				21550.
496	203.4				20280.
500	222.0				18724.
503	236.9				17651.
507	258.1				16335.
510	274.9				15425.
514	298.7				14306.
517	317.7				13531.
520	337.6				12807.

COMPOUND CONSTANTS:

MP	452.683 K	TC		VC		ZC
NBP	480.706 K	PC		DC		OM
MU	2.9800 DEBYE	RG				

VAPOR PRESSURE CORRELATION CONSTANTS, K AND KPA:

RPM2	RPM2	RPM2	RPM2
451. - 500. K	451. - 500. K	451. - 500. K	464. - 500. K
21 POINTS	21 POINTS	21 POINTS	12 POINTS
RMSD = 0.3340	RMSD = 0.3340	RMSD = 0.3340	RMSD = 0.3586
-0.73204088E+02	-0.73204088E+02	-0.73204088E+02	0.67655826E+03
0.90466559E+04	0.90466559E+04	0.90466559E+04	-0.11139630E+06
0.18287068E+00	0.18287068E+00	0.18287068E+00	-0.13723823E+01
-0.12508377E-03	-0.12508377E-03	-0.12508377E-03	0.94992874E-03

LIQUID DENSITY CORRELATION CONSTANTS, K AND G/ML:

FRANCIS1	FRANCIS1	FRANCIS1
453. - 474. K	453. - 474. K	453. - 474. K
3 POINTS	3 POINTS	3 POINTS
RMSD = 0.00000	RMSD = 0.00000	RMSD = 0.00000
0.12227294E+01	0.12227294E+01	0.12227294E+01
0.85964500E-03	0.85964500E-03	0.85964500E-03

SECOND VIRIAL COEFFICIENT CORRELATION CONSTANTS, ML/MOL:

LITERATURE DOCUMENTS FOR CORRELATED VAPOR PRESSURE VALUES:

21109 21948

LITERATURE DOCUMENTS FOR CORRELATED LIQUID DENSITY VALUES:

12455

LITERATURE DOCUMENTS REPORTING VIRIAL COEFFICIENT DATA:

T,K	VAPOR PRESSURE, KPA	SATURATED LIQUID VOLUME, ML/MOL	SECOND VIRIAL COEFFICIENT, ML/MOL	HEAT OF VAPORIZATION, J/MOL	SATURATED VAPOR VOLUME, ML/MOL
451	48.79				
452	50.05				
453	51.34	182.69			73356.
454	52.67	182.88			71672.
455	54.02	183.07			70030.
456	55.41	183.26			68429.
457	56.82	183.45			66869.
458	58.27	183.64			65348.
459	59.76	183.83			63865.
461	62.82	184.21			61010.
462	64.41	184.40			59636.
463	66.03	184.59			58297.
464	67.69	184.79			56991.
465	69.39	184.98			55719.
466	71.12	185.17			54478.
467	72.89	185.37			53268.
468	74.70	185.56			52089.
469	76.55	185.75			50939.
471	80.37	186.14			48726.
472	82.34	186.34			47661.
473	84.35	186.54			46623.
474	86.41	186.73			45610.
475	88.50				44623.
476	90.65				43661.
477	92.83				42723.
478	95.06				41808.
479	97.34				40916.
481	102.0				39198.
482	104.4				38372.
483	106.9				37566.
484	109.4				36779.
485	112.0				36013.
486	114.6				35265.
487	117.2				34537.
488	120.0				33826.
489	122.7				33133.
491	128.4				31797.
492	131.3				31154.
493	134.3				30527.
494	137.3				29915.
495	140.4				29319.
496	143.5				28737.
497	146.7				28169.
498	149.9				27616.
499	153.2				27075.

COMPOUND CONSTANTS:

MP   TC   VC   ZC
NBP   PC   DC   OM
MU   RG

VAPOR PRESSURE CORRELATION CONSTANTS, K AND KPA:

LIQUID DENSITY CORRELATION CONSTANTS, K AND G/ML:

FRANCIS1   FRANCIS1   FRANCIS1
291. - 359. K   291. - 359. K   291. - 359. K
  6 POINTS   6 POINTS   6 POINTS
RMSD = 0.00077   RMSD = 0.00077   RMSD = 0.00077
0.11929045E+01   0.11929045E+01   0.11929045E+01
0.71501243E-03   0.71501243E-03   0.71501243E-03
0.31916519E+02   0.31916519E+02   0.31916519E+02
0.73905195E+04   0.73905195E+04   0.73905195E+04

SECOND VIRIAL COEFFICIENT CORRELATION CONSTANTS, ML/MOL:

LITERATURE DOCUMENTS FOR CORRELATED VAPOR PRESSURE VALUES:

LITERATURE DOCUMENTS FOR CORRELATED LIQUID DENSITY VALUES:

   9923 14204

LITERATURE DOCUMENTS REPORTING VIRIAL COEFFICIENT DATA:

T,K	VAPOR PRESSURE, KPA	SATURATED LIQUID VOLUME, ML/MOL	SECOND VIRIAL COEFFICIENT, ML/MOL	HEAT OF VAPORIZATION, J/MOL	SATURATED VAPOR VOLUME, ML/MOL
291		155.29			
292		155.40			
294		155.63			
295		155.74			
297		155.97			
298		156.09			
300		156.32			
301		156.43			
303		156.66			
304		156.78			
306		157.01			
307		157.12			
309		157.36			
311		157.59			
312		157.71			
314		157.94			
315		158.06			
317		158.29			
318		158.41			
320		158.65			
321		158.77			
323		159.00			
324		159.12			
326		159.36			
328		159.60			
329		159.72			
331		159.96			
332		160.08			
334		160.32			
335		160.44			
337		160.68			
338		160.81			
340		161.05			
341		161.17			
343		161.42			
345		161.66			
346		161.78			
348		162.03			
349		162.15			
351		162.40			
352		162.53			
354		162.77			
355		162.90			
357		163.15			
358		163.27			

735

COMPOUND CONSTANTS:

MP	TC	VC	ZC
NBP	PC	DC	OM
MU	RG		

VAPOR PRESSURE CORRELATION CONSTANTS, K AND KPA:

RPM2	RPM2	RPM2
449. - 501. K	449. - 501. K	449. - 501. K
7 POINTS	7 POINTS	7 POINTS
RMSD = 0.0102	RMSD = 0.0102	RMSD = 0.0102
-0.43389210E+02	-0.43389210E+02	-0.43389210E+02
0.30307622E+04	0.30307622E+04	0.30307622E+04
0.12861987E+00	0.12861987E+00	0.12861987E+00
-0.92236652E-04	-0.92236652E-04	-0.92236652E-04

LIQUID DENSITY CORRELATION CONSTANTS, K AND G/ML:

FRANCIS1	FRANCIS1	FRANCIS1
301. - 412. K	301. - 412. K	301. - 412. K
9 POINTS	9 POINTS	9 POINTS
RMSD = 0.00012	RMSD = 0.00012	RMSD = 0.00012
0.11690245E+01	0.11690245E+01	0.11690245E+01
0.72760601E-03	0.72760601E-03	0.72760601E-03
0.35297394E+01	0.35297394E+01	0.35297394E+01
0.16314060E+04	0.16314060E+04	0.16314060E+04

SECOND VIRIAL COEFFICIENT CORRELATION CONSTANTS, ML/MOL:

LITERATURE DOCUMENTS FOR CORRELATED VAPOR PRESSURE VALUES:

40284

LITERATURE DOCUMENTS FOR CORRELATED LIQUID DENSITY VALUES:

40284

LITERATURE DOCUMENTS REPORTING VIRIAL COEFFICIENT DATA:

T,K	VAPOR PRESSURE, KPA	SATURATED LIQUID VOLUME, ML/MOL	SECOND VIRIAL COEFFICIENT, ML/MOL	HEAT OF VAPORIZATION, J/MOL	SATURATED VAPOR VOLUME, ML/MOL
301		177.63			
305		178.18			
310		178.87			
314		179.42			
319		180.13			
323		180.69			
328		181.40			
332		181.97			
337		182.69			
341		183.27			
346		184.01			
350		184.59			
355		185.34			
360		186.08			
364		186.69			
369		187.45			
373		188.06			
378		188.83			
382		189.45			
387		190.23			
391		190.86			
396		191.65			
400		192.29			
405		193.10			
410		193.91			
450	12.77				293067.
455	14.86				254610.
460	17.24				221838.
464	19.38				199105.
469	22.36				174403.
473	25.02				157210.
478	28.70				138463.
482	31.97				125370.
487	36.46				111044.
491	40.42				101004.
496	45.83				89982.
500	50.56				82229.

COMPOUND CONSTANTS:

MP	335.550 K	TC	VC	ZC
NBP	549.670 K	PC	DC	OM
MU	2.7500 DEBYE	RG		

VAPOR PRESSURE CORRELATION CONSTANTS, K AND KPA:

VAPRES-2	VAPRES-2	VAPRES-2	RPM2
408. - 564. K	408. - 556. K	408. - 564. K	520. - 564. K
23 POINTS	22 POINTS	23 POINTS	6 POINTS
RMSD = 0.0190	RMSD = 0.0165	RMSD = 0.0190	RMSD = 0.0248
-0.30334436E+03	-0.35752726E+03	-0.30334436E+03	0.12036858E+02
-0.12188523E+03	0.13802298E+04	-0.12188523E+03	-0.58199290E+04
-0.14965503E+00	-0.16995624E+00	-0.14965503E+00	0.10284643E-01
0.56469159E-04	0.63613684E-04	0.56469159E-04	-0.82201750E-05
0.59179741E+02	0.68760896E+02	0.59179741E+02	

LIQUID DENSITY CORRELATION CONSTANTS, K AND G/ML:

FRANCIS1	FRANCIS1	FRANCIS1
343. - 416. K	343. - 416. K	343. - 416. K
6 POINTS	6 POINTS	6 POINTS
RMSD = 0.00017	RMSD = 0.00017	RMSD = 0.00017
0.11714888E+01	0.11714888E+01	0.11714888E+01
0.66504418E-03	0.66504418E-03	0.66504418E-03
0.63866241E+02	0.63866241E+02	0.63866241E+02
0.22372144E+04	0.22372144E+04	0.22372144E+04

SECOND VIRIAL COEFFICIENT CORRELATION CONSTANTS, ML/MOL:

LITERATURE DOCUMENTS FOR CORRELATED VAPOR PRESSURE VALUES:

40284

LITERATURE DOCUMENTS FOR CORRELATED LIQUID DENSITY VALUES:

40284

LITERATURE DOCUMENTS REPORTING VIRIAL COEFFICIENT DATA:

T,K	VAPOR PRESSURE, KPA	SATURATED LIQUID VOLUME, ML/MOL	SECOND VIRIAL COEFFICIENT, ML/MOL	HEAT OF VAPORIZATION, J/MOL	SATURATED VAPOR VOLUME, ML/MOL
343		200.41			
348		201.16			
353		201.93			
358		202.69			
363		203.47			
368		204.24			
373		205.03			
378		205.82			
383		206.62			
388		207.42			
393		208.23			
398		209.05			
403		209.87			
408	1.557	210.70		58407.	2178281.
413	1.917	211.54		58132.	1790929.
418	2.347				1481010.
423	2.856				1231582.
428	3.456				1029694.
433	4.160				865384.
438	4.982				730946.
443	5.937				620383.
448	7.041				529003.
453	8.312				453113.
458	9.769				389794.
463	11.43				336727.
468	13.32				292057.
473	15.47				254297.
478	17.88				222247.
483	20.60				194937.
488	23.65				171577.
493	27.05				151521.
498	30.84				134239.
503	35.06				119297.
508	39.72				106334.
513	44.87				95051.
518	50.55				85198.
523	56.79				76570.
528	63.63				68989.
533	71.12				62311.
538	79.30				56411.
543	88.21				51184.
548	97.90				46542.
553	108.4				42407.
558	119.8				38716.
563	132.2				35413.

COMPOUND CONSTANTS:

MP		TC		VC		ZC
NBP	354.703 K	PC		DC		OM
MU	3.0000 DEBYE	RG				

    VAPOR PRESSURE CORRELATION CONSTANTS, K AND KPA:

RPM2	RPM2	RPM2
300. - 355. K	300. - 355. K	300. - 355. K
6 POINTS	6 POINTS	6 POINTS
RMSD = 0.6422	RMSD = 0.6422	RMSD = 0.6422
-0.25859342E+03	-0.25859342E+03	-0.25859342E+03
0.25691015E+05	0.25691015E+05	0.25691015E+05
0.84404370E+00	0.84404370E+00	0.84404370E+00
-0.86319912E-03	-0.86319912E-03	-0.86319912E-03

    LIQUID DENSITY CORRELATION CONSTANTS, K AND G/ML:

    SECOND VIRIAL COEFFICIENT CORRELATION CONSTANTS, ML/MOL:

LITERATURE DOCUMENTS FOR CORRELATED VAPOR PRESSURE VALUES:

   2219   5999 10531 18774

LITERATURE DOCUMENTS FOR CORRELATED LIQUID DENSITY VALUES:

LITERATURE DOCUMENTS REPORTING VIRIAL COEFFICIENT DATA:

T,K	VAPOR PRESSURE, KPA	SATURATED LIQUID VOLUME, ML/MOL	SECOND VIRIAL COEFFICIENT, ML/MOL	HEAT OF VAPORIZATION, J/MOL	SATURATED VAPOR VOLUME, ML/MOL
300	13.05				191194.
301	13.59				184170.
302	14.16				177375.
303	14.75				170805.
305	16.02				158326.
306	16.69				152409.
307	17.40				146703.
308	18.14				141201.
310	19.71				130795.
311	20.54				125880.
312	21.41				121151.
313	22.32				116602.
315	24.24				108026.
316	25.27				103988.
317	26.33				100109.
318	27.43				96386.
320	29.77				89382.
321	31.00				86091.
322	32.28				82936.
323	33.61			·	79910.
325	36.40				74230.
326	37.87				71566.
327	39.40				69014.
328	40.97				66569.
330	44.26				61986.
331	45.99				59839.
332	47.77				57785.
333	49.60				55818.
335	53.43				52134.
336	55.42				50411.
337	57.46				48762.
338	59.56				47185.
340	63.91				44235.
341	66.16				42855.
342	68.46				41536.
343	70.81				40276.
345	75.65				37919.
346	78.14				36818.
347	80.67				35766.
348	83.24				34762.
350	88.49				32886.
351	91.17				32011.
352	93.88				31176.
353	96.61				30379.
355	102.2				28894.

COMPOUND CONSTANTS:

MP   219.580 K   TC   VC   ZC
NBP 370.904 K   PC   DC   OM
MU   2.7700 DEBYE   RG

   VAPOR PRESSURE CORRELATION CONSTANTS, K AND KPA:

   RPM2   RPM2   RPM2
   311. - 371. K   311. - 371. K   311. - 371. K
     5 POINTS   5 POINTS   5 POINTS
   RMSD = 0.1654   RMSD = 0.1654   RMSD = 0.1654
   -0.11112349E+03   -0.11112349E+03   -0.11112349E+03
    0.10130898E+05   0.10130898E+05   0.10130898E+05
    0.37424566E+00   0.37424566E+00   0.37424566E+00
   -0.36622482E-03   -0.36622482E-03   -0.36622482E-03

   LIQUID DENSITY CORRELATION CONSTANTS, K AND G/ML:

   FRANCIS1   FRANCIS1   FRANCIS1
   293. - 299. K   293. - 299. K   293. - 299. K
     3 POINTS   3 POINTS   3 POINTS
   RMSD = 0.00000   RMSD = 0.00000   RMSD = 0.00000
   0.11309113E+01   0.11309113E+01   0.11309113E+01
   0.95017100E-03   0.95017100E-03   0.95017100E-03

   SECOND VIRIAL COEFFICIENT CORRELATION CONSTANTS, ML/MOL:

LITERATURE DOCUMENTS FOR CORRELATED VAPOR PRESSURE VALUES:

   1361 23278

LITERATURE DOCUMENTS FOR CORRELATED LIQUID DENSITY VALUES:

   1360 23270

LITERATURE DOCUMENTS REPORTING VIRIAL COEFFICIENT DATA:

T,K	VAPOR PRESSURE, KPA	SATURATED LIQUID VOLUME, ML/MOL	SECOND VIRIAL COEFFICIENT, ML/MOL	HEAT OF VAPORIZATION, J/MOL	SATURATED VAPOR VOLUME, ML/MOL
293		98.67			
294		98.78			
296		99.00			
298		99.22			
312	11.73				221141.
314	12.75				204799.
316	13.85				189717.
317	14.43				182619.
319	15.67				169255.
321	17.01				156931.
323	18.45				145568.
324	19.21				140224.
326	20.82				130168.
328	22.56				120900.
330	24.42				112359.
331	25.40				108343.
333	27.47				100787.
335	29.69				93823.
337	32.06				87405.
339	34.59				81489.
340	35.92				78707.
342	38.70				73471.
344	41.67				68642.
346	44.82				64189.
347	46.46				62094.
349	49.90				58149.
351	53.54				54508.
353	57.38				51147.
355	61.43				48045.
356	63.54				46583.
358	67.91				43829.
360	72.50				41285.
362	77.31				38933.
363	79.79				37825.
365	84.92				35735.
367	90.27				33802.
369	95.84				32013.
370	98.69				31170.

COMPOUND CONSTANTS:

MP		TC		VC		ZC
NBP	402.144 K	PC		DC		OM
MU		RG				

VAPOR PRESSURE CORRELATION CONSTANTS, K AND KPA:

RPM2	RPM2	RPM2	RPM2
401. - 561. K	401. - 561. K	401. - 561. K	401. - 561. K
13 POINTS	13 POINTS	13 POINTS	13 POINTS
RMSD = 1.5160	RMSD = 1.5160	RMSD = 1.5160	RMSD = 1.5160
0.12562638E+02	0.12562638E+02	0.12562638E+02	0.12562638E+02
-0.42869180E+04	-0.42869180E+04	-0.42869180E+04	-0.42869180E+04
0.11158413E-01	0.11158413E-01	0.11158413E-01	0.11158413E-01
-0.10948396E-04	-0.10948396E-04	-0.10948396E-04	-0.10948396E-04

LIQUID DENSITY CORRELATION CONSTANTS, K AND G/ML:

SECOND VIRIAL COEFFICIENT CORRELATION CONSTANTS, ML/MOL:

LITERATURE DOCUMENTS FOR CORRELATED VAPOR PRESSURE VALUES:

5067

LITERATURE DOCUMENTS FOR CORRELATED LIQUID DENSITY VALUES:

LITERATURE DOCUMENTS REPORTING VIRIAL COEFFICIENT DATA:

T,K	VAPOR PRESSURE, KPA	SATURATED LIQUID VOLUME, ML/MOL	SECOND VIRIAL COEFFICIENT, ML/MOL	HEAT OF VAPORIZATION, J/MOL	SATURATED VAPOR VOLUME, ML/MOL
401	98.11				33983.
404	107.0				31403.
408	119.8				28322.
411	130.2				26251.
415	145.2				23769.
419	161.5				21569.
422	174.7				20082.
426	193.7				18290.
430	214.2				16692.
433	230.7				15607.
437	254.2				14293.
440	273.1				13397.
444	299.9				12310.
448	328.7				11331.
451	351.7				10662.
455	384.3				9845.
459	419.1				9107.
462	446.7				8599.
466	485.7				7977.
470	527.2				7413.
473	559.9				7023.
477	606.0				6545.
480	642.3				6213.
484	693.2				5805.
488	746.9				5432.
491	789.2				5173.
495	848.1				4853.
499	910.1				4559.
502	958.6				4354.
506	1026.				4100.
510	1097.				3867.
513	1152.				3703.
517	1228.				3500.
520	1288.				3358.
524	1370.				3181.
528	1455.				3017.
531	1521.				2902.
535	1613.				2758.
539	1708.				2624.
542	1781.				2531.
546	1881.				2413.
550	1985.				2303.
553	2065.				2226.
557	2175.				2129.
560	2259.				2061.

745

COMPOUND CONSTANTS:

MP	TC	VC	ZC
NBP 402.959 K	PC	DC	OM
MU 3.2400 DEBYE	RG		

VAPOR PRESSURE CORRELATION CONSTANTS, K AND KPA:

RPM2	RPM2	RPM2
287. - 404. K	287. - 404. K	287. - 404. K
13 POINTS	13 POINTS	13 POINTS
RMSD = 0.1369	RMSD = 0.1369	RMSD = 0.1369
0.40684782E+01	0.40684782E+01	0.40684782E+01
-0.37767853E+04	-0.37767853E+04	-0.37767853E+04
0.45759886E-01	0.45759886E-01	0.45759886E-01
-0.52451646E-04	-0.52451646E-04	-0.52451646E-04

LIQUID DENSITY CORRELATION CONSTANTS, K AND G/ML:

FRANCIS1	FRANCIS1	FRANCIS1
293. - 394. K	293. - 394. K	293. - 394. K
6 POINTS	6 POINTS	6 POINTS
RMSD = 0.00004	RMSD = 0.00004	RMSD = 0.00004
0.11541748E+01	0.11541748E+01	0.11541748E+01
0.92861406E-03	0.92861406E-03	0.92861406E-03
0.12769001E+02	0.12769001E+02	0.12769001E+02
0.83510986E+03	0.83510986E+03	0.83510986E+03

SECOND VIRIAL COEFFICIENT CORRELATION CONSTANTS, ML/MOL:

LITERATURE DOCUMENTS FOR CORRELATED VAPOR PRESSURE VALUES:

11773 13236

LITERATURE DOCUMENTS FOR CORRELATED LIQUID DENSITY VALUES:

11773

LITERATURE DOCUMENTS REPORTING VIRIAL COEFFICIENT DATA:

T,K	VAPOR PRESSURE, KPA	SATURATED LIQUID VOLUME, ML/MOL	SECOND VIRIAL COEFFICIENT, ML/MOL	HEAT OF VAPORIZATION, J/MOL	SATURATED VAPOR VOLUME, ML/MOL
287	0.7570				3152107.
289	0.8554				2809062.
292	1.024				2370396.
294	1.153	114.45		42121.	2120917.
297	1.372	114.84		42109.	1800309.
300	1.627	115.23		42091.	1533427.
302	1.819	115.49		42074.	1380449.
305	2.145	115.89		42044.	1182368.
308	2.521	116.29		42005.	1016001.
310	2.802	116.56		41975.	919933.
313	3.275	116.97		41924.	794658.
316	3.816	117.38		41864.	688554.
318	4.218	117.65		41820.	626860.
321	4.889	118.07		41747.	545873.
324	5.651	118.49		41665.	476735.
326	6.213	118.77		41606.	436273.
329	7.145	119.19		41510.	382824.
332	8.195	119.62		41406.	336855.
334	8.964	119.90		41331.	309788.
337	10.23	120.33		41211.	273820.
340	11.65	120.77		41082.	242669.
342	12.68	121.06		40991.	224220.
345	14.37	121.50		40846.	199568.
348	16.25	121.95		40691.	178076.
350	17.61	122.25		40583.	165278.
353	19.82	122.70		40412.	148086.
356	22.25	123.15		40231.	133004.
358	24.01	123.46		40105.	123977.
361	26.85	123.92		39906.	111791.
364	29.95	124.38		39698.	101037.
366	32.18	124.69		39553.	94569.
369	35.76	125.16		39326.	85796.
372	39.65	125.64		39089.	78011.
374	42.42	125.96		38924.	73307.
377	46.85	126.44		38669.	66899.
380	51.64	126.92		38401.	61182.
382	55.03	127.25		38217.	57713.
385	60.43	127.74		37931.	52967.
388	66.22	128.24		37633.	48713.
390	70.31	128.57		37427.	46121.
393	76.77	129.08		37110.	42561.
396	83.66				39354.
398	88.50				37393.
401	96.11				34690.
403	101.4				33033.

COMPOUND CONSTANTS:

MP		TC		VC		ZC
NBP	394.648 K	PC		DC		OM
MU		RG				

    VAPOR PRESSURE CORRELATION CONSTANTS, K AND KPA:

RPM2	RPM2	RPM2
273. - 395. K	273. - 395. K	273. - 395. K
4 POINTS	4 POINTS	4 POINTS
RMSD = 0.0005	RMSD = 0.0005	RMSD = 0.0005
0.41406877E+02	0.41406877E+02	0.41406877E+02
-0.79886545E+04	-0.79886545E+04	-0.79886545E+04
-0.61649738E-01	-0.61649738E-01	-0.61649738E-01
0.49977846E-04	0.49977846E-04	0.49977846E-04

    LIQUID DENSITY CORRELATION CONSTANTS, K AND G/ML:

    SECOND VIRIAL COEFFICIENT CORRELATION CONSTANTS, ML/MOL:

LITERATURE DOCUMENTS FOR CORRELATED VAPOR PRESSURE VALUES:

   13236

LITERATURE DOCUMENTS FOR CORRELATED LIQUID DENSITY VALUES:

LITERATURE DOCUMENTS REPORTING VIRIAL COEFFICIENT DATA:

T,K	VAPOR PRESSURE, KPA	SATURATED LIQUID VOLUME, ML/MOL	SECOND VIRIAL COEFFICIENT, ML/MOL	HEAT OF VAPORIZATION, J/MOL	SATURATED VAPOR VOLUME, ML/MOL
273	0.3824				5935071.
275	0.4418				5175232.
278	0.5458				4234584.
281	0.6705				3484634.
284	0.8190				2883223.
286	0.9330				2548601.
289	1.130				2127105.
292	1.361				1784127.
295	1.631				1503608.
297	1.836				1344997.
300	2.184				1142140.
303	2.586				974083.
306	3.050				834228.
309	3.582				717332.
311	3.978				650056.
314	4.641				562551.
317	5.394				488608.
320	6.247				425883.
322	6.876				389344.
325	7.918				341268.
328	9.088				300089.
331	10.40				264695.
333	11.35				243856.
336	12.92				216155.
339	14.67				192140.
342	16.60				171257.
345	18.74				153044.
347	20.29				142191.
350	22.81				127599.
353	25.57				114778.
356	28.60				103482.
358	30.78				96696.
361	34.30				87498.
364	38.14				79341.
367	42.33				72090.
370	46.87				65628.
372	50.12				61710.
375	55.33				56353.
378	60.97				51551.
381	67.06				47239.
383	71.39				44608.
386	78.30				40987.
389	85.75				37718.
392	93.76				34761.
394	99.43				32946.

COMPOUND CONSTANTS:

MP   TC   VC   ZC
NBP   PC   DC   OM
MU   RG

VAPOR PRESSURE CORRELATION CONSTANTS, K AND KPA:

LIQUID DENSITY CORRELATION CONSTANTS, K AND G/ML:

FRANCIS1   FRANCIS1   FRANCIS1
277. - 374. K   277. - 374. K   277. - 374. K
 20 POINTS   20 POINTS   20 POINTS
RMSD = 0.00014   RMSD = 0.00014   RMSD = 0.00014
0.12116728E+01   0.12116728E+01   0.12116728E+01
0.71956799E-03   0.71956799E-03   0.71956799E-03
0.64043417E+01   0.64043417E+01   0.64043417E+01
0.51158325E+03   0.51158325E+03   0.51158325E+03

SECOND VIRIAL COEFFICIENT CORRELATION CONSTANTS, ML/MOL:

LITERATURE DOCUMENTS FOR CORRELATED VAPOR PRESSURE VALUES:

LITERATURE DOCUMENTS FOR CORRELATED LIQUID DENSITY VALUES:

 13846

LITERATURE DOCUMENTS REPORTING VIRIAL COEFFICIENT DATA:

T,K	VAPOR PRESSURE, KPA	SATURATED LIQUID VOLUME, ML/MOL	SECOND VIRIAL COEFFICIENT, ML/MOL	HEAT OF VAPORIZATION, J/MOL	SATURATED VAPOR VOLUME, ML/MOL
277		142.31			
279		142.55			
281		142.80			
283		143.04			
285		143.29			
288		143.66			
290		143.91			
292		144.16			
294		144.42			
296		144.67			
299		145.06			
301		145.32			
303		145.58			
305		145.84			
307		146.10			
310		146.51			
312		146.78			
314		147.05			
316		147.32			
318		147.60			
321		148.01			
323		148.29			
325		148.58			
327		148.86			
329		149.15			
332		149.59			
334		149.88			
336		150.18			
338		150.48			
340		150.78			
343		151.24			
345		151.55			
347		151.86			
349		152.18			
351		152.50			
354		152.99			
356		153.31			
358		153.65			
360		153.98			
362		154.32			
365		154.84			
367		155.19			
369		155.54			
371		155.90			
373		156.27			

COMPOUND CONSTANTS:

MP	301.141 K	TC	VC	ZC
NBP	470.391 K	PC	DC	OM
MU	2.3800 DEBYE	RG		

VAPOR PRESSURE CORRELATION CONSTANTS, K AND KPA:

LIQUID DENSITY CORRELATION CONSTANTS, K AND G/ML:

FRANCIS1	FRANCIS1	FRANCIS1
303. - 363. K	303. - 363. K	303. - 363. K
4 POINTS	4 POINTS	4 POINTS
RMSD = 0.00001	RMSD = 0.00001	RMSD = 0.00001
0.11017246E+01	0.11017246E+01	0.11017246E+01
0.64413319E-03	0.64413319E-03	0.64413319E-03
0.50127020E+01	0.50127020E+01	0.50127020E+01
0.49241553E+03	0.49241553E+03	0.49241553E+03

SECOND VIRIAL COEFFICIENT CORRELATION CONSTANTS, ML/MOL:

LITERATURE DOCUMENTS FOR CORRELATED VAPOR PRESSURE VALUES:

LITERATURE DOCUMENTS FOR CORRELATED LIQUID DENSITY VALUES:

14195

LITERATURE DOCUMENTS REPORTING VIRIAL COEFFICIENT DATA:

T,K	VAPOR PRESSURE, KPA	SATURATED LIQUID VOLUME, ML/MOL	SECOND VIRIAL COEFFICIENT, ML/MOL	HEAT OF VAPORIZATION, J/MOL	SATURATED VAPOR VOLUME, ML/MOL
303		157.04			
304		157.18			
305		157.32			
307		157.60			
308		157.75			
309		157.89			
311		158.18			
312		158.32			
313		158.47			
315		158.76			
316		158.90			
317		159.05			
319		159.35			
320		159.50			
322		159.80			
323		159.95			
324		160.10			
326		160.41			
327		160.56			
328		160.72			
330		161.03			
331		161.18			
332		161.34			
334		161.66			
335		161.82			
337		162.14			
338		162.30			
339		162.47			
341		162.80			
342		162.96			
343		163.13			
345		163.47			
346		163.63			
347		163.81			
349		164.15			
350		164.32			
352		164.67			
353		164.85			
354		165.03			
356		165.39			
357		165.57			
358		165.75			
360		166.12			
361		166.31			
362		166.49			

COMPOUND CONSTANTS:

MP   388.655 K   TC   VC   ZC
NBP   PC   DC   OM
MU   0.6800 DEBYE   RG

VAPOR PRESSURE CORRELATION CONSTANTS, K AND KPA:

RIEDEL   RIEDEL   RIEDEL
392. - 402. K   392. - 402. K   392. - 402. K
  8 POINTS   8 POINTS   8 POINTS
RMSD = 0.0256   RMSD = 0.0256   RMSD = 0.0256
 0.46007514E+04   0.46007514E+04   0.46007514E+04
-0.67240180E+03   -0.67240180E+03   -0.67240180E+03
-0.23440132E+06   -0.23440132E+06   -0.23440132E+06
 0.41021702E-14   0.41021702E-14   0.41021702E-14

LIQUID DENSITY CORRELATION CONSTANTS, K AND G/ML:

FRANCIS1   FRANCIS1   FRANCIS1
395. - 434. K   395. - 434. K   395. - 434. K
  4 POINTS   4 POINTS   4 POINTS
RMSD = 0.00017   RMSD = 0.00017   RMSD = 0.00017
0.15105515E+01   0.15105515E+01   0.15105515E+01
0.97980653E-03   0.97980653E-03   0.97980653E-03
0.19811020E+02   0.19811020E+02   0.19811020E+02
0.90835181E+03   0.90835181E+03   0.90835181E+03

SECOND VIRIAL COEFFICIENT CORRELATION CONSTANTS, ML/MOL:

LITERATURE DOCUMENTS FOR CORRELATED VAPOR PRESSURE VALUES:

  20670

LITERATURE DOCUMENTS FOR CORRELATED LIQUID DENSITY VALUES:

  14192

LITERATURE DOCUMENTS REPORTING VIRIAL COEFFICIENT DATA:

T,K	VAPOR PRESSURE, KPA	SATURATED LIQUID VOLUME, ML/MOL	SECOND VIRIAL COEFFICIENT, ML/MOL	HEAT OF VAPORIZATION, J/MOL	SATURATED VAPOR VOLUME, ML/MOL
392	13.28				245490.
393	13.79				237013.
394	14.31				228928.
395	14.85	99.63		47617.	221198.
396	15.40	99.73		47508.	213792.
397	15.97	99.83			206681.
398	16.56	99.93			199839.
399	17.17	100.02			193243.
400	17.80	100.12			186870.
401	18.45	100.22			180701.
402	19.13	100.32			174718.
403		100.42			
404		100.51			
405		100.61			
406		100.71			
407		100.81			
408		100.91			
409		101.01			
410		101.11			
411		101.21			
412		101.31			
413		101.41			
414		101.51			
415		101.62			
416		101.72			
417		101.82			
418		101.92			
419		102.02			
420		102.13			
421		102.23			
422		102.33			
423		102.43			
424		102.54			
425		102.64			
426		102.74			
427		102.85			
428		102.95			
429		103.06			
430		103.16			
431		103.27			
432		103.37			
433		103.48			
434		103.58			

COMPOUND CONSTANTS:

MP   340.141 K   TC   VC   ZC
NBP   PC   DC   OM
MU   RG

VAPOR PRESSURE CORRELATION CONSTANTS, K AND KPA:

LIQUID DENSITY CORRELATION CONSTANTS, K AND G/ML:

FRANCIS1   FRANCIS1   FRANCIS1
348. - 401. K   348. - 401. K   348. - 401. K
  5 POINTS   5 POINTS   5 POINTS
RMSD = 0.00042   RMSD = 0.00042   RMSD = 0.00042
0.13968458E+01   0.13968458E+01   0.13968458E+01
0.88775204E-03   0.88775204E-03   0.88775204E-03
0.16219463E+01   0.16219463E+01   0.16219463E+01
0.61806470E+03   0.61806470E+03   0.61806470E+03

SECOND VIRIAL COEFFICIENT CORRELATION CONSTANTS, ML/MOL:

LITERATURE DOCUMENTS FOR CORRELATED VAPOR PRESSURE VALUES:

LITERATURE DOCUMENTS FOR CORRELATED LIQUID DENSITY VALUES:

  14192

LITERATURE DOCUMENTS REPORTING VIRIAL COEFFICIENT DATA:

T,K	VAPOR PRESSURE, KPA	SATURATED LIQUID VOLUME, ML/MOL	SECOND VIRIAL COEFFICIENT, ML/MOL	HEAT OF VAPORIZATION, J/MOL	SATURATED VAPOR VOLUME, ML/MOL
348		112.88			
349		112.97			
350		113.07			
351		113.16			
352		113.26			
354		113.45			
355		113.55			
356		113.64			
357		113.74			
358		113.84			
360		114.03			
361		114.13			
362		114.22			
363		114.32			
364		114.42			
366		114.62			
367		114.71			
368		114.81			
369		114.91			
370		115.01			
372		115.21			
373		115.31			
374		115.41			
375		115.51			
376		115.61			
378		115.81			
379		115.91			
380		116.01			
381		116.11			
382		116.21			
384		116.42			
385		116.52			
386		116.62			
387		116.72			
388		116.83			
390		117.03			
391		117.13			
392		117.24			
393		117.34			
394		117.44			
396		117.65			
397		117.76			
398		117.86			
399		117.97			
400		118.07			

COMPOUND CONSTANTS:

MP		TC		VC		ZC
NBP	502.086 K	PC		DC		OM
MU		RG				

VAPOR PRESSURE CORRELATION CONSTANTS, K AND KPA:

RPM2	RPM2	RPM2	RPM2
369. - 521. K	369. - 521. K	369. - 521. K	490. - 521. K
16 POINTS	16 POINTS	16 POINTS	4 POINTS
RMSD = 0.0003	RMSD = 0.0003	RMSD = 0.0003	RMSD = 0.0002
0.17009724E+02	0.17009724E+02	0.17009724E+02	0.18340928E+02
-0.62063641E+04	-0.62063641E+04	-0.62063641E+04	-0.64288041E+04
-0.10441452E-03	-0.10441452E-03	-0.10441452E-03	-0.27587607E-02
0.87991144E-07	0.87991144E-07	0.87991144E-07	0.18513966E-05

LIQUID DENSITY CORRELATION CONSTANTS, K AND G/ML:

SECOND VIRIAL COEFFICIENT CORRELATION CONSTANTS, ML/MOL:

LITERATURE DOCUMENTS FOR CORRELATED VAPOR PRESSURE VALUES:

6619

LITERATURE DOCUMENTS FOR CORRELATED LIQUID DENSITY VALUES:

LITERATURE DOCUMENTS REPORTING VIRIAL COEFFICIENT DATA:

T,K	VAPOR PRESSURE, KPA	SATURATED LIQUID VOLUME, ML/MOL	SECOND VIRIAL COEFFICIENT, ML/MOL	HEAT OF VAPORIZATION, J/MOL	SATURATED VAPOR VOLUME, ML/MOL
369	1.178				2604576.
372	1.349				2292962.
375	1.541				2022880.
379	1.835				1717057.
382	2.087				1521967.
386	2.469				1299792.
389	2.795				1157260.
393	3.287				994058.
396	3.704				888795.
400	4.332				767646.
403	4.862				689108.
406	5.448				619627.
410	6.323				539108.
413	7.058				486552.
417	8.151				425370.
420	9.064				385257.
424	10.42				338358.
427	11.55				307478.
431	13.21				271225.
434	14.59				247258.
438	16.63				219010.
441	18.31				200263.
444	20.13				183351.
448	22.81				163313.
451	25.01				149945.
455	28.22				134050.
458	30.86				123410.
462	34.69				110717.
465	37.83				102192.
469	42.39				91989.
472	46.11				85115.
476	51.49				76863.
479	55.87				71287.
482	60.56				66179.
486	67.32				60023.
489	72.80				55846.
493	80.69				50797.
496	87.07				47362.
500	96.23				43199.
503	103.6				40359.
507	114.2				36909.
510	122.7				34549.
514	134.9				31676.
517	144.7				29706.
520	155.1				27880.

$$C(13)H(20)O(1)$$

COMPOUND CONSTANTS:

MP		TC		VC		ZC
NBP		PC		DC		OM
MU	3.5000 DEBYE	RG				

VAPOR PRESSURE CORRELATION CONSTANTS, K AND KPA:

RPM2	RPM2	RPM2
291. - 330. K	291. - 330. K	291. - 330. K
10 POINTS	10 POINTS	10 POINTS
RMSD = 0.0003	RMSD = 0.0003	RMSD = 0.0003
-0.19575553E+04	-0.19575553E+04	-0.19575553E+04
0.19631481E+06	0.19631481E+06	0.19631481E+06
0.63736451E+01	0.63736451E+01	0.63736451E+01
-0.68379550E-02	-0.68379550E-02	-0.68379550E-02

LIQUID DENSITY CORRELATION CONSTANTS, K AND G/ML:

SECOND VIRIAL COEFFICIENT CORRELATION CONSTANTS, ML/MOL:

LITERATURE DOCUMENTS FOR CORRELATED VAPOR PRESSURE VALUES:

5725

LITERATURE DOCUMENTS FOR CORRELATED LIQUID DENSITY VALUES:

LITERATURE DOCUMENTS REPORTING VIRIAL COEFFICIENT DATA:

T,K	VAPOR PRESSURE, KPA	SATURATED LIQUID VOLUME, ML/MOL	SECOND VIRIAL COEFFICIENT, ML/MOL	HEAT OF VAPORIZATION, J/MOL	SATURATED VAPOR VOLUME, ML/MOL
293		116.86			
297		117.25			
301		117.64			
305	0.05903	118.04		59001.	0.4296E+08
309	0.07978	118.44		58978.	0.3220E+08
313	0.1070	118.84		58892.	0.2433E+08
317	0.1423	119.25		58748.	0.1853E+08
322	0.2010	119.76		58491.	0.1332E+08
326	0.2626	120.18		58229.	0.1032E+08
330	0.3405	120.59		57920.	8058952.
334	0.4381	121.01		57570.	6339250.
338	0.5594	121.43		57182.	5023541.
342	0.7091	121.86		56759.	4009946.
347	0.9441	122.40		56187.	3055936.
351	1.178	122.83		55701.	2478239.
355	1.459	123.27		55193.	2023400.
359	1.795	123.71		54667.	1662968.
363	2.194	124.16		54127.	1375533.
367	2.665	124.61		53577.	1144881.
372	3.370	125.17		52881.	917897.
376	4.038	125.63		52322.	774196.
380	4.812	126.09		51766.	656640.
384	5.702	126.56		51215.	559928.
388	6.722	127.03		50675.	479925.
392	7.884	127.50		50149.	413387.
396	9.203	127.98		49641.	357757.
401	11.09	128.59		49037.	300512.
405	12.82	129.07		48583.	262634.
409	14.76	129.57		48159.	230450.
413	16.92	130.07		47770.	202976.
417	19.32	130.57		47419.	179413.
421	22.00	131.07		47109.	159114.
426	25.75	131.72		46788.	137536.
430	29.12	132.23		46587.	122793.
434	32.83	132.76		46441.	109916.
438	36.92	133.29		46353.	98626.
442	41.44	133.82		46327.	88687.
446	46.41	134.36			79906.
451	53.33	135.04			70311.
455	59.50	135.60			63578.
459	66.30	136.16			57562.
463	73.79	136.72			52169.
467	82.05	137.29			47321.
471	91.18	137.87			42950.
475	101.3	138.46			38998.

COMPOUND CONSTANTS:

MP	291.753 K	TC	VC	ZC
NBP	490.914 K	PC	DC	OM
MU	2.8800 DEBYE	RG		

VAPOR PRESSURE CORRELATION CONSTANTS, K AND KPA:

RPM2	RPM2	RPM2
405. - 491. K	405. - 491. K	405. - 491. K
7 POINTS	7 POINTS	7 POINTS
RMSD = 0.0570	RMSD = 0.0570	RMSD = 0.0570
-0.31931183E+02	-0.31931183E+02	-0.31931183E+02
0.76175152E+03	0.76175152E+03	0.76175152E+03
0.11630691E+00	0.11630691E+00	0.11630691E+00
-0.91697854E-04	-0.91697854E-04	-0.91697854E-04

LIQUID DENSITY CORRELATION CONSTANTS, K AND G/ML:

FRANCIS1	FRANCIS1	FRANCIS1
293. - 359. K	293. - 359. K	293. - 359. K
4 POINTS	4 POINTS	4 POINTS
RMSD = 0.00011	RMSD = 0.00011	RMSD = 0.00011
0.12283163E+01	0.12283163E+01	0.12283163E+01
0.72272122E-03	0.72272122E-03	0.72272122E-03
0.87706625E+00	0.87706625E+00	0.87706625E+00
0.43193555E+03	0.43193555E+03	0.43193555E+03

SECOND VIRIAL COEFFICIENT CORRELATION CONSTANTS, ML/MOL:

LITERATURE DOCUMENTS FOR CORRELATED VAPOR PRESSURE VALUES:

1361 3396 18395

LITERATURE DOCUMENTS FOR CORRELATED LIQUID DENSITY VALUES:

14417

LITERATURE DOCUMENTS REPORTING VIRIAL COEFFICIENT DATA:

T,K	VAPOR PRESSURE, KPA	SATURATED LIQUID VOLUME, ML/MOL	SECOND VIRIAL COEFFICIENT, ML/MOL	HEAT OF VAPORIZATION, J/MOL	SATURATED VAPOR VOLUME, ML/MOL
293		132.82			
297		133.22			
302		133.74			
306		134.15			
311		134.68			
315		135.10			
320		135.64			
324		136.08			
329		136.63			
333		137.09			
338		137.66			
342		138.13			
347		138.73			
351		139.22			
356		139.84			
405	7.488				449724.
410	9.009				378409.
414	10.41				330512.
419	12.44				280057.
423	14.30				245991.
428	16.95				209920.
432	19.37				185438.
437	22.80				159381.
441	25.89				141604.
446	30.25				122587.
450	34.15				109546.
455	39.60				95526.
459	44.45				85863.
464	51.15				75424.
468	57.06				68194.
473	65.17				60346.
477	72.26				54886.
482	81.90				48931.
486	90.26				44768.
491	101.5				40210.

COMPOUND CONSTANTS:

MP	257.759 K	TC		VC		ZC	
NBP		PC		DC		OM	
MU	2.7200 DEBYE	RG					

VAPOR PRESSURE CORRELATION CONSTANTS, K AND KPA:

LIQUID DENSITY CORRELATION CONSTANTS, K AND G/ML:

FRANCIS1	FRANCIS1	FRANCIS1
293. - 359. K	293. - 359. K	293. - 359. K
7 POINTS	7 POINTS	7 POINTS
RMSD = 0.00028	RMSD = 0.00028	RMSD = 0.00028
0.12354736E+01	0.12354736E+01	0.12354736E+01
0.75112213E-03	0.75112213E-03	0.75112213E-03
0.29371872E+01	0.29371872E+01	0.29371872E+01
0.51115088E+03	0.51115088E+03	0.51115088E+03

SECOND VIRIAL COEFFICIENT CORRELATION CONSTANTS, ML/MOL:

LITERATURE DOCUMENTS FOR CORRELATED VAPOR PRESSURE VALUES:

LITERATURE DOCUMENTS FOR CORRELATED LIQUID DENSITY VALUES:

  14417 40688 41319

LITERATURE DOCUMENTS REPORTING VIRIAL COEFFICIENT DATA:

T,K	VAPOR PRESSURE, KPA	SATURATED LIQUID VOLUME, ML/MOL	SECOND VIRIAL COEFFICIENT, ML/MOL	HEAT OF VAPORIZATION, J/MOL	SATURATED VAPOR VOLUME, ML/MOL
293		133.92			
294		134.03			
296		134.25			
297		134.36			
299		134.57			
300		134.69			
302		134.91			
303		135.02			
305		135.24			
306		135.35			
308		135.58			
309		135.69			
311		135.92			
312		136.03			
314		136.26			
315		136.37			
317		136.60			
318		136.72			
320		136.95			
321		137.07			
323		137.30			
324		137.42			
326		137.65			
327		137.77			
329		138.01			
330		138.13			
332		138.37			
333		138.49			
335		138.73			
336		138.85			
338		139.10			
339		139.22			
341		139.47			
342		139.59			
344		139.84			
345		139.96			
347		140.21			
348		140.34			
350		140.59			
351		140.72			
353		140.98			
354		141.11			
356		141.37			
357		141.50			
359		141.76			

COMPOUND CONSTANTS:

MP		TC		VC		ZC
NBP	497.474 K	PC		DC		OM
MU	3.2300 DEBYE	RG				

VAPOR PRESSURE CORRELATION CONSTANTS, K AND KPA:

RPM2	RPM2	RPM2	RPM2
369. - 521. K	369. - 511. K	369. - 521. K	480. - 521. K
17 POINTS	16 POINTS	17 POINTS	5 POINTS
RMSD = 0.0085	RMSD = 0.0070	RMSD = 0.0085	RMSD = 0.0002
0.17973910E+02	0.18114459E+02	0.17973910E+02	0.17958869E+02
-0.63745665E+04	-0.63943280E+04	-0.63745665E+04	-0.63834928E+04
-0.17194104E-02	-0.20508118E-02	-0.17194104E-02	-0.15442724E-02
0.12669434E-05	0.15259788E-05	0.12669434E-05	0.10486322E-05

LIQUID DENSITY CORRELATION CONSTANTS, K AND G/ML:

SECOND VIRIAL COEFFICIENT CORRELATION CONSTANTS, ML/MOL:

LITERATURE DOCUMENTS FOR CORRELATED VAPOR PRESSURE VALUES:

6619 14453

LITERATURE DOCUMENTS FOR CORRELATED LIQUID DENSITY VALUES:

LITERATURE DOCUMENTS REPORTING VIRIAL COEFFICIENT DATA:

T,K	VAPOR PRESSURE, KPA	SATURATED LIQUID VOLUME, ML/MOL	SECOND VIRIAL COEFFICIENT, ML/MOL	HEAT OF VAPORIZATION, J/MOL	SATURATED VAPOR VOLUME, ML/MOL
369	1.267				2421345.
372	1.453				2128560.
375	1.663				1875189.
379	1.983				1588784.
382	2.258				1406394.
386	2.677				1199021.
389	3.033				1066198.
393	3.574				914348.
396	4.032				816555.
400	4.723				704166.
403	5.307				631412.
406	5.952				567124.
410	6.919				492722.
413	7.730				444224.
417	8.940				387836.
420	9.951				350913.
424	11.45				307794.
427	12.70				279438.
431	14.56				246185.
434	16.09				224226.
438	18.36				198373.
441	20.23				181234.
444	22.27				165785.
448	25.25				147501.
451	27.71				135314.
455	31.31				120839.
458	34.26				111157.
462	38.56				99619.
465	42.08				91876.
469	47.20				82619.
472	51.38				76386.
476	57.43				68912.
479	62.36				63864.
482	67.64				59245.
486	75.27				53681.
489	81.46				49909.
493	90.38				45353.
496	97.59				42256.
500	108.0				38505.
503	116.3				35948.
507	128.3				32844.
510	138.0				30723.
514	151.9				28141.
517	163.0				26373.
520	174.8				24735.

COMPOUND CONSTANTS:

MP		TC	VC	ZC
NBP		PC	DC	OM
MU	3.4100 DEBYE	RG		

VAPOR PRESSURE CORRELATION CONSTANTS, K AND KPA:

LIQUID DENSITY CORRELATION CONSTANTS, K AND G/ML:

FRANCIS1	FRANCIS1	FRANCIS1
313. - 354. K	313. - 354. K	313. - 354. K
9 POINTS	9 POINTS	9 POINTS
RMSD = 0.00022	RMSD = 0.00022	RMSD = 0.00022
0.13698235E+01	0.13698235E+01	0.13698235E+01
0.83868415E-03	0.83868415E-03	0.83868415E-03
0.96535120E+01	0.96535120E+01	0.96535120E+01
0.10811353E+04	0.10811353E+04	0.10811353E+04

SECOND VIRIAL COEFFICIENT CORRELATION CONSTANTS, ML/MOL:

LITERATURE DOCUMENTS FOR CORRELATED VAPOR PRESSURE VALUES:

LITERATURE DOCUMENTS FOR CORRELATED LIQUID DENSITY VALUES:

182

LITERATURE DOCUMENTS REPORTING VIRIAL COEFFICIENT DATA:

T,K	VAPOR PRESSURE, KPA	SATURATED LIQUID VOLUME, ML/MOL	SECOND VIRIAL COEFFICIENT, ML/MOL	HEAT OF VAPORIZATION, J/MOL	SATURATED VAPOR VOLUME, ML/MOL
313		120.72			
314		120.82			
315		120.91			
316		121.01			
317		121.10			
318		121.20			
319		121.29			
320		121.39			
321		121.48			
322		121.58			
323		121.67			
324		121.77			
325		121.87			
326		121.96			
327		122.06			
328		122.16			
329		122.25			
330		122.35			
331		122.45			
332		122.54			
333		122.64			
334		122.74			
335		122.84			
336		122.93			
337		123.03			
338		123.13			
339		123.23			
340		123.33			
341		123.42			
342		123.52			
343		123.62			
344		123.72			
345		123.82			
346		123.92			
347		124.02			
348		124.12			
349		124.22			
350		124.32			
351		124.42			
352		124.52			
353		124.62			
354		124.72			

COMPOUND CONSTANTS:

MP   TC   VC   ZC
NBP   502.000 K   PC   DC   OM
MU   2.8300 DEBYE   RG

VAPOR PRESSURE CORRELATION CONSTANTS, K AND KPA:

LIQUID DENSITY CORRELATION CONSTANTS, K AND G/ML:

FRANCIS1   FRANCIS1   FRANCIS1
291. - 360. K   291. - 360. K   291. - 360. K
   7 POINTS   7 POINTS   7 POINTS
RMSD = 0.00027   RMSD = 0.00027   RMSD = 0.00027
0.11910629E+01   0.11910629E+01   0.11910629E+01
0.65790350E-03   0.65790350E-03   0.65790350E-03
0.14866219E+01   0.14866219E+01   0.14866219E+01
0.44262402E+03   0.44262402E+03   0.44262402E+03

SECOND VIRIAL COEFFICIENT CORRELATION CONSTANTS, ML/MOL:

LITERATURE DOCUMENTS FOR CORRELATED VAPOR PRESSURE VALUES:

LITERATURE DOCUMENTS FOR CORRELATED LIQUID DENSITY VALUES:

   5988 14417 40489

LITERATURE DOCUMENTS REPORTING VIRIAL COEFFICIENT DATA:

T,K	VAPOR PRESSURE, KPA	SATURATED LIQUID VOLUME, ML/MOL	SECOND VIRIAL COEFFICIENT, ML/MOL	HEAT OF VAPORIZATION, J/MOL	SATURATED VAPOR VOLUME, ML/MOL
291		149.73			
292		149.84			
294		150.06			
295		150.17			
297		150.39			
298		150.50			
300		150.73			
301		150.84			
303		151.06			
305		151.29			
306		151.40			
308		151.63			
309		151.75			
311		151.98			
312		152.10			
314		152.33			
316		152.56			
317		152.68			
319		152.92			
320		153.04			
322		153.28			
323		153.40			
325		153.64			
327		153.89			
328		154.01			
330		154.26			
331		154.38			
333		154.63			
334		154.76			
336		155.01			
338		155.27			
339		155.40			
341		155.66			
342		155.79			
344		156.06			
345		156.19			
347		156.46			
349		156.73			
350		156.87			
352		157.15			
353		157.29			
355		157.57			
356		157.72			
358		158.01			
359		158.15			

COMPOUND CONSTANTS:

MP   TC   VC   ZC
NBP   PC   DC   OM
MU   RG

VAPOR PRESSURE CORRELATION CONSTANTS, K AND KPA:

LIQUID DENSITY CORRELATION CONSTANTS, K AND G/ML:

FRANCIS1   FRANCIS1   FRANCIS1
293. - 360. K   293. - 360. K   293. - 360. K
  4 POINTS   4 POINTS   4 POINTS
RMSD = 0.00002   RMSD = 0.00002   RMSD = 0.00002
0.12143669E+01   0.12143669E+01   0.12143669E+01
0.77111414E-03   0.77111414E-03   0.77111414E-03
0.56137331E-01   0.56137331E-01   0.56137331E-01
0.37699731E+03   0.37699731E+03   0.37699731E+03

SECOND VIRIAL COEFFICIENT CORRELATION CONSTANTS, ML/MOL:

LITERATURE DOCUMENTS FOR CORRELATED VAPOR PRESSURE VALUES:

LITERATURE DOCUMENTS FOR CORRELATED LIQUID DENSITY VALUES:

  14417

LITERATURE DOCUMENTS REPORTING VIRIAL COEFFICIENT DATA:

T,K	VAPOR PRESSURE, KPA	SATURATED LIQUID VOLUME, ML/MOL	SECOND VIRIAL COEFFICIENT, ML/MOL	HEAT OF VAPORIZATION, J/MOL	SATURATED VAPOR VOLUME, ML/MOL
293		150.04			
294		150.16			
296		150.40			
297		150.52			
299		150.75			
300		150.87			
302		151.11			
303		151.23			
305		151.48			
306		151.60			
308		151.84			
309		151.96			
311		152.21			
312		152.33			
314		152.58			
315		152.70			
317		152.95			
318		153.07			
320		153.32			
321		153.45			
323		153.70			
324		153.82			
326		154.08			
328		154.33			
329		154.46			
331		154.72			
332		154.85			
334		155.10			
335		155.24			
337		155.50			
338		155.63			
340		155.89			
341		156.03			
343		156.30			
344		156.43			
346		156.71			
347		156.84			
349		157.12			
350		157.26			
352		157.55			
353		157.69			
355		157.99			
356		158.14			
358		158.45			
359		158.61			

COMPOUND CONSTANTS:

MP	TC	VC	ZC
NBP	PC	DC	OM
MU	RG		

VAPOR PRESSURE CORRELATION CONSTANTS, K AND KPA:

LIQUID DENSITY CORRELATION CONSTANTS, K AND G/ML:

FRANCIS1	FRANCIS1	FRANCIS1
298. - 361. K	298. - 361. K	298. - 361. K
4 POINTS	4 POINTS	4 POINTS
RMSD = 0.00000	RMSD = 0.00000	RMSD = 0.00000
0.11962500E+01	0.11962500E+01	0.11962500E+01
0.70684613E-03	0.70684613E-03	0.70684613E-03
0.24943739E+00	0.24943739E+00	0.24943739E+00
0.39578760E+03	0.39578760E+03	0.39578760E+03

SECOND VIRIAL COEFFICIENT CORRELATION CONSTANTS, ML/MOL:

LITERATURE DOCUMENTS FOR CORRELATED VAPOR PRESSURE VALUES:

LITERATURE DOCUMENTS FOR CORRELATED LIQUID DENSITY VALUES:

  14417 40688

LITERATURE DOCUMENTS REPORTING VIRIAL COEFFICIENT DATA:

T,K	VAPOR PRESSURE, KPA	SATURATED LIQUID VOLUME, ML/MOL	SECOND VIRIAL COEFFICIENT, ML/MOL	HEAT OF VAPORIZATION, J/MOL	SATURATED VAPOR VOLUME, ML/MOL
298		150.76			
299		150.87			
300		150.98			
302		151.21			
303		151.32			
305		151.55			
306		151.67			
308		151.90			
309		152.01			
310		152.13			
312		152.36			
313		152.47			
315		152.71			
316		152.83			
318		153.06			
319		153.18			
320		153.30			
322		153.54			
323		153.66			
325		153.90			
326		154.02			
328		154.26			
329		154.39			
330		154.51			
332		154.76			
333		154.88			
335		155.13			
336		155.26			
338		155.51			
339		155.64			
340		155.77			
342		156.03			
343		156.16			
345		156.42			
346		156.55			
348		156.82			
349		156.96			
350		157.09			
352		157.37			
353		157.51			
355		157.80			
356		157.94			
358		158.24			
359		158.39			
360		158.54			

COMPOUND CONSTANTS:

MP   331.640 K   TC   VC   ZC
NBP   PC   DC   OM
MU   RG

VAPOR PRESSURE CORRELATION CONSTANTS, K AND KPA:

LIQUID DENSITY CORRELATION CONSTANTS, K AND G/ML:

FRANCIS1   FRANCIS1   FRANCIS1
333. - 374. K   333. - 374. K   333. - 374. K
  8 POINTS   8 POINTS   8 POINTS
RMSD = 0.00092   RMSD = 0.00092   RMSD = 0.00092
0.13859816E+01   0.13859816E+01   0.13859816E+01
0.94468961E-03   0.94468961E-03   0.94468961E-03
0.49999952E-02   0.49999952E-02   0.49999952E-02
0.49347217E+03   0.49347217E+03   0.49347217E+03

SECOND VIRIAL COEFFICIENT CORRELATION CONSTANTS, ML/MOL:

LITERATURE DOCUMENTS FOR CORRELATED VAPOR PRESSURE VALUES:

LITERATURE DOCUMENTS FOR CORRELATED LIQUID DENSITY VALUES:

   13839

LITERATURE DOCUMENTS REPORTING VIRIAL COEFFICIENT DATA:

T,K	VAPOR PRESSURE, KPA	SATURATED LIQUID VOLUME, ML/MOL	SECOND VIRIAL COEFFICIENT, ML/MOL	HEAT OF VAPORIZATION, J/MOL	SATURATED VAPOR VOLUME, ML/MOL
333		151.38			
334		151.52			
335		151.65			
336		151.79			
337		151.92			
338		152.05			
339		152.19			
340		152.32			
341		152.46			
342		152.60			
343		152.73			
344		152.87			
345		153.00			
346		153.14			
347		153.28			
348		153.41			
349		153.55			
350		153.69			
351		153.83			
352		153.96			
353		154.10			
354		154.24			
355		154.38			
356		154.52			
357		154.66			
358		154.80			
359		154.94			
360		155.08			
361		155.22			
362		155.36			
363		155.50			
364		155.64			
365		155.78			
366		155.92			
367		156.06			
368		156.21			
369		156.35			
370		156.49			
371		156.63			
372		156.78			
373		156.92			
374		157.06			

COMPOUND CONSTANTS:

MP	TC	VC	ZC
NBP	PC	DC	OM
MU	RG		

    VAPOR PRESSURE CORRELATION CONSTANTS, K AND KPA:

RPM2	RPM2	RPM2
375. - 535. K	375. - 535. K	375. - 535. K
20 POINTS	20 POINTS	20 POINTS
RMSD = 0.0044	RMSD = 0.0044	RMSD = 0.0044
0.99367506E+02	0.99367506E+02	0.99367506E+02
-0.19759721E+05	-0.19759721E+05	-0.19759721E+05
-0.17211686E+00	-0.17211686E+00	-0.17211686E+00
0.11966357E-03	0.11966357E-03	0.11966357E-03

    LIQUID DENSITY CORRELATION CONSTANTS, K AND G/ML:

    SECOND VIRIAL COEFFICIENT CORRELATION CONSTANTS, ML/MOL:

LITERATURE DOCUMENTS FOR CORRELATED VAPOR PRESSURE VALUES:

  8739 15685

LITERATURE DOCUMENTS FOR CORRELATED LIQUID DENSITY VALUES:

LITERATURE DOCUMENTS REPORTING VIRIAL COEFFICIENT DATA:

T,K	VAPOR PRESSURE, KPA	SATURATED LIQUID VOLUME, ML/MOL	SECOND VIRIAL COEFFICIENT, ML/MOL	HEAT OF VAPORIZATION, J/MOL	SATURATED VAPOR VOLUME, ML/MOL
375	0.3530				8831957.
378	0.4193				7494583.
382	0.5239				6061886.
385	0.6161				5195437.
389	0.7599				4256099.
393	0.9308				3510413.
396	1.079				3051224.
400	1.307				2544842.
404	1.573				2135170.
407	1.801				1878805.
411	2.147				1591761.
414	2.440				1410454.
418	2.883				1205650.
422	3.388				1035480.
425	3.814				926549.
429	4.448				801934.
433	5.166				696925.
436	5.764				628900.
440	6.649				550192.
444	7.643				483027.
447	8.465				439050.
451	9.673				387646.
454	10.67				353779.
458	12.13				313959.
462	13.75				279367.
465	15.08				256372.
469	17.02				229104.
473	19.17				205193.
476	20.92				189171.
480	23.47				170028.
484	26.28				153100.
487	28.58				141676.
491	31.91				127935.
494	34.62				118625.
498	38.56				107382.
502	42.89				97324.
505	46.41				90470.
509	51.52				82148.
513	57.13				74660.
516	61.70				69531.
520	68.33				63277.
524	75.61				57619.
527	81.55				53729.
531	90.16				48966.
534	97.19				45683.

COMPOUND CONSTANTS:

MP   TC   VC   ZC
NBP   PC   DC   OM
MU   RG

   VAPOR PRESSURE CORRELATION CONSTANTS, K AND KPA:

   RPM2   RPM2   RPM2
   292. - 417. K   292. - 417. K   292. - 417. K
    11 POINTS   11 POINTS   11 POINTS
   RMSD = 0.0265   RMSD = 0.0265   RMSD = 0.0265
   0.25265859E+02   0.25265859E+02   0.25265859E+02
   -0.47381032E+04   -0.47381032E+04   -0.47381032E+04
   -0.49092128E-01   -0.49092128E-01   -0.49092128E-01
   0.47445496E-04   0.47445496E-04   0.47445496E-04

   LIQUID DENSITY CORRELATION CONSTANTS, K AND G/ML:

   SECOND VIRIAL COEFFICIENT CORRELATION CONSTANTS, ML/MOL:

LITERATURE DOCUMENTS FOR CORRELATED VAPOR PRESSURE VALUES:

   10983

LITERATURE DOCUMENTS FOR CORRELATED LIQUID DENSITY VALUES:

LITERATURE DOCUMENTS REPORTING VIRIAL COEFFICIENT DATA:

T,K	VAPOR PRESSURE, KPA	SATURATED LIQUID VOLUME, ML/MOL	SECOND VIRIAL COEFFICIENT, ML/MOL	HEAT OF VAPORIZATION, J/MOL	SATURATED VAPOR VOLUME, ML/MOL
292	0.2865				8474134.
294	0.3066				7972827.
297	0.3387				7290405.
300	0.3733				6681744.
303	0.4105				6137415.
306	0.4504				5649361.
309	0.4931				5210667.
311	0.5232				4942655.
314	0.5708				4573418.
317	0.6217				4239431.
320	0.6759				3936653.
323	0.7334				3661577.
326	0.7946				3411144.
328	0.8374				3256546.
331	0.9049				3041416.
334	0.9763				2844521.
337	1.052				2663981.
340	1.132				2498131.
343	1.216				2345509.
345	1.275				2250464.
348	1.367				2117081.
351	1.464				1993786.
354	1.566				1879637.
357	1.673				1773788.
360	1.786				1675492.
363	1.905				1584075.
365	1.988				1526653.
368	2.117				1445395.
371	2.252				1369546.
374	2.394				1298651.
377	2.544				1232303.
380	2.700				1170134.
382	2.809				1130843.
385	2.978				1074914.
388	3.155				1022344.
391	3.342				972876.
394	3.537				926277.
397	3.741				882336.
399	3.883				854423.
402	4.104				814490.
405	4.335				776738.
408	4.578				741012.
411	4.832				707176.
414	5.099				675102.
416	5.284				654639.

COMPOUND CONSTANTS:

MP   248.914 K   TC   VC   ZC
NBP   PC   DC   OM
MU   RG

    VAPOR PRESSURE CORRELATION CONSTANTS, K AND KPA:

RPM2	RPM2	RPM2
294. - 368. K	294. - 368. K	294. - 368. K
8 POINTS	8 POINTS	8 POINTS
RMSD = 0.0302	RMSD = 0.0302	RMSD = 0.0302
-0.83424914E+02	-0.83424914E+02	-0.83424914E+02
0.52538877E+04	0.52538877E+04	0.52538877E+04
0.31324636E+00	0.31324636E+00	0.31324636E+00
-0.33243093E-03	-0.33243093E-03	-0.33243093E-03

LIQUID DENSITY CORRELATION CONSTANTS, K AND G/ML:

SECOND VIRIAL COEFFICIENT CORRELATION CONSTANTS, ML/MOL:

LITERATURE DOCUMENTS FOR CORRELATED VAPOR PRESSURE VALUES:

  10983

LITERATURE DOCUMENTS FOR CORRELATED LIQUID DENSITY VALUES:

LITERATURE DOCUMENTS REPORTING VIRIAL COEFFICIENT DATA:

T,K	VAPOR PRESSURE, KPA	SATURATED LIQUID VOLUME, ML/MOL	SECOND VIRIAL COEFFICIENT, ML/MOL	HEAT OF VAPORIZATION, J/MOL	SATURATED VAPOR VOLUME, ML/MOL
294	0.1115				0.2193E+08
295	0.1180				0.2079E+08
297	0.1321				0.1869E+08
299	0.1477				0.1683E+08
300	0.1562				0.1597E+08
302	0.1744				0.1440E+08
304	0.1945				0.1300E+08
305	0.2053				0.1235E+08
307	0.2286				0.1117E+08
309	0.2542				0.1011E+08
310	0.2679				9620889.
312	0.2974				8723843.
314	0.3296				7920210.
315	0.3469				7550156.
317	0.3838				6867689.
319	0.4240				6255036.
320	0.4455				5972481.
322	0.4912				5450564.
324	0.5408				4981034.
325	0.5672				4764131.
327	0.6232				4362830.
329	0.6837				4001008.
330	0.7157				3833578.
332	0.7835				3523292.
334	0.8563				3242897.
336	0.9346				2989284.
337	0.9757				2871645.
339	1.062				2653113.
341	1.155				2455003.
342	1.203				2362950.
344	1.305				2191661.
346	1.413				2036029.
347	1.469				1963592.
349	1.587				1828583.
351	1.711				1705646.
352	1.776				1648334.
354	1.910				1541348.
356	2.050				1443729.
357	2.123				1398151.
359	2.273				1312946.
361	2.430				1235056.
362	2.511				1198639.
364	2.677				1130475.
366	2.849				1068063.
367	2.937				1038851.

COMPOUND CONSTANTS:

MP   303.640 K   TC   VC   ZC
NBP   PC   DC   OM
MU   RG

   VAPOR PRESSURE CORRELATION CONSTANTS, K AND KPA:

   RPM2   RPM2   RPM2
   323. - 418. K   323. - 418. K   323. - 418. K
     5 POINTS   5 POINTS   5 POINTS
   RMSD = 0.0133   RMSD = 0.0133   RMSD = 0.0133
   -0.99897398E+02   -0.99897398E+02   -0.99897398E+02
    0.12670804E+05   0.12670804E+05   0.12670804E+05
    0.23088809E+00   0.23088809E+00   0.23088809E+00
   -0.14890088E-03   -0.14890088E-03   -0.14890088E-03

   LIQUID DENSITY CORRELATION CONSTANTS, K AND G/ML:

   SECOND VIRIAL COEFFICIENT CORRELATION CONSTANTS, ML/MOL:

LITERATURE DOCUMENTS FOR CORRELATED VAPOR PRESSURE VALUES:

  11006

LITERATURE DOCUMENTS FOR CORRELATED LIQUID DENSITY VALUES:

LITERATURE DOCUMENTS REPORTING VIRIAL COEFFICIENT DATA:

T,K	VAPOR PRESSURE, KPA	SATURATED LIQUID VOLUME, ML/MOL	SECOND VIRIAL COEFFICIENT, ML/MOL	HEAT OF VAPORIZATION, J/MOL	SATURATED VAPOR VOLUME, ML/MOL
323	0.1966				0.1366E+08
325	0.2020				0.1338E+08
327	0.2080				0.1307E+08
329	0.2145				0.1275E+08
331	0.2216				0.1242E+08
333	0.2293				0.1207E+08
335	0.2376				0.1172E+08
338	0.2514				0.1118E+08
340	0.2615				0.1081E+08
342	0.2724				0.1044E+08
344	0.2841				0.1007E+08
346	0.2967				9696731.
348	0.3102				9327017.
351	0.3325				8778271.
353	0.3487				8417943.
355	0.3661				8063164.
357	0.3847				7714797.
359	0.4048				7373601.
361	0.4263				7040235.
364	0.4616				6556071.
366	0.4873				6244544.
368	0.5149				5942431.
370	0.5445				5650004.
372	0.5762				5367461.
374	0.6103				5094935.
376	0.6469				4832499.
379	0.7069				4457790.
381	0.7506				4220555.
383	0.7974				3993272.
385	0.8478				3775809.
387	0.9018				3568004.
389	0.9598				3369666.
392	1.055				3089436.
394	1.124				2913784.
396	1.199				2746764.
398	1.279				2588113.
400	1.364				2437546.
402	1.457				2294781.
405	1.608				2094624.
407	1.718				1970088.
409	1.836				1852333.
411	1.963				1741069.
413	2.099				1636010.
415	2.245				1536873.
417	2.402				1443382.

COMPOUND CONSTANTS:

MP   386.655 K   TC   VC   ZC
NBP   PC   DC   OM
MU   2.7400 DEBYE   RG

   VAPOR PRESSURE CORRELATION CONSTANTS, K AND KPA:

   RPM2   RPM2   RPM2
   388. - 431. K   388. - 431. K   388. - 431. K
     8 POINTS   8 POINTS   8 POINTS
   RMSD = 0.0260   RMSD = 0.0260   RMSD = 0.0260
    0.66819317E+04   0.66819317E+04   0.66819317E+04
   -0.92881114E+06   -0.92881114E+06   -0.92881114E+06
   -0.16059749E+02   -0.16059749E+02   -0.16059749E+02
    0.12896429E-01   0.12896429E-01   0.12896429E-01

   LIQUID DENSITY CORRELATION CONSTANTS, K AND G/ML:

   SECOND VIRIAL COEFFICIENT CORRELATION CONSTANTS, ML/MOL:

LITERATURE DOCUMENTS FOR CORRELATED VAPOR PRESSURE VALUES:

   11006

LITERATURE DOCUMENTS FOR CORRELATED LIQUID DENSITY VALUES:

LITERATURE DOCUMENTS REPORTING VIRIAL COEFFICIENT DATA:

T,K	VAPOR PRESSURE, KPA	SATURATED LIQUID VOLUME, ML/MOL	SECOND VIRIAL COEFFICIENT, ML/MOL	HEAT OF VAPORIZATION, J/MOL	SATURATED VAPOR VOLUME, ML/MOL
388	0.1991				0.1620E+08
389	0.2233				0.1449E+08
390	0.2489				0.1303E+08
391	0.2760				0.1178E+08
392	0.3045				0.1071E+08
393	0.3341				9779254.
394	0.3649				8976584.
395	0.3968				8277560.
396	0.4295				7666187.
397	0.4630				7129245.
398	0.4972				6655748.
399	0.5319				6236519.
400	0.5672				5863866.
401	0.6028				5531295.
402	0.6387				5233315.
403	0.6748				4965249.
404	0.7112				4723097.
405	0.7477				4503430.
406	0.7844				4303284.
407	0.8213				4120094.
408	0.8584				3951632.
409	0.8958				3795946.
410	0.9336				3651329.
411	0.9718				3516277.
412	1.011				3389467.
413	1.050				3269726.
414	1.091				3156018.
415	1.132				3047422.
416	1.175				2943130.
417	1.220				2842419.
418	1.266				2744661.
419	1.315				2649300.
420	1.366				2555861.
421	1.421				2463928.
422	1.478				2373157.
423	1.540				2283263.
424	1.607				2194020.
425	1.678				2105256.
426	1.756				2016855.
427	1.841				1928748.
428	1.933				1840920.
429	2.034				1753397.
430	2.146				1666249.
431	2.269				1579581.

COMPOUND CONSTANTS:

MP   264.155 K   TC   VC   ZC
NBP   PC   DC   OM
MU   RG

VAPOR PRESSURE CORRELATION CONSTANTS, K AND KPA:

LIQUID DENSITY CORRELATION CONSTANTS, K AND G/ML:

FRANCIS1   FRANCIS1   FRANCIS1
292. - 355. K   292. - 355. K   292. - 355. K
    9 POINTS   9 POINTS   9 POINTS
RMSD = 0.00011   RMSD = 0.00011   RMSD = 0.00011
0.12204685E+01   0.12204685E+01   0.12204685E+01
0.79455716E-03   0.79455716E-03   0.79455716E-03
0.14701186E+02   0.14701186E+02   0.14701186E+02
0.12922126E+04   0.12922126E+04   0.12922126E+04

SECOND VIRIAL COEFFICIENT CORRELATION CONSTANTS, ML/MOL:

LITERATURE DOCUMENTS FOR CORRELATED VAPOR PRESSURE VALUES:

LITERATURE DOCUMENTS FOR CORRELATED LIQUID DENSITY VALUES:

   6711 13407 40489

LITERATURE DOCUMENTS REPORTING VIRIAL COEFFICIENT DATA:

T,K	VAPOR PRESSURE, KPA	SATURATED LIQUID VOLUME, ML/MOL	SECOND VIRIAL COEFFICIENT, ML/MOL	HEAT OF VAPORIZATION, J/MOL	SATURATED VAPOR VOLUME ML/MOL
292		166.60			
293		166.74			
294		166.88			
296		167.16			
297		167.30			
299		167.58			
300		167.72			
302		168.00			
303		168.14			
304		168.28			
306		168.56			
307		168.71			
309		168.99			
310		169.13			
312		169.42			
313		169.56			
314		169.71			
316		169.99			
317		170.14			
319		170.43			
320		170.57			
322		170.87			
323		171.01			
324		171.16			
326		171.45			
327		171.60			
329		171.89			
330		172.04			
332		172.34			
333		172.48			
334		172.63			
336		172.93			
337		173.08			
339		173.38			
340		173.53			
342		173.83			
343		173.98			
344		174.14			
346		174.44			
347		174.59			
349		174.90			
350		175.05			
352		175.36			
353		175.51			
354		175.66			

COMPOUND CONSTANTS:

MP   TC   VC   ZC
NBP   PC   DC   OM
MU   RG

VAPOR PRESSURE CORRELATION CONSTANTS, K AND KPA:

LIQUID DENSITY CORRELATION CONSTANTS, K AND G/ML:

FRANCIS1   FRANCIS1   FRANCIS1
293. - 360. K   293. - 360. K   293. - 360. K
   4 POINTS   4 POINTS   4 POINTS
RMSD = 0.00025   RMSD = 0.00025   RMSD = 0.00025
0.12095985E+01   0.12095985E+01   0.12095985E+01
0.79942029E-03   0.79942029E-03   0.79942029E-03
0.79448156E+01   0.79448156E+01   0.79448156E+01
0.24433069E+04   0.24433069E+04   0.24433069E+04

SECOND VIRIAL COEFFICIENT CORRELATION CONSTANTS, ML/MOL:

LITERATURE DOCUMENTS FOR CORRELATED VAPOR PRESSURE VALUES:

LITERATURE DOCUMENTS FOR CORRELATED LIQUID DENSITY VALUES:

   14417

LITERATURE DOCUMENTS REPORTING VIRIAL COEFFICIENT DATA:

T,K	VAPOR PRESSURE, KPA	SATURATED LIQUID VOLUME, ML/MOL	SECOND VIRIAL COEFFICIENT, ML/MOL	HEAT OF VAPORIZATION, J/MOL	SATURATED VAPOR VOLUME, ML/MOL
293		166.96			
294		167.10			
296		167.37			
297		167.51			
299		167.79			
300		167.93			
302		168.21			
303		168.35			
305		168.63			
306		168.77			
308		169.05			
309		169.19			
311		169.48			
312		169.62			
314		169.90			
315		170.04			
317		170.33			
318		170.47			
320		170.76			
321		170.91			
323		171.19			
324		171.34			
326		171.63			
328		171.92			
329		172.07			
331		172.36			
332		172.51			
334		172.80			
335		172.95			
337		173.25			
338		173.39			
340		173.69			
341		173.84			
343		174.14			
344		174.29			
346		174.59			
347		174.74			
349		175.04			
350		175.19			
352		175.50			
353		175.65			
355		175.96			
356		176.11			
358		176.42			
359		176.57			

795

COMPOUND CONSTANTS:

MP	300.141 K	TC	VC	ZC
NBP	538.400 K	PC	DC	OM
MU		RG		

VAPOR PRESSURE CORRELATION CONSTANTS, K AND KPA:

LIQUID DENSITY CORRELATION CONSTANTS, K AND G/ML:

FRANCIS1	FRANCIS1	FRANCIS1
299. - 354. K	299. - 354. K	299. - 354. K
8 POINTS	8 POINTS	8 POINTS
RMSD = 0.00013	RMSD = 0.00013	RMSD = 0.00013
0.11945877E+01	0.11945877E+01	0.11945877E+01
0.78531238E-03	0.78531238E-03	0.78531238E-03
0.15419426E+02	0.15419426E+02	0.15419426E+02
0.65498828E+04	0.65498828E+04	0.65498828E+04

SECOND VIRIAL COEFFICIENT CORRELATION CONSTANTS, ML/MOL:

LITERATURE DOCUMENTS FOR CORRELATED VAPOR PRESSURE VALUES:

LITERATURE DOCUMENTS FOR CORRELATED LIQUID DENSITY VALUES:

6711

LITERATURE DOCUMENTS REPORTING VIRIAL COEFFICIENT DATA:

T,K	VAPOR PRESSURE, KPA	SATURATED LIQUID VOLUME, ML/MOL	SECOND VIRIAL COEFFICIENT, ML/MOL	HEAT OF VAPORIZATION, J/MOL	SATURATED VAPOR VOLUME, ML/MOL
299		184.12			
300		184.27			
301		184.42			
302		184.57			
304		184.88			
305		185.03			
306		185.18			
307		185.33			
309		185.64			
310		185.79			
311		185.95			
312		186.10			
314		186.41			
315		186.57			
316		186.72			
317		186.88			
319		187.19			
320		187.35			
321		187.50			
322		187.66			
324		187.97			
325		188.13			
326		188.29			
327		188.45			
329		188.77			
330		188.92			
331		189.08			
332		189.24			
334		189.56			
335		189.72			
336		189.88			
337		190.04			
339		190.37			
340		190.53			
341		190.69			
342		190.85			
344		191.18			
345		191.34			
346		191.50			
347		191.67			
349		192.00			
350		192.16			
351		192.33			
352		192.49			
354		192.82			

COMPOUND CONSTANTS:

MP   TC   VC   ZC
NBP   568.721 K   PC   DC   OM
MU   RG

VAPOR PRESSURE CORRELATION CONSTANTS, K AND KPA:

LIQUID DENSITY CORRELATION CONSTANTS, K AND G/ML:

FRANCIS1   FRANCIS1   FRANCIS1
293. - 359. K   293. - 359. K   293. - 359. K
  4 POINTS   4 POINTS   4 POINTS
RMSD = 0.00026   RMSD = 0.00026   RMSD = 0.00026
0.13363056E+01   0.13363056E+01   0.13363056E+01
0.72387094E-03   0.72387094E-03   0.72387094E-03
0.21761551E+02   0.21761551E+02   0.21761551E+02
0.45346445E+04   0.45346445E+04   0.45346445E+04

SECOND VIRIAL COEFFICIENT CORRELATION CONSTANTS, ML/MOL:

LITERATURE DOCUMENTS FOR CORRELATED VAPOR PRESSURE VALUES:

LITERATURE DOCUMENTS FOR CORRELATED LIQUID DENSITY VALUES:

  14713

LITERATURE DOCUMENTS REPORTING VIRIAL COEFFICIENT DATA:

T,K	VAPOR PRESSURE, KPA	SATURATED LIQUID VOLUME, ML/MOL	SECOND VIRIAL COEFFICIENT, ML/MOL	HEAT OF VAPORIZATION, J/MOL	SATURATED VAPOR VOLUME, ML/MOL
293		152.10			
294		152.20			
296		152.39			
297		152.49			
299		152.69			
300		152.79			
302		152.99			
303		153.09			
305		153.29			
306		153.39			
308		153.59			
309		153.69			
311		153.89			
312		153.99			
314		154.20			
315		154.30			
317		154.50			
318		154.60			
320		154.81			
321		154.91			
323		155.11			
324		155.22			
326		155.42			
327		155.52			
329		155.73			
330		155.83			
332		156.04			
333		156.15			
335		156.35			
336		156.46			
338		156.67			
339		156.77			
341		156.98			
342		157.09			
344		157.30			
345		157.40			
347		157.61			
348		157.72			
350		157.93			
351		158.04			
353		158.25			
354		158.36			
356		158.57			
357		158.68			
359		158.89			

COMPOUND CONSTANTS:

MP 290.143 K	TC	VC	ZC
NBP 544.516 K	PC	DC	OM
MU	RG		

VAPOR PRESSURE CORRELATION CONSTANTS, K AND KPA:

LIQUID DENSITY CORRELATION CONSTANTS, K AND G/ML:

FRANCIS1	FRANCIS1	FRANCIS1
294. - 354. K	294. - 354. K	294. - 354. K
4 POINTS	4 POINTS	4 POINTS
RMSD = 0.00013	RMSD = 0.00013	RMSD = 0.00013
0.11599350E+01	0.11599350E+01	0.11599350E+01
0.66308654E-03	0.66308654E-03	0.66308654E-03
0.26666517E+01	0.26666517E+01	0.26666517E+01
0.49937061E+03	0.49937061E+03	0.49937061E+03

SECOND VIRIAL COEFFICIENT CORRELATION CONSTANTS, ML/MOL:

LITERATURE DOCUMENTS FOR CORRELATED VAPOR PRESSURE VALUES:

LITERATURE DOCUMENTS FOR CORRELATED LIQUID DENSITY VALUES:

   6711

LITERATURE DOCUMENTS REPORTING VIRIAL COEFFICIENT DATA:

T,K	VAPOR PRESSURE, KPA	SATURATED LIQUID VOLUME, ML/MOL	SECOND VIRIAL COEFFICIENT, ML/MOL	HEAT OF VAPORIZATION, J/MOL	SATURATED VAPOR VOLUME, ML/MOL
294		199.88			
295		200.03			
296		200.18			
298		200.49			
299		200.65			
300		200.80			
302		201.11			
303		201.27			
304		201.42			
306		201.74			
307		201.89			
308		202.05			
310		202.37			
311		202.53			
313		202.85			
314		203.01			
315		203.17			
317		203.49			
318		203.65			
319		203.81			
321		204.14			
322		204.30			
323		204.47			
325		204.80			
326		204.96			
328		205.30			
329		205.46			
330		205.63			
332		205.97			
333		206.14			
334		206.31			
336		206.65			
337		206.82			
338		206.99			
340		207.34			
341		207.51			
343		207.86			
344		208.04			
345		208.21			
347		208.57			
348		208.75			
349		208.92			
351		209.28			
352		209.46			
353		209.65			

COMPOUND CONSTANTS:

MP		TC		VC		ZC
NBP	579.223 K	PC		DC		OM
MU		RG				

VAPOR PRESSURE CORRELATION CONSTANTS, K AND KPA:

LIQUID DENSITY CORRELATION CONSTANTS, K AND G/ML:

FRANCIS1	FRANCIS1	FRANCIS1
293. - 359. K	293. - 359. K	293. - 359. K
4 POINTS	4 POINTS	4 POINTS
RMSD = 0.00002	RMSD = 0.00002	RMSD = 0.00002
0.13082275E+01	0.13082275E+01	0.13082275E+01
0.72633754E-03	0.72633754E-03	0.72633754E-03
0.51078349E-02	0.51078349E-02	0.51078349E-02
0.38099902E+03	0.38099902E+03	0.38099902E+03

SECOND VIRIAL COEFFICIENT CORRELATION CONSTANTS, ML/MOL:

LITERATURE DOCUMENTS FOR CORRELATED VAPOR PRESSURE VALUES:

LITERATURE DOCUMENTS FOR CORRELATED LIQUID DENSITY VALUES:

14713

LITERATURE DOCUMENTS REPORTING VIRIAL COEFFICIENT DATA:

T,K	VAPOR PRESSURE, KPA	SATURATED LIQUID VOLUME, ML/MOL	SECOND VIRIAL COEFFICIENT, ML/MOL	HEAT OF VAPORIZATION, J/MOL	SATURATED VAPOR VOLUME, ML/MOL
293		168.20			
294		168.31			
296		168.53			
297		168.65			
299		168.87			
300		168.98			
302		169.21			
303		169.32			
305		169.55			
306		169.66			
308		169.89			
309		170.00			
311		170.23			
312		170.35			
314		170.58			
315		170.69			
317		170.92			
318		171.04			
320		171.27			
321		171.39			
323		171.62			
324		171.73			
326		171.97			
327		172.08			
329		172.32			
330		172.44			
332		172.67			
333		172.79			
335		173.03			
336		173.14			
338		173.38			
339		173.50			
341		173.74			
342		173.86			
344		174.10			
345		174.22			
347		174.46			
348		174.58			
350		174.82			
351		174.95			
353		175.19			
354		175.31			
356		175.56			
357		175.68			
359		175.93			

COMPOUND CONSTANTS:

MP	320.989 K	TC		VC		ZC
NBP	579.097 K	PC		DC		OM
MU	2.9800 DEBYE	RG				

VAPOR PRESSURE CORRELATION CONSTANTS, K AND KPA:

VAPRES-2	VAPRES-2	VAPRES-2	VAPRES-2
329. - 622. K	329. - 598. K	329. - 622. K	560. - 622. K
45 POINTS	41 POINTS	45 POINTS	26 POINTS
RMSD = 0.3010	RMSD = 0.2040	RMSD = 0.3010	RMSD = 0.1995
-0.73382832E+03	-0.99581025E+03	-0.73382832E+03	0.16527152E+05
0.11307120E+05	0.18157724E+05	0.11307120E+05	-0.57617584E+06
-0.29425735E+00	-0.39969474E+00	-0.29425735E+00	0.45684349E+01
0.10027145E-03	0.13929486E-03	0.10027145E-03	-0.12489337E-02
0.13451238E+03	0.18137657E+03	0.13451238E+03	-0.27909133E+04

LIQUID DENSITY CORRELATION CONSTANTS, K AND G/ML:

FRANCIS1	FRANCIS1	FRANCIS1
328. - 524. K	328. - 524. K	328. - 524. K
16 POINTS	16 POINTS	16 POINTS
RMSD = 0.00021	RMSD = 0.00021	RMSD = 0.00021
0.13401823E+01	0.13401823E+01	0.13401823E+01
0.75588422E-03	0.75588422E-03	0.75588422E-03
0.45055351E+01	0.45055351E+01	0.45055351E+01
0.69762964E+03	0.69762964E+03	0.69762964E+03

SECOND VIRIAL COEFFICIENT CORRELATION CONSTANTS, ML/MOL:

LITERATURE DOCUMENTS FOR CORRELATED VAPOR PRESSURE VALUES:

    1361   4651   6664   7481 40222

LITERATURE DOCUMENTS FOR CORRELATED LIQUID DENSITY VALUES:

    182 20318

LITERATURE DOCUMENTS REPORTING VIRIAL COEFFICIENT DATA:

T,K	VAPOR PRESSURE, KPA	SATURATED LIQUID VOLUME, ML/MOL	SECOND VIRIAL COEFFICIENT, ML/MOL	HEAT OF VAPORIZATION, J/MOL	SATURATED VAPOR VOLUME, ML/MOL
328		168.71			
334	0.004515	169.46			0.6151E+09
341	0.007512	170.33			0.3774E+09
348	0.01226	171.22			0.2360E+09
354	0.01839	171.99			0.1600E+09
361	0.02904	172.90			0.1034E+09
368	0.04506	173.82			0.6790E+08
374	0.06481	174.61			0.4798E+08
381	0.09756	175.56			0.3247E+08
388	0.1446	176.51			0.2231E+08
394	0.2002	177.34			0.1636E+08
401	0.2887	178.32			0.1155E+08
408	0.4105	179.31			8263514.
414	0.5492	180.17			6267516.
421	0.7620	181.19			4593436.
428	1.044	182.23			3407954.
434	1.355	183.13			2663427.
441	1.816	184.20			2018900.
448	2.408	185.28			1547104.
454	3.040	186.22			1241813.
461	3.952	187.34			969920.
468	5.087	188.48			764902.
474	6.269	189.47			628646.
481	7.932	190.65			504178.
488	9.950	191.85			407802.
495	12.38	193.07			332535.
501	14.83	194.15			280909.
508	18.18	195.43			232311.
515	22.13	196.73			193484.
521	26.05	197.88			166302.
528	31.31				140211.
535	37.40				118939.
541	43.35				103772.
548	51.22				88958.
555	60.20				76659.
561	68.85				67745.
568	80.18				58902.
575	92.94				51440.
581	105.1				45952.
588	120.9				40431.
595	138.6				35703.
601	155.3				32180.
608	176.8				28591.
615	200.7				25477.
621	223.2				23129.

COMPOUND CONSTANTS:

MP	TC	VC	ZC
NBP	PC	DC	OM
MU	RG		

VAPOR PRESSURE CORRELATION CONSTANTS, K AND KPA:

LIQUID DENSITY CORRELATION CONSTANTS, K AND G/ML:

FRANCIS1	FRANCIS1	FRANCIS1
293. - 359. K	293. - 359. K	293. - 359. K
4 POINTS	4 POINTS	4 POINTS
RMSD = 0.00039	RMSD = 0.00039	RMSD = 0.00039
0.12868528E+01	0.12868528E+01	0.12868528E+01
0.70454692E-03	0.70454692E-03	0.70454692E-03
0.12738742E+02	0.12738742E+02	0.12738742E+02
0.17073696E+04	0.17073696E+04	0.17073696E+04

SECOND VIRIAL COEFFICIENT CORRELATION CONSTANTS, ML/MOL:

LITERATURE DOCUMENTS FOR CORRELATED VAPOR PRESSURE VALUES:

LITERATURE DOCUMENTS FOR CORRELATED LIQUID DENSITY VALUES:

14713

LITERATURE DOCUMENTS REPORTING VIRIAL COEFFICIENT DATA:

T,K	VAPOR PRESSURE, KPA	SATURATED LIQUID VOLUME, ML/MOL	SECOND VIRIAL COEFFICIENT, ML/MOL	HEAT OF VAPORIZATION, J/MOL	SATURATED VAPOR VOLUME, ML/MOL
293		185.05			
294		185.17			
296		185.42			
297		185.54			
299		185.79			
300		185.91			
302		186.16			
303		186.29			
305		186.53			
306		186.66			
308		186.91			
309		187.03			
311		187.29			
312		187.41			
314		187.66			
315		187.79			
317		188.04			
318		188.17			
320		188.43			
321		188.55			
323		188.81			
324		188.94			
326		189.19			
327		189.32			
329		189.58			
330		189.71			
332		189.97			
333		190.10			
335		190.36			
336		190.49			
338		190.75			
339		190.88			
341		191.14			
342		191.27			
344		191.53			
345		191.66			
347		191.93			
348		192.06			
350		192.33			
351		192.46			
353		192.72			
354		192.86			
356		193.13			
357		193.26			
359		193.53			

COMPOUND CONSTANTS:

MP	368.148 K	TC		VC		ZC	
NBP	620.160 K	PC		DC		OM	
MU	3.7100 DEBYE	RG					

VAPOR PRESSURE CORRELATION CONSTANTS, K AND KPA:

LIQUID DENSITY CORRELATION CONSTANTS, K AND G/ML:

FRANCIS1	FRANCIS1	FRANCIS1
403. - 461. K	403. - 461. K	403. - 461. K
7 POINTS	7 POINTS	7 POINTS
RMSD = 0.00072	RMSD = 0.00072	RMSD = 0.00072
0.14260750E+01	0.14260750E+01	0.14260750E+01
0.73180185E-03	0.73180185E-03	0.73180185E-03
0.32127676E+01	0.32127676E+01	0.32127676E+01
0.16007200E+04	0.16007200E+04	0.16007200E+04

SECOND VIRIAL COEFFICIENT CORRELATION CONSTANTS, ML/MOL:

LITERATURE DOCUMENTS FOR CORRELATED VAPOR PRESSURE VALUES:

LITERATURE DOCUMENTS FOR CORRELATED LIQUID DENSITY VALUES:

14192

LITERATURE DOCUMENTS REPORTING VIRIAL COEFFICIENT DATA:

T,K	VAPOR PRESSURE, KPA	SATURATED LIQUID VOLUME, ML/MOL	SECOND VIRIAL COEFFICIENT, ML/MOL	HEAT OF VAPORIZATION, J/MOL	SATURATED VAPOR VOLUME, ML/MOL
403		186.30			
404		186.42			
405		186.54			
406		186.66			
408		186.91			
409		187.03			
410		187.15			
412		187.39			
413		187.52			
414		187.64			
416		187.89			
417		188.01			
418		188.13			
420		188.38			
421		188.50			
422		188.63			
424		188.88			
425		189.00			
426		189.13			
428		189.38			
429		189.50			
430		189.63			
431		189.75			
433		190.01			
434		190.13			
435		190.26			
437		190.51			
438		190.64			
439		190.76			
441		191.02			
442		191.15			
443		191.27			
445		191.53			
446		191.66			
447		191.79			
449		192.04			
450		192.17			
451		192.30			
453		192.56			
454		192.69			
455		192.82			
457		193.08			
458		193.21			
459		193.34			
460		193.47			

COMPOUND CONSTANTS:

MP   558.019 K   TC   VC   ZC
NBP   649.984 K   PC   DC   OM
MU   RG

    VAPOR PRESSURE CORRELATION CONSTANTS, K AND KPA:

    RIEDEL   RIEDEL   RIEDEL   RIEDEL
    558. - 651. K   558. - 651. K   558. - 651. K   630. - 651. K
     26 POINTS   26 POINTS   26 POINTS   7 POINTS
    RMSD = 0.3105   RMSD = 0.3105   RMSD = 0.3105   RMSD = 0.1840
    0.36652016E+03   0.36652016E+03   0.36652016E+03   0.17529189E+05
   -0.48575010E+02   -0.48575010E+02   -0.48575010E+02   -0.24029397E+04
   -0.32112042E+05   -0.32112042E+05   -0.32112042E+05   -0.13170764E+07
    0.28048558E-16   0.28048558E-16   0.28048558E-16   0.86815023E-15

    LIQUID DENSITY CORRELATION CONSTANTS, K AND G/ML:

    FRANCIS1   FRANCIS1   FRANCIS1
    565. - 576. K   565. - 576. K   565. - 576. K
     2 POINTS   2 POINTS   2 POINTS
    RMSD = 0.00000   RMSD = 0.00000   RMSD = 0.00000
    0.15605904E+01   0.15605904E+01   0.15605904E+01
    0.87257900E-03   0.87257900E-03   0.87257900E-03

    SECOND VIRIAL COEFFICIENT CORRELATION CONSTANTS, ML/MOL:

LITERATURE DOCUMENTS FOR CORRELATED VAPOR PRESSURE VALUES:

   6817 17342

LITERATURE DOCUMENTS FOR CORRELATED LIQUID DENSITY VALUES:

   20318

LITERATURE DOCUMENTS REPORTING VIRIAL COEFFICIENT DATA:

T,K	VAPOR PRESSURE, KPA	SATURATED LIQUID VOLUME, ML/MOL	SECOND VIRIAL COEFFICIENT, ML/MOL	HEAT OF VAPORIZATION, J/MOL	SATURATED VAPOR VOLUME, ML/MOL
558	13.63				
560	14.33				324812.
562	15.06				310199.
564	15.82				296387.
566	16.61	195.19		64394.	283322.
568	17.43	195.51		64236.	270956.
570	18.28	195.84		64091.	259243.
572	19.17	196.16		63960.	248141.
574	20.09	196.48		63843.	237611.
577	21.53				222809.
579	22.54				213557.
581	23.59				204761.
583	24.68				196394.
585	25.81				188430.
587	26.99				180845.
589	28.21				173616.
591	29.47				166723.
593	30.79				160146.
596	32.85				150835.
598	34.30				144973.
600	35.79				139368.
602	37.35				134007.
604	38.97				128877.
606	40.64				123965.
608	42.39				119259.
610	44.20				114748.
612	46.08				110423.
615	49.04				104262.
617	51.12				100359.
619	53.27				96611.
621	55.51				93009.
623	57.85				89547.
625	60.27				86218.
627	62.80				83015.
629	65.43				79932.
631	68.17				76964.
634	72.49				72716.
636	75.53				70012.
638	78.70				67407.
640	82.00				64894.
642	85.44				62472.
644	89.04				60134.
646	92.80				57879.
648	96.72				55702.
650	100.8				53600.

COMPOUND CONSTANTS:

MP   287.144 K   TC   VC   ZC
NBP   PC   DC   OM
MU   RG

VAPOR PRESSURE CORRELATION CONSTANTS, K AND KPA:

LIQUID DENSITY CORRELATION CONSTANTS, K AND G/ML:

FRANCIS1   FRANCIS1   FRANCIS1
293. - 354. K   293. - 354. K   293. - 354. K
   8 POINTS   8 POINTS   8 POINTS
RMSD = 0.00010   RMSD = 0.00010   RMSD = 0.00010
0.11673403E+01   0.11673403E+01   0.11673403E+01
0.70211850E-03   0.70211850E-03   0.70211850E-03
0.19182373E+02   0.19182373E+02   0.19182373E+02
0.99052905E+03   0.99052905E+03   0.99052905E+03

SECOND VIRIAL COEFFICIENT CORRELATION CONSTANTS, ML/MOL:

LITERATURE DOCUMENTS FOR CORRELATED VAPOR PRESSURE VALUES:

LITERATURE DOCUMENTS FOR CORRELATED LIQUID DENSITY VALUES:

    6711

LITERATURE DOCUMENTS REPORTING VIRIAL COEFFICIENT DATA:

T,K	VAPOR PRESSURE, KPA	SATURATED LIQUID VOLUME, ML/MOL	SECOND VIRIAL COEFFICIENT, ML/MOL	HEAT OF VAPORIZATION, J/MOL	SATURATED VAPOR VOLUME, ML/MOL
293		233.74			
294		233.92			
295		234.11			
297		234.48			
298		234.67			
299		234.86			
301		235.23			
302		235.42			
304		235.80			
305		235.99			
306		236.18			
308		236.56			
309		236.75			
311		237.13			
312		237.32			
313		237.51			
315		237.90			
316		238.09			
317		238.29			
319		238.67			
320		238.87			
322		239.26			
323		239.45			
324		239.65			
326		240.04			
327		240.24			
329		240.63			
330		240.83			
331		241.03			
333		241.43			
334		241.63			
335		241.83			
337		242.23			
338		242.43			
340		242.83			
341		243.03			
342		243.24			
344		243.64			
345		243.85			
347		244.25			
348		244.46			
349		244.66			
351		245.08			
352		245.28			
353		245.49			

COMPOUND CONSTANTS:

MP	TC	VC	ZC
NBP	PC	DC	OM
MU	RG		

VAPOR PRESSURE CORRELATION CONSTANTS, K AND KPA:

LIQUID DENSITY CORRELATION CONSTANTS, K AND G/ML:

FRANCIS1	FRANCIS1	FRANCIS1
293. - 359. K	293. - 359. K	293. - 359. K
4 POINTS	4 POINTS	4 POINTS
RMSD = 0.00011	RMSD = 0.00011	RMSD = 0.00011
0.12601871E+01	0.12601871E+01	0.12601871E+01
0.69377758E-03	0.69377758E-03	0.69377758E-03
0.52163324E+01	0.52163324E+01	0.52163324E+01
0.12629890E+04	0.12629890E+04	0.12629890E+04

SECOND VIRIAL COEFFICIENT CORRELATION CONSTANTS, ML/MOL:

LITERATURE DOCUMENTS FOR CORRELATED VAPOR PRESSURE VALUES:

LITERATURE DOCUMENTS FOR CORRELATED LIQUID DENSITY VALUES:

14713

LITERATURE DOCUMENTS REPORTING VIRIAL COEFFICIENT DATA:

T,K	VAPOR PRESSURE, KPA	SATURATED LIQUID VOLUME, ML/MOL	SECOND VIRIAL COEFFICIENT, ML/MOL	HEAT OF VAPORIZATION, J/MOL	SATURATED VAPOR VOLUME, ML/MOL
293		201.89			
294		202.02			
296		202.29			
297		202.43			
299		202.70			
300		202.83			
302		203.10			
303		203.24			
305		203.51			
306		203.65			
308		203.92			
309		204.06			
311		204.33			
312		204.47			
314		204.75			
315		204.89			
317		205.16			
318		205.30			
320		205.58			
321		205.72			
323		206.00			
324		206.14			
326		206.42			
327		206.56			
329		206.84			
330		206.98			
332		207.27			
333		207.41			
335		207.69			
336		207.83			
338		208.12			
339		208.26			
341		208.55			
342		208.69			
344		208.98			
345		209.12			
347		209.41			
348		209.56			
350		209.85			
351		209.99			
353		210.28			
354		210.43			
356		210.72			
357		210.87			
359		211.16			

COMPOUND CONSTANTS:

MP	307.640 K	TC		VC		ZC
NBP	604.154 K	PC		DC		OM
MU	2.8100 DEBYE	RG				

VAPOR PRESSURE CORRELATION CONSTANTS, K AND KPA:

RIEDEL	RIEDEL	RIEDEL
503. - 604. K	503. - 604. K	503. - 604. K
39 POINTS	39 POINTS	39 POINTS
RMSD = 0.1583	RMSD = 0.1583	RMSD = 0.1583
0.73424260E+02	0.73424260E+02	0.73424260E+02
-0.76018739E+01	-0.76018739E+01	-0.76018739E+01
-0.12094587E+05	-0.12094587E+05	-0.12094587E+05
-0.21751014E-17	-0.21751014E-17	-0.21751014E-17

LIQUID DENSITY CORRELATION CONSTANTS, K AND G/ML:

SECOND VIRIAL COEFFICIENT CORRELATION CONSTANTS, ML/MOL:

LITERATURE DOCUMENTS FOR CORRELATED VAPOR PRESSURE VALUES:

13834

LITERATURE DOCUMENTS FOR CORRELATED LIQUID DENSITY VALUES:

LITERATURE DOCUMENTS REPORTING VIRIAL COEFFICIENT DATA:

T,K	VAPOR PRESSURE, KPA	SATURATED LIQUID VOLUME, ML/MOL	SECOND VIRIAL COEFFICIENT, ML/MOL	HEAT OF VAPORIZATION, J/MOL	SATURATED VAPOR VOLUME, ML/MOL
503	7.815				535175.
505	8.333				503886.
507	8.880				474727.
509	9.456				447535.
512	10.38				410114.
514	11.04				387216.
516	11.73				365816.
519	12.83				336287.
521	13.61				318169.
523	14.44				301201.
525	15.30				285301.
528	16.67				263288.
530	17.65				249738.
532	18.66				237015.
535	20.28				219358.
537	21.42				208463.
539	22.61				198213.
542	24.50				183955.
544	25.83				175137.
546	27.21				166826.
548	28.66				158989.
551	30.94				148056.
553	32.55				141275.
555	34.21				134870.
558	36.85				125916.
560	38.69				120350.
562	40.60				115084.
564	42.59				110099.
567	45.72				103112.
569	47.90				98758.
571	50.17				94630.
574	53.72				88832.
576	56.20				85213.
578	58.77				81776.
581	62.79				76940.
583	65.58				73915.
585	68.47				71038.
587	71.46				68302.
590	76.12				64443.
592	79.36				62024.
594	82.70				59720.
597	87.91				56464.
599	91.52				54420.
601	95.23				52470.
603	99.06				50609.

COMPOUND CONSTANTS:

MP   TC   VC   ZC
NBP   PC   DC   OM
MU   RG

VAPOR PRESSURE CORRELATION CONSTANTS, K AND KPA:

LIQUID DENSITY CORRELATION CONSTANTS, K AND G/ML:

FRANCIS1   FRANCIS1   FRANCIS1
293. - 359. K   293. - 359. K   293. - 359. K
  4 POINTS   4 POINTS   4 POINTS
RMSD = 0.00024   RMSD = 0.00024   RMSD = 0.00024
0.12418633E+01   0.12418633E+01   0.12418633E+01
0.70035667E-03   0.70035667E-03   0.70035667E-03
0.49999952E-02   0.49999952E-02   0.49999952E-02
0.80786211E+04   0.80786211E+04   0.80786211E+04

SECOND VIRIAL COEFFICIENT CORRELATION CONSTANTS, ML/MOL:

LITERATURE DOCUMENTS FOR CORRELATED VAPOR PRESSURE VALUES:

LITERATURE DOCUMENTS FOR CORRELATED LIQUID DENSITY VALUES:

  14713

LITERATURE DOCUMENTS REPORTING VIRIAL COEFFICIENT DATA:

T,K	VAPOR PRESSURE, KPA	SATURATED LIQUID VOLUME, ML/MOL	SECOND VIRIAL COEFFICIENT, ML/MOL	HEAT OF VAPORIZATION, J/MOL	SATURATED VAPOR VOLUME, ML/MOL
293		218.32			
294		218.46			
296		218.76			
297		218.91			
299		219.20			
300		219.35			
302		219.65			
303		219.80			
305		220.10			
306		220.25			
308		220.55			
309		220.70			
311		221.00			
312		221.15			
314		221.46			
315		221.61			
317		221.91			
318		222.07			
320		222.37			
321		222.52			
323		222.83			
324		222.99			
326		223.29			
327		223.45			
329		223.76			
330		223.91			
332		224.22			
333		224.38			
335		224.69			
336		224.85			
338		225.16			
339		225.32			
341		225.63			
342		225.79			
344		226.11			
345		226.26			
347		226.58			
348		226.74			
350		227.06			
351		227.22			
353		227.54			
354		227.70			
356		228.02			
357		228.18			
359		228.50			

COMPOUND CONSTANTS:

MP	TC	VC	ZC
NBP	PC	DC	OM
MU	RG		

VAPOR PRESSURE CORRELATION CONSTANTS, K AND KPA:

LIQUID DENSITY CORRELATION CONSTANTS, K AND G/ML:

FRANCIS1	FRANCIS1	FRANCIS1
293. - 359. K	293. - 359. K	293. - 359. K
4 POINTS	4 POINTS	4 POINTS
RMSD = 0.00005	RMSD = 0.00005	RMSD = 0.00005
0.12268534E+01	0.12268534E+01	0.12268534E+01
0.67655882E-03	0.67655882E-03	0.67655882E-03
0.18401814E+01	0.18401814E+01	0.18401814E+01
0.69340601E+03	0.69340601E+03	0.69340601E+03

SECOND VIRIAL COEFFICIENT CORRELATION CONSTANTS, ML/MOL:

LITERATURE DOCUMENTS FOR CORRELATED VAPOR PRESSURE VALUES:

LITERATURE DOCUMENTS FOR CORRELATED LIQUID DENSITY VALUES:

14713

LITERATURE DOCUMENTS REPORTING VIRIAL COEFFICIENT DATA:

T,K	VAPOR PRESSURE, KPA	SATURATED LIQUID VOLUME, ML/MOL	SECOND VIRIAL COEFFICIENT, ML/MOL	HEAT OF VAPORIZATION, J/MOL	SATURATED VAPOR VOLUME, ML/MOL
293		234.71			
294		234.86			
296		235.18			
297		235.34			
299		235.66			
300		235.81			
302		236.13			
303		236.29			
305		236.61			
306		236.77			
308		237.10			
309		237.26			
311		237.58			
312		237.74			
314		238.07			
315		238.23			
317		238.56			
318		238.72			
320		239.05			
321		239.21			
323		239.54			
324		239.70			
326		240.04			
327		240.20			
329		240.53			
330		240.70			
332		241.03			
333		241.20			
335		241.53			
336		241.70			
338		242.04			
339		242.21			
341		242.55			
342		242.71			
344		243.05			
345		243.22			
347		243.57			
348		243.74			
350		244.08			
351		244.25			
353		244.59			
354		244.77			
356		245.11			
357		245.29			
359		245.63			

COMPOUND CONSTANTS:

MP	447.181 K	TC		VC		ZC
NBP		PC		DC		OM
MU	3.4100 DEBYE	RG				

VAPOR PRESSURE CORRELATION CONSTANTS, K AND KPA:

RPM2	RPM2	RPM2
456. - 549. K	456. - 549. K	456. - 549. K
9 POINTS	9 POINTS	9 POINTS
RMSD = 0.0126	RMSD = 0.0126	RMSD = 0.0126
0.54175963E+03	0.54175963E+03	0.54175963E+03
-0.10033771E+06	-0.10033771E+06	-0.10033771E+06
-0.10049879E+01	-0.10049879E+01	-0.10049879E+01
0.63820178E-03	0.63820178E-03	0.63820178E-03

LIQUID DENSITY CORRELATION CONSTANTS, K AND G/ML:

SECOND VIRIAL COEFFICIENT CORRELATION CONSTANTS, ML/MOL:

LITERATURE DOCUMENTS FOR CORRELATED VAPOR PRESSURE VALUES:

11052

LITERATURE DOCUMENTS FOR CORRELATED LIQUID DENSITY VALUES:

LITERATURE DOCUMENTS REPORTING VIRIAL COEFFICIENT DATA:

T,K	VAPOR PRESSURE, KPA	SATURATED LIQUID VOLUME, ML/MOL	SECOND VIRIAL COEFFICIENT, ML/MOL	HEAT OF VAPORIZATION, J/MOL	SATURATED VAPOR VOLUME, ML/MOL
456	0.02131				0.1779E+09
458	0.02397				0.1589E+09
460	0.02687				0.1423E+09
462	0.03003				0.1279E+09
464	0.03346				0.1153E+09
466	0.03717				0.1042E+09
468	0.04117				0.9451E+08
470	0.04548				0.8592E+08
472	0.05012				0.7831E+08
475	0.05769				0.6846E+08
477	0.06318				0.6277E+08
479	0.06903				0.5769E+08
481	0.07526				0.5314E+08
483	0.08187				0.4905E+08
485	0.08889				0.4537E+08
487	0.09632				0.4204E+08
489	0.1042				0.3902E+08
491	0.1125				0.3629E+08
494	0.1258				0.3265E+08
496	0.1353				0.3048E+08
498	0.1453				0.2850E+08
500	0.1558				0.2669E+08
502	0.1668				0.2502E+08
504	0.1784				0.2349E+08
506	0.1906				0.2207E+08
508	0.2034				0.2076E+08
510	0.2169				0.1955E+08
513	0.2383				0.1790E+08
515	0.2535				0.1689E+08
517	0.2694				0.1596E+08
519	0.2861				0.1508E+08
521	0.3037				0.1426E+08
523	0.3222				0.1350E+08
525	0.3416				0.1278E+08
527	0.3620				0.1210E+08
529	0.3836				0.1147E+08
532	0.4180				0.1058E+08
534	0.4426				0.1003E+08
536	0.4685				9512956.
538	0.4958				9021377.
540	0.5248				8555503.
542	0.5554				8113513.
544	0.5879				7693758.
546	0.6223				7294748.
548	0.6589				6915135.

COMPOUND CONSTANTS:

MP   317.639 K   TC   VC   ZC
NBP   PC   DC   OM
MU   RG

VAPOR PRESSURE CORRELATION CONSTANTS, K AND KPA:

LIQUID DENSITY CORRELATION CONSTANTS, K AND G/ML:

FRANCIS1   FRANCIS1   FRANCIS1
319. - 357. K   319. - 357. K   319. - 357. K
  5 POINTS   5 POINTS   5 POINTS
RMSD = 0.00015   RMSD = 0.00015   RMSD = 0.00015
0.11383848E+01   0.11383848E+01   0.11383848E+01
0.66913106E-03   0.66913106E-03   0.66913106E-03
0.49246109E+02   0.49246109E+02   0.49246109E+02
0.24330369E+04   0.24330369E+04   0.24330369E+04

SECOND VIRIAL COEFFICIENT CORRELATION CONSTANTS, ML/MOL:

LITERATURE DOCUMENTS FOR CORRELATED VAPOR PRESSURE VALUES:

LITERATURE DOCUMENTS FOR CORRELATED LIQUID DENSITY VALUES:

   6711

LITERATURE DOCUMENTS REPORTING VIRIAL COEFFICIENT DATA:

T,K	VAPOR PRESSURE, KPA	SATURATED LIQUID VOLUME, ML/MOL	SECOND VIRIAL COEFFICIENT, ML/MOL	HEAT OF VAPORIZATION, J/MOL	SATURATED VAPOR VOLUME, ML/MOL
319		288.83			
320		289.05			
321		289.27			
322		289.48			
323		289.70			
324		289.92			
325		290.14			
326		290.36			
327		290.58			
328		290.80			
329		291.02			
330		291.25			
331		291.47			
332		291.69			
333		291.91			
334		292.13			
335		292.36			
336		292.58			
337		292.81			
338		293.03			
339		293.25			
340		293.48			
341		293.70			
342		293.93			
343		294.16			
344		294.38			
345		294.61			
346		294.84			
347		295.06			
348		295.29			
349		295.52			
350		295.75			
351		295.98			
352		296.20			
353		296.43			
354		296.66			
355		296.89			
356		297.12			
357		297.36			

COMPOUND CONSTANTS:

MP   TC   VC   ZC
NBP   PC   DC   OM
MU   RG

VAPOR PRESSURE CORRELATION CONSTANTS, K AND KPA:

LIQUID DENSITY CORRELATION CONSTANTS, K AND G/ML:

FRANCIS1   FRANCIS1   FRANCIS1
293. - 359. K   293. - 359. K   293. - 359. K
  4 POINTS   4 POINTS   4 POINTS
RMSD = 0.00009   RMSD = 0.00009   RMSD = 0.00009
0.12137480E+01   0.12137480E+01   0.12137480E+01
0.67674927E-03   0.67674927E-03   0.67674927E-03
0.45696878E+01   0.45696878E+01   0.45696878E+01
0.12091201E+04   0.12091201E+04   0.12091201E+04

SECOND VIRIAL COEFFICIENT CORRELATION CONSTANTS, ML/MOL:

LITERATURE DOCUMENTS FOR CORRELATED VAPOR PRESSURE VALUES:

LITERATURE DOCUMENTS FOR CORRELATED LIQUID DENSITY VALUES:

  14713

LITERATURE DOCUMENTS REPORTING VIRIAL COEFFICIENT DATA:

T,K	VAPOR PRESSURE, KPA	SATURATED LIQUID VOLUME, ML/MOL	SECOND VIRIAL COEFFICIENT, ML/MOL	HEAT OF VAPORIZATION, J/MOL	SATURATED VAPOR VOLUME, ML/MOL
293		251.73			
294		251.90			
296		252.25			
297		252.42			
299		252.76			
300		252.93			
302		253.27			
303		253.45			
305		253.79			
306		253.96			
308		254.31			
309		254.48			
311		254.83			
312		255.01			
314		255.36			
315		255.53			
317		255.88			
318		256.06			
320		256.41			
321		256.59			
323		256.94			
324		257.12			
326		257.47			
327		257.65			
329		258.01			
330		258.19			
332		258.54			
333		258.72			
335		259.08			
336		259.26			
338		259.63			
339		259.81			
341		260.17			
342		260.35			
344		260.72			
345		260.90			
347		261.26			
348		261.45			
350		261.82			
351		262.00			
353		262.37			
354		262.55			
356		262.92			
357		263.11			
359		263.48			

COMPOUND CONSTANTS:

MP   325.639 K   TC   VC   ZC
NBP   PC   DC   OM
MU   RG

VAPOR PRESSURE CORRELATION CONSTANTS, K AND KPA:

LIQUID DENSITY CORRELATION CONSTANTS, K AND G/ML:

FRANCIS1   FRANCIS1   FRANCIS1
328. - 355. K   328. - 355. K   328. - 355. K
  5 POINTS   5 POINTS   5 POINTS
RMSD = 0.00014   RMSD = 0.00014   RMSD = 0.00014
0.11290541E+01   0.11290541E+01   0.11290541E+01
0.66905282E-03   0.66905282E-03   0.66905282E-03
0.51053818E+02   0.51053818E+02   0.51053818E+02
0.27560354E+04   0.27560354E+04   0.27560354E+04

SECOND VIRIAL COEFFICIENT CORRELATION CONSTANTS, ML/MOL:

LITERATURE DOCUMENTS FOR CORRELATED VAPOR PRESSURE VALUES:

LITERATURE DOCUMENTS FOR CORRELATED LIQUID DENSITY VALUES:

    6711

LITERATURE DOCUMENTS REPORTING VIRIAL COEFFICIENT DATA:

T,K	VAPOR PRESSURE, KPA	SATURATED LIQUID VOLUME, ML/MOL	SECOND VIRIAL COEFFICIENT, ML/MOL	HEAT OF VAPORIZATION, J/MOL	SATURATED VAPOR VOLUME, ML/MOL
328		324.64			
329		324.89			
330		325.14			
331		325.39			
332		325.64			
333		325.89			
334		326.14			
335		326.39			
336		326.64			
337		326.89			
338		327.14			
339		327.39			
340		327.64			
341		327.90			
342		328.15			
343		328.40			
344		328.66			
345		328.91			
346		329.16			
347		329.42			
348		329.67			
349		329.93			
350		330.19			
351		330.44			
352		330.70			
353		330.96			
354		331.21			
355		331.47			

COMPOUND CONSTANTS:

MP   332.140 K   TC   VC   ZC
NBP   PC   DC   OM
MU   RG

    VAPOR PRESSURE CORRELATION CONSTANTS, K AND KPA:

RPM2   RPM2   RPM2
443. - 525. K   443. - 525. K   443. - 525. K
  2 POINTS   2 POINTS   2 POINTS
RMSD = 0.0000   RMSD = 0.0000   RMSD = 0.0000
0.28106180E+02   0.28106180E+02   0.28106180E+02
-0.14368583E+05   -0.14368583E+05   -0.14368583E+05

    LIQUID DENSITY CORRELATION CONSTANTS, K AND G/ML:

FRANCIS1   FRANCIS1   FRANCIS1
353. - 574. K   353. - 574. K   353. - 574. K
  5 POINTS   5 POINTS   5 POINTS
RMSD = 0.00064   RMSD = 0.00064   RMSD = 0.00064
0.11021729E+01   0.11021729E+01   0.11021729E+01
0.68568555E-03   0.68568555E-03   0.68568555E-03
0.27912140E+00   0.27912140E+00   0.27912140E+00
0.60188306E+03   0.60188306E+03   0.60188306E+03

    SECOND VIRIAL COEFFICIENT CORRELATION CONSTANTS, ML/MOL:

LITERATURE DOCUMENTS FOR CORRELATED VAPOR PRESSURE VALUES:

  8226

LITERATURE DOCUMENTS FOR CORRELATED LIQUID DENSITY VALUES:

  8125

LITERATURE DOCUMENTS REPORTING VIRIAL COEFFICIENT DATA:

T,K	VAPOR PRESSURE, KPA	SATURATED LIQUID VOLUME, ML/MOL	SECOND VIRIAL COEFFICIENT, ML/MOL	HEAT OF VAPORIZATION, J/MOL	SATURATED VAPOR VOLUME, ML/MOL
353		368.48			
358		369.97			
363		371.47			
368		372.98			
373		374.50			
378		376.04			
383		377.59			
388		379.15			
393		380.73			
398		382.32			
403		383.93			
408		385.55			
413		387.18			
418		388.83			
423		390.50			
428		392.18			
433		393.88			
438		395.59			
443	0.01319	397.32			0.2793E+09
448	0.01894	399.06			0.1967E+09
453	0.02698	400.83			0.1396E+09
458	0.03815	402.61			0.9982E+08
463	0.05353	404.41			0.7191E+08
468	0.07458	406.23			0.5218E+08
473	0.1032	408.06			0.3812E+08
478	0.1418	409.92			0.2804E+08
483	0.1935	411.80			0.2075E+08
488	0.2625	413.70			0.1546E+08
493	0.3538	415.62			0.1159E+08
498	0.4741	417.57			8734066.
503	0.6315	419.54			6622124.
508	0.8366	421.54			5048788.
513	1.102	423.57			3870040.
518	1.444	425.63			2982038.
523	1.883	427.73			2309489.
528		429.86			
533		432.03			
538		434.25			
543		436.52			
548		438.87			
553		441.29			
558		443.81			
563		446.47			
568		449.31			
573		452.43			

COMPOUND CONSTANTS:

MP		TC	VC	ZC
NBP	391.656 K	PC	DC	OM
MU	2.3800 DEBYE	RG		

VAPOR PRESSURE CORRELATION CONSTANTS, K AND KPA:

RPM2	RPM2	RPM2
303. - 392. K	303. - 392. K	303. - 392. K
4 POINTS	4 POINTS	4 POINTS
RMSD = 0.0005	RMSD = 0.0005	RMSD = 0.0005
0.63297187E+02	0.63297187E+02	0.63297187E+02
-0.10289174E+05	-0.10289174E+05	-0.10289174E+05
-0.13078063E+00	-0.13078063E+00	-0.13078063E+00
0.12264536E-03	0.12264536E-03	0.12264536E-03

LIQUID DENSITY CORRELATION CONSTANTS, K AND G/ML:

SECOND VIRIAL COEFFICIENT CORRELATION CONSTANTS, ML/MOL:

LITERATURE DOCUMENTS FOR CORRELATED VAPOR PRESSURE VALUES:

22644

LITERATURE DOCUMENTS FOR CORRELATED LIQUID DENSITY VALUES:

LITERATURE DOCUMENTS REPORTING VIRIAL COEFFICIENT DATA:

T,K	VAPOR PRESSURE, KPA	SATURATED LIQUID VOLUME, ML/MOL	SECOND VIRIAL COEFFICIENT, ML/MOL	HEAT OF VAPORIZATION, J/MOL	SATURATED VAPOR VOLUME, ML/MOL
303	2.646				952213.
305	2.954				858442.
307	3.292				775359.
309	3.662				701594.
311	4.066				635969.
313	4.507				577472.
315	4.986				525228.
317	5.508				478481.
319	6.075				436577.
321	6.690				398946.
323	7.356				365095.
325	8.076				334593.
327	8.854				307061.
329	9.694				282172.
331	10.60				259636.
333	11.57				239199.
335	12.62				220637.
337	13.75				203755.
339	14.96				188379.
341	16.26				174354.
343	17.65				161544.
345	19.14				149829.
347	20.74				139101.
349	22.45				129265.
351	24.27				120235.
353	26.22				111935.
355	28.30				104298.
357	30.52				97262.
359	32.88				90773.
361	35.40				84783.
363	38.09				79245.
365	40.94				74123.
367	43.98				69378.
369	47.21				64980.
371	50.65				60899.
373	54.31				57108.
375	58.19				53584.
377	62.31				50306.
379	66.69				47253.
381	71.33				44407.
383	76.27				41753.
385	81.50				39276.
387	87.06				36961.
389	92.95				34798.
391	99.19				32773.

COMPOUND CONSTANTS:

MP		TC		VC		ZC
NBP	391.656 K	PC		DC		OM
MU	1.9700 DEBYE	RG				

VAPOR PRESSURE CORRELATION CONSTANTS, K AND KPA:

RPM2	RPM2	RPM2
292. - 392. K	292. - 392. K	292. - 392. K
4 POINTS	4 POINTS	4 POINTS
RMSD = 0.0004	RMSD = 0.0004	RMSD = 0.0004
0.26169771E+02	0.26169771E+02	0.26169771E+02
-0.55232119E+04	-0.55232119E+04	-0.55232119E+04
-0.28809006E-01	-0.28809006E-01	-0.28809006E-01
0.24994206E-04	0.24994206E-04	0.24994206E-04

LIQUID DENSITY CORRELATION CONSTANTS, K AND G/ML:

SECOND VIRIAL COEFFICIENT CORRELATION CONSTANTS, ML/MOL:

LITERATURE DOCUMENTS FOR CORRELATED VAPOR PRESSURE VALUES:

22644

LITERATURE DOCUMENTS FOR CORRELATED LIQUID DENSITY VALUES:

LITERATURE DOCUMENTS REPORTING VIRIAL COEFFICIENT DATA:

T,K	VAPOR PRESSURE, KPA	SATURATED LIQUID VOLUME, ML/MOL	SECOND VIRIAL COEFFICIENT, ML/MOL	HEAT OF VAPORIZATION, J/MOL	SATURATED VAPOR VOLUME, ML/MOL
292	2.647				917090.
294	2.927				835209.
296	3.231				761775.
298	3.561				695811.
301	4.109				609038.
303	4.513				558251.
305	4.949				512388.
307	5.420				470912.
310	6.197				415903.
312	6.765				383446.
314	7.376				353951.
316	8.032				327112.
319	9.107				291253.
321	9.887				269937.
323	10.72				250457.
326	12.08				224289.
328	13.07				208653.
330	14.12				194303.
332	15.24				181121.
335	17.06				163300.
337	18.36				152585.
339	19.75				142704.
341	21.22				133583.
344	23.60				121182.
346	25.31				113682.
348	27.11				106736.
351	30.01				97252.
353	32.08				91492.
355	34.27				86141.
357	36.57				81163.
360	40.27				74334.
362	42.90				70167.
364	45.66				66279.
366	48.57				62650.
369	53.22				57648.
371	56.51				54582.
373	59.97				51711.
376	65.48				47742.
378	69.38				45300.
380	73.46				43009.
382	77.74				40858.
385	84.53				37870.
387	89.32				36026.
389	94.32				34290.
391	99.56				32654.

COMPOUND CONSTANTS:

MP   316.139 K   TC   VC   ZC
NBP 445.185 K   PC   DC   OM
MU   RG

    VAPOR PRESSURE CORRELATION CONSTANTS, K AND KPA:

RPM2   RPM2   RPM2
348. - 446. K   348. - 446. K   348. - 446. K
   5 POINTS   5 POINTS   5 POINTS
RMSD = 0.0338   RMSD = 0.0338   RMSD = 0.0338
   0.43814304E+01   0.43814304E+01   0.43814304E+01
  -0.42660432E+04   -0.42660432E+04   -0.42660432E+04
   0.37510628E-01   0.37510628E-01   0.37510628E-01
  -0.34712350E-04   -0.34712350E-04   -0.34712350E-04

    LIQUID DENSITY CORRELATION CONSTANTS, K AND G/ML:

    SECOND VIRIAL COEFFICIENT CORRELATION CONSTANTS, ML/MOL:

LITERATURE DOCUMENTS FOR CORRELATED VAPOR PRESSURE VALUES:

  22644

LITERATURE DOCUMENTS FOR CORRELATED LIQUID DENSITY VALUES:

LITERATURE DOCUMENTS REPORTING VIRIAL COEFFICIENT DATA:

T,K	VAPOR PRESSURE, KPA	SATURATED LIQUID VOLUME, ML/MOL	SECOND VIRIAL COEFFICIENT, ML/MOL	HEAT OF VAPORIZATION, J/MOL	SATURATED VAPOR VOLUME, ML/MOL
348	2.645				1094017.
350	2.913				998976.
352	3.205				913140.
354	3.523				835539.
356	3.868				765308.
359	4.441				672134.
361	4.863				617169.
363	5.321				567250.
365	5.815				521871.
368	6.632				461335.
370	7.231				425429.
372	7.877				392679.
374	8.572				362780.
376	9.319				335461.
379	10.55				298788.
381	11.44				276899.
383	12.40				256836.
385	13.43				238431.
388	15.10				213606.
390	16.32				198717.
392	17.62				185019.
394	19.00				172406.
396	20.48				160783.
399	22.88				145018.
401	24.60				135511.
403	26.44				126726.
405	28.39				118602.
408	31.55				107537.
410	33.81				100835.
412	36.20				94622.
414	38.74				88858.
417	42.82				80974.
419	45.73				76180.
421	48.81				71721.
423	52.05				67572.
425	55.47				63707.
428	60.94				58397.
430	64.82				55153.
432	68.91				52125.
434	73.20				49297.
437	80.04				45397.
439	84.87				43006.
441	89.94				40767.
443	95.25				38671.
445	100.8				36706.

COMPOUND CONSTANTS:

MP   140.141 K   TC   410.665 K   VC   ZC
NBP 280.948 K   PC   2.878 MPA   DC   OM 0.3449
MU   RG   4.0101 ANGSTROM

VAPOR PRESSURE CORRELATION CONSTANTS, K AND KPA:

RIEDEL	RIEDEL	VAPRES-2	RIEDEL
232. - 411. K	232. - 287. K	232. - 345. K	273. - 411. K
17 POINTS	7 POINTS	11 POINTS	13 POINTS
RMSD = 0.7267	RMSD = 0.0991	RMSD = 0.3514	RMSD = 0.7664
0.52021099E+02	0.51924067E+02	0.48887711E+03	0.51987333E+02
-0.55057334E+01	-0.54678495E+01	-0.12792830E+05	-0.54992005E+01
-0.46044849E+04	-0.46277611E+04	0.28424310E+00	-0.46055390E+04
0.60184321E-16	-0.94190148E-17	-0.15548657E-03	0.59499181E-16
		-0.89800244E+02	

LIQUID DENSITY CORRELATION CONSTANTS, K AND G/ML:

SECOND VIRIAL COEFFICIENT CORRELATION CONSTANTS, ML/MOL:

LITERATURE DOCUMENTS FOR CORRELATED VAPOR PRESSURE VALUES:

   10015

LITERATURE DOCUMENTS FOR CORRELATED LIQUID DENSITY VALUES:

LITERATURE DOCUMENTS REPORTING VIRIAL COEFFICIENT DATA:

T,K	VAPOR PRESSURE, KPA	SATURATED LIQUID VOLUME, ML/MOL	SECOND VIRIAL COEFFICIENT, ML/MOL	HEAT OF VAPORIZATION, J/MOL	SATURATED VAPOR VOLUME, ML/MOL
232	8.983				213406.
236	11.46				169971.
240	14.47				136604.
244	18.12				110725.
248	22.49				90472.
252	27.69				74485.
256	33.84				61763.
260	41.04				51559.
264	49.43				43315.
268	59.15				36608.
272	70.33				31115.
276	83.13				26587.
280	97.70				22832.
284	114.2				19700.
288	132.8				17074.
293	159.3				14362.
297	183.3				12560.
301	209.9				11025.
305	239.5				9710.
309	272.1				8579.
313	307.9				7602.
317	347.2				6755.
321	390.2				6017.
325	437.1				5373.
329	488.1				4808.
333	543.4				4310.
337	603.2				3871.
341	667.8				3482.
345	737.5				3136.
349	812.4				2828.
354	913.9				2487.
358	1002.				2246.
362	1096.				2029.
366	1196.				1832.
370	1303.				1654.
374	1418.				1492.
378	1540.				1344.
382	1670.				1208.
386	1808.				1083.
390	1956.				966.8
394	2112.				857.9
398	2279.				754.5
402	2456.				654.1
406	2644.				551.5
410	2845.				420.0

COMPOUND CONSTANTS:

MP 147.698 K	TC 357.245 K	VC 329.08 ML/MOL	ZC 0.3139
NBP 245.805 K	PC 2.841 MPA	DC 0.5045 G/ML	OM 0.3649
MU	RG 3.8076 ANGSTROM		

VAPOR PRESSURE CORRELATION CONSTANTS, K AND KPA:

WAGNER	WAGNER	WAGNER	WAGNER
194. - 358. K	194. - 265. K	194. - 304. K	225. - 358. K
35 POINTS	15 POINTS	22 POINTS	29 POINTS
RMSD = 2.9460	RMSD = 0.2163	RMSD = 1.2410	RMSD = 3.1480
-0.76569653E+01	-0.68017611E+01	-0.74864472E+01	-0.77804023E+01
0.12828020E+01	-0.90258908E+00	0.82054154E+00	0.17061565E+01
-0.39738249E+01	0.74037148E-01	-0.29970338E+01	-0.55119078E+01
-0.72826939E+01	-0.18709122E+02	-0.10487798E+02	0.36491593E+01

LIQUID DENSITY CORRELATION CONSTANTS, K AND G/ML:

FRANCIS1	FRANCIS1	FRANCIS1
210. - 295. K	210. - 295. K	210. - 295. K
12 POINTS	12 POINTS	12 POINTS
RMSD = 0.00142	RMSD = 0.00142	RMSD = 0.00142
0.23480034E+01	0.23480034E+01	0.23480034E+01
0.25871226E-02	0.25871226E-02	0.25871226E-02
0.31269958E+02	0.31269958E+02	0.31269958E+02
0.42212915E+03	0.42212915E+03	0.42212915E+03

SECOND VIRIAL COEFFICIENT CORRELATION CONSTANTS, ML/MOL:

LITERATURE DOCUMENTS FOR CORRELATED VAPOR PRESSURE VALUES:

7465 10015 15034 16255 23080

LITERATURE DOCUMENTS FOR CORRELATED LIQUID DENSITY VALUES:

7465

LITERATURE DOCUMENTS REPORTING VIRIAL COEFFICIENT DATA:

T,K	VAPOR PRESSURE, KPA	SATURATED LIQUID VOLUME, ML/MOL	SECOND VIRIAL COEFFICIENT, ML/MOL	HEAT OF VAPORIZATION, J/MOL	SATURATED VAPOR VOLUME, ML/MOL
194	4.130				389298.
197	5.283				308796.
201	7.228				230023.
205	9.732				173983.
208	12.05				142433.
212	15.82	100.58		24545.	110350.
216	20.50	101.39		24212.	86553.
220	26.24	102.22		23883.	68666.
223	31.35	102.86		23638.	58124.
227	39.37	103.73		23313.	46946.
231	48.96	104.63		22989.	38270.
234	57.29	105.32		22746.	33018.
238	70.10	106.26		22420.	27309.
242	85.07	107.23		22093.	22757.
246	102.4	108.23		21763.	19096.
249	117.2	109.00		21514.	16813.
253	139.3	110.05		21177.	14262.
257	164.6	111.15		20834.	12167.
261	193.2	112.28		20486.	10433.
264	217.1	113.15		20220.	9325.
268	252.3	114.36		19858.	8061.
272	291.8	115.61		19487.	6997.
275	324.3	116.59		19203.	6308.
279	371.8	117.94		18815.	5510.
283	424.5	119.35		18415.	4829.
287	482.5	120.82		18002.	4243.
290	529.8	121.98		17684.	3858.
294	598.2	123.59		17245.	3405.
298	672.9				3011.
302	754.5				2667.
305	820.4				2437.
309	914.8				2163.
313	1017.				1921.
316	1099.				1758.
320	1217.				1561.
324	1343.				1385.
328	1480.				1228.
331	1589.				1120.
335	1744.				987.3
339	1910.				866.4
343	2089.				754.6
346	2231.				675.3
350	2434.				572.7
354	2652.				466.8
357	2826.				356.7

COMPOUND CONSTANTS:

MP   TC   VC   ZC
NBP   427.659 K   PC   DC   OM
MU   2.3500 DEBYE   RG

VAPOR PRESSURE CORRELATION CONSTANTS, K AND KPA:

RPM2   RPM2   RPM2
322. - 428. K   322. - 428. K   322. - 428. K
 9 POINTS   9 POINTS   9 POINTS
RMSD = 0.0807   RMSD = 0.0807   RMSD = 0.0807
 0.68041358E+02   0.68041358E+02   0.68041358E+02
-0.12469551E+05   -0.12469551E+05   -0.12469551E+05
-0.12079499E+00   -0.12079499E+00   -0.12079499E+00
 0.95103685E-04   0.95103685E-04   0.95103685E-04

LIQUID DENSITY CORRELATION CONSTANTS, K AND G/ML:

SECOND VIRIAL COEFFICIENT CORRELATION CONSTANTS, ML/MOL:

LITERATURE DOCUMENTS FOR CORRELATED VAPOR PRESSURE VALUES:

  14415

LITERATURE DOCUMENTS FOR CORRELATED LIQUID DENSITY VALUES:

LITERATURE DOCUMENTS REPORTING VIRIAL COEFFICIENT DATA:

T,K	VAPOR PRESSURE, KPA	SATURATED LIQUID VOLUME, ML/MOL	SECOND VIRIAL COEFFICIENT, ML/MOL	HEAT OF VAPORIZATION, J/MOL	SATURATED VAPOR VOLUME, ML/MOL
322	1.324				2021821.
324	1.493				1803736.
326	1.681				1612616.
329	1.999				1368531.
331	2.238				1229787.
334	2.641				1051420.
336	2.943				949364.
338	3.272				858787.
341	3.824				741335.
343	4.234				673553.
346	4.916				585150.
348	5.420				533840.
350	5.966				487797.
353	6.869				427290.
355	7.531				391907.
358	8.623				345175.
360	9.421				317709.
362	10.28				292823.
365	11.68				259736.
367	12.71				240162.
370	14.37				214021.
372	15.58				198489.
374	16.87				184295.
377	18.97				165228.
379	20.48				153834.
382	22.94				138467.
384	24.70				129248.
387	27.56				116769.
389	29.60				109254.
391	31.77				102316.
394	35.27				92879.
396	37.77				87169.
399	41.79				79376.
401	44.67				74646.
403	47.70				70251.
406	52.56				64227.
408	56.02				60554.
411	61.56				55506.
413	65.51				52420.
415	69.66				49537.
418	76.29				45558.
420	80.99				43116.
423	88.50				39739.
425	93.83				37660.
427	99.42				35709.

COMPOUND CONSTANTS:

MP		TC	VC	ZC
NBP	409.159 K	PC	DC	OM
MU		RG		

    VAPOR PRESSURE CORRELATION CONSTANTS, K AND KPA:

RPM2	RPM2	RPM2
306. - 410. K	306. - 410. K	306. - 410. K
8 POINTS	8 POINTS	8 POINTS
RMSD = 0.0902	RMSD = 0.0902	RMSD = 0.0902
0.11101456E+03	0.11101456E+03	0.11101456E+03
-0.16741801E+05	-0.16741801E+05	-0.16741801E+05
-0.25145797E+00	-0.25145797E+00	-0.25145797E+00
0.22344785E-03	0.22344785E-03	0.22344785E-03

    LIQUID DENSITY CORRELATION CONSTANTS, K AND G/ML:

    SECOND VIRIAL COEFFICIENT CORRELATION CONSTANTS, ML/MOL:

LITERATURE DOCUMENTS FOR CORRELATED VAPOR PRESSURE VALUES:

  14415

LITERATURE DOCUMENTS FOR CORRELATED LIQUID DENSITY VALUES:

LITERATURE DOCUMENTS REPORTING VIRIAL COEFFICIENT DATA:

T,K	VAPOR PRESSURE, KPA	SATURATED LIQUID VOLUME, ML/MOL	SECOND VIRIAL COEFFICIENT, ML/MOL	HEAT OF VAPORIZATION, J/MOL	SATURATED VAPOR VOLUME, ML/MOL
306	1.322				1923994.
308	1.501				1706030.
310	1.699				1516931.
313	2.036				1278186.
315	2.290				1143929.
317	2.568				1026294.
320	3.037				876003.
322	3.387				790489.
324	3.768				714883.
327	4.405				617241.
329	4.875				561086.
331	5.385				511029.
334	6.230				445740.
336	6.850				407825.
339	7.872				358051.
341	8.619				328958.
343	9.421				302700.
346	10.74				267937.
348	11.69				247448.
350	12.72				228837.
353	14.39				204005.
355	15.60				189259.
357	16.89				175784.
360	18.98				157680.
362	20.50				146852.
365	22.95				132237.
367	24.72				123456.
369	26.59				115361.
372	29.64				104368.
374	31.82				97724.
376	34.14				91571.
379	37.89				83170.
381	40.58				78065.
383	43.43				73318.
386	48.04				66803.
388	51.35				62826.
391	56.69				57350.
393	60.51				53997.
395	64.57				50860.
398	71.13				46523.
400	75.83				43857.
402	80.82				41355.
405	88.88				37884.
407	94.67				35744.
409	100.8				33730.

COMPOUND CONSTANTS:

MP		TC		VC		ZC	
NBP	388.742 K	PC		DC		OM	
MU		RG					

VAPOR PRESSURE CORRELATION CONSTANTS, K AND KPA:

RPM2	RPM2	RPM2
309. - 391. K	309. - 391. K	309. - 391. K
17 POINTS	17 POINTS	17 POINTS
RMSD = 0.0002	RMSD = 0.0002	RMSD = 0.0002
0.16738657E+02	0.16738657E+02	0.16738657E+02
-0.46777777E+04	-0.46777777E+04	-0.46777777E+04
-0.33828305E-03	-0.33828305E-03	-0.33828305E-03
0.29316526E-06	0.29316526E-06	0.29316526E-06

LIQUID DENSITY CORRELATION CONSTANTS, K AND G/ML:

SECOND VIRIAL COEFFICIENT CORRELATION CONSTANTS, ML/MOL:

LITERATURE DOCUMENTS FOR CORRELATED VAPOR PRESSURE VALUES:

15556

LITERATURE DOCUMENTS FOR CORRELATED LIQUID DENSITY VALUES:

LITERATURE DOCUMENTS REPORTING VIRIAL COEFFICIENT DATA:

T,K	VAPOR PRESSURE, KPA	SATURATED LIQUID VOLUME, ML/MOL	SECOND VIRIAL COEFFICIENT, ML/MOL	HEAT OF VAPORIZATION, J/MOL	SATURATED VAPOR VOLUME, ML/MOL
309	4.589				559840.
310	4.818				534967.
312	5.306				488930.
314	5.835				447386.
316	6.411				409850.
318	7.034				375893.
320	7.709				345136.
322	8.439				317245.
323	8.826				304280.
325	9.646				280142.
327	10.53				258190.
329	11.48				238203.
331	12.51				219986.
333	13.61				203364.
335	14.80				188181.
336	15.43				181085.
338	16.75				167805.
340	18.16				155645.
342	19.68				144497.
344	21.30				134269.
346	23.04				124875.
348	24.89				116239.
349	25.87				112183.
351	27.91				104558.
353	30.09				97532.
355	32.42				91052.
357	34.89				85071.
359	37.52				79546.
361	40.32				74437.
363	43.30				69710.
364	44.85				67479.
366	48.10				63264.
368	51.55				59356.
370	55.20				55729.
372	59.07				52361.
374	63.16				49231.
376	67.49				46320.
377	69.75				44941.
379	74.45				42325.
381	79.42				39889.
383	84.65				37617.
385	90.18				35496.
387	96.00				33517.
389	102.1				31667.
390	105.3				30788.

COMPOUND CONSTANTS:

MP	TC	VC	ZC
NBP	PC	DC	OM
MU	RG		

    VAPOR PRESSURE CORRELATION CONSTANTS, K AND KPA:

RPM2	RPM2	RPM2
272. - 333. K	272. - 333. K	272. - 333. K
13 POINTS	13 POINTS	13 POINTS
RMSD = 0.0000	RMSD = 0.0000	RMSD = 0.0000
0.17581976E+02	0.17581976E+02	0.17581976E+02
-0.54557925E+04	-0.54557925E+04	-0.54557925E+04
-0.54307219E-03	-0.54307219E-03	-0.54307219E-03
0.48989525E-06	0.48989525E-06	0.48989525E-06

    LIQUID DENSITY CORRELATION CONSTANTS, K AND G/ML:

    SECOND VIRIAL COEFFICIENT CORRELATION CONSTANTS, ML/MOL:

LITERATURE DOCUMENTS FOR CORRELATED VAPOR PRESSURE VALUES:

  13274

LITERATURE DOCUMENTS FOR CORRELATED LIQUID DENSITY VALUES:

LITERATURE DOCUMENTS REPORTING VIRIAL COEFFICIENT DATA:

T,K	VAPOR PRESSURE, KPA	SATURATED LIQUID VOLUME, ML/MOL	SECOND VIRIAL COEFFICIENT, ML/MOL	HEAT OF VAPORIZATION, J/MOL	SATURATED VAPOR VOLUME, ML/MOL
272	0.07520				0.3007E+08
273	0.08091				0.2805E+08
274	0.08701				0.2618E+08
276	0.1005				0.2284E+08
277	0.1079				0.2135E+08
278	0.1158				0.1997E+08
280	0.1331				0.1749E+08
281	0.1426				0.1638E+08
283	0.1635				0.1439E+08
284	0.1749				0.1350E+08
285	0.1871				0.1267E+08
287	0.2137				0.1117E+08
288	0.2282				0.1049E+08
290	0.2599				9276714.
291	0.2772				8728197.
292	0.2955				8215639.
294	0.3354				7288327.
295	0.3571				6869000.
296	0.3800				6476472.
298	0.4298				5764459.
299	0.4568				5441642.
301	0.5155				4854979.
302	0.5472				4588488.
303	0.5807				4338293.
305	0.6532				3882499.
306	0.6923				3674948.
308	0.7769				3296175.
309	0.8226				3123391.
310	0.8706				2960727.
312	0.9741				2663193.
313	1.030				2527158.
314	1.088				2398898.
316	1.214				2163784.
317	1.282				2056051.
319	1.427				1858249.
320	1.505				1767469.
321	1.587				1681665.
323	1.762				1523806.
324	1.856				1451207.
326	2.057				1317443.
327	2.165				1255834.
328	2.277				1197466.
330	2.518				1089719.
331	2.646				1039996.
332	2.780				992831.

COMPOUND CONSTANTS:

MP		TC		VC		ZC
NBP	490.027 K	PC		DC		OM
MU		RG				

VAPOR PRESSURE CORRELATION CONSTANTS, K AND KPA:

RPM2	RPM2	RPM2
377. - 488. K	377. - 488. K	377. - 488. K
4 POINTS	4 POINTS	4 POINTS
RMSD = 0.0021	RMSD = 0.0021	RMSD = 0.0021
-0.53089226E+02	-0.53089226E+02	-0.53089226E+02
0.31022259E+04	0.31022259E+04	0.31022259E+04
0.17381281E+00	0.17381281E+00	0.17381281E+00
-0.14074304E-03	-0.14074304E-03	-0.14074304E-03

LIQUID DENSITY CORRELATION CONSTANTS, K AND G/ML:

SECOND VIRIAL COEFFICIENT CORRELATION CONSTANTS, ML/MOL:

LITERATURE DOCUMENTS FOR CORRELATED VAPOR PRESSURE VALUES:

22755

LITERATURE DOCUMENTS FOR CORRELATED LIQUID DENSITY VALUES:

LITERATURE DOCUMENTS REPORTING VIRIAL COEFFICIENT DATA:

T,K	VAPOR PRESSURE, KPA	SATURATED LIQUID VOLUME, ML/MOL	SECOND VIRIAL COEFFICIENT, ML/MOL	HEAT OF VAPORIZATION, J/MOL	SATURATED VAPOR VOLUME, ML/MOL
377	1.941				1614827.
379	2.127				1481596.
382	2.436				1303646.
384	2.665				1198027.
387	3.045				1056735.
389	3.325				972739.
392	3.789				860190.
394	4.130				793169.
397	4.694				703217.
399	5.107				649561.
402	5.788				577424.
404	6.287				534320.
407	7.105				476269.
409	7.702				441520.
412	8.680				394640.
414	9.391				366527.
417	10.55				328533.
419	11.40				305707.
422	12.77				274803.
424	13.76				256202.
427	15.37				230973.
429	16.53				215761.
432	18.41				195089.
435	20.47				176715.
437	21.94				165603.
440	24.31				150459.
442	26.01				141285.
445	28.74				128758.
447	30.68				121155.
450	33.78				110756.
452	35.99				104433.
455	39.50				95768.
457	41.99				90491.
460	45.94				83247.
462	48.73				78826.
465	53.14				72748.
467	56.25				69033.
470	61.14				63917.
472	64.56				60784.
475	69.95				56463.
477	73.70				53814.
480	79.57				50153.
482	83.66				47905.
485	90.02				44795.
487	94.42				42882.

COMPOUND CONSTANTS:

MP   291.543 K   TC   VC   ZC
NBP 510.208 K   PC   DC   OM
MU   2.3400 DEBYE   RG

   VAPOR PRESSURE CORRELATION CONSTANTS, K AND KPA:

   RPM2   RPM2   RPM2
   415. - 511. K   415. - 511. K   415. - 511. K
     7 POINTS   7 POINTS   7 POINTS
   RMSD = 0.0428   RMSD = 0.0428   RMSD = 0.0428
   0.63746621E+02   0.63746621E+02   0.63746621E+02
   -0.13829575E+05   -0.13829575E+05   -0.13829575E+05
   -0.95899972E-01   -0.95899972E-01   -0.95899972E-01
   0.64946645E-04   0.64946645E-04   0.64946645E-04

   LIQUID DENSITY CORRELATION CONSTANTS, K AND G/ML:

   FRANCIS1   FRANCIS1   FRANCIS1
   293. - 299. K   293. - 299. K   293. - 299. K
     2 POINTS   2 POINTS   2 POINTS
   RMSD = 0.00000   RMSD = 0.00000   RMSD = 0.00000
   0.14689927E+01   0.14689927E+01   0.14689927E+01
   0.94420500E-03   0.94420500E-03   0.94420500E-03

   SECOND VIRIAL COEFFICIENT CORRELATION CONSTANTS, ML/MOL:

LITERATURE DOCUMENTS FOR CORRELATED VAPOR PRESSURE VALUES:

   1361

LITERATURE DOCUMENTS FOR CORRELATED LIQUID DENSITY VALUES:

   1360

LITERATURE DOCUMENTS REPORTING VIRIAL COEFFICIENT DATA:

T,K	VAPOR PRESSURE, KPA	SATURATED LIQUID VOLUME, ML/MOL	SECOND VIRIAL COEFFICIENT, ML/MOL	HEAT OF VAPORIZATION, J/MOL	SATURATED VAPOR VOLUME, ML/MOL
293		129.66			
297		130.07			
416	6.344				545229.
421	7.649				457615.
426	9.168				386345.
431	10.93				327993.
436	12.95				279922.
441	15.27				240085.
446	17.92				206885.
451	20.94				179066.
456	24.36				155636.
461	28.22				135803.
466	32.58				118936.
471	37.47				104526.
476	42.94				92162.
481	49.06				81509.
486	55.89				72294.
491	63.50				64292.
496	71.95				57319.
501	81.33				51220.
506	91.72				45869.
510	100.8				42054.

COMPOUND CONSTANTS:

MP   TC   VC   ZC
NBP   PC   DC   OM
MU   RG

VAPOR PRESSURE CORRELATION CONSTANTS, K AND KPA:

LIQUID DENSITY CORRELATION CONSTANTS, K AND G/ML:

FRANCIS1   FRANCIS1   FRANCIS1
296. - 354. K   296. - 354. K   296. - 354. K
  4 POINTS   4 POINTS   4 POINTS
RMSD = 0.00018   RMSD = 0.00018   RMSD = 0.00018
0.16353292E+01   0.16353292E+01   0.16353292E+01
0.10515784E-02   0.10515784E-02   0.10515784E-02
0.53056593E+01   0.53056593E+01   0.53056593E+01
0.85100244E+03   0.85100244E+03   0.85100244E+03

SECOND VIRIAL COEFFICIENT CORRELATION CONSTANTS, ML/MOL:

LITERATURE DOCUMENTS FOR CORRELATED VAPOR PRESSURE VALUES:

LITERATURE DOCUMENTS FOR CORRELATED LIQUID DENSITY VALUES:

  14195

LITERATURE DOCUMENTS REPORTING VIRIAL COEFFICIENT DATA:

T,K	VAPOR PRESSURE, KPA	SATURATED LIQUID VOLUME, ML/MOL	SECOND VIRIAL COEFFICIENT, ML/MOL	HEAT OF VAPORIZATION, J/MOL	SATURATED VAPOR VOLUME, ML/MOL
296		188.00			
297		188.16			
298		188.31			
299		188.46			
301		188.77			
302		188.93			
303		189.08			
305		189.39			
306		189.55			
307		189.70			
309		190.01			
310		190.17			
311		190.33			
313		190.64			
314		190.80			
315		190.96			
317		191.27			
318		191.43			
319		191.59			
321		191.91			
322		192.07			
323		192.23			
324		192.39			
326		192.71			
327		192.87			
328		193.03			
330		193.35			
331		193.52			
332		193.68			
334		194.00			
335		194.17			
336		194.33			
338		194.66			
339		194.82			
340		194.99			
342		195.32			
343		195.49			
344		195.65			
346		195.98			
347		196.15			
348		196.32			
350		196.65			
351		196.82			
352		196.99			
353		197.16			

C(16)H(19)BR(1)O(2)

COMPOUND CONSTANTS:

MP	TC	VC	ZC
NBP	PC	DC	OM
MU	RG		

VAPOR PRESSURE CORRELATION CONSTANTS, K AND KPA:

LIQUID DENSITY CORRELATION CONSTANTS, K AND G/ML:

FRANCIS1	FRANCIS1	FRANCIS1
367. - 427. K	367. - 427. K	367. - 427. K
4 POINTS	4 POINTS	4 POINTS
RMSD = 0.00414	RMSD = 0.00414	RMSD = 0.00414
0.15772915E+01	0.15772915E+01	0.15772915E+01
0.82694599E-03	0.82694599E-03	0.82694599E-03
0.15674782E+03	0.15674782E+03	0.15674782E+03
0.31303765E+04	0.31303765E+04	0.31303765E+04

SECOND VIRIAL COEFFICIENT CORRELATION CONSTANTS, ML/MOL:

LITERATURE DOCUMENTS FOR CORRELATED VAPOR PRESSURE VALUES:

LITERATURE DOCUMENTS FOR CORRELATED LIQUID DENSITY VALUES:

   14195

LITERATURE DOCUMENTS REPORTING VIRIAL COEFFICIENT DATA:

T,K	VAPOR PRESSURE, KPA	SATURATED LIQUID VOLUME, ML/MOL	SECOND VIRIAL COEFFICIENT, ML/MOL	HEAT OF VAPORIZATION, J/MOL	SATURATED VAPOR VOLUME, ML/MOL
367		265.58			
368		265.76			
369		265.95			
371		266.32			
372		266.51			
373		266.69			
375		267.07			
376		267.25			
377		267.44			
379		267.82			
380		268.00			
381		268.19			
383		268.57			
384		268.76			
386		269.14			
387		269.33			
388		269.52			
390		269.90			
391		270.09			
392		270.28			
394		270.67			
395		270.86			
396		271.05			
398		271.44			
399		271.63			
401		272.02			
402		272.21			
403		272.41			
405		272.80			
406		272.99			
407		273.19			
409		273.58			
410		273.78			
411		273.97			
413		274.37			
414		274.57			
416		274.96			
417		275.16			
418		275.36			
420		275.76			
421		275.96			
422		276.16			
424		276.56			
425		276.76			
426		276.96			

COMPOUND CONSTANTS:

MP	TC	VC	ZC
NBP	PC	DC	OM
MU	RG		

VAPOR PRESSURE CORRELATION CONSTANTS, K AND KPA:

LIQUID DENSITY CORRELATION CONSTANTS, K AND G/ML:

FRANCIS1	FRANCIS1	FRANCIS1
346. - 409. K	346. - 409. K	346. - 409. K
4 POINTS	4 POINTS	4 POINTS
RMSD = 0.00082	RMSD = 0.00082	RMSD = 0.00082
0.17867718E+01	0.17867718E+01	0.17867718E+01
0.96267578E-03	0.96267578E-03	0.96267578E-03
0.19923569E+02	0.19923569E+02	0.19923569E+02
0.15690449E+04	0.15690449E+04	0.15690449E+04

SECOND VIRIAL COEFFICIENT CORRELATION CONSTANTS, ML/MOL:

LITERATURE DOCUMENTS FOR CORRELATED VAPOR PRESSURE VALUES:

LITERATURE DOCUMENTS FOR CORRELATED LIQUID DENSITY VALUES:

14195

LITERATURE DOCUMENTS REPORTING VIRIAL COEFFICIENT DATA:

T,K	VAPOR PRESSURE, KPA	SATURATED LIQUID VOLUME, ML/MOL	SECOND VIRIAL COEFFICIENT, ML/MOL	HEAT OF VAPORIZATION, J/MOL	SATURATED VAPOR VOLUME, ML/MOL
346		279.76			
347		279.95			
348		280.14			
350		280.52			
351		280.71			
353		281.10			
354		281.29			
356		281.67			
357		281.86			
358		282.06			
360		282.44			
361		282.64			
363		283.03			
364		283.22			
366		283.61			
367		283.81			
368		284.00			
370		284.40			
371		284.59			
373		284.99			
374		285.18			
376		285.58			
377		285.78			
378		285.98			
380		286.37			
381		286.57			
383		286.97			
384		287.17			
386		287.57			
387		287.77			
388		287.98			
390		288.38			
391		288.58			
393		288.99			
394		289.19			
396		289.60			
397		289.80			
398		290.01			
400		290.41			
401		290.62			
403		291.03			
404		291.24			
406		291.65			
407		291.86			
408		292.06			

C(16)H(17)BR(1)O(3)

COMPOUND CONSTANTS:

MP	TC	VC	ZC
NBP	PC	DC	OM
MU	RG		

VAPOR PRESSURE CORRELATION CONSTANTS, K AND KPA:

LIQUID DENSITY CORRELATION CONSTANTS, K AND G/ML:

FRANCIS1	FRANCIS1	FRANCIS1
367. - 423. K	367. - 423. K	367. - 423. K
4 POINTS	4 POINTS	4 POINTS
RMSD = 0.00085	RMSD = 0.00085	RMSD = 0.00085
0.15926456E+01	0.15926456E+01	0.15926456E+01
0.86395908E-03	0.86395908E-03	0.86395908E-03
0.50203278E+02	0.50203278E+02	0.50203278E+02
0.44107930E+04	0.44107930E+04	0.44107930E+04

SECOND VIRIAL COEFFICIENT CORRELATION CONSTANTS, ML/MOL:

LITERATURE DOCUMENTS FOR CORRELATED VAPOR PRESSURE VALUES:

LITERATURE DOCUMENTS FOR CORRELATED LIQUID DENSITY VALUES:

14195

LITERATURE DOCUMENTS REPORTING VIRIAL COEFFICIENT DATA:

T,K	VAPOR PRESSURE, KPA	SATURATED LIQUID VOLUME, ML/MOL	SECOND VIRIAL COEFFICIENT, ML/MOL	HEAT OF VAPORIZATION, J/MOL	SATURATED VAPOR VOLUME, ML/MOL
367		266.96			
368		267.14			
369		267.33			
370		267.51			
372		267.88			
373		268.06			
374		268.25			
375		268.43			
377		268.81			
378		268.99			
379		269.18			
380		269.36			
382		269.74			
383		269.92			
384		270.11			
386		270.49			
387		270.68			
388		270.86			
389		271.05			
391		271.43			
392		271.62			
393		271.81			
394		272.00			
396		272.38			
397		272.57			
398		272.76			
400		273.15			
401		273.34			
402		273.53			
403		273.72			
405		274.11			
406		274.30			
407		274.50			
408		274.69			
410		275.08			
411		275.27			
412		275.47			
414		275.86			
415		276.06			
416		276.25			
417		276.45			
419		276.84			
420		277.04			
421		277.24			
422		277.43			

COMPOUND CONSTANTS:

MP		TC		VC		ZC
NBP	416.551 K	PC		DC		OM
MU		RG				

   VAPOR PRESSURE CORRELATION CONSTANTS, K AND KPA:

RPM2	RPM2	RPM2
372. - 416. K	372. - 416. K	372. - 416. K
13 POINTS	13 POINTS	13 POINTS
RMSD = 0.8504	RMSD = 0.8504	RMSD = 0.8504
-0.37836491E+03	-0.37836491E+03	-0.37836491E+03
0.46708999E+05	0.46708999E+05	0.46708999E+05
0.10059766E+01	0.10059766E+01	0.10059766E+01
-0.85404877E-03	-0.85404877E-03	-0.85404877E-03

   LIQUID DENSITY CORRELATION CONSTANTS, K AND G/ML:

   SECOND VIRIAL COEFFICIENT CORRELATION CONSTANTS, ML/MOL:

LITERATURE DOCUMENTS FOR CORRELATED VAPOR PRESSURE VALUES:

   11752

LITERATURE DOCUMENTS FOR CORRELATED LIQUID DENSITY VALUES:

LITERATURE DOCUMENTS REPORTING VIRIAL COEFFICIENT DATA:

T,K	VAPOR PRESSURE, KPA	SATURATED LIQUID VOLUME, ML/MOL	SECOND VIRIAL COEFFICIENT, ML/MOL	HEAT OF VAPORIZATION, J/MOL	SATURATED VAPOR VOLUME, ML/MOL
372	25.37				121919.
373	26.22				118269.
374	27.11				114716.
375	28.02				111261.
376	28.97				107902.
377	29.96				104639.
378	30.97				101470.
379	32.03				98394.
380	33.11				95410.
381	34.24				92517.
382	35.40				89712.
383	36.60				86995.
384	37.85				84363.
385	39.13				81815.
386	40.45				79349.
387	41.81				76963.
388	43.21				74656.
389	44.66				72425.
390	46.15				70269.
391	47.68				68185.
392	49.25				66173.
393	50.87				64230.
394	52.54				62354.
395	54.25				60543.
396	56.00				58796.
397	57.80				57111.
398	59.64				55485.
399	61.53				53918.
400	63.46				52406.
401	65.44				50950.
402	67.46				49547.
403	69.52				48195.
404	71.63				46893.
405	73.78				45639.
406	75.97				44431.
407	78.21				43269.
408	80.48				42151.
409	82.79				41074.
410	85.14				40039.
411	87.52				39044.
412	89.94				38087.
413	92.39				37166.
414	94.87				36282.
415	97.38				35433.
416	99.92				34617.

COMPOUND CONSTANTS:

MP		TC		VC		ZC
NBP	462.775 K	PC		DC		OM
MU		RG				

VAPOR PRESSURE CORRELATION CONSTANTS, K AND KPA:

RPM2	RPM2	RPM2
382. - 463. K	382. - 463. K	382. - 463. K
5 POINTS	5 POINTS	5 POINTS
RMSD = 0.0078	RMSD = 0.0078	RMSD = 0.0078
-0.36593305E+02	-0.36593305E+02	-0.36593305E+02
0.11972196E+04	0.11972196E+04	0.11972196E+04
0.13428023E+00	0.13428023E+00	0.13428023E+00
-0.10980973E-03	-0.10980973E-03	-0.10980973E-03

LIQUID DENSITY CORRELATION CONSTANTS, K AND G/ML:

FRANCIS1	FRANCIS1	FRANCIS1
293. - 299. K	293. - 299. K	293. - 299. K
2 POINTS	2 POINTS	2 POINTS
RMSD = 0.00000	RMSD = 0.00000	RMSD = 0.00000
0.12291625E+01	0.12291625E+01	0.12291625E+01
0.82216200E-03	0.82216200E-03	0.82216200E-03

SECOND VIRIAL COEFFICIENT CORRELATION CONSTANTS, ML/MOL:

LITERATURE DOCUMENTS FOR CORRELATED VAPOR PRESSURE VALUES:

1361

LITERATURE DOCUMENTS FOR CORRELATED LIQUID DENSITY VALUES:

1360

LITERATURE DOCUMENTS REPORTING VIRIAL COEFFICIENT DATA:

T,K	VAPOR PRESSURE, KPA	SATURATED LIQUID VOLUME, ML/MOL	SECOND VIRIAL COEFFICIENT, ML/MOL	HEAT OF VAPORIZATION, J/MOL	SATURATED VAPOR VOLUME, ML/MOL
293		103.35			
296		103.60			
385	6.943				461063.
389	8.190				394926.
393	9.633				339209.
397	11.30				292164.
401	13.21				252349.
405	15.41				218576.
408	17.25				196612.
412	20.01				171149.
416	23.15				149412.
420	26.70				130813.
424	30.69				114862.
428	35.18				101152.
432	40.20				89340.
435	44.35				81553.
439	50.41				72404.
443	57.13				64473.
447	64.54				57583.
451	72.69				51583.
455	81.62				46349.
459	91.36				41772.
462	99.22				38715.

COMPOUND CONSTANTS:

MP		TC		VC		ZC	
NBP	416.404 K	PC		DC		OM	
MU		RG					

VAPOR PRESSURE CORRELATION CONSTANTS, K AND KPA:

RPM2	RPM2	RPM2
317. - 419. K	317. - 419. K	317. - 419. K
14 POINTS	14 POINTS	14 POINTS
RMSD = 0.1870	RMSD = 0.1870	RMSD = 0.1870
0.20595620E+02	0.20595620E+02	0.20595620E+02
-0.56476721E+04	-0.56476721E+04	-0.56476721E+04
-0.66797188E-02	-0.66797188E-02	-0.66797188E-02
0.21173870E-05	0.21173870E-05	0.21173870E-05

LIQUID DENSITY CORRELATION CONSTANTS, K AND G/ML:

SECOND VIRIAL COEFFICIENT CORRELATION CONSTANTS, ML/MOL:

LITERATURE DOCUMENTS FOR CORRELATED VAPOR PRESSURE VALUES:

7045

LITERATURE DOCUMENTS FOR CORRELATED LIQUID DENSITY VALUES:

LITERATURE DOCUMENTS REPORTING VIRIAL COEFFICIENT DATA:

T,K	VAPOR PRESSURE, KPA	SATURATED LIQUID VOLUME, ML/MOL	SECOND VIRIAL COEFFICIENT, ML/MOL	HEAT OF VAPORIZATION, J/MOL	SATURATED VAPOR VOLUME, ML/MOL
317	2.399				1098794.
319	2.654				999470.
321	2.932				910339.
323	3.235				830244.
326	3.739				724858.
328	4.112				663180.
330	4.517				607495.
333	5.187				533810.
335	5.679				490441.
337	6.211				451114.
340	7.089				398796.
342	7.730				367840.
344	8.421				339653.
347	9.554				301967.
349	10.38				279557.
351	11.26				259072.
354	12.71				231556.
356	13.76				215116.
358	14.88				200034.
361	16.70				179685.
363	18.02				167474.
365	19.42				156233.
367	20.92				145875.
370	23.34				131825.
372	25.07				123349.
374	26.92				115515.
377	29.90				104846.
379	32.03				98385.
381	34.29				92394.
384	37.92				84207.
386	40.51				79231.
388	43.24				74605.
391	47.63				68259.
393	50.75				64389.
395	54.03				60781.
398	59.29				55817.
400	63.01				52779.
402	66.93				49940.
405	73.17				46021.
407	77.58				43616.
409	82.21				41363.
412	89.57				38243.
414	94.77				36323.
416	100.2				34520.
418	105.9				32825.

COMPOUND CONSTANTS:

MP		TC		VC		ZC	
NBP	458.142 K	PC		DC		OM	
MU		RG					

VAPOR PRESSURE CORRELATION CONSTANTS, K AND KPA:

RPM2	RPM2	RPM2
317. - 459. K	317. - 459. K	317. - 459. K
8 POINTS	8 POINTS	8 POINTS
RMSD = 0.0858	RMSD = 0.0858	RMSD = 0.0858
0.41970324E+02	0.41970324E+02	0.41970324E+02
-0.99808633E+04	-0.99808633E+04	-0.99808633E+04
-0.52226934E-01	-0.52226934E-01	-0.52226934E-01
0.39833865E-04	0.39833865E-04	0.39833865E-04

LIQUID DENSITY CORRELATION CONSTANTS, K AND G/ML:

SECOND VIRIAL COEFFICIENT CORRELATION CONSTANTS, ML/MOL:

LITERATURE DOCUMENTS FOR CORRELATED VAPOR PRESSURE VALUES:

1146

LITERATURE DOCUMENTS FOR CORRELATED LIQUID DENSITY VALUES:

LITERATURE DOCUMENTS REPORTING VIRIAL COEFFICIENT DATA:

T,K	VAPOR PRESSURE, KPA	SATURATED LIQUID VOLUME, ML/MOL	SECOND VIRIAL COEFFICIENT, ML/MOL	HEAT OF VAPORIZATION, J/MOL	SATURATED VAPOR VOLUME, ML/MOL
317	0.1264				0.2085E+08
320	0.1567				0.1698E+08
323	0.1933				0.1390E+08
326	0.2373				0.1142E+08
329	0.2901				9430436.
333	0.3766				7352547.
336	0.4558				6129649.
339	0.5494				5130321.
342	0.6597				4310333.
346	0.8371				3436807.
349	0.9965				2911999.
352	1.182				2475822.
355	1.398				2111999.
358	1.647				1807464.
362	2.040				1475743.
365	2.386				1272015.
368	2.783				1099609.
371	3.236				953256.
375	3.940				791310.
378	4.553				690295.
381	5.247				603711.
384	6.032				529292.
387	6.917				465159.
391	8.272				392984.
394	9.435				347206.
397	10.74				307425.
400	12.19				272774.
404	14.40				233305.
407	16.27				207973.
410	18.35				185743.
413	20.66				166194.
417	24.13				143687.
420	27.05				129089.
423	30.28				116166.
426	33.83				104705.
429	37.74				94522.
433	43.55				82661.
436	48.41				74880.
439	53.74				67926.
442	59.56				61702.
446	68.18				54392.
449	75.33				49557.
452	83.13				45207.
455	91.63				41288.
458	100.9				37752.

COMPOUND CONSTANTS:

MP	TC	VC	ZC
NBP	PC	DC	OM
MU	RG		

VAPOR PRESSURE CORRELATION CONSTANTS, K AND KPA:

RPM2	RPM2	RPM2
277. - 333. K	277. - 333. K	277. - 333. K
11 POINTS	11 POINTS	11 POINTS
RMSD = 0.0000	RMSD = 0.0000	RMSD = 0.0000
0.31404936E+02	0.31404936E+02	0.31404936E+02
-0.87815700E+04	-0.87815700E+04	-0.87815700E+04
-0.32014182E-01	-0.32014182E-01	-0.32014182E-01
0.34129310E-04	0.34129310E-04	0.34129310E-04

LIQUID DENSITY CORRELATION CONSTANTS, K AND G/ML:

SECOND VIRIAL COEFFICIENT CORRELATION CONSTANTS, ML/MOL:

LITERATURE DOCUMENTS FOR CORRELATED VAPOR PRESSURE VALUES:

  13274

LITERATURE DOCUMENTS FOR CORRELATED LIQUID DENSITY VALUES:

LITERATURE DOCUMENTS REPORTING VIRIAL COEFFICIENT DATA:

T,K	VAPOR PRESSURE, KPA	SATURATED LIQUID VOLUME, ML/MOL	SECOND VIRIAL COEFFICIENT, ML/MOL	HEAT OF VAPORIZATION, J/MOL	SATURATED VAPOR VOLUME, ML/MOL
277	0.001435				0.1605E+10
278	0.001587				0.1456E+10
279	0.001755				0.1322E+10
280	0.001938				0.1201E+10
282	0.002360				0.9937E+09
283	0.002601				0.9048E+09
284	0.002864				0.8243E+09
285	0.003153				0.7516E+09
287	0.003811				0.6261E+09
288	0.004186				0.5720E+09
289	0.004595				0.5229E+09
290	0.005041				0.4783E+09
292	0.006054				0.4010E+09
293	0.006628				0.3676E+09
294	0.007252				0.3371E+09
296	0.008665				0.2840E+09
297	0.009463				0.2609E+09
298	0.01033				0.2399E+09
299	0.01127				0.2207E+09
301	0.01338				0.1870E+09
302	0.01457				0.1723E+09
303	0.01586				0.1589E+09
304	0.01725				0.1465E+09
306	0.02037				0.1249E+09
307	0.02212				0.1154E+09
308	0.02401				0.1067E+09
310	0.02823				0.9129E+08
311	0.03059				0.8452E+08
312	0.03313				0.7830E+08
313	0.03586				0.7256E+08
315	0.04196				0.6242E+08
316	0.04535				0.5793E+08
317	0.04899				0.5380E+08
318	0.05290				0.4998E+08
320	0.06160				0.4319E+08
321	0.06642				0.4018E+08
322	0.07158				0.3740E+08
324	0.08303				0.3244E+08
325	0.08937				0.3024E+08
326	0.09615				0.2819E+08
327	0.1034				0.2630E+08
329	0.1194				0.2291E+08
330	0.1282				0.2140E+08
331	0.1377				0.1999E+08
332	0.1477				0.1869E+08

COMPOUND CONSTANTS:

MP	226.177 K	TC		VC		ZC
NBP	441.076 K	PC		DC		OM
MU	3.2400 DEBYE	RG				

VAPOR PRESSURE CORRELATION CONSTANTS, K AND KPA:

RPM2	RPM2	RPM2
310. - 442. K	310. - 442. K	310. - 442. K
15 POINTS	15 POINTS	15 POINTS
RMSD = 0.0022	RMSD = 0.0022	RMSD = 0.0022
0.21864140E+02	0.21864140E+02	0.21864140E+02
-0.64159922E+04	-0.64159922E+04	-0.64159922E+04
-0.77320691E-02	-0.77320691E-02	-0.77320691E-02
0.36538030E-05	0.36538030E-05	0.36538030E-05

LIQUID DENSITY CORRELATION CONSTANTS, K AND G/ML:

FRANCIS1	FRANCIS1	FRANCIS1
293. - 394. K	293. - 394. K	293. - 394. K
6 POINTS	6 POINTS	6 POINTS
RMSD = 0.00018	RMSD = 0.00018	RMSD = 0.00018
0.12066355E+01	0.12066355E+01	0.12066355E+01
0.88522420E-03	0.88522420E-03	0.88522420E-03
0.25476007E+01	0.25476007E+01	0.25476007E+01
0.59638232E+03	0.59638232E+03	0.59638232E+03

SECOND VIRIAL COEFFICIENT CORRELATION CONSTANTS, ML/MOL:

LITERATURE DOCUMENTS FOR CORRELATED VAPOR PRESSURE VALUES:

   1146   6962

LITERATURE DOCUMENTS FOR CORRELATED LIQUID DENSITY VALUES:

   11773

LITERATURE DOCUMENTS REPORTING VIRIAL COEFFICIENT DATA:

T,K	VAPOR PRESSURE, KPA	SATURATED LIQUID VOLUME, ML/MOL	SECOND VIRIAL COEFFICIENT, ML/MOL	HEAT OF VAPORIZATION, J/MOL	SATURATED VAPOR VOLUME, ML/MOL
293		123.72			
296		124.09			
299		124.45			
303		124.94			
306		125.31			
309		125.68			
313	0.4984	126.18		48909.	5221858.
316	0.5956	126.56		48841.	4410945.
320	0.7513	127.07		48751.	3541335.
323	0.8906	127.45		48683.	3015484.
326	1.052	127.84		48616.	2576141.
330	1.307	128.36		48525.	2098602.
333	1.533	128.75		48456.	1805962.
337	1.887	129.28		48365.	1485094.
340	2.197	129.68		48297.	1286762.
343	2.551	130.09		48228.	1118037.
347	3.099	130.63		48136.	930892.
350	3.575	131.04		48067.	813888.
353	4.114	131.45		47998.	713410.
357	4.941	132.00		47906.	600779.
360	5.652	132.42		47836.	529623.
364	6.736	132.99		47743.	449311.
367	7.662	133.42		47674.	398227.
370	8.697	133.85		47604.	353729.
374	10.26	134.43		47510.	303044.
377	11.59	134.87		47440.	270513.
381	13.58	135.46		47346.	233236.
384	15.26	135.91		47275.	209169.
387	17.12	136.36		47204.	187945.
391	19.89	136.96		47109.	163430.
394	22.21	137.42		47038.	147478.
397	24.76				133318.
401	28.54				116841.
404	31.68				106041.
408	36.31				93414.
411	40.16				85098.
414	44.33				77642.
418	50.46				68868.
421	55.52				63052.
425	62.90				56180.
428	68.96				51605.
431	75.49				47467.
435	85.00				42549.
438	92.77				39257.
441	101.1				36266.

COMPOUND CONSTANTS:

MP 467.190 K	TC	VC	ZC
NBP 723.129 K	PC	DC	OM
MU	RG		

    VAPOR PRESSURE CORRELATION CONSTANTS, K AND KPA:

RPM2	RPM2	RPM2
474. - 724. K	474. - 724. K	474. - 724. K
7 POINTS	7 POINTS	7 POINTS
RMSD = 0.0390	RMSD = 0.0390	RMSD = 0.0390
-0.22362756E+01	-0.22362756E+01	-0.22362756E+01
-0.64738203E+04	-0.64738203E+04	-0.64738203E+04
0.43962829E-01	0.43962829E-01	0.43962829E-01
-0.30566467E-04	-0.30566467E-04	-0.30566467E-04

    LIQUID DENSITY CORRELATION CONSTANTS, K AND G/ML:

    SECOND VIRIAL COEFFICIENT CORRELATION CONSTANTS, ML/MOL:

LITERATURE DOCUMENTS FOR CORRELATED VAPOR PRESSURE VALUES:

   11052

LITERATURE DOCUMENTS FOR CORRELATED LIQUID DENSITY VALUES:

LITERATURE DOCUMENTS REPORTING VIRIAL COEFFICIENT DATA:

T,K	VAPOR PRESSURE, KPA	SATURATED LIQUID VOLUME, ML/MOL	SECOND VIRIAL COEFFICIENT, ML/MOL	HEAT OF VAPORIZATION, J/MOL	SATURATED VAPOR VOLUME, ML/MOL
474	0.1461				0.2697E+08
479	0.1815				0.2194E+08
485	0.2340				0.1723E+08
491	0.2998				0.1361E+08
496	0.3669				0.1124E+08
502	0.4649				8977292.
508	0.5857				7211810.
513	0.7068				6034747.
519	0.8811				4897682.
525	1.092				3996393.
530	1.301				3387065.
536	1.597				2790483.
542	1.950				2310741.
547	2.294				1982168.
553	2.776				1656408.
559	3.342				1390902.
564	3.886				1206839.
570	4.636				1022202.
576	5.506				869810.
581	6.331				762962.
587	7.456				654598.
593	8.740				564103.
598	9.944				499977.
604	11.56				434273.
610	13.39				378797.
616	15.44				331785.
621	17.32				298030.
627	19.82				263000.
633	22.59				233018.
638	25.10				211298.
644	28.40				188563.
650	31.99				168925.
655	35.23				154582.
661	39.41				139449.
667	43.92				126269.
672	47.93				116570.
678	53.05				106265.
684	58.50				97223.
689	63.28				90525.
695	69.32				83365.
701	75.65				77042.
706	81.15				72332.
712	88.00				67272.
718	95.09				62780.
723	101.2				59420.

COMPOUND CONSTANTS:

MP	TC	VC	ZC
NBP	PC	DC	OM
MU	RG		

    VAPOR PRESSURE CORRELATION CONSTANTS, K AND KPA:

RPM2	RPM2	RPM2
410. - 496. K	410. - 496. K	410. - 496. K
22 POINTS	22 POINTS	22 POINTS
RMSD = 0.0010	RMSD = 0.0010	RMSD = 0.0010
0.18832730E+03	0.18832730E+03	0.18832730E+03
-0.38592193E+05	-0.38592193E+05	-0.38592193E+05
-0.36089778E+00	-0.36089778E+00	-0.36089778E+00
0.26967276E-03	0.26967276E-03	0.26967276E-03

    LIQUID DENSITY CORRELATION CONSTANTS, K AND G/ML:

    SECOND VIRIAL COEFFICIENT CORRELATION CONSTANTS, ML/MOL:

LITERATURE DOCUMENTS FOR CORRELATED VAPOR PRESSURE VALUES:

  10550

LITERATURE DOCUMENTS FOR CORRELATED LIQUID DENSITY VALUES:

LITERATURE DOCUMENTS REPORTING VIRIAL COEFFICIENT DATA:

T,K	VAPOR PRESSURE, KPA	SATURATED LIQUID VOLUME, ML/MOL	SECOND VIRIAL COEFFICIENT, ML/MOL	HEAT OF VAPORIZATION, J/MOL	SATURATED VAPOR VOLUME, ML/MOL
410	0.0002169				0.1572E+11
411	0.0002372				0.1441E+11
413	0.0002832				0.1212E+11
415	0.0003374				0.1023E+11
417	0.0004012				0.8643E+10
419	0.0004759				0.7320E+10
421	0.0005634				0.6213E+10
423	0.0006657				0.5283E+10
425	0.0007850				0.4501E+10
427	0.0009241				0.3842E+10
429	0.001086				0.3285E+10
431	0.001273				0.2814E+10
433	0.001491				0.2414E+10
435	0.001743				0.2075E+10
437	0.002034				0.1786E+10
439	0.002371				0.1540E+10
441	0.002759				0.1329E+10
443	0.003205				0.1149E+10
445	0.003719				0.9949E+09
447	0.004309				0.8624E+09
449	0.004987				0.7486E+09
451	0.005764				0.6506E+09
452	0.006194				0.6068E+09
454	0.007146				0.5283E+09
456	0.008235				0.4604E+09
458	0.009479				0.4017E+09
460	0.01090				0.3509E+09
462	0.01252				0.3068E+09
464	0.01437				0.2685E+09
466	0.01648				0.2352E+09
468	0.01887				0.2062E+09
470	0.02160				0.1809E+09
472	0.02470				0.1589E+09
474	0.02823				0.1396E+09
476	0.03223				0.1228E+09
478	0.03678				0.1080E+09
480	0.04194				0.9515E+08
482	0.04780				0.8384E+08
484	0.05444				0.7392E+08
486	0.06196				0.6521E+08
488	0.07049				0.5756E+08
490	0.08016				0.5083E+08
492	0.09110				0.4490E+08
494	0.1035				0.3968E+08
495	0.1103				0.3731E+08

COMPOUND CONSTANTS:

MP	TC	VC	ZC
NBP	PC	DC	OM
MU	RG		

VAPOR PRESSURE CORRELATION CONSTANTS, K AND KPA:

RPM2	RPM2	RPM2
373. - 457. K	373. - 457. K	373. - 457. K
20 POINTS	20 POINTS	20 POINTS
RMSD = 0.0005	RMSD = 0.0005	RMSD = 0.0005
-0.35975370E+00	-0.35975370E+00	-0.35975370E+00
-0.10541719E+05	-0.10541719E+05	-0.10541719E+05
0.76920902E-01	0.76920902E-01	0.76920902E-01
-0.68264351E-04	-0.68264351E-04	-0.68264351E-04

LIQUID DENSITY CORRELATION CONSTANTS, K AND G/ML:

SECOND VIRIAL COEFFICIENT CORRELATION CONSTANTS, ML/MOL:

LITERATURE DOCUMENTS FOR CORRELATED VAPOR PRESSURE VALUES:

10550

LITERATURE DOCUMENTS FOR CORRELATED LIQUID DENSITY VALUES:

LITERATURE DOCUMENTS REPORTING VIRIAL COEFFICIENT DATA:

T,K	VAPOR PRESSURE, KPA	SATURATED LIQUID VOLUME, ML/MOL	SECOND VIRIAL COEFFICIENT, ML/MOL	HEAT OF VAPORIZATION, J/MOL	SATURATED VAPOR VOLUME, ML/MOL
373	0.0000805				0.3855E+11
374	0.0000891				0.3492E+11
376	0.0001089				0.2870E+11
378	0.0001329				0.2364E+11
380	0.0001619				0.1952E+11
382	0.0001968				0.1614E+11
384	0.0002387				0.1338E+11
386	0.0002889				0.1111E+11
388	0.0003489				0.9245E+10
390	0.0004207				0.7708E+10
392	0.0005062				0.6439E+10
393	0.0005548				0.5890E+10
395	0.0006656				0.4934E+10
397	0.0007970				0.4142E+10
399	0.0009525				0.3483E+10
401	0.001136				0.2934E+10
403	0.001353				0.2477E+10
405	0.001608				0.2094E+10
407	0.001908				0.1774E+10
409	0.002259				0.1505E+10
411	0.002670				0.1280E+10
413	0.003151				0.1090E+10
414	0.003421				0.1006E+10
416	0.004026				0.8590E+09
418	0.004731				0.7346E+09
420	0.005549				0.6293E+09
422	0.006498				0.5400E+09
424	0.007596				0.4641E+09
426	0.008865				0.3995E+09
428	0.01033				0.3445E+09
430	0.01202				0.2975E+09
432	0.01396				0.2574E+09
434	0.01618				0.2230E+09
435	0.01742				0.2077E+09
437	0.02015				0.1803E+09
439	0.02327				0.1568E+09
441	0.02684				0.1366E+09
443	0.03091				0.1192E+09
445	0.03553				0.1041E+09
447	0.04079				0.9110E+08
449	0.04676				0.7983E+08
451	0.05353				0.7005E+08
453	0.06118				0.6156E+08
455	0.06982				0.5418E+08
456	0.07455				0.5085E+08

COMPOUND CONSTANTS:

MP	TC	VC	ZC
NBP	PC	DC	OM
MU	RG		

VAPOR PRESSURE CORRELATION CONSTANTS, K AND KPA:

LIQUID DENSITY CORRELATION CONSTANTS, K AND G/ML:

FRANCIS1	FRANCIS1	FRANCIS1
289. - 352. K	289. - 352. K	289. - 352. K
4 POINTS	4 POINTS	4 POINTS
RMSD = 0.00021	RMSD = 0.00021	RMSD = 0.00021
0.12822104E+01	0.12822104E+01	0.12822104E+01
0.88562258E-03	0.88562258E-03	0.88562258E-03
0.85805788E+01	0.85805788E+01	0.85805788E+01
0.38810552E+04	0.38810552E+04	0.38810552E+04

SECOND VIRIAL COEFFICIENT CORRELATION CONSTANTS, ML/MOL:

LITERATURE DOCUMENTS FOR CORRELATED VAPOR PRESSURE VALUES:

LITERATURE DOCUMENTS FOR CORRELATED LIQUID DENSITY VALUES:

  14195

LITERATURE DOCUMENTS REPORTING VIRIAL COEFFICIENT DATA:

T,K	VAPOR PRESSURE, KPA	SATURATED LIQUID VOLUME, ML/MOL	SECOND VIRIAL COEFFICIENT, ML/MOL	HEAT OF VAPORIZATION, J/MOL	SATURATED VAPOR VOLUME, ML/MOL
289		191.67			
290		191.84			
291		192.00			
293		192.34			
294		192.50			
296		192.84			
297		193.01			
299		193.34			
300		193.51			
301		193.68			
303		194.02			
304		194.19			
306		194.53			
307		194.70			
309		195.05			
310		195.22			
311		195.39			
313		195.74			
314		195.91			
316		196.26			
317		196.43			
319		196.78			
320		196.95			
321		197.13			
323		197.48			
324		197.66			
326		198.01			
327		198.19			
329		198.54			
330		198.72			
331		198.90			
333		199.26			
334		199.44			
336		199.80			
337		199.98			
339		200.34			
340		200.52			
341		200.70			
343		201.07			
344		201.25			
346		201.62			
347		201.80			
349		202.17			
350		202.35			
351		202.54			

C(16)H(18)O(3)

COMPOUND CONSTANTS:

MP   TC   VC   ZC
NBP   PC   DC   OM
MU   RG

VAPOR PRESSURE CORRELATION CONSTANTS, K AND KPA:

LIQUID DENSITY CORRELATION CONSTANTS, K AND G/ML:

FRANCIS1   FRANCIS1   FRANCIS1
348. - 407. K   348. - 407. K   348. - 407. K
  4 POINTS   4 POINTS   4 POINTS
RMSD = 0.00029   RMSD = 0.00029   RMSD = 0.00029
0.13531923E+01   0.13531923E+01   0.13531923E+01
0.77704038E-03   0.77704038E-03   0.77704038E-03
0.37081345E+02   0.37081345E+02   0.37081345E+02
0.17556663E+04   0.17556663E+04   0.17556663E+04

SECOND VIRIAL COEFFICIENT CORRELATION CONSTANTS, ML/MOL:

LITERATURE DOCUMENTS FOR CORRELATED VAPOR PRESSURE VALUES:

LITERATURE DOCUMENTS FOR CORRELATED LIQUID DENSITY VALUES:

   14195

LITERATURE DOCUMENTS REPORTING VIRIAL COEFFICIENT DATA:

T,K	VAPOR PRESSURE, KPA	SATURATED LIQUID VOLUME, ML/MOL	SECOND VIRIAL COEFFICIENT, ML/MOL	HEAT OF VAPORIZATION, J/MOL	SATURATED VAPOR VOLUME, ML/MOL
348		244.52			
349		244.70			
350		244.88			
352		245.25			
353		245.44			
354		245.63			
356		246.00			
357		246.18			
358		246.37			
360		246.75			
361		246.93			
362		247.12			
364		247.50			
365		247.69			
366		247.88			
368		248.26			
369		248.45			
370		248.64			
372		249.02			
373		249.21			
374		249.40			
376		249.79			
377		249.98			
378		250.17			
380		250.56			
381		250.75			
382		250.95			
384		251.34			
385		251.53			
386		251.73			
388		252.12			
389		252.31			
390		252.51			
392		252.90			
393		253.10			
394		253.30			
396		253.70			
397		253.89			
398		254.09			
400		254.49			
401		254.69			
402		254.89			
404		255.29			
405		255.50			
406		255.70			